COMPACT
WORLD
ATLAS

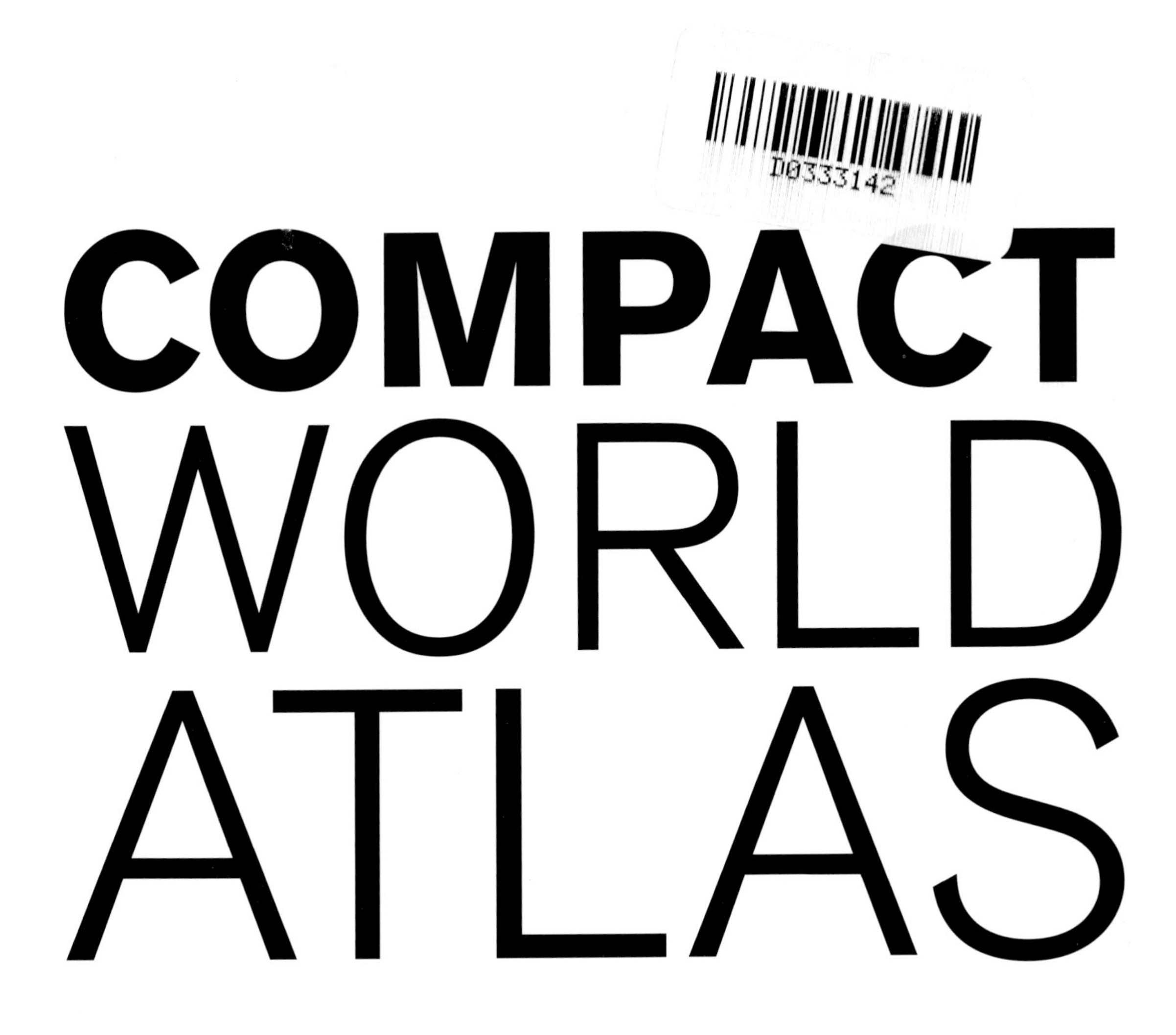

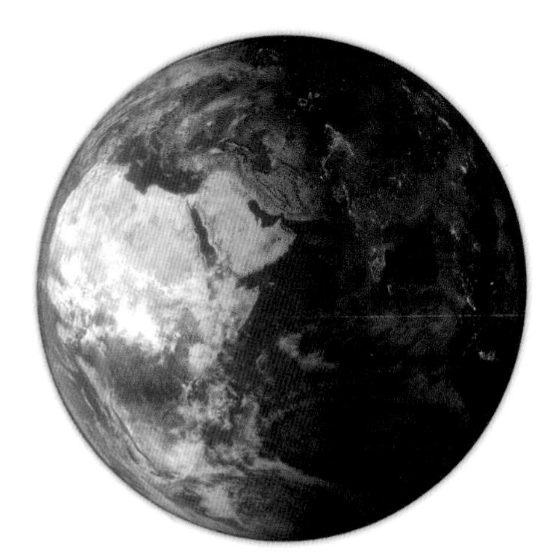

DK | Penguin

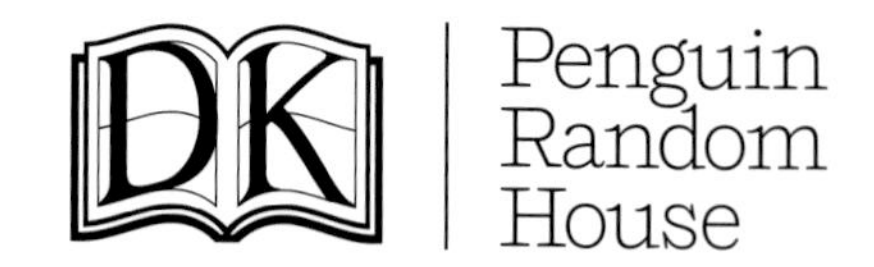

FOR THE SEVENTH EDITION

SENIOR CARTOGRAPHIC EDITOR Simon Mumford

JACKET DESIGNER Dhirendra Singh   JACKET DESIGN DEVELOPMENT Sophia MTT

PRODUCER (PRE PRODUCTION) Jacqueline Street   SENIOR PRODUCER Angela Graef

PUBLISHER
Andrew Macintyre

PUBLISHING DIRECTOR
Jonathan Metcalf

ASSOCIATE PUBLISHING DIRECTOR
Liz Wheeler

ART DIRECTOR
Karen Self

First published in Great Britain in 2001
by Dorling Kindersley Limited
80 Strand, London WC2R 0RL

Reprinted with revisions 2002. Second edition 2003. Reprinted with revisions 2004.
Third edition 2005. Fourth edition 2009. Fifth edition 2012. Sixth edition 2015. Seventh edition 2018

Copyright © 2001, 2002, 2003, 2004, 2005, 2009, 2012, 2015, 2018
Dorling Kindersley Limited
A Penguin Random House Company

10 9 8 7 6 5 4 3 2 1
001–308089–May/2018

A CIP catalogue record for this book is available from the British Library
ISBN 978-0-2413-1764-8

Printed and bound in Hong Kong

A WORLD OF IDEAS:
SEE ALL THERE IS TO KNOW

www.dk.com

Key to map symbols

Physical features

Elevation

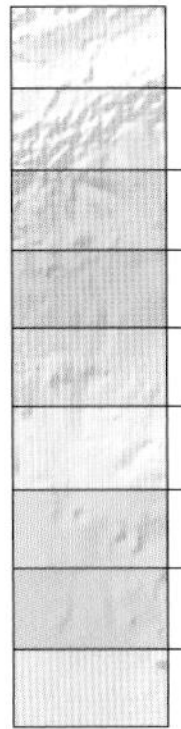

- 6000m/19,686ft
- 4000m/13,124ft
- 3000m/9843ft
- 2000m/6562ft
- 1,000m/3281ft
- 500m/1640ft
- 250m/820ft
- 0
- Below sea level

△   Mountain

▽   Depression

⌂   Volcano

)(Pass/tunnel

Sandy desert

Drainage features

Major perennial river

Minor perennial river

Seasonal river

Canal

Waterfall

Perennial lake

Seasonal lake

Wetland

Ice features

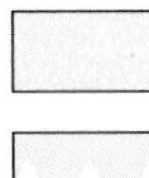

Permanent ice cap/ice shelf

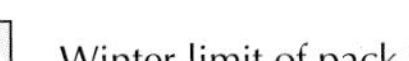

Winter limit of pack ice

Summer limit of pack ice

Borders

Full international border

Disputed de facto border

Territorial claim border

×—×—×   Cease-fire line

Undefined boundary

Internal administrative boundary

Communications

Major road

Minor road

Railway

✈   International airport

Settlements

◉   Above 500,000

◉   100,000 to 500,000

○   50,000 to 100,000

○   Below 50,000

●   National capital

●   Internal administrative capital

Miscellaneous features

+   Site of interest

⎍⎍⎍   Ancient wall

Graticule features

Line of latitude/longitude/Equator

Tropic/Polar circle

25°   Degrees of latitude/longitude

Names

Physical features

Andes
Sahara | Landscape features
Ardennes

Land's End | Headland

Mont Blanc 4,807m | Elevation/volcano/pass

Blue Nile | River/canal/waterfall

Ross Ice Shelf | Ice feature

PACIFIC OCEAN
Sulu Sea | Sea features
Palk Strait

Chile Rise | Undersea feature

Regions

FRANCE | Country

BERMUDA (to UK) | Dependent territory

KANSAS | Administrative region

Dordogne | Cultural region

Settlements

PARIS | Capital city

SAN JUAN | Dependent territory capital city

Chicago
Kettering | Other settlements
Burke

Inset map symbols

Urban area

City

Park

▪   Place of interest

□   Suburb/district

COMPACT WORLD ATLAS

Contents

The World's Regions

North & Central America

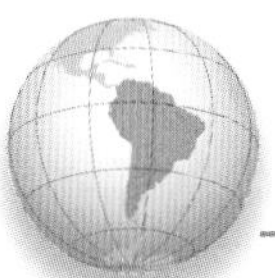

South America

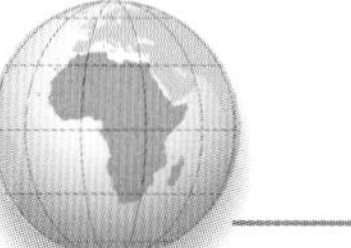

Africa

Europe

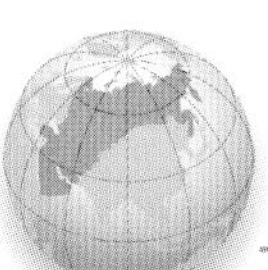

North & West Asia

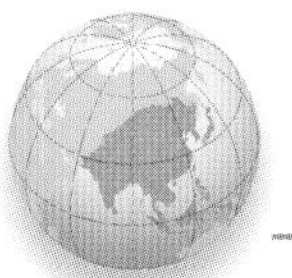

Australasia & Oceania

South & East Asia

Index – Gazetteer

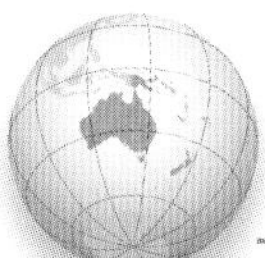

The Political World

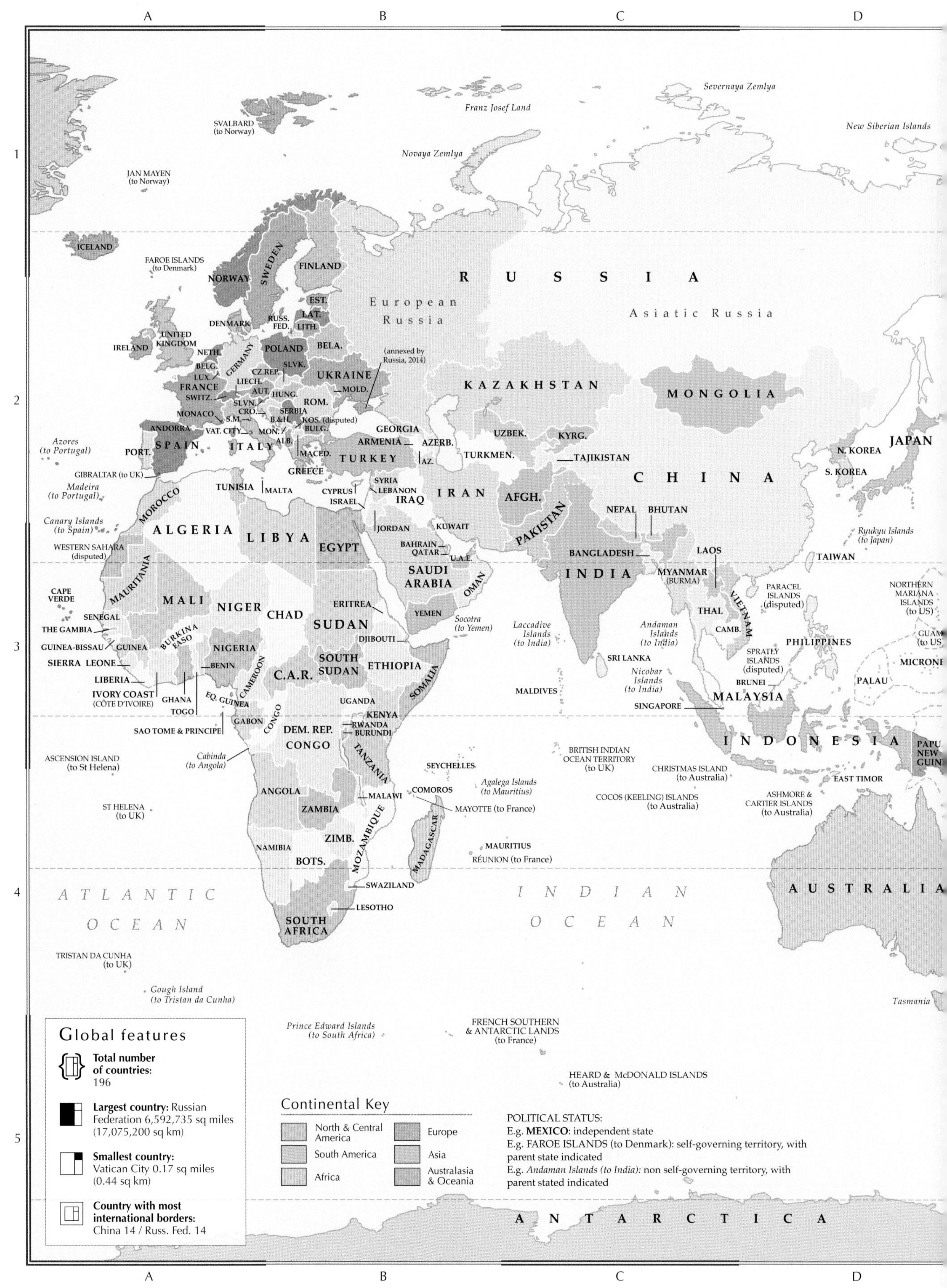

A   B   C   D

1

Severnaya Zemlya
Franz Josef Land
New Siberian Islands

SVALBARD
(to Norway)

JAN MAYEN
(to Norway)

Novaya Zemlya

ICELAND

FAROE ISLANDS
(to Denmark)

NORWAY   SWEDEN   FINLAND

R U S S I A

European
Russia

Asiatic Russia

EST.
LAT.
RUSS.
FED.
LITH.

IRELAND   UNITED
KINGDOM   DENMARK   BELA.

(annexed by
Russia, 2014)

NETH.   GERMANY   POLAND
BELG.   CZ.REP.   SLVK.
LUX.   LIECH.   UKRAINE
FRANCE   AUT. HUNG.   MOLD.
SWITZ.   SLVN.   ROM.
MONACO   CRO.   SERBIA
S.M.   B.&H.   KOS. (disputed)
ANDORRA   VAT. CITY   MON.   BULG.
ALB.   GEORGIA

K A Z A K H S T A N

M O N G O L I A

J A P A N

Azores
(to Portugal)

SPAIN   ITALY   MACED.   ARMENIA   AZERB.
GREECE   TURKEY   AZ.
PORT.

UZBEK.   KYRG.
TURKMEN.   TAJIKISTAN

C H I N A

N. KOREA
S. KOREA

GIBRALTAR (to UK)
Madeira
(to Portugal)

TUNISIA   MALTA   CYPRUS   SYRIA
ISRAEL   LEBANON
IRAQ

I R A N   AFGH.

NEPAL   BHUTAN

Ryukyu Islands
(to Japan)

Canary Islands
(to Spain)

MOROCCO

JORDAN   KUWAIT

PAKISTAN

WESTERN SAHARA
(disputed)

ALGERIA   LIBYA   EGYPT

BAHRAIN
QATAR   U.A.E.

BANGLADESH   LAOS   TAIWAN

CAPE
VERDE

MAURITANIA   MALI   NIGER   CHAD

SAUDI
ARABIA   OMAN

YEMEN

Socotra
(to Yemen)

I N D I A   MYANMAR
(BURMA)

THAI.

PARACEL
ISLANDS
(disputed)

NORTHERN
MARIANA
ISLANDS
(to US)

ERITREA

Laccadive
Islands
(to India)

Andaman
Islands
(to India)

VIETNAM

GUAM
(to US)

SENEGAL
THE GAMBIA
GUINEA-BISSAU   GUINEA
SIERRA LEONE
LIBERIA   IVORY COAST
(CÔTE D'IVOIRE)

BURKINA
FASO   NIGERIA
BENIN
GHANA
TOGO   CAMEROON
EQ. GUINEA   GABON

SUDAN   DJIBOUTI

SOUTH
SUDAN   ETHIOPIA

SOMALIA

C.A.R.

UGANDA

CAMB.

SPRATLY
ISLANDS
(disputed)

PHILIPPINES

MICRONE

SRI LANKA

Nicobar
Islands
(to India)

BRUNEI

PALAU

MALDIVES   SINGAPORE   MALAYSIA

SAO TOME & PRINCIPE

CONGO   DEM. REP.
CONGO

KENYA
RWANDA
BURUNDI

I N D O N E S I A

PAPU
NEW
GUIN

ASCENSION ISLAND
(to St Helena)

Cabinda
(to Angola)

TANZANIA

BRITISH INDIAN
OCEAN TERRITORY
(to UK)

CHRISTMAS ISLAND
(to Australia)

EAST TIMOR

ASHMORE &
CARTIER ISLANDS
(to Australia)

SEYCHELLES

Agalega Islands
(to Mauritius)

COCOS (KEELING) ISLANDS
(to Australia)

ANGOLA   MALAWI   COMOROS
ZAMBIA   MOZAMBIQUE   MAYOTTE (to France)

ST HELENA
(to UK)

ZIMB.   MADAGASCAR

MAURITIUS
RÉUNION (to France)

NAMIBIA   BOTS.

SWAZILAND

LESOTHO

A T L A N T I C
O C E A N

I N D I A N
O C E A N

A U S T R A L I A

SOUTH
AFRICA

TRISTAN DA CUNHA
(to UK)

Tasmania

Gough Island
(to Tristan da Cunha)

Prince Edward Islands
(to South Africa)

FRENCH SOUTHERN
& ANTARCTIC LANDS
(to France)

HEARD & McDONALD ISLANDS
(to Australia)

Global features

**Total number
of countries:**
196

Largest country: Russian
Federation 6,592,735 sq miles
(17,075,200 sq km)

Smallest country:
Vatican City 0.17 sq miles
(0.44 sq km)

**Country with most
international borders:**
China 14 / Russ. Fed. 14

Continental Key

North & Central
America

South America

Africa

Europe

Asia

Australasia
& Oceania

POLITICAL STATUS:
E.g. **MEXICO**: independent state
E.g. FAROE ISLANDS (to Denmark): self-governing territory, with
parent state indicated
E.g. *Andaman Islands (to India):* non self-governing territory, with
parent stated indicated

A N T A R C T I C A

A   B   C   D

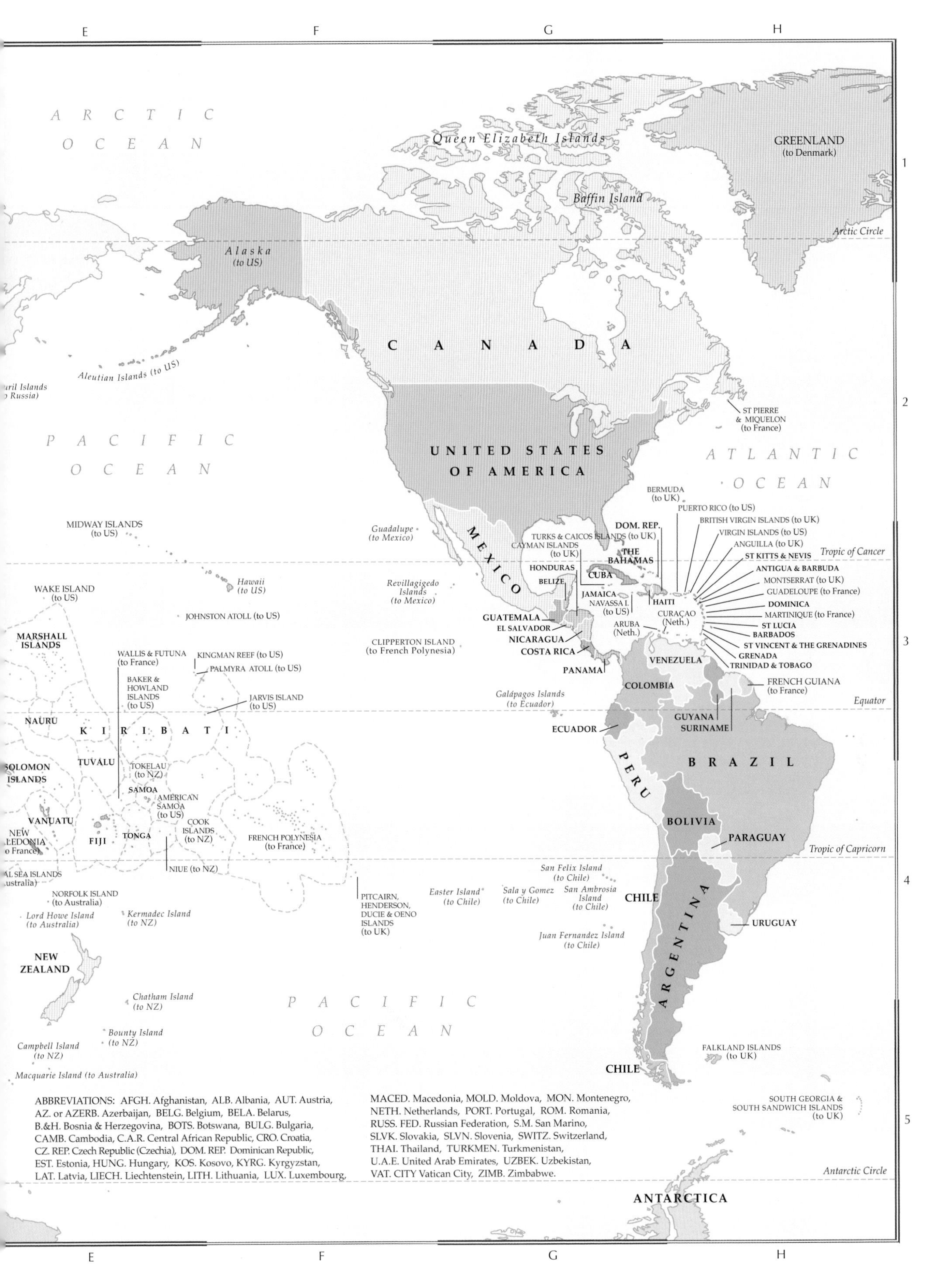

E   F   G   H

A R C T I C

O C E A N

Queen Elizabeth Islands

GREENLAND
(to Denmark)

Baffin Island

1

Arctic Circle

Alaska
(to US)

C A N A D A

Aleutian Islands (to US)

uril Islands
o Russia)

2

P A C I F I C

O C E A N

UNITED STATES
OF AMERICA

A T L A N T I C

O C E A N

ST PIERRE
& MIQUELON
(to France)

BERMUDA
(to UK)

PUERTO RICO (to US)

MIDWAY ISLANDS
(to US)

Guadalupe
(to Mexico)

DOM. REP.

BRITISH VIRGIN ISLANDS (to UK)

VIRGIN ISLANDS (to US)

TURKS & CAICOS ISLANDS (to UK)

ANGUILLA (to UK)

CAYMAN ISLANDS
(to UK)

ST KITTS & NEVIS

Tropic of Cancer

THE
BAHAMAS

ANTIGUA & BARBUDA

HONDURAS

CUBA

MONTSERRAT (to UK)

WAKE ISLAND
(to US)

Hawaii
(to US)

Revillagigedo
Islands
(to Mexico)

BELIZE

GUADELOUPE (to France)

JAMAICA

DOMINICA

NAVASSA I.

MARTINIQUE (to France)

JOHNSTON ATOLL (to US)

(to US)

HAITI

MARSHALL
ISLANDS

GUATEMALA

CURAÇAO

ST LUCIA

(Neth.)

BARBADOS

EL SALVADOR

ARUBA

WALLIS & FUTUNA
(to France)

KINGMAN REEF (to US)

NICARAGUA

(Neth.)

ST VINCENT & THE GRENADINES

CLIPPERTON ISLAND
(to French Polynesia)

COSTA RICA

GRENADA

PALMYRA ATOLL (to US)

PANAMA

VENEZUELA

TRINIDAD & TOBAGO

3

BAKER &
HOWLAND
ISLANDS
(to US)

JARVIS ISLAND
(to US)

COLOMBIA

FRENCH GUIANA
(to France)

NAURU

Galápagos Islands
(to Ecuador)

GUYANA

Equator

K I R I B A T I

ECUADOR

SURINAME

SOLOMON
ISLANDS

TUVALU

TOKELAU
(to NZ)

P E R U

B R A Z I L

SAMOA

AMERICAN
SAMOA
(to US)

BOLIVIA

VANUATU

COOK
ISLANDS
(to NZ)

FRENCH POLYNESIA
(to France)

PARAGUAY

Tropic of Capricorn

NEW
LEDONIA
o France)

FIJI

TONGA

AL SEA ISLANDS
ustralia)

NIUE (to NZ)

San Felix Island
(to Chile)

4

NORFOLK ISLAND
(to Australia)

Easter Island
(to Chile)

Sala y Gomez
(to Chile)

San Ambrosia
Island
(to Chile)

CHILE

Lord Howe Island
(to Australia)

Kermadec Island
(to NZ)

PITCAIRN,
HENDERSON,
DUCIE & OENO
ISLANDS
(to UK)

A R G E N T I N A

URUGUAY

Juan Fernandez Island
(to Chile)

NEW
ZEALAND

Chatham Island
(to NZ)

P A C I F I C

FALKLAND ISLANDS
(to UK)

Campbell Island
(to NZ)

Bounty Island
(to NZ)

O C E A N

CHILE

Macquarie Island (to Australia)

SOUTH GEORGIA &
SOUTH SANDWICH ISLANDS
(to UK)

5

ABBREVIATIONS: AFGH. Afghanistan, ALB. Albania, AUT. Austria,
AZ. or AZERB. Azerbaijan, BELG. Belgium, BELA. Belarus,
B.&H. Bosnia & Herzegovina, BOTS. Botswana, BULG. Bulgaria,
CAMB. Cambodia, C.A.R. Central African Republic, CRO. Croatia,
CZ. REP. Czech Republic (Czechia), DOM. REP. Dominican Republic,
EST. Estonia, HUNG. Hungary, KOS. Kosovo, KYRG. Kyrgyzstan,
LAT. Latvia, LIECH. Liechtenstein, LITH. Lithuania, LUX. Luxembourg,

MACED. Macedonia, MOLD. Moldova, MON. Montenegro,
NETH. Netherlands, PORT. Portugal, ROM. Romania,
RUSS. FED. Russian Federation, S.M. San Marino,
SLVK. Slovakia, SLVN. Slovenia, SWITZ. Switzerland,
THAI. Thailand, TURKMEN. Turkmenistan,
U.A.E. United Arab Emirates, UZBEK. Uzbekistan,
VAT. CITY Vatican City, ZIMB. Zimbabwe.

Antarctic Circle

ANTARCTICA

E   F   G   H

The Physical World

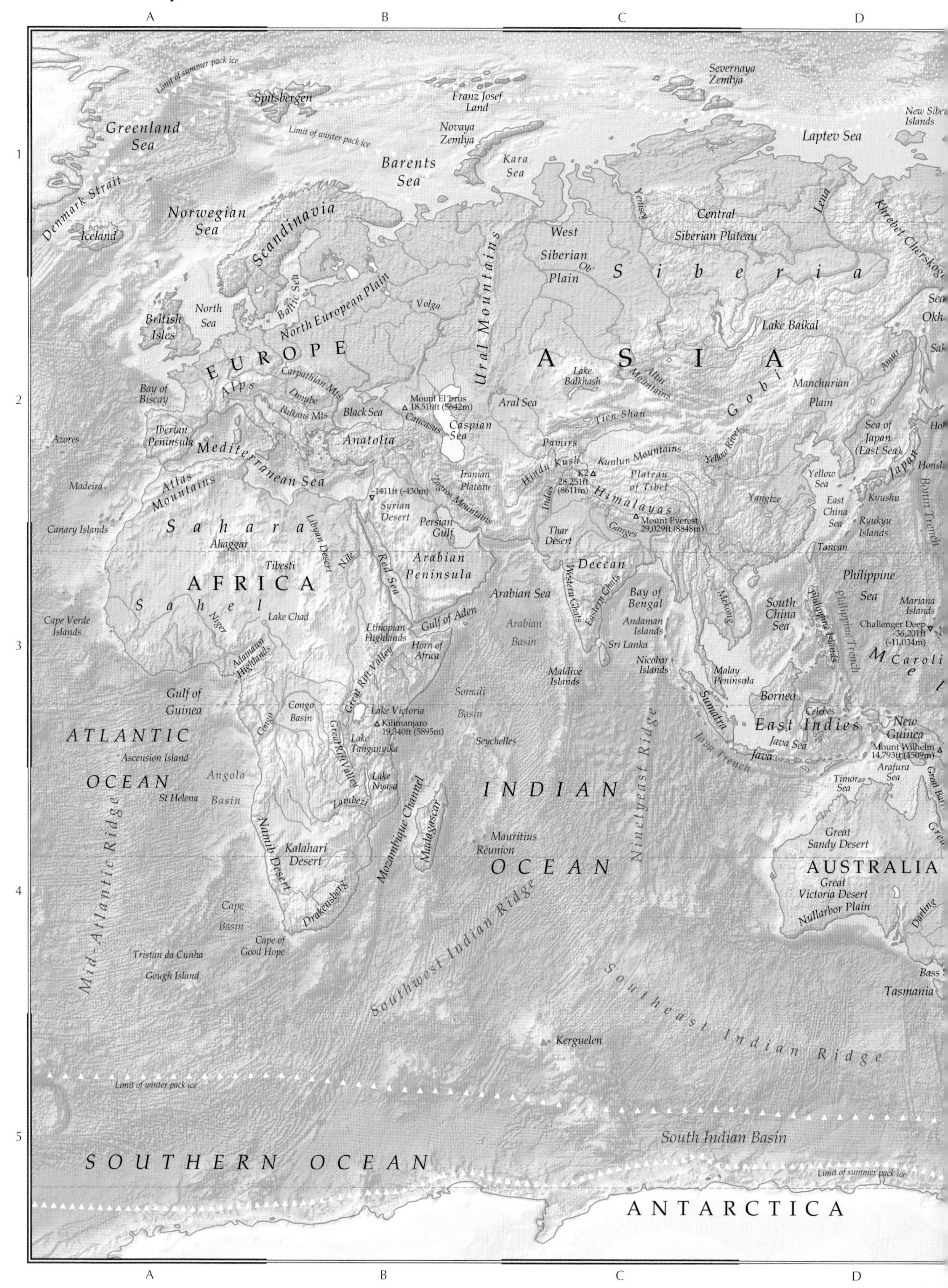

Greenland Sea

Spitsbergen

Limit of summer pack ice

Franz Josef Land

Severnaya Zemlya

New Sibe Islands

1

Denmark Strait

Iceland

Norwegian Sea

Scandinavia

Limit of winter pack ice

Novaya Zemlya

Barents Sea

Kara Sea

Laptev Sea

British Isles

North Sea

Baltic Sea

EUROPE

North European Plain

West Siberian Plain

Ob'

Yenisey

Central Siberian Plateau

Lena

Khrebet Cherskogo

Siberia

Sea Okh

2

Bay of Biscay

Alps

Carpathian Mts

Danube

Balkans Mts

Iberian Peninsula

Mediterranean Sea

Azores

Madeira

Atlas Mountains

Canary Islands

Sahara

Black Sea

Caucasus

Anatolia

Mount El'brus
△18,510ft (5642m)

Caspian Sea

−1411ft (−430m)

Syrian Desert

Zagros Mountains

Iranian Plateau

Volga

Ural Mountains

Aral Sea

Lake Balkhash

Pamirs

ASIA

Altai Mountains

Tien Shan

Hindu Kush

Kunlun Mountains

K2 △ 28,251ft (8611m)

Indus

Plateau of Tibet

Himalayas

△ Mount Everest 29,029ft (8848m)

Ganges

Lake Baikal

Amur

Gobi

Manchurian Plain

Yellow River

Yangtze

Sea of Japan (East Sea)

Yellow Sea

East China Sea

Japan

Honsh

Kyushu

Ryukyu Islands

Taiwan

Hol

Bonin Trench

Sak

3

Sahel

AFRICA

Niger

Tibesti

Ahaggar

Libyan Desert

Nile

Red Sea

Lake Chad

Adamawa Highlands

Ethiopian Highlands

Gulf of Aden

Horn of Africa

Arabian Peninsula

Persian Gulf

Arabian Sea

Thar Desert

Deccan

Western Ghats

Eastern Ghats

Bay of Bengal

Andaman Islands

Nicobar Islands

Sri Lanka

Maldive Islands

Mekong

South China Sea

Philippine Sea

Mariana Islands

Challenger Deep −36,201ft (−11,034m)

Philippine Trench

Philippine Islands

M Caroli e

Cape Verde Islands

Gulf of Guinea

Congo

Congo Basin

Great Rift Valley

Lake Victoria

△ Kilimanjaro 19,340ft (5895m)

Lake Tanganyika

Somali Basin

Seychelles

Malay Peninsula

Borneo

Celebes

Java Sea

East Indies

New Guinea

Mount Wilhelm 14,793ft (4509m)

ATLANTIC

Ascension Island

St Helena

Angola Basin

Namib Desert

Great Rift Valley

Lake Nyasa

Zambezi

Mozambique Channel

Madagascar

INDIAN

Sumatra

Java Trench

Java

Timor Sea

Arafura Sea

Great Bar

Grea

4

OCEAN

Mid-Atlantic Ridge

Tristan da Cunha

Gough Island

Cape Basin

Kalahari Desert

Drakensberg

Cape of Good Hope

Ninetyeast Ridge

Mauritius
Réunion

OCEAN

Great Sandy Desert

AUSTRALIA

Great Victoria Desert

Nullarbor Plain

Darling

Bass

Tasmania

△ Kerguelen

Southwest Indian Ridge

Southeast Indian Ridge

Limit of winter pack ice

5

SOUTHERN OCEAN

South Indian Basin

Limit of summer pack ice

ANTARCTICA

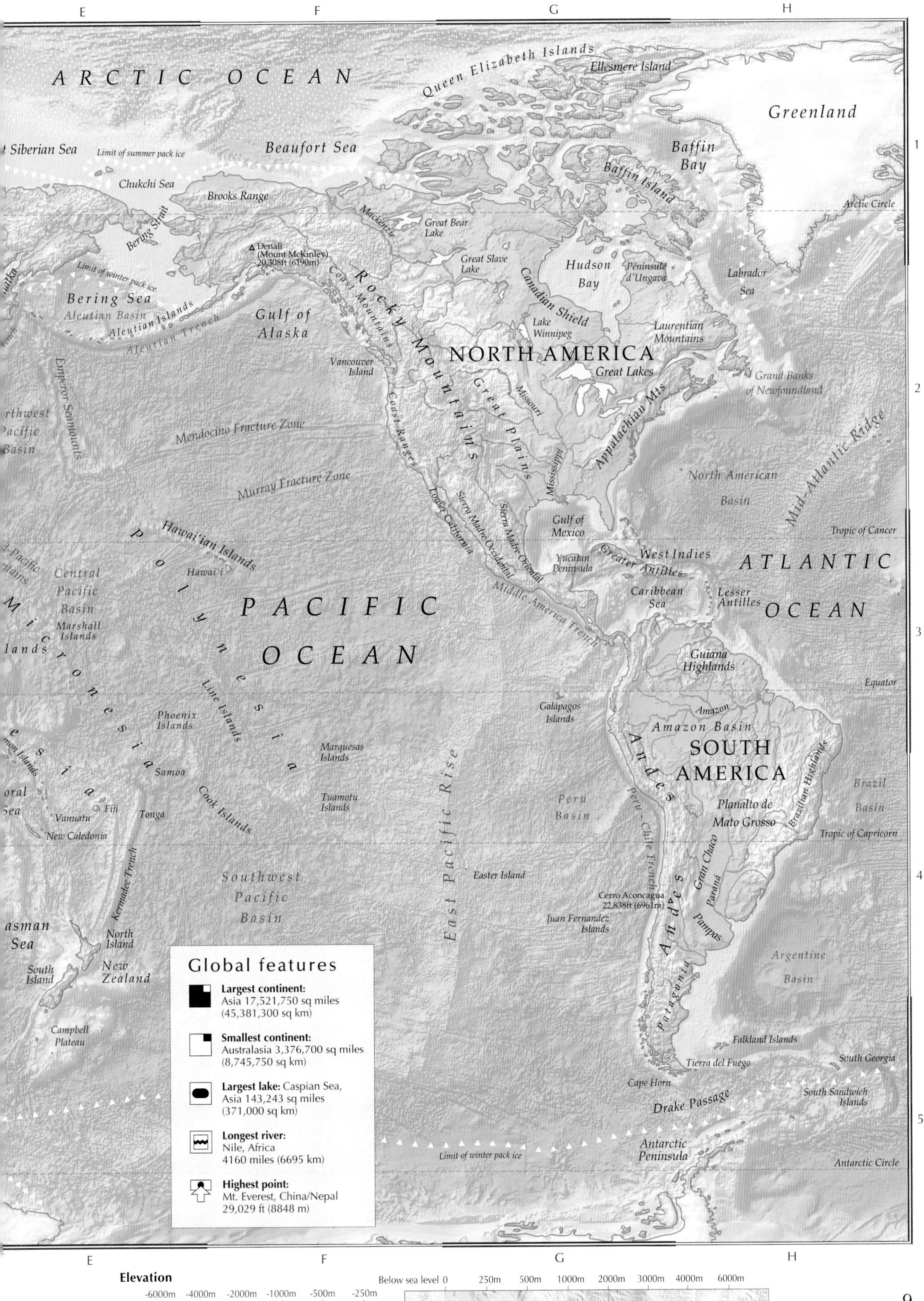

Standard Time Zones

The numbers at the top of the map indicate how many hours each time zone is ahead or behind Coordinated Universal Time (UTC). The row of clocks indicate the time in each zone when it is 12:00 noon UTC.

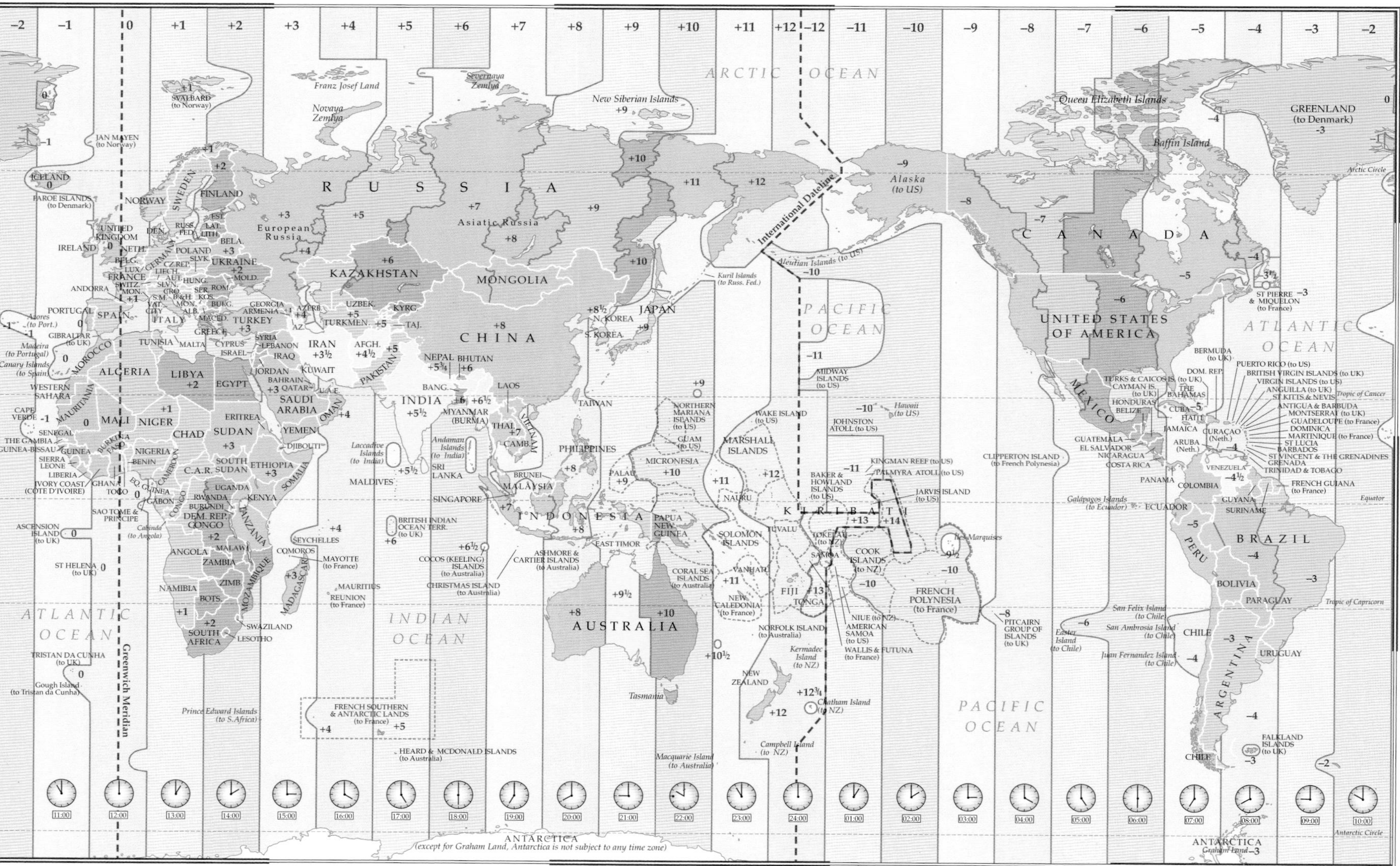

TIME ZONES

Because Earth is a rotating sphere, the Sun shines on only half of its surface at any one time. Thus, it is simultaneously morning, evening, and night time in different parts of the world. Because of these disparities, each country or part of a country adheres to a local time. A region of the Earth's surface within which a single local time is used is called a time zone.

COORDINATED UNIVERSAL TIME (UTC)

Coordinated Universal Time (UTC) is a reference by which the local time in each time zone is set. UTC is a successor to, and closely approximates, Greenwich Mean Time (GMT). However, UTC is based on an atomic clock, whereas GMT is determined by the Sun's position in the sky relative to the 0° longitudinal meridian, which runs through Greenwich, UK.

THE INTERNATIONAL DATELINE

The International Dateline is an imaginary line from pole to pole that roughly corresponds to the 180° longitudinal meridian. It is an arbitrary marker between calendar days. The dateline is needed because of the use of local times around the world rather than a single universal time.

The
WORLD
ATLAS

THE MAPS IN THIS ATLAS ARE ARRANGED CONTINENT BY CONTINENT, STARTING FROM THE
INTERNATIONAL DATE LINE, AND MOVING EASTWARD. THE MAPS PROVIDE A UNIQUE VIEW
OF TODAY'S WORLD, COMBINING TRADITIONAL CARTOGRAPHIC TECHNIQUES WITH THE
LATEST REMOTE-SENSED AND DIGITAL TECHNOLOGY.

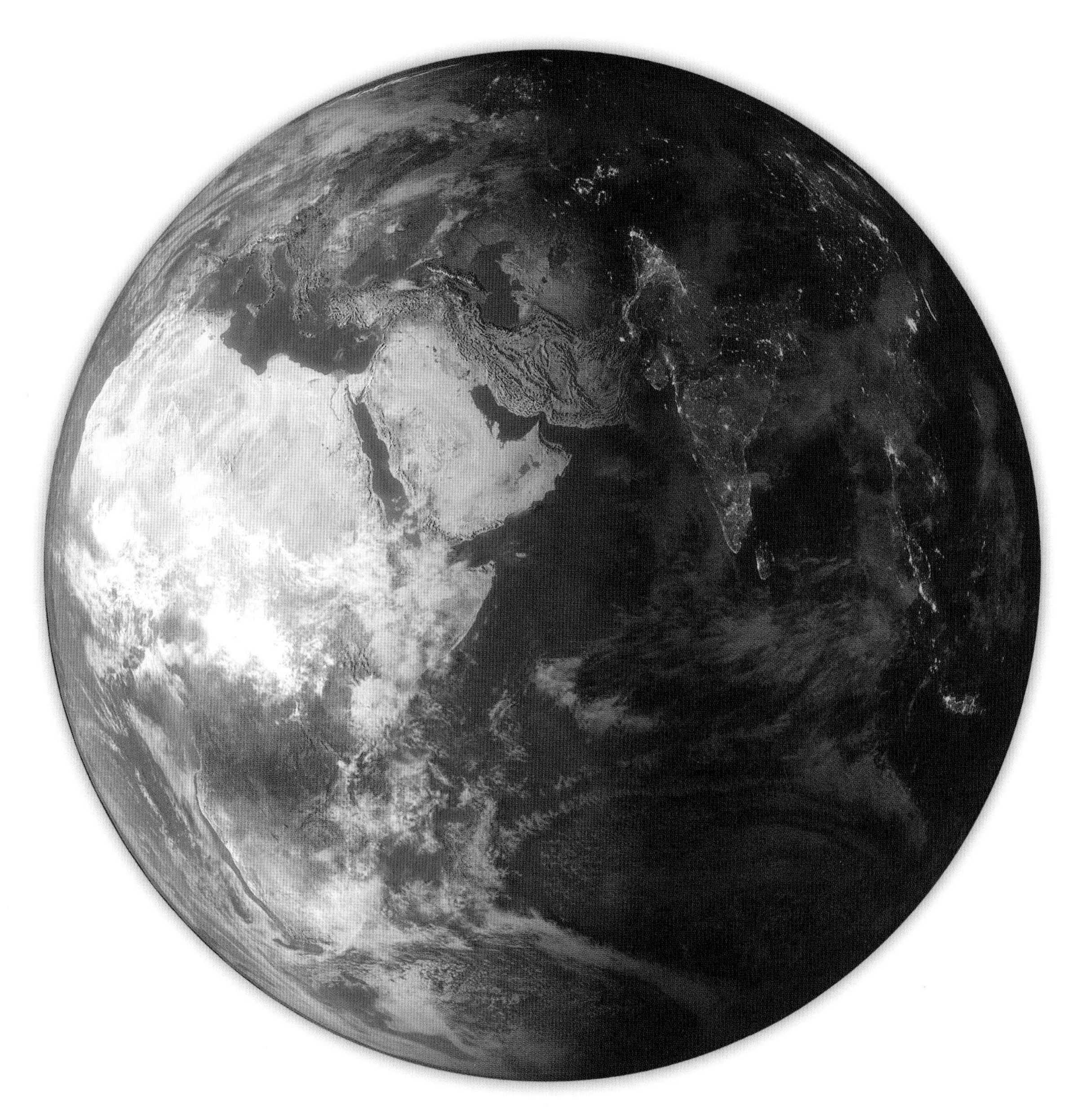

North & Central America

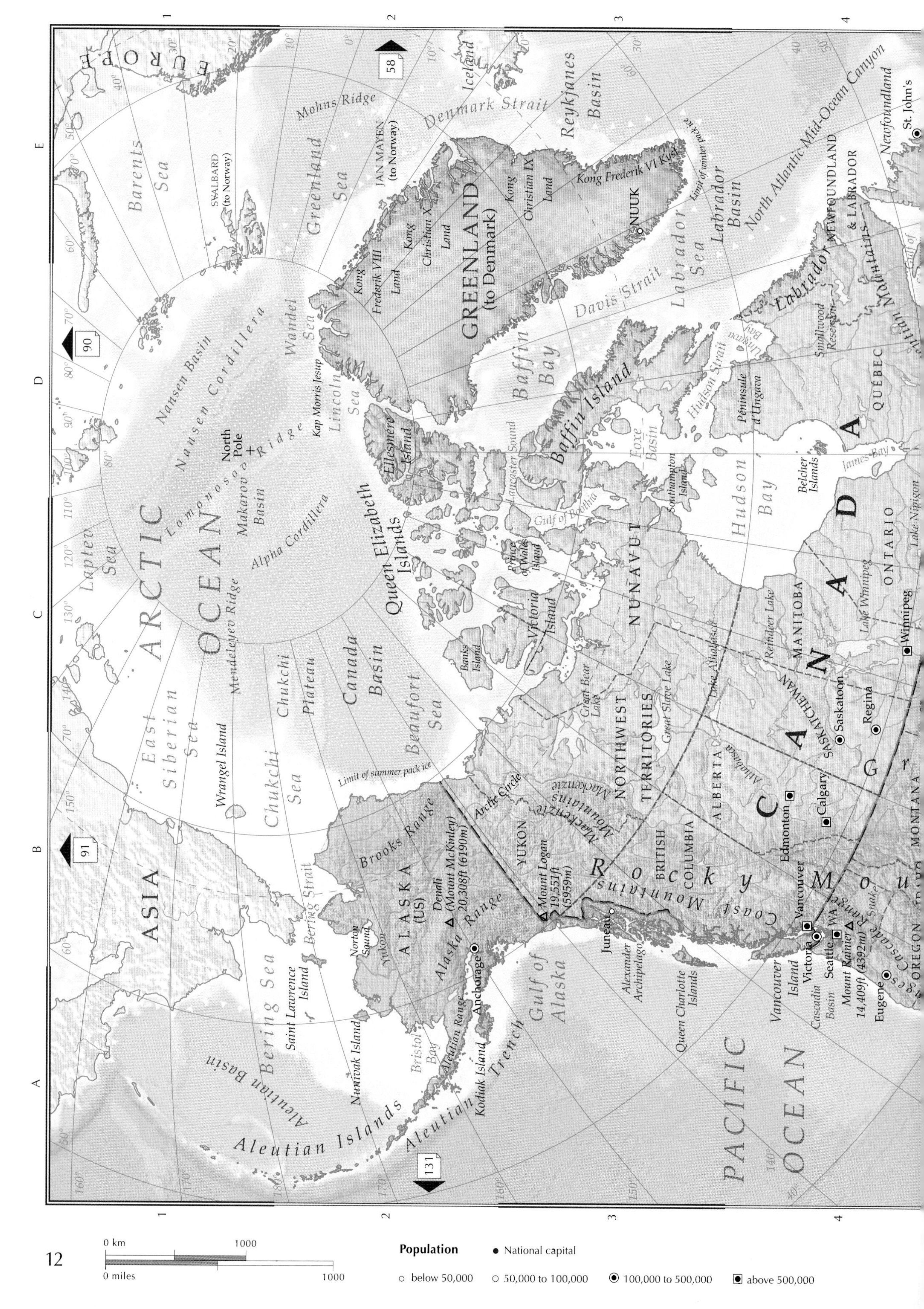

EUROPE

Barents Sea

SVALBARD (to Norway)

Greenland Sea

Mohns Ridge

JAN MAYEN (to Norway)

Denmark Strait

Iceland

Reykjanes Basin

North Atlantic Mid-Ocean Canyon

St. John's

Newfoundland

90

Kong Frederik VI Kyst

Kong Christian IX Land

Kong Christian X Land

Kong Frederik VIII Land

GREENLAND (to Denmark)

NUUK

Limit of winter pack ice

Labrador Basin

NEWFOUNDLAND & LABRADOR

Nansen Basin

Nansen Cordillera

Wandel Sea

Kap Morris Jesup

Lincoln Sea

Davis Strait

Labrador Sea

Smallwood Reservoir

QUEBEC

Ungava Bay

Lake Nipigon

North Pole

Makarov Basin

Lomonosov Ridge

Mendeleyev Ridge

Alpha Cordillera

Ellesmere Island

Queen Elizabeth Islands

Baffin Bay

Baffin Island

Péninsule d'Ungava

Lancaster Sound

Foxe Basin

Belcher Islands

James Bay

ONTARIO

ARCTIC OCEAN

Laptev Sea

Chukchi Plateau

Canada Basin

Prince of Wales Island

Victoria Island

Gulf of Boothia

Southampton Island

Hudson Bay

East Siberian Sea

Chukchi Sea

Banks Island

NUNAVUT

Great Bear Lake

Lake Athabasca

Reindeer Lake

MANITOBA

Saskatoon

Winnipeg

Lake Winnipeg

91

Wrangel Island

Beaufort Sea

Great Slave Lake

SASKATCHEWAN

Regina

C A N A D A

Limit of summer pack ice

Arctic Circle

Mackenzie Mountains

NORTHWEST TERRITORIES

Mackenzie

ALBERTA

Edmonton

Calgary

Bering Strait

Brooks Range

YUKON

Mount Logan 19,551ft (5959m)

Rocky Mountains

BRITISH COLUMBIA

MONTANA

ASIA

Saint Lawrence Island

Norton Sound

ALASKA (US)

Denali (Mount McKinley) 20,308ft (6190m)

Alaska Range

Yukon

Juneau

Coast Mountains

Vancouver

Seattle WA

Mount Rainier 14,409ft (4392m)

Snake R.

Cascade Range

Bering Sea

Nunivak Island

Bristol Bay

Anchorage

Aleutian Range

Kodiak Island

Gulf of Alaska

Alexander Archipelago

Queen Charlotte Islands

Vancouver Island

Victoria

Cascadia Basin

Eugene

OREGON

Aleutian Basin

Aleutian Islands

Aleutian Trench

PACIFIC OCEAN

131

0 km   1000

0 miles   1000

Population   ● National capital

○ below 50,000   ○ 50,000 to 100,000   ◉ 100,000 to 500,000   ◼ above 500,000

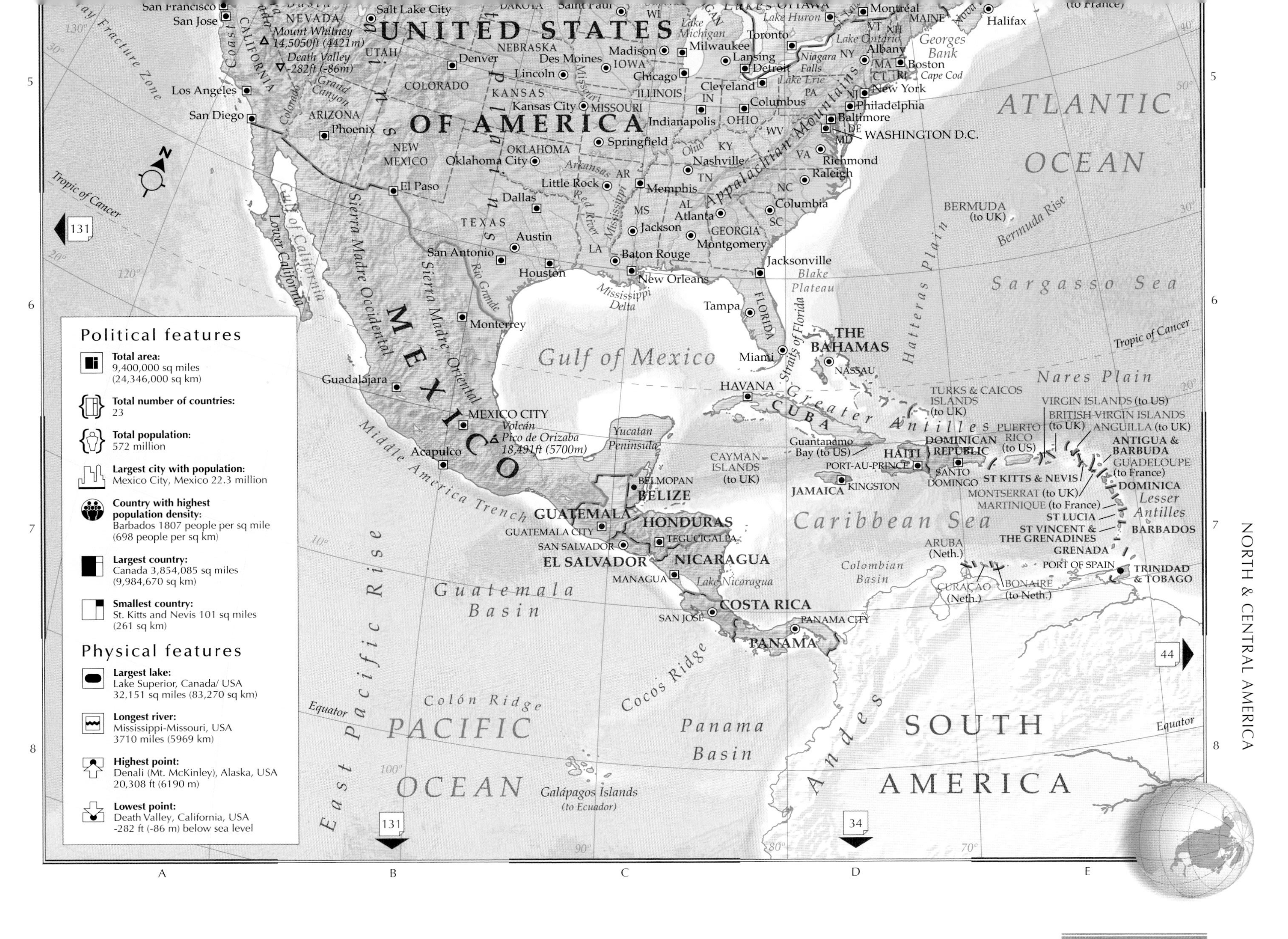

Political features

Total area:
9,400,000 sq miles
(24,346,000 sq km)

Total number of countries:
23

Total population:
572 million

Largest city with population:
Mexico City, Mexico 22.3 million

Country with highest population density:
Barbados 1807 people per sq mile
(698 people per sq km)

Largest country:
Canada 3,854,085 sq miles
(9,984,670 sq km)

Smallest country:
St. Kitts and Nevis 101 sq miles
(261 sq km)

Physical features

Largest lake:
Lake Superior, Canada/ USA
32,151 sq miles (83,270 sq km)

Longest river:
Mississippi-Missouri, USA
3710 miles (5969 km)

Highest point:
Denali (Mt. McKinley), Alaska, USA
20,308 ft (6190 m)

Lowest point:
Death Valley, California, USA
-282 ft (-86 m) below sea level

Western Canada & Alaska

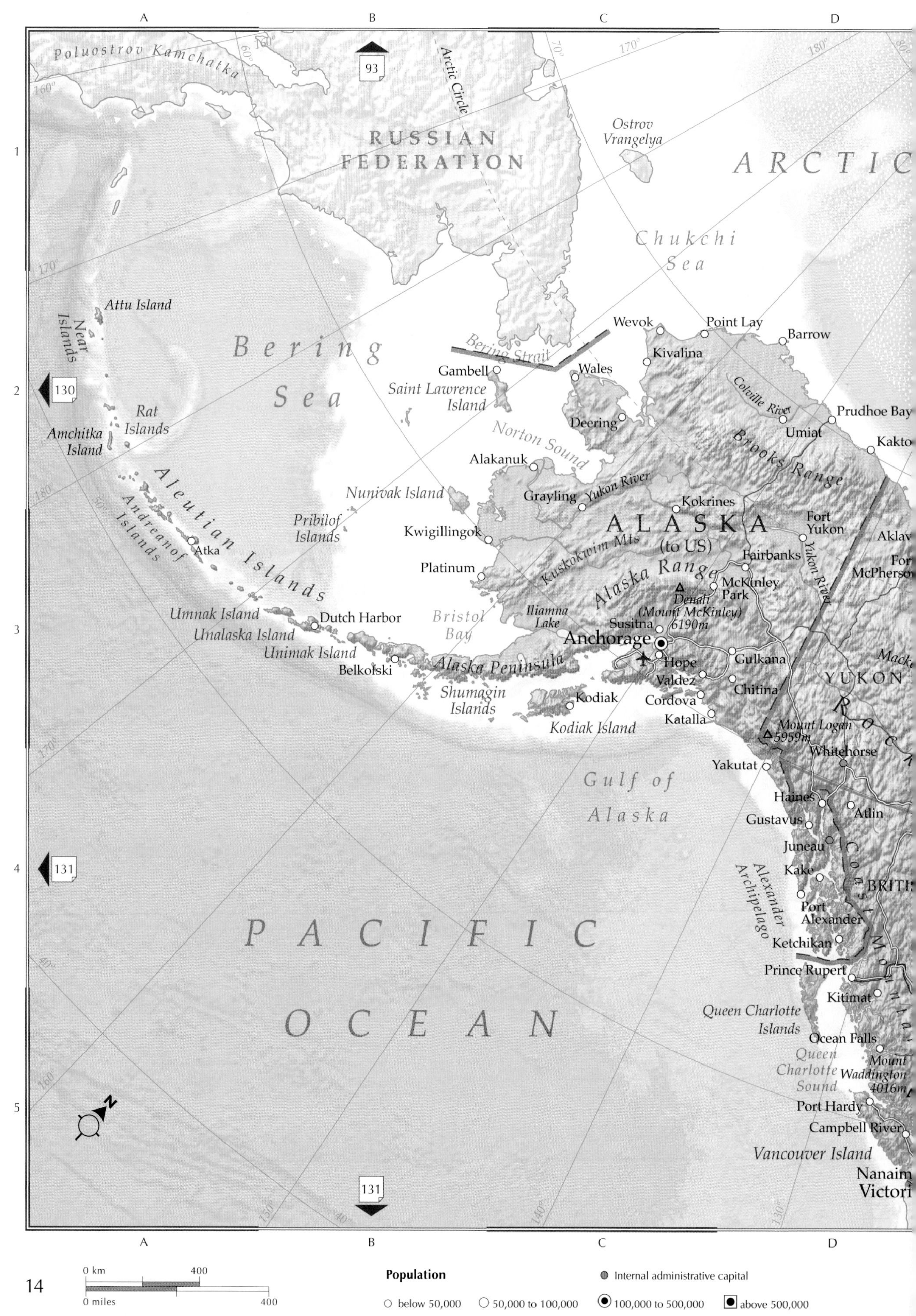

poluostrov Kamchatka

93

RUSSIAN
FEDERATION

Arctic Circle

Ostrov
Vrangelya

A R C T I C

Chukchi
Sea

Wevok   Point Lay   Barrow

Kivalina

Bering Strait

Gambell   Wales

Prudhoe Bay

Umiat

Brooks Range

Kakto

Attu Island

Near
Islands

Bering

Sea

Saint Lawrence
Island

Deering

Norton Sound

Colville River

130

Rat
Islands

Alakanuk

Grayling   Yukon River

Kokrines

A L A S K A
(to US)

Fort
Yukon

Aklav

For
McPherson

Amchitka
Island

Nunivak Island

Kwigillingok

Kuskokwim Mts

Fairbanks

Aleutian Islands

Andreanof
Islands

Atka

Pribilof
Islands

Platinum

Alaska Range

McKinley
Park

Yukon Rive

Denali
(Mount McKinley)
6190m

Umnak Island

Dutch Harbor

Bristol
Bay

Iliamna
Lake

Susitna

Anchorage

Hope   Gulkana

Mack

Unalaska Island

Unimak Island

Belkofski

Alaska Peninsula

Valdez

Chitina

YUKON

Shumagin
Islands

Kodiak

Cordova

Katalla

Mount Logan
5959m

Whitehorse

Kodiak Island

Gulf of
Alaska

Yakutat

Haines

Atlin

Gustavus

Juneau

131

Kake

Alexander
Archipelago

BRITI

P A C I F I C

Port
Alexander

Ketchikan

Prince Rupert

O C E A N

Kitimat

Queen Charlotte
Islands

Ocean Falls

Queen
Charlotte
Sound

Mount
Waddington
4016m

Port Hardy

Campbell River

Vancouver Island

Nanaim
Victori

131

0 km   400

0 miles   400

Population

○ below 50,000   ◯ 50,000 to 100,000   ◉ 100,000 to 500,000   ■ above 500,000

● Internal administrative capital

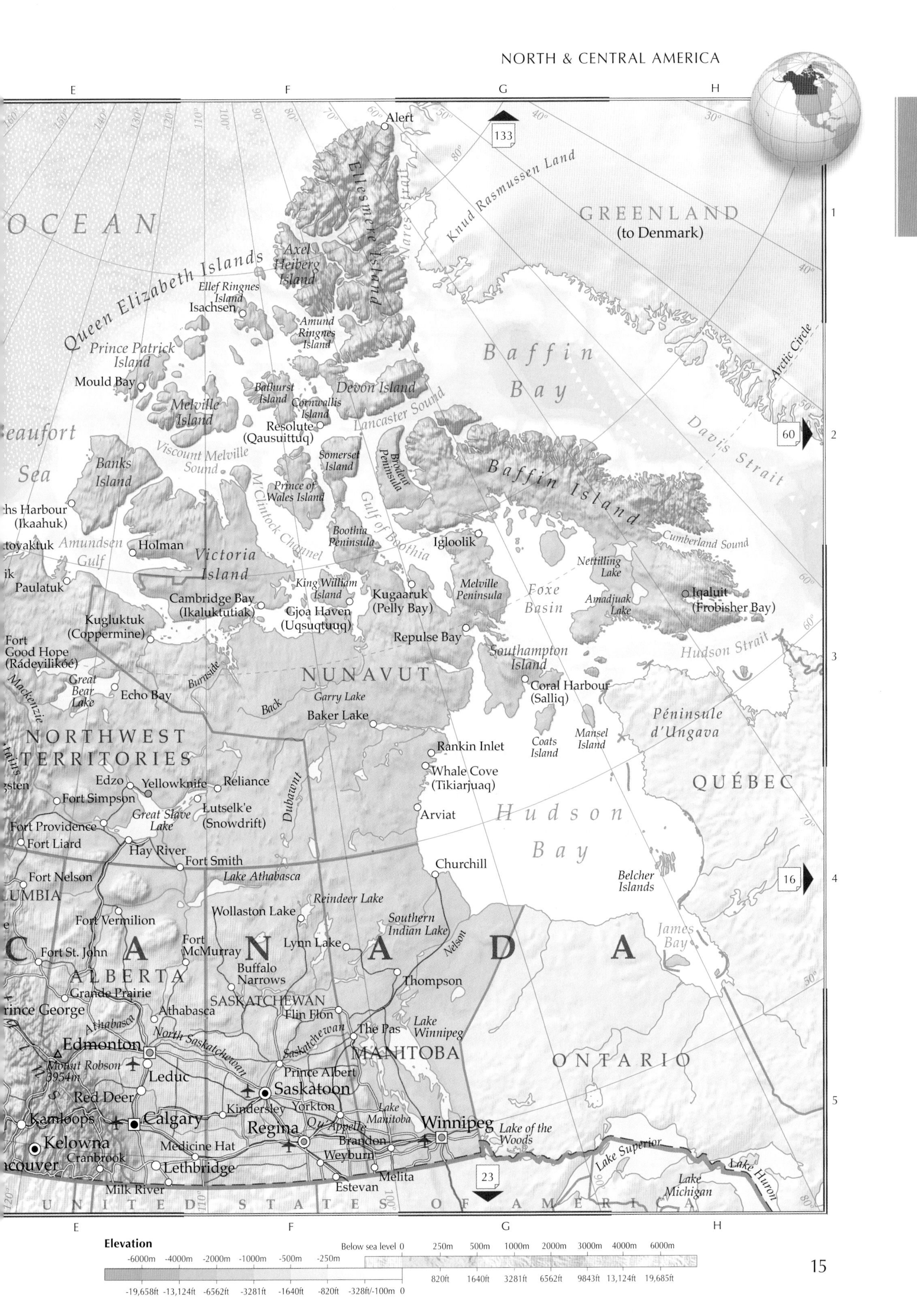

Alert

133

Knud Rasmussen Land

GREENLAND
(to Denmark)

Arctic Circle

OCEAN

Queen Elizabeth Islands

Ellesmere Island

Axel Heiberg Island

Ellef Ringnes Island
Isachsen

Amund Ringnes Island

Prince Patrick Island

Mould Bay

Melville Island

Bathurst Island

Cornwallis Island

Devon Island

Baffin Bay

Beaufort Sea

Banks Island

Viscount Melville Sound

M'Clintock Channel

Resolute (Qausuittuq)

Somerset Island

Prince of Wales Island

Lancaster Sound

Brodeur Peninsula

Davis Strait

60

hs Harbour (Ikaahuk)

toyaktuk

Amundsen Gulf

Holman

ik

Paulatuk

Victoria Island

Gulf of Boothia

Boothia Peninsula

Baffin Island

Cumberland Sound

Fort Good Hope (Rádeyilikóé)

Great Bear Lake

Echo Bay

Kugluktuk (Coppermine)

Cambridge Bay (Ikaluktutiak)

King William Island

Gjoa Haven (Uqsuqtuuq)

Kugaaruk (Pelly Bay)

Igloolik

Melville Peninsula

Nettilling Lake

Amadjuak Lake

Iqaluit (Frobisher Bay)

Foxe Basin

Repulse Bay

Burnside

NUNAVUT

Back

Garry Lake

Baker Lake

Southampton Island

Coral Harbour (Salliq)

Péninsule d'Ungava

NORTHWEST TERRITORIES

sten

Edzo

Yellowknife

Reliance

Fort Simpson

Great Slave Lake

Lutselk'e (Snowdrift)

Dubaunt

Rankin Inlet

Whale Cove (Tikiarjuaq)

Arviat

Coats Island

Mansel Island

QUÉBEC

Fort Providence

Fort Liard

Hay River

Fort Smith

Lake Athabasca

Churchill

Hudson Bay

Belcher Islands

16

Fort Nelson

LUMBIA

Fort Vermilion

Wollaston Lake

Reindeer Lake

Southern Indian Lake

Nelson

James Bay

CANADA

Fort St. John

ALBERTA

Grande Prairie

rince George

Athabasca

SASKATCHEWAN

Fort McMurray

Buffalo Narrows

Lynn Lake

Thompson

ONTARIO

Athabasca

North Saskatchewan

Flin Flon

Lake Winnipeg

Edmonton

Mount Robson 3954m

Leduc

Red Deer

Saskatchewan

Prince Albert

The Pas

MANITOBA

Kamloops

Calgary

Kindersley

Saskatoon

Yorkton

Lake Manitoba

Kelowna

Medicine Hat

Regina

Qu'Appelle

Lake Winnipeg

Winnipeg

Lake of the Woods

Lake Superior

ancouver

Cranbrook

Lethbridge

Brandon

Weyburn

Melita

Lake Huron

Milk River

UNITED

STATES

Estevan

23

OF

AMERICA

Lake Michigan

Elevation

Below sea level 0   250m   500m   1000m   2000m   3000m   6000m

-6000m -4000m -2000m -1000m -500m -250m

-19,658ft -13,124ft -6562ft -3281ft -1640ft -820ft -328ft/-100m 0

820ft   1640ft   3281ft   6562ft   9843ft 13,124ft 19,685ft

Eastern Canada

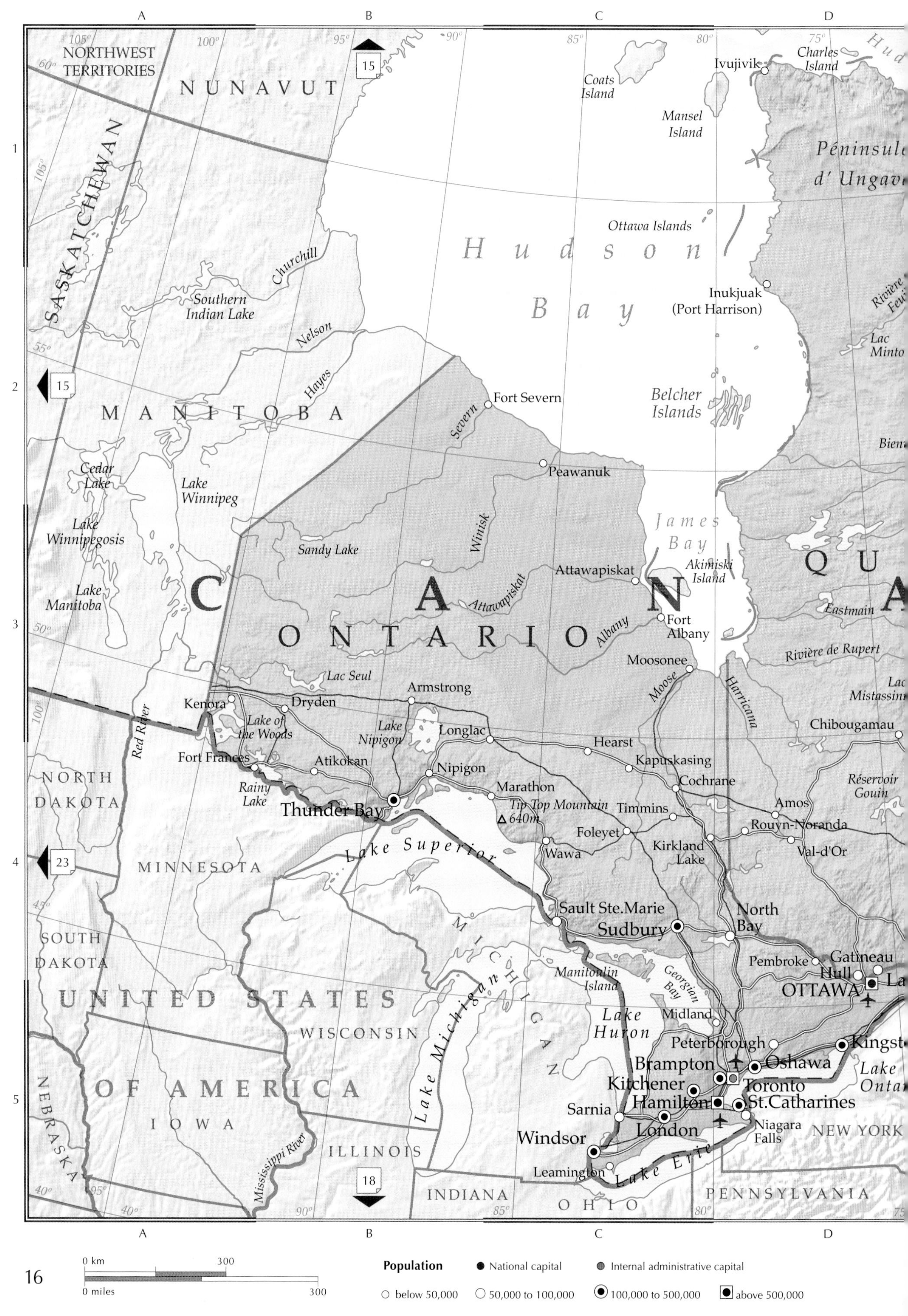

NORTHWEST TERRITORIES

NUNAVUT

SASKATCHEWAN

MANITOBA

Churchill

Southern Indian Lake

Nelson

Hayes

Cedar Lake

Lake Winnipeg

Lake Winnipegosis

Lake Manitoba

Sandy Lake

Hudson Bay

Coats Island

Mansel Island

Ivujivik

Charles Island

Péninsule d'Ungave

Ottawa Islands

Inukjuak (Port Harrison)

Rivière Feui

Lac Minto

Fort Severn

Belcher Islands

Peawanuk

James Bay

Bien

Severn

Winisk

Akimiski Island

Attawapiskat

Attawapiskat

Q U

Fort Albany

Albany

Eastmain

C   A   N

O N T A R I O

Moosonee

Moose

Rivière de Rupert

Lac Mistassin

Chibougamau

Réservoir Gouin

Lac Seul

Armstrong

Kenora

Dryden

Lake of the Woods

Lake Nipigon

Longlac

Hearst

Kapuskasing

Cochrane

Harricana

Red River

Fort Frances

Atikokan

Nipigon

Marathon

Tip Top Mountain ▲640m

Timmins

Amos

Rouyn-Noranda

NORTH DAKOTA

Rainy Lake

Thunder Bay

Lake Superior

Foleyet

Wawa

Kirkland Lake

Val-d'Or

MINNESOTA

Sault Ste.Marie

Sudbury

North Bay

SOUTH DAKOTA

M I C H I G A N

Manitoulin Island

Georgian Bay

Pembroke

Gatineau

Hull

OTTAWA

La

UNITED   STATES

WISCONSIN

Lake Michigan

Lake Huron

Midland

Peterborough

Kingst

NEBRASKA

OF   AMERICA

IOWA

Brampton

Kitchener

Hamilton

Sarnia

London

Windsor

Oshawa

Toronto

St.Catharines

Niagara Falls

Lake Onta

NEW YORK

ILLINOIS

Leamington

Lake Erie

INDIANA

OHIO

PENNSYLVANIA

0 km   300

0 miles   300

Population   ● National capital   ● Internal administrative capital

○ below 50,000   ○ 50,000 to 100,000   ◉ 100,000 to 500,000   ▣ above 500,000

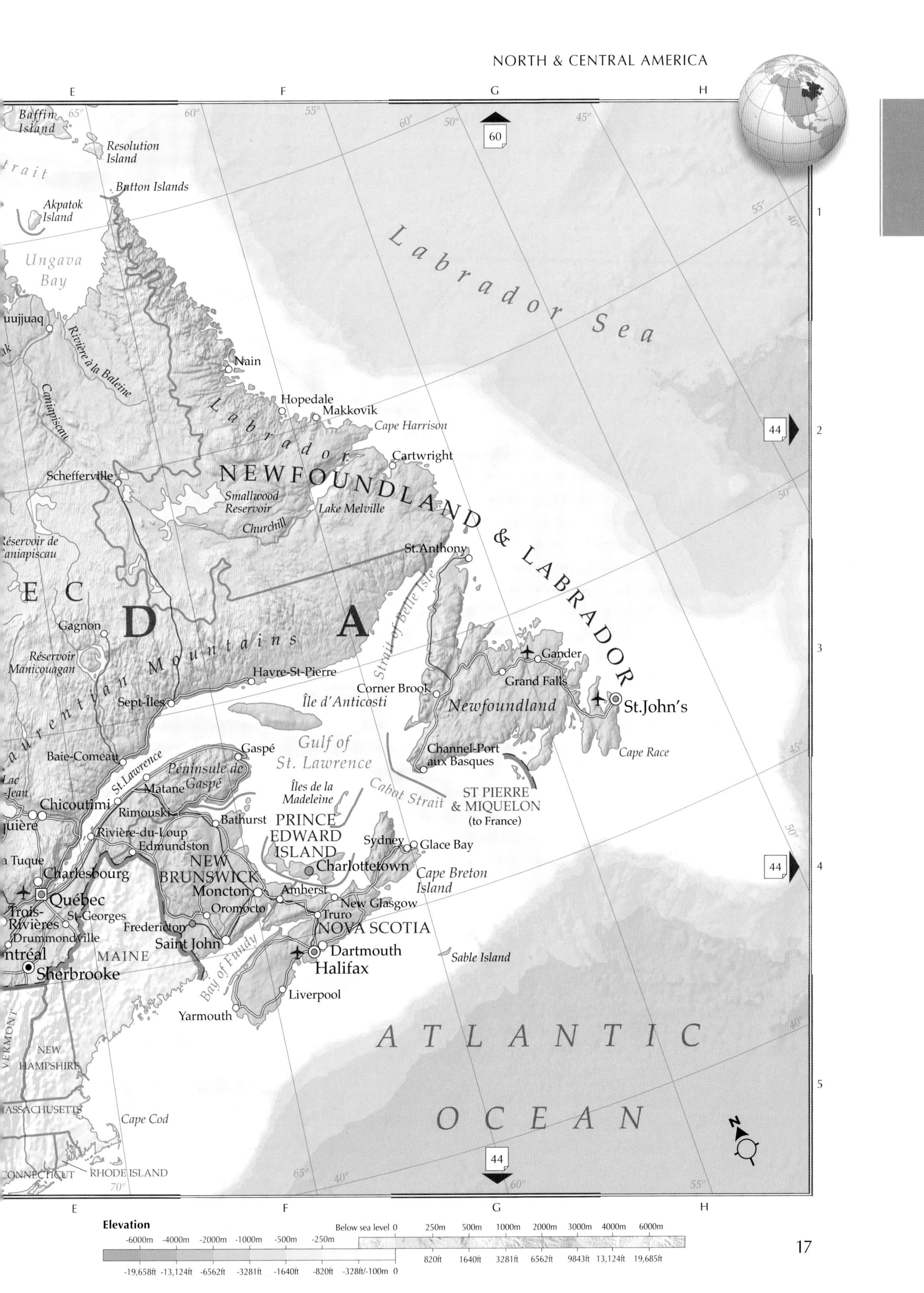

Baffin Island

Resolution Island

Button Islands

Akpatok Island

Ungava Bay

uujjuaq

Rivière à la Baleine

Caniapiscau

Labrador Sea

Nain

Hopedale

Makkovik

Cape Harrison

Cartwright

Scheffervile

N E W F O U N D L A N D

Smallwood Reservoir

Lake Melville

Churchill

St.Anthony

& L A B R A D O R

Réservoir de aniapiscau

E   C

D

A

Gagnon

Réservoir Manicouagan

Strait of Belle Isle

Laurentian Mountains

Havre-St-Pierre

Corner Brook

Newfoundland

Gander

Grand Falls

St.John's

Sept-Îles

Île d'Anticosti

Cape Race

Baie-Comeau

St.Lawrence

Gulf of St. Lawrence

Channel-Port aux Basques

Lac Jean

Gaspé

Îles de la Madeleine

Cabot Strait

ST PIERRE & MIQUELON (to France)

Chicoutimi

Matane

Péninsule de Gaspé

uière

Rimouski

Rivière-du-Loup

Bathurst

PRINCE EDWARD ISLAND

Sydney

Glace Bay

Edmundston

Charlottetown

a Tuque

Charlesbourg

NEW BRUNSWICK

Cape Breton Island

Québec

Moncton

Oromocto

Amherst

New Glasgow

Trois-Rivières

St-Georges

Truro

NOVA SCOTIA

Drummondville

Fredericton

Sable Island

ntréal

MAINE

Saint John

Bay of Fundy

Dartmouth

Sherbrooke

Halifax

Liverpool

Yarmouth

VERMONT

NEW HAMPSHIRE

A T L A N T I C

ASSACHUSETTS

Cape Cod

O C E A N

CONNECTICUT

RHODE ISLAND

E   F   G   H

60

44

44

44

1   2   3   4   5

65° 60° 55° 60° 50° 45° 40°

55°

50°

45°

50°

40°

55°

70° 65° 60°

N

Elevation

-6000m -4000m -2000m -1000m -500m -250m   Below sea level 0   250m 500m 1000m 2000m 3000m 4000m 6000m

-19,658ft -13,124ft -6562ft -3281ft -1640ft -820ft -328ft/-100m 0   820ft 1640ft 3281ft 6562ft 9843ft 13,124ft 19,685ft

17

USA: The Northeast

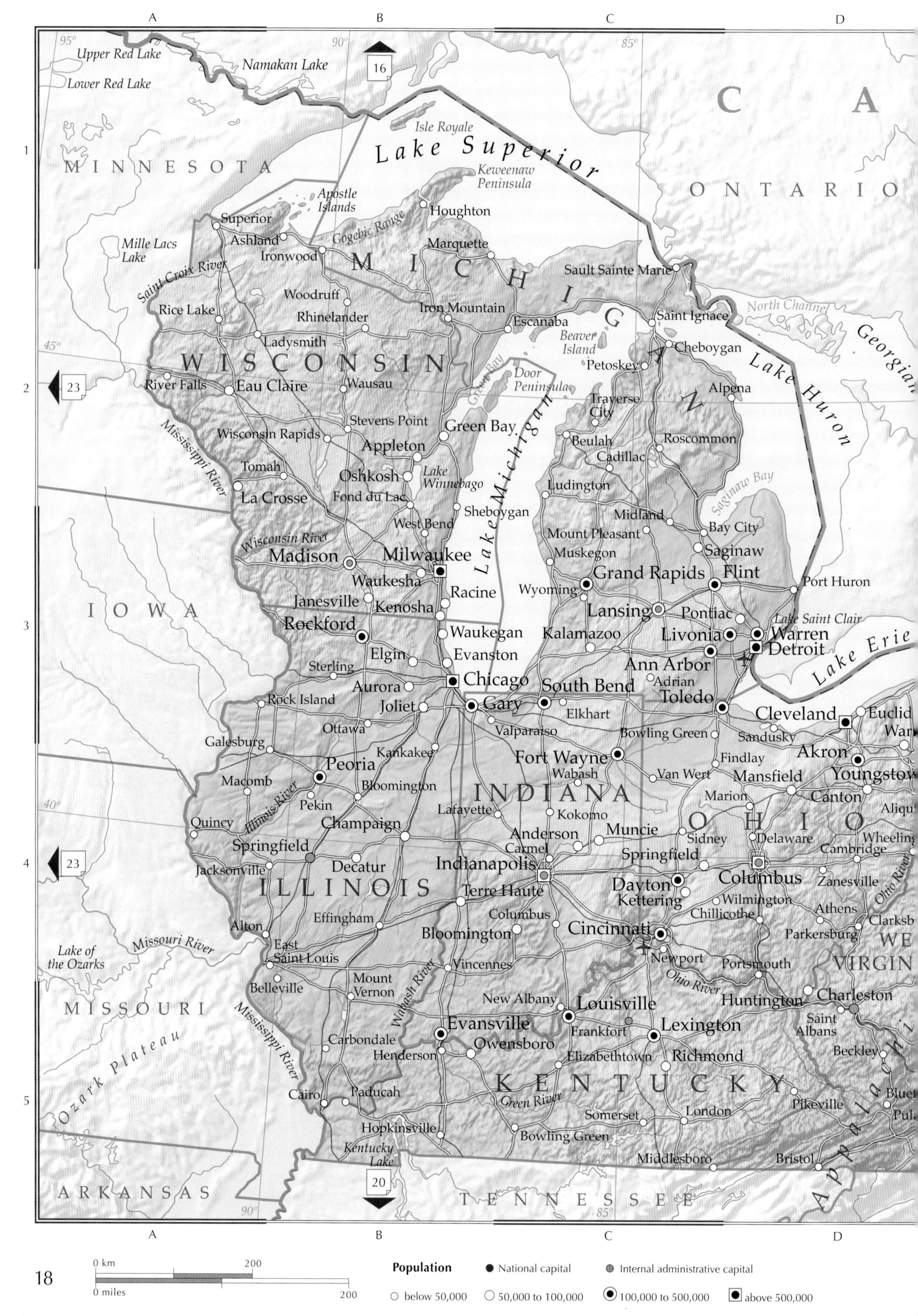

0 km   200
0 miles   200

Population   ● National capital   ● Internal administrative capital
○ below 50,000   ○ 50,000 to 100,000   ◉ 100,000 to 500,000   ■ above 500,000

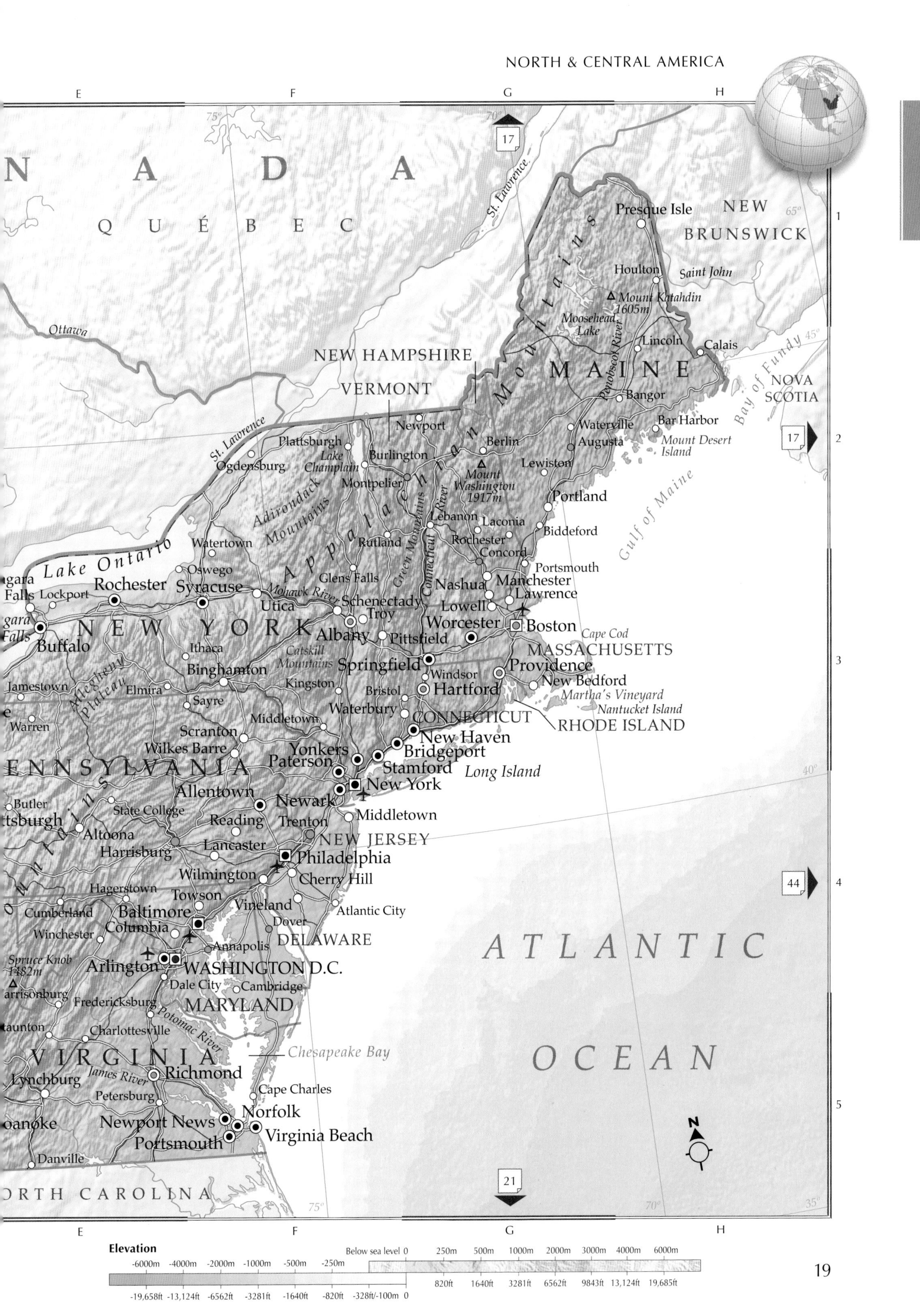

E   F   G   H

17

1

N A D A

Q U É B E C

Ottawa

NEW BRUNSWICK

Presque Isle

Houlton   *Saint John*

△ *Mount Katahdin*
1605m

Moosehead Lake

Lincoln   Calais

NOVA SCOTIA

Bay of Fundy

65°

45°

17

2

St. Lawrence

NEW HAMPSHIRE

VERMONT

Newport

Plattsburgh

Lake Champlain

Burlington

Ogdensburg

Montpelier

St. Lawrence

Adirondack Mountains

Berlin

Mount Washington 1917m

Lebanon

Laconia

Rochester

Concord

Portsmouth

Waterville

Augusta

Mount Desert Island

Bangor

Portland

Biddeford

Gulf of Maine

M A I N E

Appalachian Mountains

Green Mountains

Connecticut River

Watertown

Oswego

Rutland

Glens Falls

Nashua   Manchester

Lawrence

Niagara Falls   Lockport

Rochester   Syracuse

Mohawk River

Utica

Schenectady   Troy

Albany

Pittsfield

Lowell

Worcester

Boston

Cape Cod

3

Niagara Falls

Buffalo

N E W   Y O R K

Ithaca

Binghamton

Catskill Mountains

Kingston

Springfield

Windsor

Bristol

Hartford

MASSACHUSETTS

Providence

New Bedford

Martha's Vineyard

Nantucket Island

Jamestown

Allegheny Plateau

Elmira

Sayre

Waterbury

CONNECTICUT

New Haven

RHODE ISLAND

Warren

Scranton

Middletown

Bridgeport

E N N S Y L V A N I A

Wilkes Barre

Paterson

Yonkers

Stamford

New York

Long Island

40°

Butler

State College

Allentown

Newark

Middletown

tsburgh   Altoona

Reading

Trenton

Harrisburg

Lancaster

NEW JERSEY

44

4

Hagerstown

Wilmington

Philadelphia

Cherry Hill

Cumberland

Towson

Vineland

Atlantic City

Baltimore

Columbia

Dover

Winchester

Annapolis

DELAWARE

Spruce Knob 1482m △

Arlington

WASHINGTON D.C.

Dale City   Cambridge

arrisonburg

Fredericksburg

MARYLAND

Potomac River

A T L A N T I C

taunton

Charlottesville

Chesapeake Bay

V I R G I N I A

O C E A N

Lynchburg

James River

Richmond

Cape Charles

oanoke

Petersburg

Newport News

Norfolk

Portsmouth

Virginia Beach

N

Danville

5

RTH C A R O L I N A

21

75°   70°   35°

E   F   G   H

Elevation

| | | | | | | | Below sea level 0 | 250m | 500m | 1000m | 2000m | 3000m | 4000m | 6000m |

-6000m   -4000m   -2000m   -1000m   -500m   -250m

820ft   1640ft   3281ft   6562ft   9843ft   13,124ft   19,685ft

-19,658ft   -13,124ft   -6562ft   -3281ft   -1640ft   -820ft   -328ft/-100m   0

USA: The Southeast

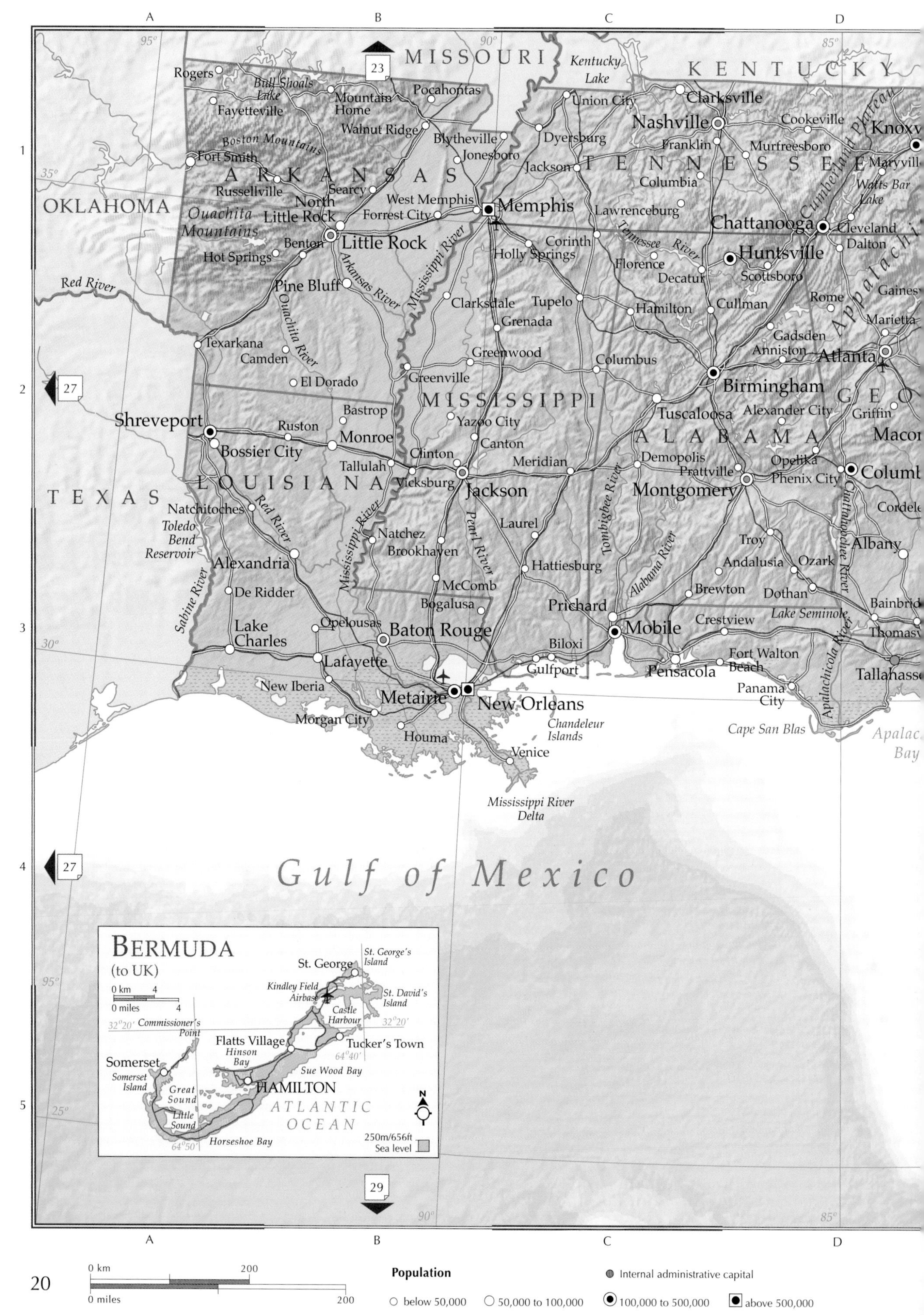

MISSOURI

KENTUCKY

Rogers
Fayetteville
Bull Shoals Lake
Mountain Home
Walnut Ridge
Pocahontas
Kentucky Lake
Union City
Clarksville
Nashville
Cookeville
Knoxv

Boston Mountains
Blytheville
Jonesboro
Dyersburg
Franklin
Murfreesboro
Maryville

Fort Smith
ARKANSAS
Jackson
TENNESSEE
Columbia
Watts Bar Lake

OKLAHOMA
Russellville
Searcy
West Memphis
Memphis
Lawrenceburg
Chattanooga
Cleveland

North Little Rock
Forrest City
Corinth
Holly Springs
Tennessee River
Huntsville
Dalton

Ouachita Mountains
Little Rock
Benton
Clarksdale
Florence
Decatur
Scottsboro
Rome
Gaines

Hot Springs
Tupelo
Hamilton
Cullman
Gadsden
Marietta

Pine Bluff
Grenada
Columbus
Anniston
Atlanta

Texarkana
Greenwood
Birmingham
GEO

Camden
Ouachita River
Greenville
MISSISSIPPI
Tuscaloosa
Alexander City
Griffin

El Dorado
Bastrop
Yazoo City
ALABAMA
Demopolis
Opelika
Macon

Shreveport
Ruston
Monroe
Canton
Meridian
Prattville
Phenix City
Columb

Bossier City
Clinton
LOUISIANA
Tallulah
Vicksburg
Jackson
Laurel
Montgomery
Cordele

TEXAS
Natchitoches
Red River
Natchez
Brookhaven
Hattiesburg
Troy
Andalusia
Ozark
Albany

Toledo Bend Reservoir
Alexandria
McComb
Tombigbee River
Brewton
Dothan
Bainbrid

De Ridder
Bogalusa
Prichard
Crestview
Lake Seminole
Thomas

Sabine River
Lake Charles
Opelousas
Baton Rouge
Biloxi
Mobile
Pensacola
Fort Walton Beach

Lafayette
Gulfport
Panama City
Tallahasse

New Iberia
Metairie
New Orleans
Cape San Blas
Apalac Bay

Morgan City
Houma
Chandeleur Islands

Venice

Mississippi River Delta

Gulf of Mexico

BERMUDA
(to UK)

0 km 4
0 miles 4

Commissioner's Point
St. George's Island
St. George
Kindley Field Airbase
St. David's Island
Castle Harbour
Flatts Village
Tucker's Town
Hinson Bay
Sue Wood Bay
Somerset
Somerset Island
Great Sound
HAMILTON
Little Sound
Horseshoe Bay

ATLANTIC OCEAN

250m/656ft Sea level

0 km 200
0 miles 200

Population

○ below 50,000   ○ 50,000 to 100,000   ◉ 100,000 to 500,000   ● above 500,000

● Internal administrative capital

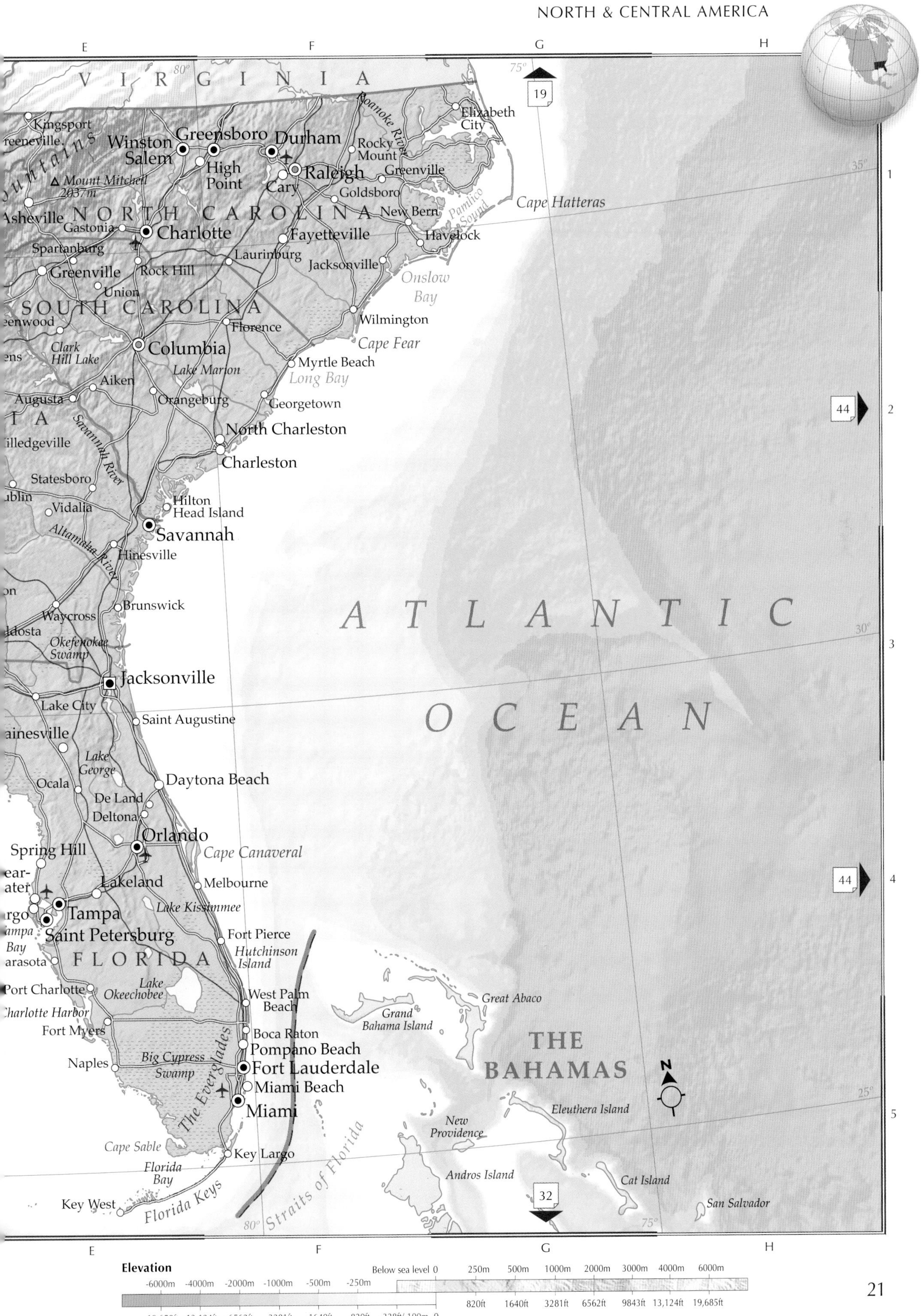

E   **F**   **G**   **H**

19

VIRGINIA

Kingsport
reenville
Winston
Salem
Greensboro
Durham
Rocky
Mount
Elizabeth
City
High
Point
Cary
Raleigh
Greenville
Mount Mitchell
2037m
NORTH CAROLINA
Goldsboro
Asheville
New Bern
Cape Hatteras
Gastonia
Charlotte
Fayetteville
Havelock
Spartanburg
Rock Hill
Laurinburg
Jacksonville
Greenville
Union
Onslow
Bay
SOUTH CAROLINA
eenwood
Florence
Wilmington
Cape Fear
ens
Clark
Hill Lake
Columbia
Myrtle Beach
Long Bay
Augusta
Aiken
Lake Marion
Orangeburg
Georgetown   **44**
illedgeville
North Charleston
TA
Statesboro
Charleston
Vidalia
Hilton
Head Island
ublin
Savannah

Altamaha River   Hinesville

Brunswick

Waycross
Okefenokee
Swamp
ldosta

ATLANTIC   **30°**

Jacksonville
Lake City
Saint Augustine
ainesville

Lake
George
Ocala

OCEAN

De Land
Deltona
Daytona Beach

Orlando
Cape Canaveral
Spring Hill
ear-
ater
Lakeland
Melbourne   **44**
rgo
Tampa
Lake Kissimmee
Saint Petersburg
ampa
Bay
arasota
FLORIDA
Fort Pierce
Hutchinson
Island
Port Charlotte
Lake
Okeechobee
West Palm
Beach
Great Abaco
Charlotte Harbor
Fort Myers
Boca Raton
Grand
Bahama Island
Big Cypress
Swamp
Pompano Beach
Naples
Fort Lauderdale
THE
BAHAMAS
Miami Beach
Miami
Eleuthera Island   **25°**
Cape Sable
Key Largo
New
Providence
Florida
Bay
Andros Island
Cat Island
Key West
Florida Keys   Straits of Florida   **32**   San Salvador

80°   **75°**

E   **F**   **G**   **H**

Elevation

							Below sea level 0	250m	500m	1000m	2000m	3000m	4000m	6000m
-6000m	-4000m	-2000m	-1000m	-500m	-250m									

820ft   1640ft   3281ft   6562ft   9843ft   13,124ft   19,685ft

-19,658ft   -13,124ft   -6562ft   -3281ft   -1640ft   -820ft   -328ft/-100m   0

USA: Central States

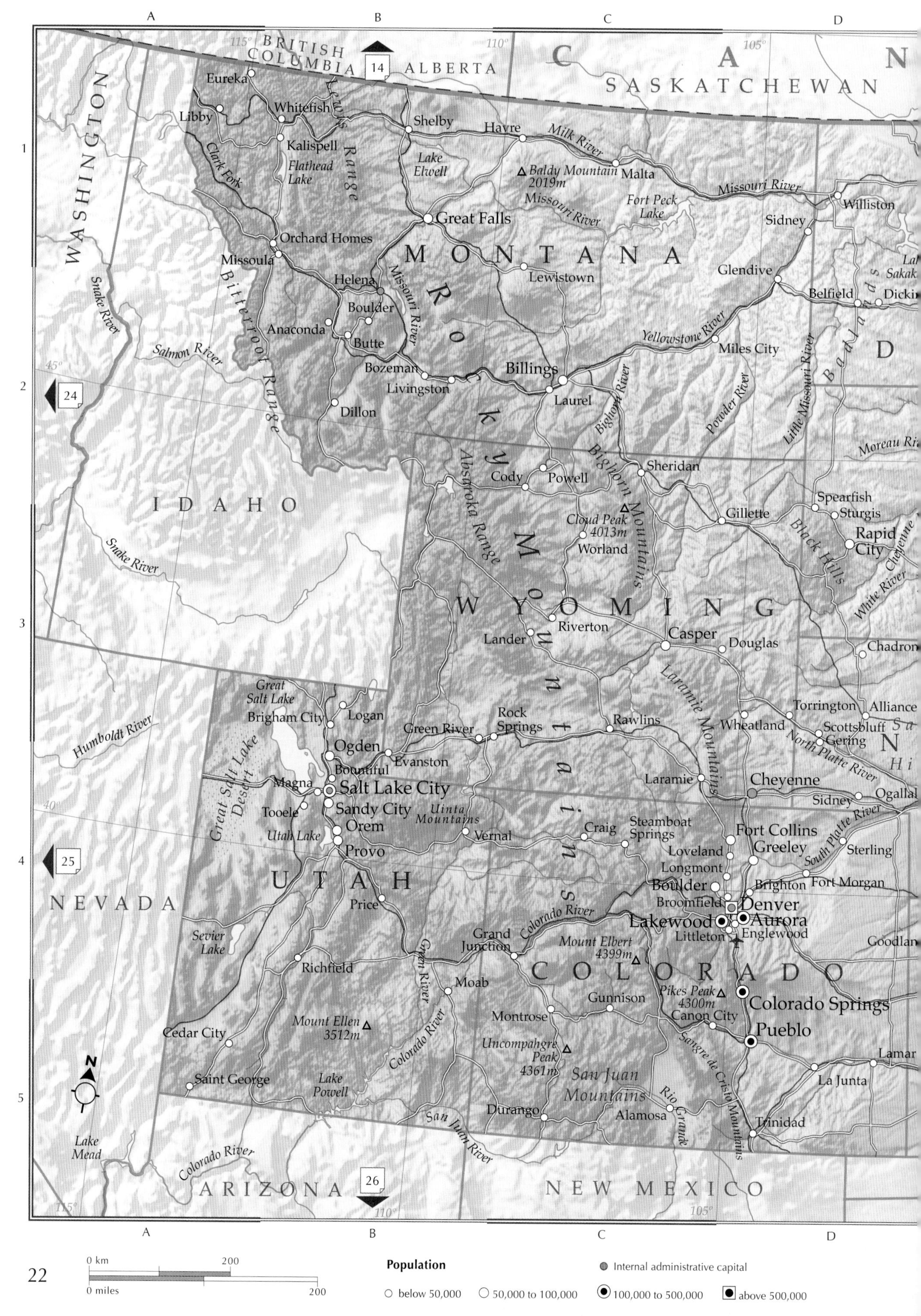

0 km   200

0 miles   200

Population

○ below 50,000   ○ 50,000 to 100,000   ◉ 100,000 to 500,000   ◼ above 500,000

● Internal administrative capital

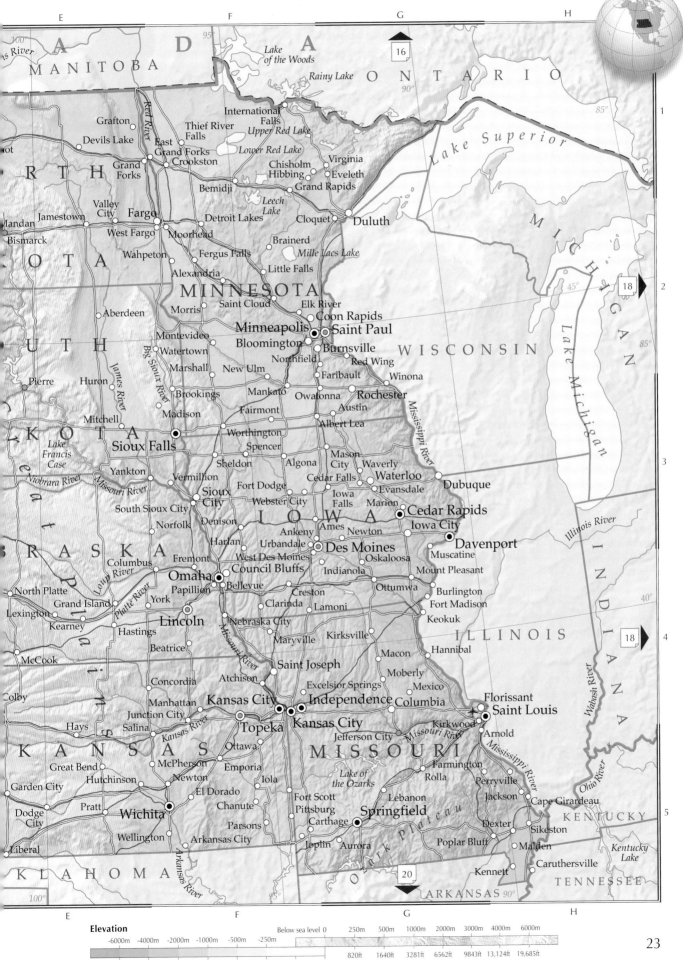

Elevation

| | | | | | Below sea level 0 | 250m | 500m | 1000m | 2000m | 3000m | 4000m | 6000m |
| -6000m | -4000m | -2000m | -1000m | -500m | -250m | | | | | | | |

820ft   1640ft   3281ft   6562ft   9843ft   13,124ft   19,685ft

-19,658ft   -13,124ft   -6562ft   -3281ft   -1640ft   -820ft   -328ft/-100m   0

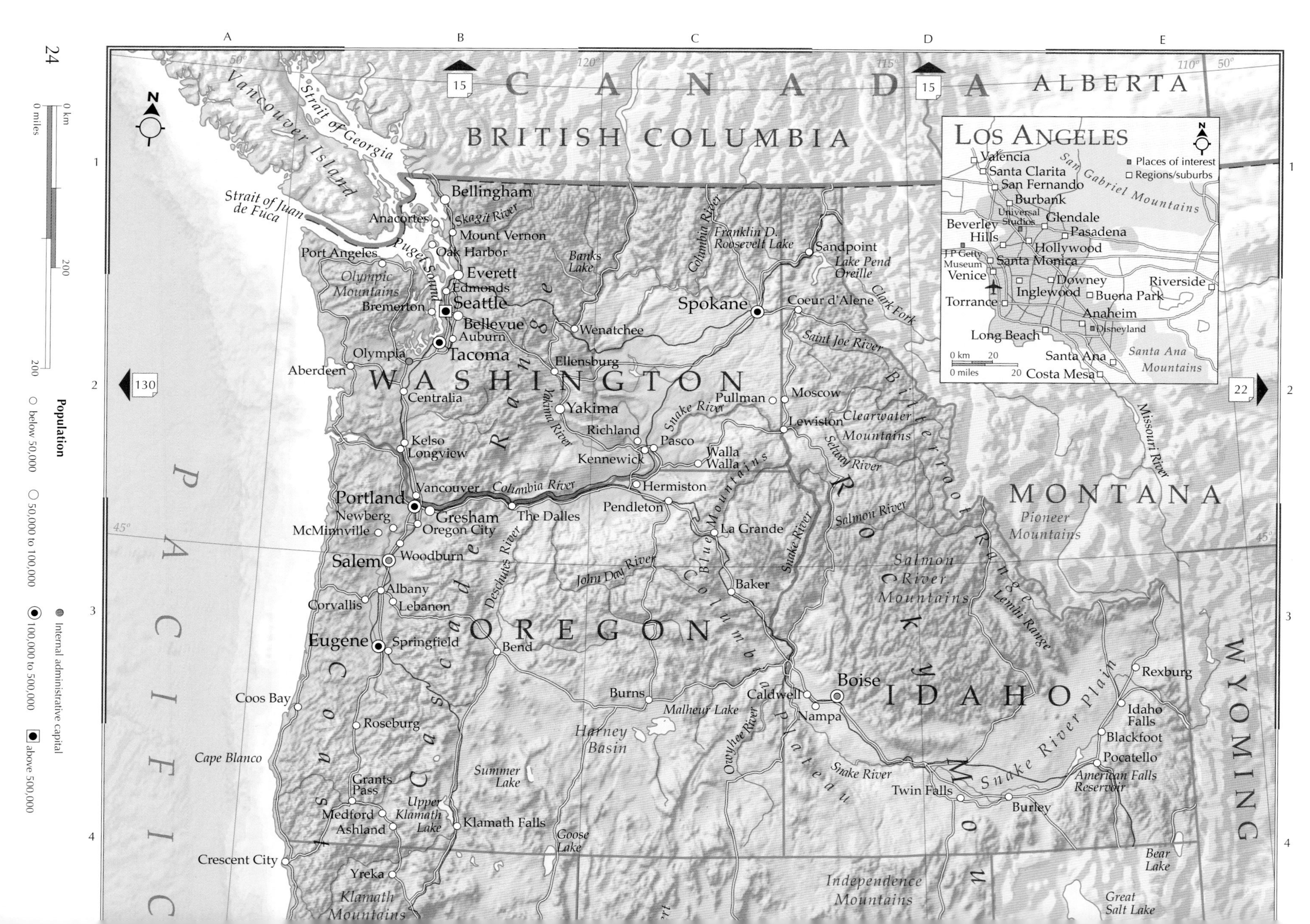

LOS ANGELES

- ■ Places of interest
- □ Regions/suburbs

Valencia
Santa Clarita
San Fernando
San Gabriel Mountains
Glendale
Pasadena
Burbank
Hollywood
Universal Studios
Beverley Hills
Santa Monica
J P Getty Museum
Venice
Torrance
Inglewood
Downey
Buena Park
Anaheim
Disneyland
Santa Ana
Santa Ana Mountains
Long Beach
Costa Mesa

0 km   20
0 miles   20

WYOMING

MONTANA

IDAHO

ROCKY Mountains

Bitterroot Mountains

Pioneer Mountains

Lemhi Range

Salmon River Mountains

Independence Mountains

CANADA

ALBERTA

BRITISH COLUMBIA

Missouri River

Rexburg
Idaho Falls
Blackfoot
Pocatello
American Falls Reservoir
Burley
Bear Lake
Great Salt Lake

Snake River Plain

Boise
Nampa
Caldwell
Twin Falls
Snake River

Sandpoint
Lake Pend Oreille
Clark Fork
Coeur d'Alene
Saint Joe River
Franklin D. Roosevelt Lake
Columbia River
Spokane

Moscow
Pullman
Lewiston
Clearwater Mountains
Selway River
Salmon River
La Grande
Baker
Owyhee River

Walla Walla
Pasco
Hermiston
Blue Mountains
Pendleton
John Day River

WASHINGTON

OREGON

Columbia Plateau

Wenatchee
Ellensburg
Yakima
Yakima River
Richland
Kennewick

Banks Lake

Bellingham
Skagit River
Mount Vernon
Oak Harbor
Everett
Edmonds
Seattle
Bellevue
Auburn
Tacoma
Centralia
Kelso
Longview
Vancouver
Gresham
Oregon City
The Dalles
Deschutes River
Bend

Anacortes
Puget Sound
Port Angeles
Bremerton
Olympia
Olympic Mountains
Aberdeen

Strait of Georgia
Vancouver Island
Strait of Juan de Fuca

Portland
Newberg
McMinnville
Salem
Woodburn
Albany
Lebanon
Corvallis
Eugene
Springfield

Coast Ranges

Roseburg
Grants Pass
Medford
Ashland
Yreka
Crescent City
Klamath Mountains

Burns
Harney Basin

Malheur Lake
Summer Lake
Upper Klamath Lake
Klamath Falls
Goose Lake

Coos Bay
Cape Blanco

PACIFIC

0 km   200
0 miles   200

Population

○ below 50,000   ○ 50,000 to 100,000   ◉ 100,000 to 500,000   ■ above 500,000

● Internal administrative capital

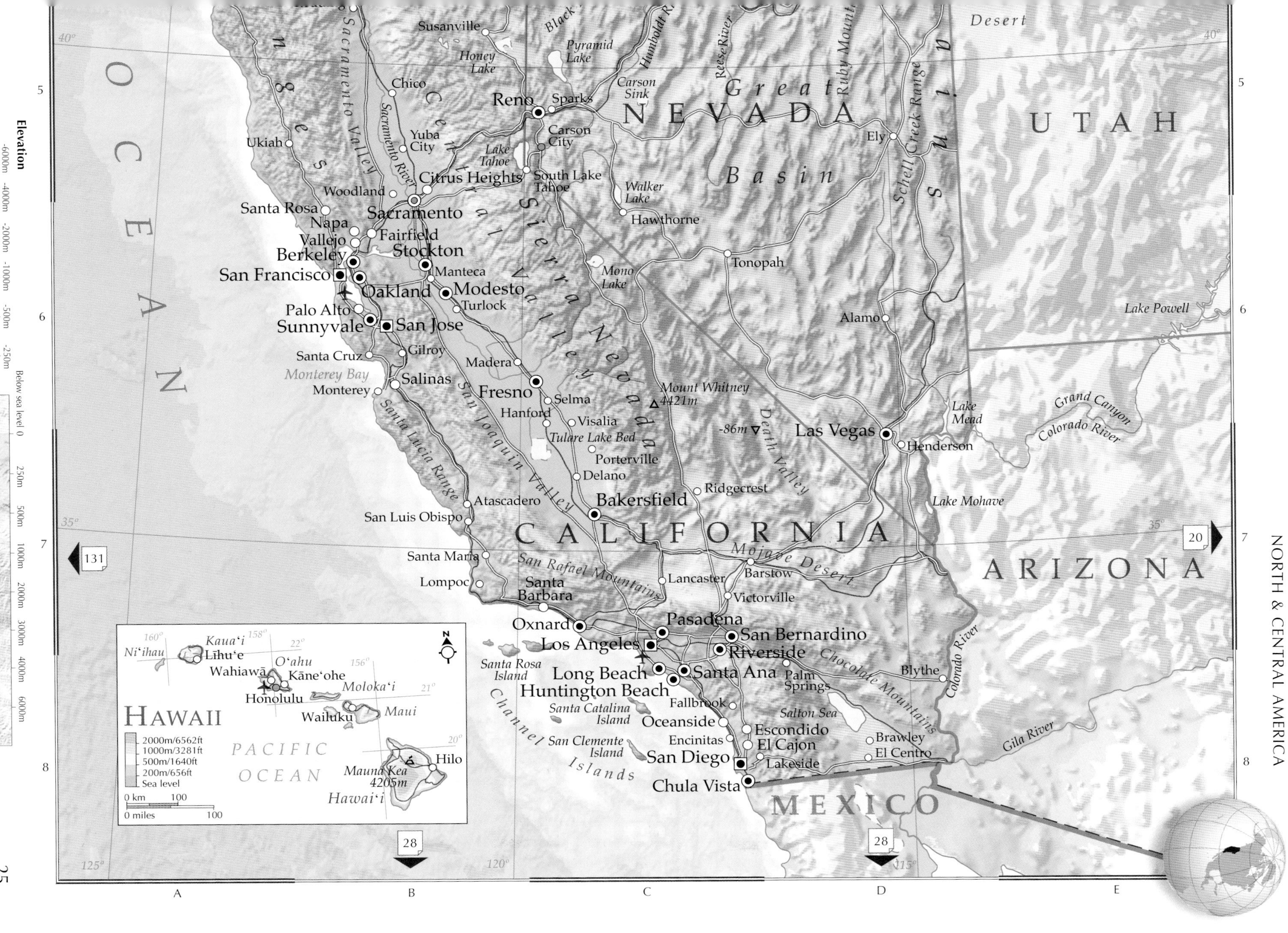

20

131

28

28

UTAH

Desert

NEVADA

Great Basin

ains

Schell Creek Range

Ruby Mountains

Ely

Reese River

Humboldt R.

Black R.

Pyramid Lake

Honey Lake

Susanville

Sparks

Reno

Carson Sink

Carson City

Walker Lake

Hawthorne

Tonopah

Alamo

Lake Powell

Grand Canyon

Colorado River

ARIZONA

Lake Mead

Lake Mohave

Henderson

Las Vegas

Death Valley

-86m

Mojave Desert

Barstow

Victorville

Mount Whitney
4421m

Ridgecrest

Lancaster

Colorado River

Gila River

MEXICO

Blythe

Chocolate Mountains

Brawley

El Centro

Salton Sea

Palm Springs

San Bernardino

Riverside

Escondido

El Cajon

Lakeside

San Diego

Chula Vista

Santa Ana

Fallbrook

Oceanside

Encinitas

Pasadena

Los Angeles

Long Beach

Huntington Beach

Santa Catalina Island

San Clemente Island

Channel Islands

CALIFORNIA

San Rafael Mountains

Oxnard

Santa Barbara

Santa Rosa Island

Bakersfield

Porterville

Delano

Tulare Lake Bed

Visalia

Selma

Hanford

Madera

Fresno

San Joaquin Valley

Atascadero

Santa Lucia Range

Santa Maria

Lompoc

San Luis Obispo

Sierra Nevada

Mono Lake

Central Valley

Modesto

Turlock

Manteca

Stockton

San Jose

Sunnyvale

Palo Alto

Gilroy

Salinas

Santa Cruz

Monterey

Monterey Bay

South Lake Tahoe

Lake Tahoe

Citrus Heights

Sacramento

Fairfield

Oakland

Berkeley

San Francisco

Napa

Vallejo

Santa Rosa

Woodland

Yuba City

Sacramento River

Chico

Sacramento Valley

Ukiah

nges

Kauaʻi

Niʻihau

Lihuʻe

Oʻahu

Wahiawā

Kāneʻohe

Honolulu

Molokaʻi

Wailuku

Maui

Mauna Kea
4205m

Hilo

Hawaiʻi

PACIFIC OCEAN

HAWAII

2000m/6562ft
1000m/3281ft
500m/1640ft
200m/656ft
Sea level

0 km   100
0 miles   100

OCEAN

O C E A N

25

Elevation

| | | | | | | | | | Below sea level 0 | 250m | 500m | 1000m | 2000m | 3000m | 4000m | 6000m |

-6000m   -4000m   -2000m   -1000m   -500m   -250m

-19,658ft   -13,124ft   -6562ft   -3281ft   -1640ft   -820ft   -328ft/-100m   0   820ft   1640ft   3281ft   6562ft   9843ft   13,124ft   19,685ft

USA: The Southwest

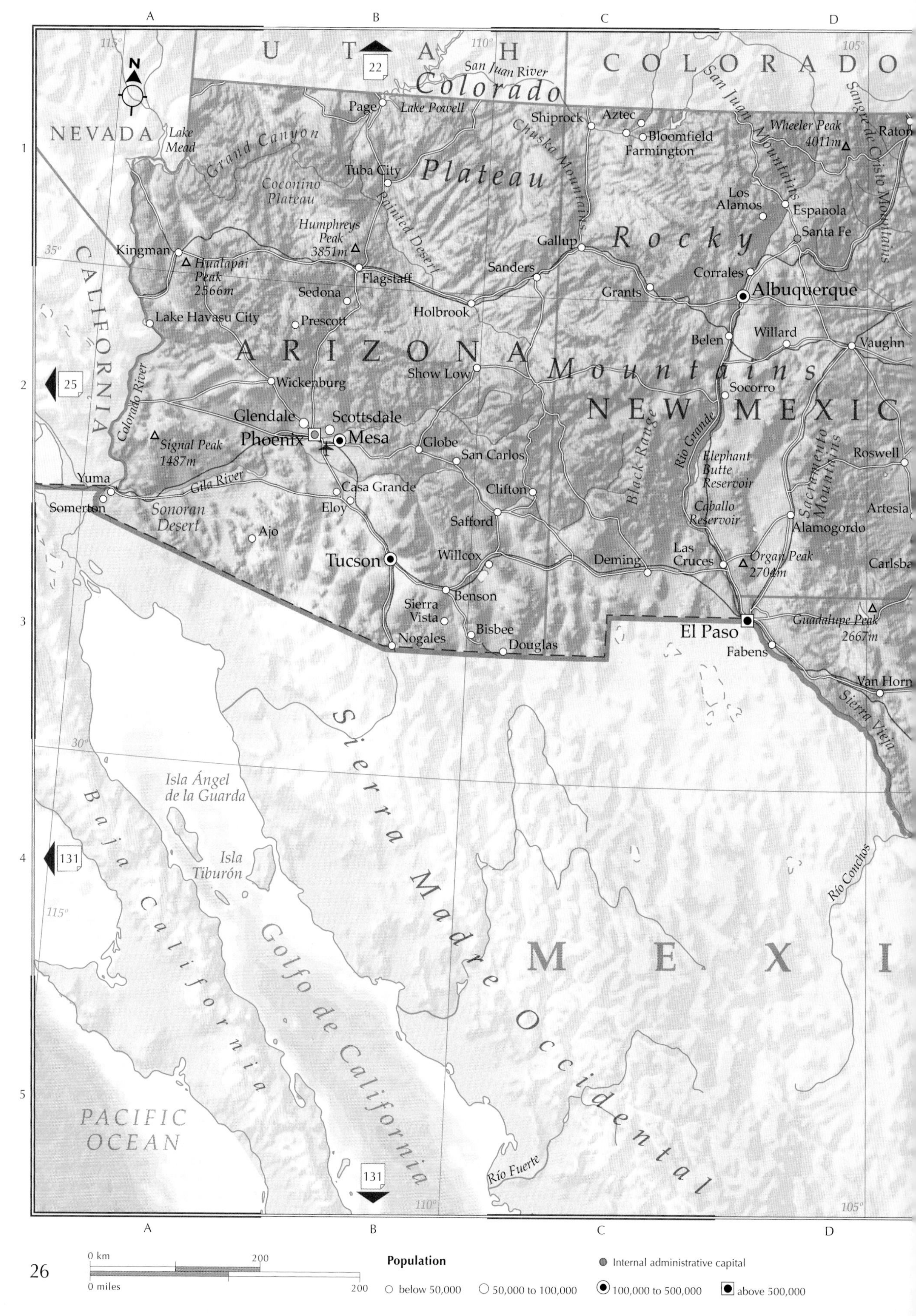

0 km
200

0 miles
200

Population

○ below 50,000   ○ 50,000 to 100,000   ◉ 100,000 to 500,000   ■ above 500,000

● Internal administrative capital

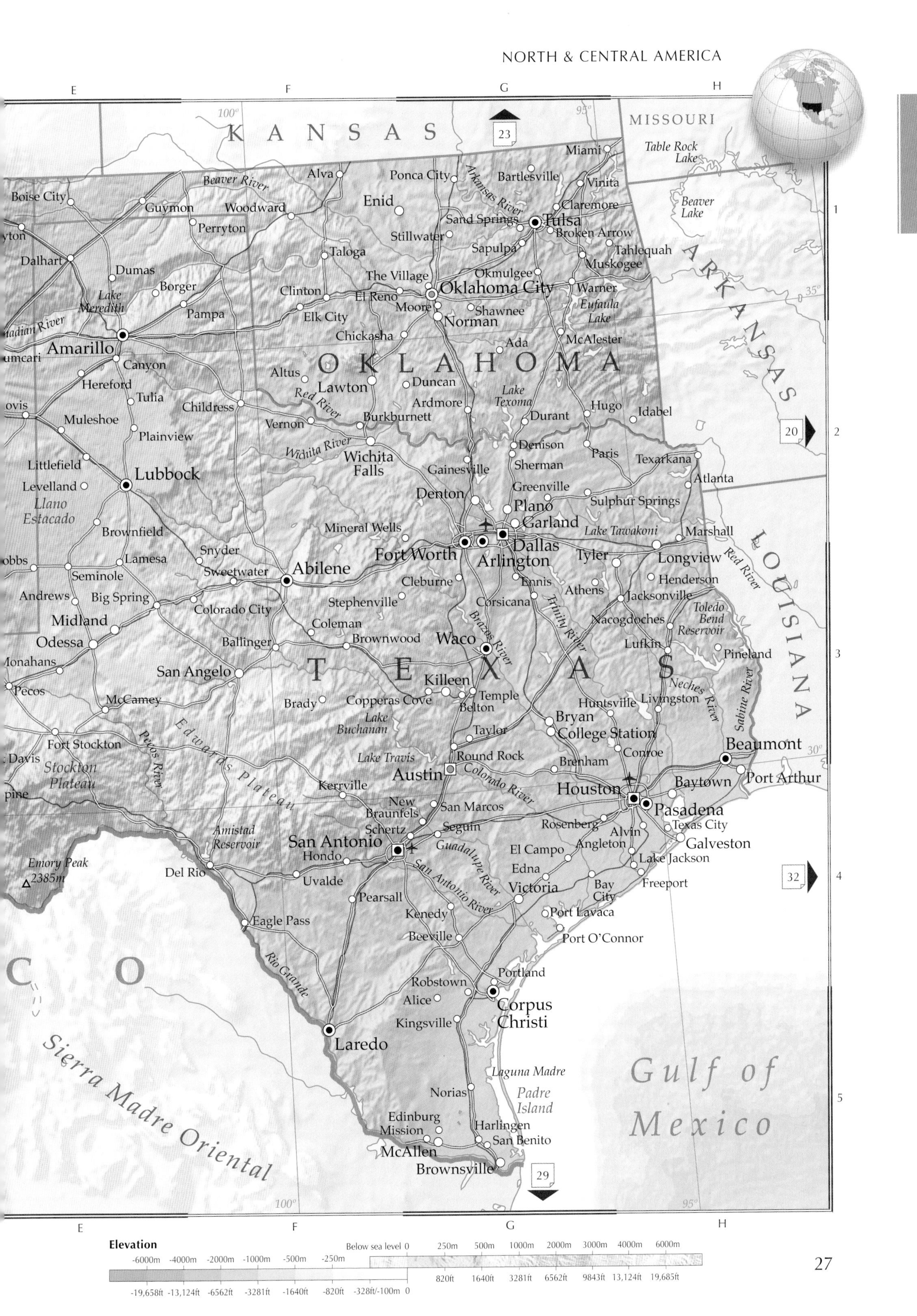

E · F · G · H

MISSOURI

KANSAS

100° · 23 · *95°*

Table Rock
Lake

Boise City · Alva · Ponca City · Bartlesville · Miami · Vinita · Beaver
Lake · 1

Guymon · Woodward · *Beaver River* · Enid · Sand Springs · Claremore
yton · Perryton · Stillwater · Tulsa · Broken Arrow · Tahlequah
Dalhart · Dumas · Taloga · The Village · Okmulgee · Sapulpa · Muskogee · *ARKANSAS*
Borger · Clinton · El Reno · Oklahoma City · Warner · 35° · 20 · 2
*Lake
Meredith* · Pampa · Elk City · Moore · Shawnee · *Eufaula
Lake* · Idabel
adian River · Canyon · Altus · Lawton · Chickasha · Norman · Ada · McAlester
Amarillo · Hereford · Duncan · *Lake
Texoma* · Hugo
umcari · Tulia · Childress · Vernon · Ardmore · Durant · Denison · Paris · Texarkana
ovis · Muleshoe · Plainview · *Red River* · Burkburnett · *LOUISIANA*
Littlefield · *Wichita River* · Wichita
Falls · Gainesville · Sherman · Greenville · Atlanta
Levelland · Lubbock · Denton · Plano · Sulphur Springs · *Red River*
*Llano
Estacado* · Mineral Wells · Garland · *Lake Tawakoni* · Marshall
obbs · Brownfield · Fort Worth · Dallas · Tyler · Longview
Snyder · Arlington · Ennis · Henderson
Lamesa · Sweetwater · Abilene · Cleburne · Athens · Jacksonville
Seminole · Stephenville · Corsicana · Nacogdoches · *Toledo
Bend
Reservoir*
Andrews · Big Spring · Colorado City · Coleman · *Brazos River* · Lufkin · Pineland · 3
Midland · Ballinger · Brownwood · Waco · *Trinity River* · *Sabine River* · *Neches River*
Odessa · T · E · X · A · S
onahans · San Angelo · Brady · Killeen · Temple · Huntsville · Livingston
Pecos · McCamey · Copperas Cove · Belton · Bryan · Beaumont · 30°
Davis · Fort Stockton · *Lake
Buchanan* · Taylor · College Station · Port Arthur
pine · *Stockton
Plateau* · *Pecos River* · *Lake Travis* · Round Rock · Brenham · Conroe · Baytown
Edwards Plateau · Austin · *Colorado River* · Houston · Pasadena
Kerrville · New
Braunfels · San Marcos · Rosenberg · Alvin · Texas City · 4
*Amistad
Reservoir* · Schertz · Seguin · El Campo · Angleton · Galveston · 32
Emory Peak · San Antonio · *Guadalupe River* · Edna · Bay
City · Lake Jackson
△2385m · Hondo · Victoria · Freeport
Del Rio · Uvalde · *San Antonio River* · Port Lavaca
Pearsall · Kenedy · Port O'Connor
Rio Grande · Eagle Pass · Beeville
C · O · Portland
Robstown · *Laguna Madre*
Alice · Corpus
Christi · *Padre
Island* · 5
Kingsville
Sierra Madre Oriental · Laredo · Norias · **Gulf of**
Edinburg · **Mexico**
Mission · Harlingen
McAllen · San Benito
Brownsville · 29
100° · *95°*

E · F · G · H

Elevation

-6000m -4000m -2000m -1000m -500m -250m   Below sea level 0   250m 500m 1000m 2000m 3000m 4000m 6000m

-19,658ft -13,124ft -6562ft -3281ft -1640ft -820ft -328ft/-100m 0   820ft 1640ft 3281ft 6562ft 9843ft 13,124ft 19,685ft

Mexico

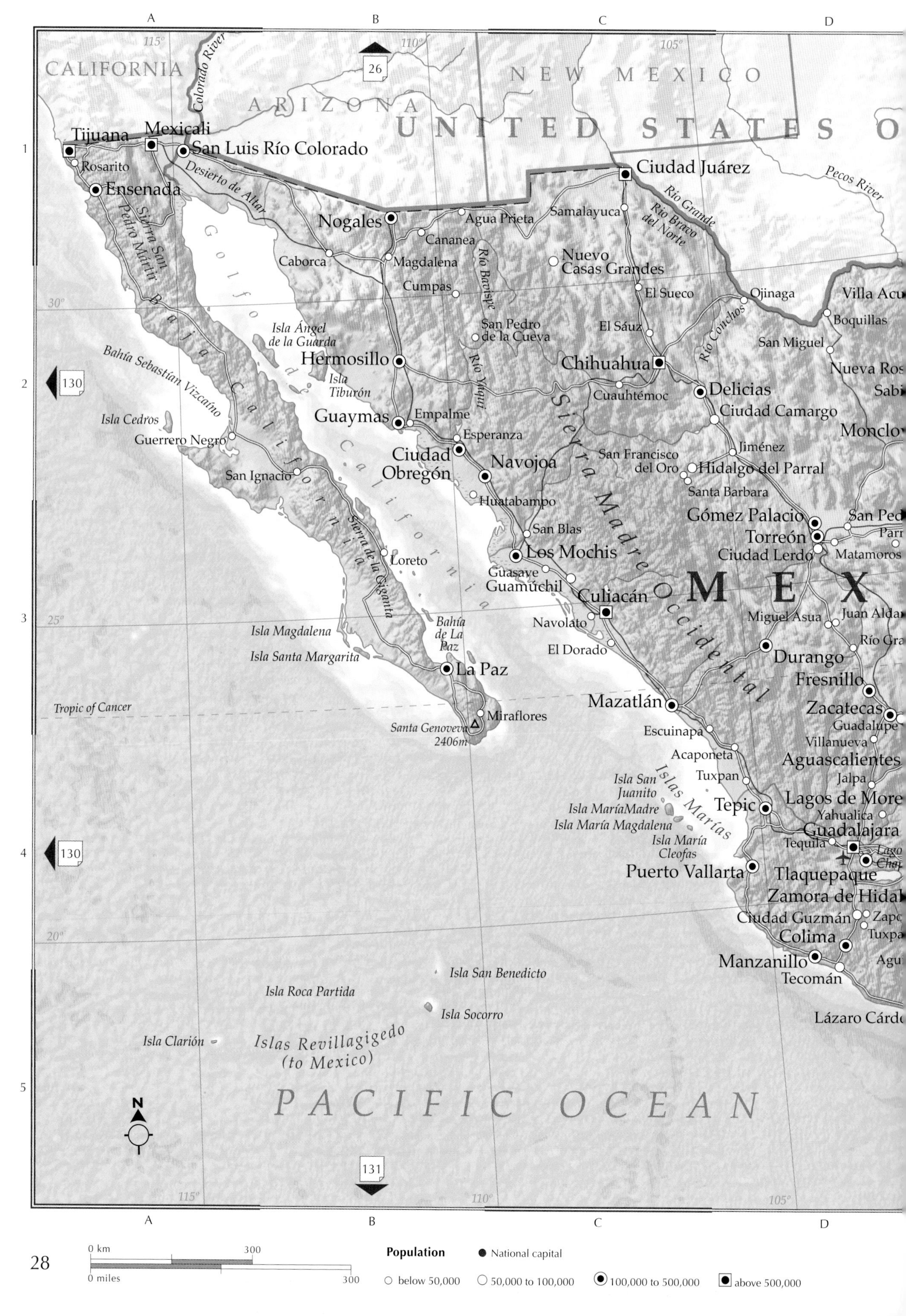

CALIFORNIA

ARIZONA

NEW MEXICO

UNITED STATES O

Colorado River

Pecos River

Tijuana
Rosarito
Ensenada
Mexicali
San Luis Río Colorado
Ciudad Juárez
Río Grande
del Norte
Desierto de Altar
Nogales
Agua Prieta
Samalayuca
Cananea
Caborca
Magdalena
Nuevo
Casas Grandes
Ojinaga
Villa Acu
Cumpas
El Sueco
Boquillas
San Pedro
de la Cueva
El Sáuz
San Miguel
Nueva Ros
Sabi
Isla Ángel
de la Guarda
Río Bavispe
Río Yaqui
Chihuahua
Río Conchos
Monclo
Hermosillo
Isla
Tiburón
Cuauhtémoc
Delicias
Ciudad Camargo
Guaymas
Empalme
San Francisco
del Oro
Jiménez
Esperanza
Navojoa
Hidalgo del Parral
Ciudad
Obregón
Santa Barbara
Huatabampo
Gómez Palacio
San Ped
San Blas
Torreón
Parr
Los Mochis
Ciudad Lerdo
Matamoros
Loreto
Guasave
Guamúchil
Culiacán
M E X
Navolato
Miguel Asua
Juan Alda
Bahía de La
Paz
El Dorado
Río Gra
Durango
La Paz
Fresnillo
Zacatecas
Miraflores
Mazatlán
Guadalupe
Santa Genoveva
2406m
Escuinapa
Villanueva
Aguascalientes
Acaponeta
Jalpa
Tuxpan
Isla San
Juanito
Tepic
Lagos de More
Isla MaríaMadre
Yahualica
Isla María Magdalena
Guadalajara
Isla María
Cleofas
Tequila
Lago
Cha
Puerto Vallarta
Tlaquepaque
Zamora de Hida
Ciudad Guzmán
Zapc
Colima
Tuxpa
Manzanillo
Agu
Tecomán
Isla San Benedicto
Isla Roca Partida
Lázaro Cárde
Isla Socorro
Isla Clarión
Islas Revillagigedo
(to Mexico)

Golfo de California
Baja California
Sierra San Pedro Mártir
Bahía Sebastián Vizcaíno
Isla Cedros
Guerrero Negro
San Ignacio
Sierra de la Giganta
Isla Magdalena
Isla Santa Margarita

Sierra Madre Occidental

Islas Marías

Tropic of Cancer

PACIFIC OCEAN

N

115°
110°
105°

30°
25°
20°

26
130
130
131

0 km   300
0 miles   300

Population   ● National capital

○ below 50,000   ◎ 50,000 to 100,000   ◉ 100,000 to 500,000   ▣ above 500,000

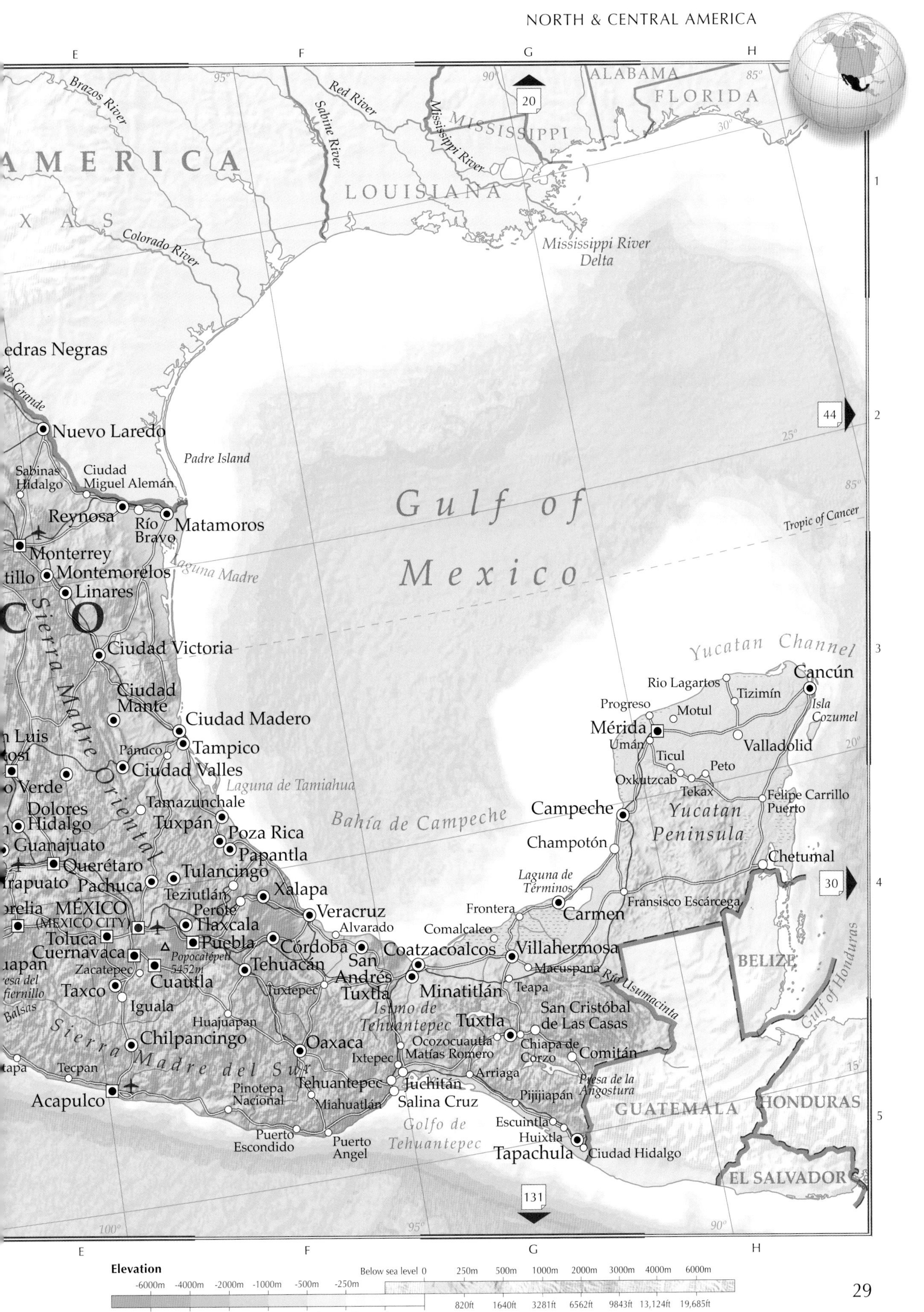

E   F   G   H

Brazos River

95°

Red River

Sabine River

90°

20

ALABAMA

FLORIDA

MISSISSIPPI

Mississippi River

30°

85°

1

A M E R I C A

L O U I S I A N A

T E X A S

Colorado River

Mississippi River Delta

dras Negras

Río Grande

44

2

Nuevo Laredo

Padre Island

G u l f   o f

25°

85°

Sabinas
Hidalgo

Ciudad
Miguel Alemán

M e x i c o

Reynosa

Río
Bravo

Matamoros

Tropic of Cancer

Monterrey

Laguna Madre

illo
Montemorelos

Linares

Sierra Madre Oriental

Yucatan Channel

3

C O

Ciudad Victoria

Rio Lagartos

Cancún

Ciudad
Mante

Progreso

Tizimín

Isla
Cozumel

n Luis
osí

Ciudad Madero

Motul

Mérida

Valladolid

20°

Pánuco
Tampico

Umán

Ticul

Ciudad Valles

Laguna de Tamiahua

Peto

o Verde

Tamazunchale

Bahía de Campeche

Campeche

Oxkutzcab
Tekax

Felipe Carrillo
Puerto

Dolores
Hidalgo

Túxpán

Poza Rica

Champotón

*Yucatan
Peninsula*

Guanajuato

Papantla

Chetumal

Querétaro

Tulancingo

*Laguna de
Términos*

30

4

rapuato
Pachuca

Teziutlán

Xalapa

Fransisco Escárcega

elia
MÉXICO

Perote

Veracruz

Frontera

Carmen

(MEXICO CITY)

Tlaxcala

Alvarado

Comalcalco

BELIZE

Toluca
Cuernavaca

Puebla

Córdoba

Coatzacoalcos

Villahermosa

Gulf of Honduras

uapan

Popocatépetl
5452m

Tehuacán

San
Andrés

Macuspana

Río Usumacinta

sa del
fiernillo

Zacatepec

Cuautla

Tuxtepec

Tuxtla

Minatitlán

Teapa

Taxco

Iguala

Huajuapan

*Istmo de
Tehuantepec*

Tuxtla

San Cristóbal
de Las Casas

Balsas

Sierra Madre del Sur

Chilpancingo

Oaxaca

Ocozocuautla
Matías Romero

Chiapa de
Corzo

Comitán

Ixtepec

Arriaga

15°

apa

Tecpan

Tehuantepec

Juchitán

*Presa de la
Angostura*

HONDURAS

Acapulco

Pinotepa
Nacional

Miahuatlán

Salina Cruz

Pijijiapán

GUATEMALA

5

Puerto
Escondido

Puerto
Angel

*Golfo de
Tehuantepec*

Escuintla

Huixtla

Tapachula

Ciudad Hidalgo

EL SALVADOR

131

100°

95°

90°

E   F   G   H

Elevation

					Below sea level 0	250m	500m	1000m	2000m	3000m	4000m	6000m
-6000m	-4000m	-2000m	-1000m	-500m	-250m							

-19,658ft -13,124ft -6562ft -3281ft -1640ft -820ft -328ft/-100m 0   820ft 1640ft 3281ft 6562ft 9843ft 13,124ft 19,685ft

Central America

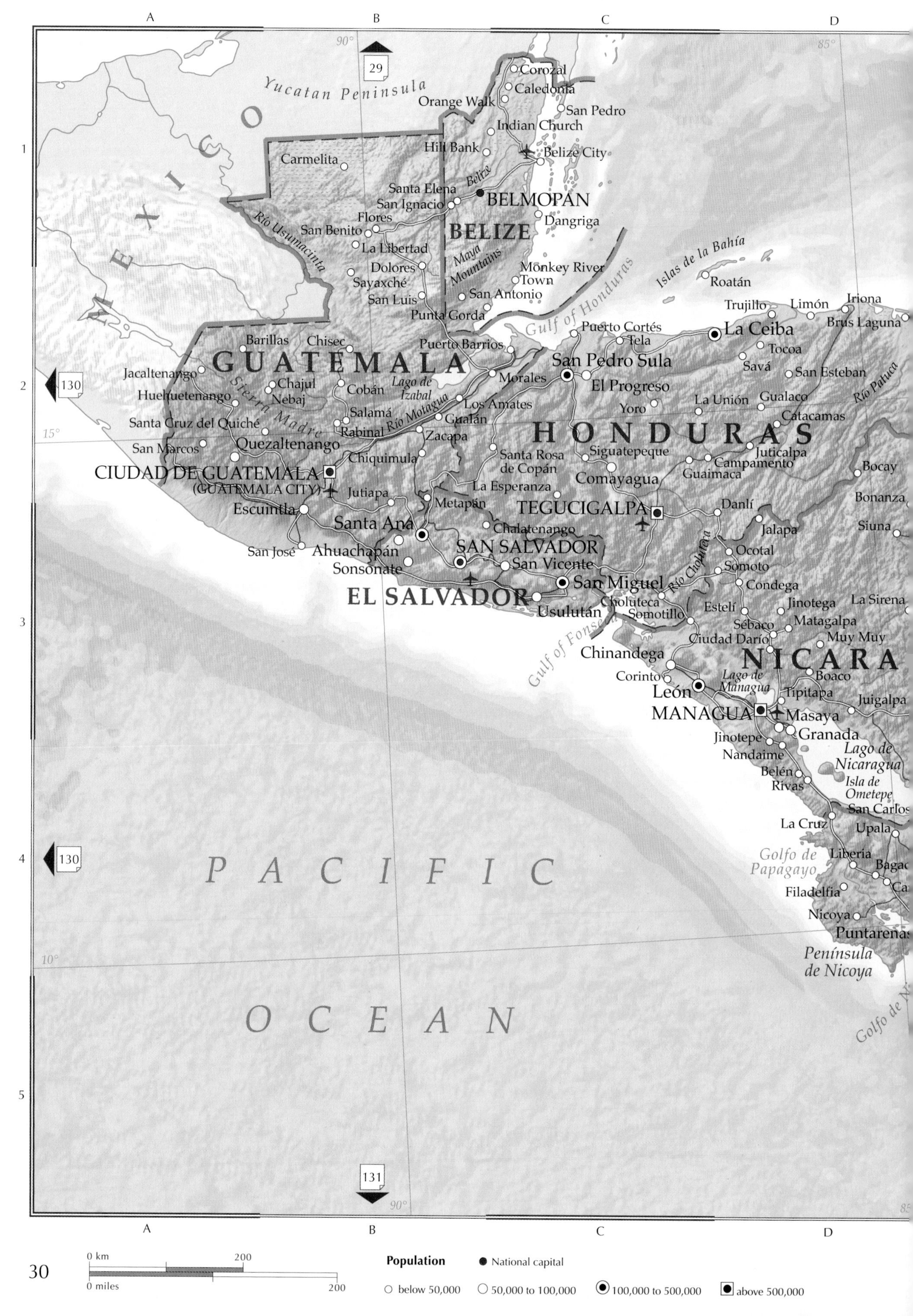

MEXICO

Yucatan Peninsula

Corozal
Caledonia
Orange Walk
San Pedro
Indian Church
Hill Bank
Belize City
Carmelita
Santa Elena
San Ignacio
Flores
San Benito
La Libertad
Dolores
Sayaxché
San Luis
BELMOPAN
Dangriga
BELIZE
Maya Mountains
Monkey River Town
San Antonio
Punta Gorda

Río Usumacinta

Barillas
Chisec
Jacaltenango
Chajul
Nebaj
Cobán
Huehuetenango
Santa Cruz del Quiché
Salamá
San Marcos
Quezaltenango
Rabinal
CIUDAD DE GUATEMALA
(GUATEMALA CITY)
Escuintla
Santa Ana
Ahuachapán
Sonsonate
San José
Jutiapa
Metapán
Chiquimula
GUATEMALA
Lago de Izabal
Río Motagua
Los Amates
Gualán
Zacapa
Sierra Madre

Puerto Barrios
Morales
Santa Rosa de Copán
La Esperanza
Chalatenango
SAN SALVADOR
San Vicente
EL SALVADOR
Usulután
San Miguel

Gulf of Honduras
Islas de la Bahía
Roatán
Puerto Cortés
Tela
La Ceiba
Trujillo
Limón
Iriona
Brus Laguna
Tocoa
Savá
San Esteban
San Pedro Sula
El Progreso
Yoro
La Unión
Gualaco
Catacamas
Comayagua
Siguatepeque
Guaimaca
Campamento
Juticalpa
HONDURAS
Bocay
Bonanza
TEGUCIGALPA
Danlí
Jalapa
Ocotal
Somoto
Condega
Estelí
Siuna
Choluteca
Somotillo
Jinotega
La Sirena
Sébaco
Matagalpa
Muy Muy
Ciudad Darío
Chinandega
NICARA
Corinto
Boaco
Juigalpa
Tipitapa
León
Lago de Managua
MANAGUA
Masaya
Jinotepe
Granada
Nandaime
Lago de Nicaragua
Belén
Rivas
Isla de Ometepe
San Carlos
La Cruz
Upala
Golfo de Papagayo
Liberia
Bagac
Filadelfia
Ca
Nicoya
Puntarenas
Península de Nicoya

Gulf of Fonseca
Río Choluteca

PACIFIC

OCEAN

Golfo de N

0 km   200
0 miles   200

Population   ● National capital
○ below 50,000   ○ 50,000 to 100,000   ◉ 100,000 to 500,000   ■ above 500,000

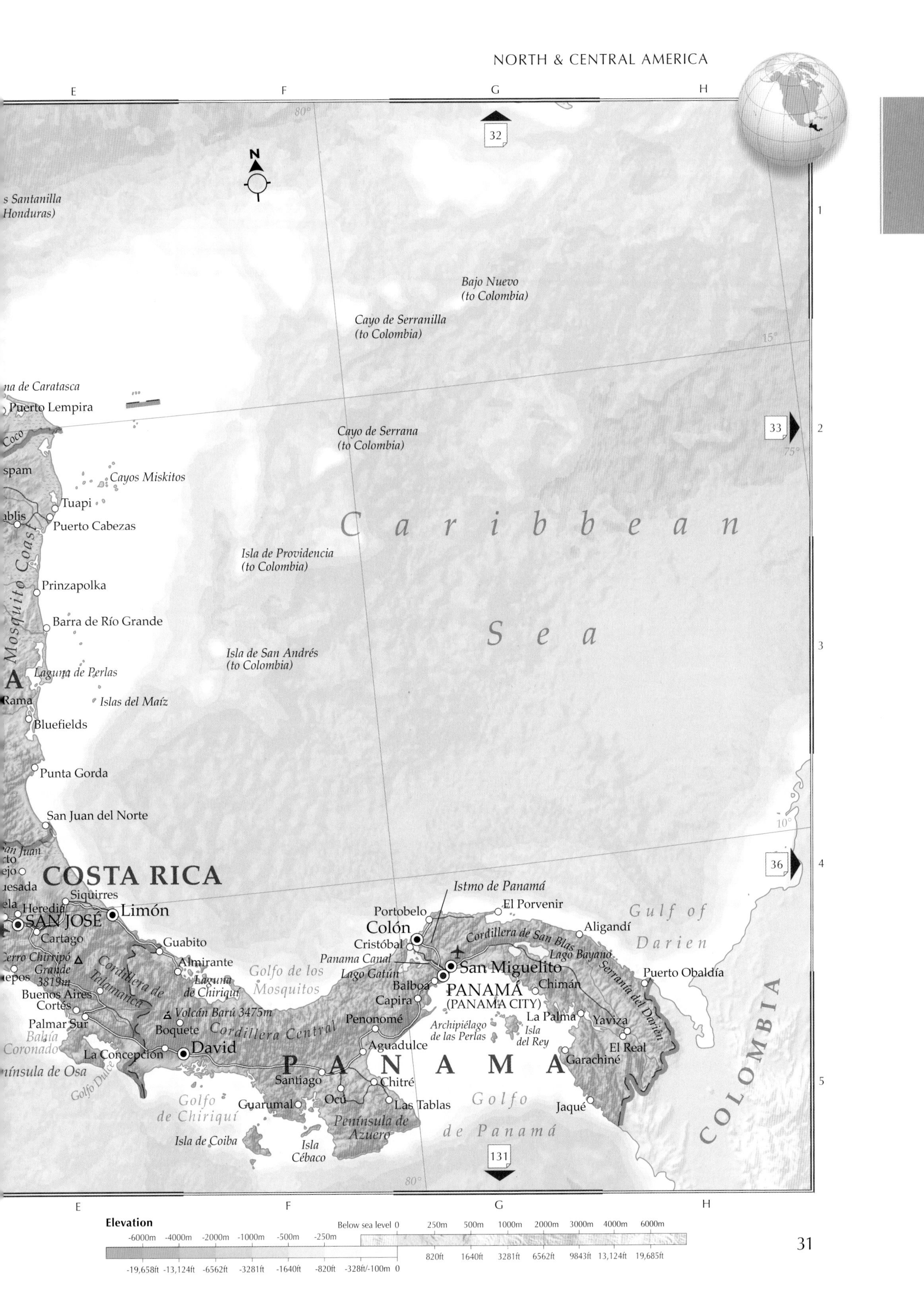

E   F   G   H

32

1

s Santanilla
Honduras)

Bajo Nuevo
(to Colombia)

15°

Cayo de Serranilla
(to Colombia)

na de Caratasca
Puerto Lempira

33   75°   2

Cayo de Serrana
(to Colombia)

Coco

spam

Cayos Miskitos

Tuapi

ablis

Puerto Cabezas

C a r i b b e a n

Prinzapolka

Isla de Providencia
(to Colombia)

S e a

Mosquito Coast

Barra de Río Grande

3

Laguna de Perlas

A

Isla de San Andrés
(to Colombia)

Rama

Islas del Maíz

Bluefields

Punta Gorda

10°

San Juan del Norte

36   4

an Juan
to
ejo

COSTA RICA

Istmo de Panamá

El Porvenir

Gulf of

esada
ela

Siquirres
Heredia

Portobelo

Aligandí

Darien

SAN JOSÉ

Limón

Colón

Cordillera de San Blas

Cartago

Cristóbal

Lago Bayano

Cerro Chirripó
Grande
3819m

Guabito

Almirante

Laguna
de Chiriquí

*Golfo de los
Mosquitos*

Panama Canal
Lago Gatún

Balboa

San Miguelito

Serranía del Darién

Puerto Obaldía

Cordillera
de
Talamanca

PANAMÁ
(PANAMA CITY)

Chimán

COLOMBIA

epos
Buenos Aires
Cortés

Capira

La Palma

Yaviza

Palmar Sur

Volcán Barú 3475m

Penonomé

Archipiélago
de las Perlas

Isla
del Rey

El Real

Bahía
Coronado

Boquete

La Concepción

David

Cordillera Central

Aguadulce

P A N A M A

Garachiné

nínsula de Osa

Santiago

Ocú

Chitré

*Golfo
de Panamá*

5

*Golfo
de Chiriquí*

Guarumal

Las Tablas

Jaqué

Golfo Dulce

Isla de Coiba

*Península de
Azuero*

Isla
Cébaco

131

E   F   G   H

Elevation

-6000m -4000m -2000m -1000m -500m   -250m   Below sea level 0   250m 500m 1000m 2000m 3000m 4000m 6000m

-19,658ft -13,124ft -6562ft -3281ft -1640ft -820ft -328ft/-100m 0   820ft 1640ft 3281ft 6562ft 9843ft 13,124ft 19,685ft

The Caribbean

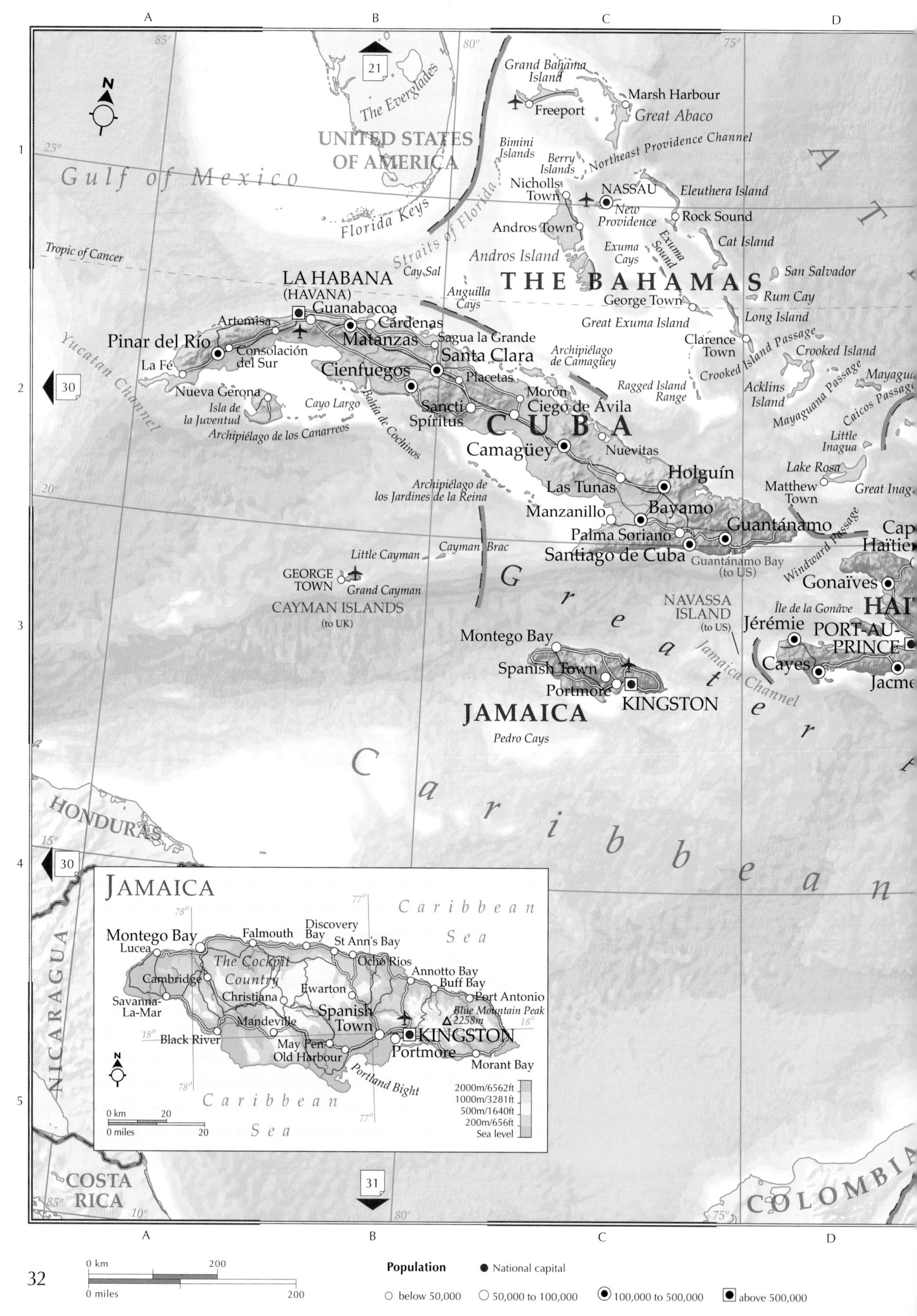

A B C D

21

N

Grand Bahama Island
Marsh Harbour
Freeport
Great Abaco
UNITED STATES OF AMERICA
Gulf of Mexico
The Everglades
Bimini Islands
Berry Islands
Northeast Providence Channel
Nicholls Town
NASSAU
Eleuthera Island
Florida Keys
Straits of Florida
Andros Town
New Providence
Rock Sound
Cat Island
Tropic of Cancer
Andros Island
Cay Sal
Anguilla Cays
THE BAHAMAS
San Salvador
LA HABANA (HAVANA)
Guanabacoa
George Town
Rum Cay
Long Island
Artemisa
Cárdenas
Great Exuma Island
Exuma Cays
Exuma Sound
Pinar del Río
Matanzas
Sagua la Grande
Clarence Town
Crooked Island
Consolación del Sur
Santa Clara
Archipiélago de Camagüey
Crooked Island Passage
La Fé
Cienfuegos
Placetas
Ragged Island Range
Acklins Island
Caicos Passage
Nueva Gerona
Sancti Spíritus
Morón
Ciego de Ávila
CUBA
Little Inagua
Isla de la Juventud
Cayo Largo
Bahía de Cochinos
Camagüey
Nuevitas
Lake Rosa
Archipiélago de los Canarreos
Holguín
Matthew Town
Great Inag
Archipiélago de los Jardines de la Reina
Las Tunas
Bayamo
Manzanillo
Guantánamo
Cap
Palma Soriano
Haïtie
Santiago de Cuba
Guantánamo Bay (to US)
Gonaïves
Cayman Brac
Little Cayman
NAVASSA ISLAND (to US)
Jérémie
HAÏ
PORT-AU-PRINCE
GEORGE TOWN
Grand Cayman
Île de la Gonâve
CAYMAN ISLANDS (to UK)
Montego Bay
Windward Passage
Cayes
Jacmel
Jamaica Channel
Spanish Town
Portmore
KINGSTON
JAMAICA
Pedro Cays
HONDURAS
Caribbean
NICARAGUA

JAMAICA

Montego Bay
Lucea
Falmouth
Discovery Bay
St Ann's Bay
Caribbean Sea
The Cockpit Country
Ocho Rios
Annotto Bay
Buff Bay
Cambridge
Christiana
Ewarton
Port Antonio
Savanna-La-Mar
Mandeville
Spanish Town
Blue Mountain Peak △2258m
Black River
May Pen
Old Harbour
KINGSTON
Portmore
Morant Bay
Portland Bight
N
Caribbean Sea

0 km   20
0 miles   20

2000m/6562ft
1000m/3281ft
500m/1640ft
200m/656ft
Sea level

COSTA RICA

31

COLOMBIA

0 km   200
0 miles   200

Population   ● National capital

○ below 50,000   ○ 50,000 to 100,000   ◉ 100,000 to 500,000   ◼ above 500,000

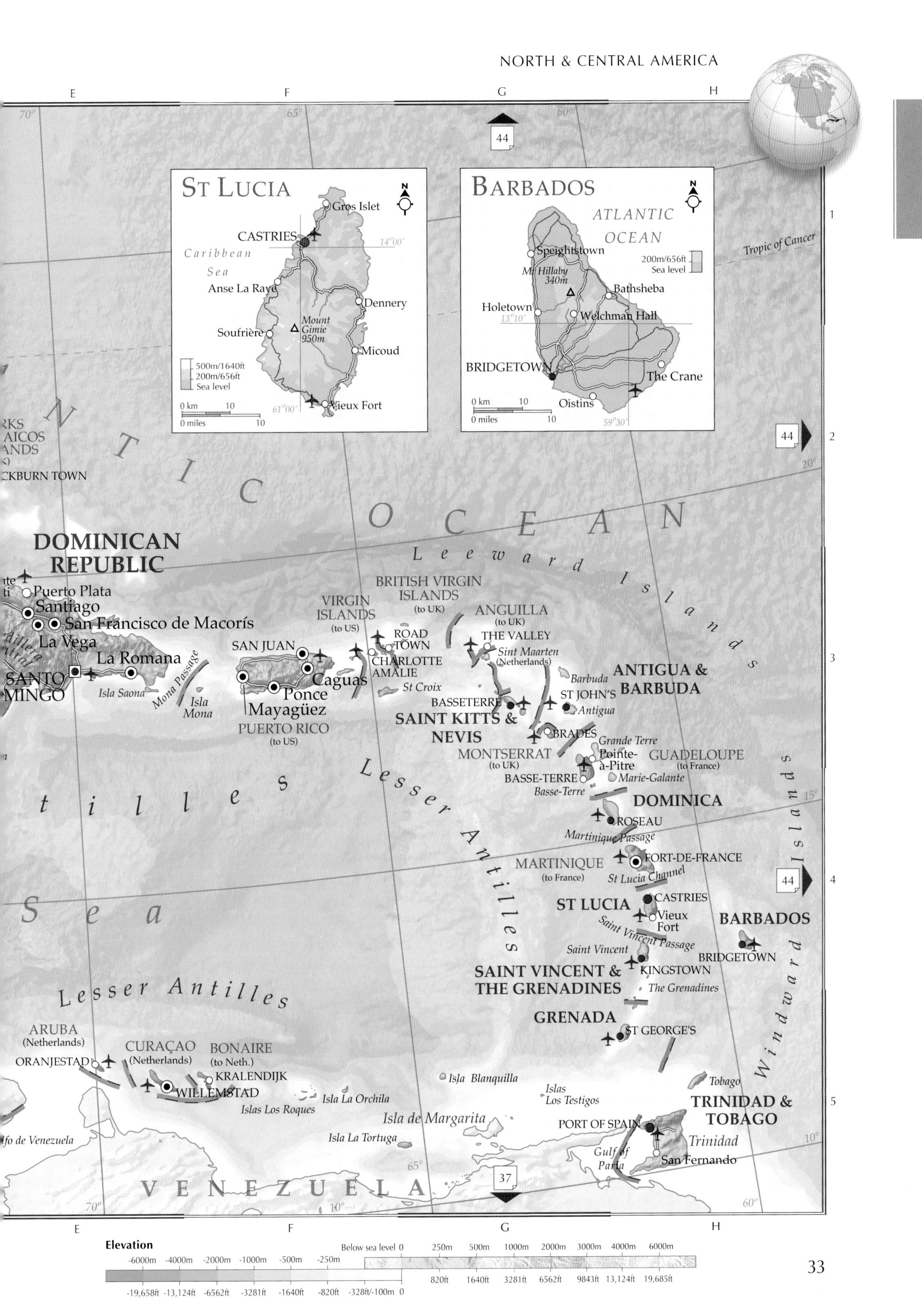

44

St Lucia

N

Gros Islet

CASTRIES

Caribbean Sea

14°00'

Anse La Raye

Dennery

Soufrière

△ *Mount Gimie 950m*

Micoud

500m/1640ft
200m/656ft
Sea level

0 km · · · · 10
0 miles · · · · 10

61°00'

Vieux Fort

Barbados

N

ATLANTIC OCEAN

Speightstown

200m/656ft
Sea level

Mt Hillaby 340m △

Bathsheba

Holetown

13°10'

Welchman Hall

BRIDGETOWN

The Crane

0 km · · · · 10
0 miles · · · · 10

Oistins

59°30'

44

RKS
AICOS
ANDS
CKBURN TOWN

DOMINICAN REPUBLIC

Puerto Plata

Santiago

San Francisco de Macorís

La Vega

SANTO MINGO

La Romana

Isla Saona

Mona Passage

Isla Mona

SAN JUAN

Caguas

Ponce

Mayagüez

PUERTO RICO (to US)

A N T I L L E S

O C E A N

Leeward Islands

VIRGIN ISLANDS (to US)

BRITISH VIRGIN ISLANDS (to UK)

ROAD TOWN

CHARLOTTE AMALIE

St Croix

ANGUILLA (to UK)

THE VALLEY

Sint Maarten (Netherlands)

Barbuda

ANTIGUA & BARBUDA

ST JOHN'S

Antigua

BASSETERRE

SAINT KITTS & NEVIS

BRADES

MONTSERRAT (to UK)

Pointe-à-Pitre

BASSE-TERRE

Basse-Terre

Grande Terre

GUADELOUPE (to France)

Marie-Galante

DOMINICA

ROSEAU

Martinique Passage

MARTINIQUE (to France)

FORT-DE-FRANCE

St Lucia Channel

ST LUCIA

CASTRIES

Vieux Fort

Saint Vincent Passage

Saint Vincent

BARBADOS

BRIDGETOWN

SAINT VINCENT & THE GRENADINES

KINGSTOWN

The Grenadines

GRENADA

ST GEORGE'S

L e s s e r   A n t i l l e s

S e a

ARUBA (Netherlands)

ORANJESTAD

CURAÇAO (Netherlands)

BONAIRE (to Neth.)

KRALENDIJK

WILLEMSTAD

Islas Los Roques

Isla La Orchila

fo de Venezuela

Isla La Tortuga

Isla de Margarita

Isla Blanquilla

Islas Los Testigos

Tobago

TRINIDAD & TOBAGO

PORT OF SPAIN

Trinidad

Gulf of Paria

San Fernando

V E N E Z U E L A

Windward Islands

Tropic of Cancer

44

15°

44

37

70°

65°

60°

20°

10°

Elevation

-6000m -4000m -2000m -1000m -500m -250m Below sea level 0 250m 500m 1000m 2000m 3000m 4000m 6000m

-19,658ft -13,124ft -6562ft -3281ft -1640ft -820ft -328ft/-100m 0 820ft 1640ft 3281ft 6562ft 9843ft 13,124ft 19,685ft

South America

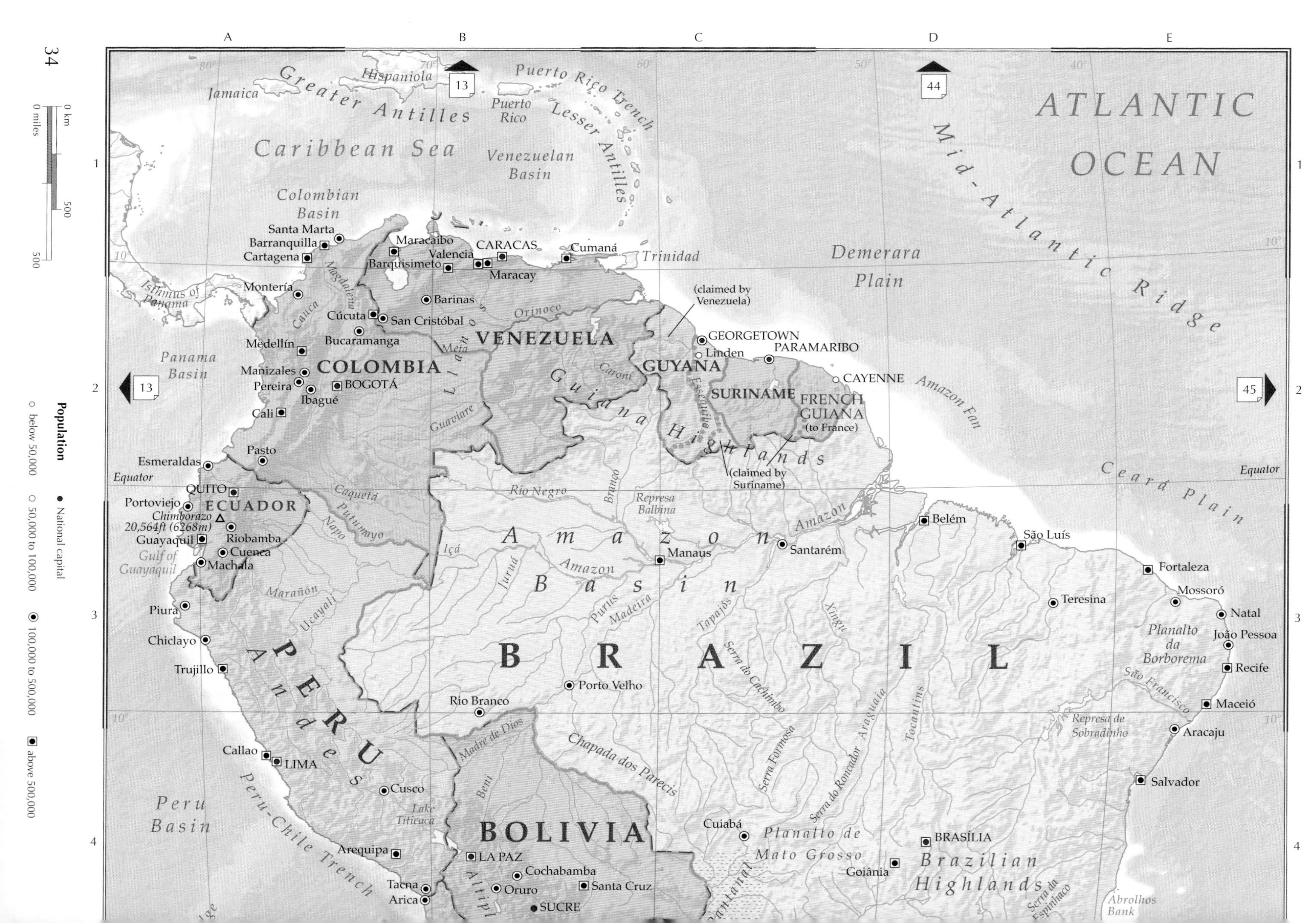

Population   • National capital

○ below 50,000   ⊙ 50,000 to 100,000   ◉ 100,000 to 500,000   ■ above 500,000

0 km   500
0 miles   500

ATLANTIC

OCEAN

Mid-Atlantic Ridge

Demerara Plain

Ceará Plain

Equator

Caribbean Sea

Greater Antilles

Jamaica

Hispaniola

Puerto Rico

Puerto Rico Trench

Lesser Antilles

Trinidad

Venezuelan Basin

Colombian Basin

Panama Basin

Isthmus of Panama

Peru Basin

Peru-Chile Trench

Abrolhos Bank

VENEZUELA

CARACAS

Maracay

Valencia

Barquisimeto

Barinas

Maracaibo

Cumaná

San Cristóbal

COLOMBIA

BOGOTÁ

Cúcuta

Bucaramanga

Ibagué

Medellín

Manizales

Pereira

Cali

Pasto

Santa Marta

Barranquilla

Cartagena

Montería

Magdalena

Cauca

Orinoco

Meta

Caroní

Guaviare

Caquetá

Putumayo

Napo

Marañón

Ucayali

ECUADOR

QUITO

Portoviejo

Guayaquil

Machala

Cuenca

Riobamba

Chimborazo
20,564ft (6268m) △

Esmeraldas

Gulf of Guayaquil

Equator

GUYANA

GEORGETOWN

Linden

(claimed by Venezuela)

Essequibo

SURINAME

PARAMARIBO

FRENCH GUIANA
(to France)

CAYENNE

Guiana Highlands

(claimed by Suriname)

Amazon Fan

Belém

Santarém

Manaus

Amazon

Rio Negro

Branco

Içá

Japurá

Juruá

Purus

Madeira

Tapajós

Xingu

Tocantins

Araguaia

BRAZIL

BRASÍLIA

Goiânia

São Luís

Teresina

Fortaleza

Mossoró

Natal

João Pessoa

Recife

Maceió

Aracaju

Salvador

Planalto da Borborema

São Francisco

Represa de Sobradinho

Brazilian Highlands

Serra do Espinhaço

Serra do Cachimbo

Serra Formosa

Serra do Roncador

Planalto de Mato Grosso

Chapada dos Parecis

Cuiabá

Porto Velho

Rio Branco

Represa Balbina

Amazon Basin

PERU

LIMA

Callao

Trujillo

Chiclayo

Piura

Cusco

Arequipa

Tacna

Arica

Andes

Lake Titicaca

BOLIVIA

LA PAZ

SUCRE

Cochabamba

Oruro

Santa Cruz

Altiplano

Beni

Mamoré

Madre de Dios

Equator

10°

40°

50°

60°

70°

80°

0°

10°

20°

A   B   C   D   E

1   2   3   4

44   45   13

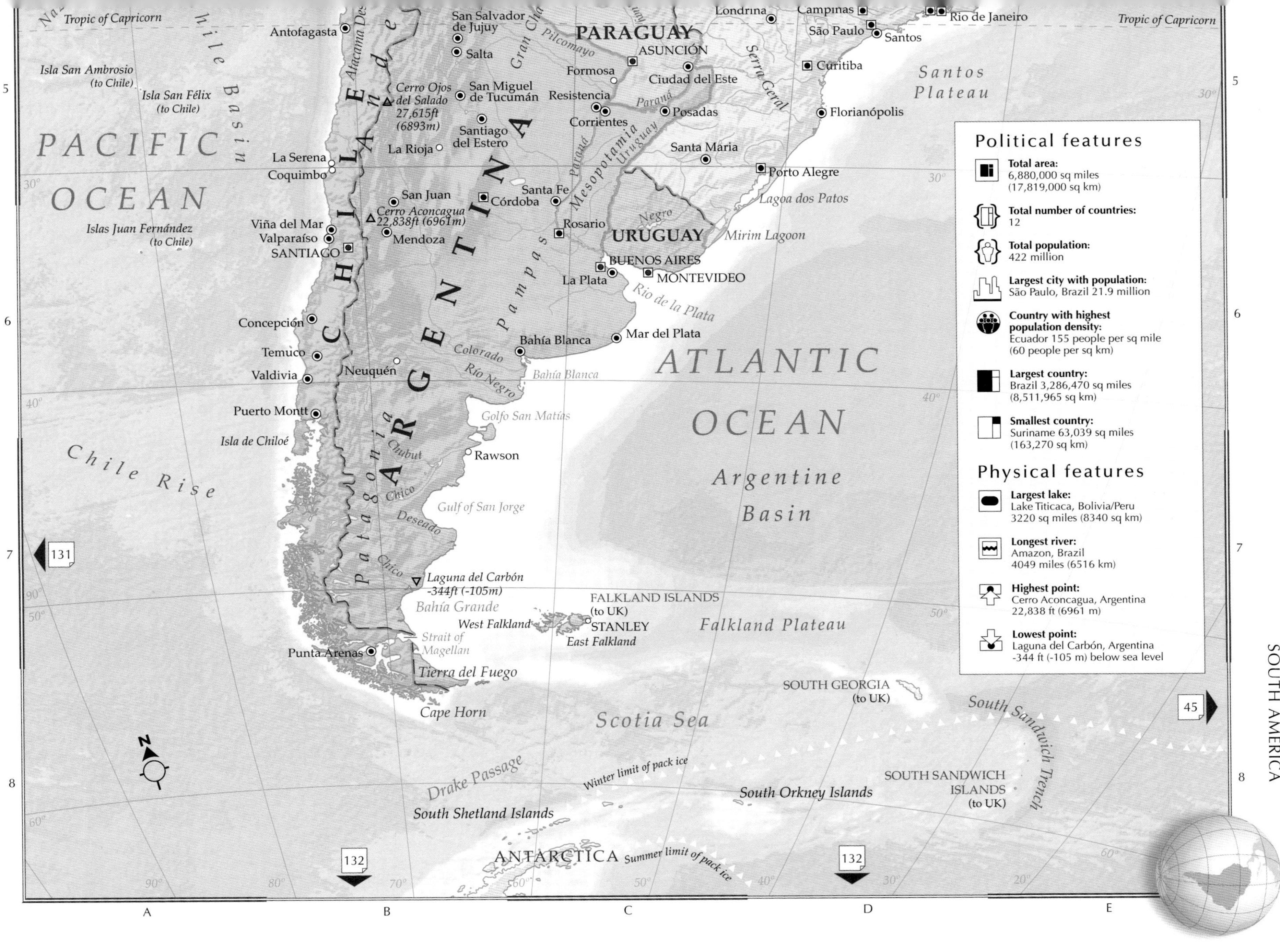

Political features

Total area:
6,880,000 sq miles
(17,819,000 sq km)

Total number of countries:
12

Total population:
422 million

Largest city with population:
São Paulo, Brazil 21.9 million

Country with highest population density:
Ecuador 155 people per sq mile
(60 people per sq km)

Largest country:
Brazil 3,286,470 sq miles
(8,511,965 sq km)

Smallest country:
Suriname 63,039 sq miles
(163,270 sq km)

Physical features

Largest lake:
Lake Titicaca, Bolivia/Peru
3220 sq miles (8340 sq km)

Longest river:
Amazon, Brazil
4049 miles (6516 km)

Highest point:
Cerro Aconcagua, Argentina
22,838 ft (6961 m)

Lowest point:
Laguna del Carbón, Argentina
-344 ft (-105 m) below sea level

Northern South America

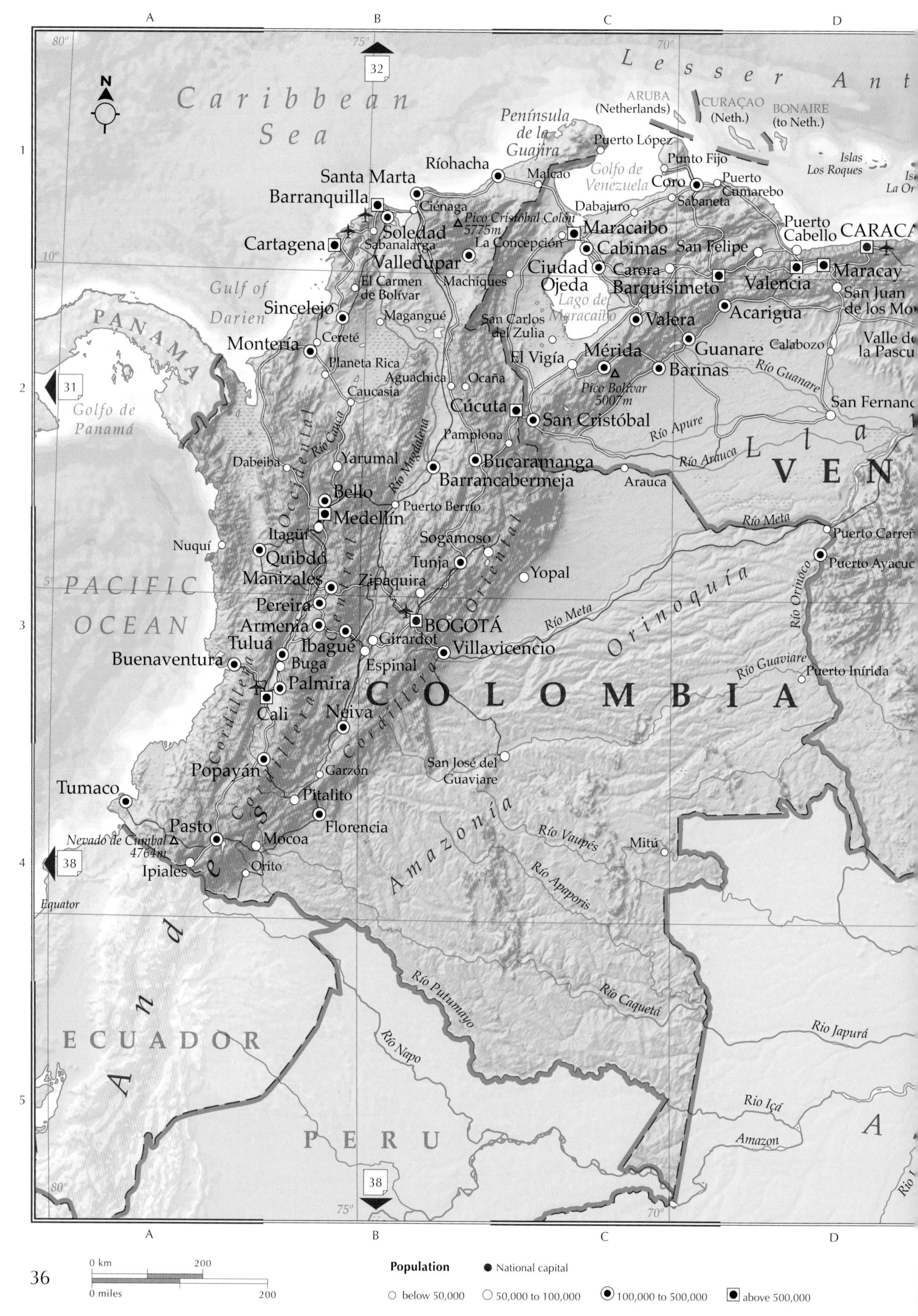

Caribbean Sea

ARUBA (Netherlands)   CURAÇAO (Neth.)   BONAIRE (to Neth.)

Lesser Ant

Islas Los Roques

La Or   *Is*

Península de la Guajira

Puerto López
Punto Fijo
Ríohacha
Coro
Puerto Cumarebo
Maicao
Golfo de Venezuela
Sabaneta
Santa Marta
Barranquilla
Ciénaga
Dabajuro
Puerto Cabello
CARACA
Soledad
Sabanalarga
△ Pico Cristóbal Colón 5775m
Maracaibo
San Felipe
Cartagena
Valledupar
La Concepción
Cabimas
Maracay
El Carmen de Bolívar
Machiques
Ciudad Ojeda
Carora
Barquisimeto
Valencia
San Juan de los Mo
Sincelejo
Gulf of Darien
Magangué
San Carlos del Zulia
Lago de Maracaibo
Valera
Acarigua
Valle de la Pascu
Montería
Cereté
El Vigía
Mérida
Guanare
Calabozo
PANAMA
Planeta Rica
Aguachica
Ocaña
△ Pico Bolívar 5007m
Barinas
Caucasia
Golfo de Panamá
Cúcuta
San Cristóbal
Río Guanare
San Fernan
Dabeiba
Yarumal
Pamplona
Río Apure
Río Arauca
L   a
Bucaramanga
Arauca
V E N
Bello
Barrancabermeja
Medellín
Puerto Berrío
Río Meta
Itagüí
Sogamoso
Puerto Carre
Quibdó
Tunja
Orinoquía
Puerto Ayacuc
Nuquí
PACIFIC OCEAN
Manizales
Zipaquira
Yopal
Río Orinoco
Pereira
Armenia
BOGOTÁ
Río Meta
Tuluá
Ibagué
Girardot
Villavicencio
Buenaventura
Buga
Espinal
Puerto Inírida
Palmira
Río Guaviare
Cali
Neiva
C O L O M B I A
Popayán
Garzón
San José del Guaviare
Tumaco
Pitalito
Amazonía
Pasto
Mocoa
Florencia
Río Vaupés
Mitú
△ Nevado de Cumbal 4764m
Orito
Ipiales
Equator
ECUADOR
An d e s
Cordillera Occidental
Cordillera Central
Cordillera Oriental
Río Cauca
Río Magdalena
Río Putumayo
Río Napo
Río Caquetá
Río Japurá
Rio Içá
Amazon
PERU
A
Rio I

0 km   200
0 miles   200

Population   ● National capital

○ below 50,000   ◎ 50,000 to 100,000   ◉ 100,000 to 500,000   ■ above 500,000

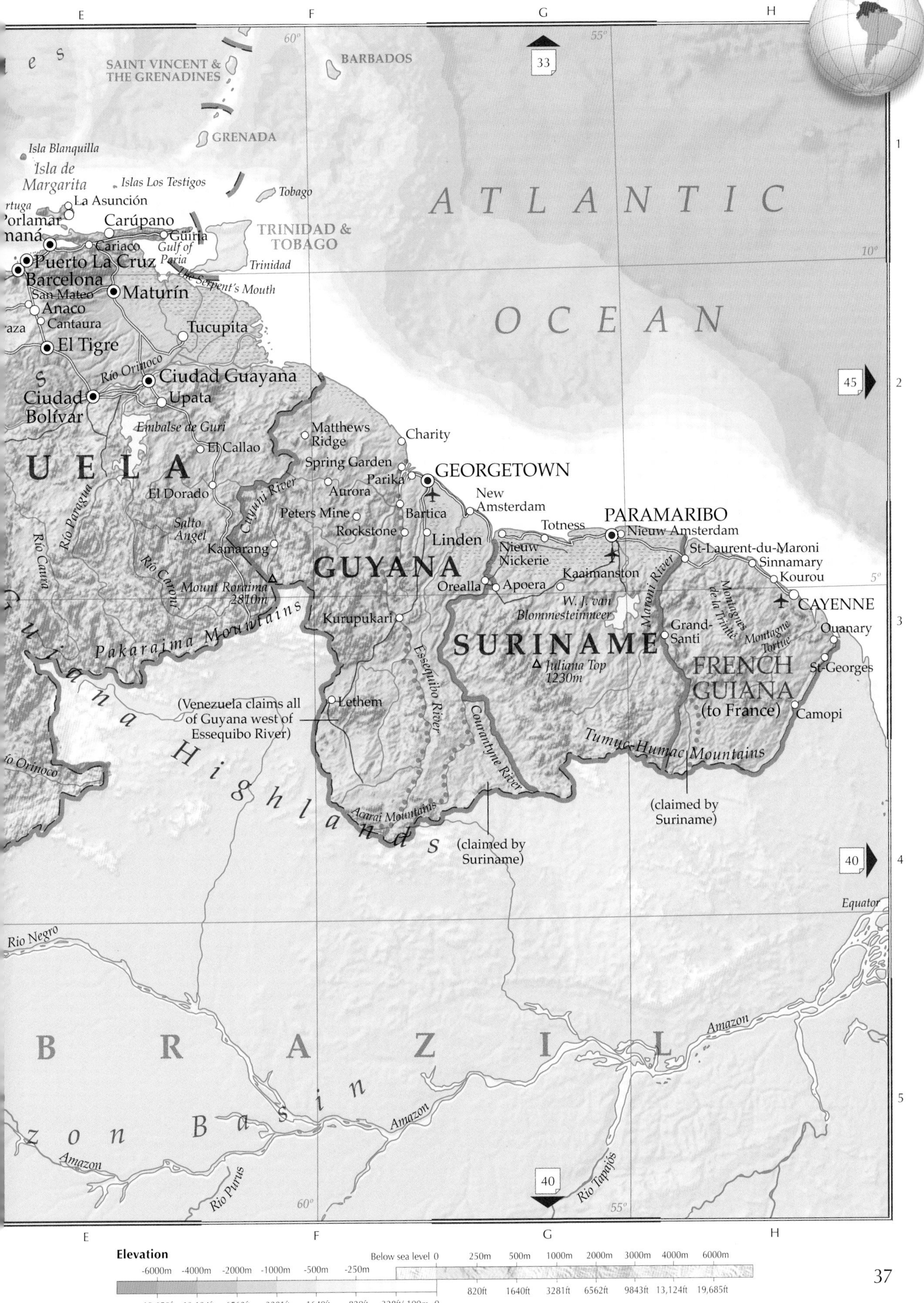

E · F · G · H

33

1

Isla Blanquilla
Isla de
Margarita
rtuga   Islas Los Testigos
Porlamar   La Asunción   Tobago
nana   TRINIDAD &
              Carúpano   TOBAGO
         Cariaco Guiria
    Puerto La Cruz Gulf of   Trinidad
    Barcelona   Paria
San Mateo   The Serpent's Mouth
Anaco   Maturín
aza Cantaura
El Tigre   Tucupita

Río Orinoco
        Ciudad Guayana   45 2
Ciudad   Upata
Bolívar
Embalse de Guri
U E L A   Matthews Charity
              El Callao   Ridge
                        Spring Garden
         El Dorado   Aurora   Parika   GEORGETOWN
Salto   Peters Mine   Bartica   New
Ángel   Kamarang   Rockstone   Linden   Amsterdam   PARAMARIBO
                              Nieuw   Nieuw Amsterdam
Mount Roraima △   GUYANA   Nickerie   St-Laurent-du-Maroni
2810m   Totness   Sinnamary
Pakaraima Mountains   Orealla Apoera   Kaaimanston   Kourou 5°
                              W. J. van   CAYENNE 3
              Kurupukari   Blommesteinmeer   Grand-   Ouanary
                        SURINAME   Santi   FRENCH
                        △ Juliana Top   GUIANA   St-Georges
                        1230m   Montagne   (to France)
Río Caura   Tortue   Camopi
(Venezuela claims all   Lethem
of Guyana west of
Essequibo River)   Tumuc-Humac Mountains

H i g h l a n d s   Acarai Mountains   (claimed by
o Orinoco   Suriname)

                        (claimed by   40 4
                        Suriname)
                                        Equator
Río Negro

B R A Z I L   Amazon   5
    zon   Basin
                    Amazon
Amazon
            Río Purus   Río Tapajós
                        40

60°   60°   55°

Elevation

Below sea level 0   250m   500m   1000m   2000m   3000m   4000m   6000m

-6000m -4000m -2000m -1000m -500m -250m

-19,658ft -13,124ft -6562ft -3281ft -1640ft -820ft -328ft/-100m 0   820ft   1640ft   3281ft   6562ft   9843ft 13,124ft 19,685ft

Western South America

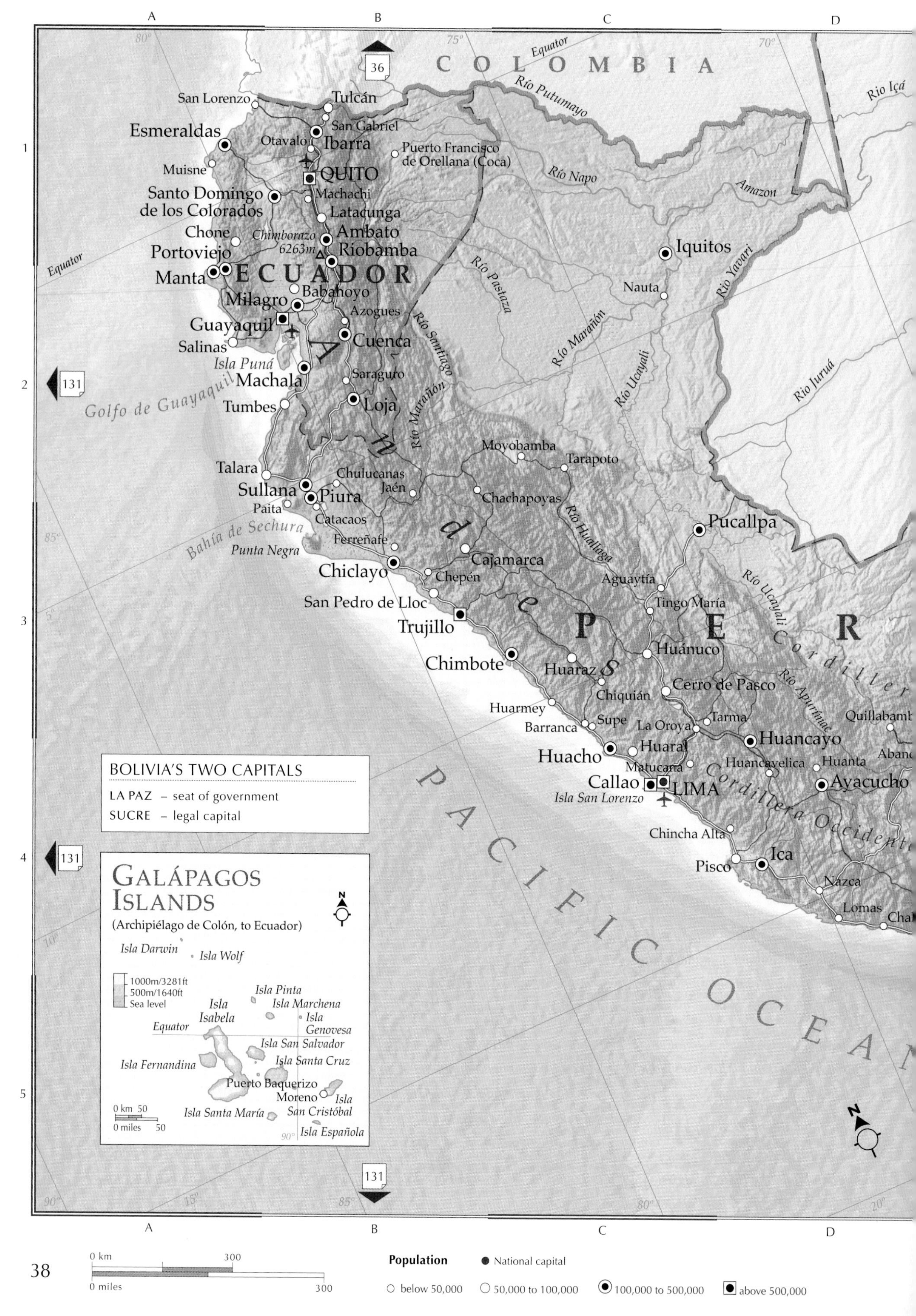

COLOMBIA

San Lorenzo
Esmeraldas
Tulcán
San Gabriel
Otavalo
Ibarra
Muisne
Puerto Francisco
de Orellana (Coca)
QUITO
Machachi
Santo Domingo
de los Colorados
Latacunga
Chone
Chimborazo
6263m
Ambato
Portoviejo
Riobamba
Manta
ECUADOR
Milagro
Babahoyo
Azogues
Guayaquil
Salinas
Cuenca
Isla Puná
Saraguro
Machala
Loja
Golfo de Guayaquil
Tumbes

Iquitos
Nauta

Río Putumayo
Equator
Río Napo
Amazon
Río Içá
Río Pastaza
Río Santiago
Río Marañón
Río Yavari
Río Juruá

Talara
Sullana
Piura
Paita
Catacaos
Chulucanas
Jaén
Ferreñafe
Chiclayo
Chepén
San Pedro de Lloc
Trujillo
Chimbote
Huaraz
Chiquián
Huarmey
Barranca
Supe
La Oroya
Huacho
Huaral
Matucana
Callao
LIMA
Isla San Lorenzo
Chincha Alta
Pisco
Ica
Nazca
Lomas

Moyobamba
Tarapoto
Chachapoyas
Cajamarca
Pucallpa
Aguaytía
Tingo María
Huánuco
Cerro de Pasco
Tarma
Huancayo
Huancavelica
Huanta
Ayacucho
Quillabamba
Aband

Bahía de Sechura
Punta Negra

PERU
Río Huallaga
Río Ucayali
Río Apurímac
Cordillera
Cordillera Occidental

PACIFIC OCEAN

BOLIVIA'S TWO CAPITALS

LA PAZ – seat of government

SUCRE – legal capital

GALÁPAGOS ISLANDS

(Archipiélago de Colón, to Ecuador)

Isla Darwin
Isla Wolf

1000m/3281ft
500m/1640ft
Sea level

Isla Pinta
Isla Marchena
Isla
Isabela
Isla
Genovesa
Equator
Isla San Salvador
Isla Fernandina
Isla Santa Cruz
Puerto Baquerizo
Moreno
Isla
Isla Santa María
San Cristóbal
Isla Española

0 km 50
0 miles 50

0 km 300
0 miles 300

Population   ● National capital

○ below 50,000   ○ 50,000 to 100,000   ◉ 100,000 to 500,000   ◼ above 500,000

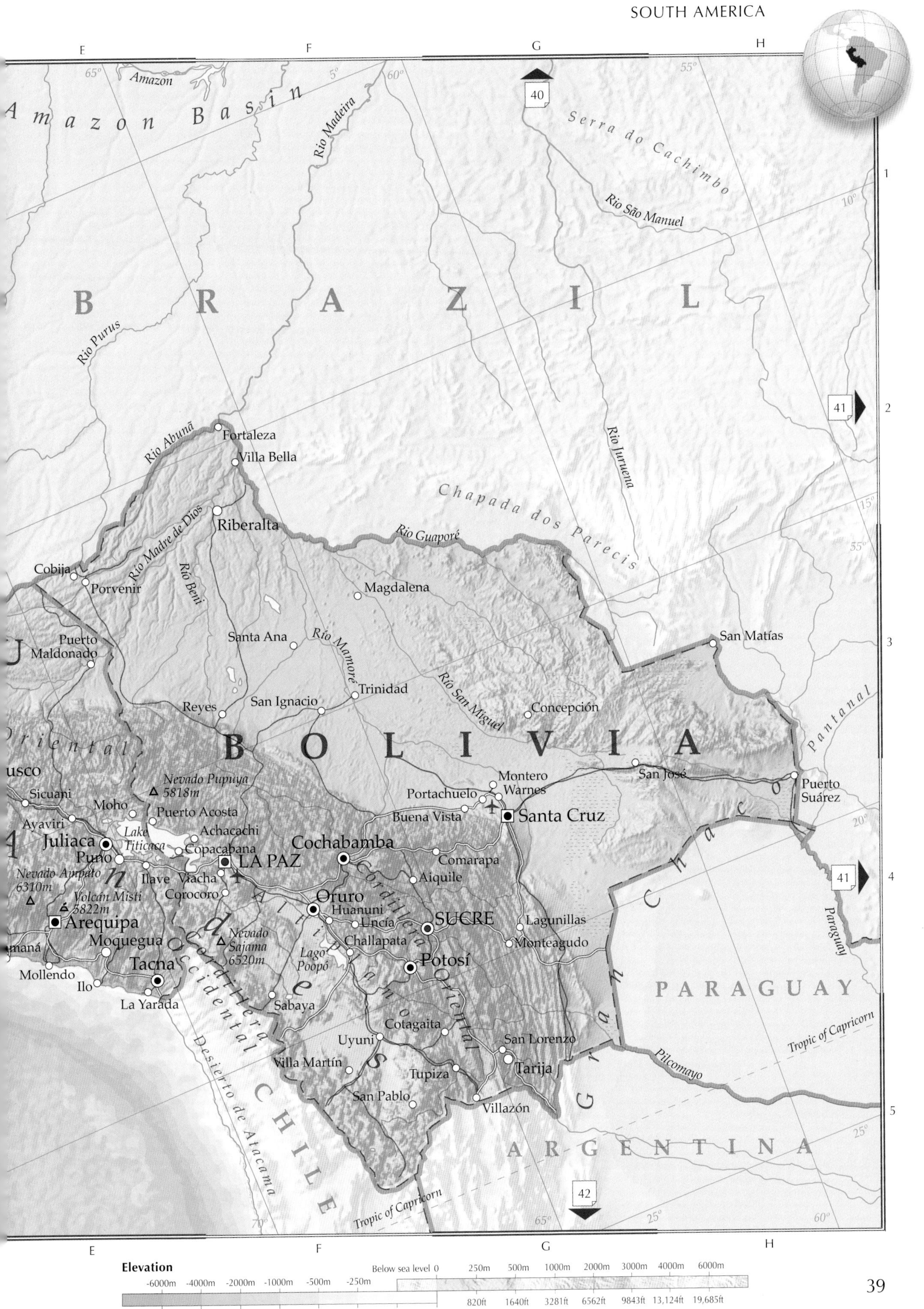

E   65°   *Amazon*   F   5°   60°   G   55°   H

40

A m a z o n   B a s i n

Serra do Cachimbo

Rio Madeira

1

Rio São Manuel

10°

B   R   A   Z   I   L

Rio Purus

41 ▶

2

Rio Juruena

Rio Abunã

Fortaleza
Villa Bella

C h a p a d a   d o s   P a r e c i s

15°

Riberalta

Rio Guaporé

55°

Cobija
Porvenir

Magdalena

Río Madre de Dios

Rio Beni

San Matías

3

J

Puerto
Maldonado

Santa Ana

Río Mamoré

Trinidad

Concepción

Pantanal

Reyes   San Ignacio

Río San Miguel

Puerto
Suárez

Oriental

B   O   L   I   V   I   A

20°

usco
Sicuani

△ *Nevado Pupuya*
5818m

Montero
Warnes

San José

Moho   Puerto Acosta

Portachuelo

Ayaviri
Juliaca   *Lake Titicaca*

Achacachi
Copacabana

Cochabamba

Buena Vista

● Santa Cruz

41 ▶

4

Puno
Nevado Ampato
6310m △

Ilave   Viacha
Corocoro

■ LA PAZ

Comarapa
Aiquile

△ *Volcán Misti*
5822m

Oruro ●
Huanuni

Lagunillas

● Arequipa

Uncía
Challapata

SUCRE ■

Monteagudo

aná

Moquegua

*Nevado
Sajama*
6520m △

*Lago
Poopó*

Potosí ●

P   A   R   A   G   U   A   Y

Mollendo
Ilo   Tacna ●

Sabaya

Tropic of Capricorn

La Yarada

Cotagaita

San Lorenzo

Pilcomayo

Desierto de Atacama

Uyuni

Villa Martín

Tarija ●

25°

San Pablo

Tupiza

C   H   I   L   E

Villazón

5

42

Tropic of Capricorn

70°   65°   25°   60°

A   R   G   E   N   T   I   N   A

25°

E   F   G   H

Elevation

Below sea level 0   250m   500m   1000m   2000m   3000m   4000m   6000m

-6000m   -4000m   -2000m   -1000m   -500m   -250m

820ft   1640ft   3281ft   6562ft   9843ft   13,124ft   19,685ft

-19,658ft   -13,124ft   -6562ft   -3281ft   -1640ft   -820ft   -328ft/-100m   0

Brazil

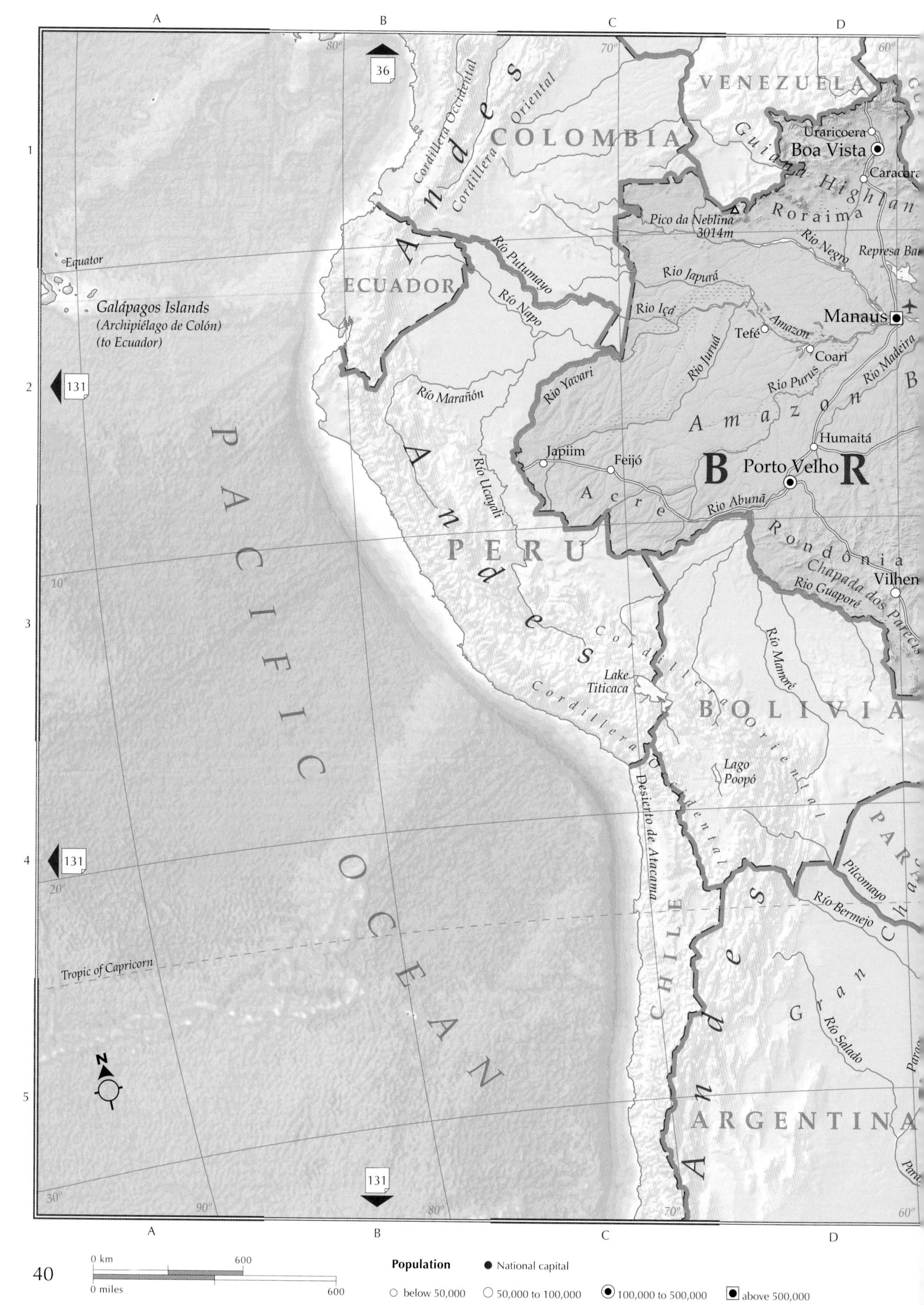

80°

36

VENEZUELA

COLOMBIA

Cordillera Occidental

Cordillera Oriental

Guiana Highlan

Uraricoera

Boa Vista

Caracora

1

70°

60°

Pico da Neblina
3014m

Roraima

Rio Negro

Represa Ba

Río Putumayo

ECUADOR

Río Napo

Rio Japurá

Rio Içá

Amazon

Manaus

Tefé

Coari

Rio Juruá

Río Marañón

Río Yavari

Rio Purus

Rio Madeira

Equator

Galápagos Islands
(Archipiélago de Colón)
(to Ecuador)

131

Amazon

2

A   n   d   e   s

Japiim

Feijó

Humaitá

B   R

Porto Velho

Rio Abunã

Acre

Río Ucayali

A

PERU

Rondônia

Chapada dos Parecis

Rio Guaporé

Vilhen

10°

3

P A C I F I C

Cordillera

Lake
Titicaca

Río Mamoré

BOLIVIA

Cordillera Oriental

Lago
Poopó

O C E A N

Desierto de Atacama

Cordillera Occidental

PARA

131

Pilcomayo

4

20°

Río Bermejo

Parag

Tropic of Capricorn

CHILE

A n d e s

G   r   a   n

C h

G

Río Salado

N

ARGENTINA

Para

5

30°

90°

80°

70°

60°

0 km   600

0 miles   600

Population   ● National capital

○ below 50,000   ○ 50,000 to 100,000   ◉ 100,000 to 500,000   ▣ above 500,000

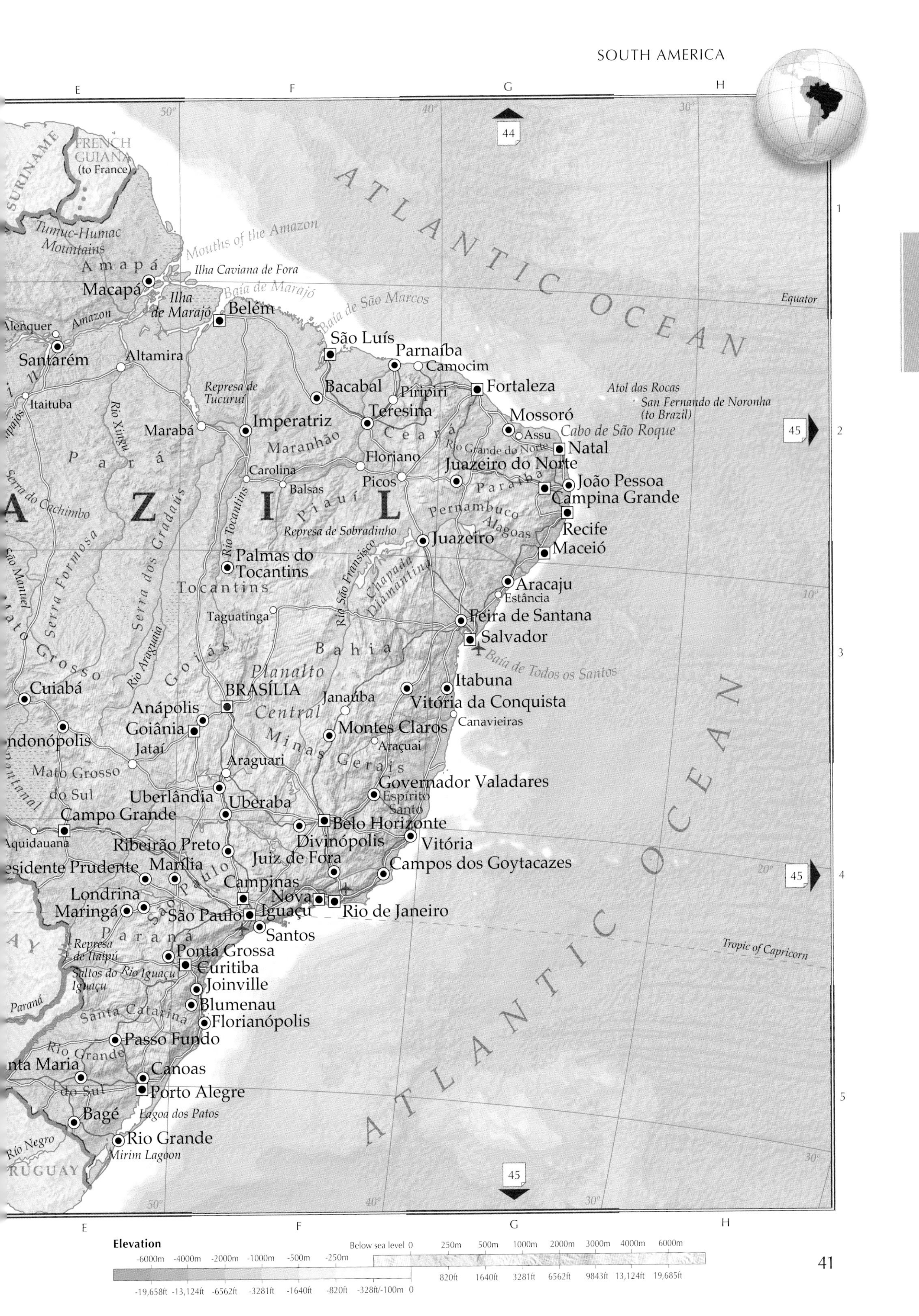

ATLANTIC OCEAN

ATLANTIC OCEAN

44

45

45

45

E   F   G   H

SURINAME

FRENCH GUIANA
(to France)

Tumuc-Humac
Mountains

Mouths of the Amazon

Ilha Caviana de Fora

Amapá

Macapá

Ilha
de Marajó

Baía de Marajó

Belém

Baía de São Marcos

Equator

Alenquer

Amazon

São Luís

Parnaíba

Camocim

Santarém

Altamira

Bacabal

Piripiri

Fortaleza

Atol das Rocas

San Fernando de Noronha
(to Brazil)

Itaituba

Represa de
Tucuruí

Teresina

Mossoró

Assu

Cabo de São Roque

Rio Xingu

Imperatriz

Marabá

Maranhão

Ceará

Floriano

Rio Grande do Norte

Natal

Pará

Carolina

Picos

Juazeiro do Norte

Paraíba

João Pessoa

Campina Grande

Serra do Cachimbo

Balsas

Piauí

Pernambuco

Recife

Serra dos Gradaús

BRAZIL

Represa de Sobradinho

Alagoas

Maceió

Serra Formosa

Rio Tocantins

Palmas do
Tocantins

Rio São Francisco

Juazeiro

São Manuel

Tocantins

Chapada
Diamantina

Aracaju

Estância

Mato Grosso

Taguatinga

Feira de Santana

Cuiabá

Goiás

Bahia

Salvador

Baía de Todos os Santos

Anápolis

BRASÍLIA

Planalto
Central

Janaúba

Itabuna

Vitória da Conquista

Goiânia

Canavieiras

Rondonópolis

Jataí

Minas

Montes Claros

Araçuaí

Mato Grosso
do Sul

Araguari

Gerais

Governador Valadares

Uberlândia

Uberaba

Espírito
Santo

Campo Grande

Belo Horizonte

Aquidauana

Ribeirão Preto

Divinópolis

Vitória

Presidente Prudente

Marília

Juiz de Fora

Campos dos Goytacazes

Londrina

São Paulo

Campinas

Nova

Maringá

São Paulo

Iguaçu

Rio de Janeiro

Paraná

Santos

Tropic of Capricorn

Represa
de Itaipú

Saltos do Rio Iguaçu

Ponta Grossa

Curitiba

Paraná

Joinville

Santa Catarina

Blumenau

Florianópolis

Passo Fundo

Rio Grande

Canoas

do Sul

Porto Alegre

Santa Maria

Bagé

Lagoa dos Patos

Rio Negro

Rio Grande

URUGUAY

Mirim Lagoon

50°   40°   30°

E   F   G   H

1

2

3

4

5

Elevation

| Below sea level | 0 | 250m | 500m | 1000m | 2000m | 3000m | 4000m | 6000m |

-6000m -4000m -2000m -1000m -500m -250m

-19,658ft -13,124ft -6562ft -3281ft -1640ft -820ft -328ft/-100m 0

820ft 1640ft 3281ft 6562ft 9843ft 13,124ft 19,685ft

Southern South America

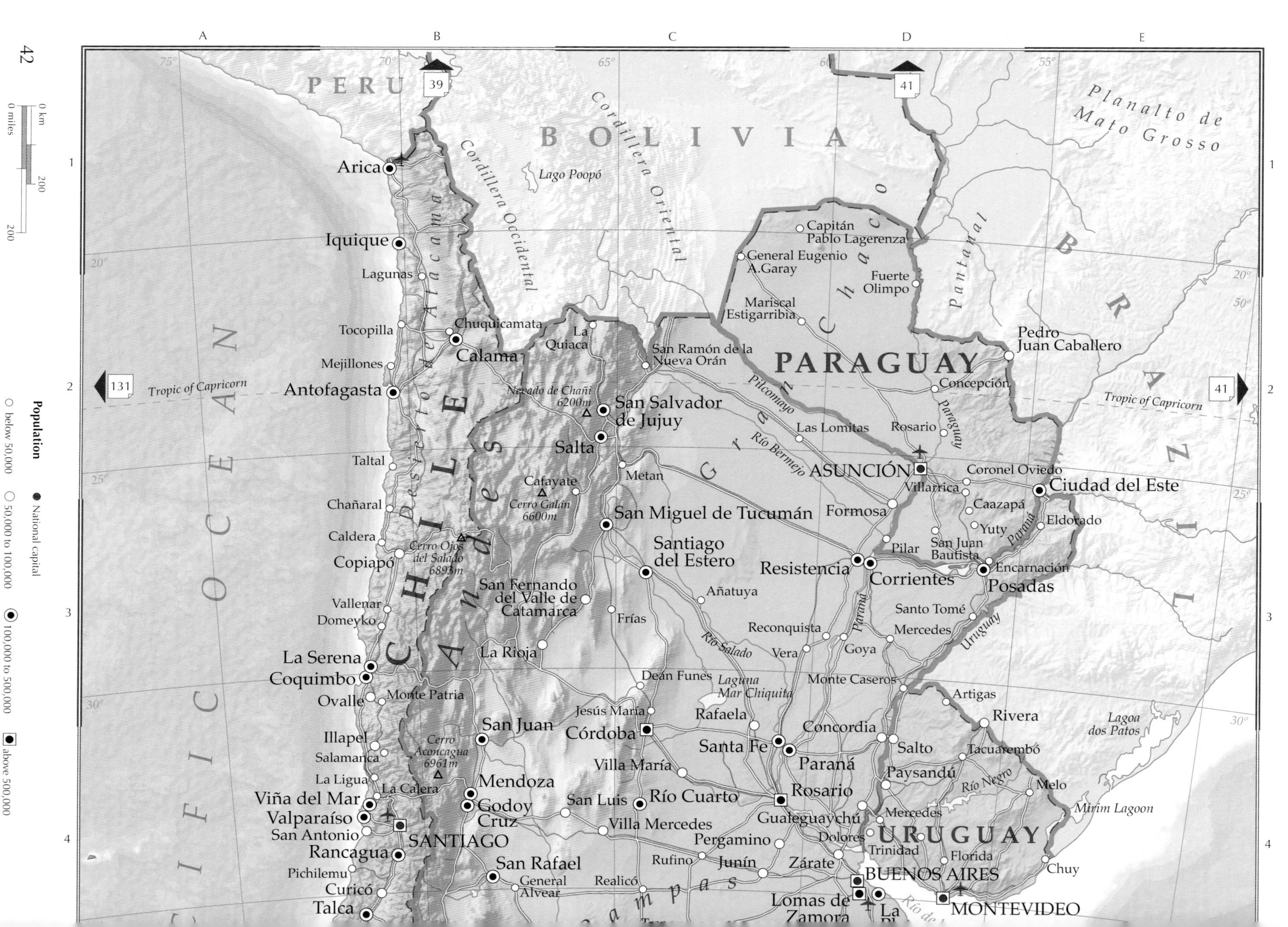

Population   ● National capital

○ below 50,000   ○ 50,000 to 100,000   ● 100,000 to 500,000   ■ above 500,000

0 km   200
0 miles   200

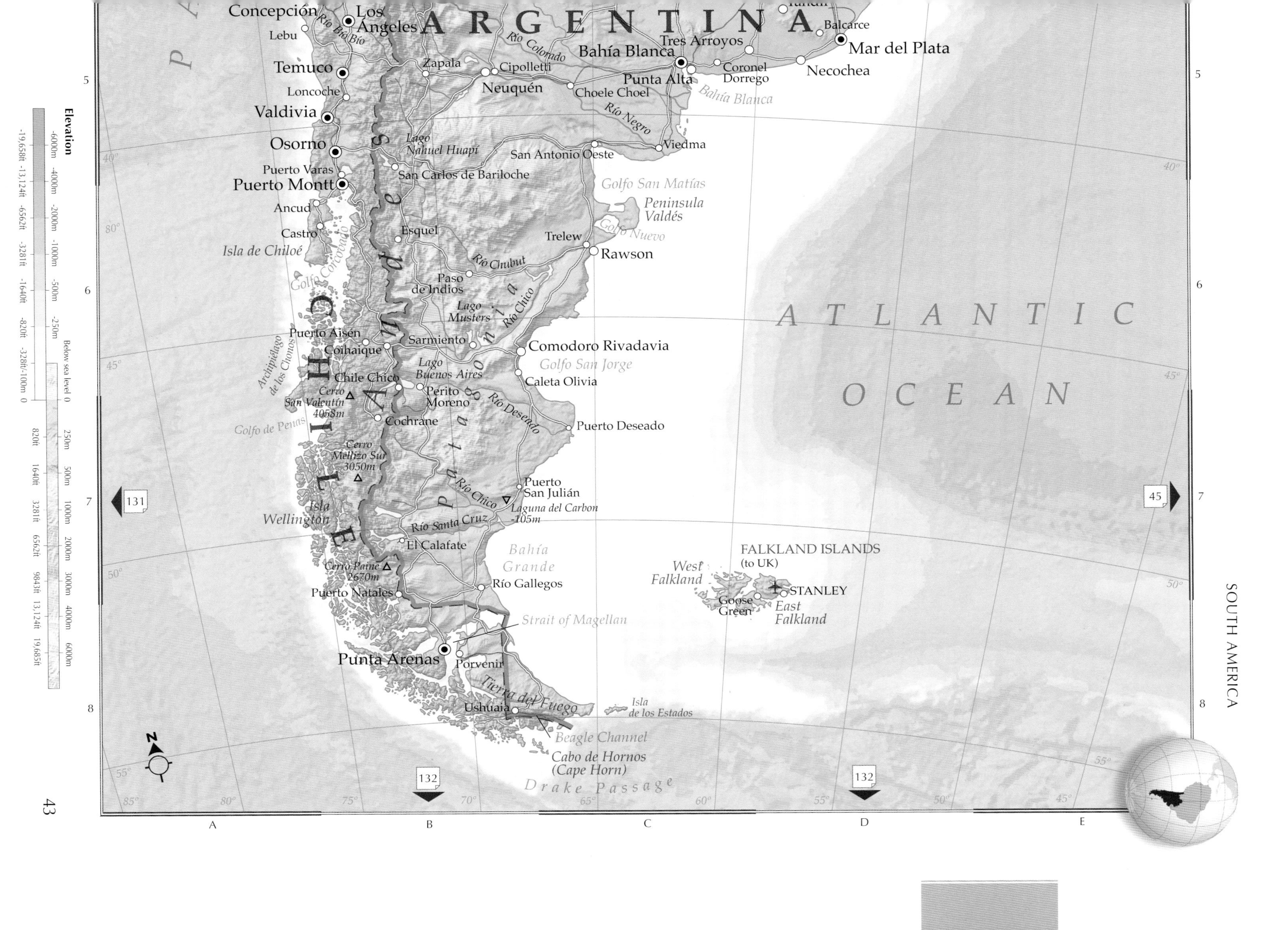

ATLANTIC

OCEAN

PACIFIC

ARGENTINA

CHILE

Mar del Plata
Balcarce
Necochea
Coronel
Dorrego
Tres Arroyos
Bahía Blanca
Punta Alta
Choele Choel
Cipolletti
Neuquén
Zapala
Los Ángeles
Concepción
Lebu
Temuco
Loncoche
Valdivia
Osorno
Puerto Varas
Puerto Montt
Ancud
Castro
Viedma
San Antonio Oeste
Trelew
Rawson
San Carlos de Bariloche
Esquel
Paso de Indios
Sarmiento
Coyhaique
Puerto Aisén
Chile Chico
Cochrane
Comodoro Rivadavia
Caleta Olivia
Puerto Deseado
Puerto San Julián
Laguna del Carbón
-105m
Río Gallegos
El Calafate
Puerto Natales
Punta Arenas
Porvenir
Ushuaia

Bahía Blanca
Río Colorado
Río Negro
Golfo San Matías
Península Valdés
Golfo Nuevo
Río Chubut
Lago Musters
Lago Buenos Aires
Perito Moreno
Río Deseado
Golfo San Jorge
Río Santa Cruz
Río Chico
Bahía Grande
Strait of Magellan
Tierra del Fuego
Beagle Channel
Cabo de Hornos
(Cape Horn)
Drake Passage
Isla de los Estados

FALKLAND ISLANDS
(to UK)
West Falkland
East Falkland
STANLEY
Goose Green

Río Bío Bío
Lago Nahuel Huapi
Cerro Tronador
Corcovado
Golfo Corcovado
Isla de Chiloé
Archipiélago de los Chonos
Golfo de Penas
Isla Wellington
Cerro San Valentín
4058m
Cerro Murallón
3050m
Cerro Fitz Roy
3375m
Río Chico

45
132
132
131

Z

43

The Atlantic Ocean

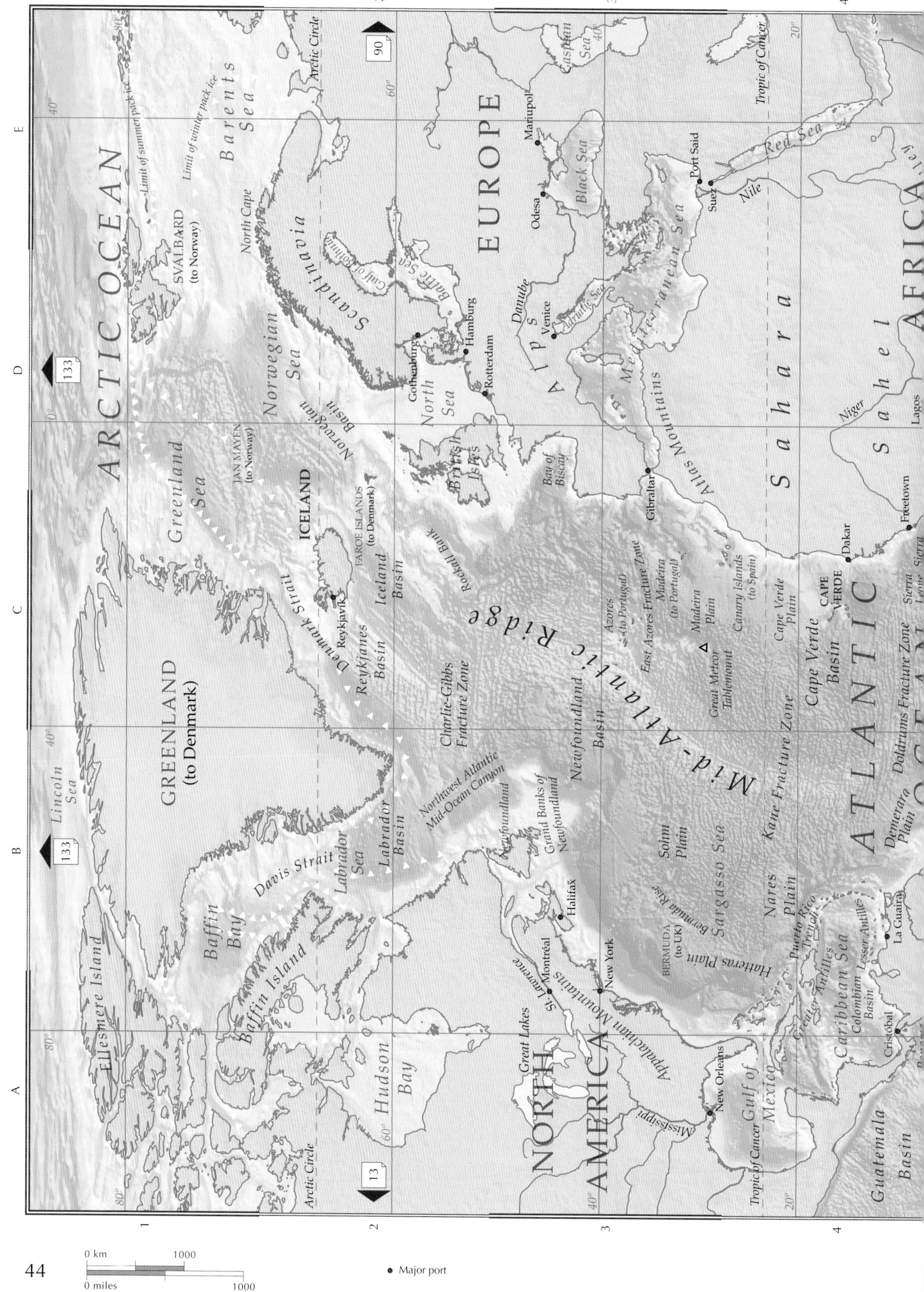

ARCTIC OCEAN

Limit of summer pack ice
Limit of winter pack ice

Barents Sea

Arctic Circle

North Cape

SVALBARD (to Norway)

EUROPE

Mariupol

Black Sea

Caspian Sea

Tropic of Cancer

Red Sea

Port Said

Odesa

Suez

Nile

Scandinavia

Gulf of Bothnia

Baltic Sea

Danube

Venice

Alps

Adriatic Sea

Mediterranean Sea

AFRICA

Norwegian Sea

Gothenburg

Hamburg

Rotterdam

North Sea

British Isles

Bay of Biscay

Atlas Mountains

Gibraltar

Sahara

Sahel

Niger

Lagos

Norwegian Basin

JAN MAYEN (to Norway)

Greenland Sea

ICELAND

FAROE ISLANDS (to Denmark)

Reykjavik

Denmark Strait

Reykjanes Basin

Iceland Basin

Rockall Bank

Charlie-Gibbs Fracture Zone

Azores (to Portugal)

East Azores Fracture Zone

Madeira (to Portugal)

Madeira Plain

Great Meteor Tablemount

Canary Islands (to Spain)

Mid-Atlantic Ridge

Cape Verde Plain

Cape Verde

CAPE VERDE

Dakar

Freetown

Sierra Leone Sierra

ARCTIC

GREENLAND (to Denmark)

Lincoln Sea

Ellesmere Island

Baffin Bay

Baffin Island

Davis Strait

Labrador Sea

Labrador Basin

Northwest Atlantic Mid-Ocean Canyon

Newfoundland

Grand Banks of Newfoundland

Newfoundland Basin

Sohm Plain

Cape Verde Basin

Kane Fracture Zone

Nares Plain

Doldrums Fracture Zone

Demerara Plain

ATLANTIC OCEAN

Hudson Bay

Great Lakes

St. Lawrence

Montréal

New York

Appalachian Mountains

Halifax

BERMUDA (to UK)

Bermuda Rise

Hatteras Plain

Sargasso Sea

Puerto Rico Trench

Greater Antilles

Lesser Antilles

La Guaira

NORTH AMERICA

New Orleans

Gulf of Mexico

Mississippi

Tropic of Cancer

Cristóbal

Caribbean Sea

Colombian Basin

Guatemala Basin

Arctic Circle

90

133

133

13

0 km 1000
0 miles 1000

● Major port

44

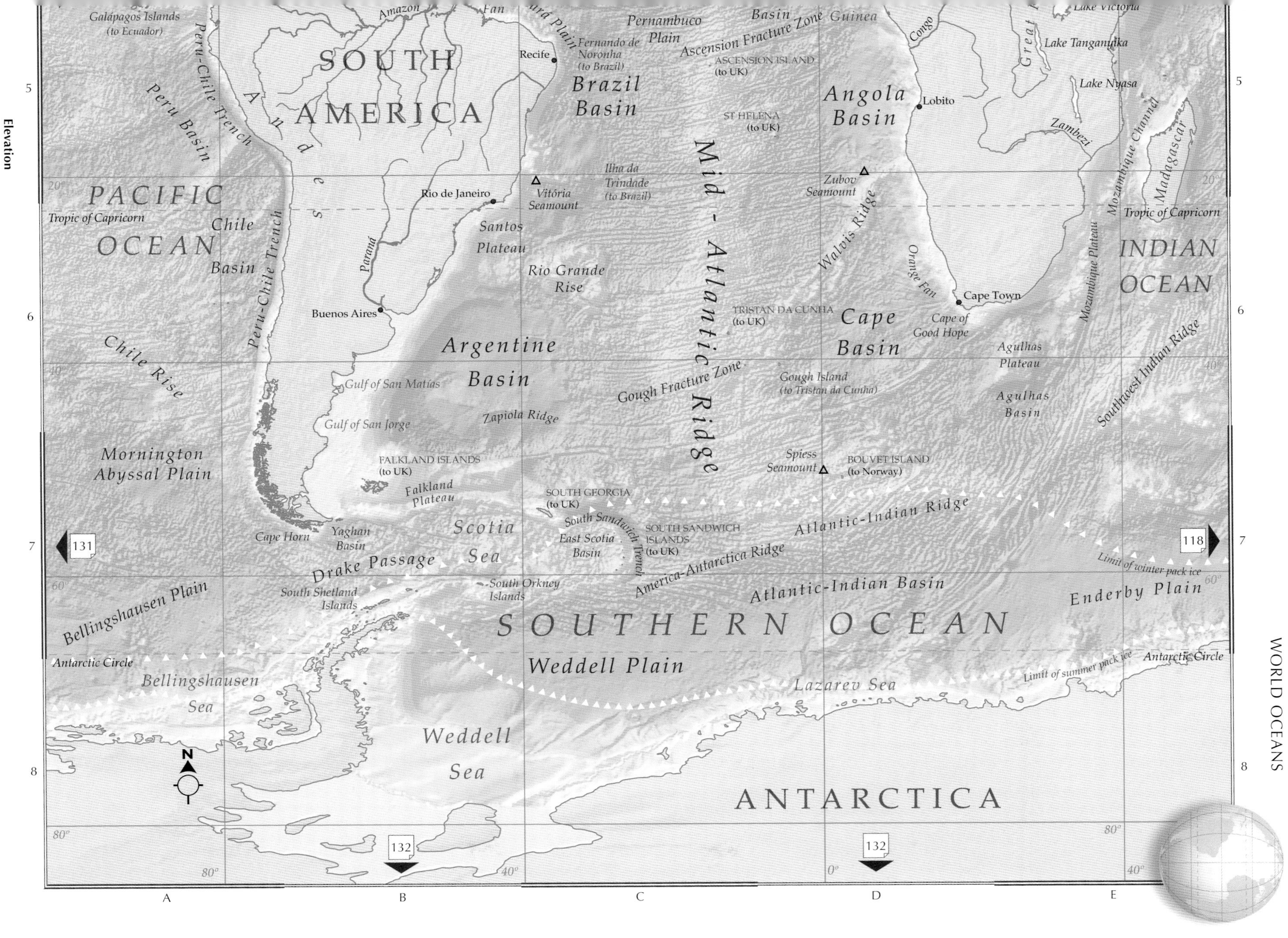
Galápagos Islands (to Ecuador)
Peru-Chile Trench
Peru Basin
SOUTH AMERICA
Andes
Amazon
Recife
Pernambuco Plain
Fernando de Noronha (to Brazil)
Brazil Basin
Basin
Ascension Fracture Zone
Guinea
ASCENSION ISLAND (to UK)
ST HELENA (to UK)
Angola Basin
Lobito
Congo
Lake Tanganyika
Lake Nyasa
Zambezi
Mozambique Channel
Madagascar
Tropic of Capricorn
PACIFIC OCEAN
Chile Basin
Rio de Janeiro
Vitória Seamount
Ilha da Trindade (to Brazil)
Santos Plateau
Rio Grande Rise
Paraná
Mid - Atlantic Ridge
Zubov Seamount
Walvis Ridge
Orange Fan
Cape Town
Cape of Good Hope
Mozambique Plateau
INDIAN OCEAN
Buenos Aires
TRISTAN DA CUNHA (to UK)
Cape Basin
Chile Rise
Argentine Basin
Gulf of San Matías
Gulf of San Jorge
Zapiola Ridge
Gough Fracture Zone
Gough Island (to Tristan da Cunha)
Agulhas Plateau
Agulhas Basin
Southwest Indian Ridge
Mornington Abyssal Plain
FALKLAND ISLANDS (to UK)
Falkland Plateau
Spiess Seamount
BOUVET ISLAND (to Norway)
SOUTH GEORGIA (to UK)
South Sandwich Trench
SOUTH SANDWICH ISLANDS (to UK)
East Scotia Basin
Atlantic-Indian Ridge
Cape Horn
Yaghan Basin
Scotia Sea
Drake Passage
America-Antarctica Ridge
Limit of winter pack ice
South Shetland Islands
South Orkney Islands
Atlantic-Indian Basin
Enderby Plain
Bellingshausen Plain
SOUTHERN OCEAN
Antarctic Circle
Weddell Plain
Limit of summer pack ice
Bellingshausen Sea
Lazarev Sea
Weddell Sea
ANTARCTICA
131
118
132
A B C D E

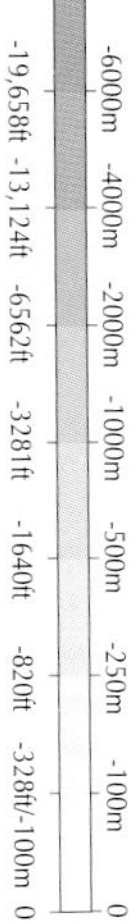
Elevation
-6000m -4000m -2000m -1000m -500m -250m -100m 0
-19,658ft -13,124ft -6562ft -3281ft -1640ft -820ft -328ft/-100m 0

Africa

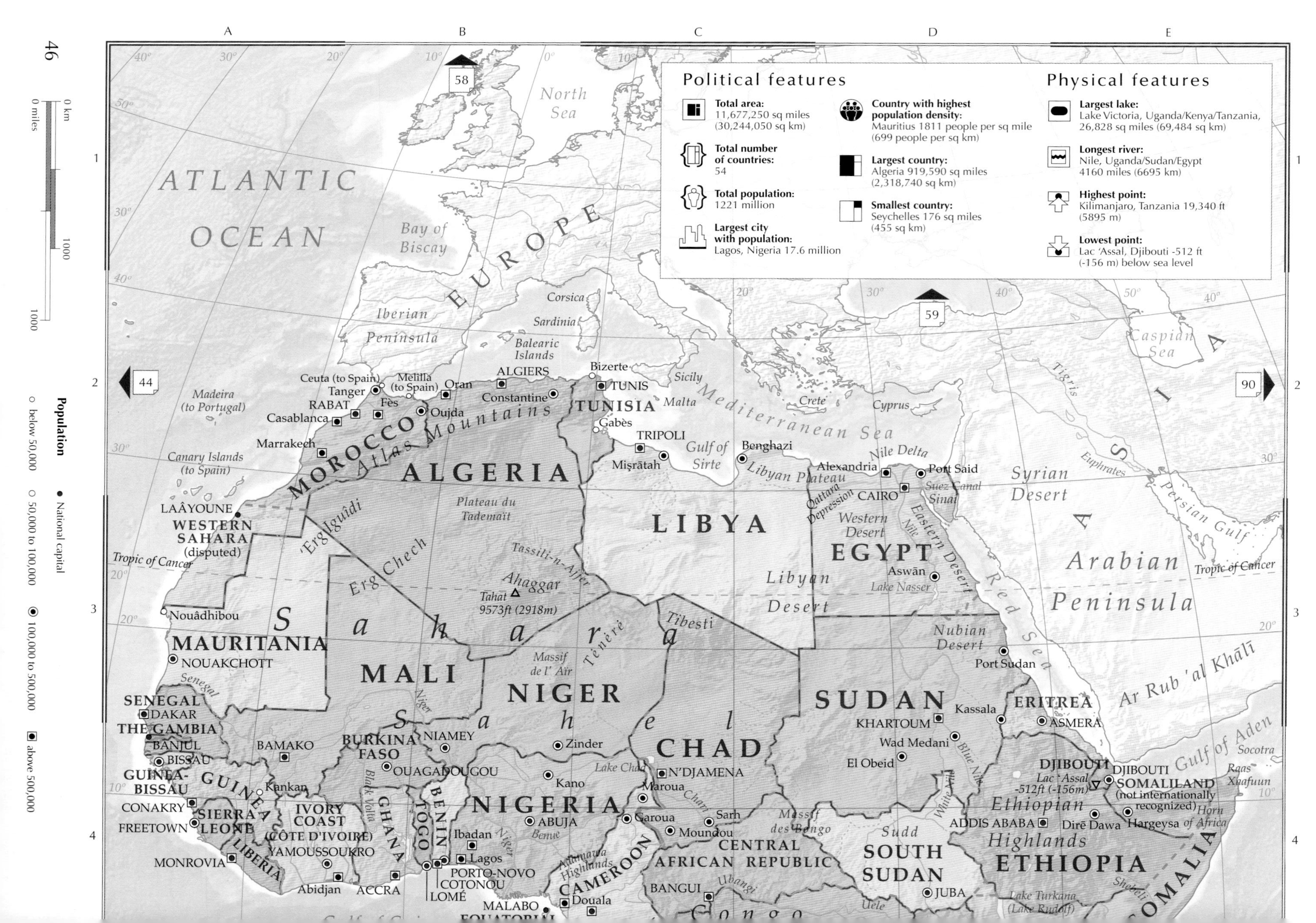

Political features

Total area:
11,677,250 sq miles
(30,244,050 sq km)

Total number of countries:
54

Total population:
1221 million

Largest city with population:
Lagos, Nigeria 17.6 million

Country with highest population density:
Mauritius 1811 people per sq mile
(699 people per sq km)

Largest country:
Algeria 919,590 sq miles
(2,318,740 sq km)

Smallest country:
Seychelles 176 sq miles
(455 sq km)

Physical features

Largest lake:
Lake Victoria, Uganda/Kenya/Tanzania,
26,828 sq miles (69,484 sq km)

Longest river:
Nile, Uganda/Sudan/Egypt
4160 miles (6695 km)

Highest point:
Kilimanjaro, Tanzania 19,340 ft
(5895 m)

Lowest point:
Lac 'Assal, Djibouti -512 ft
(-156 m) below sea level

Population   • National capital

○ below 50,000   ◎ 50,000 to 100,000   ◉ 100,000 to 500,000   ◼ above 500,000

0 km   1000

0 miles   1000

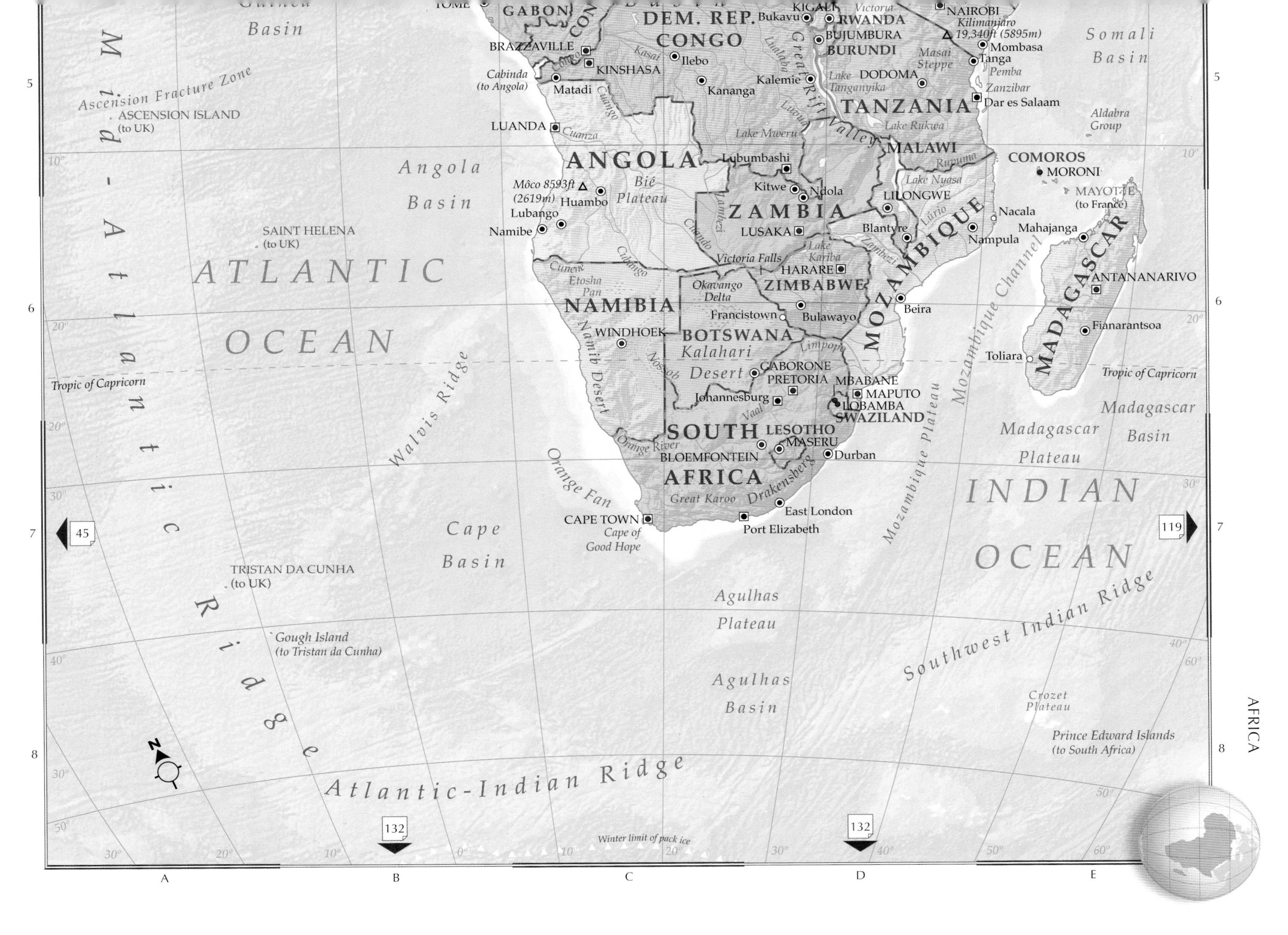

Somali Basin

Aldabra Group

COMOROS
MORONI
MAYOTTE (to France)

ANTANANARIVO
Fianarantsoa
Mahajanga
Nacala
Nampula
Toliara

MADAGASCAR

Madagascar Basin

Tropic of Capricorn

Mozambique Channel

Madagascar Plateau

INDIAN
OCEAN

Mozambique Plateau

Southwest Indian Ridge

Crozet Plateau

Prince Edward Islands (to South Africa)

NAIROBI
Kilimanjaro 19,340ft (5895m)
Mombasa
Tanga
Pemba
Zanzibar
Dar es Salaam
KIGALI
RWANDA
Bukavu
BUJUMBURA
BURUNDI
DODOMA

Masai Steppe

TANZANIA
MALAWI
LILONGWE
Blantyre
Beira

Lake Victoria
Lake Nyasa
Ruvuma
Lake Tanganyika
Great Rift Valley
Lukuga
Lake Rukwa

DEM. REP. CONGO
Kalemie
Lualaba
Lubumbashi
Kitwe
Ndola
Lake Mweru
Kananga
Luvua
Ilebo
Kasai

Agulhas Plateau
Agulhas Basin

ZAMBIA
LUSAKA
HARARE
ZIMBABWE
Bulawayo
Francistown

Lake Kariba
Zambezi
Victoria Falls
Okavango Delta
Kalahari Desert
Cuando
Cubango

MOZAMBIQUE
MAPUTO
MBABANE
SWAZILAND
LOBAMBA
MASERU
LESOTHO
Durban
East London
Port Elizabeth

Limpopo

ANGOLA
Huambo
Bié Plateau
Cuango
Cuanza
KINSHASA
MATADI
BRAZZAVILLE
Cabinda (to Angola)
LUANDA

GABON
CONGO

Môco 8593ft (2619m)
Lubango
Namibe

Etosha Pan
Cunene
Cubango
Okavango
Nossob

NAMIBIA
WINDHOEK
Namib Desert

BOTSWANA
GABORONE
PRETORIA
Johannesburg
BLOEMFONTEIN

SOUTH AFRICA

Great Karoo
Drakensberg
CAPE TOWN
Cape of Good Hope

Orange River
Orange Fan

Cape Basin

ATLANTIC
OCEAN

Angola Basin

SAINT HELENA (to UK)

Ascension Fracture Zone
ASCENSION ISLAND (to UK)

Walvis Ridge

TRISTAN DA CUNHA (to UK)
Gough Island (to Tristan da Cunha)

Mid-Atlantic Ridge

Atlantic-Indian Ridge

Winter limit of pack ice

Tropic of Capricorn

N

119
132
132
45

47

Northwest Africa

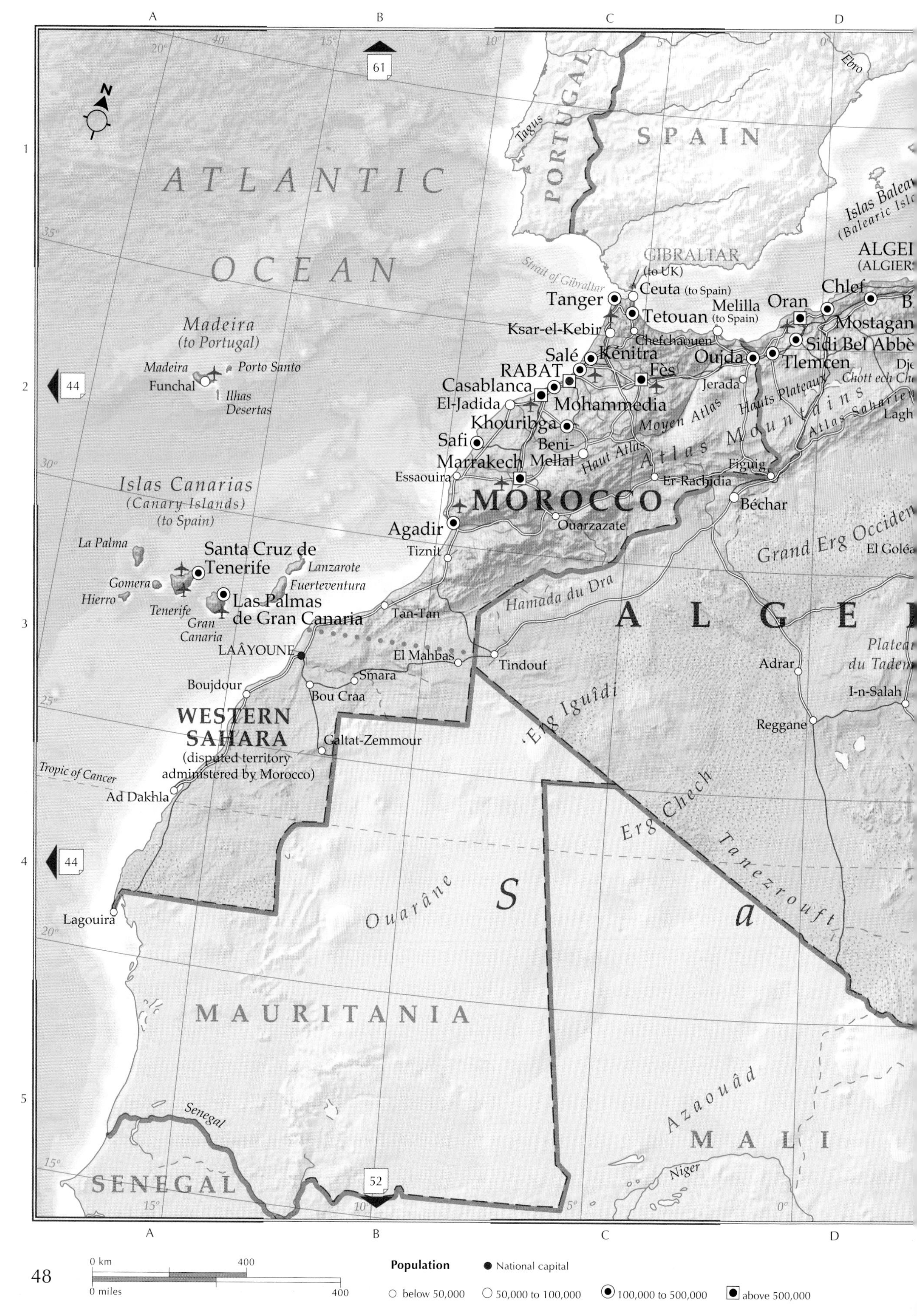

ATLANTIC

OCEAN

Madeira
(to Portugal)

Madeira *Porto Santo*
Funchal *Ilhas*
Desertas

Islas Canarias
(Canary Islands)
(to Spain)

La Palma
Santa Cruz de
Tenerife *Lanzarote*
Gomera *Fuerteventura*
Hierro *Gran*
Tenerife *Canaria*
Las Palmas
de Gran Canaria

LAÂYOUNE
Boujdour
Bou Craa
WESTERN
SAHARA
(disputed territory
administered by Morocco)
Tropic of Cancer
Ad Dakhla
Galtat-Zemmour

Lagouira

MAURITANIA

Senegal

SENEGAL

PORTUGAL SPAIN
Tagus *Ebro*

Islas Balear
(Balearic Isla

GIBRALTAR ALGIE
(to UK) (ALGIER)
Ceuta (to Spain)
Tanger Melilla Oran Chlef
Tetouan (to Spain) Mostagan
Ksar-el-Kebir Chefchaouen Sidi Bel Abbè
Salé Kénitra Oujda Tlemcen
RABAT Fès Jerada *Chott ech Che*
Casablanca Mohammedia *Moyen Atlas* *Hauts Plateaux* Lagh
El-Jadida *Atlas Saharien*
Khouribga *Haut Atlas* Figuig
Safi Beni- *Atlas Mountains*
Marrakech Mellal Er-Rachidia
Essaouira Béchar
MOROCCO *Grand Erg Occiden*
Agadir Ouarzazate El Golé
Tiznit
Hamada du Dra ALGE
Tan-Tan
Plateau
El Mahbas *du Tade*
Smara Tindouf Adrar
I-n-Salah
'Erg Iguîdi Reggane

Erg Chech *Tanezrouft*
S
a
Ouarâne

Azaouâd
MALI
Niger

0 km 400
0 miles 400

Population ● National capital
○ below 50,000 ◉ 50,000 to 100,000 ◉ 100,000 to 500,000 ■ above 500,000

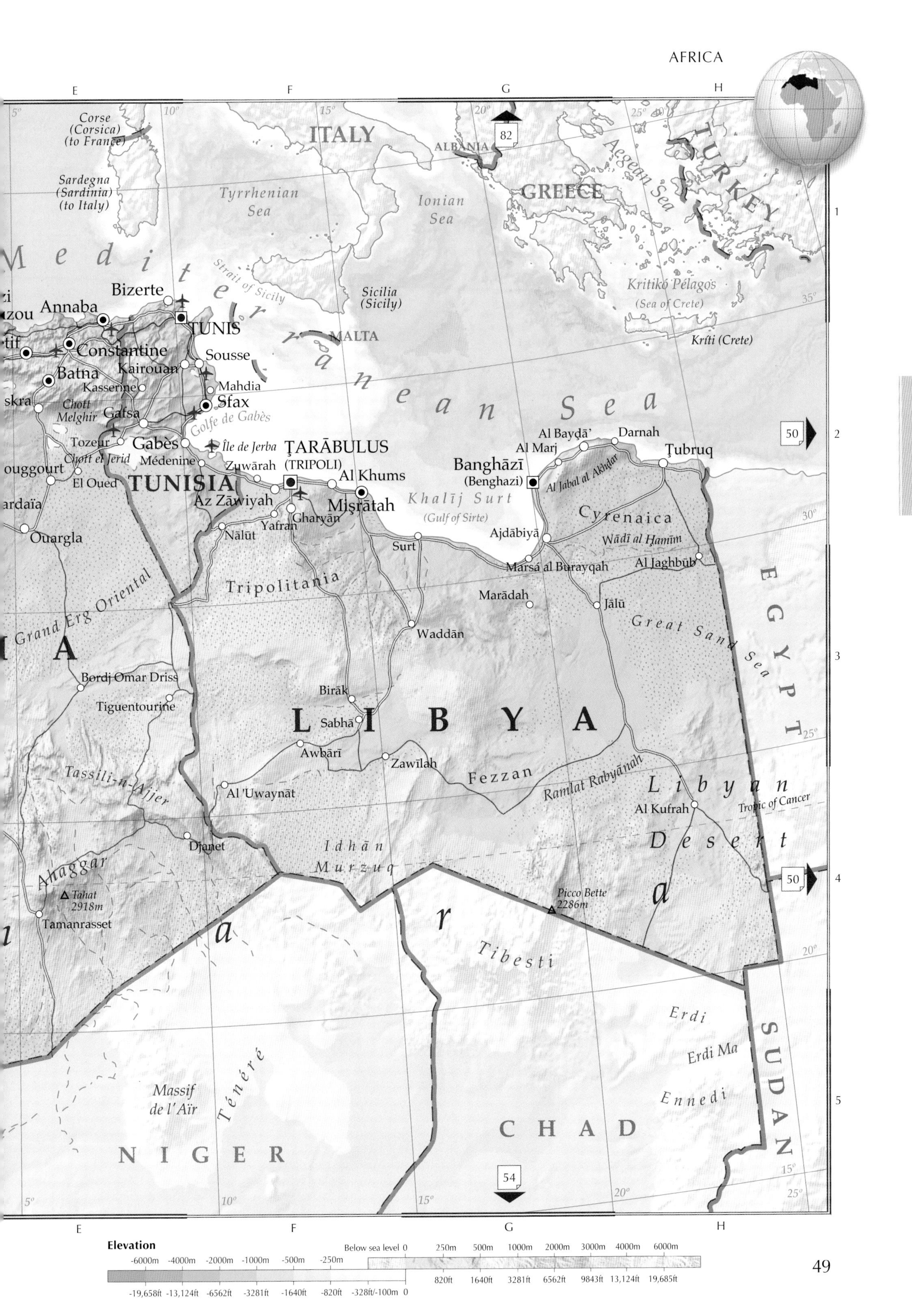

E   10°   F   15°   G   20°   25°   H

ITALY

ALBANIA   82

GREECE

Corse (Corsica) (to France)

Sardegna (Sardinia) (to Italy)

Tyrrhenian Sea

Aegean Sea

TURKEY

Ionian Sea

Kritikó Pélagos (Sea of Crete)

Strait of Sicily

Sicilia (Sicily)

MALTA

35°

Kríti (Crete)

zou   Annaba   Bizerte

tif   Constantine   TUNIS

Batna   Kairouan   Sousse

skra   Kasserine   Mahdia

Chott Melghir   Gafsa   Sfax

Tozeur   *Golfe de Gabès*

ouggourt   Gabès   *Île de Jerba*

Chott el Jerid   Médenine   ṬARĀBULUS (TRIPOLI)   Al Bayḍāʾ   Darnah

ardaïa   Zuwārah   Al Marj   Ṭubruq

TUNISIA   Al Khums   Banghāzī (Benghazi)   Al Jabal al Akhḍar

Ouargla   Az Zāwiyah   Miṣrātah

El Oued   Naʹlūt   Yafran   Gharyān   Surt   *Khalīj Surt (Gulf of Sirte)*   Ajdābiyā   *Cyrenaica*   50   2

30°

Marsá al Burayqah   Al Jaghbūb   E

Grand Erg Oriental   *Tripolitania*   Marādah   Jālū   *Wādī al Ḥamīm*   G

Waddān   *Great Sand Sea*   Y

Bordj Omar Driss   P   3

Tiguentourine   Birāk   T

Tassili-n-Ajjer   Sabhā   **L I B Y A**   25°

Awbārī   Zawīlah   *Fezzan*   *Ramlat Rabyānah*   *Libyan*

Al ʾUwaynāt   Al Kufrah   *Tropic of Cancer*   50   4

Djanet   *Idhān*   *Desert*

Ahaggar   *Murzuq*   20°

△ Tahat 2918m   Picco Bette 2286m △   *Tibesti*   a

Tamanrasset   r

S   5

Massif de l'Aïr   *Ténéré*   *Erdi*   U

Erdi Ma   D

Ennedi   A

N I G E R   **C H A D**   N

54   15°

5°   10°   15°   20°   25°

E   F   G   H

49

Elevation

-6000m   -4000m   -2000m   -1000m   -500m   -250m   Below sea level 0   250m   500m   1000m   2000m   3000m   4000m   6000m

-19,658ft   -13,124ft   -6562ft   -3281ft   -1640ft   -820ft   -328ft/-100m   0   820ft   1640ft   3281ft   6562ft   9843ft   13,124ft   19,685ft

Northeast Africa

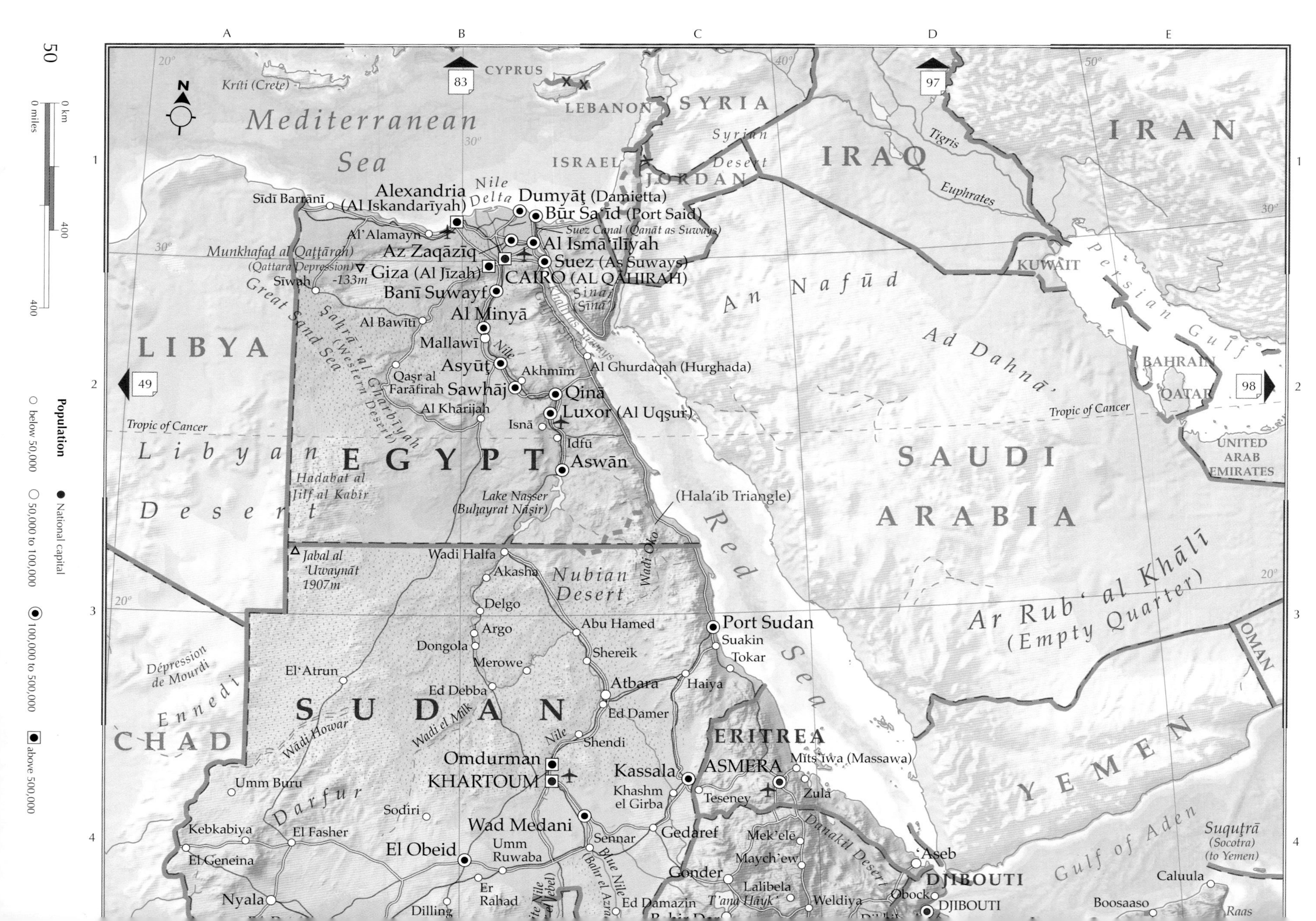

IRAN

IRAQ

SYRIA

LEBANON

CYPRUS

ISRAEL

JORDAN

KUWAIT

BAHRAIN

QATAR

UNITED ARAB EMIRATES

OMAN

SAUDI ARABIA

YEMEN

DJIBOUTI

ERITREA

LIBYA

EGYPT

SUDAN

CHAD

Mediterranean Sea

Kríti (Crete)

Persian Gulf

Red Sea

Gulf of Aden

Suquṭrā (Socotra) (to Yemen)

Tigris

Euphrates

Syrian Desert

An Nafūd

Ad Dahnā'

Ar Rub' al Khālī (Empty Quarter)

Nile Delta

Tropic of Cancer

Danakil Desert

Nubian Desert

Libyan Desert

Darfur

Ṣaḥrā' al Gharbīyah (Western Desert)

Great Sand Sea

Lake Nasser (Buḥayrat Nāṣir)

Gulf of Suez (Khalīj as Suways)

Gulf of Aqaba (Khalīj al 'Aqabah)

Sinai (Sīnā')

Suez Canal (Qanāt as Suways)

Blue Nile (Bahr el Azraq)

White Nile (Bahr el Jebel)

Munkhafaḍ al Qaṭṭārah (Qattara Depression) –133m

Ennedi

Dépression de Mourdi

Cities and places

Sidī Barrāni
Al Alamayn
Alexandria (Al Iskandarīyah)
Dumyāṭ (Damietta)
Būr Sa'īd (Port Said)
Al Ismā'īlīyah
Suez (As Suways)
CAIRO (AL QĀHIRAH)
Giza (Al Jīzah)
Az Zaqāzīq
Banī Suwayf
Al Minyā
Al Bawīti
Mallawī
Asyūṭ
Akhmīm
Qaṣr al Farāfirah
Sawhāj
Al Khārijah
Qinā
Luxor (Al Uqṣur)
Isnā
Idfū
Aswān
Siwah
Jabal al 'Uwaynāt 1907m
Hadabat al Jilf al Kabīr

Port Sudan
Suakin
Tokar
Haiya
Abu Hamed
Akasha
Wadi Halfa
Delgo
Argo
Merowe
Dongola
Ed Debba
Shereik
Atbara
Ed Damer
Shendi
Khashm el Girba
Kassala
Gedaref
Sennar
Wad Medani
Umm Ruwaba
Er Rahad
El Obeid
Omdurman
KHARTOUM
Sodiri
El'Atrun
Umm Burru
Kebkabiya
El Fasher
El Geneina
Dilling
Nyala
Wadi el Milk
Wadi el Melik
Wadi Howar

Mits'iwa (Massawa)
Zula
ASMERA
Teseney
Mek'elē
Maych'ew
Tanu Hayk'
Lalibela
Weldiya
Gonder
Ed Damazin
Raas
Calula
Boosaaso
Djibouti
DJIBOUTI
Obock
Asseb

Wadi Oko
(Hala'ib Triangle)

References

83
97
98
49

0 km   400
0 miles   400

Population

● National capital

○ below 50,000

◯ 50,000 to 100,000

◉ 100,000 to 500,000

▣ above 500,000

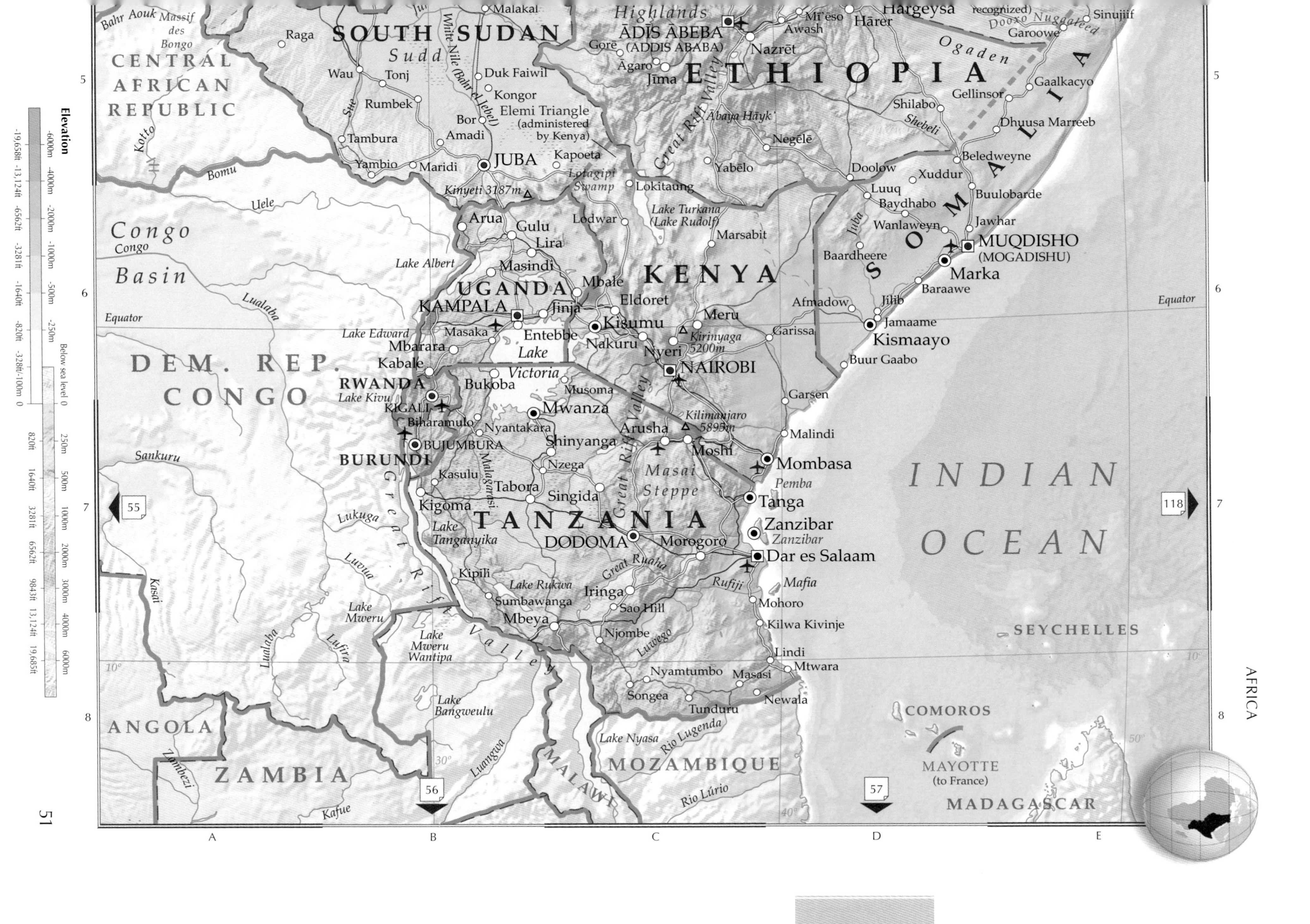

118

E

D

SEYCHELLES

COMOROS

MAYOTTE
(to France)

57

MADAGASCAR

INDIAN
OCEAN

not recognized)
Dooso Nugaaleed Sinujiif
Garoowe Gaalkacyo

O g a d e n

Hargeysa
Härer
Mi'eso Härer
Awash

ETHIOPIA

Nazret
ĀDĪS ĀBEBA
(ADDIS ABABA)

Highlands

Gore Jima
Agaro

Abaya Hāyk'

Negēlē

Yabelo

Gellinsor
Shilabo
Shebeli

Dhuusa Marreeb
Beledweyne
Buulobarde
Jawhar

Xuddur
Doolow
Luuq
Baydhabo
Wanlaweyn

MUQDISHO
(MOGADISHU)
Marka
Baraawe

SOMALIA

Juba
Baardheere
Jilib
Jamaame
Kismaayo
Buur Gaabo

Afmadow
Garissa
Garsen
Malindi

Great Rift Valley

Mombasa
Pemba
Tanga
Zanzibar
Zanzibar
Dar es Salaam
Mafia
Mohoro
Kilwa Kivinje
Lindi
Mtwara
Newala

Marsabit
Meru
Lake Turkana
(Lake Rudolf)
Lokitaung

KENYA

Eldoret
Nakuru
Nyeri
Kirinyaga
5200m
NAIROBI
Kilimanjaro
5895m
Moshi
Arusha

Masai
Steppe

Morogoro
Rufiji
Great Ruaha

Iringa
Sao Hill
Niombe

Lugenga

Lake Nyasa / Lake Malawi

MOZAMBIQUE

MALAWI

Rio Lúrio

Nyamtumbo
Songea
Tundura
Masasi

Lwangwa

Mbale
Kisumu
Nakuru
Entebbe
Musoma
Mwanza
Shinyanga
Nzega
Singida
Tabora
DODOMA
Kilosa

Kampala
Masaka
Bukoba
Nyantakara
Malagarasi

UGANDA

TANZANIA

Mbeya
Sumbawanga
Kipili
Lake Rukwa

Lotagipi
Swamp
Lokichokio
Lodwar
Lira
Masindi
Arua
Gulu

JUBA
Kinyeti 3187m

Elemi Triangle
(administered
by Kenya)
Kapoeta

SOUTH
SUDAN

Sudd

White Nile (Bahr el Jebel)

Bor
Amadi
Maridi
Yambio
Tambura

Malakal
Duk Faiwil
Kongor

Raga
Wau
Tonj
Rumbek

Bahr Aouk
Massif des Bongo

CENTRAL
AFRICAN
REPUBLIC

Kotto
Uele
Bomu

Congo
Basin

Lualaba
Congo

DEM. REP.
CONGO

Sankuru
Kasai

Equator

Lake Albert
Lake Edward
Lake Kivu
Kabale
Mbarara

RWANDA
KIGALI
Biharamulo
Kasulu
Kigoma

BURUNDI
BUJUMBURA

Lake
Victoria
Jinja
Bukoba

Lukuga
Lake
Tanganyika

Luvua
Lake Mweru

Lufira

Lake Mweru
Wantipa

Lake
Bangweulu

Luangwa

ZAMBIA

ANGOLA

Zambezi
Kafue
Lualaba

Great Rift Valley

55

56

10°   50°   40°   30°   20°   10°

Equator

5   6   7   8

Elevation

Below sea level 0   250m   500m   1000m   2000m   3000m   4000m   6000m

-6000m -4000m -2000m -1000m -500m -250m

-19,658ft -13,124ft -6562ft -3281ft -1640ft -820ft -328ft/-100m 0   820ft   1640ft   3281ft   6562ft   9843ft 13,124ft 19,685ft

West Africa

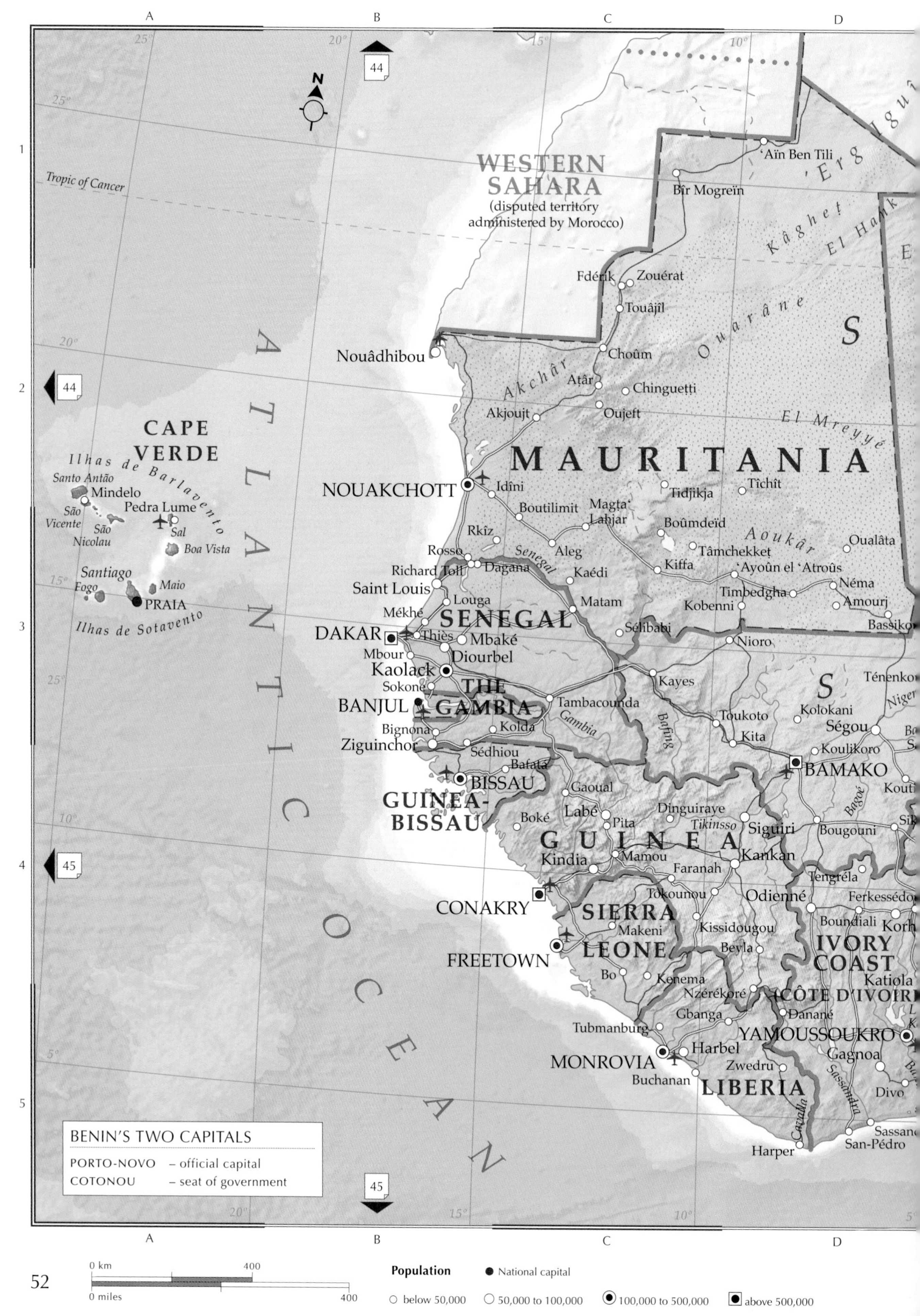

A | B | C | D

WESTERN
SAHARA
(disputed territory
administered by Morocco)

'Aïn Ben Tili
Bîr Mogreïn

Erg

Kâghet

El Hank

Fdérik • Zouérat
Touâjil

Ouarâne

S

Nouâdhibou

Akchâr

Atâr • Chinguetti

El Mreyyé

Akjoujt • Oujeft

MAURITANIA

NOUAKCHOTT • Idîni

Boutilimit

Tidjikja

Tîchît

CAPE
VERDE

Ilhas de Barlavento

Santo Antão
Mindelo
São
Vicente Pedra Lume
São Sal
Nicolau Boa Vista

Rkîz

Rosso
Richard Toll • Dagana

Senegal

Aleg

Magta
Lahjar

Boûmdeïd

Aoukâr

Kaédi
Kiffa • Tâmchekket

Oualâta

'Ayoûn el 'Atroûs

Néma

Amourj

Santiago
Fogo Maio

PRAIA

Ilhas de Sotavento

Saint Louis

Mékhé

DAKAR • Thiès
Mbour
Kaolack
Sokone

SENEGAL

Mbaké
Diourbel

Louga

Matam

Sélibabi

Kobenni

Timbedgha

Nioro

Bassiko

THE
BANJUL GAMBIA

Bignona
Ziguinchor

Sédhiou

Kolda

Gambia

Tambacounda

Kayes

Toukoto

Bafing

Kita

Ténenkou

Niger

Kolokani

Ségou
Koulikoro

S

BISSAU

Bafatá

Gaoual

GUINEA-
BISSAU

Boké

Labé

Dinguiraye

Pita

Tikinsso

Siguiri

Bougouni

Bagoé

Sik

Kouti

BAMAKO

CONAKRY

GUINEA

Kindia

Mamou

Faranah

Tokounou

Kankan

Odienné

Tengréla

Ferkéssédo

Korh

FREETOWN

SIERRA
LEONE

Makeni

Kissidougou

Beyla

Boundiali

IVORY
COAST

Katiola

Bo

Kenema

Nzérékoré

CÔTE D'IVOIR

Gbanga

Tubmanburg

MONROVIA Harbel

Buchanan

Danané

YAMOUSSOUKRO

Gagnoa

Zwedru

LIBERIA

Sassandra

Divo

Cavalla

Harper

Sassand
San-Pédro

BENIN'S TWO CAPITALS

PORTO-NOVO – official capital
COTONOU – seat of government

A | B | C | D

0 km   400
0 miles   400

Population • National capital

○ below 50,000   ◯ 50,000 to 100,000   ◉ 100,000 to 500,000   ▣ above 500,000

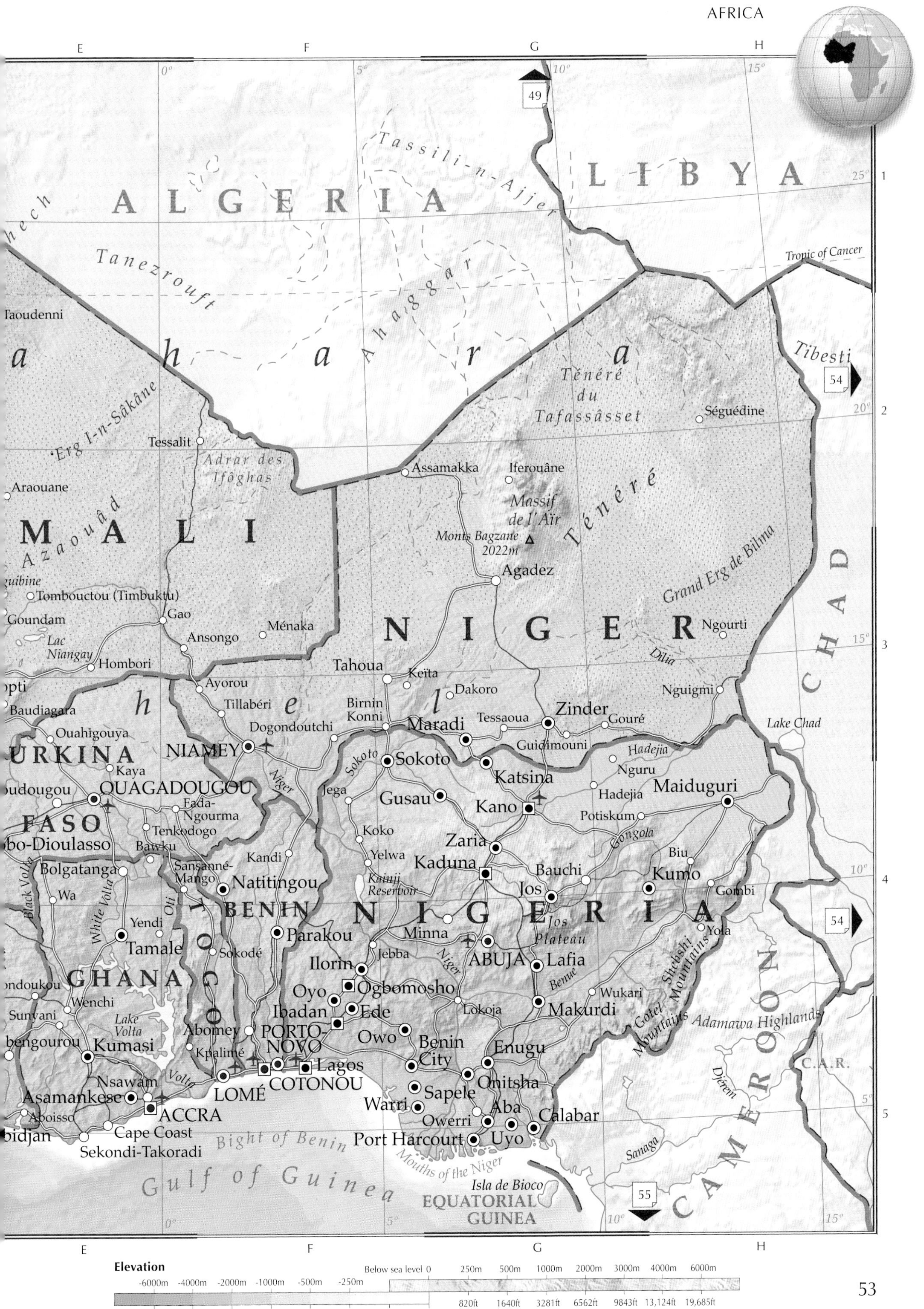

ALGERIA

LIBYA

Tassili-n-Ajjer

Tanezrouft

Taoudenni

25° 1

Tropic of Cancer

S   a   h   a   r   a

Ahaggar

Ténéré du Tafassâsset

Séguédine

Tibesti

54

'Erg I-n-Sâkâne

Tessalit

Adrar des Ifôghas

20° 2

Araouane

Assamakka   Iferouâne

Ténéré

M   A   L   I

Massif de l'Aïr

Azaouâd

Monts Bagzane △ 2022m

Grand Erg de Bilma

Aguibine

Agadez

C   H   A   D

Tombouctou (Timbuktu)

Gao

N   I   G   E   R

Ngourti

Goundam

Ansongo   Ménaka

Dîlia

Lac Niangay

15° 3

Hombori

Tahoua   Keïta

Dakoro

Nguigmi

Lake Chad

opti

h   Ayorou   *e*

Tillabéri

Birnin Konni   Maradi   Tessaoua   Zinder   Gouré

Baudiagara

Dogondoutchi   *l*

Hadejia

Ouahigouya   NIAMEY   Sokoto   Guidimouni   Nguru

URKINA   Kaya   *Sokoto*   Katsina   Hadejia   Maiduguri

oudougou   OUAGADOUGOU   Jega   Gusau   Kano

FASO   Fada-Ngourma   Koko   Zaria   Potiskum   *Gongola*   Biu

bo-Dioulasso   Tenkodogo   Yelwa   Kaduna   Bauchi   Kumo   Gombi

Bolgatanga   Bawku   Kandi   Sansanné-   *Kainji*   Jos   10° 4

Wa   Mango   Natitingou   *Reservoir*   NIGERIA   54

Yendi   BENIN   Minna   *Jos Plateau*   Yola

Tamale   Parakou   *Niger*   ABUJA   Lafia   *Shebshi Mountains*

GHANA   Sokodé   Ilorin   Jebba   *Benue*   Wukari   *Gotel Mountains*   *Adamawa Highlands*

ondoukou   Oyo   Ogbomosho   Lokoja   Makurdi   C.A.R.

Wenchi   Abomey   Ibadan   Ede   Owo   Benin   *Déyem*

Sunyani   Kpalimé   PORTO-   City   Enugu   5°

bengourou   Kumasi   NOVO   Owo   Onitsha

Nsawam   LOMÉ   COTONOU   Lagos   Sapele   Aba   Calabar

Asamankese   ACCRA   Warri   Owerri   Uyo

Aboisso   Cape Coast   *Bight of Benin*   Port Harcourt   Uyo

bidjan   Sekondi-Takoradi   *Mouths of the Niger*   *Sanaga*   CAMEROON

Gulf of Guinea   *Isla de Bioco*   55   15°

EQUATORIAL   10°

GUINEA

E   F   G   H

Central Africa

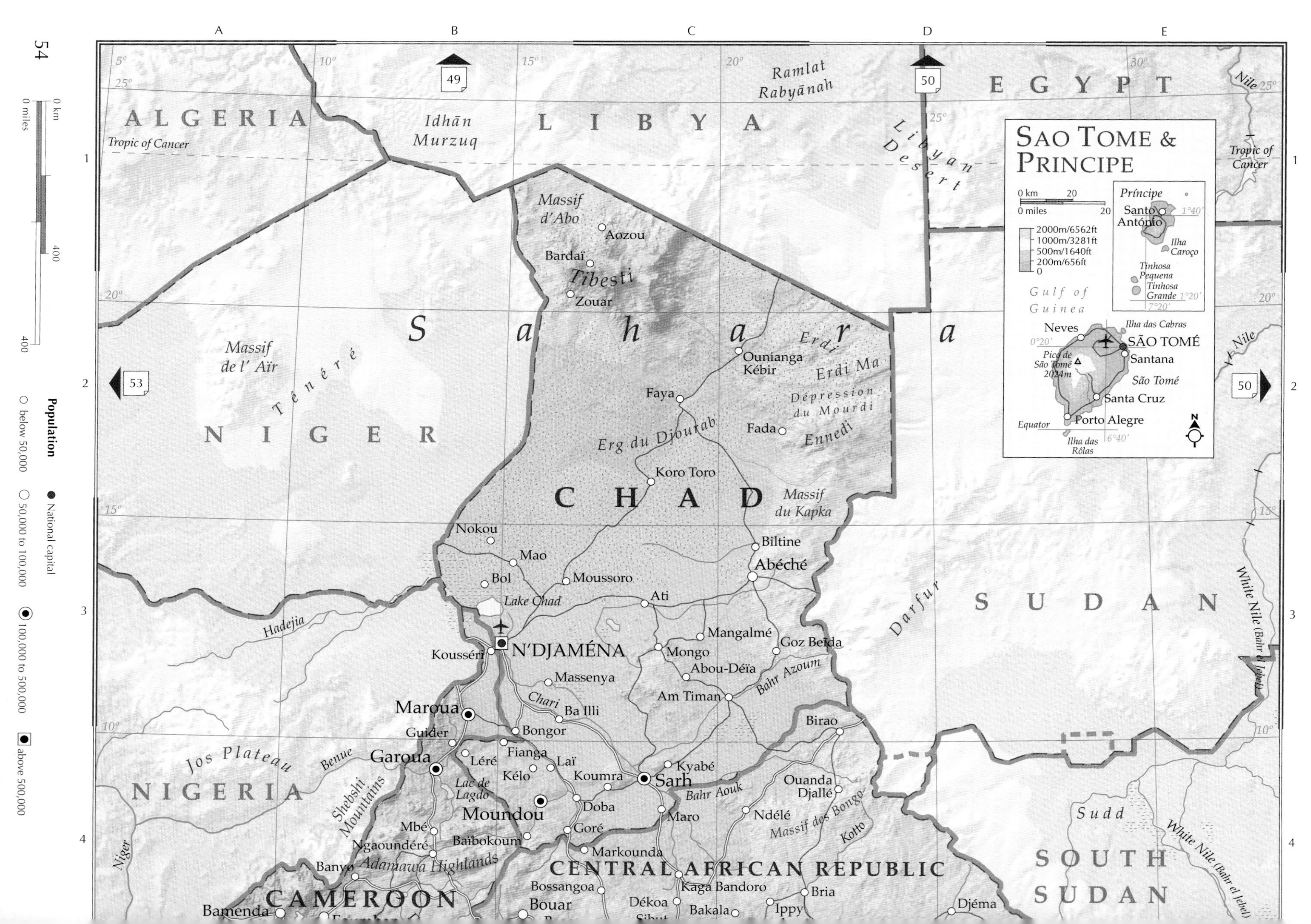

SÃO TOMÉ & PRINCIPE

Príncipe
Santo António
Ilha Caroço
Tinhosa Pequena
Tinhosa Grande

0 km   20
0 miles   20

2000m/6562ft
1000m/3281ft
500m/1640ft
200m/656ft
0

Ilha das Cabras
SÃO TOMÉ
Santana
São Tomé
Santa Cruz
Neves
Porto Alegre
Pico de São Tomé 2024m
Ilha das Rôlas
Equator

Gulf of Guinea

EGYPT

Nile

Tropic of Cancer

Libyan Desert

Ramlat Rabyānah

LIBYA

Idhān Murzuq

ALGERIA

Tropic of Cancer

NIGER

Massif de l' Aïr

Ténéré

SUDAN

Darfur

SOUTH SUDAN

White Nile (Bahr el Jebel)

White Nile (Bahr el Jebel)

Sudd

Djéma

CENTRAL AFRICAN REPUBLIC

Birao
Ouanda Djallé
Ndélé
Bria
Ippy
Kaga Bandoro
Bakala
Bamingui
Dékoa
Bossangoa
Bouar
Bozoum

Kotto
Massif des Bongo

SAHARA

CHAD

Erdi
Erdi Ma
Dépression du Mourdi
Ennedi
Massif du Kapka
Ounianga Kébir
Fada
Biltine
Abéché
Goz Beïda
Mangalmé
Abou-Déïa
Mongo
Am Timan
Ati
Koro Toro
Faya

Erg du Djourab

Bahr Azoum
Bahr Aouk

Sarh
Kyabé
Maro
Koumra
Doba
Goré
Moundou

Moussoro
Massenya
N'DJAMÉNA
Chari
Ba Illi
Bongor
Fianga
Kélo
Léré
Lac de Léré
Lai
Baïbokoum
Markounda

Mao
Bol
Lake Chad
Nokou
Koussèri
Maroua
Guider
Garoua
Mbé
Ngaoundéré
Banyo
Adamawa Highlands

Massif d'Abo
Aozou
Bardaï
Tibesti
Zouar

Shebshi Mountains
Jos Plateau
Benue

NIGERIA
Hadejia
Niger
Bamenda

CAMEROON

Population

● National capital

○ below 50,000
◎ 50,000 to 100,000
◉ 100,000 to 500,000
▣ above 500,000

0 km   400
0 miles   400

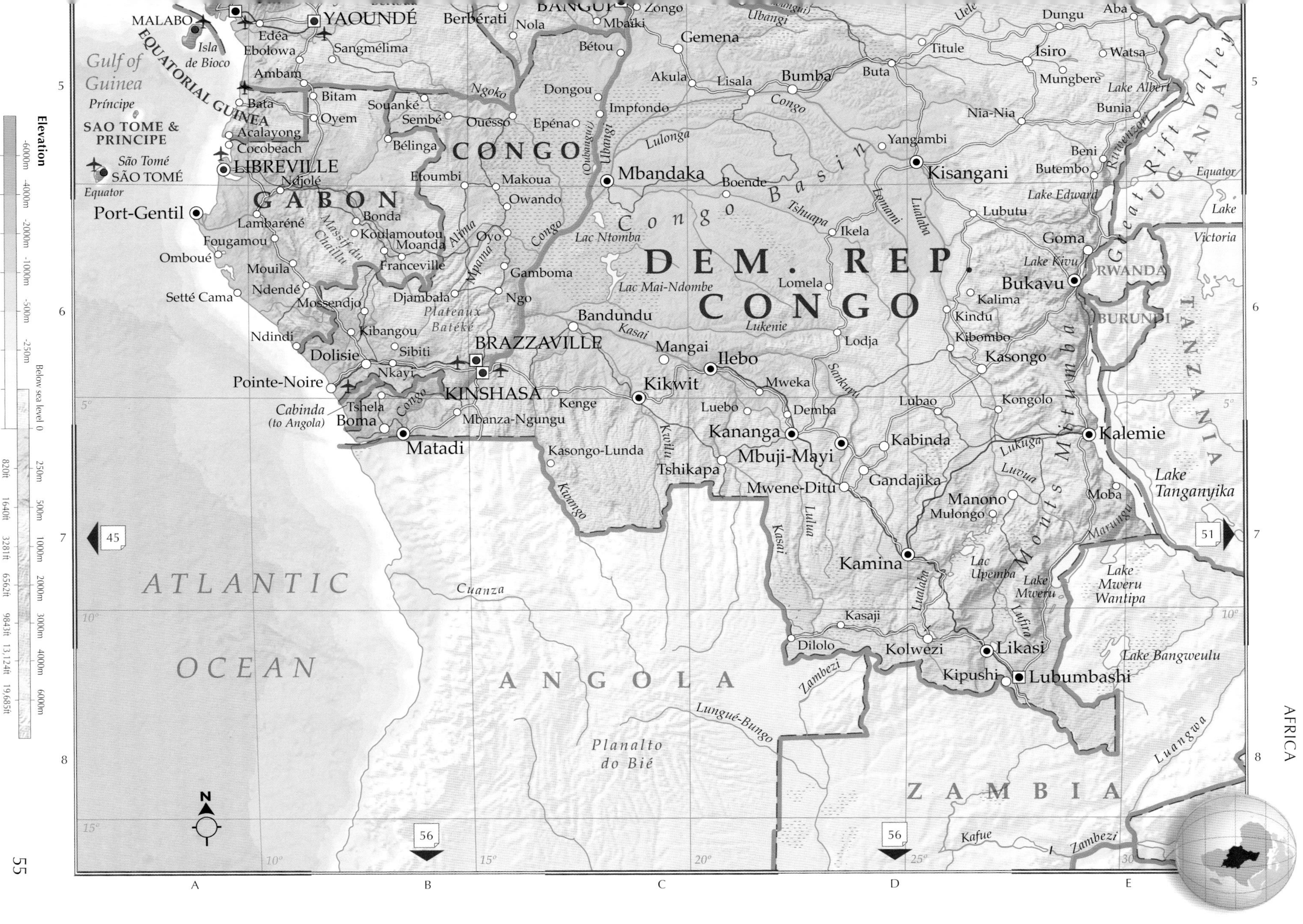

Elevation

-6000m	-4000m	-2000m	-1000m	-500m	-250m	Below sea level 0	250m	500m	1000m	2000m	3000m	4000m	6000m
-19,658ft	-13,124ft	-6562ft	-3281ft	-1640ft	-820ft	-328ft/-100m 0		820ft	1640ft	3281ft	6562ft	9843ft	13,124ft	19,685ft

Southern Africa

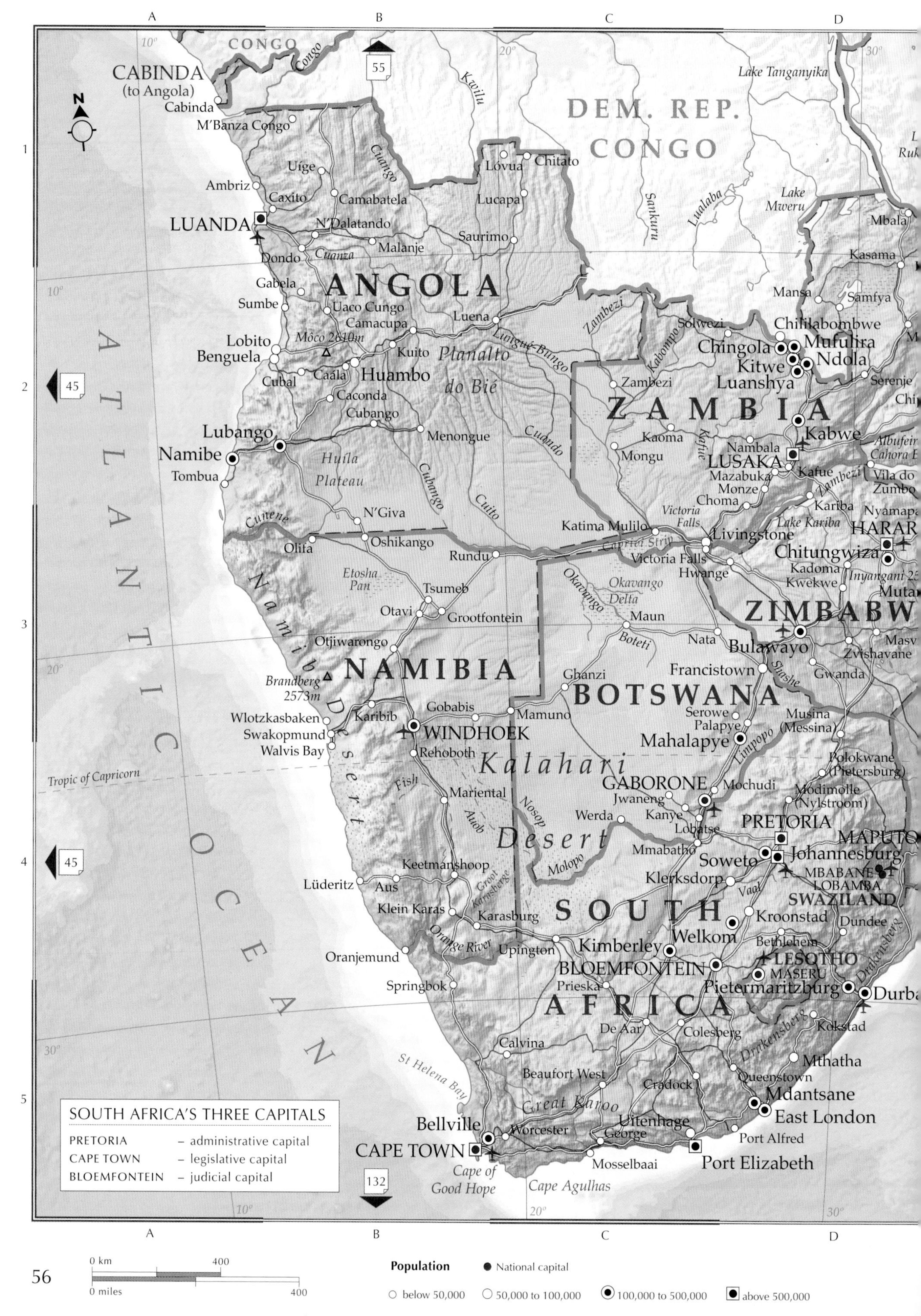

CABINDA (to Angola)
Cabinda
M'Banza Congo
CONGO
Congo
Congo
Kwilu

DEM. REP. CONGO
Lake Tanganyika
Lake Mweru
Lualaba
Sankuru
Mbala
Kasama

Uíge
Ambriz
Caxito
Camabatela
Lovua
Chitato
Lucapa
Mansa
Samfya

LUANDA
N'Dalatando
Saurimo
Solwezi
Chililabombwe
Mufulira
Serenje
Chi

Dondo
Cuanza
Malanje
Luena
Zambezi
Chingola
Ndola
Kitwe
Luanshya

Gabela
Sumbe
Uaco Cungo
Camacupa
Kaoma
Nambala
Kabwe
Albufeir
Cahora

ANGOLA
Lingué-Bungo
ZAMBIA
Mongu
LUSAKA
Vila do
Zúmbo

Lobito
Benguela
Môco 2610m
Kuito
Planalto do Bié
Mazabuka
Monze
Kafue
Kariba
Nyamapa

Cubal
Caála
Huambo
Cuito
Choma
HARAR

Caconda
Cubango
Menongue
Cuando
Victoria Falls
Livingstone
Chitungwiza

Lubango
Namibe
Tombua
Huíla Plateau
Cubango
Katima Mulilo
Caprivi Strip
Victoria Falls
Hwange
Kadoma
Kwekwe
Inyangani 2
Muta

N'Giva
Olita
Oshikango
Rundu
Okavango
Okavango Delta
Maun
Nata
Bulawayo
ZIMBABW

Etosha Pan
Tsumeb
Grootfontein
Boteti
Francistown
Zvishavane
Gwanda
Masv

Otavi
Otjiwarongo
Ghanzi
BOTSWANA
Serowe
Palapye
Musina (Messina)

Brandberg 2573m
NAMIBIA
Gobabis
Mamuno
Mahalapye
Polokwane (Pietersburg)

Wlotzkasbaken
Swakopmund
Walvis Bay
Karibib
WINDHOEK
Rehoboth
Kalahari
Modimolle (Nylstroom)

Tropic of Capricorn
Fish
Mariental
Desert
Jwaneng
Werda
GABORONE
Mochudi
Kanye
Lobatse
PRETORIA
MAPUTO

Auob
Nosop
Mmabatho
Johannesburg
MBABANE
LOBAMBA
SWAZILAND

Lüderitz
Aus
Klein Karas
Groot Karasberge
Molopo
SOUTH
Klerksdorp
Vaal
Kroonstad
Dundee

Oranjemund
Karasburg
Kimberley
Welkom
Bethlehem
LESOTHO

Springbok
Orange River
Upington
BLOEMFONTEIN
MASERU
Pietermaritzburg
Durba

Prieska
AFRICA
Kokstad

Calvina
De Aar
Colesberg
Mthatha

Beaufort West
Cradock
Queenstown
Mdantsane

St Helena Bay
Great Karoo
George
Uitenhage
East London

Bellville
Worcester
Port Alfred

CAPE TOWN
Cape of Good Hope
Mosselbaai
Cape Agulhas
Port Elizabeth

ATLANTIC OCEAN
Namib Desert

SOUTH AFRICA'S THREE CAPITALS

PRETORIA — administrative capital
CAPE TOWN — legislative capital
BLOEMFONTEIN — judicial capital

0 km   400
0 miles   400

Population   ● National capital

○ below 50,000   ○ 50,000 to 100,000   ◉ 100,000 to 500,000   ◼ above 500,000

45   55   132

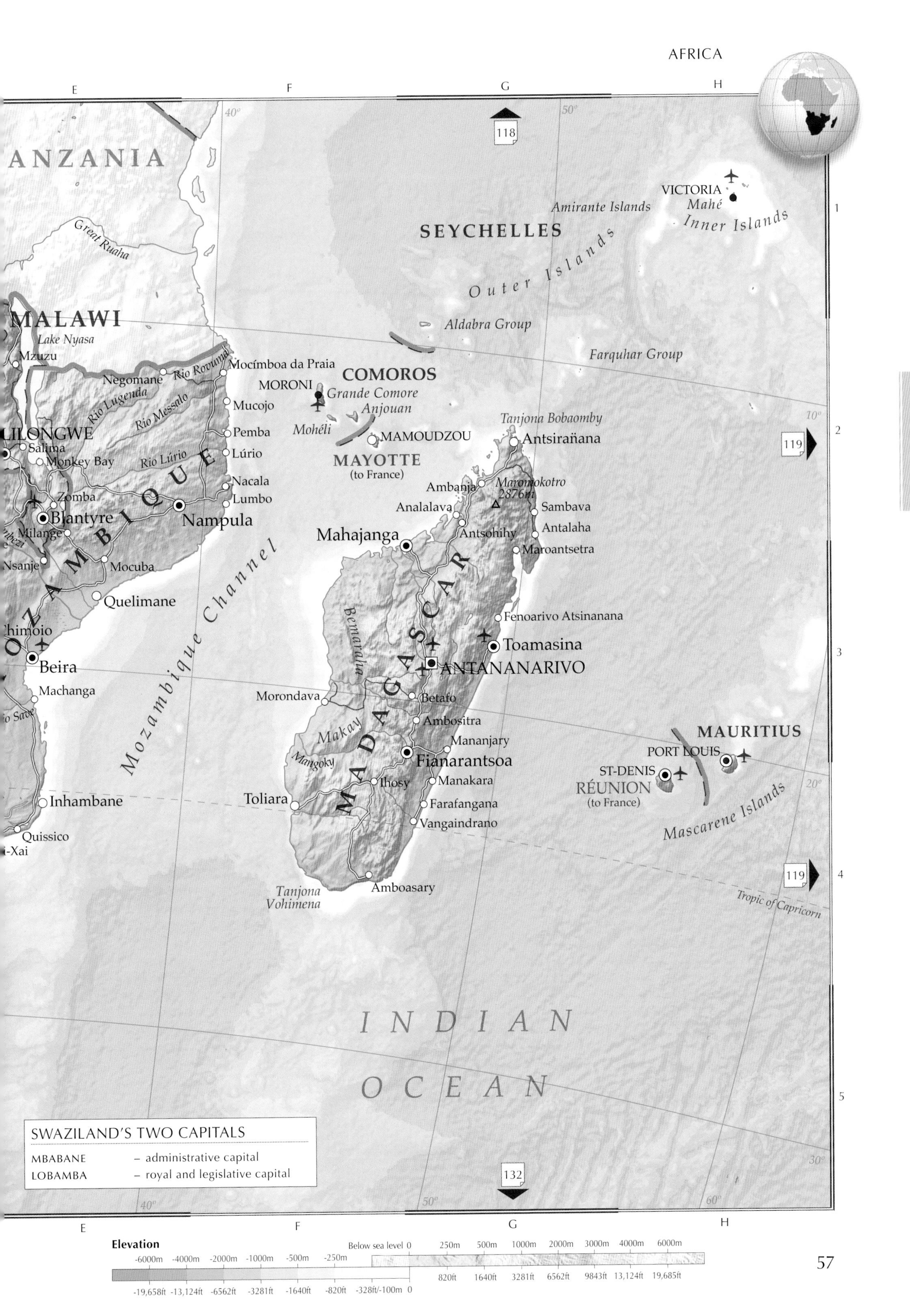

E   F   G   H

118

40°

50°

TANZANIA

Great Ruaha

SEYCHELLES

Amirante Islands

VICTORIA

Mahé

Inner Islands

1

Outer Islands

MALAWI

Lake Nyasa

Mzuzu

Negomane

Rio Rovuma

Mocímboa da Praia

Aldabra Group

Farquhar Group

COMOROS

MORONI

Grande Comore

Anjouan

10°

2

Rio Lugenda

Rio Messalo

LILONGWE

Salima

Monkey Bay

Rio Lúrio

Mucojo

Pemba

Mohéli

MAMOUDZOU

MAYOTTE

(to France)

Tanjona Bobaomby

Antsirañana

119

Zomba

Lúrio

Ambanja

Maromokotro
2876m

Blantyre

Milange

Nacala

Analalava

△

Sambava

Nampula

Lumbo

Antsohihy

Antalaha

Mocuba

Mahajanga

Maroantsetra

MOZAMBIQUE

Quelimane

Bemaraha

Fenoarivo Atsinanana

3

Chimoio

Beira

Mozambique Channel

MADAGASCAR

Toamasina

ANTANANARIVO

Machanga

Morondava

Betafo

Rio Save

Makay

Ambositra

MAURITIUS

Inhambane

Mangoky

Mananjary

PORT LOUIS

Fianarantsoa

ST-DENIS

Quissico

Ihosy

Manakara

RÉUNION

(to France)

20°

Xai-Xai

Toliara

Farafangana

Mascarene Islands

Vangaindrano

Amboasary

Tanjona
Vohimena

119

4

Tropic of Capricorn

I N D I A N

O C E A N

5

SWAZILAND'S TWO CAPITALS

MBABANE   – administrative capital

LOBAMBA   – royal and legislative capital

132

40°   50°   60°

30°

E   F   G   H

Elevation

						Below sea level 0	250m	500m	1000m	2000m	3000m	4000m	6000m

-6000m   -4000m   -2000m   -1000m   -500m   -250m

820ft   1640ft   3281ft   6562ft   9843ft   13,124ft   19,685ft

-19,658ft   -13,124ft   -6562ft   -3281ft   -1640ft   -820ft   -328ft/-100m   0

Europe

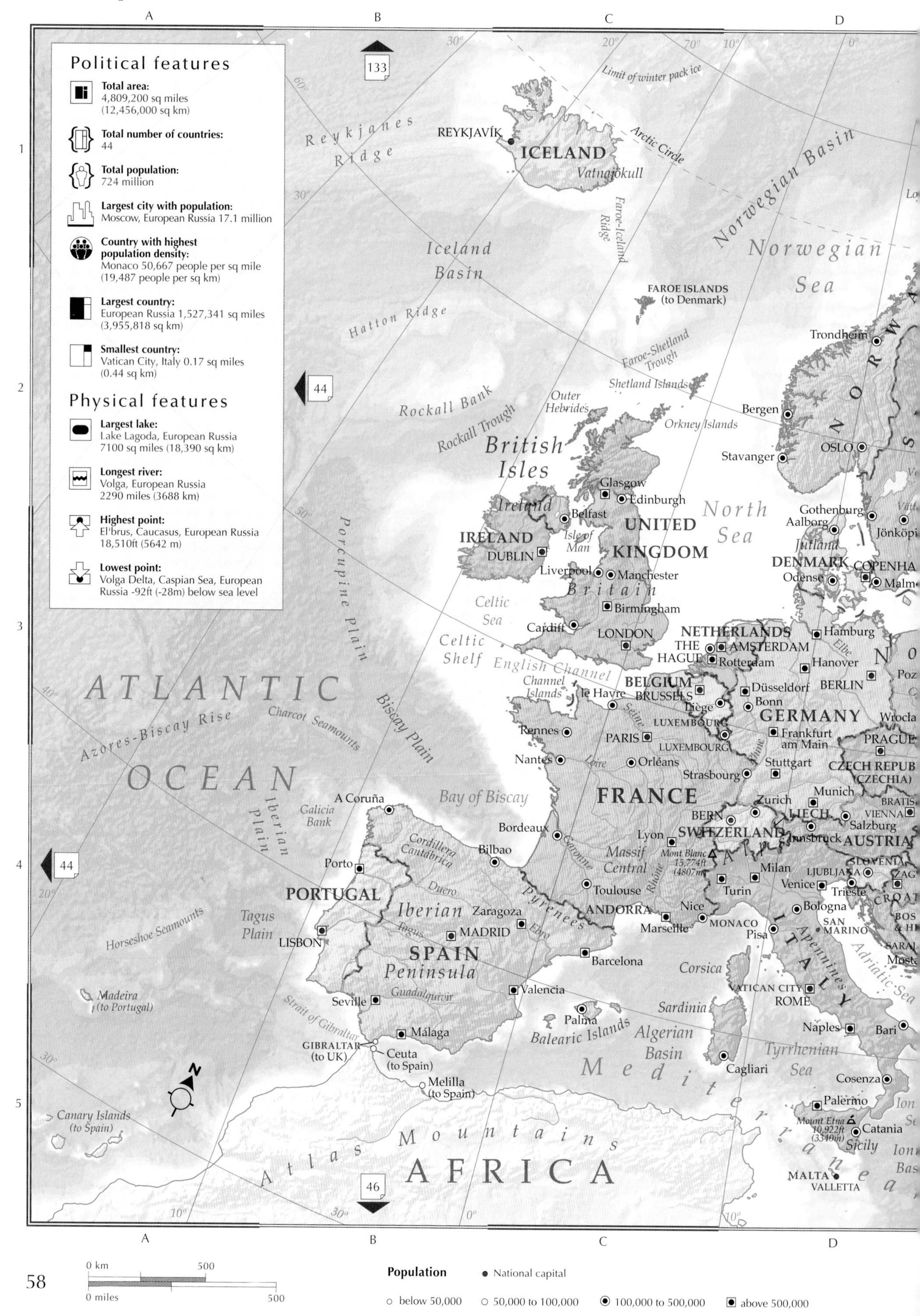

133

Political features

Total area:
4,809,200 sq miles
(12,456,000 sq km)

Total number of countries:
44

Total population:
724 million

Largest city with population:
Moscow, European Russia 17.1 million

Country with highest population density:
Monaco 50,667 people per sq mile
(19,487 people per sq km)

Largest country:
European Russia 1,527,341 sq miles
(3,955,818 sq km)

Smallest country:
Vatican City, Italy 0.17 sq miles
(0.44 sq km)

Physical features

Largest lake:
Lake Lagoda, European Russia
7100 sq miles (18,390 sq km)

Longest river:
Volga, European Russia
2290 miles (3688 km)

Highest point:
El'brus, Caucasus, European Russia
18,510ft (5642 m)

Lowest point:
Volga Delta, Caspian Sea, European
Russia -92ft (-28m) below sea level

0 km   500
0 miles   500

Population   ● National capital

○ below 50,000   ◎ 50,000 to 100,000   ◉ 100,000 to 500,000   ■ above 500,000

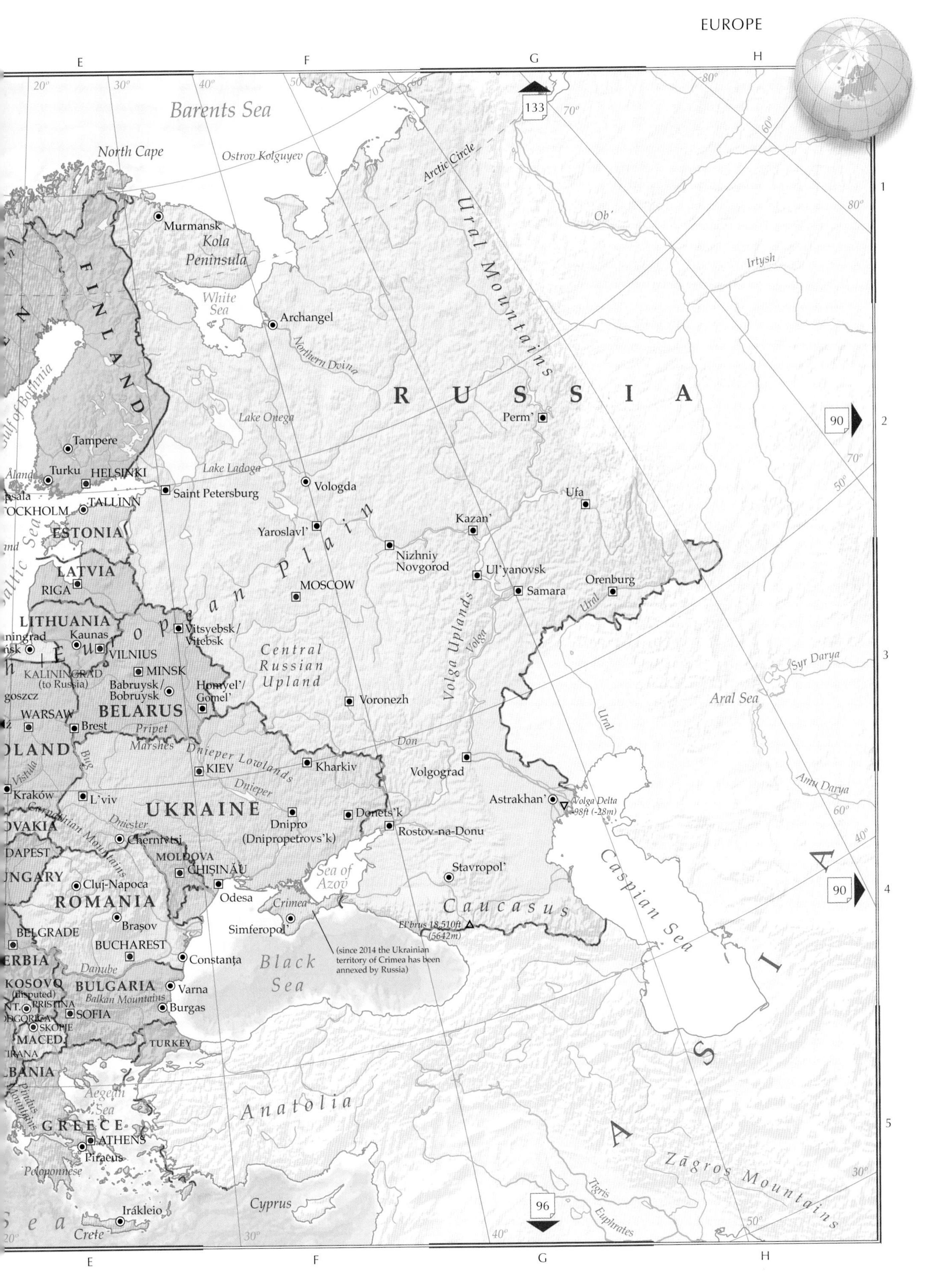

Barents Sea

North Cape

Ostrov Kolguyev

Arctic Circle

133

Murmansk

Kola Peninsula

Ob'

Irtysh

White Sea

Archangel

Northern Dvina

R U S S I A

Ural Mountains

FINLAND

Lake Onega

Perm'

90

Tampere

Turku

HELSINKI

Lake Ladoga

Saint Petersburg

Vologda

Ufa

TALLINN

STOCKHOLM

Yaroslavl'

Kazan'

Gulf of Bothnia

Åland

ESTONIA

Nizhniy Novgorod

Ul'yanovsk

Orenburg

Uppsala

LATVIA

MOSCOW

Samara

Ural

RIGA

Volga Uplands

Volga

LITHUANIA

Vitsyebsk/ Vitebsk

Syr Darya

Kaliningrad

Kaunas

Central Russian Upland

Aral Sea

Minsk

VILNIUS

Bydgoszcz

KALININGRAD (to Russia)

Babruysk/ Bobruysk

Homyel'/ Gomel'

Voronezh

Ural

Anu Darya

WARSAW

BELARUS

Brest

Pripet Marshes

Don

POLAND

Bug

Dnieper Lowlands

Kharkiv

Volgograd

Vistula

KIEV

Dnieper

Astrakhan'

Volga Delta 98ft (-28m)

Kraków

L'viv

UKRAINE

Donets'k

SLOVAKIA

Carpathian Mountains

Dniester

Dnipro (Dnipropetrovs'k)

Rostov-na-Donu

Ural

BUDAPEST

Chernivtsi

Caspian Sea

HUNGARY

MOLDOVA

Stavropol'

90

Cluj-Napoca

CHIŞINĂU

Sea of Azov

ROMANIA

Odesa

Crimea

Braşov

Simferopol'

Caucasus

BELGRADE

BUCHAREST

Constanţa

El'brus 18,510ft (5642m)

SERBIA

Danube

(since 2014 the Ukrainian territory of Crimea has been annexed by Russia)

A

KOSOVO (disputed)

BULGARIA

Varna

Black Sea

PODGORICA

PRISTINA

SOFIA

Burgas

Balkan Mountains

SKOPJE

MACED.

TURKEY

TIRANA

ALBANIA

Aegean Sea

Anatolia

Pindus Mountains

GREECE

ATHENS

Piraeus

Zāgros Mountains

Peloponnese

Sea

Irákleio

Cyprus

96

Crete

Tigris

Euphrates

The North Atlantic

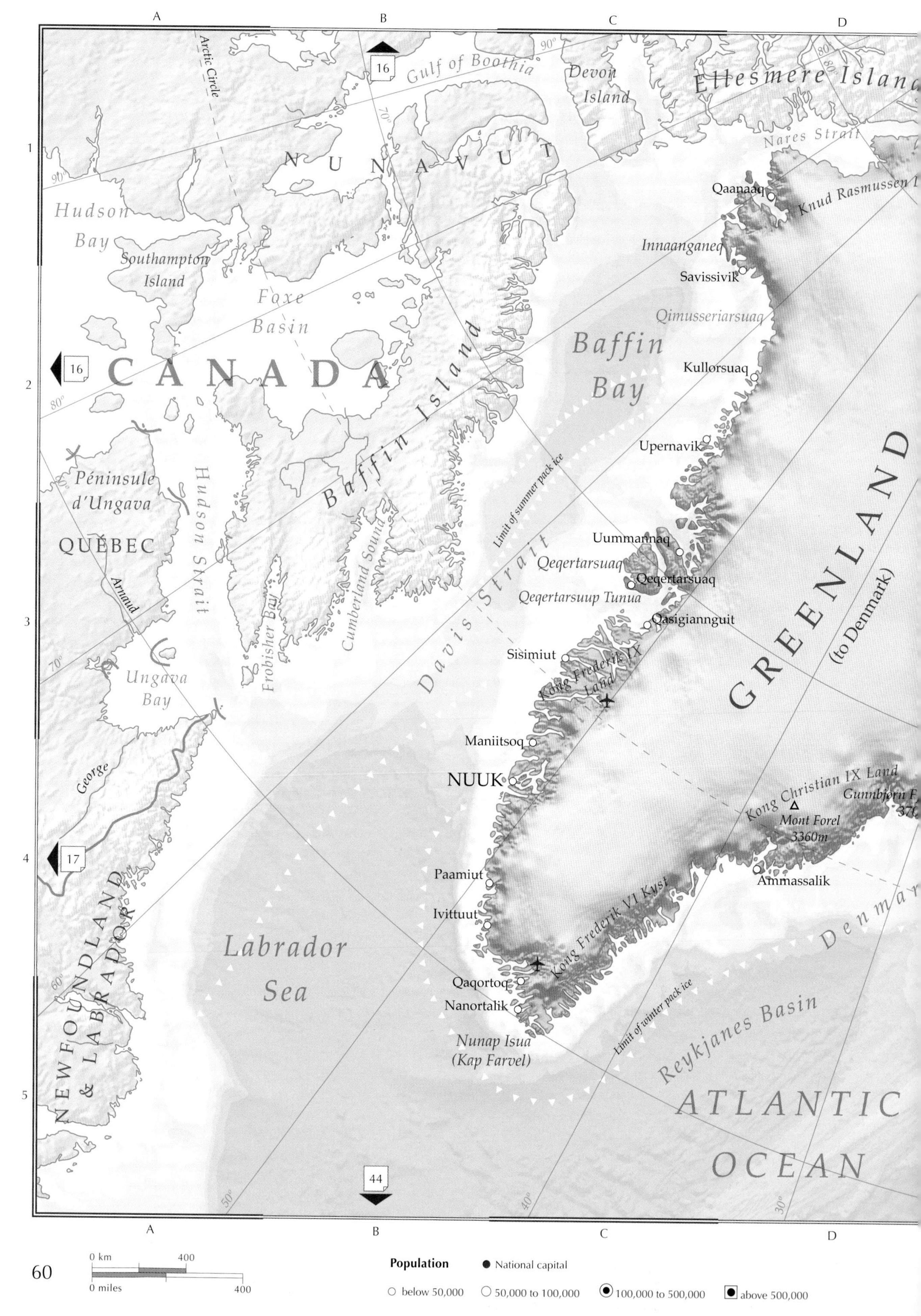

A | B | C | D

1

16

Gulf of Boothia

Devon Island

Ellesmere Island

Nares Strait

Arctic Circle

90°

70°

80°

Hudson Bay

Southampton Island

Foxe Basin

N U N A V U T

Qaanaaq

Knud Rasmussen I

Innaanganeq

Savissivik

Qimusseriarsuaq

Baffin Bay

Kullorsuaq

2

16

80°

C A N A D A

Baffin Island

Upernavik

G R E E N L A N D (to Denmark)

Péninsule d'Ungava

QUÉBEC

Hudson Strait

Arnaud

Frobisher Bay

Cumberland Sound

Limit of summer pack ice

Davis Strait

Uummannaq

Qeqertarsuaq

Qeqertarsuaq

Qeqertarsuup Tunua

3

70°

Ungava Bay

Qasigianguit

Sisimiut

Kong Frederik IX Land

George

Maniitsoq

NUUK

Kong Christian IX Land

Mont Forel 3360m

Gunnbjorn F 37

4

17

N E W F O U N D L A N D & L A B R A D O R

Paamiut

Ivittuut

Kong Frederik VI Kyst

Ammassalik

Denmar

Labrador Sea

Qaqortoq

Nanortalik

Limit of winter pack ice

Reykjanes Basin

5

60°

Nunap Isua (Kap Farvel)

ATLANTIC

50°

44

40°

OCEAN

30°

A | B | C | D

60

0 km   400

0 miles   400

Population ● National capital

○ below 50,000   ○ 50,000 to 100,000   ◉ 100,000 to 500,000   ■ above 500,000

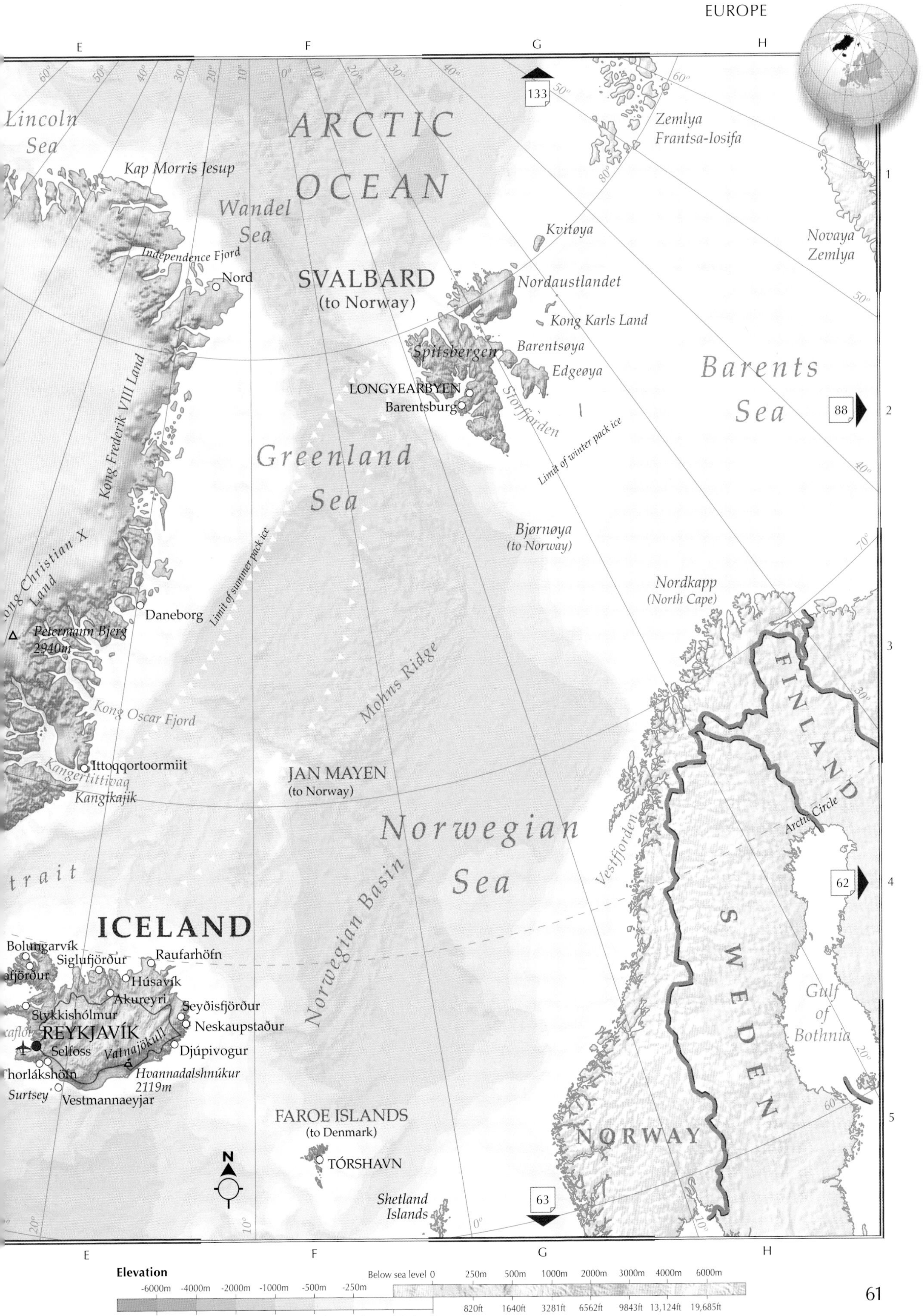

E   F   G   H

133

88

62

63

Lincoln
Sea

Kap Morris Jesup

A R C T I C
O C E A N

Zemlya
Frantsa-Iosifa

1

Wandel
Sea

Kvitøya

Novaya
Zemlya

Independence Fjord

Nord

SVALBARD
(to Norway)

Nordaustlandet

Kong Karls Land

Barentsøya

Spitsbergen

Edgeøya

LONGYEARBYEN

Barentsburg

B a r e n t s
S e a

2

Storfjorden

Greenland
Sea

Limit of winter pack ice

Bjørnøya
(to Norway)

Kong Frederik VIII Land

Nordkapp
(North Cape)

Limit of summer pack ice

Kong Christian X
Land

Daneborg

△ *Petermann Bjerg*
2940m

F I N L A N D

3

Mohns Ridge

Kong Oscar Fjord

Ittoqqortoormiit

Kangertittivaq
Kangikajik

JAN MAYEN
(to Norway)

Arctic Circle

Vestfjorden

trait

N o r w e g i a n
S e a

Norwegian Basin

S
W
E
D
E
N

Gulf
of
Bothnia

4

ICELAND

Bolungarvík
Siglufjörður   Raufarhöfn
afjörður   Húsavík
Akureyri
Stykkishólmur   Seyðisfjörður
caflói   Neskaupstaður
REYKJAVÍK
Selfoss   *Vatnajökull*   Djúpivogur
Thorlákshöfn   △ *Hvannadalshnúkur*
2119m
Surtsey   Vestmannaeyjar

FAROE ISLANDS
(to Denmark)

▲ TÓRSHAVN

Shetland
Islands

N O R W A Y

5

N

E   F   G   H

Elevation

						Below sea level 0	250m	500m	1000m	2000m	3000m	4000m	6000m
-6000m	-4000m	-2000m	-1000m	-500m	-250m								

							820ft	1640ft	3281ft	6562ft	9843ft	13,124ft	19,685ft
-19,658ft	-13,124ft	-6562ft	-3281ft	-1640ft	-820ft	-328ft/-100m	0						

Scandinavia & Finland

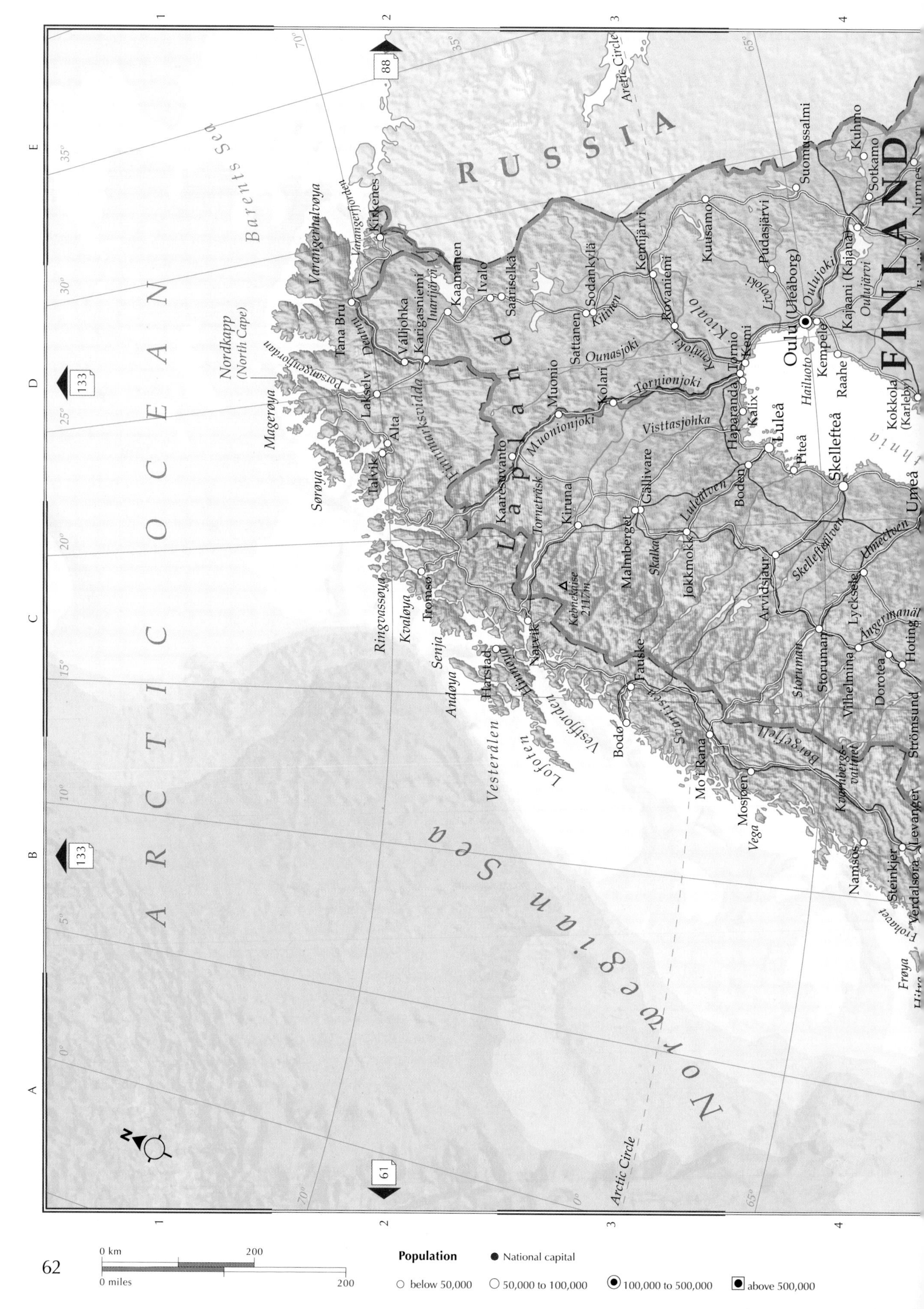

Population

● National capital

○ below 50,000   ○ 50,000 to 100,000   ◉ 100,000 to 500,000   ■ above 500,000

0 km   200

0 miles   200

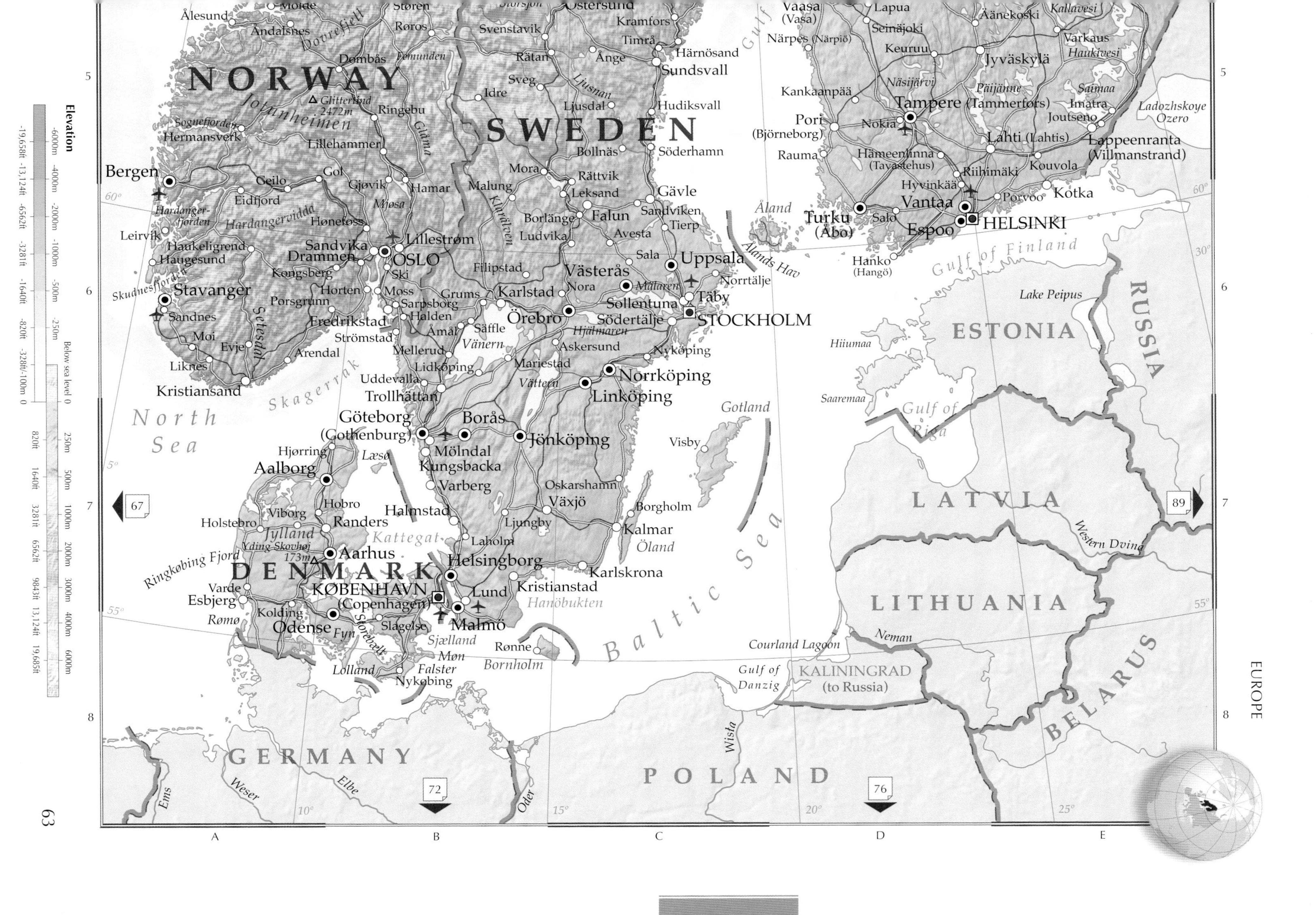

Elevation

-6000m	-4000m	-2000m	-1000m	-500m	-250m	Below sea level 0	250m	500m	1000m	2000m	3000m	4000m	6000m

| -19,658ft | -13,124ft | -6562ft | -3281ft | -1640ft | -820ft | -328ft/-100m 0 | | 820ft | 1640ft | 3281ft | 6562ft | 9843ft | 13,124ft | 19,685ft |

The Low Countries

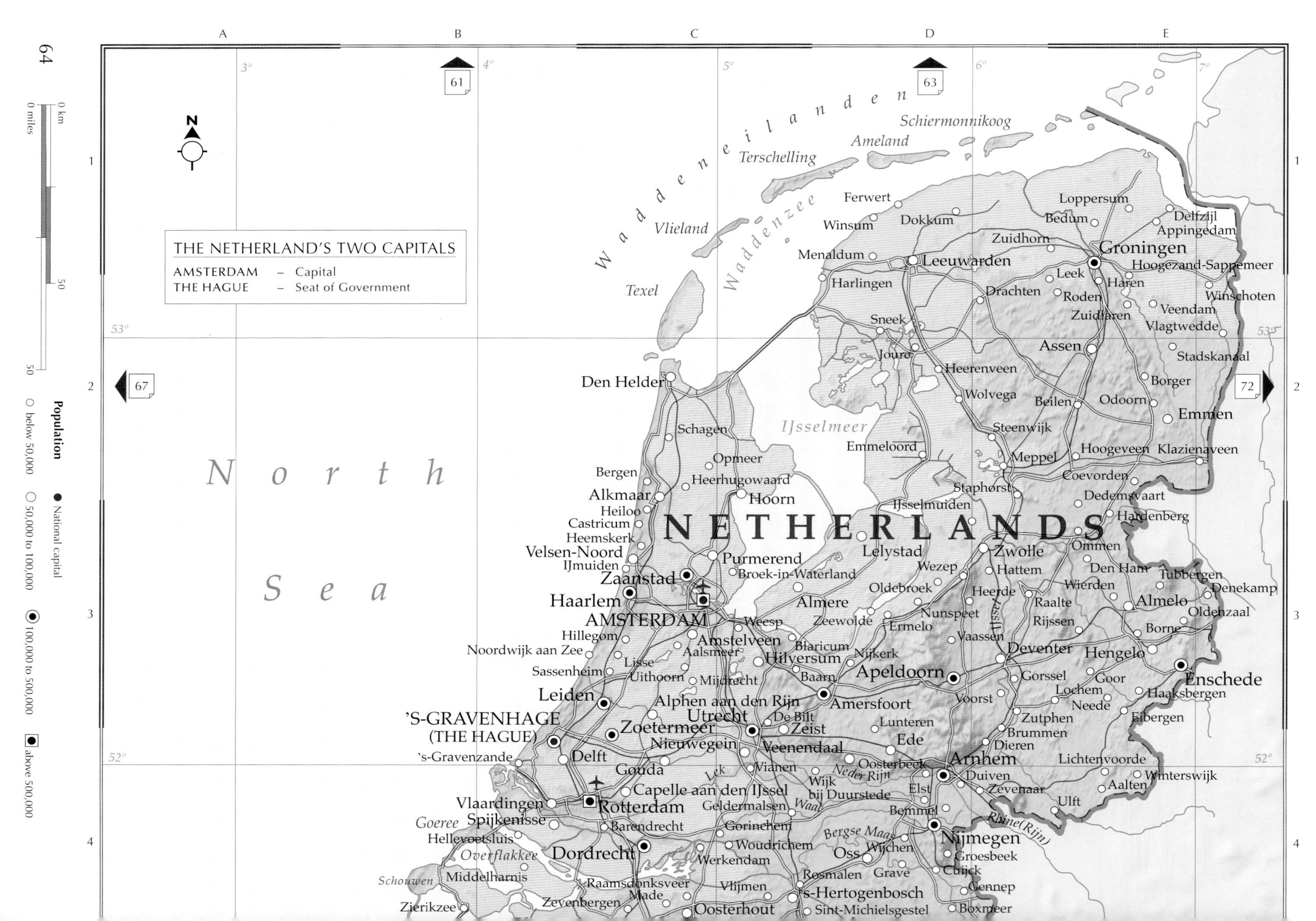

THE NETHERLAND'S TWO CAPITALS

AMSTERDAM — Capital
THE HAGUE — Seat of Government

Schiermonnikoog
Ameland
Terschelling
Vlieland
Texel

W a d d e n   e i l a n d e n

W a d d e n z e e

IJsselmeer

N o r t h   S e a

Deltzijl
Appingedam
Winschoten
Hoogezand-Sappemeer
Veendam
Vlagtwedde
Stadskanaal
Emmen
Borger
Klazienaveen
Hoogeveen
Odoorn
Coevorden
Dedemsvaart
Hardenberg
Denekamp
Enschede
Oldenzaal
Tubbergen
Den Ham
Almelo
Borne
Haaksbergen
Wintersvijk
Aalten
Ulft
Lichtenvoorde
Zevenaar
Groenlo
Eibergen
Haaksbergen
Hengelo
Goor
Needg
Lochem
Gorssel
Ruurlo
Brummen
Dieren
Zutphen
Duiven
Arnhem
Elst
Nijmegen
Groesbeek
Cuijk
Bemmel
Gennep
Boxmeer

Loppersum
Bedum
Winsum
Zuidhorn
Haren
Leek
Roden
Groningen
Assen
Beilen
Meppel
Steenwijk
Staphorst
Wolvega
Heerenveen
Zuidlaren

Rhine (Rijn)

Ferwert
Dokkum
Winsum
Menaldum
Leeuwarden
Harlingen
Sneek
Joure
Emmeloord
Lelystad
Zwolle
Hattem
Wezep
Heerde
Raalte
Rijssen
Wierden
Ommen
Deventer
Vaassen
Epe
Nunspeet
Ermelo
Apeldoorn
Voorst
Harderwijk
Barneveld
Ede
Lunteren
Oosterbeek
Bennekom
Wageningen
Rosmalen
Oss
Wijchen
Grave
Sint-Michielsgestel
's-Hertogenbosch
Oosterhout

Den Helder
Bergen
Alkmaar
Heiloo
Castricum
Heemskerk
Velsen-Noord
IJmuiden
Haarlem
Hillegom
Noordwijk aan Zee
Sassenheim
Lisse
Leiden
'S-GRAVENHAGE
(THE HAGUE)
's-Gravenzande
Delft
Vlaardingen
Spijkenisse
Hellevoetsluis
Goeree
Middelharnis
Zevenbergen
Raamsdonksveer
Zierikzee
Overflakkee
Schouwen
Made
Werkendam
Vlijmen
Woudrichem
Geertruidenberg

Schagen
Opmeer
Heerhugowaard
Hoorn
Purmerend
Broek-in-Waterland
Zaanstad
AMSTERDAM
Amstelveen
Aalsmeer
Uithoorn
Mijdrecht
Weesp
Almere
Zeewolde
Baarn
Hilversum
Bilthoven
Blaricum
Nijkerk
Amersfoort
Soest
Zeist
De Bilt
Utrecht
Veenendaal
Wijk bij Duurstede
Geldermalsen
Zaltbommel
Tiel
Culemborg
IJk
Nieuwegein
Vianen
Capelle aan den/IJssel
Gorinchem
Rotterdam
Dordrecht
Barendrecht
Gouda
Alphen aan den Rijn
Zoetermeer
Nieuwegein

NETHERLANDS

Nederrijn
Waal
Maas
Bergse Maas
IJssel

0 km   50
0 miles   50

Population   ● National capital

○ below 50,000   ○ 50,000 to 100,000   ◉ 100,000 to 500,000   ■ above 500,000

N

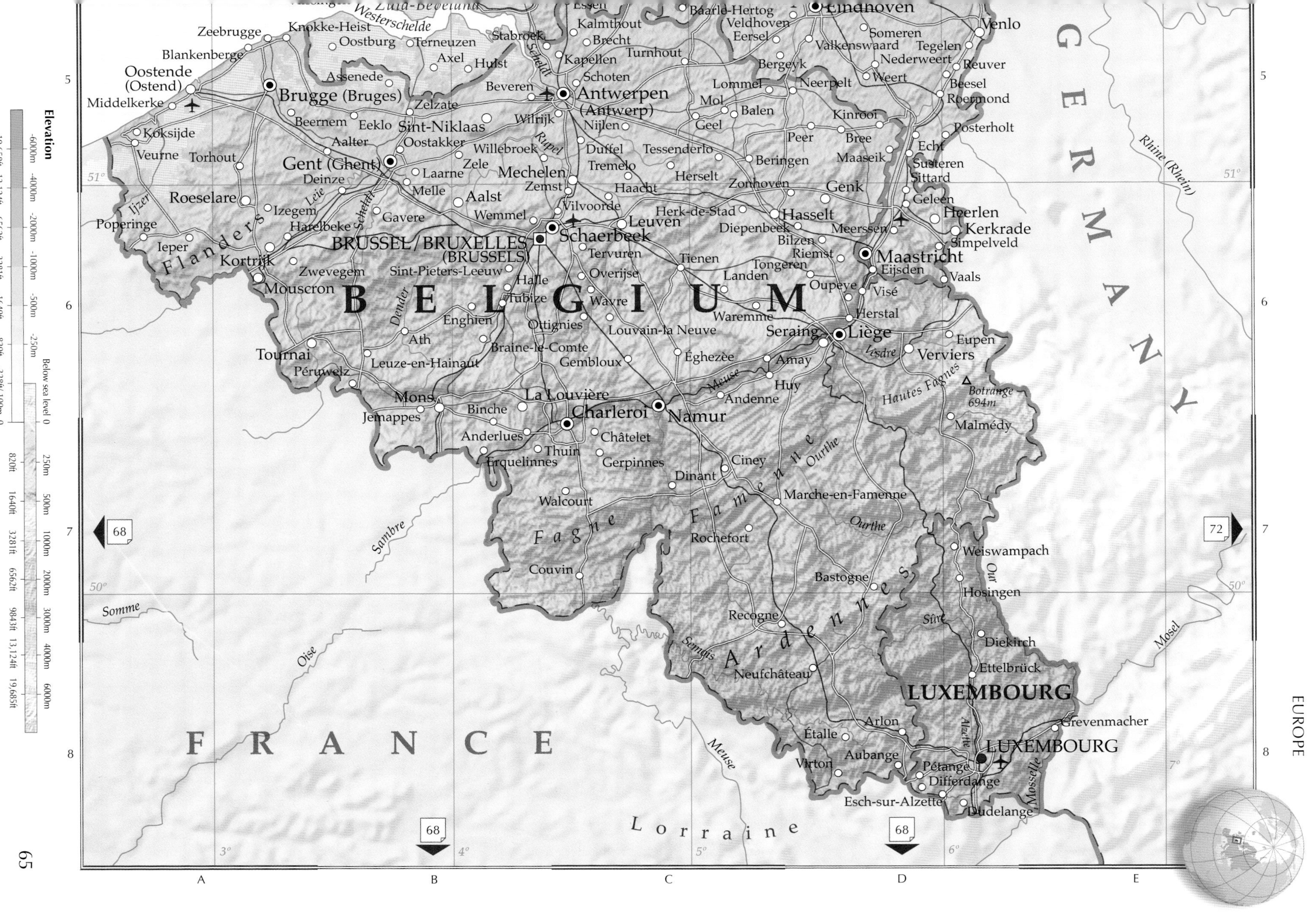

GERMANY

Rhine (Rhein)

Mosel

LUXEMBOURG

LUXEMBOURG

Grevenmacher

Mosselle

Alzette

Diekirch

Ettelbrück

Weiswampach

Hosingen

Our

Bastogne

Marche-en-Famenne

Arlon

Neuchâteau

Étalle

Virton

Aubange

Pétange

Differdange

Dudelange

Esch-sur-Alzette

Lorraine

FRANCE

Venlo

Reuver

Beesel

Roermond

Posterholt

Bree

Maaseik

Genk

Herleen

Kerkrade

Simpelveld

Vaals

Eupen

Verviers

Malmédy

Botrange
694m

Hautes Fagnes

Vesdre

Herstal

Liège

Seraing

Oupeye

Visé

Maastricht

Eijsden

Meerssen

Riemst

Bilzen

Tongeren

Landen

Waremme

Amay

Huy

Andenne

Ciney

Rochefort

Dinant

Namur

Gembloux

Recogne

Meuse

Ourthe

M a s s i f   d e   l ' O u r t h e

A r d e n n e s

F a g n e

F a m e

Somme

Oise

Sambre

Sambre

BELGIUM

BRUSSEL/BRUXELLES
(BRUSSELS)

Schaerbeek

Mechelen

Antwerpen
(Antwerp)

Leuven

Tienen

Hasselt

Herk-de-Stad

Diepenbeek

Beringen

Peer

Neerpelt

Lommel

Mol

Geel

Turnhout

Brecht

Schoten

Kalmthout

Kapellen

Essen

Stabroek

Beveren

Sint-Niklaas

Gent (Ghent)

Brugge (Bruges)

Zeebrugge

Blankenberge

Oostende
(Ostend)

Middelkerke

Koksijde

Veurne

Torhout

Roeselare

Ieper

Poperinge

Kortrijk

Mouscron

Tournai

Péruwelz

Leuze-en-Hainaut

Ath

Enghien

Mons

Jemappes

Binche

La Louvière

Charleroi

Châtelet

Gerpinnes

Thuin

Anderlues

Frameries

Walcourt

Couvin

Fagne

Elevation

-6000m   -4000m   -2000m   -1000m   -500m   -250m   Below sea level 0   250m   500m   1000m   2000m   3000m   4000m   6000m

-19,658ft -13,124ft -6562ft -3281ft -1640ft -820ft -328ft/-100m 0   820ft   1640ft   3281ft   6562ft   9843ft 13,124ft 19,685ft

65

The British Isles

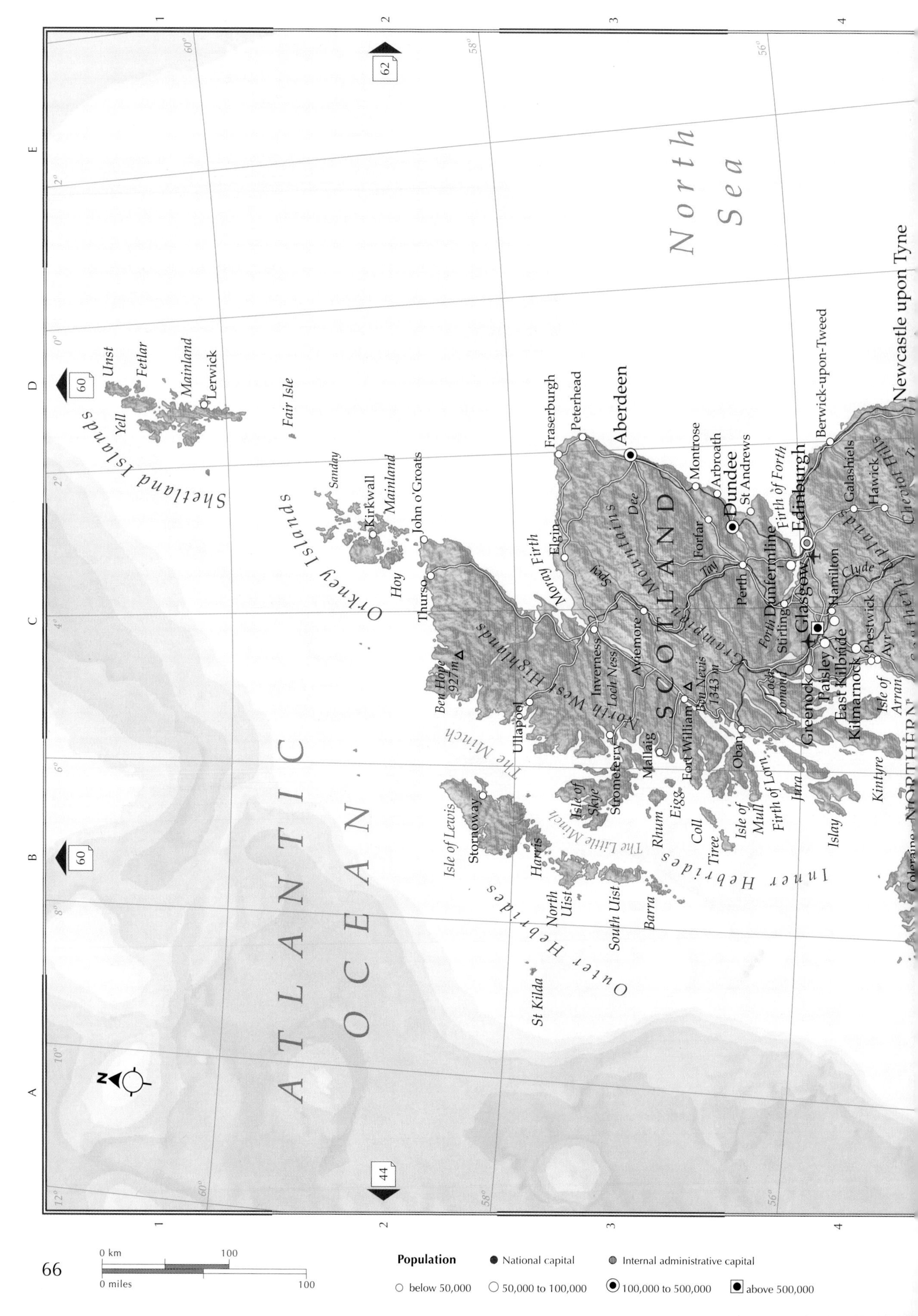

North Sea

Newcastle upon Tyne

ATLANTIC OCEAN

Shetland Islands
Unst
Yell
Fetlar
Mainland
Lerwick

Fair Isle

Sanday
Orkney Islands
Kirkwall
Mainland
Hoy
John o'Groats

Thurso
Ben Hope 927m △

The Minch
Ullapool

North West Highlands

Isle of Lewis
Stornoway
Harris

The Little Minch

North Uist
South Uist
Barra

St Kilda

Outer Hebrides

Isle of Skye
Stromeferry
Rhum
Eigg
Coll
Tiree

Isle of Mull

Firth of Lorn

Jura

Islay

Inner Hebrides

Kintyre
Isle of Arran

Mallaig
Fort William
Ben Nevis 1343 m △
Oban

NORTHERN
Coleraine

Moray Firth
Elgin
Spey
Inverness
Loch Ness
Aviemore

Fraserburgh
Peterhead
Aberdeen

SCOTLAND
Grampian Mountains
Dee

Montrose
Arbroath
Dundee
St Andrews
Firth of Forth

Forfar
Tay
Perth

Loch Lomond
Stirling
Forth
Dunfermline
Edinburgh

Galashiels
Pentland Hills
Hawick
Cheviot Hills
Berwick-upon-Tweed

Greenock
Paisley
Glasgow
Clyde
Hamilton
East Kilbride
Kilmarnock
Prestwick
Ayr

Southern Uplands

0 km   100
0 miles   100

Population   ● National capital   ● Internal administrative capital

○ below 50,000   ○ 50,000 to 100,000   ◉ 100,000 to 500,000   ▣ above 500,000

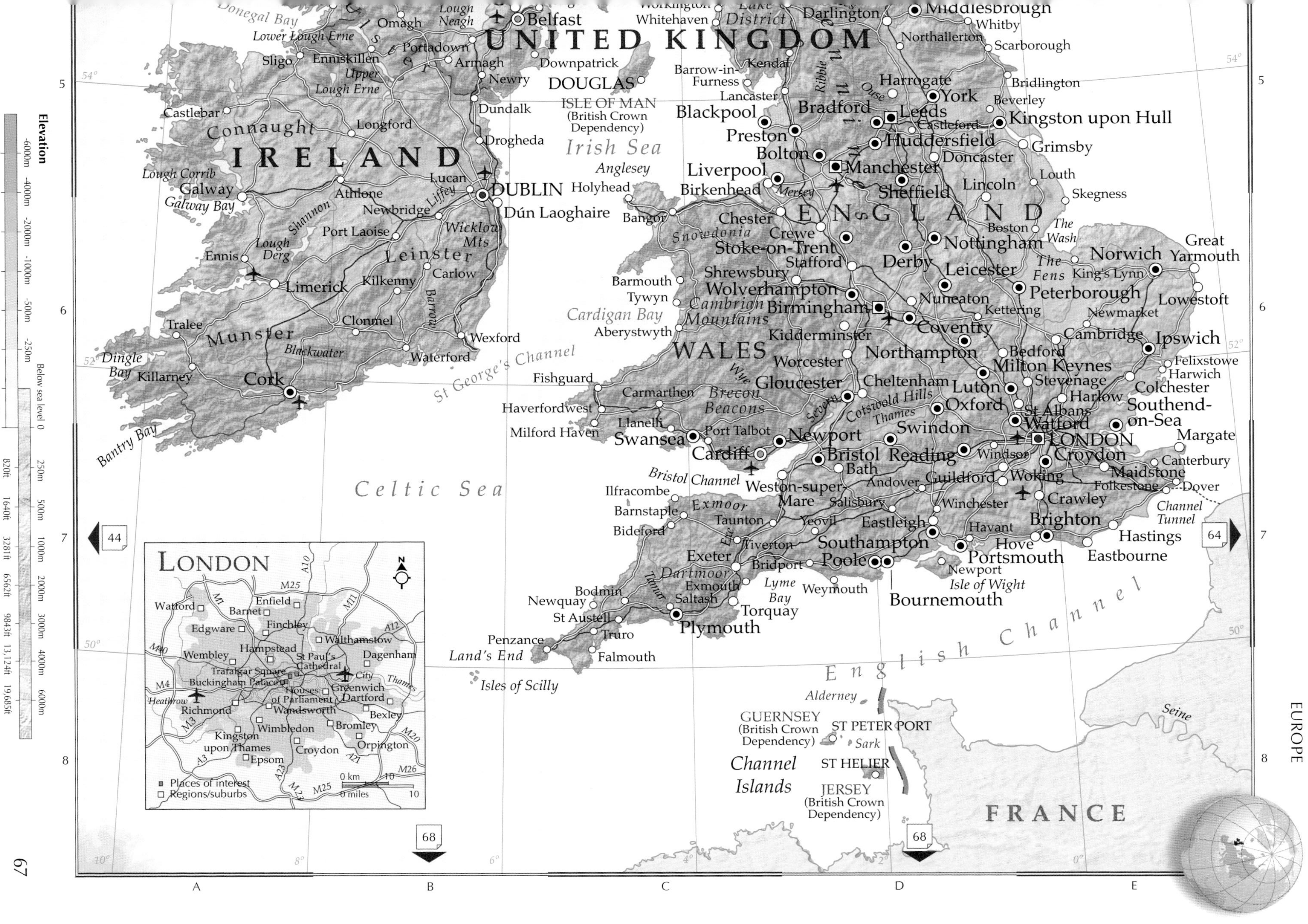

Elevation

| -6000m | -4000m | -2000m | -1000m | -500m | -250m | Below sea level 0 | 250m | 500m | 1000m | 2000m | 3000m | 4000m | 6000m |

| -19,658ft | -13,124ft | -6562ft | -3281ft | -1640ft | -820ft | -328ft/-100m 0 | 820ft | 1640ft | 3281ft | 6562ft | 9843ft | 13,124ft | 19,685ft |

France, Andorra & Monaco

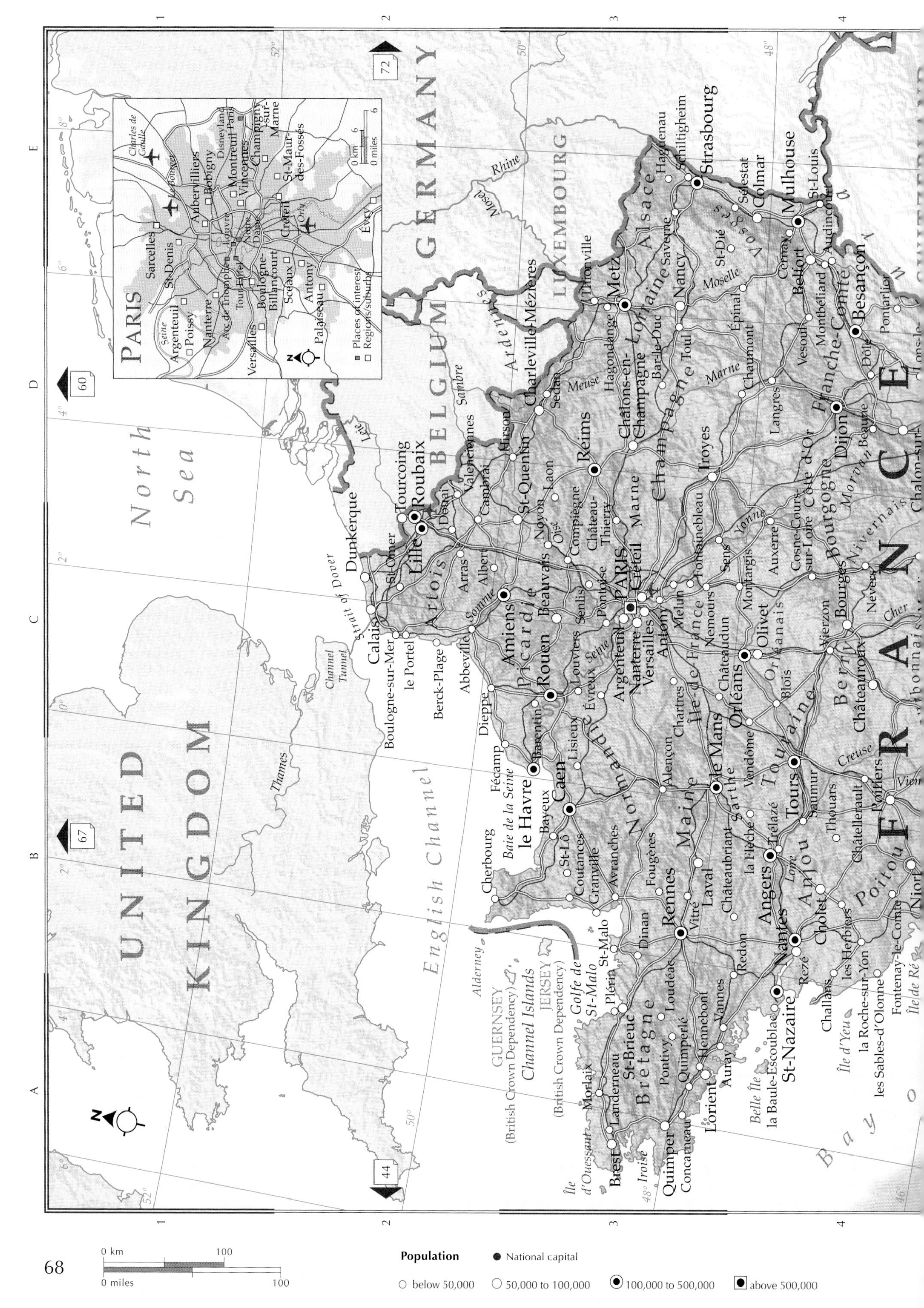

PARIS

Charles de Gaulle
Disneyland
Sarcelles
St-Denis
Aubervilliers
Bobigny
Montreuil
Vincennes
Champigny-sur-Marne
St-Maur-des-Fossés
Nanterre
Le Louvre
Tour Eiffel
Arc de Triomphe
Boulogne-Billancourt
Sceaux
Antony
Créteil
Orly
Evry
Versailles
Palaiseau
Seine
Argenteuil
Poissy

Places of interest
Regions/suburbs

GERMANY
BELGIUM
LUXEMBOURG

UNITED KINGDOM

North Sea

English Channel

Rhine
Mosel
Moselle

Dunkerque
Tourcoing
Roubaix
Calais
St-Omer
Lille
Douai
Valenciennes
le Portel
Boulogne-sur-Mer
Berck-Plage
Arras
Albert
St-Quentin
Cambrai
Noyon
Laon
Mézières
Charleville-Mézières
Sedan
Hirson
Oise
Compiègne
Château-Thierry
Senlis
Pontoise
Reims
Châlons-en-Champagne
Épernay
Thionville
Metz
Hagondange
Bar-le-Duc
Toul
Nancy
St-Dié
Haguenau
Schiltigheim
Strasbourg
Sélestat
Colmar
Mulhouse
St-Louis
Audincourt
Belfort
Montbéliard
Vesoul
Besançon
Dole
Pontarlier
Langres
Chaumont
Épinal
Saverne
Verne
Troyes
Sens
Montargis
Auxerre
Cosne-Cours-sur-Loire
Nevers
Dijon
Beaune
Chalon-sur-Saône

Dieppe
Fécamp
le Havre
Rouen
Louviers
Évreux
Amiens
Beauvais
Abbeville
Somme
Seine
Paris
Argenteuil
Nanterre
Versailles
Antony
Melun
Nemours
Fontainebleau
Étampes
Château-Thierry

Cherbourg
Baie de la Seine
Bayeux
Caen
St-Lô
Coutances
Granville
Avranches
Lisieux
Alençon
Chartres
Le Mans
Châteaudun
Orléans
Olivet
Blois
Vendôme
Tours
Vierzon
Bourges
Châteauroux
Poitiers

Alderney
GUERNSEY
(British Crown Dependency)
Channel Islands
JERSEY
(British Crown Dependency)
Golfe de St-Malo
St-Malo
Dinan
Plérin
St-Brieuc
Landerneau
Morlaix
Bretagne
Pontivy
Loudéac
Quimperlé
Vannes
Auray
Lorient
Quimper
Concarneau
île d'Ouessant
île d'Iroise
Brest

Rennes
Vitré
Fougères
Laval
Châteaubriant
Redon
la Flèche
Sablé
Angers
Cholet
les Herbiers
Nantes
Rezé
St-Nazaire
la Baule-Escoublac
Challans
la Roche-sur-Yon
Fontenay-le-Comte
Niort
les Sables-d'Olonne
île d'Yeu
Belle Île
île de Ré

Loire
Cher
Creuse
Vienne
Sarthe
Maine
Anjou
Poitou
Touraine
Berry
Normandie

Bay of

68

Population
● National capital

○ below 50,000
◯ 50,000 to 100,000
◉ 100,000 to 500,000
■ above 500,000

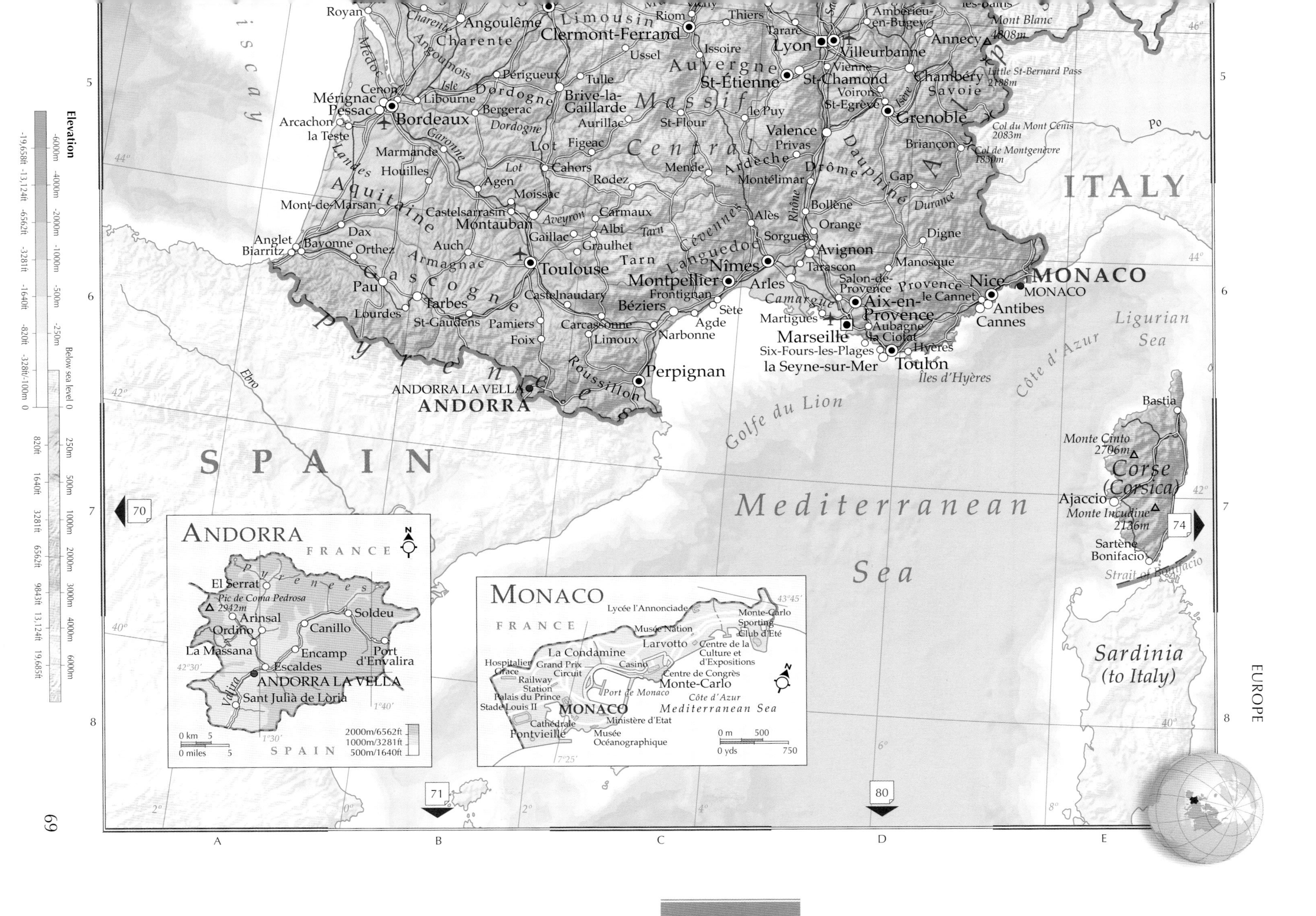

ITALY

MONACO

Mont Blanc
4808m

Col du Mont Cenis
2083m

Col de Montgenèvre
1850m

Ligurian Sea

Corse (Corsica)

Bastia

Monte Cinto
2706m△

Ajaccio
Monte Incudine
2136m△

Sartène
Bonifacio

Strait of Bonifacio

Sardinia
(to Italy)

Annecy
Savoie
Grenoble
Chambéry
Ambérieu-en-Bugey
Vienne
St-Chamond
Voiron
St-Égrève
Briançon
Digne

Lyon
Villeurbanne
Rive-de-Gier
Dauphiné
Gap
Durance

Dignes

Monaco
MONACO
Antibes
Nice
le Cannet
Cannes
Hyères
Toulon
Iles d'Hyères
la Seyne-sur-Mer
Six-Fours-les-Plages

Côte d'Azur

Aix-en-Provence
Aubagne
la Ciotat
Marseille
Martigues
Salon-de-Provence
Manosque
Provence

St-Étienne
le Puy
Roanne
Thiers
Tarare
Valence
Privas
Ardèche
Montélimar
Drôme
Rhône
Bollène
Orange
Avignon
Tarascon
Sorgues
Arles
Nîmes
Camargue

Mediterranean Sea

Golfe du Lion

Clermont-Ferrand
Riom
Issoire
Ussel
Auvergne
St-Flour
Mende
Massif Central
Aurillac
Rodez
Millau
Cévennes

Sète
Agde
Narbonne
Frontignan
Béziers
Montpellier
Languedoc
Limoux
Carcassonne
Tarn
Perpignan
Roussillon

Angoulême
Charente
Tulle
Brive-la-Gaillarde
Périgueux
Dordogne
Limousin
Bergerac
Dordogne
Figeac
Cahors
Gaillac
Albi
Carmaux
Aveyron
Montauban
Moissac
Castelsarrasin
Gramat

Toulouse
Castelnaudary
Pamiers
Foix
Gascogne
Auch
Ariège

ANDORRA LA VELLA
ANDORRA

Royan
Médoc
Cenon
Mérignac
Pessac
Arcachon
la Teste
Landes
Marmande
Libourne
Isle
Angoumois
Bordeaux
Garonne
Aquitaine
Agen
Lot
Nérac
Houilles

Mont-de-Marsan
Dax
Orthez
Pau
Lourdes
Tarbes
St-Gaudens
Landes
Pyrénées

Anglet
Biarritz
Bayonne

Biscay

SPAIN

Ebro

Po

MONACO

FRANCE

Monte-Carlo
Sporting Club d'Eté
Larvotto
Musée National
Lycée l'Annonciade
Centre de la Culture et d'Expositions
Centre de Congrès
La Condamine
Casino
Monte-Carlo
Grand Prix Circuit
Hospitalier Princesse Grace
Railway Station
Port de Monaco
Palais du Prince
Stade Louis II
Fontvieille
Cathédrale
Ministère d'Etat
Musée Océanographique
MONACO
Côte d'Azur
Mediterranean Sea

0 m   500
0 yds   750

43°45'

7°25'

ANDORRA

FRANCE

El Serrat
Soldeu
Pic de Coma Pedrosa
2942m△
Arinsal
Ordino
Canillo
La Massana
Port d'Envalira
Encamp
Escaldes
ANDORRA LA VELLA
Sant Julià de Lòria
Valira
Pyrénées

SPAIN

2000m/6562ft
1000m/3281ft
500m/1640ft

0 km   5
0 miles   5

1°40'
1°30'
42°30'

Elevation

						Below sea level 0	250m	500m	1000m	2000m	3000m	4000m	6000m
-6000m	-4000m	-2000m	-1000m	-500m	-250m								
-19,658ft	-13,124ft	-6562ft	-3281ft	-1640ft	-820ft	-328ft/-100m 0		820ft	1640ft	3281ft	6562ft	9843ft	13,124ft 19,685ft

Spain & Portugal

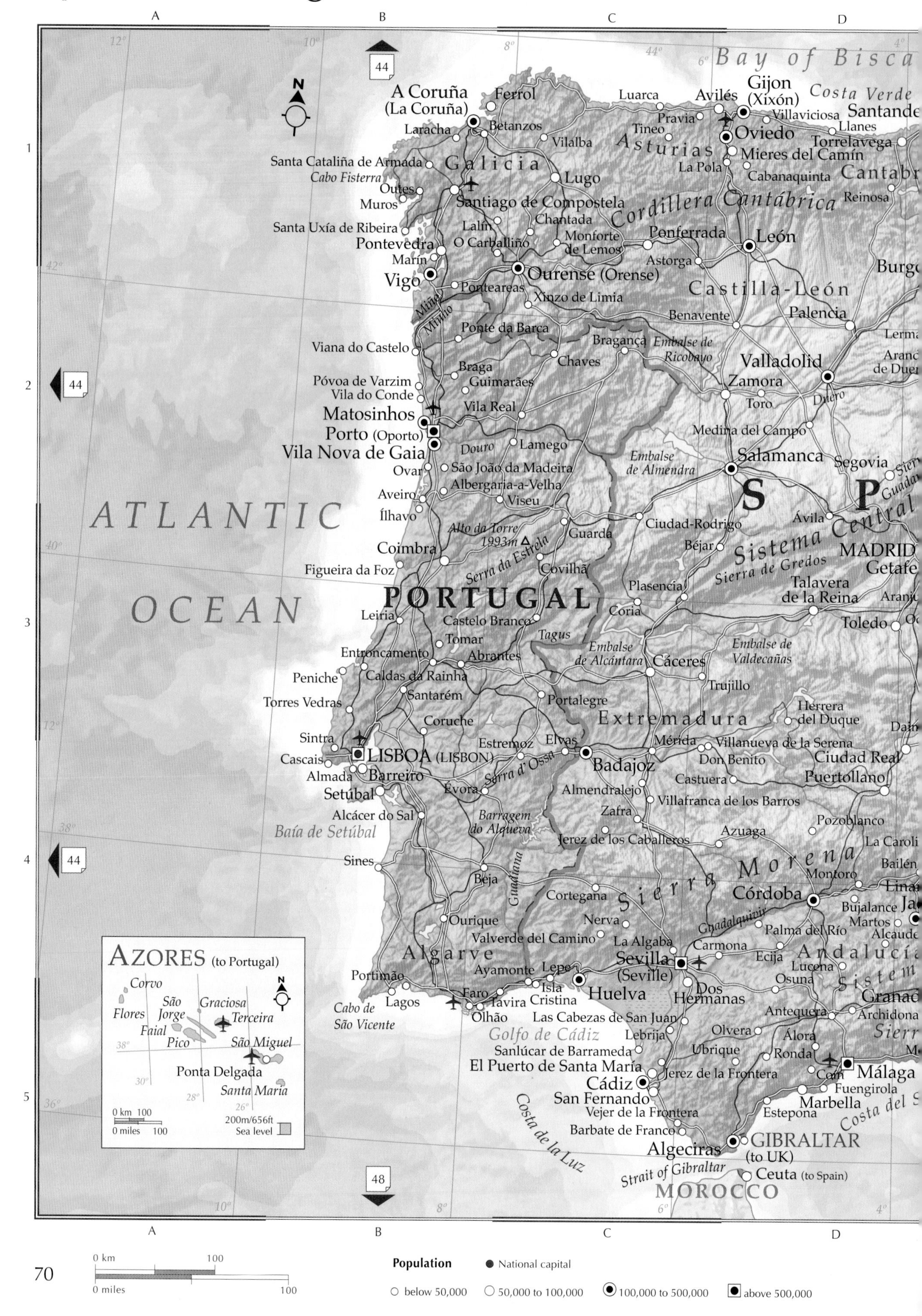

A Coruña (La Coruña)
Ferrol
Luarca
Avilés
Gijon (Xixón)
Costa Verde
Bay of Biscay
Santander
Betanzos
Pravia
Villaviciosa
Llanes
Laracha
Vilalba
Tineo
Oviedo
Torrelavega
Santa Cataliña de Armada
Galicia
Asturias
Mieres del Camín
Cabanaquinta
Cantabr
Cabo Fisterra
Lugo
La Pola
Cordillera Cantábrica
Outes
Santiago de Compostela
Chantada
Reinosa
Muros
Lalín
Monforte de Lemos
Ponferrada
León
Santa Uxía de Ribeira
O Carballiño
Astorga
Pontevedra
Ourense (Orense)
Benavente
Castilla-León
Palencia
Marín
Ponteareas
Xinzo de Limia
Lerma
Vigo
Miño Miño
Bragança
Embalse de Ricobayo
Valladolid
Arand de Duer
Viana do Castelo
Ponte da Barca
Chaves
Zamora
Póvoa de Varzim
Braga
Guimarães
Toro
Duero
Vila do Conde
Vila Real
Medina del Campo
Matosinhos
Porto (Oporto)
Lamego
Embalse de Almendra
Salamanca
Segovia
Sierr
Vila Nova de Gaia
Douro
São João da Madeira
Guadar
Ovar
Albergaria-a-Velha
S
P
Central
Aveiro
Viseu
Ílhavo
Ciudad-Rodrigo
Ávila
MADRID
Alto da Torre 1993m
Guarda
Béjar
Getafe
Coimbra
Sistema
Figueira da Foz
Serra da Estrela
Covilhã
Plasencia
Sierra de Gredos
Talavera de la Reina
Aranj
PORTUGAL
Coria
Toledo
Leiria
Castelo Branco
Embalse de Valdecañas
Castelo Branco
Embalse de Alcántara
Cáceres
Tomar
Tagus
Entroncamento
Abrantes
Peniche
Caldas da Rainha
Trujillo
Herrera del Duque
Torres Vedras
Santarém
Portalegre
Extremadura
Coruche
Sintra
Estremoz
Elvas
Mérida
Villanueva de la Serena
Cascais
LISBOA (LISBON)
Badajoz
Don Benito
Ciudad Real
Almada
Barreiro
Évora
Castuera
Puertollano
Setúbal
Serra d'Ossa
Almendralejo
Villafranca de los Barros
Alcácer do Sal
Zafra
Pozoblanco
La Caroli
Baía de Setúbal
Barragem do Alqueva
Jerez de los Caballeros
Azuaga
Sines
Beja
Morena
Bailén
Guadiana
Montoro
Linar
Ourique
Cortegana
Nerva
Córdoba
Bujalance
Jae
Martos
Algarve
Valverde del Camino
La Algaba
Guadalquivir
Palma del Río
Alcaude
Portimão
Ayamonte
Lepe
Carmona
Écija
Andalucía
Lagos
Faro
Tavira
Isla Cristina
Sevilla (Seville)
Lucena
Sistem
Cabo de São Vicente
Olhão
Las Cabezas de San Juan
Huelva
Dos Hermanas
Osuna
Granac
Golfo de Cádiz
Lebrija
Antequera
Archidona
Sierr
Sanlúcar de Barrameda
Olvera
Álora
El Puerto de Santa María
Jerez de la Frontera
Ubrique
Ronda
Coín
Málaga
Cádiz
Fuengirola
San Fernando
Vejer de la Frontera
Estepona
Marbella
Costa del
Barbate de Franco
Costa de la Luz
Algeciras
GIBRALTAR (to UK)
Strait of Gibraltar
Ceuta (to Spain)
MOROCCO

ATLANTIC

OCEAN

AZORES (to Portugal)
Corvo
São Jorge
Graciosa
Flores
Faial
Pico
Terceira
São Miguel
Ponta Delgada
Santa Maria
0 km 100
0 miles 100
200m/656ft
Sea level

Population

● National capital
○ below 50,000
◯ 50,000 to 100,000
◉ 100,000 to 500,000
◼ above 500,000

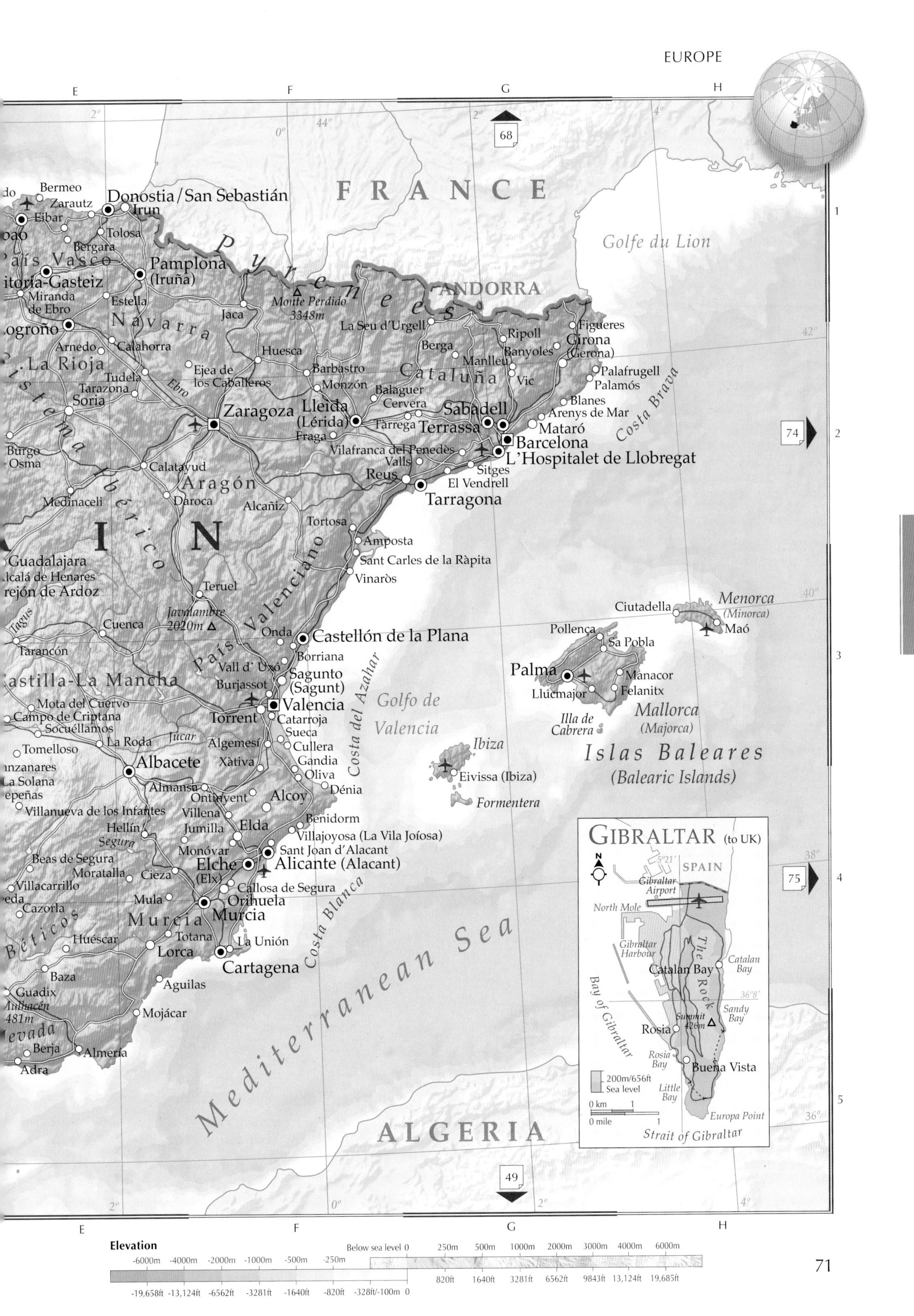

E   2°   F   0°   44°   G   2°   H   4°

68

F R A N C E

Golfe du Lion

do
Bermeo
Zarautz
Donostia / San Sebastián
Eibar
Irun
ao
Tolosa
Bergara
País Vasco
Pamplona
(Iruña)
toria-Gasteiz
Miranda
de Ebro
Navarra
Estella
Jaca
△ Monte Perdido
3348m
La Seu d'Urgell
ANDORRA
Ripoll
Figueres
42°
ogroño
Arnedo
Calahorra
La Rioja
Ejea de
los Caballeros
Huesca
Barbastro
Berga
Banyoles
Girona
(Gerona)
Palafrugell
Palamós
Tudela
Tarazona
Soria
Monzón
Balaguer
Cervera
Cataluña
Manlleu
Vic
Blanes
74
Burgo
Osma
Zaragoza
Lleida
(Lérida)
Tàrrega
Terrassa
Sabadell
Arenys de Mar
Mataró
Costa Brava
Fraga
Barcelona
L'Hospitalet de Llobregat
Calatayud
Aragón
Daroca
Vilafranca del Penedès
Valls
Sitges
Reus
Medinaceli
Alcañiz
Tortosa
El Vendrell
Tarragona
Guadalajara
lcalá de Henares
rejón de Ardoz
Teruel
Amposta
Sant Carles de la Ràpita
Vinaròs
N
Javalambre
2020m △
Valenciano
Ciutadella
Menorca
(Minorca)
40°
Cuenca
Onda
Castellón de la Plana
Pollença
Sa Pobla
Maó
astilla-La Mancha
Tarancón
Borriana
Sagunto
(Sagunt)
País
Palma
Manacor
Felanitx
Mota del Cuervo
Campo de Criptana
Socuéllamos
Burjassot
Valencia
Vall d'Uxó
Catarroja
Sueca
Golfo de
Valencia
Lluemajor
Mallorca
(Majorca)
Torrent
La Roda
Júcar
Algemesí
Cullera
Illa de
Cabrera
Tomelloso
Xàtiva
Gandia
Oliva
Ibiza
Islas Baleares
(Balearic Islands)
anzanares
La Solana
epeñas
Albacete
Almansa
Ontinyent
Villena
Alcoy
Dénia
Eivissa (Ibiza)
Villanueva de los Infantes
Hellín
Jumilla
Elda
Benidorm
Formentera
Beas de Segura
Moratalla
Segura
Monóvar
Elche
(Elx)
Villajoyosa (La Vila Joíosa)
Sant Joan d'Alacant
Alicante (Alacant)
Villacarrillo
eda
Mula
Cieza
Callosa de Segura
Orihuela
Costa Blanca
Cazorla
Murcia
Murcia
38°
75
Béticos
Huéscar
Totana
La Unión
Lorca
Cartagena
Baza
Aguilas
Guadix
lhacén
481m
Mojácar
evada
Berja
Almería
Adra

Costa del Azahar

Mediterranean Sea

A L G E R I A

49

GIBRALTAR (to UK)

N
SPAIN
5°21'
*Gibraltar
Airport*
North Mole
*Gibraltar
Harbour*
The Rock
Catalan Bay
*Catalan
Bay*
Bay of Gibraltar
36°8'
Rosia
Summit
426m △
*Sandy
Bay*
*Rosia
Bay*
Buena Vista
*Little
Bay*
200m/656ft
Sea level
Europa Point
0 km   1
0 mile   1
Strait of Gibraltar

E   2°   F   0°   G   2°   H   4°

Elevation

| | | | | | | | Below sea level 0 | 250m | 500m | 1000m | 2000m | 3000m | 4000m | 6000m |

-6000m   -4000m   -2000m   -1000m   -500m   -250m

-19,658ft   -13,124ft   -6562ft   -3281ft   -1640ft   -820ft   -328ft/-100m 0

820ft   1640ft   3281ft   6562ft   9843ft   13,124ft   19,685ft

Germany & The Alpine States

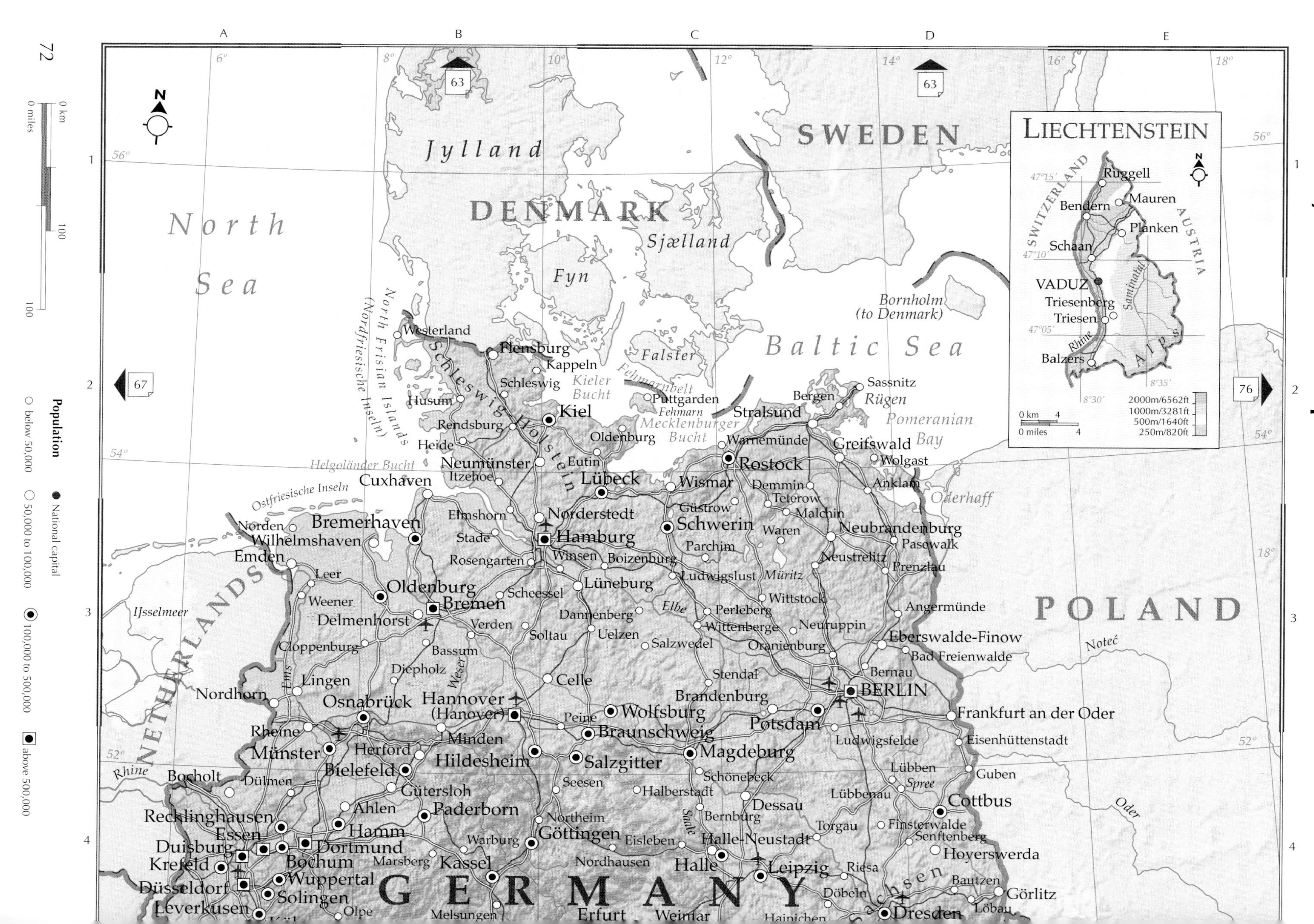

Population ● National capital

○ below 50,000   ○ 50,000 to 100,000   ◉ 100,000 to 500,000   ■ above 500,000

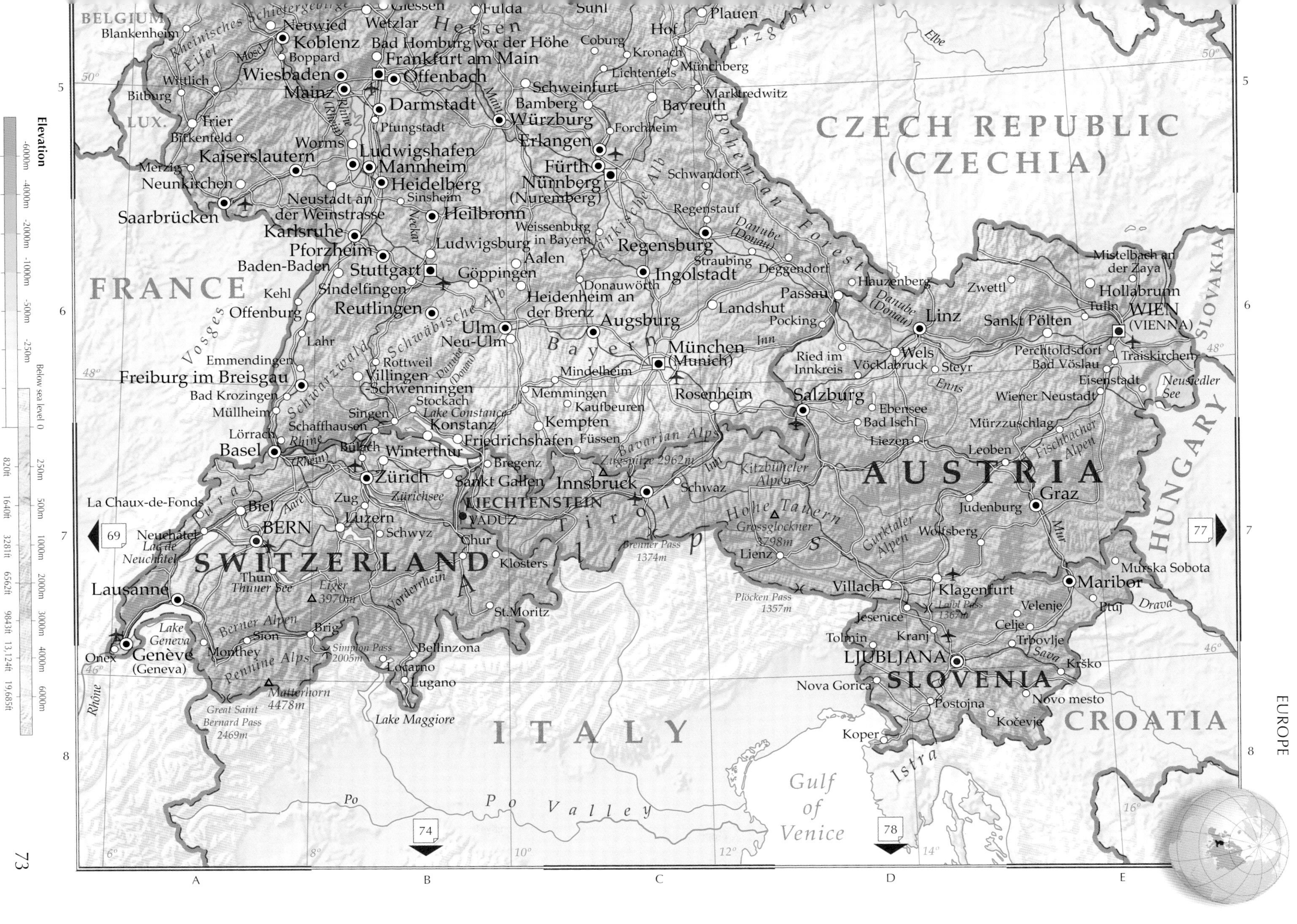

SLOVAKIA

HUNGARY

CZECH REPUBLIC (CZECHIA)

CROATIA

AUSTRIA

SLOVENIA

ITALY

SWITZERLAND

FRANCE

BELGIUM

LUX.

Wien (Vienna)
Mistelbach an der Zaya
Hollabrunn
Traiskirchen
Perchtoldsdorf
Bad Vöslau
Eisenstadt
Neusiedler See
Murska Sobota
Graz
Maribor
Ptuj
Drava
Krško
Novo mesto
Ljubljana
Velenje
Celje
Trbovlje
Sava
Ljubno Pass 1367m
Klagenfurt
Kranj
Jesenice
Wolfsberg
Judenburg
Mürzzuschlag
Leoben
Sankt Pölten
Zwettl
Mur
Linz
Wels
Danube (Donau)
Enns
Steyr
Hauzenberg
Passau
Ried im Innkreis
Vöcklabruck
Ebensee
Bad Ischl
Salzburg
Deggendorf
Pocking
Inn
Regenstauf
Straubing
Landshut
Rosenheim
Kitzbüheler Alpen
Hohe Tauern
Grossglockner 3798m
Nova Gorica
Tolmin
Villach
Lienz
Plöcken Pass 1357m
Koper
Gulf of Venice
Istra
Deggendorf
Regensburg
Ingolstadt
Donauwörth
Heidenheim an der Brenz
Augsburg
München (Munich)
Schwandorf
Alb
Forchheim
Bamberg
Würzburg
Schweinfurt
Fürth
Nürnberg (Nuremberg)
Erlangen
Weissenburg in Bayern
Aalen
Göppingen
Ulm
Neu-Ulm
Memmingen
Kempten
Kaufbeuren
Füssen
Mindelheim
Schwaz
Karwendel 2962m
Innsbruck
Tirol
Brenner Pass 1374m
Timmelsjoch 2509m
Po Valley
Po
Plauen
Hof
Sunl
Suhl
Coburg
Lichtenfels
Bayreuth
Kronach
Markredwitz
Mühlberg
München
Marktredwitz
Bohemian Forest
Elbe
Erzgebirge
Fichtelgebirge
Heilbronn
Ludwigsburg
Stuttgart
Sindelfingen
Reutlingen
Tübingen
Schwäbische Alb
Rottweil
Villingen-Schwenningen
Schramberg
Stockach
Singen
Konstanz
Lake Constance
Friedrichshafen
Bregenz
Sankt Gallen
Winterthur
Zürich
Vaduz
LIECHTENSTEIN
Chur
Klosters
St.Moritz
Ofenpass 2149m
Bernina Pass 2330m
Bellinzona
Locarno
Lugano
Lake Maggiore
Lake Lugano
Offenburg
Lahr
Kehl
Baden-Baden
Pforzheim
Freiburg im Breisgau
Emmendingen
Bad Krozingen
Müllheim
Lörrach
Basel
Rhine
Schaffhausen
Waldshut
Bülach
Baden
Zug
Luzern
Schwyz
Zürichsee
Berner Alpen
Interlaken
Thuner See
Thun
Brienzer Alpen
Penninne Alps
Simplon Pass 2005m
Brig
Sion
Monthey
Matterhorn 4478m
Great Saint Bernard Pass 2469m
Martigny
Eiger 3970m
BERN
Biel
Neuchâtel
Lake Neuchâtel
La Chaux-de-Fonds
Lausanne
Lake Geneva
Lac Léman
Geneva (Genève)
Onex
Rhône
Frankfurt am Main
Offenbach
Darmstadt
Pfungstadt
Wiesbaden
Mainz
Worms
Mannheim
Ludwigshafen
Heidelberg
Neustadt an der Weinstrasse
Karlsruhe
Kaiserslautern
Neunkirchen
Saarbrücken
Neustadt
Trier
Bitburg
Merzig
Birkenfeld
Wittlich
Blankenheim
Neuwied
Koblenz
Boppard
Wetzlar
Giessen
Fulda
Bad Homburg vor der Höhe
Hessen
Eifel
Rheinisches Schiefergebirge
Mosel
Rhine (Rhein)
Main
Neckar
Lech
Bayern
Vosges
Schwarzwald (Black Forest)
Aare
Säntis
Reuss
Iller
Donau (Danube)
Schliersee

Elevation

Below sea level 0   250m   500m   1000m   2000m   3000m   4000m   6000m

-6000m -4000m -2000m -1000m -500m -250m

820ft   1640ft   3281ft   6562ft   9843ft 13,124ft 19,685ft

-19,658ft -13,124ft -6562ft -3281ft -1640ft -820ft -328ft/-100m 0

Italy

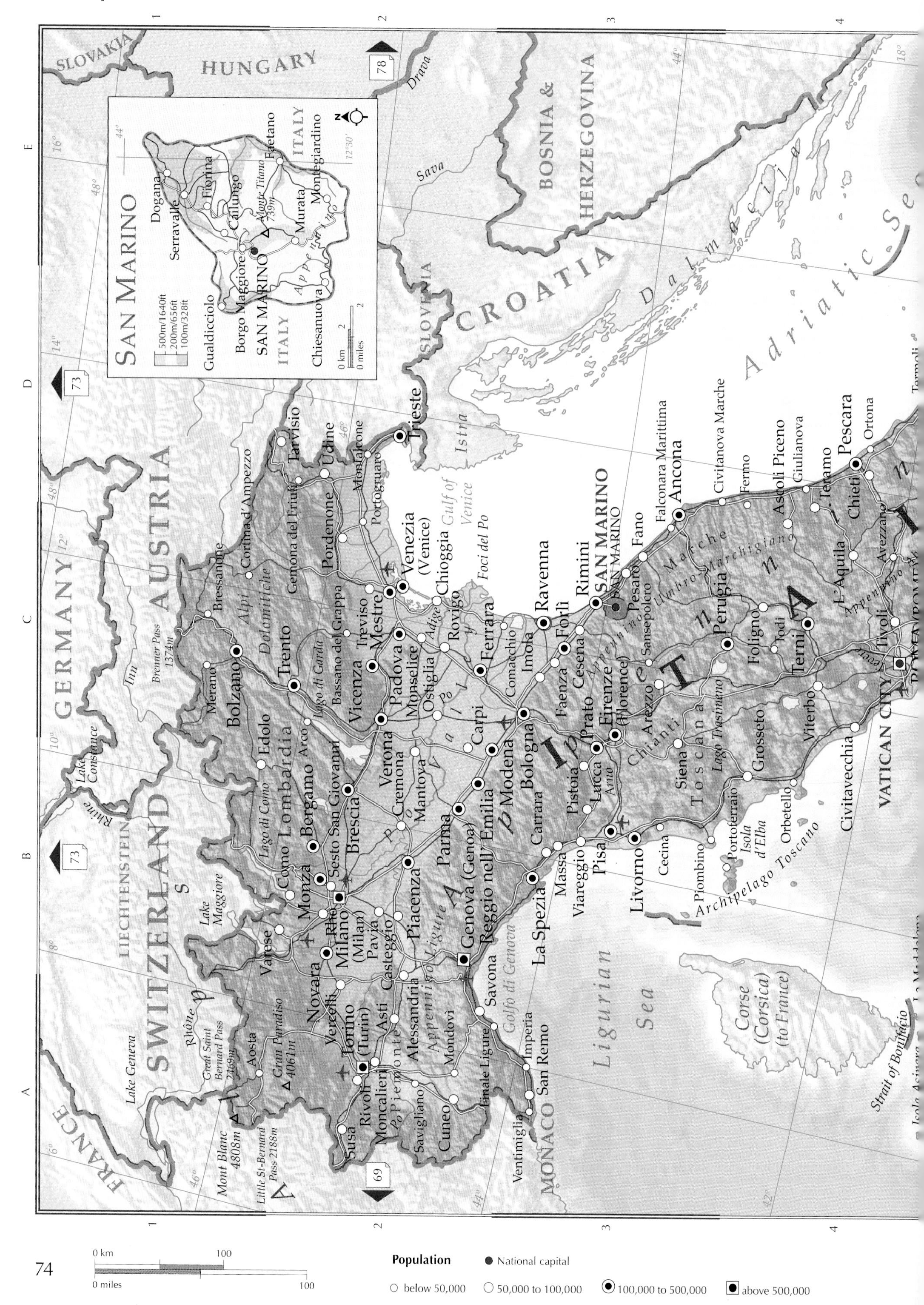

Population   ● National capital

○ below 50,000   ○ 50,000 to 100,000   ◉ 100,000 to 500,000   ■ above 500,000

SAN MARINO

Dogana
Serravalle
Fiorina
Cailungo
Fietano
Murata
Montegiardino
Monte Titano
739m
Gualdicciolo
Borgo Maggiore
SAN MARINO
Chiesanuova
ITALY
ITALY

500m/1640ft
200m/656ft
100m/328ft

0 km   2
0 miles   2

SLOVAKIA
HUNGARY
Drava
Sava
CROATIA
BOSNIA &
HERZEGOVINA
Dalmacija
Adriatic Sea

FRANCE
GERMANY
LIECHTENSTEIN
SWITZERLAND
AUSTRIA
SLOVENIA
Lake Geneva
Lake Constance
Rhône
Rhine
Inn
Bremer Pass 1374m
Mont Blanc 4808m
Great Saint Bernard Pass 2469m
Little St-Bernard Pass 2188m
Gran Paradiso 4061m
Lake Maggiore
Lago di Como
Merano
Bolzano
Bressanone
Alpi
Dolomitiche
Cortina d'Ampezzo
Tarvisio
Udine
Trieste
Istra
Gulf of Venice
Monfalcone
Portogruaro
Trento
Edolo
Arco
Lago di Garda
Gemona del Friuli
Pordenone
Treviso
Mestre
Venezia (Venice)
Chioggia
Bassano del Grappa
Vicenza
Padova
Monselice
Ostiglia
Adige
Rovigo
Foci del Po
Varese
Como
Bergamo
Brescia
Verona
Mantova
Po
Ferrara
Comacchio
Ravenna
Monza
Sesto San Giovanni
Cremona
Parma
Modena
Bologna
Imola
Faenza
Forlì
Cesena
Rimini
SAN MARINO
Falconara Marittima
Ancona
Civitanova Marche
Milano (Milan)
Pavia
Piacenza
Reggio nell'Emilia
Carpi
Po
Fano
Pesaro
Fermo
Ascoli Piceno
Giulianova
Teramo
Pescara
Ortona
Chieti
Novara
Vercelli
Casteggio
Genova (Genoa)
La Spezia
Carrara
Massa
Pistoia
Prato
Firenze (Florence)
Arezzo
Sansepolcro
Perugia
Foligno
Todi
Terni
L'Aquila
Avezzano
Tivoli
ROME
VATICAN CITY
Torino (Turin)
Rivoli
Moncalieri
Asti
Alessandria
Mondovì
Savona
Finale Ligure
Imperia
San Remo
Ventimiglia
MONACO
Golfo di Genova
Ligurian Sea
Viareggio
Lucca
Arno
Pisa
Livorno
Cecina
Piombino
Portoferraio
Isola d'Elba
Archipelago Toscano
Grosseto
Orbetello
Siena
Lago Trasimeno
Viterbo
Civitavecchia
Corse (Corsica) (to France)
Strait of Bonifacio
Cuneo
Savigliano
Susa
Rivoli
Appennino Ligure
Piemonte
Lombardia
Veneto
Marche
Toscana
Chianti
Umbro-Marchigiano
Umbria
ITALIA
Appennino
Appennino Abruzzese

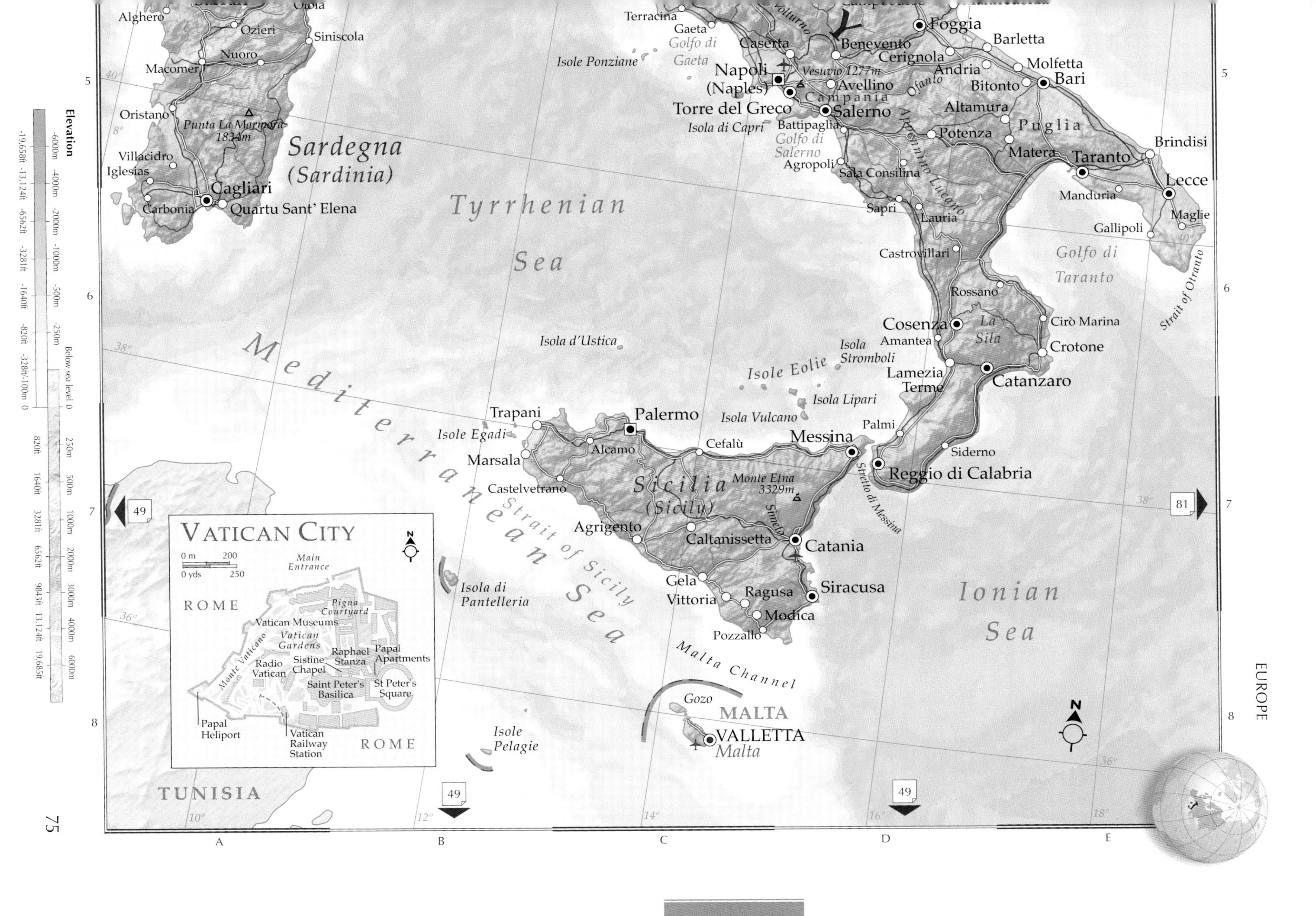

Brindisi
Lecce
Maglie
Gallipoli
Strait of Otranto
Golfo di Taranto
Manduria
Taranto
Matera
Bari
Molfetta
Bitonto
Puglia
Altamura
Potenza
Ciró Marina
Crotone
Rossano
La Sila
Catanzaro
Appennino Lucano
Lauria
Castrovillari
Cosenza
Amantea
Sapri
Sala Consilina
Lamezia Terme
Isola Stromboli
Isola Liparì
Isole Eolie
Isola Vulcano
Siderno
Palmi
Reggio di Calabria
Stretto di Messina
Messina
Catania
Siracusa
Monte Etna 3329m
Sineto
Caltanissetta
Ragusa
Modica
Enna
Cefalù
Pozzallo
Palermo
Sicilia (Sicily)
Gela
Vittoria
Alcamo
Agrigento
Trapani
Isole Egadi
Marsala
Castelvetrano
Strait of Sicily
Isola di Pantelleria
Isole Pelagie
Gozo
MALTA
VALLETTA
Malta
Malta Channel

Barletta
Molfetta
Andria
Cerignola
Foggia
Benevento
Avellino
Caserta
Salerno
Campania
Battipaglia
Golfo di Salerno
Agropoli
Vesuvio 1277m
Napoli (Naples)
Torre del Greco
Isola di Capri
Gaeta
Golfo di Gaeta
Terracina
Isole Ponziane
Volturno
Ofanto

Tyrrhenian Sea
Isola d'Ustica
Mediterranean Sea

Siniscola
Ozieri
Nuoro
Macomer
Oristano
Villacidro
Iglesias
Carbonia
Sardegna (Sardinia)
Punta La Marmora 1834m
Cagliari
Quartu Sant' Elena
Alghero

Ionian Sea

81
49
49
49

VATICAN CITY
ROME
Pigna Courtyard
Vatican Museums
Raphael Stanza
Papal Apartments
St Peter's Square
Sistine Chapel
Radio Vatican
Vatican Gardens
Main Entrance
Monte Vaticano
Saint Peter's Basilica
Vatican Railway Station
Papal Heliport
ROME
0 m   200
0 yds   250

TUNISIA

Elevation

| Below sea level 0 | 250m | 500m | 1000m | 2000m | 3000m | 4000m | 6000m |

| -6000m | -4000m | -2000m | -1000m | -500m | -250m |
| -19,658ft | -13,124ft | -6562ft | -3281ft | -1640ft | -820ft | -328ft/-100m 0 |

| 820ft | 1640ft | 3281ft | 6562ft | 9843ft | 13,124ft | 19,685ft |

Central Europe

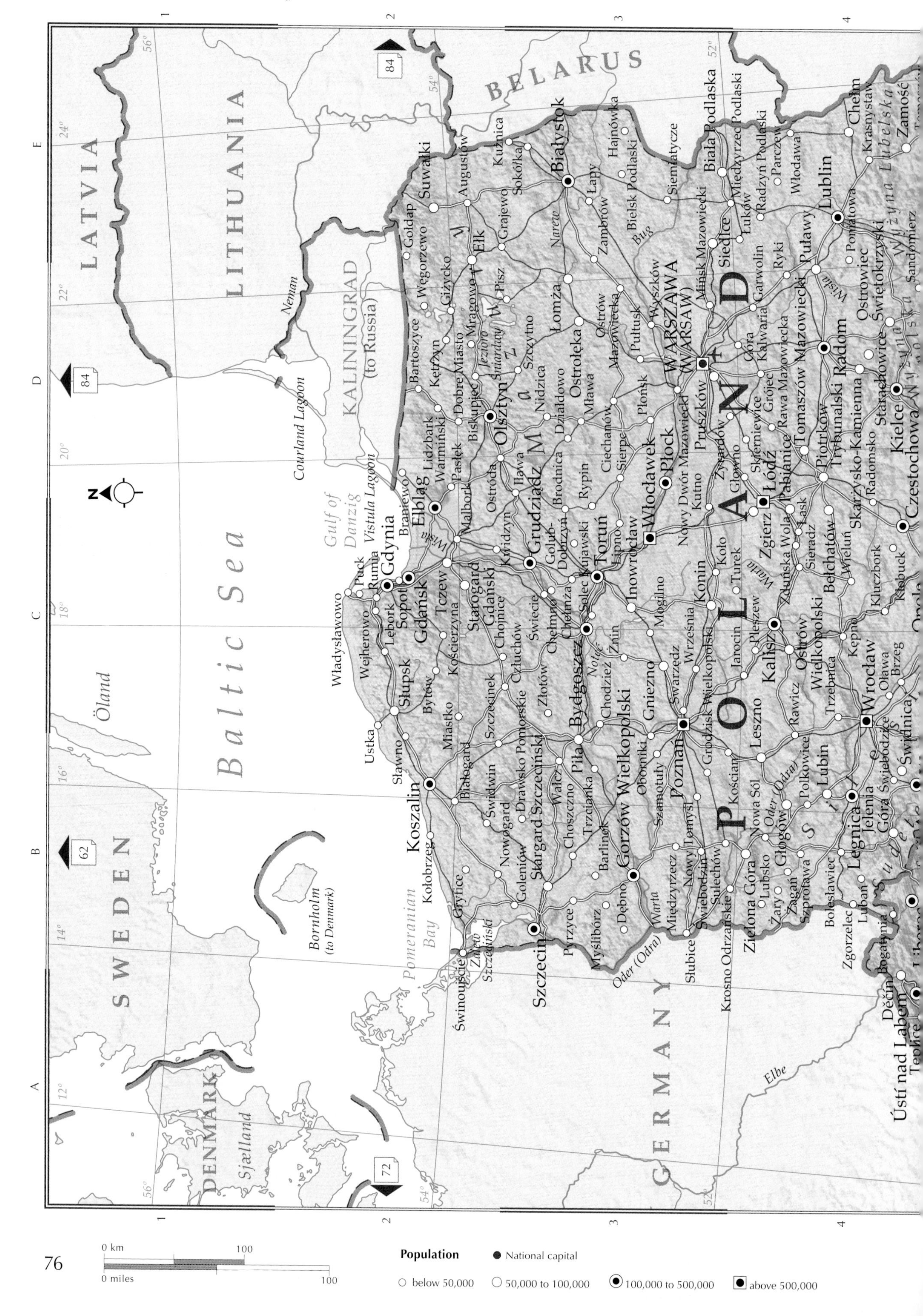

SWEDEN

Öland

Baltic Sea

LATVIA

LITHUANIA

Neman

KALININGRAD
(to Russia)

Courland Lagoon

BELARUS

DENMARK

Sjælland

Bornholm
(to Denmark)

Pomeranian Bay

Gulf of Danzig

Vistula Lagoon

GERMANY

POLAND

Świnoujście
Zalew Szczeciński
Szczecin
Pyrzyce
Myślibórz
Gryfice
Kołobrzeg
Goleniów
Nowogard
Stargard Szczeciński
Barlinek
Dębno
Kostrzyn
Krosno Odrzańskie
Słubice
Sulechów
Świebodzin
Nowy Tomyśl
Międzyrzecz
Gorzów Wielkopolski
Zielona Góra
Nowa Sól
Lubsko
Żary
Żagań
Szprotawa
Głogów
Polkowice
Lubin
Bolesławiec
Zgorzelec
Lubań
Bogatynia

Ustka
Słupsk
Sławno
Koszalin
Białogard
Świdwin
Drawsko Pomorskie
Szczecinek
Miastko
Bytów
Człuchów
Złotów
Piła
Wałcz
Trzcianka
Chodzież
Czarnków
Oborniki
Szamotuły
Poznań
Kościan
Leszno
Góra
Rawicz
Trzebnica
Oleśnica

Władysławowo
Wejherowo
Rumia
Puck
Gdynia
Sopot
Lębork
Gdańsk
Tczew
Starogard Gdański
Kościerzyna
Chojnice
Świecie
Bydgoszcz
Gniezno
Września
Swarzędz
Środa Wielkopolska
Jarocin
Pleszew
Kalisz
Ostrów Wielkopolski
Krotoszyn
Kępno
Ostrzeszów
Wieluń
Kluczbork
Opole
Brzeg
Oława
Wrocław
Świdnica
Legnica
Jelenia Góra
Góra Śląska

Braniewo
Elbląg
Malbork
Kwidzyn
Grudziądz
Chełmno
Świecie
Grudziądz M
Toruń
Inowrocław
Mogilno
Żnin
Szubin
Konin
Koło
Turek
Łódź
Zgierz
Pabianice
Zduńska Wola
Sieradz
Łask
Bełchatów
Wieruszów
Radomsko
Częstochowa

Frombork
Pasłęk
Lidzbark Warmiński
Orneta
Dobre Miasto
Bartoszyce
Kętrzyn
Olsztyn
Ostróda
Iława
Nidzica
Działdowo
Brodnica
Golub-Dobrzyń
Rypin
Lipno
Ciechanów
Sierpc
Włocławek
Płock
Kutno
Łęczyca
Głowno
Skierniewice
Rawa Mazowiecka
Tomaszów Mazowiecki
Piotrków Trybunalski
Skarżysko-Kamienna
Kielce
Starachowice
Ostrowiec Świętokrzyski
Sandomierz

Goldap
Gołdap
Gorzewo
Węgorzewo
Giżycko
Jezioro Śniardwy
Mrągowo
Pisz
Ełk
Szczytno
Mława
Nidzica
Pułtusk
Płońsk
Nowy Dwór Mazowiecki
Nasielsk
Wyszków
Legionowo
WARSZAWA (WARSAW)
Pruszków
Żyrardów
Grójec
Garwolin
Kozienice
Radom
Puławy
Dęblin
Ryki
Łuków

Suwałki
Augustów
Kuźnica
Sokółka
Białystok
Grajewo
Łomża
Zambrów
Łapy
Bielsk Podlaski
Hajnówka
Mińsk Mazowiecki
Siedlce
Węgrów
Sokołów Podlaski
Międzyrzec Podlaski
Biała Podlaska
Radzyń Podlaski
Parczew
Łuków
Włodawa

Lublin
Poniatowa
Kraśnik
Świdnik
Chełm
Krasnystaw
Zamość
Puławy

Narew
Bug
Wisła
Warta
Noteć
Oder (Odra)
Elbe

Déčin
Ústí nad Labem
Teplice

0 km		100
0 miles		100

Population   ● National capital

○ below 50,000   ◯ 50,000 to 100,000   ◉ 100,000 to 500,000   ■ above 500,000

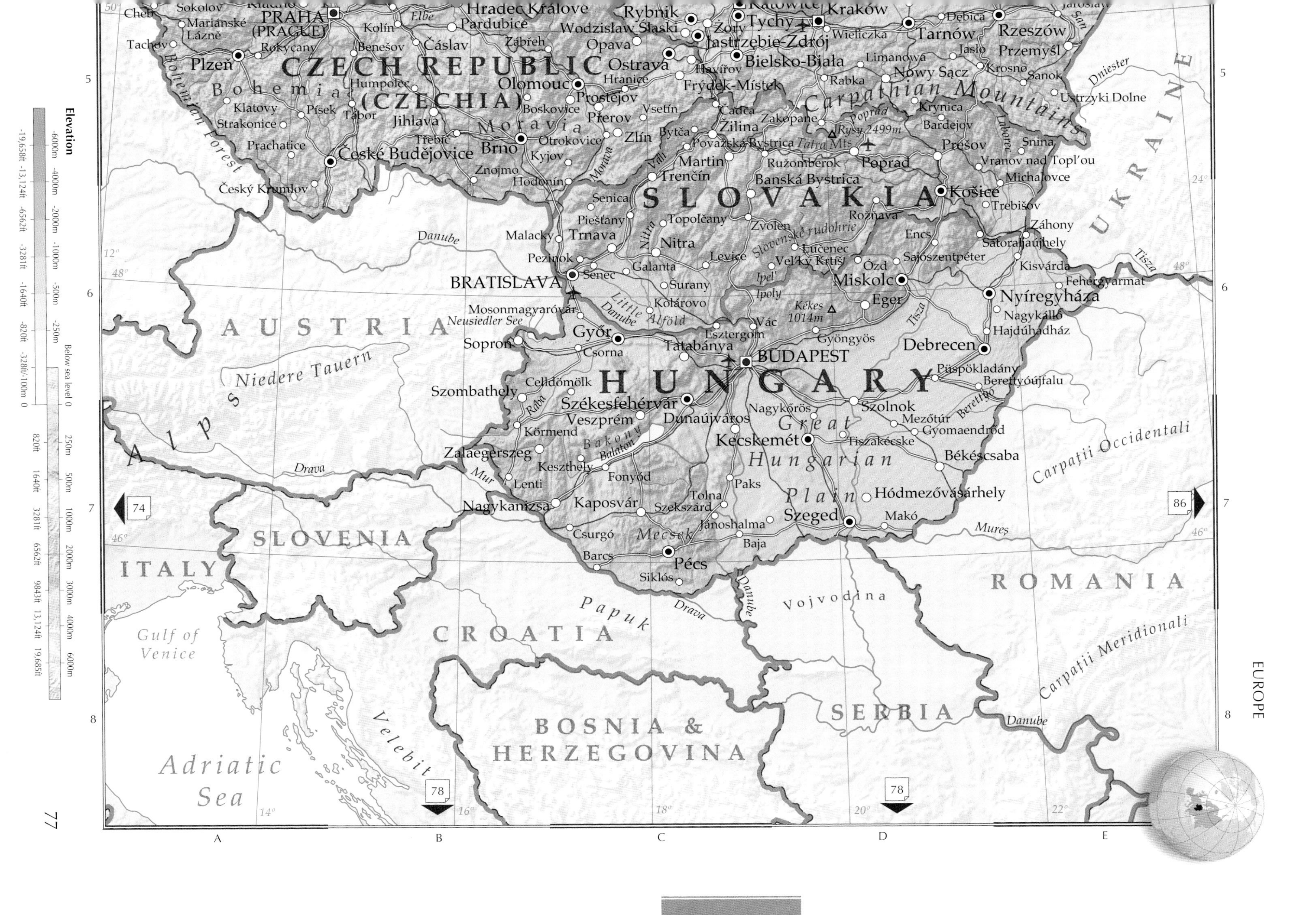

Elevation

						Below sea level 0	250m	500m	1000m	2000m	3000m	4000m	6000m

-6000m -4000m -2000m -1000m -500m -250m

820ft 1640ft 3281ft 6562ft 9843ft 13,124ft 19,685ft

-19,658ft -13,124ft -6562ft -3281ft -1640ft -820ft -328ft/-100m 0

Southeast Europe

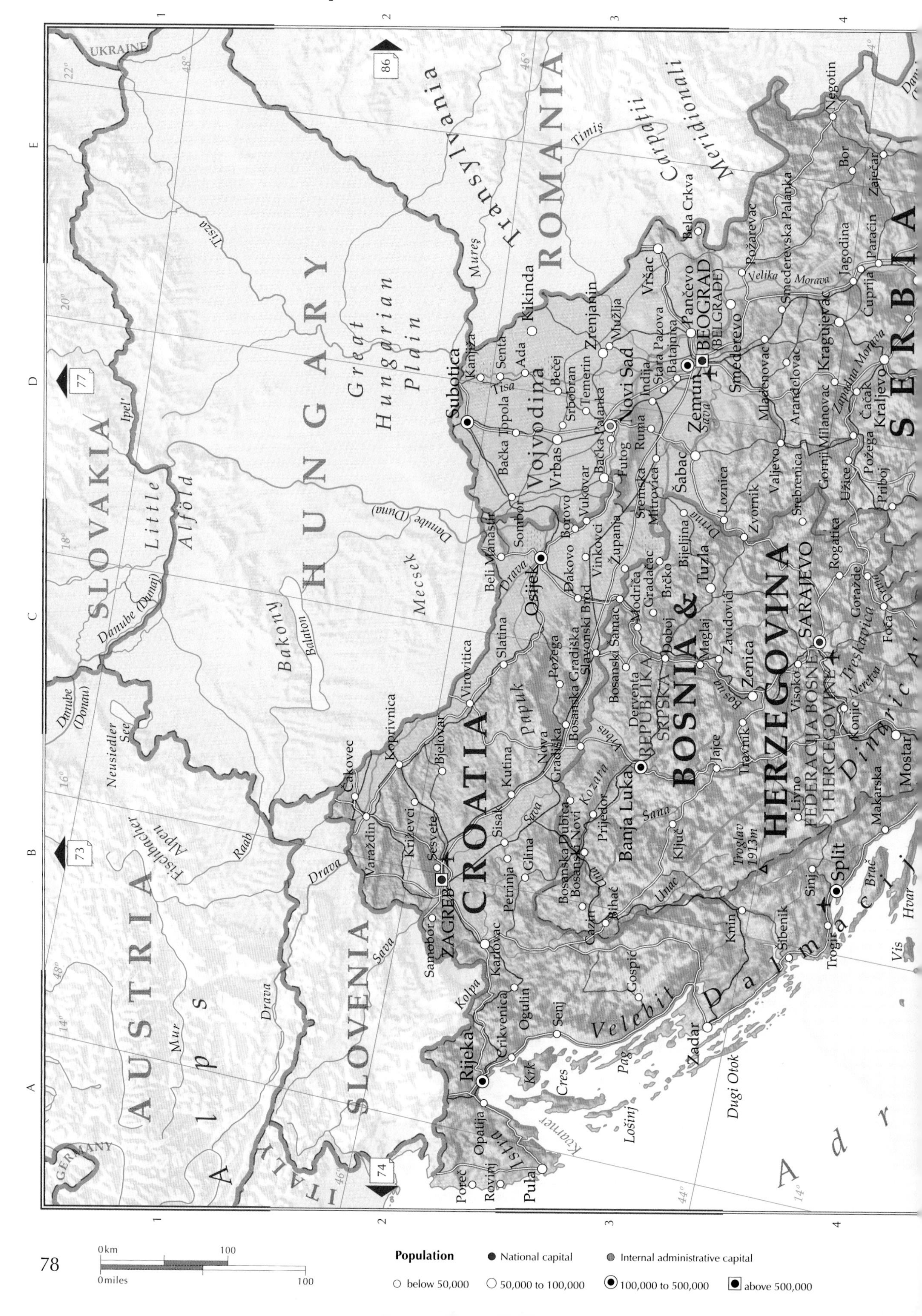

UKRAINE

SLOVAKIA

AUSTRIA

A l p s

Fischbacher Alpen

GERMANY

ITALY

SLOVENIA

HUNGARY

Great Hungarian Plain

Little Alföld

Bakony

Mecsek

Danube (Duna)

Balaton

Neusiedler See

Tisza

Ipel

Danube (Donau)

Raab

Drava

Mur

Drava

Sava

Kolpa

Istria

Transylvania

ROMANIA

Carpaṭii Meridionali

Timiş

Mureş

CROATIA

Kupa

Papuk

BOSNIA & HERZEGOVINA

REPUBLIKA SRPSKA

FEDERACIJA BOSNE I HERCEGOVINE

SERBIA

Voivodina

Dinaric Alps

Velebit

P a l m a t i j

A d r i a

Cities and places

SLOVAKIA / Danube (Dunaj) — Tisza

Čakovec, Koprivnica, Varaždin, Križevci, Sesvete, Zagreb, Samobor, Karlovac, Petrinja, Glina, Sisak, Kutina, Nova Gradiška, Slatina, Požega, Virovitica, Bjelovar

Subotica, Kanjiža, Senta, Ada, Bečej, Srbobran, Bačka Topola, Tisa, Sombor, Beli Manastir, Osijek, Đakovo, Vinkovci, Županja, Bačka Palanka, Temerin, Indija, Ruma, Stara Pazova, Batajnica, Novi Sad, Futog, Sremska Mitrovica, Šabac, Loznica, Zvornik

Kikinda, Zrenjanin, Mužlja, Pančevo, Vršac, Bela Crkva, BEOGRAD (BELGRADE), Zemun, Smederevo, Velika Morava, Smederevska Palanka, Mladenovac, Aranđelovac, Kragujevac, Gornji Milanovac, Čačak, Kraljevo, Zapadna Morava, Cuprija, Paraćin, Jagodina, Požarevac, Negotin, Bor, Zaječar

Požega, Priboj, Užice, Rogatika, Gorazde, Foča, Treskavica, Konjic, Neretva, SARAJEVO, Visoko, Zenica, Zavidovići, Maglaj, Tuzla, Bijeljina, Brčko, Gradačac, Modriča, Bosanski Šamac, Doboj, Derventa, Bosanski Brod, Slavonski Brod, Bosanska Gradiška, Bosanska Dubica, Bosanski Novi, Prijedor, Banja Luka, Jajce, Travnik, Ključ, Sana, Kozara, Una, Sana, Bihać, Cazin, Una, Unac, Knin, Sinj, Livno, Troglav 1913m, Gospić

Ogulin, Senj, Crikvenica, Krk, Cres, Lošinj, Rab, Pag, Zadar, Dugi Otok, Šibenik, Trogir, Split, Brač, Hvar, Vis, Makarska, Mostar

Rijeka, Opatija, Poreč, Rovinj, Pula

Bos. na

Vrbas

Drina

Sava

Zala

Kvarner

Velika Morava

78

Population

○ below 50,000
○ 50,000 to 100,000
◉ 100,000 to 500,000
■ above 500,000

● National capital
● Internal administrative capital

0 km ———— 100
0 miles ———— 100

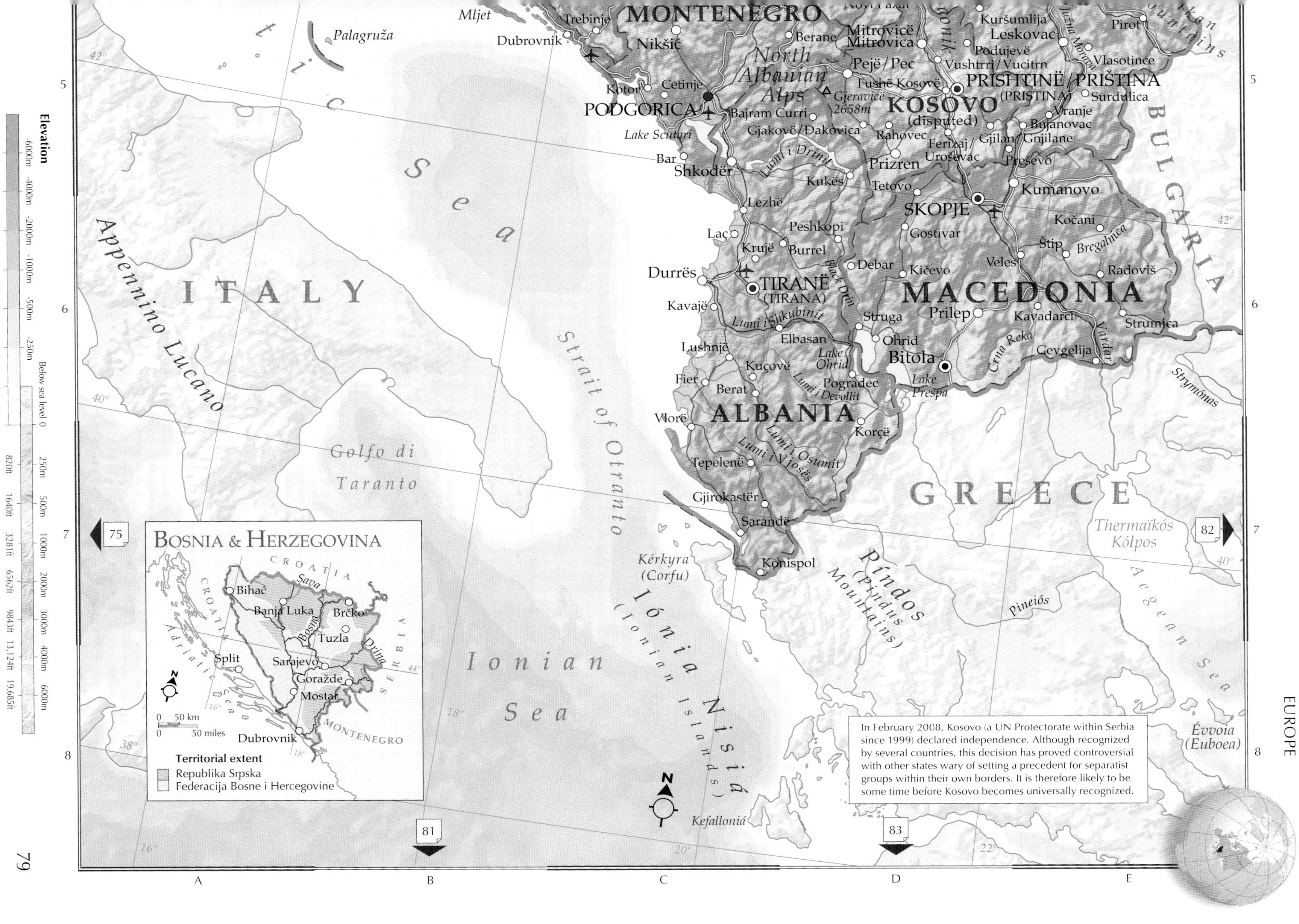

In February 2008, Kosovo (a UN Protectorate within Serbia since 1999) declared independence. Although recognized by several countries, this decision has proved controversial with other states wary of setting a precedent for separatist groups within their own borders. It is therefore likely to be some time before Kosovo becomes universally recognized.

Elevation

-6000m -4000m -2000m -1000m -500m -250m   Below sea level 0   250m 500m 1000m 2000m 3000m 4000m 6000m

-19,658ft -13,124ft -6562ft -3281ft -1640ft -820ft -328ft/-100m 0   820ft 1640ft 3281ft 6562ft 9843ft 13,124ft 19,685ft

BOSNIA & HERZEGOVINA

Territorial extent
Republika Srpska
Federacija Bosne i Hercegovine

The Mediterranean

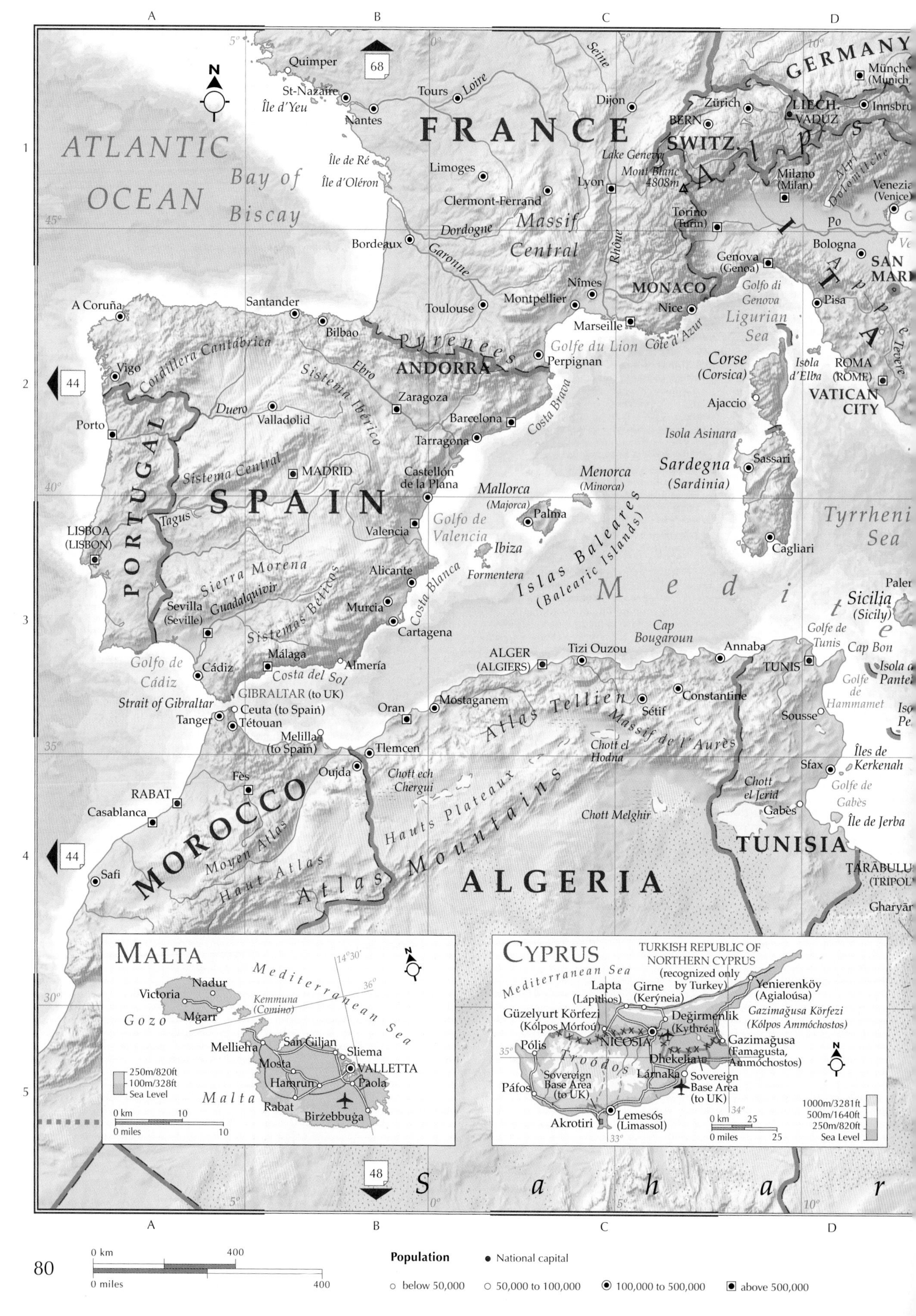

ATLANTIC OCEAN

Bay of Biscay

FRANCE

GERMANY

Quimper
St-Nazaire
Île d'Yeu
Nantes
Tours
Loire
Dijon
Zürich
BERN
SWITZ.
München (Munich)
LIECH.
VADUZ
Innsbru

Limoges
Clermont-Ferrand
Lyon
Lake Geneva
Mont Blanc
4808m
Milano (Milan)
Venezia (Venice)

Bordeaux
Dordogne
Garonne
Massif Central
Rhône
Torino (Turin)
Po
Bologna
Ve

Santander
Bilbao
Toulouse
Montpellier
Nîmes
MONACO
Nice
Genova (Genoa)
Golfo di Genova
Pisa
San Marino
ROMA (ROME)

A Coruña
Cordillera Cantábrica
Pyrenees
ANDORRA
Marseille
Perpignan
Côte d'Azur
Golfe du Lion
Ligurian Sea
Corse (Corsica)
Isola d'Elba
VATICAN CITY

Vigo
Duero
Valladolid
Sistema Ibérico
Zaragoza
Barcelona
Tarragona
Costa Brava
Ajaccio
Isola Asinara
Sassari
Sardegna (Sardinia)
Tyrrheni Sea

Porto
PORTUGAL
Sistema Central
MADRID
SPAIN
Castellón de la Plana
Valencia
Mallorca (Majorca)
Palma
Menorca (Minorca)
Golfo de Valencia
Ibiza
Formentera
Islas Baleares (Balearic Islands)
Cagliari
Medi

LISBOA (LISBON)
Tagus
Sierra Morena
Guadalquivir
Sistemas Béticos
Alicante
Murcia
Costa Blanca

Sevilla (Seville)
Málaga
Almería
Cartagena
Cap Bougaroun
Tizi Ouzou
Annaba
Golfe de Tunis
Cap Bon
Sicilia (Sicily)
Paler
Isola di Pante

Golfo de Cádiz
Cádiz
Costa del Sol
GIBRALTAR (to UK)
Ceuta (to Spain)
ALGER (ALGIERS)
Oran
Mostaganem
Constantine
Sétif
TUNIS
Golfe de Hammamet
Sousse
Isola Pe

Strait of Gibraltar
Tanger
Tétouan
Melilla (to Spain)
Tlemcen
Atlas Tellien
Massif de l'Aurès
Chott el Hodna
Sfax
Îles de Kerkenah
Golfe de Gabès

Fès
Oujda
Chott ech Chergui
Chott el Jerid
Gabès
Île de Jerba

RABAT
Casablanca
MOROCCO
Moyen Atlas
Haut Atlas
Hauts Plateaux
Atlas Mountains
Chott Melghir
TUNISIA

Safi
ALGERIA
TARĀBULU (TRIPOL
Gharyān

S   a   h   a   r

MALTA

Mediterranean Sea

Nadur
Victoria
Gozo
Mgarr
Kemmuna (Comino)

Mellieha
San Giljan
Sliema
Mosta
Hamrun
VALLETTA
Paola
Rabat
Birżebbuġa
Malta

250m/820ft
100m/328ft
Sea Level

0 km   10
0 miles   10

CYPRUS

TURKISH REPUBLIC OF NORTHERN CYPRUS
(recognized only by Turkey)

Mediterranean Sea
Lapta (Lápithos)
Girne (Kerýneia)
Yenierenköy (Agialoúsa)

Güzelyurt Körfezi (Kólpos Mórfou)
Değirmenlik (Kythréa)
Gazimağusa Körfezi (Kólpos Ammóchostos)

Pólis
NICOSIA
Gazimağusa (Famagusta, Ammóchostos)

Troodos
Dhekelia
Páfos
Sovereign Base Area (to UK)
Lárnaka
Sovereign Base Area (to UK)

Akrotiri
Lemesós (Limassol)

1000m/3281ft
500m/1640ft
250m/820ft
Sea Level

0 km   25
0 miles   25

0 km   400
0 miles   400

Population   • National capital

○ below 50,000   ◎ 50,000 to 100,000   ◉ 100,000 to 500,000   ■ above 500,000

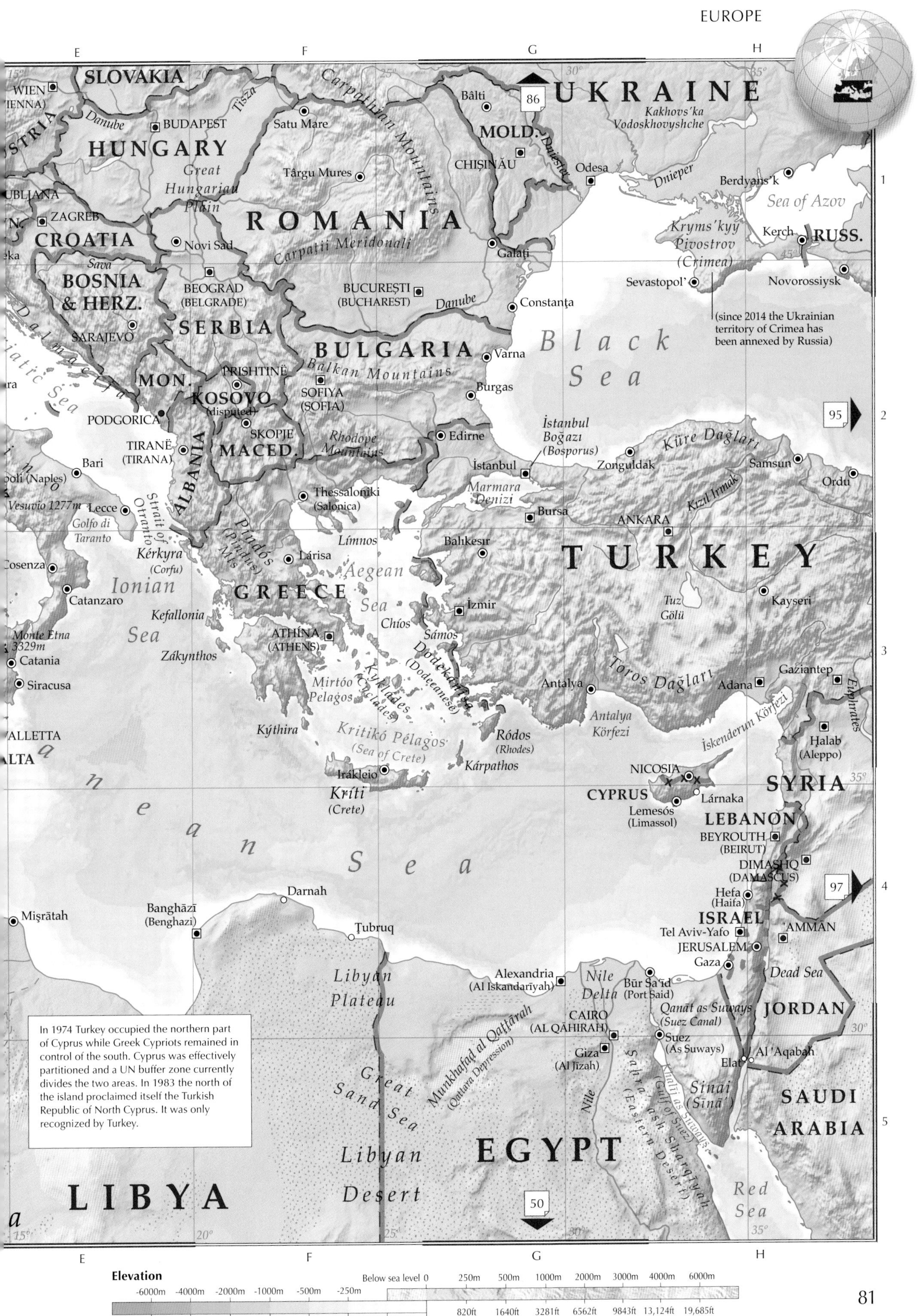

SLOVAKIA

WIEN (VIENNA)

Danube

BUDAPEST

HUNGARY

Great Hungarian Plain

Tisza

Satu Mare

Bâlti

86

UKRAINE

Kakhovs'ka Vodoskhovyshche

MOLD.

CHIŞINĂU

Odesa

Berdyans'k

Dnieper

Sea of Azov

RUSS.

Kerch

Krylm's'kyy Pivostrov (Crimea)

Novorossiysk

Sevastopol'

(since 2014 the Ukrainian territory of Crimea has been annexed by Russia)

Târgu Mures

ROMANIA

Carpații Meridonali

Galați

ZAGREB

CROATIA

Sava

Novi Sad

UBLJANA

N

BOSNIA & HERZ.

BEOGRAD (BELGRADE)

SERBIA

BUCUREŞTI (BUCHAREST)

Danube

Constanța

Black Sea

SARAJEVO

Dalmatia

adriatic Sea

PODGORICA

MON.

PRISHTINË

KOSOVO (disputed)

SKOPJE

MACED.

Balkan Mountains

SOFIYA (SOFIA)

BULGARIA

Varna

Burgas

ra

TIRANË (TIRANA)

Bari

ALBANIA

Rhodope Mountains

Edirne

İstanbul Boğazı (Bosporus)

İstanbul

Zonguldak

Küre Dağları

Samsun

Ordu

oli (Naples)

Vesuvio 1277m

Lecce

Strait of Otranto

Pindos Mountains

Thessaloníki (Salonica)

Límnos

Marmara Denizi

Bursa

ANKARA

Kızıl Irmak

Golfo di Taranto

Kérkyra (Corfu)

Larísa

Balıkesır

TURKEY

Ionian Sea

Kefallonia

GREECE

Aegean Sea

İzmir

Tuz Gölü

Kayseri

Cosenza

Catanzaro

Monte Etna 3329m

Catania

Siracusa

Zákynthos

ATHÍNA (ATHENS)

Chíos

Mirtóo Pelagos

Kyklades (Cyclades)

Sámos

Dodekanisos (Dodecanese)

Toros Dağları

Antalya

Antalya Körfezi

Adana

Gaziantep

Euphrates

İskenderun Körfezi

Halab (Aleppo)

ALLETTA

ALTA

Kýthira

Kritikó Pélagos (Sea of Crete)

Ródos (Rhodes)

Kárpathos

NICOSIA

CYPRUS

Lárnaka

Lemesós (Limassol)

SYRIA

LEBANON

a

Irákleio

Kríti (Crete)

BEYROUTH (BEIRUT)

DIMASHQ (DAMASCUS)

97

n

Darnah

Hefa (Haifa)

ISRAEL

Tel Aviv-Yafo

AMMAN

e

Banghāzī (Benghazi)

Ţubruq

a

Mişrātah

Alexandria (Al Iskandarīyah)

Nile Delta

Būr Sa'īd (Port Said)

JERUSALEM

Gaza

Dead Sea

JORDAN

S

e

a

Libyan Plateau

Qanāt as Suways (Suez Canal)

In 1974 Turkey occupied the northern part of Cyprus while Greek Cypriots remained in control of the south. Cyprus was effectively partitioned and a UN buffer zone currently divides the two areas. In 1983 the north of the island proclaimed itself the Turkish Republic of North Cyprus. It was only recognized by Turkey.

Munkhafad al Qaţţārah (Qattara Depression)

CAIRO (AL QĀHIRAH)

Suez (As Suways)

Al 'Aqabah

Elat

Giza (Al Jīzah)

Great Sand Sea

Sahra' ash Sharqīyah (Eastern Desert)

Khalīj as Suways (Gulf of Suez)

Sinai (Sīnā')

SAUDI ARABIA

LIBYA

Libyan Desert

EGYPT

50

Nile

Red Sea

Elevation

Below sea level 0   250m   500m   1000m   2000m   3000m   4000m   6000m

-6000m -4000m -2000m -1000m -500m -250m

-19,658ft -13,124ft -6562ft -3281ft -1640ft -820ft -328ft/-100m 0

820ft   1640ft   3281ft   6562ft   9843ft 13,124ft 19,685ft

Bulgaria & Greece

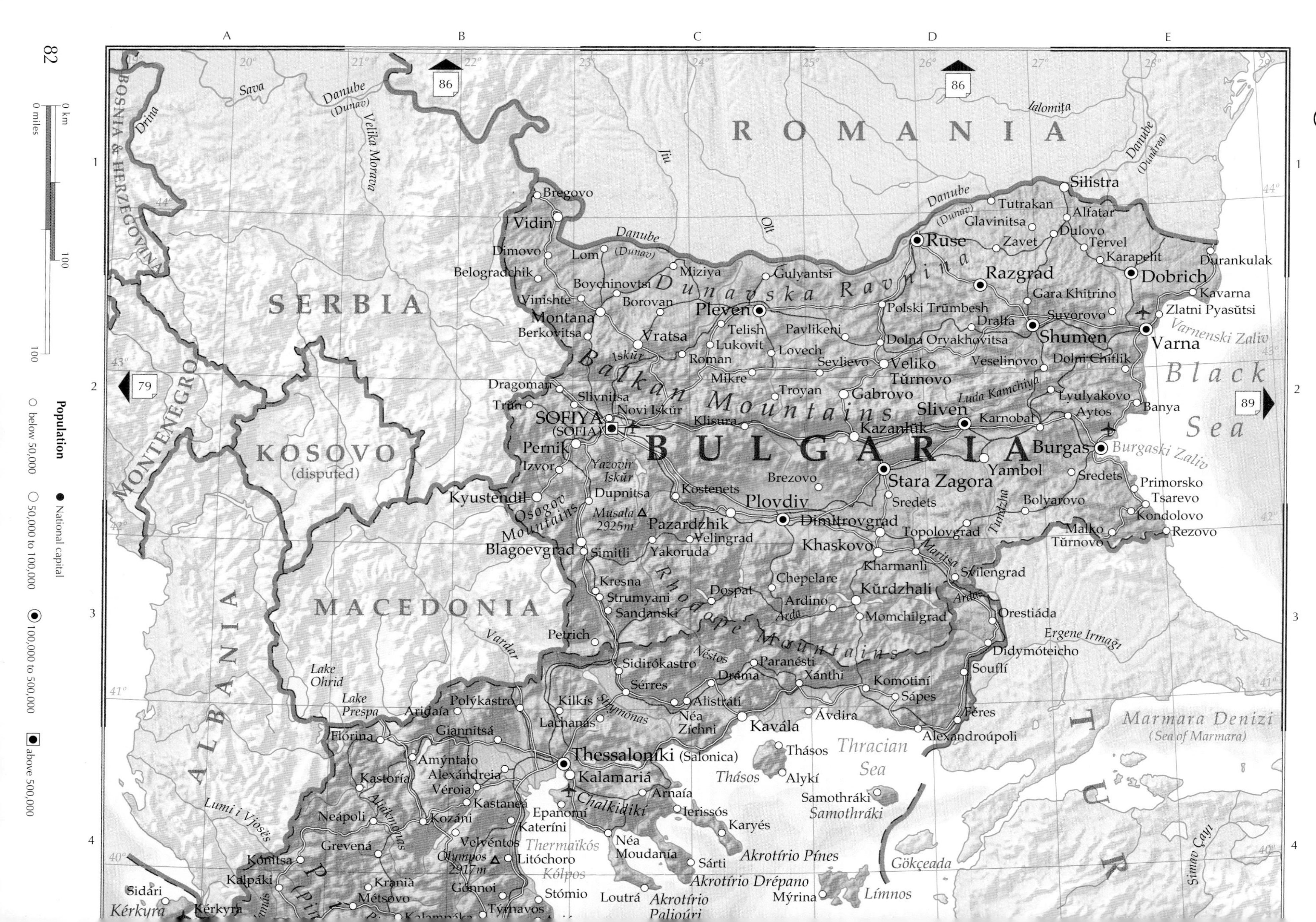

0 km   100

0 miles   100

Population   ● National capital

○ below 50,000   ◯ 50,000 to 100,000   ◉ 100,000 to 500,000   ◼ above 500,000

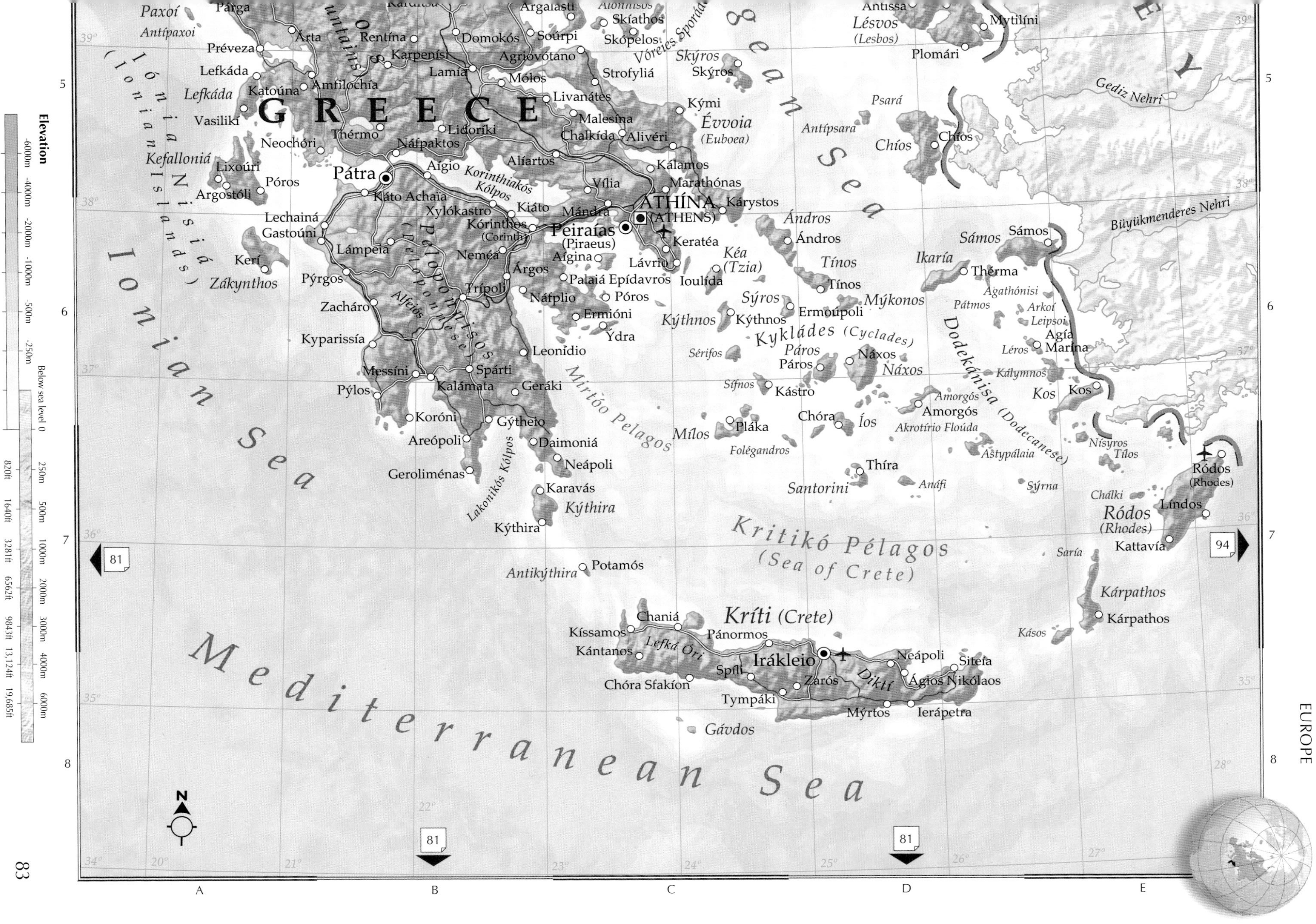

Below sea level 0   250m   500m   1000m   2000m   3000m   4000m   6000m

Elevation

-6000m   -4000m   -2000m   -1000m   -500m   -250m

-19,658ft   -13,124ft   -6562ft   -3281ft   -1640ft   -820ft   -328ft/-100m   0

820ft   1640ft   3281ft   6562ft   9843ft   13,124ft   19,685ft

The Baltic States & Belarus

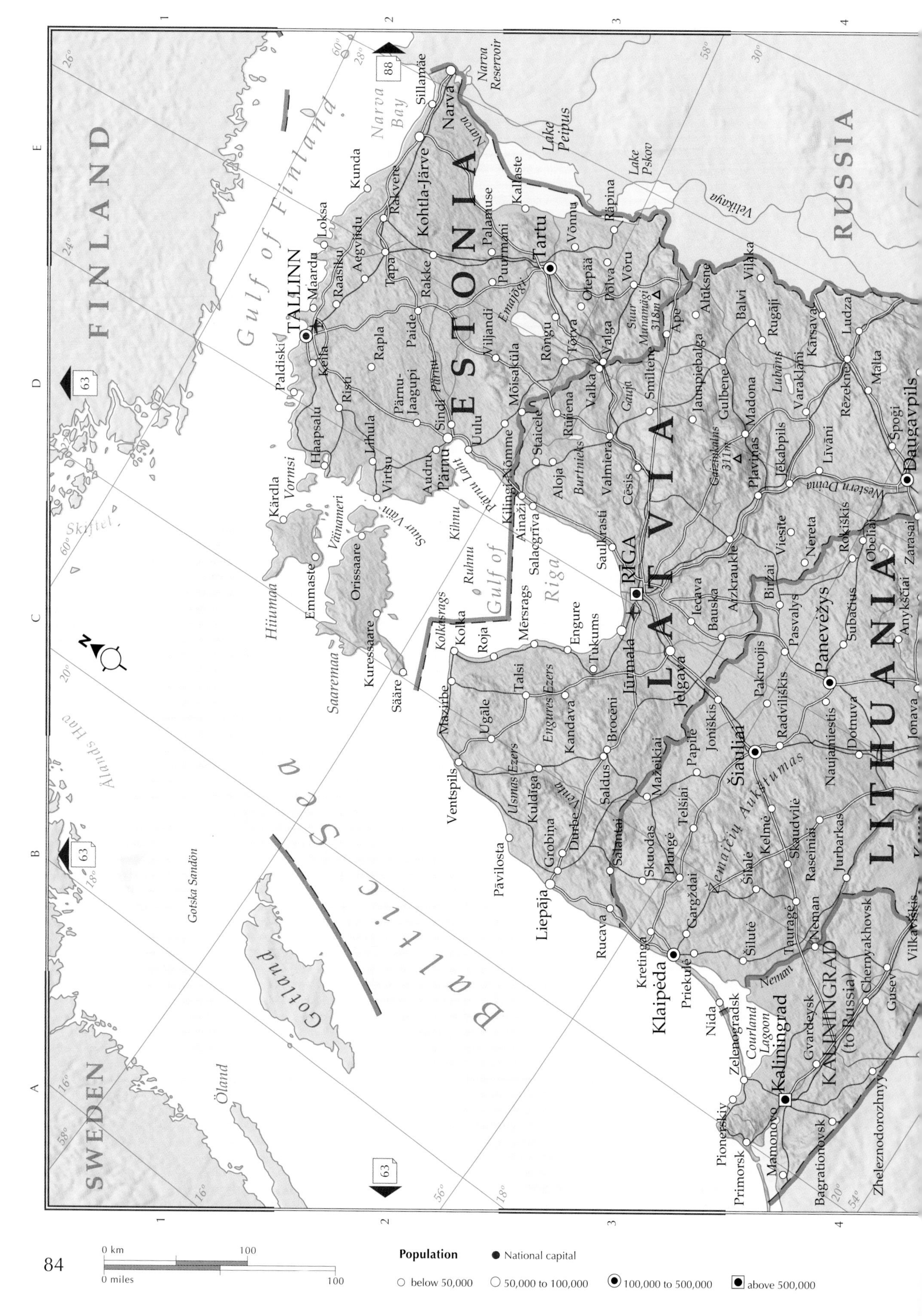

0 km   100
0 miles   100

Population   ● National capital

○ below 50,000   ◎ 50,000 to 100,000   ◉ 100,000 to 500,000   ■ above 500,000

SWEDEN
FINLAND
ESTONIA
LATVIA
LITHUANIA
RUSSIA
KALININGRAD (to Russia)

Gulf of Finland
Baltic Sea
Gulf of Riga
Narva Bay

TALLINN
Tartu
RĪGA
Panevėžys
Šiauliai
Klaipėda
Kaliningrad
Daugavpils

Paldiski
Keila
Maardu
Loksa
Kunda
Rakvere
Kohtla-Järve
Sillamäe
Narva
Aegviidu
Raasiku
Tapa
Rakke
Paide
Rapla
Haapsalu
Risti
Lihula
Pärnu-Jaagupi
Virtsu
Audru
Pärnu
Sindi
Kilingi-Nõmme
Kärdla
Vormsi
Emmaste
Orissaare
Kuressaare
Sääre
Viljandi
Mõisaküla
Rõngu
Tõrva
Valga
Valka
Rūjiena
Staicele
Ainaži
Salacgrīva
Aloja
Cēsis
Valmiera
Burtnieks
Saulkrasti
Kolkasrags
Kolka
Roja
Mērsrags
Engure
Tukums
Jūrmala
Mazirbe
Ugāle
Talsi
Kandava
Broceni
Saldus
Ventspils
Usmas Ezers
Kuldīga
Engures Ezers
Pāvilosta
Grobiņa
Durbe
Liepāja
Rucava
Kretinga
Priekule
Nida
Zelenogradsk
Pionerskiy
Primorsk
Mamonovo
Bagrationovsk
Gvardeysk
Chernyakhovsk
Gusev
Zheleznodorozhnyy
Vilkaviškis
Kaunas
Jurbarkas
Neman
Šilutė
Thaurage
Skaudvilė
Raseiniai
Kelmė
Šilalė
Telšiai
Plungė
Gargždai
Skuodas
Salantai
Mažeikiai
Papilė
Joniškiai
Radviliškis
Pakruojis
Pasvalys
Biržai
Jelgava
Bauska
Iecava
Aizkraukle
Jēkabpils
Pļaviņas
Madona
Gulbene
Jaunpiebalga
Lubāna
Varakļāni
Līvāni
Balvi
Rugāji
Alūksne
Viļaka
Kārsava
Rēzekne
Ludza
Malta
Spoģi
Obeliai
Rokiškis
Nereta
Viesīte
Subačius
Naujamiestis
Dotnuva
Jonava
Anykščiai
Zarasai
Šiauliai

Hiiumaa
Saaremaa
Vormsi
Kihnu
Ruhnu
Gotska Sandön
Gotland
Öland
Courland Lagoon
Lake Peipus
Lake Pskov
Narva Reservoir
Suur Väin
Väinameri
Skärjär.
Ålands Hav

Emajõgi
Pärnu
Gauja
Western Dvina
Velikaya
Nemunas
Neman

Žemaičių Aukštumas
Munamägi 318m
Gaiziņkalns 311m

63
63
63
88

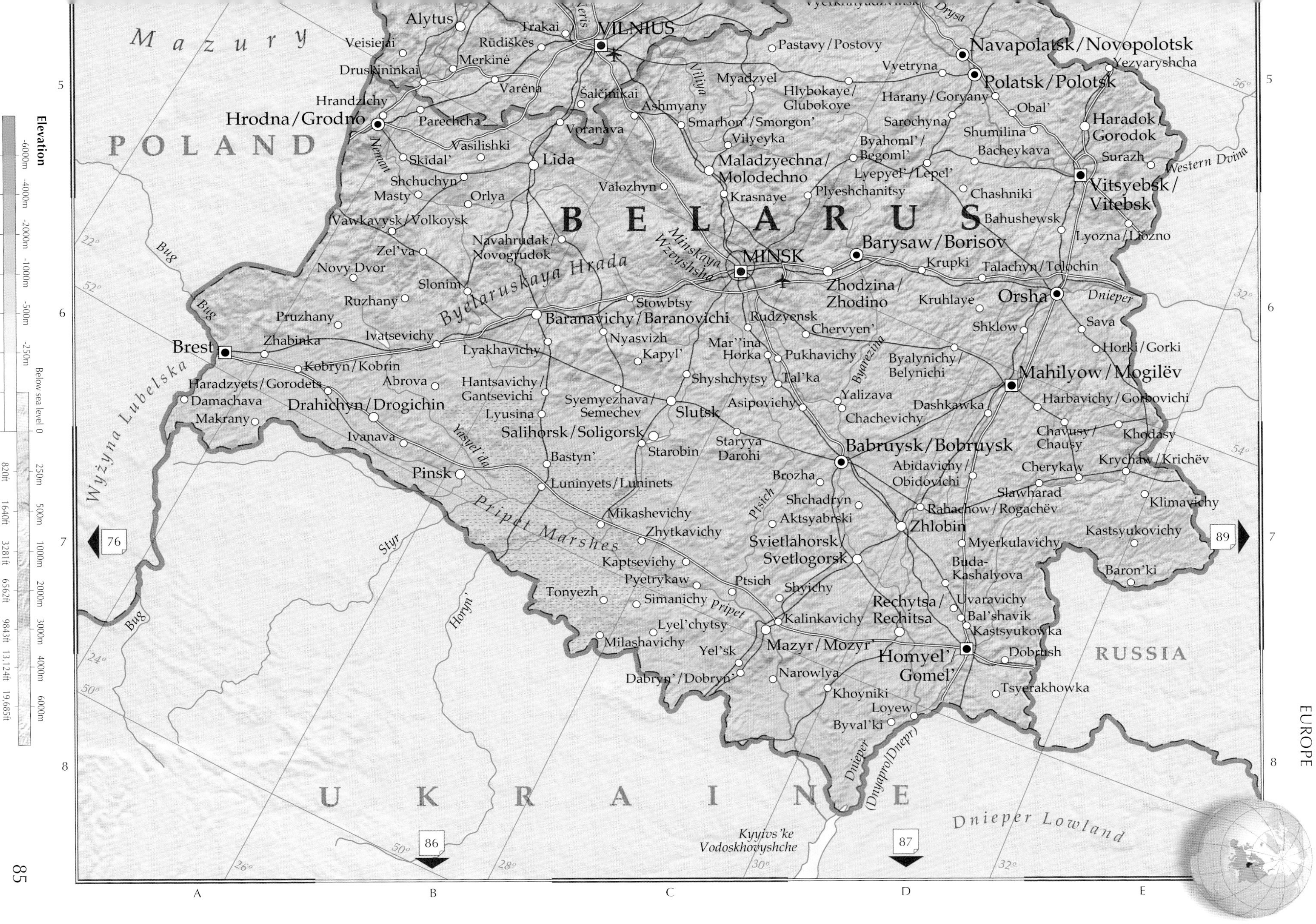

RUSSIA

BELARUS

POLAND

UKRAINE

VILNIUS

Dnieper Lowland

Prieft Marshes

Navapolatsk/Novopolotsk
Polatsk/Polotsk
Haradok/Gorodok
Vitsyebsk/Vitebsk
Surazh
Lyozna/Liozno
Horki/Gorki
Klimavichy
Baron'ki
Kastsyukovichy
Khodasy
Kryčhaŭ/Krichëv
Cherykaw
Chaúsy/Chausy
Mahilyow/Mogilëv
Dashkawka
Shklow
Talachyn/Tolochin
Orsha
Sava
Dnieper
Obal'
Shumlina
Bacheykava
Chashniki
Bahushewsk
Myerkulavichy
Tsyerakhowka
Budda-
Kashalyova
Uvaravichy
Bal'shavik
Kastsyukowka
Dobrush
Rahachow/Rogachëv
Zhlobin
Harbavichy/Gorbovichi
Slawharad
Byalynichy/Belynichi
Dashkawka
Chachevichy
Yalizava
Abidavichy/
Obidovichi
Babruysk/Bobruysk
Rechytsa/Rechitsa
Homyel'/Gomel
Kalinkavichy
Narowlya
Loyew
Byval'ki
Khoyniki
Dabryn'/Dobryn'
Milashavichy
Lyel'chytsy
Yel'sk
Tonyezh
Pyetrykaw
Simanichy
Pripet
Kaptsevichy
Zhytkavichy
Mikashevichy
Lyusina
Luninyets/Luninets
Yasyvel'da
Pinsk
Ivanava
Drahichyn/Drogichin
Kobryn/Kobrin
Zhabinka
Pruzhany
Abrova
Ivatsevichy
Ruzhany
Zel'va
Hantsavichy/
Gantsevichi
Lyakhavichy
Baranavichy/Baranovichi
Nyasvizh
Kapyl'
Syemyezhava
Semechev
Salihorsk/Soligorsk
Starobin
Bastyn'
Starryya
Darohi
Pitsich
Shatsilki
Kalinkavichy
Aktsyabrski
Shchadryn
Brozha
Tal'ka
Asipovichy
Pukhavichy
Mar'ina
Horka
Shyshchytsy
Slutsk
MINSK
Kapyl'
Nyasvizh
Stowbtsy
Stowbtsy
Rudzyensk
Byelaruskaja Hrada
Valozhyn
Navahrudak/
Novogrudok
Slonim
Vawkavysk/Volkovysk
Masty
Shchuchyn
Orlya
Skidal'
Vasilishki
Parechcha
Merkinė
Rūdiškės
Varėna
Hrandzichy
Hrodna/Grodno
Neman
Novy Dvor
Haradzyets/Gorodets
Drahichyn/Drogichin
Damachava
Makrany
Bug
Brest
Bug

Krasnaye
Maladzyechna/
Molodechno
Vilyeyka
Smarhon'/Smorgon'
Ashmyany
Salčininkai
Veranava
Lida
Veisiejai
Druskininkai
Alytus
Trakai
Lazdijai
Seris
Viliya
Myadzyel
Hlybokaye/
Glubokoye
Pastavy/Postavy
Drysa
Vyetryna
Lyepyel'/Lepel'
Byahoml'/
Begoml'
Plyeshchanitsy
Dokshytsy
Sharkawshchyna
Vyelikaya/Velikaya
Western Dvina
Yezyaryshcha
Harany/Goryany
Sarochyna
Navapolatsk/Novopolotsk
Barysaw/Borisov
Zhodzina/
Zhodino
Krupki
Kruhlaye
Byarezina
Chervyen'
Talachyn/Tolochin

Mazury
Wyżyna Lubelska
Bug
Horyn'
Styr
Kyyiŭs'ke
Vodoskhovyshche
(Dnyapro/Dnepr)

Ukraine, Moldova & Romania

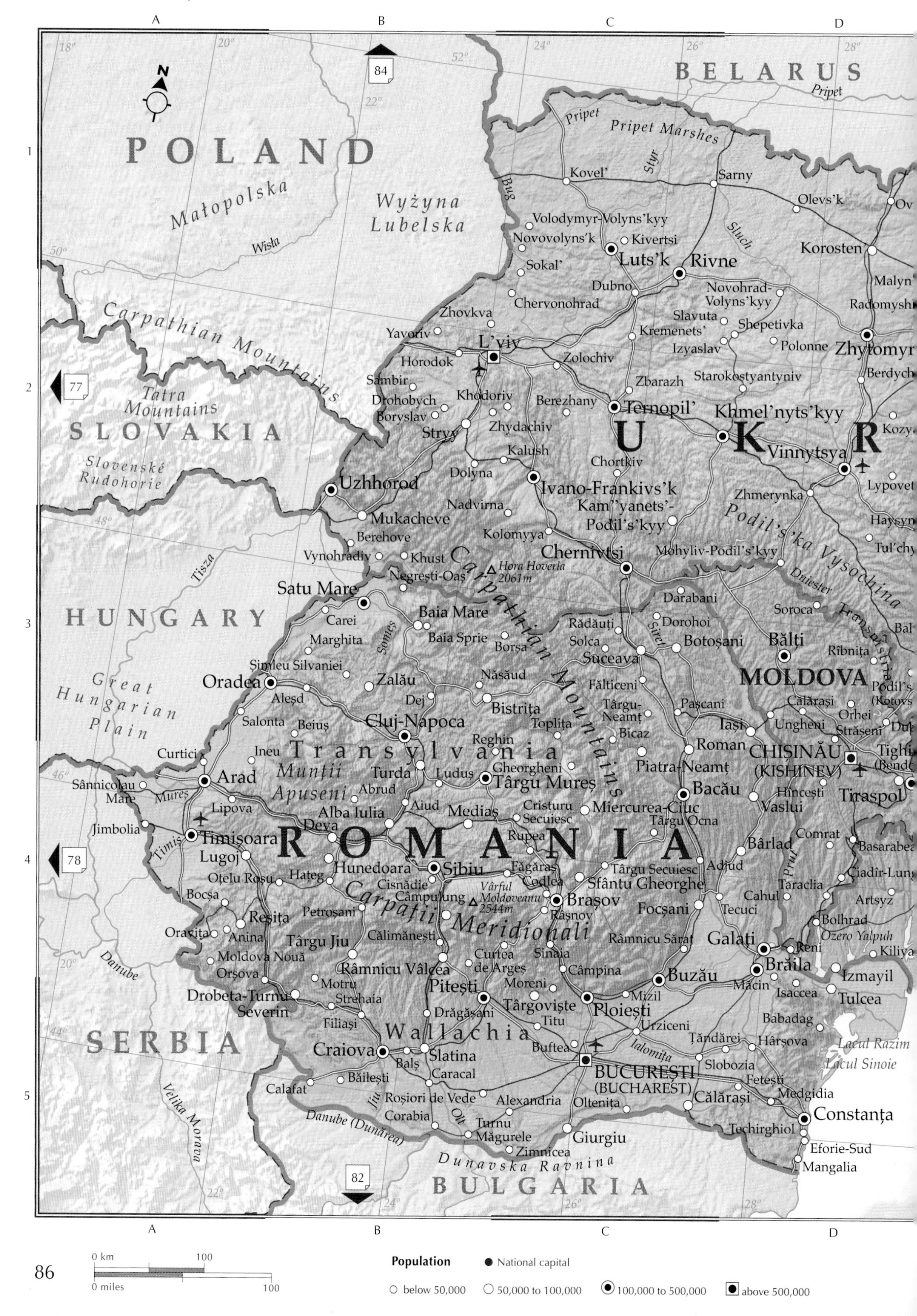

POLAND

BELARUS

Pripet

Małopolska

Wyżyna Lubelska

Wisła

Pripet Marshes

Pripet

Styr

Słuch

Kovel' Sarny

Olevs'k Ov

Volodymyr-Volyns'kyy

Novovolyns'k Kivertsi

Korosten

Sokal' Luts'k Rivne

Malyn

Dubno Novohrad- Radomyshl

Zhovkva Chervonohrad Volyns'kyy

Yavoriv L'viv Zolochiv Slavuta Shepetivka Zhytomyr

Horodok Kremenets' Polonne Berdych

Sambir Khodoriv Berezhany Zbarazh Izyaslav Starokostyantyniv

Drohobych Zhydachiv Ternopil' Khmel'nyts'kyy

Boryslav Stryy Kalush Chortkiv Kozy

Dolyna Vinnytsya

Uzhhorod Ivano-Frankivs'k Zhmerynka Lypovet

Nadvirna Kam''yanets'- *Podil's'ka Vysochyna*

Mukacheve Kolomyya Podil's'kyy Haysyn

Berehove Chernivtsi Mohyliv-Podil's'kyy Tul'chy

Vynohradiv Khust *Hora Hoverla* Dniester

Negreşti-Oaş *2061m*

Satu Mare Darabani Soroca Bal

Carei Rădăuţi Dorohoi Rîbniţa

Baia Mare Solca Botoşani Bălţi

Marghita Baia Sprie Borşa Suceava MOLDOVA

Şimleu Silvaniei Năsăud Fălticeni Paşcani Călăraşi Podil's' Kotovs

Oradea Zalău Dej Bistriţa Târgu- Roman Iaşi Ungheni Orhei Straşeni Dub

Aleşd Toplita Neamţ CHIŞINĂU Tighi

Salonta Reghin Bicaz Piatra-Neamţ (KISHINEV) (Bende

Beiuş Cluj-Napoca Gheorgheni Bacău Hînceşti

Curtici Turda Ludus Târgu Mureş Miercurea-Ciuc Vaslui Tiraspol

Sânnicolau Arad *Muntii* Abrud Aiud Medias Cristuru Târgu Ocna Bârlad Comrat Basaraba

Mare Lipova *Apuseni* Alba Iulia Secuiesc Adjud Ciadîr-Lung

Jimbolia Deva Rupea Târgu Secuiesc Tecuci Artsyz

Timişoara Hunedoara Sibiu Făgăraş Sfântu Gheorghe Cahul Bolhrad

Lugoj Cisnădie Codlea Focşani *Ozero Yalpuh*

Otelu Roşu Haţeg *Vârful* Braşov Râmnicu Sărat Galaţi Reni Kiliya

Bocşa Petroşani Câmpulung *Moldoveanu* Râşnov Reni Izmayil

Reşiţa *2544m* Câmpina Buzău Brăila

Oraviţa Anina Târgu Jiu Curtea Sinaia Mizil Măcin Tulcea

Moldova Nouă Călimăneşti de Argeş Câmpina Isaccea

Orşova Motru Petroşani Moreni Ploieşti Babadag

Drobeta-Turnu Streha ia Râmnicu Vâlcea Târgovişte Urziceni Hârşova *Lacul Razim*

Severin Filiaşi Draǧǎşani Titu Buftea Tăndărei *Lacul Sinoie*

Craiova Piteşti BUCUREŞTI Slobozia

Balş Slatina (BUCHAREST) Feteşti Medgidia

Calafat Băileşti Caracal Oltenita Călăraşi Constanţa

Roşiori de Vede Alexandria Techirghiol

Corabia Turnu Giurgiu Eforie-Sud

Măgurele Zimnicea Mangalia

Danube (Dunărea) *Dunavska Ravnina*

SERBIA BULGARIA

Velika Morava

Transylvania *Carpaţii Meridionali* ROMANIA Wallachia

Tisza *Someş* *Mureş* *Timiş* *Danube* *Jiu* *Olt* *Ialomiţa* *Siret* *Prut*

HUNGARY

Great Hungarian Plain

SLOVAKIA

Carpathian Mountains *Tatra Mountains* *Slovenské Rudohorie*

UKR

0 km 100
0 miles 100

Population   ● National capital

○ below 50,000   ◐ 50,000 to 100,000   ◉ 100,000 to 500,000   ■ above 500,000

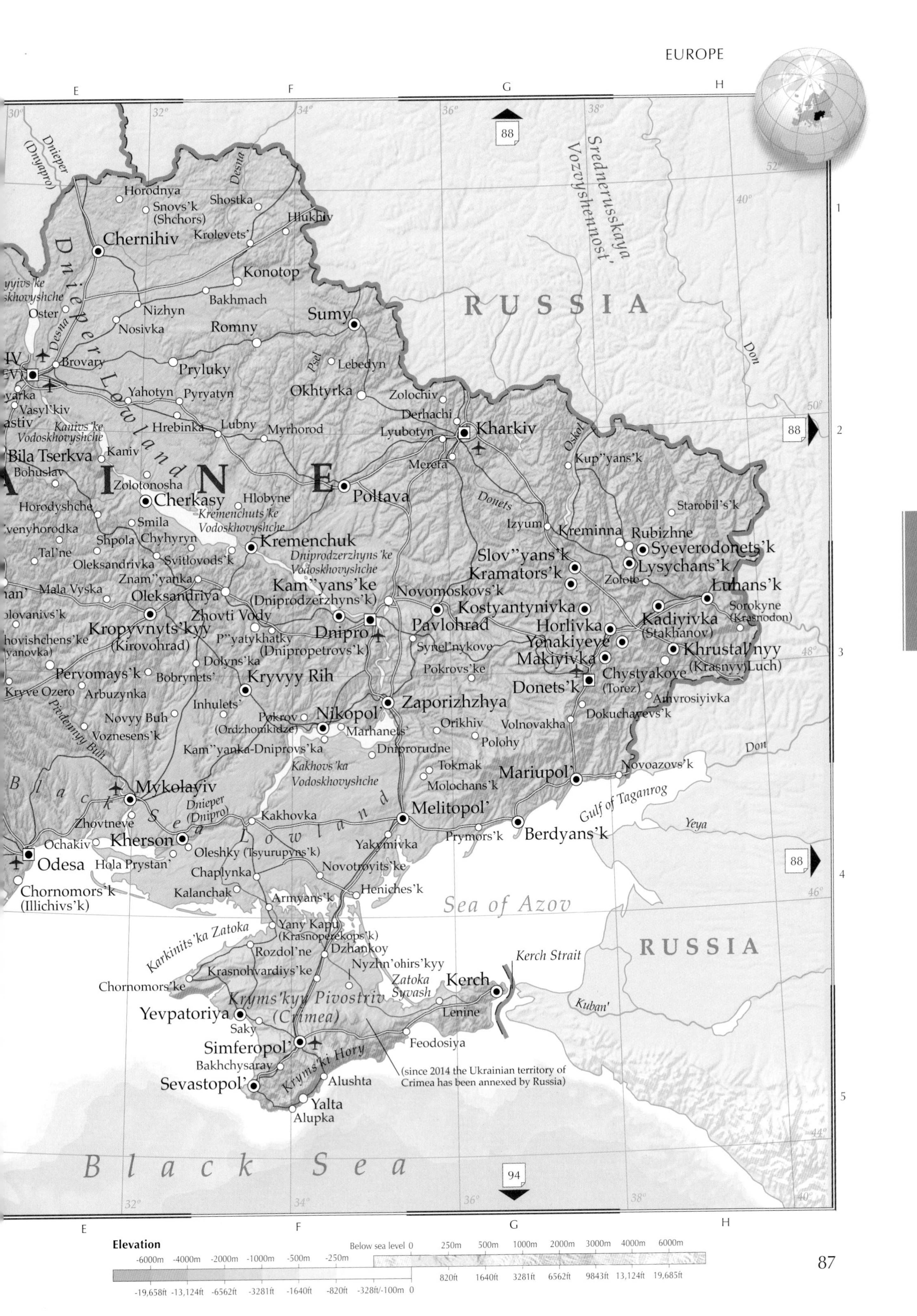

E · 32° · F · 34° · G · 36° · H · 38° · 40° · 52° · 1

RUSSIA

Horodnya
Snovs'k
(Shchors)
Shostka
Hlukhiv
Krolevets'
Chernihiv
Konotop
Bakhmach
Nizhyn
Romny
Sumy
Nosivka
Brovary
Pryluky
Lebedyn
IV
Yahotyn
Pyryatyn
Okhtyrka
Zolochiv
V
VI
yarka
Derhachi
Vasyl'kiv
Hrebinka
Lubny
Myrhorod
Lyubotyn
Kharkiv
astiv
Kaniv's'ke
Vodoskhovyshche
Bila Tserkva
Kaniv
Mereta
Kup"yans'k
Bohuslav
Zolotonosha
Donets
Starobil's'k
Horodyshche
Cherkasy
Hlobyne
Poltava
Izyum
Kreminna
venyhorodka
Kremenchuts'ke
Smila
Vodoskhovyshche
Slov"yans'k
Rubizhne
Tal'ne
Shpola
Chyhyryn
Kremenchuk
Kramators'k
Syeverodonets'k
Oleksandrivka
Svitlovods'k
Dniprodzerzhyns'ke
Novomoskovs'k
Zolote
Lysychans'k
Mala Vyska
Znam"yanka
Vodoskhovyshche
Kostyantynivka
Luhans'k
olovanivs'k
Oleksandriya
Kam"yans'ke
Horlivka
Kadiyivka
Sorokyne
Zhovti Vody
(Dniprodzerzhyns'k)
(Stakhanov)
(Krasnodon)
Kropyvnyts'kyy
P"yatykhatky
Pavlohrad
Yenakiyeye
Khrustal'nyy
(Kirovohrad)
Dnipro
Synel'nykove
Makiyivka
(Krasnyy Luch)
han
Dolyns'ka
(Dnipropetrovs'k)
Pokrovs'ke
Chystyakove
hovishchens'ke
Bobrynets'
Kryvyy Rih
Donets'k
(Torez)
yanovka)
Arbuzynka
Inhulets'
Pokrov
Zaporizhzhya
Amvrosiyivka
Kryve Ozero
Novyy Buh
(Ordzhonikidze)
Orikhiv
Dokuchayevs'k
Pervomays'k
Kam"yanka-Dniprovs'ka
Marhanets'
Volnovakha
Don
Voznesens'k
Dniprorudne
Polohy
Nikopol
Novoazovs'k
Mykolayiv
Kakhovs'ka
Tokmak
Mariupol
Vodoskhovyshche
Zhovtneve
Dnieper
Molochans'k
Gulf of Taganrog
(Dnipro)
Melitopol'
Yeya
Kakhovka
Kherson
Berdyans'k
Ochakiv
Oleshky (Tsyurupyns'k)
Yakymivka
Prymors'k
Odesa
Hola Prystan'
Novotroyits'ke
Chaplynka
Chornomors'k
Kalanchak
Heniches'k
Sea of Azov
(Illichivs'k)
Armyans'k
RUSSIA
Karkinits'ka Zatoka
Yany Kapu
(Krasnoperekops'k)
Rozdol'ne
Dzhankoy
Kerch Strait
Krasnohvardiys'ke
Nyzhn'ohirs'kyy
Kerch
Chornomors'ke
Zatoka
Syvash
Lenine
Kuban'
Yevpatoriya
Kryms'kyy Pivostriv
Saky
(Crimea)
Feodosiya
Simferopol'
Bakhchysaray
(since 2014 the Ukrainian territory of
Crimea has been annexed by Russia)
Sevastopol'
Krymski Hory
Alushta
Yalta
Alupka

B l a c k   S e a

Dnieper
(Dnyapro)
Desna
Desna
Psel
Donets
Oskol
Srednerusskaya
Vozvyshennost'
Don
A
N
E
E
Dnieper Lowland
Dnieper Lowland
Black Sea Lowland
Pridonnyy Buh
Black   Sea

88
88
88
94

E · 32° · F · 34° · G · 36° · H · 38° · 40° · 50° · 48° · 46° · 44° · 2 3 4 5

Elevation

-6000m -4000m -2000m -1000m -500m -250m   Below sea level 0   250m 500m 1000m 2000m 3000m 4000m 6000m

-19,658ft -13,124ft -6562ft -3281ft -1640ft -820ft -328ft/-100m 0   820ft 1640ft 3281ft 6562ft 9843ft 13,124ft 19,685ft

European Russia

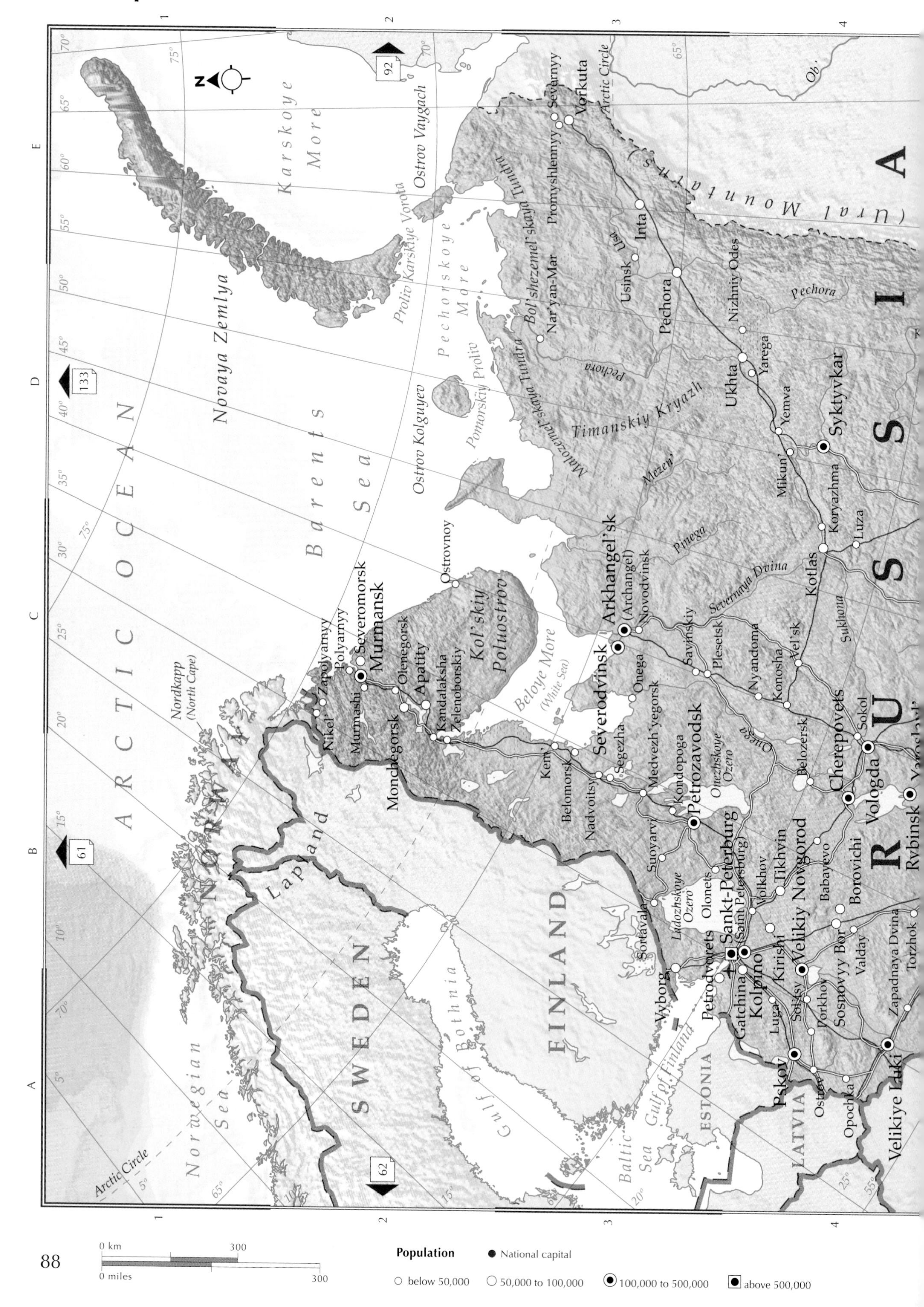

Karskoye More

Novaya Zemlya

Ostrov Vaygach

Proliv Karskiye Vorota

ARCTIC OCEAN

Barents Sea

Nordkapp (North Cape)

NORWAY

Norwegian Sea

Arctic Circle

SWEDEN

Lapland

Gulf of Bothnia

FINLAND

Baltic Sea

ESTONIA

LATVIA

Pechorskoye More

Pomorskiy Proliv

Ostrov Kolguyev

Beloye More (White Sea)

Kol'skiy Poluostrov

Ostrovnoy

Severomorsk
Murmansk
Zapolyarnyy
Polyarnyy
Niker'
Murmashi
Olenegorsk
Apatity
Monchegorsk
Kandalaksha
Zelenoborskiy

Kem'
Belomorsk
Nadvoitsy
Suoyarvi
Segezha
Medvezh'yegorsk
Kondopoga
Petrozavodsk
Ladozhskoye Ozero
Olonets
Sortavala

Onezhskoye Ozero
Onega

Severodvinsk
Arkhangel'sk (Archangel)
Novodvinsk
Savinskiy
Plesetsk
Nyandoma
Konosha
Vel'sk
Belozersk

Promyshlennyy
Severnyy
Vorkuta
Arctic Circle
Inta
Usa
Bol'shezemel'skaya Tundra
Usinsk
Nar'yan-Mar
Malozemel'skaya Tundra
Pechora
Pechora
Nizhniy Odes
Pechora
Timanskiy Kryazh
Ukhta
Yarega
Yemva
Mikun'
Mezen'
Pinega
Severnaya Dvina
Kotlas
Koryazhma
Luza
Sukhona
Sokol

(Ural Mountains)

SIBERIA

Ob'

RUSSIA

Syktyvkar

Cherepovets
Vologda
Rybinsk

Sankt-Peterburg
Saint Petersburg
Kolpino
Gatchina
Petrodvorets
Vyborg
Volkhov
Tikhvin
Babayevo
Velikiy Novgorod
Borovichi
Sospovyy Bor
Valday
Kirishi
Luga
Soltsy
Porkhov
Zapadnaya Dvina
Torzhok
Pskov
Ostrov
Opochka
Velikiye Luki

Zapolyarnyy

88

0 km   300
0 miles   300

Population   ● National capital

○ below 50,000   ○ 50,000 to 100,000   ◉ 100,000 to 500,000   ■ above 500,000

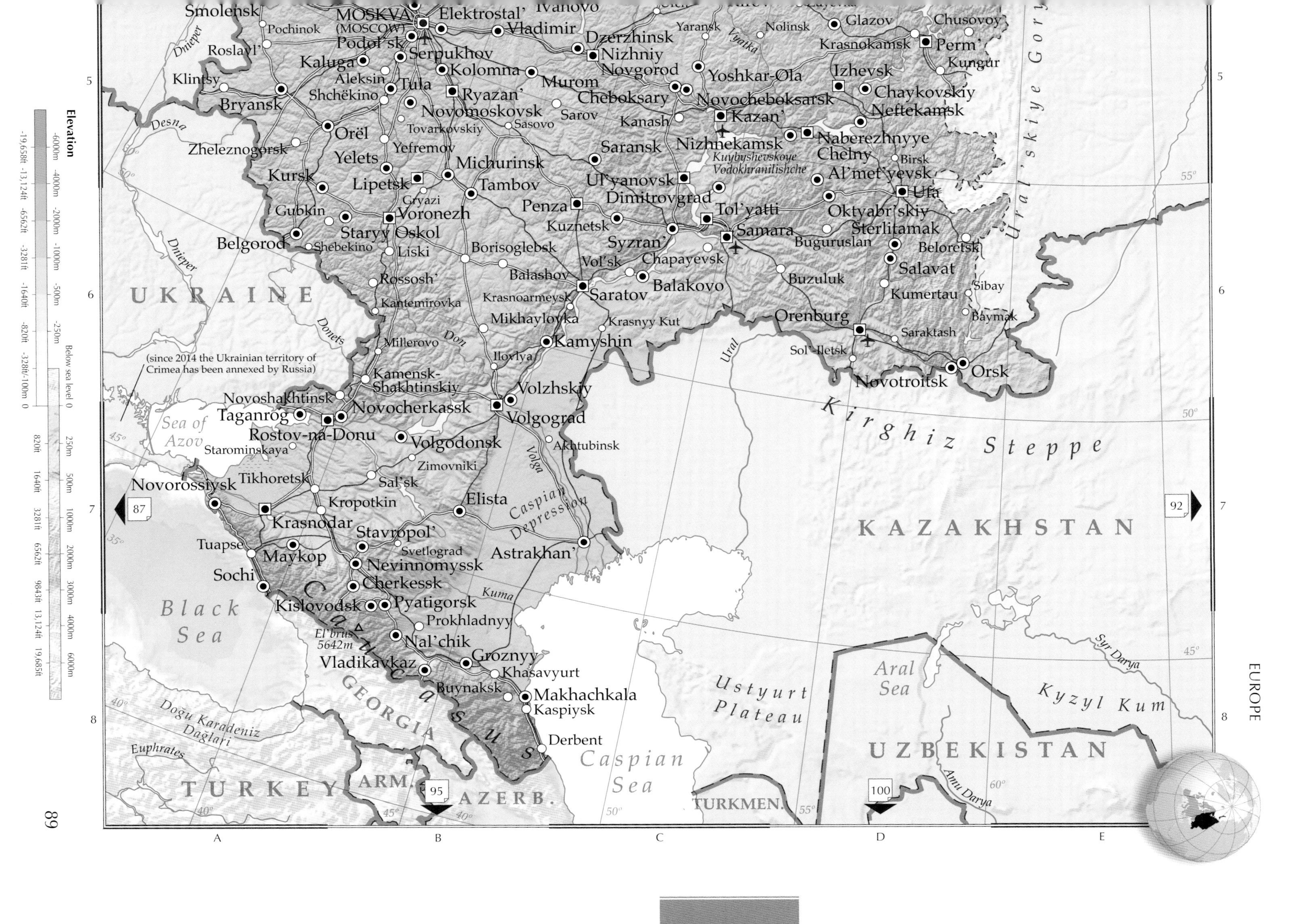

Elevation

-6000m	-4000m	-2000m	-1000m	-500m	-250m	Below sea level 0	250m	500m	1000m	2000m	3000m	4000m	6000m
-19,658ft	-13,124ft	-6562ft	-3281ft	-1640ft	-820ft	-328ft/-100m 0	820ft	1640ft	3281ft	6562ft	9843ft	13,124ft	19,685ft

North & West Asia

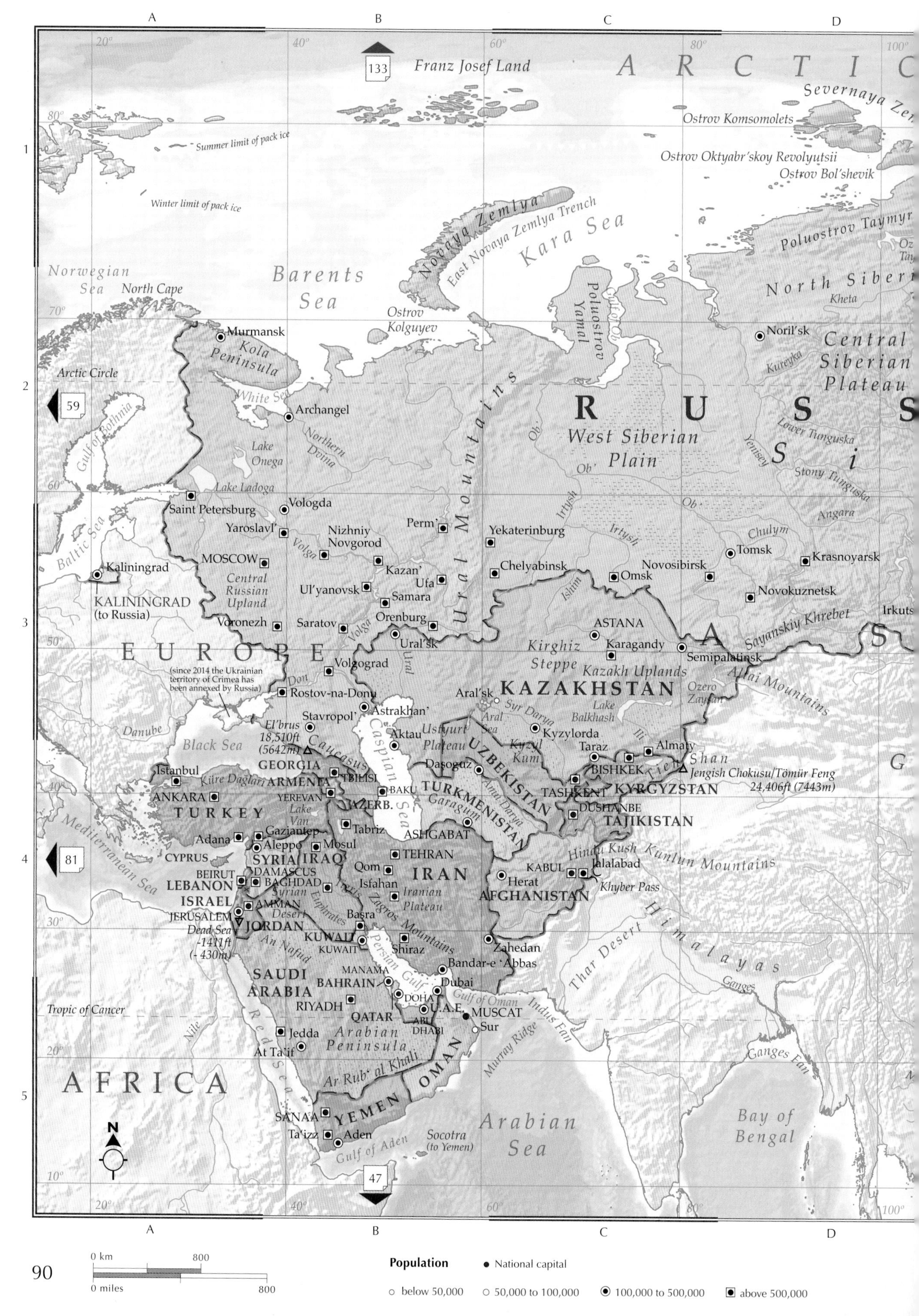

ARCTIC

Franz Josef Land

Severnaya Zen

Ostrov Komsomolets

Ostrov Oktyabr'skoy Revolyutsii
Ostrov Bol'shevik

Norwegian Sea

North Cape

Summer limit of pack ice

Winter limit of pack ice

Barents Sea

Kara Sea

North Siber

Kheta

Poluostrov Taymyr

Oz Tay

Central Siberian Plateau

Novaya Zemlya

East Novaya Zemlya Trench

Ostrov Yamal

Poluostrov Yamal

Gulf Ob

Ostrov Kolguyev

Murmansk

Kola Peninsula

White Sea

Arctic Circle

Archangel

Lake Onega

Northern Dvina

Lake Ladoga

Gulf of Bothnia

Noril'sk

Kureyka

Lower Tunguska

Yenisey

Stony Tunguska

Chulym

Tomsk

Krasnoyarsk

Irkuts

Angara

R U S S I A

West Siberian Plain

Ob

Irtysh

Ob

Irtysh

Ural Mountains

Saint Petersburg

Vologda

Yaroslavl'

Nizhniy Novgorod

Perm'

Yekaterinburg

Chelyabinsk

Omsk

Novosibirsk

Novokuznetsk

Baltic Sea

Kaliningrad

KALININGRAD (to Russia)

MOSCOW

Central Russian Upland

Volga

Kazan'

Ul'yanovsk

Samara

Ufa

Ural Mountains

Ishim

Sayanskiy Khrebet

Voronezh

Saratov

Orenburg

Ural'sk

EUROPE

Kirghiz Steppe

Karagandy

ASTANA

Semipalatinsk

Altai Mountains

(since 2014 the Ukrainian territory of Crimea has been annexed by Russia)

Volgograd

Volga

Kazakh Uplands

Ozero Zaysan

Danube

Don

Rostov-na-Donu

Astrakhan'

Aral'sk

Aral Sea

Sur Darya

KAZAKHSTAN

Lake Balkhash

Il

Black Sea

Stavropol'

El'brus 18,510ft (5642m)

Caucasus

Aktau

Ustyurt Plateau

Kyzyl Kum

Kyzylorda

Taraz

Almaty

Tien Shan

Jengish Chokusu/Tömür Feng 24,406ft (7443m)

Istanbul

Kıroe Dağları

GEORGIA

TBILISI

ARMENIA

Caspian Sea

BAKU

Dasoguz

Amu Darya

UZBEKISTAN

BISHKEK

KYRGYZSTAN

Kunlun Mountains

ANKARA

YEREVAN

AZERB.

TURKMENISTAN

TASHKENT

DUSHANBE

Lake Van

Tabriz

Garagum

ASHGABAT

TAJIKISTAN

TURKEY

Gaziantep

Adana

Aleppo

Mosul

Qom

TEHRAN

KABUL

Jalalabad

Hindu Kush

CYPRUS

SYRIA

IRAQ

Isfahan

IRAN

Herat

AFGHANISTAN

Khyber Pass

BEIRUT

DAMASCUS

BAGHDAD

Syrian Desert

Iranian Plateau

Zagros Mountains

Euphrates

LEBANON

ISRAEL

AMMAN

Basra

Zahedan

JERUSALEM

JORDAN

Dead Sea -1411ft (-430m)

KUWAIT

KUWAIT

Shiraz

Bandar-e 'Abbas

Thar Desert

Himalayas

Ganges

An Nafud

Persian Gulf

MANAMA

BAHRAIN

Dubai

Gulf of Oman

Indus Fan

SAUDI ARABIA

RIYADH

DOHA

QATAR

U.A.E.

ABU DHABI

MUSCAT

Sur

Murray Ridge

Ganges Fan

Tropic of Cancer

Jedda

Red Sea

Arabian Peninsula

Nile

At Ta'if

Ar Rub' al Khali

OMAN

Bay of Bengal

AFRICA

SANA'A

YEMEN

Arabian Sea

Ta'izz

Aden

Gulf of Aden

Socotra (to Yemen)

N

Population   ● National capital

○ below 50,000   ◎ 50,000 to 100,000   ◉ 100,000 to 500,000   ◼ above 500,000

0 km   800

0 miles   800

90

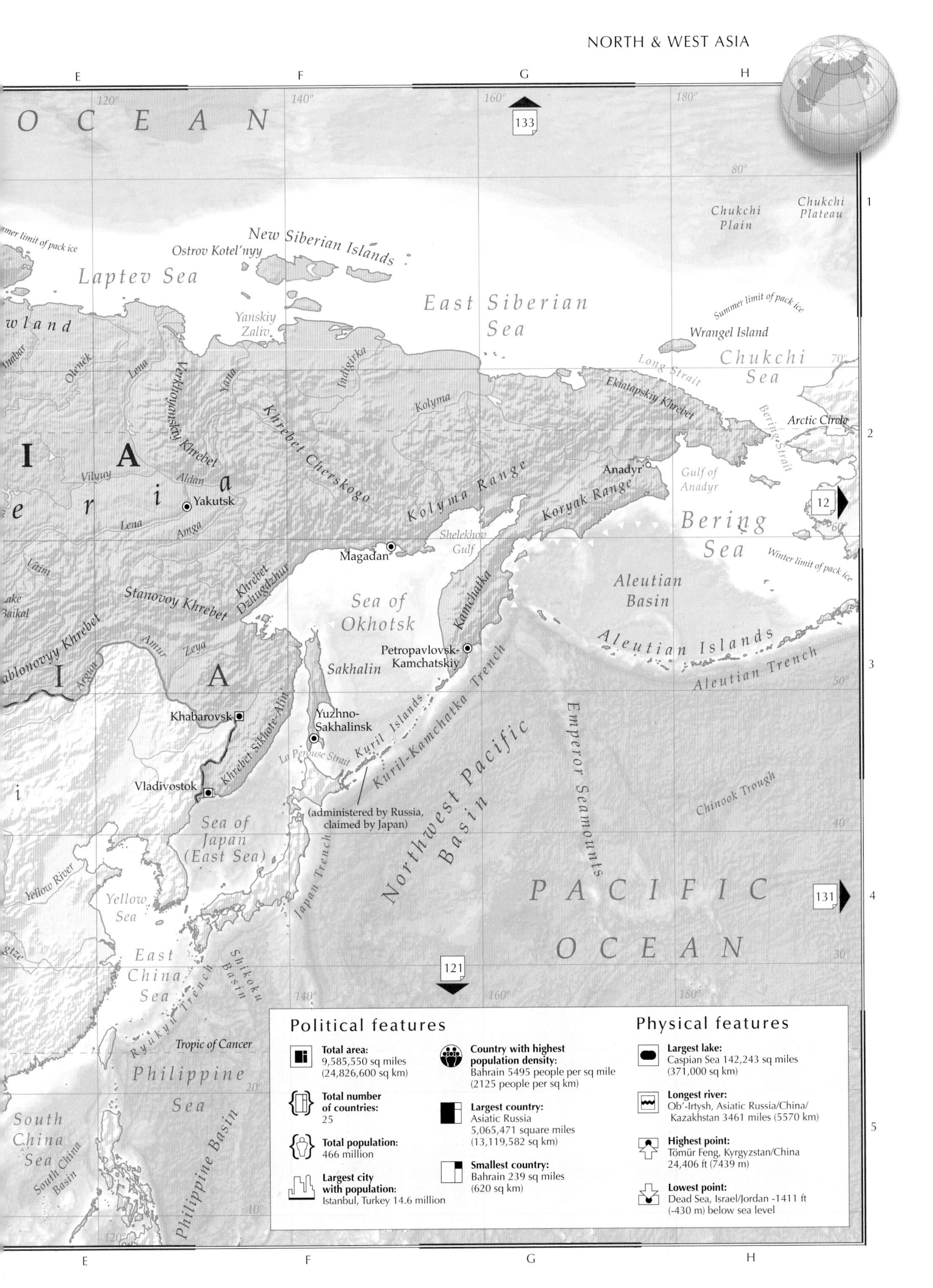

E   120°   F   140°   G   160°   H   180°

133

O C E A N

Chukchi
Plain

Chukchi
Plateau

1

80°

Summer limit of pack ice

Ostrov Kotel'nyy   New Siberian Islands

mer limit of pack ice

Laptev Sea

wland

Yanskiy
Zaliv

East Siberian
Sea

Wrangel Island

Long Strait

Chukchi
Sea   70°

Arctic Circle

Bering Strait

2

nabar

Olenek   Lena

Verkhoyanskiy Khrebet

Yana

Khrebet Cherskogo

Indigirka

Kolyma

Ekiatapskiy Khrebet

I   A

Vilyuy   Aldan   Yakutsk

Lena   Amga

Kolyma Range

Koryak Range   Anadyr

Gulf of
Anadyr

Bering
Sea   60°

12

e   r   i   a

Vatim

Khrebet Dzhugdzhur

Shelekhov
Gulf

Aleutian
Basin

Winter limit of pack ice

Magadan

Stanovoy Khrebet

Sea of
Okhotsk

Kamchatka

Aleutian Islands

ake
Baikal

Yablonovyy Khrebet   Amur   Zeya

Petropavlovsk-
Kamchatskiy

Aleutian Trench   50°

3

I   A

Argun

Sakhalin

Khabarovsk   Khrebet Sikhote-Alin'

Yuzhno-
Sakhalinsk

Kuril Islands

Kuril-Kamchatka Trench

Emperor Seamounts

Chinook Trough

i

La Perouse Strait

Vladivostok

Japan Trench

(administered by Russia,
claimed by Japan)

Northwest Pacific
Basin

40°

Sea of
Japan
(East Sea)

Yellow River

Yellow
Sea

P A C I F I C

131

4

East
China
Sea

Ryukyu Trench   Shikoku
Basin

O C E A N   30°

121

Tropic of Cancer

Philippine

20°

South
China
Sea

Sea

South China
Basin

Philippine Basin

10°

120°   140°   160°   180°

E   F   G   H

Political features

Total area:
9,585,550 sq miles
(24,826,600 sq km)

**Total number
of countries:**
25

Total population:
466 million

**Largest city
with population:**
Istanbul, Turkey 14.6 million

**Country with highest
population density:**
Bahrain 5495 people per sq mile
(2125 people per sq km)

Largest country:
Asiatic Russia
5,065,471 square miles
(13,119,582 sq km)

Smallest country:
Bahrain 239 sq miles
(620 sq km)

Physical features

Largest lake:
Caspian Sea 142,243 sq miles
(371,000 sq km)

Longest river:
Ob'-Irtysh, Asiatic Russia/China/
Kazakhstan 3461 miles (5570 km)

Highest point:
Tömür Feng, Kyrgyzstan/China
24,406 ft (7439 m)

Lowest point:
Dead Sea, Israel/Jordan -1411 ft
(-430 m) below sea level

Russia & Kazakhstan

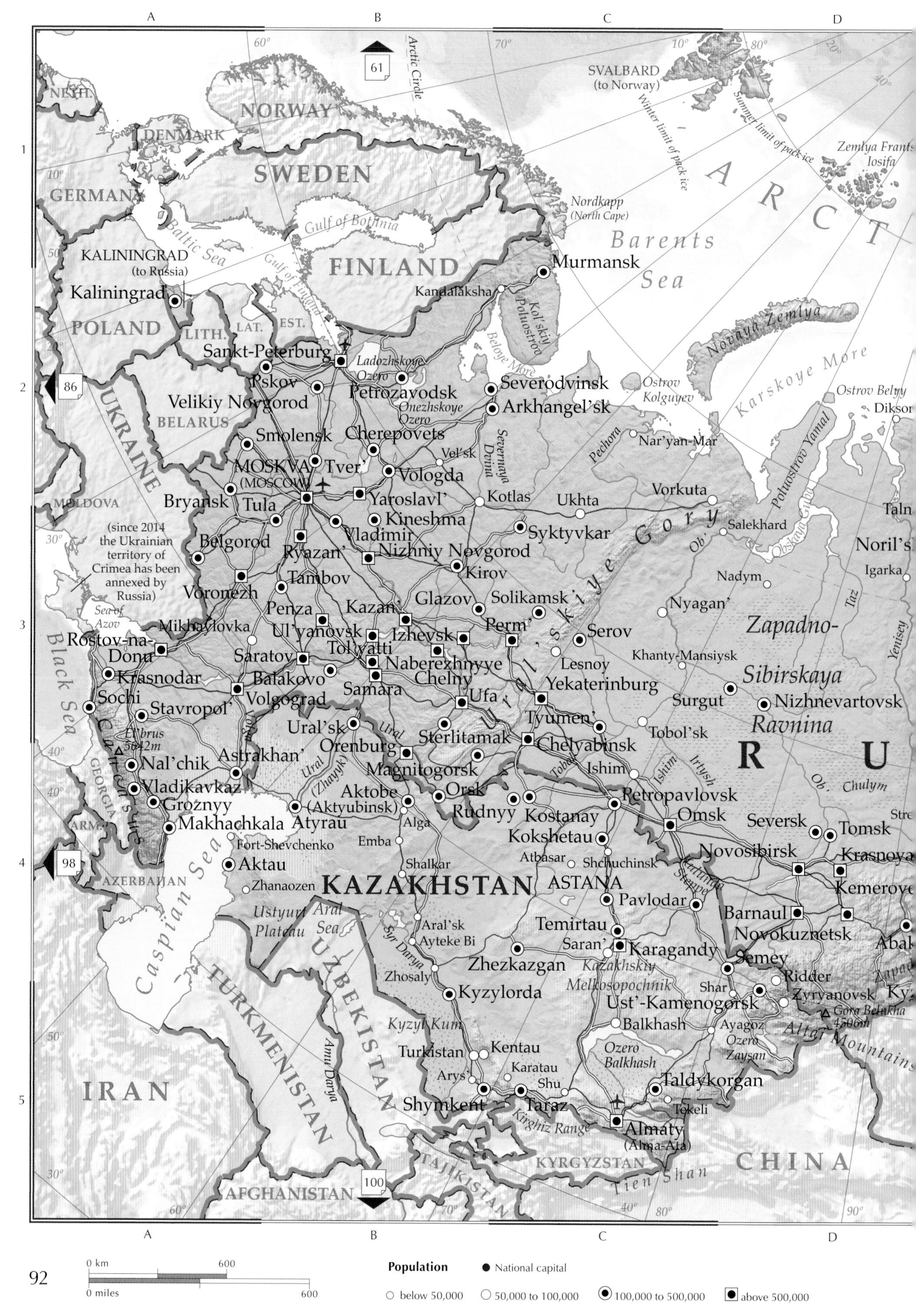

A   B   C   D

60°   70°   10°   80°

61

Arctic Circle

SVALBARD
(to Norway)

A R C T

Zemlya Frantsa Iosifa

NORWAY

SWEDEN

GERMANY

10°

DENMARK

1

Baltic Sea

Gulf of Bothnia

Nordkapp
(North Cape)

**Barents
Sea**

Murmansk

Novaya Zemlya

Karskoye More

Kandalaksha

50°

KALININGRAD
(to Russia)

Kaliningrad

POLAND

Gulf of Finland

FINLAND

Sankt-Peterburg

Ladozhskoye Ozero

Petrozavodsk

Severodvinsk

Arkhangel'sk

Kol'skiy Poluostrov

Beloye More

Ostrov Kolguyev

Ostrov Belyy

Diksor

86

LITH.

LAT.

EST.

Pskov

Velikiy Novgorod

2

BELARUS

Smolensk

Cherepovets

Onezhskoye Ozero

Nar'yan-Mar

Pechora

Poluostrov Yamal

UKRAINE

MOSKVA
(MOSCOW)

Tver'

Vologda

Vel'sk

Severnaya Dvina

Kotlas

Ukhta

Vorkuta

Taln

20°

Bryansk

Tula

Yaroslavl'

Kineshma

Vladimir

Nizhniy Novgorod

Syktyvkar

Salekhard

Ob'

Obskaya Guba

Noril's

MOLDOVA

30°

Belgorod

Ryazan'

Tambov

Kirov

Glazov

Perm'

Solikamsk

Igarka

Yenisey

(since 2014
the Ukrainian
territory of
Crimea has been
annexed by
Russia)

Voronezh

Penza

Kazan'

Izhevsk

Serov

Nyagan'

Nadym

Taz

Zapadno-

Sea of Azov

Mikhaylovka

Ul'yanovsk

U r a l ' s k i y e

Lesnoy

Khanty-Mansiysk

Sibirskaya

3

Rostov-na-
Donu

Saratov

Tol'yatti

Naberezhnyye
Chelny

Perm'

Yekaterinburg

Surgut

Nizhnevartovsk

Ravnina

Krasnodar

Black Sea

Balakovo

Samara

Ufa

G o r y

Tyumen'

Tobol'sk

R

U

Sochi

Stavropol'

Volgograd

U r a l

Sterlitamak

Chelyabinsk

Ishim

Tobol

Chulym

Caucasus

El'brus 5642m

Nal'chik

Astrakhan'

Orenburg

Ural (Zhayyk)

Ishim

Irtysh

Ob'

Vladikavkaz

Groznyy

Makhachkala

Atyrau
(Aktyubinsk)

Aktobe

Orsk

Rudnyy

Kostanay

Petropavlovsk

Omsk

Seversk

Tomsk

Stre

40°

98

Fort-Shevchenko

Alga

Kokshetau

Novosibirsk

Krasnoya

4

AZERBAIJAN

Aktau

Zhanaozen

Emba

Shalkar

KAZAKHSTAN

Atbasar

Shchuchinsk

ASTANA

Kulanda Steppe

Pavlodar

Barnaul

Novokuznetsk

Kemerovo

Abak

Caspian Sea

Ustyurt Plateau

Aral Sea

Aral'sk

Ayteke Bi

Syr Darya

Temirtau

Saran'

Karagandy

Shar

Semey

Ridder

Zapa

Ky

TURKMENISTAN

UZBEKISTAN

Zhosaly

Zhezkazgan

Kazakhskiy Melkosopochnik

Zyryanovsk

Gora Belukha 4506m

Altai Mountains

Amu Darya

Kyzylorda

Ust'-Kamenogorsk

Balkhash

Ozero Zaysan

Ayagoz

IRAN

50°

Kyzyl Kum

Turkistan

Kentau

Ozero Balkhash

Taldykorgan

Tekeli

5

TAJIKISTAN

Arys

Karatau

Shu

Zhosaly

Shymkent

Taraz

Kirghiz Range

Almaty
(Alma-Ata)

CHINA

30°

Tien Shan

KYRGYZSTAN

AFGHANISTAN

100

60°   70°   80°   90°

A   B   C   D

0 km	600
0 miles	600

Population   ● National capital

○ below 50,000   ○ 50,000 to 100,000   ◉ 100,000 to 500,000   ◼ above 500,000

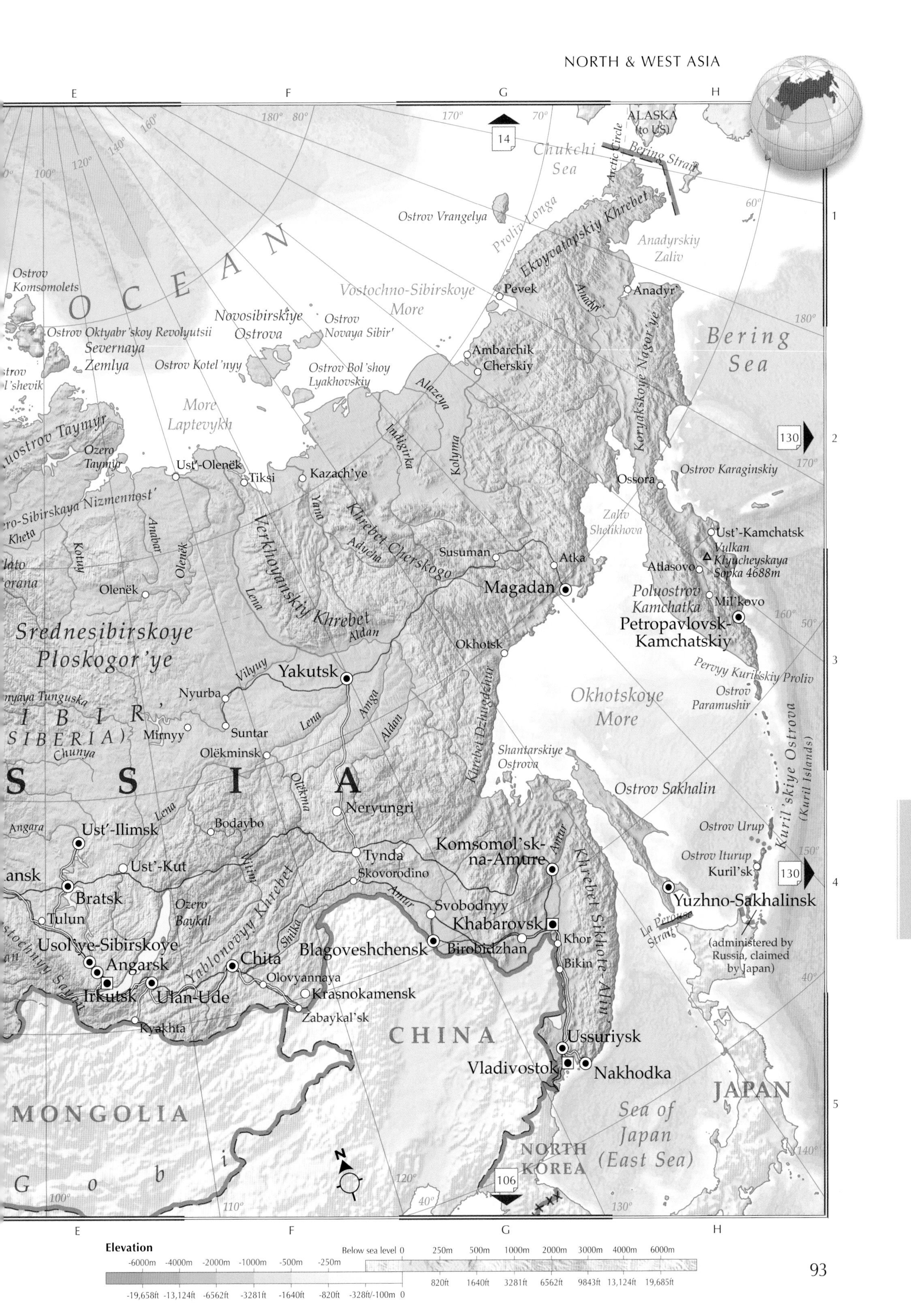

ALASKA
(to US)

Chukchi
Sea

Bering Strait

Arctic Circle

14

Ostrov Vrangelya

Proliv Longa

Vostochno-Sibirskoye
More

Ostrov
Komsomolets

OCEAN

Ekvyvatapskiy Khrebet

Anadyrskiy
Zaliv

Anadyr'

Pevek

Ostrov Oktyabr'skoy Revolyutsii

Novosibirskiye
Ostrova

Ostrov
Novaya Sibir'

Severnaya
Zemlya

Ambarchik
Cherskiy

Bering
Sea

Koryakskoye Nagor'ye

Ossora

130

Ostrov
l'shevik

Ostrov Bol'shoy
Lyakhovskiy

Ostrov Karaginskiy

More
Laptevykh

Ostrov Kotel'nyy

uostrov Taymyr

Zaliv
Shelikhova

Ust'-Olenëk   Tiksi

Kazach'ye

Ozero
Taymyr

Ossora

Ust'-Kamchatsk

Atlasovo   Vulkan
Klyucheyskaya
Sopka 4688m

o-Sibirskaya Nizmennost'

Anabar

Verkhoyanskiy Khrebet

Indigirka

Kolyma

Khrebet Cherskogo

Susuman

Atka

Poluostrov
Kamchatka

Mil'kovo

Kheta

Kotuy

Olenëk

Adycha

Yana

Alazeya

Magadan

Petropavlovsk-
Kamchatskiy

lato
orana

Olenëk

Srednesibirskoye
Ploskogor'ye

Aldan

Lena

Okhotsk

Okhotskoye
More

Pervyy Kurilskiy Proliv

Ostrov
Paramushir

Vilyuy

Yakutsk

Nyurba

Amga

Khrebet Dzhugdzhur

SIBIR'
SIBERIA)

nyaya Tunguska

Suntar

Lena

Aldan

Shantarskiye
Ostrova

Ostrov Sakhalin

Mirnyy

Chunya

Olëkminsk

SSIA

Olëkma

Kuril'skiye Ostrova
(Kuril Islands)

Ostrov Urup

Neryungri

Ostrov Iturup

Kuril'sk

Angara

Ust'-Ilimsk

Bodaybo

Vitim

Tynda

Amur

Komsomol'sk-
na-Amure

Khrebet Sikhote Alin'

130

ansk

Ust'-Kut

Skovorodino

Amur

Svobodnyy

Yuzhno-Sakhalinsk

Bratsk

Tulun

Ozero
Baykal

Yablonovyy Khrebet

Shilka

Amur

Blagoveshchensk

Birobidzhan

Khabarovsk

Khor

La Pérouse Strait

(administered by
Russia, claimed
by Japan)

Usol'ye-Sibirskoye

Angarsk

Chita

Olovyannaya

Bikin

Irkutsk

Ulan-Ude

Krasnokamensk

ostochnyy Sayan

Kyakhta

Zabaykal'sk

CHINA

Ussuriysk

Sea of
Japan
(East Sea)

JAPAN

MONGOLIA

Vladivostok

Nakhodka

G o b i

N

NORTH
KOREA

106

Elevation

-6000m -4000m -2000m -1000m -500m -250m

Below sea level 0   250m   500m   1000m   2000m   3000m   4000m   6000m

-19,658ft -13,124ft -6562ft -3281ft -1640ft -820ft -328ft/-100m 0

820ft   1640ft   3281ft   6562ft   9843ft 13,124ft 19,685ft

Turkey & The Caucasus

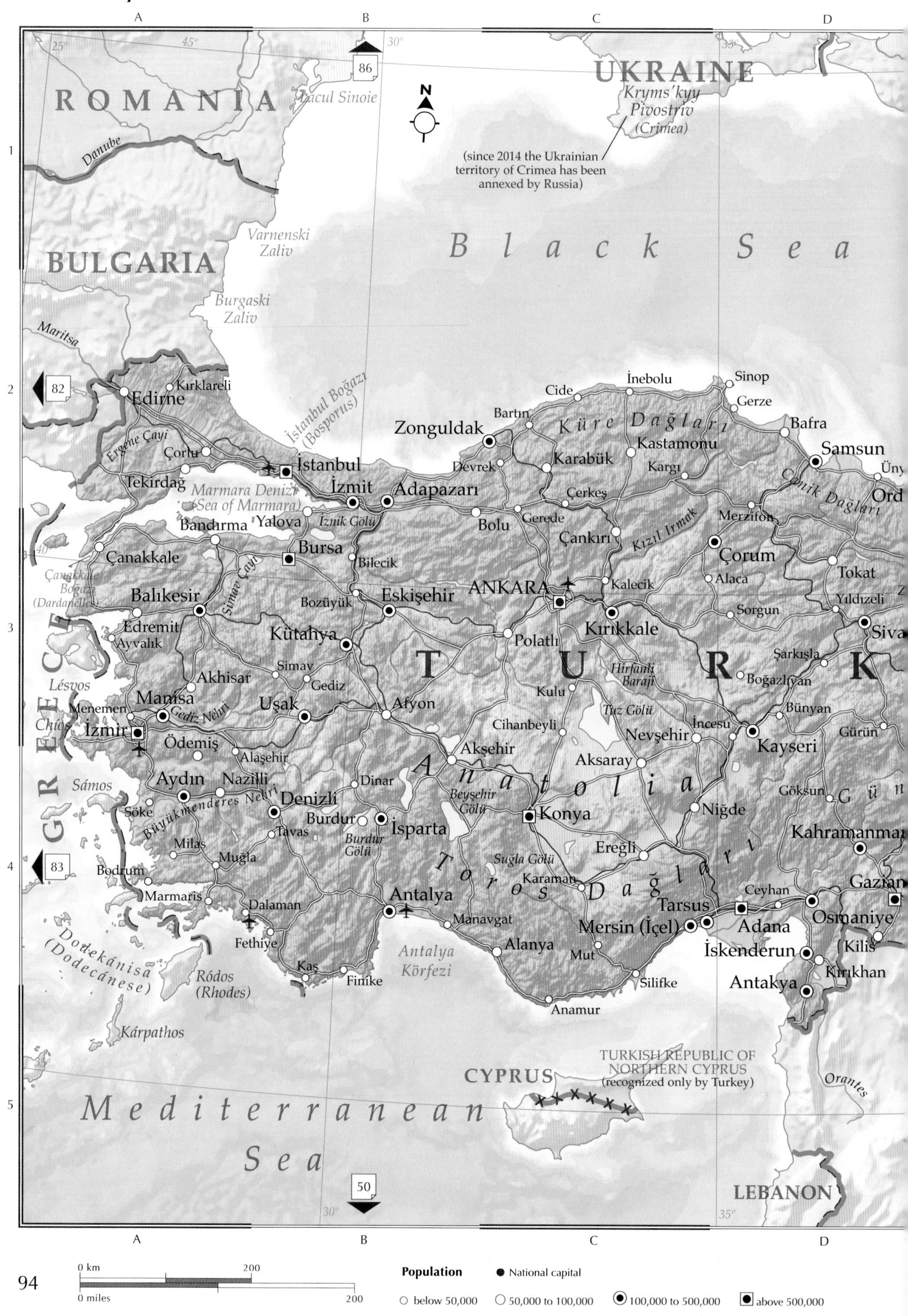

ROMANIA

Iacul Sinoie

86

N

UKRAINE

Kryms'kyy
Pivostriv
(Crimea)

(since 2014 the Ukrainian
territory of Crimea has been
annexed by Russia)

Danube

BULGARIA

Varnenski
Zaliv

Black Sea

Burgaski
Zaliv

Maritsa

82

Kırklareli
Edirne

İnebolu
Cide   Sinop
                Gerze

Ergene Çayı   Çorlu

Zonguldak   Küre Dağları   Bafra
                                Bartın   Kastamonu
Devrek Karabük Kargı   Samsun
Çerkeş   Merzifon

Tekirdağ İstanbul
          İzmit Adapazarı
Marmara Denizi   Bolu Gerede Çankırı Kızıl Irmak Çorum Ünye Ord
(Sea of Marmara)   Alaca
Bandırma Yalova İznik Gölü   Kalecik   Tokat
Çanakkale   Bursa   ANKARA Sorgun Yıldızeli
            Bilecik Eskişehir   Sivas
Balıkesir Bozüyük   Kırıkkale Şarkışla
Edremit   Polatlı   Boğazlıyan
Ayvalık Kütahya   Hirfanlı Tuz Gölü Bünyan
Lésvos   Simav   Baraji   İncesu Gürün
Chíos Akhisar Gediz   Afyon Kulu   Nevşehir Kayseri
Menemen Manisa Uşak   Cihanbeyli   Göksun
İzmir   Akşehir Aksaray Niğde
Ödemiş Alaşehir   Anatolia
Sámos Aydın Nazilli Dinar Beyşehir Konya Ereğli Kahramanma
Söke Büyükmenderes   Gölü   Karaman
Milas Denizli Burdur Isparta   Ceyhan Gaziar
    Tavas Burdur   Suğla Gölü Toros Tarsus Adana Osmaniye
Bodrum Muğla Gölü   Karaman Mersin (İçel) Kilis
Marmaris Dalaman Antalya   Manavgat Dağları İskenderun Kırıkhan
Dodekánisa   Alanya Mut   Antakya
(Dodecánese) Fethiye Kaş Finike Antalya Silifke
Ródos   Körfezi
(Rhodes)   Anamur
Kárpathos
                          TURKISH REPUBLIC OF
          CYPRUS   NORTHERN CYPRUS
                        (recognized only by Turkey)

Mediterranean
Sea

Orantes

LEBANON

50

GREECE

T U R K E Y

Population

● National capital

○ below 50,000   ○ 50,000 to 100,000   ◉ 100,000 to 500,000   ◪ above 500,000

94

0 km   200
0 miles   200

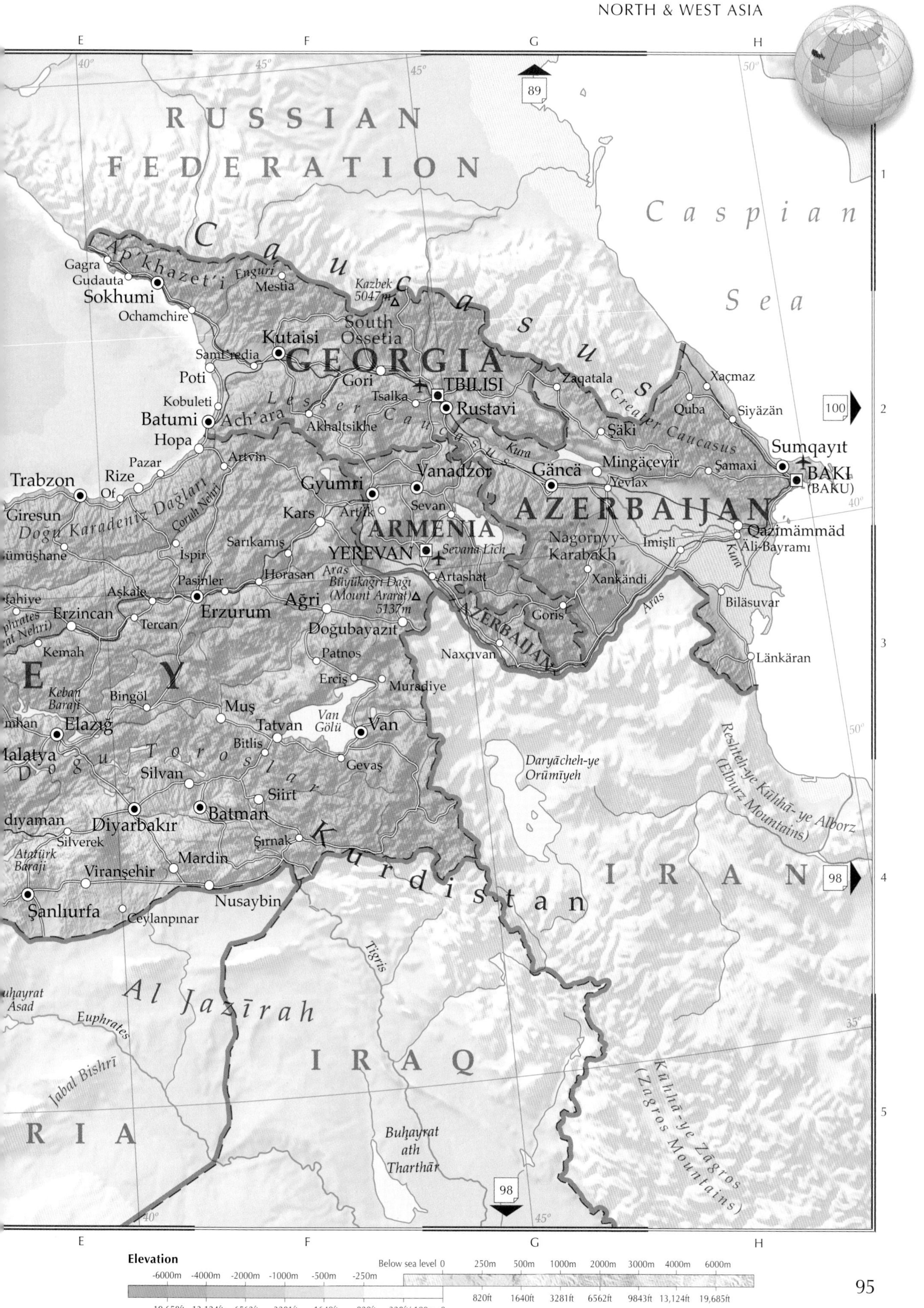

RUSSIAN

FEDERATION

Caspian

Sea

C
a
u
c
a
s
u
s

Gagra
Ap'khazet'i
Gudauta
Sokhumi
Enguri
Mestia
Ochamchire
Kazbek
5047m

South
Ossetia
Kutaisi
GEORGIA
Samtredia
Gori
Tsalka
TBILISI
Poti
Rustavi
Kobuleti
Batumi
Ach'ara
Akhaltsikhe
Hopa
Lesser Caucasus
Kura

Zaqatala
Greater Caucasus
Xaçmaz
Quba
Siyäzän

100

Şäki
Şamaxi
Sumqayıt
BAKI
(BAKU)

Trabzon
Rize
Pazar
Of
Artvin
Gyumri
Vanadzor
Gäncä
Mingäçevir
Yevlax

Giresun
Doğu Karadeniz Dağları
Çoruh Nehri
İspir
Kars
Artik
Sevan
ARMENIA
AZERBAIJAN
Qazimämmäd
Äli-Bayramı

Gümüşhane
Sarıkamış
YEREVAN
Sevana Lich
Artashat
Nagornyy-
Karabakh
İmişli
Kura

İahiye
Pasinler
Horasan
Aras
Büyükağrı Dağı
(Mount Ararat)
5137m
Xankändi
Biläsuvar

Erzincan
Aşkale
Erzurum
Ağrı
AZERBAIJAN
Goris
Aras

Euphrates
at Nehri
Tercan
Doğubayazıt
Naxçıvan
Länkäran

Kemah
Bingöl
Patnos
 Erciş

E Y
Keban
Baraji
Muş
Muradiye

İhan
Elazığ
Tatvan
Van
Gölü
Van
Daryācheh-ye
Orūmīyeh
Reshteh-ye Kūlhhā-ye Alborz
(Elburz Mountains)

Malatya
Doğu
Bitlis
Gevaş

Toroslar
Silvan
Siirt

dıyaman
Diyarbakır
Batman
Şırnak

Silverek
Mardin
I R A N

98

Atatürk
Baraji
Viranşehir
Nusaybin

Şanlıurfa
Ceylanpınar
Kurdistan

Tigris

RIA
Al Jazīrah
Euphrates

uhayrat
Asad
Buhayrat
ath
Tharthār
Kūhhā-ye Zagros
(Zagros Mountains)

Jabal Bishrī
I R A Q

R I A

89

98

98

1

2

3

4

5

E   F   G   H

Elevation

-6000m	-4000m	-2000m	-1000m	-500m	-250m	Below sea level 0	250m	500m	1000m	2000m	3000m	4000m	6000m

-19,658ft -13,124ft -6562ft -3281ft -1640ft -820ft -328ft/-100m 0   820ft 1640ft 3281ft 6562ft 9843ft 13,124ft 19,685ft

The Near East

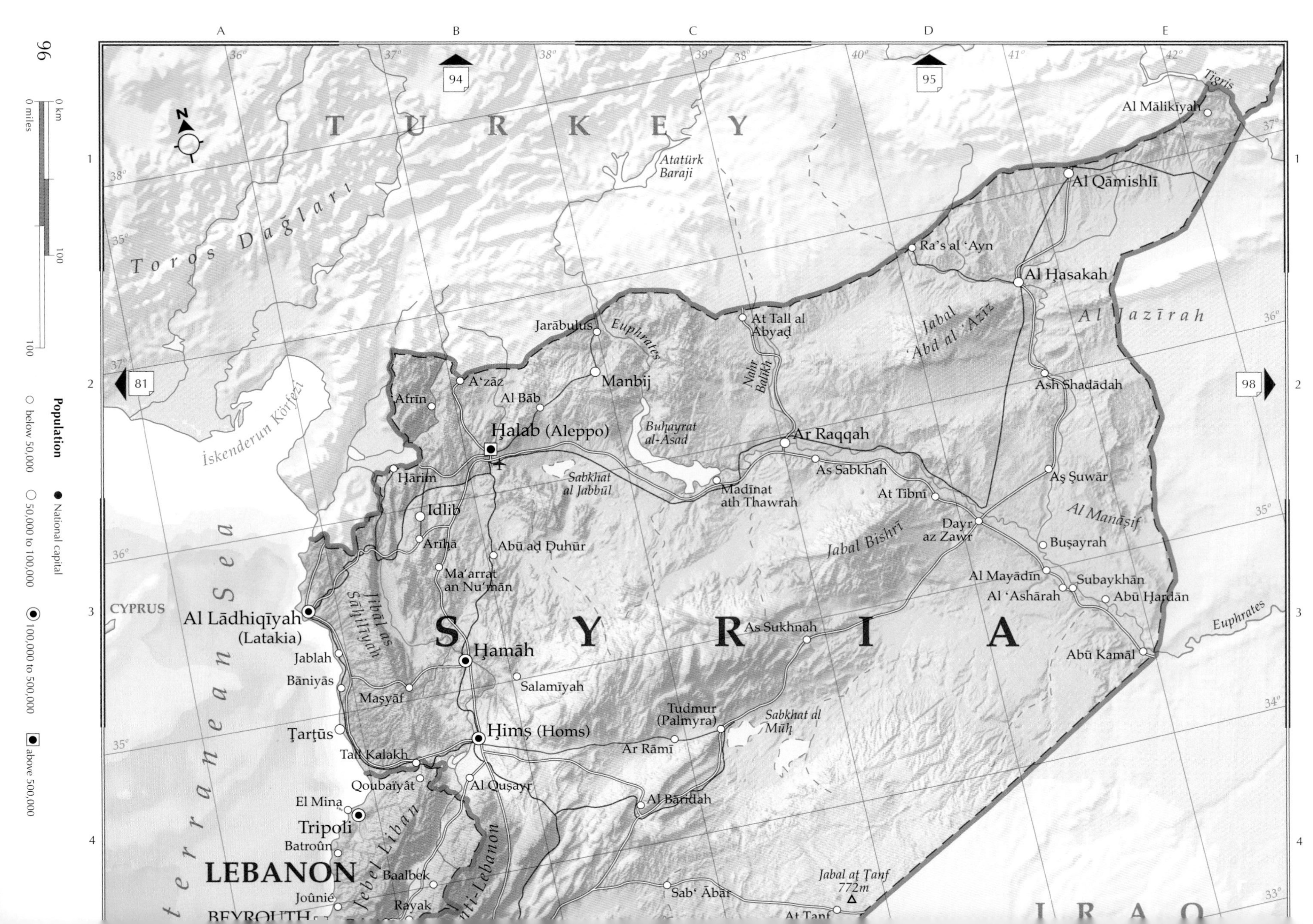

96

0 km   100
0 miles   100

Population ● National capital

○ below 50,000 ◯ 50,000 to 100,000 ◉ 100,000 to 500,000 ■ above 500,000

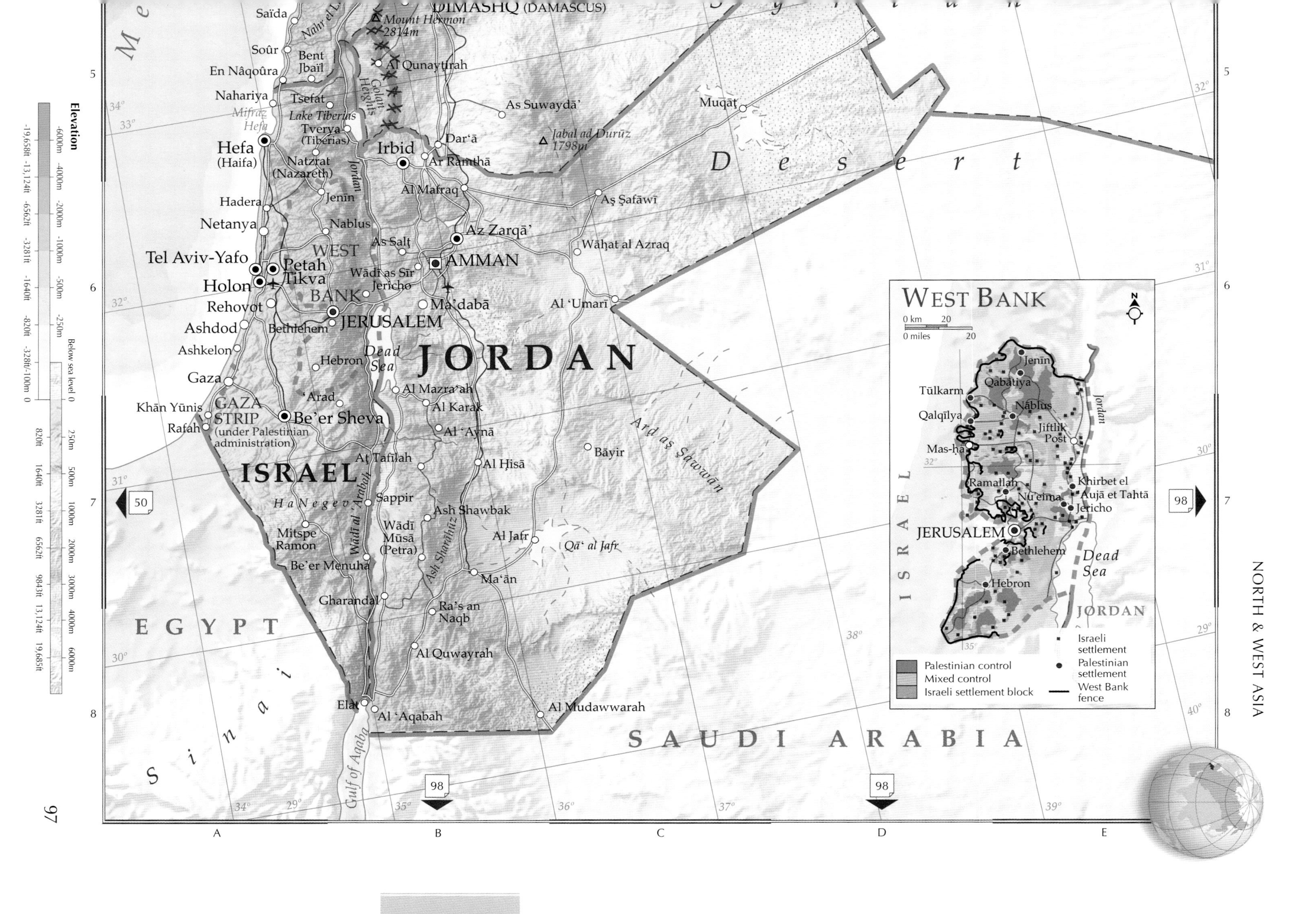
DIMASHQ (DAMASCUS)
Mount Hermon 2814m
Al Qunaytirah
Golan Heights
As Suwaydā'
Muqāṭ
Jabal ad Durūz 1798m
Desert
Saïda
Soûr
Bent Jbaïl
En Nâqoûra
Nahariya
Tsefat
Lake Tiberias
Mifraz Hefa
Tverya (Tiberias)
Hefa (Haifa)
Natzrat (Nazareth)
Irbid
Dar'ā
Ar Ramthā
Al Mafraq
Jordan
Jenīn
Hadera
Netanya
Nablus
Aş Şafāwī
Az Zarqā'
As Salṭ
Wāḥat al Azraq
WEST BANK
Tel Aviv-Yafo
Petah Tikva
Holon
Wādī as Sīr
Jericho
AMMAN
Rehovot
Ma'dabā
Al 'Umarī
Ashdod
Bethlehem
JERUSALEM
Ashkelon
Hebron
Dead Sea
JORDAN
Gaza
Al Mazra'ah
'Arad
Al Karak
Khān Yūnis
GAZA STRIP (under Palestinian administration)
Be'er Sheva
Rafah
Al 'Aynā
Ard aş Şawwān
Bāyir
Aṭ Ṭafīlah
Al Ḥisā
ISRAEL
HaNegev
Wādī al 'Arabah
Sappir
Ash Shawbak
Wādī Mūsā (Petra)
Al Jafr
Qā' al Jafr
Mitspe Ramon
Ash Sharāh
Be'er Menuha
Ma'ān
Gharandal
Ra's an Naqb
EGYPT
Al Quwayrah
Sinai
Elat
Al 'Aqabah
Al Mudawwarah
SAUDI ARABIA
Gulf of Aqaba
WEST BANK
0 km 20
0 miles 20
Jenīn
Qabāṭiya
Ṭūlkarm
Nāblus
Qalqīlya
Jiftlik Post
Jordan
Mas-ha
ISRAEL
Ramallah
Khirbet el Aujā et Taḥtā
Nu'eima
Jericho
JERUSALEM
Bethlehem
Dead Sea
Hebron
JORDAN
Israeli settlement
Palestinian settlement
West Bank fence
Palestinian control
Mixed control
Israeli settlement block
Elevation
-6000m -4000m -2000m -1000m -500m -250m
-19,658ft -13,124ft -6562ft -3281ft -1640ft -820ft -328ft/-100m 0
Below sea level 0
250m 500m 1000m 2000m 3000m 4000m 6000m
820ft 1640ft 3281ft 6562ft 9843ft 13,124ft 19,685ft

The Middle East

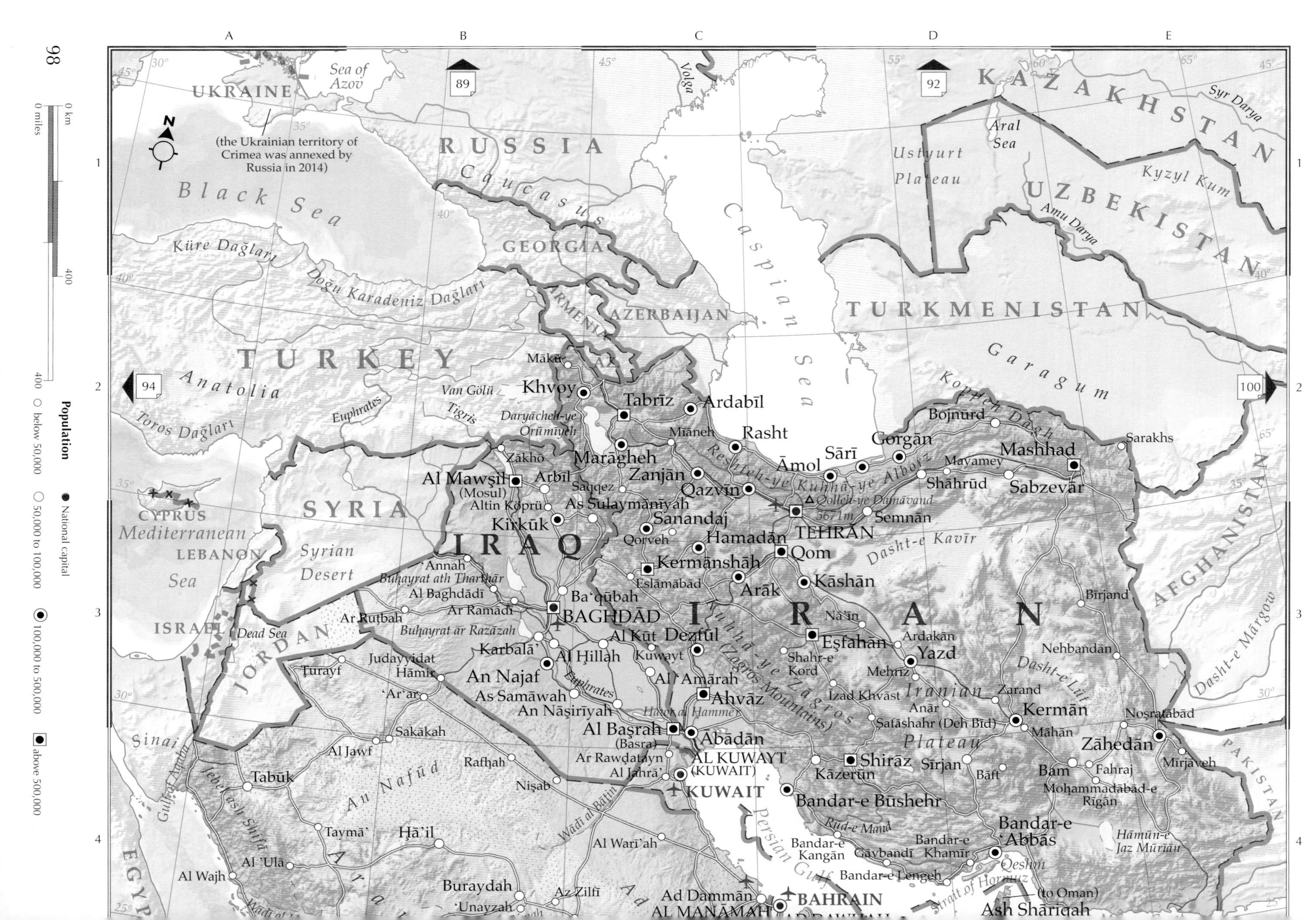

0 km   400

0 miles   400

Population ● National capital

○ below 50,000   ○ 50,000 to 100,000   ◉ 100,000 to 500,000   ■ above 500,000

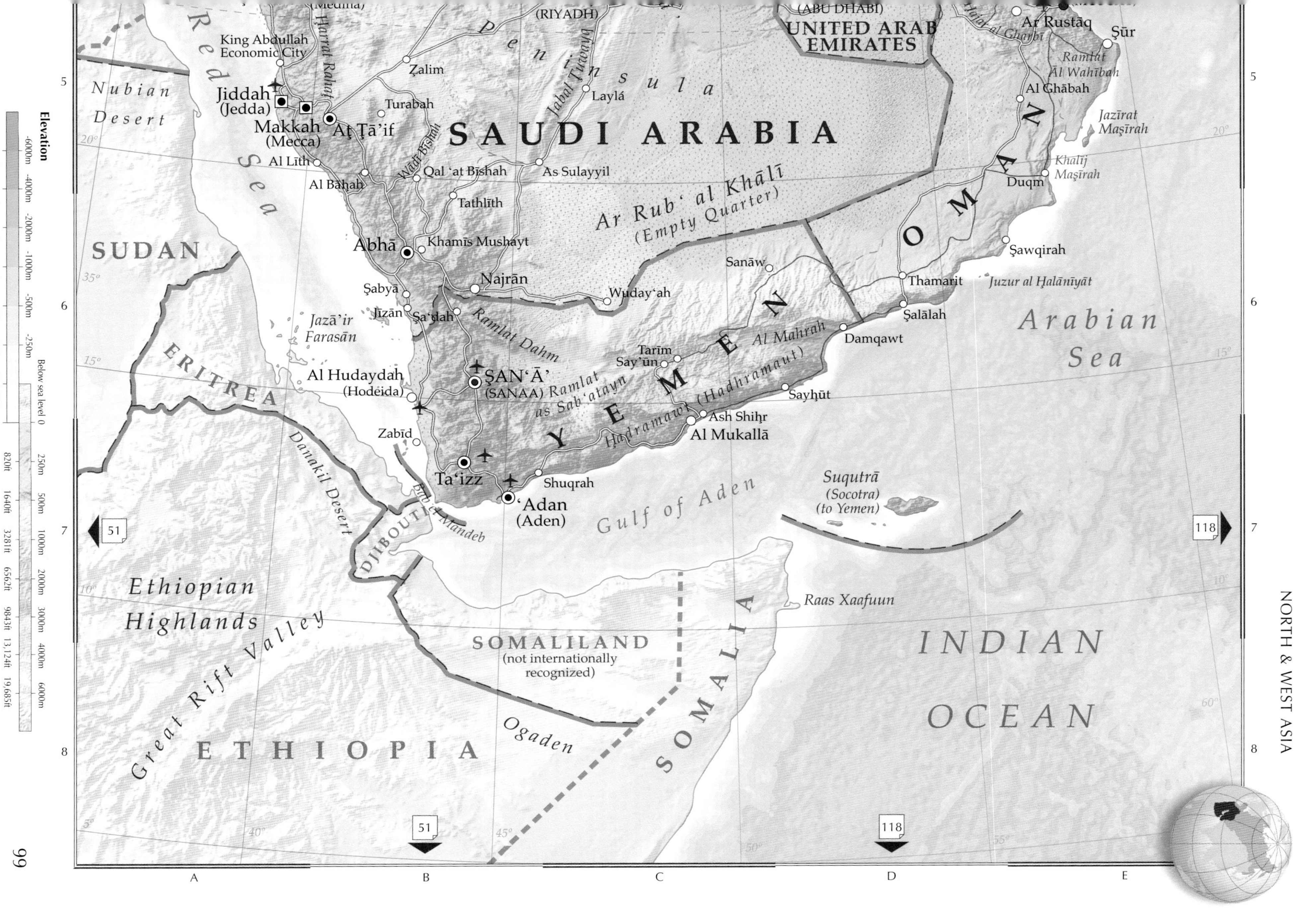

5

6

7

8

118

20°

15°

60°

55°

118

E

D

C

B

A

Şūr

Ar Rustāq

Ḩāsik al Ghubb

Ramlat

Al Wahībah

Jazīrat Maşīrah

Al Ghābah

Khalīj Maşīrah

Duqm

Şawqirah

Juzur al Ḩalānīyāt

O M A N

Thamarīt

Şalālah

Damqawt

Al Mahrah

Sayḩūt

Sanāw

Suquţrā (Socotra) (to Yemen)

Raas Xaafuun

ABU DHABI

UNITED ARAB EMIRATES

SAUDI ARABIA

Ar Rub' al Khālī (Empty Quarter)

Arabian Sea

INDIAN OCEAN

Jabal Ţuwayq

(RIYADH)

Layla

As Sulayyil

Ramlat Dahm

Ash Shiḩr

Al Mukallā

Tarīm

Sayʼūn

Ḩaḑramawt

(Ḩaḑramawt)

Y E M E N

Wudayʼah

Najrān

Ramlat as Sabʼatayn

SAN'Ā' (SANAA)

SOMALIA

SOMALILAND (not internationally recognized)

Ogaden

Gulf of Aden

Khamis Mushayṭ

Şaʼdah

Shuqrah

Adan (Aden)

Zalim

Turabah

Wadi Bishah

Qal'at Bishah

Tathlīth

Abhā

Ta'izz

Bāb el Mandeb

DJIBOUTI

Jīzān

Zabīd

Al Hudaydah (Hodeida)

At Ṭā'if

Al Bāḩah

Şabyā

Jazā'ir Farasān

ETHIOPIA

Ethiopian Highlands

Great Rift Valley

Harrat Rahat

Makkah (Mecca)

Al Lith

King Abdullah Economic City

Jiddah (Jedda)

Al Bahdb

Danakil Desert

ERITREA

SUDAN

Red Sea

Nubian Desert

Arabian Peninsula

51

45°

50°

40°

35°

15°

20°

51

118

51

Elevation

-6000m	-4000m	-2000m	-1000m	-500m	-250m	Below sea level 0	250m	500m	1000m	2000m	3000m	4000m	6000m

-19,658ft -13,124ft -6562ft -3281ft -1640ft -820ft -328ft/-100m 0

820ft 1640ft 3281ft 6562ft 9843ft 13,124ft 19,685ft

Central Asia

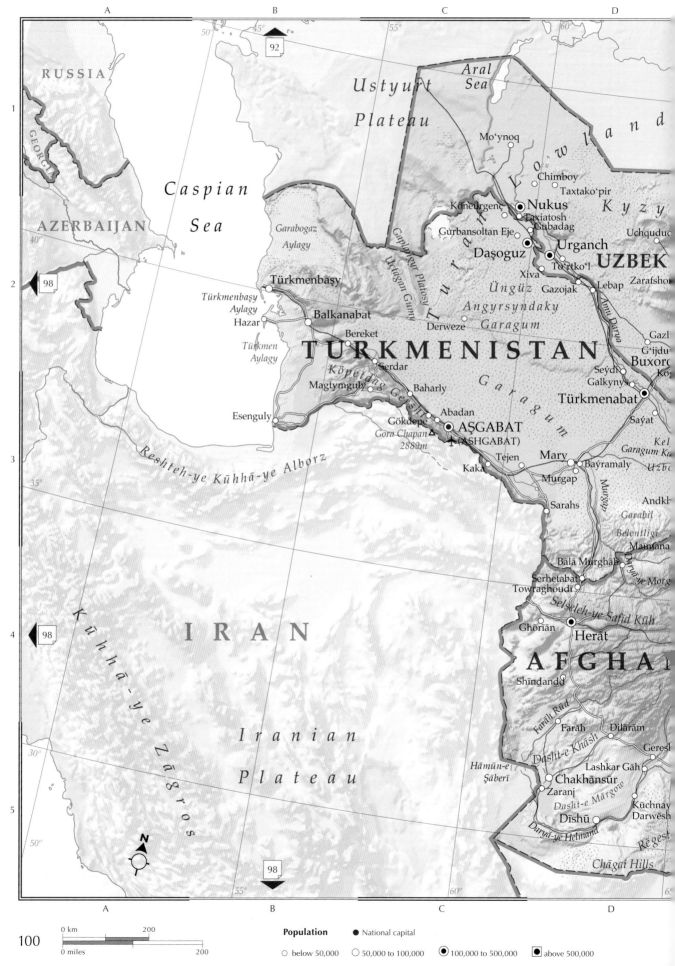

RUSSIA

GEORGIA

Caspian
Sea

AZERBAIJAN

Ustyurt
Plateau

Aral
Sea

Mo'ynoq

Turan *Lowland*

Chimboy
Taxtako'pir

Kyzy

Köneürgenç
Nukus
Taxiatosh
Gurbansoltan Eje
Cubadag
Daşoguz
Urganch
To'rtko'l
Xiva
Gazojak

Uchqudu
UZBEK
Zarafsho
Lebap

Garabogaz
Aylagy

Üngüz

Angyrsyndaky
Derweze *Garagum*

Türkmenbaşy

Türkmenbaşy
Aylagy
Hazar

Balkanabat

Bereket

TURKMENISTAN

Garagum

Gazl
G'ijdu
Buxorc
Seýdi
Galkynyş
Türkmenabat
Saýat

Türkmen
Aylagy

Köpet Serdar
Magtymguly
Baharly

Dag Gersh

Esenguly

Abadan
Gökdepe
Gora Chapan △
2889m

AŞGABAT
(ASHGABAT)

Kaka
Tejen
Murgap
Mary
Baýramaly

Garagum K
Uzbe
Kel

Reshteh-ye Kūhhā-ye Alborz

Sarahs

Murgap

Andkł

Garabil
Belentligi
Maimana

Bālā Murghāb

Daryā-ye Morg

Serhetabat
Towraghoudi

Selseleh-ye Safid Kūh

I R A N

Ghōriān
Herāt

Kūhhā-ye Zāgros

AFGHAN

Shindand

Iranian
Plateau

Farāh Rūd

Dasht-e Khāsh

Farāh
Dilārām
Geresl

Hāmūn-e
Şāberī

Lashkar Gāh
Chakhānsur
Zaranj

Dasht-e Mārgow

Kūchnay
Darwēsh

Dīshū

Daryā-ye Helmand

Rēgest

Chāgai Hills

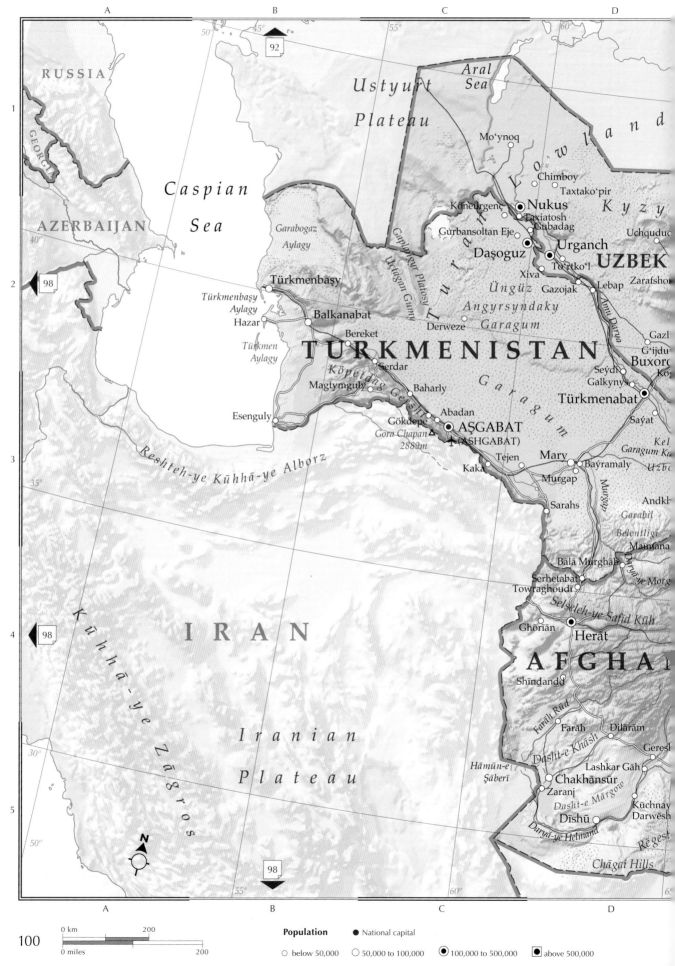

0 km   200

0 miles   200

Population   ● National capital

○ below 50,000   ◎ 50,000 to 100,000   ◉ 100,000 to 500,000   ◼ above 500,000

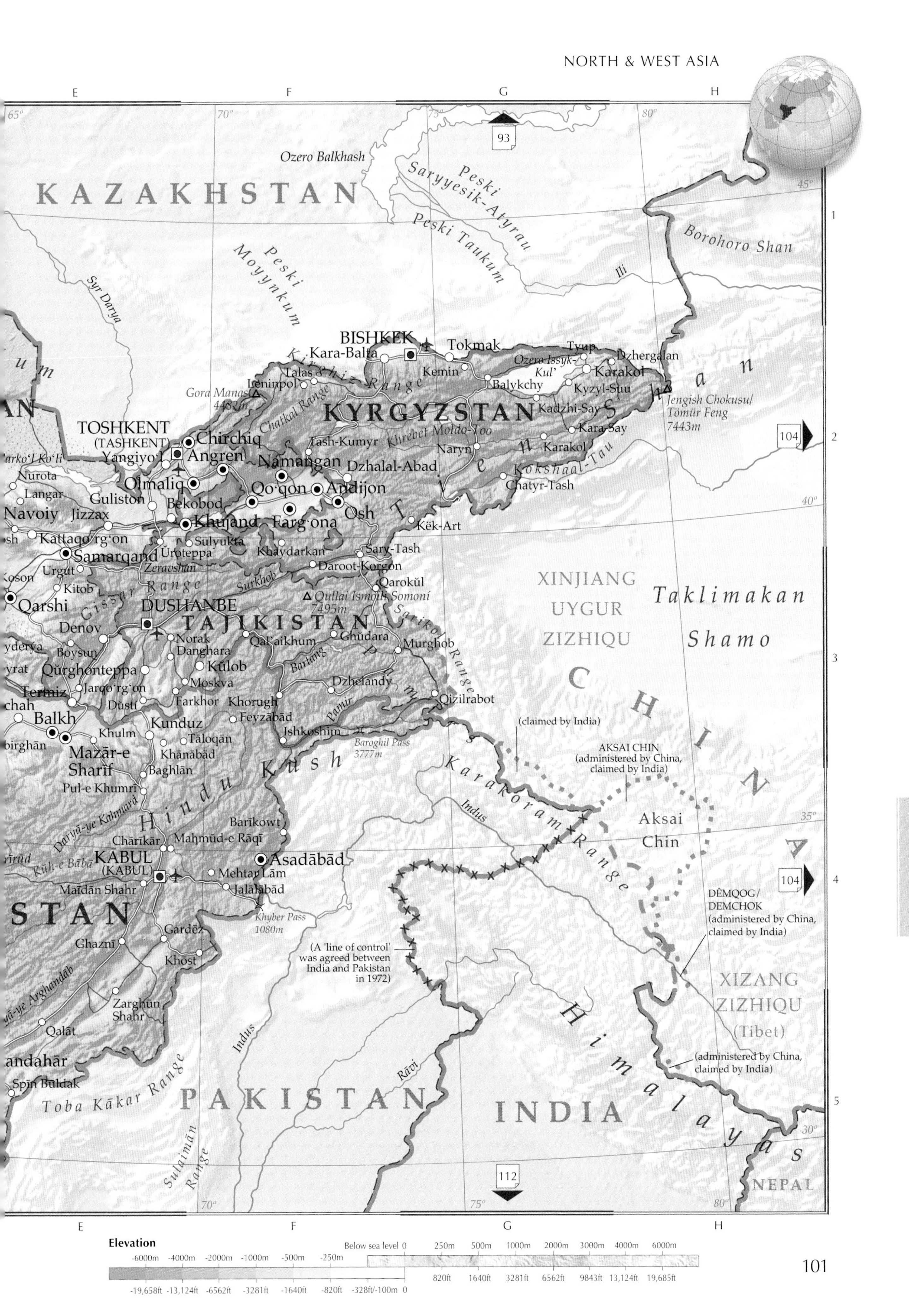

E   F   G   H

65°   70°   75°   80°   45°

Ozero Balkhash

K A Z A K H S T A N   1

Syr Darya   Peski Moyynkum   Saryyesik-Atyrau

Peski Taukum   Ili   Borohoro Shan

93

BISHKEK
Kara-Balta   Tokmak   Tyup   Dzhergalan
Talas   Kemin   Ozero Issyk-Kul'   Karakol
Gora Manas   Ireninpol   Balykchy   Kyzyl-Suu
4482m   Chatkal Range   Kadzhi-Say   Jengish Chokusu/
K I R G I Z S T A N   Tömür Feng
TOSHKENT   Kara-Say   7443m   104
(TASHKENT)   Khrebet Moldo-Too   Karakol
Yangiyo'l   Chirchiq   Tash-Kumyr   Kokshaal-Tau
Angren   Namangan   Naryn   Chatyr-Tash
Olmaliq   Qo'qon   Dzhalal-Abad   2
Nurota   Bekobod   Andijon   40°
Langar   Guliston   Farg'ona   Osh   T
Navoiy   Jizzax   Khujand   Kёk-Art
Kattaqo'rg'on   Sulyukta   Khaydarkan   Sary-Tash
Samarqand   Uroteppa   Daroot-Korgon
Urgut   Zeravshan   Qarokül
Kitob   Gissar Range   Surkhob   XINJIANG
Qarshi   Qullai Ismoili Somoni   UYGUR
Denov   DUSHANBE   7495m   ZIZHIQU   Taklimakan
TAJIKISTAN   Shamo
Boysun   Norak   Qal'aikhum   Ghüdara
yrat   Danghara   Murghob   C
Qürghonteppa   Kulob   Bartang   H
Termiz   Jarqo'rg'on   Moskva   Dzhelandy   I
chah   Dûstî   Farkhor   Khorugh   Qizilrabot   N
Balkh   Kunduz   Feyzäbäd   Pamir   (claimed by India)   A
birghän   Khulm   Taloqän   Ishköshim   AKSAI CHIN
Mazâr-e   Khānābād   Baroghil Pass   (administered by China,
Sharîf   Baghlän   3777m   claimed by India)
Pul-e Khumrî   Karakoram Range   Aksai   35°
Hindu   Kush   Indus   Chin
Chärikär   Barīkowt
KABUL   Mahmüd-e Räqî   104
(KABUL)   Asadäbäd   DÊMQOG/
Maïdän Shahr   Mehtar Läm   DEMCHOK   4
Jaläläbäd   (administered by China,
Ghaznî   Khyber Pass   claimed by India)
STAN   1080m
Gardëz   (A 'line of control'   XIZANG
Khôst   was agreed between   ZIZHIQU
India and Pakistan   (Tibet)
Zarghün   in 1972)
Shahr   Indus
Qalät   Indus   Rävi   (administered by China,
andahär   claimed by India)
Spin Buldak   Toba Käkar Range   P A K I S T A N   I N D I A   H
                                                                        i
                                                                        m
                                                                        a   5
Sulaimän   l   30°
Range   a
112   y
a
s   NEPAL

E   F   G   H

South & East Asia

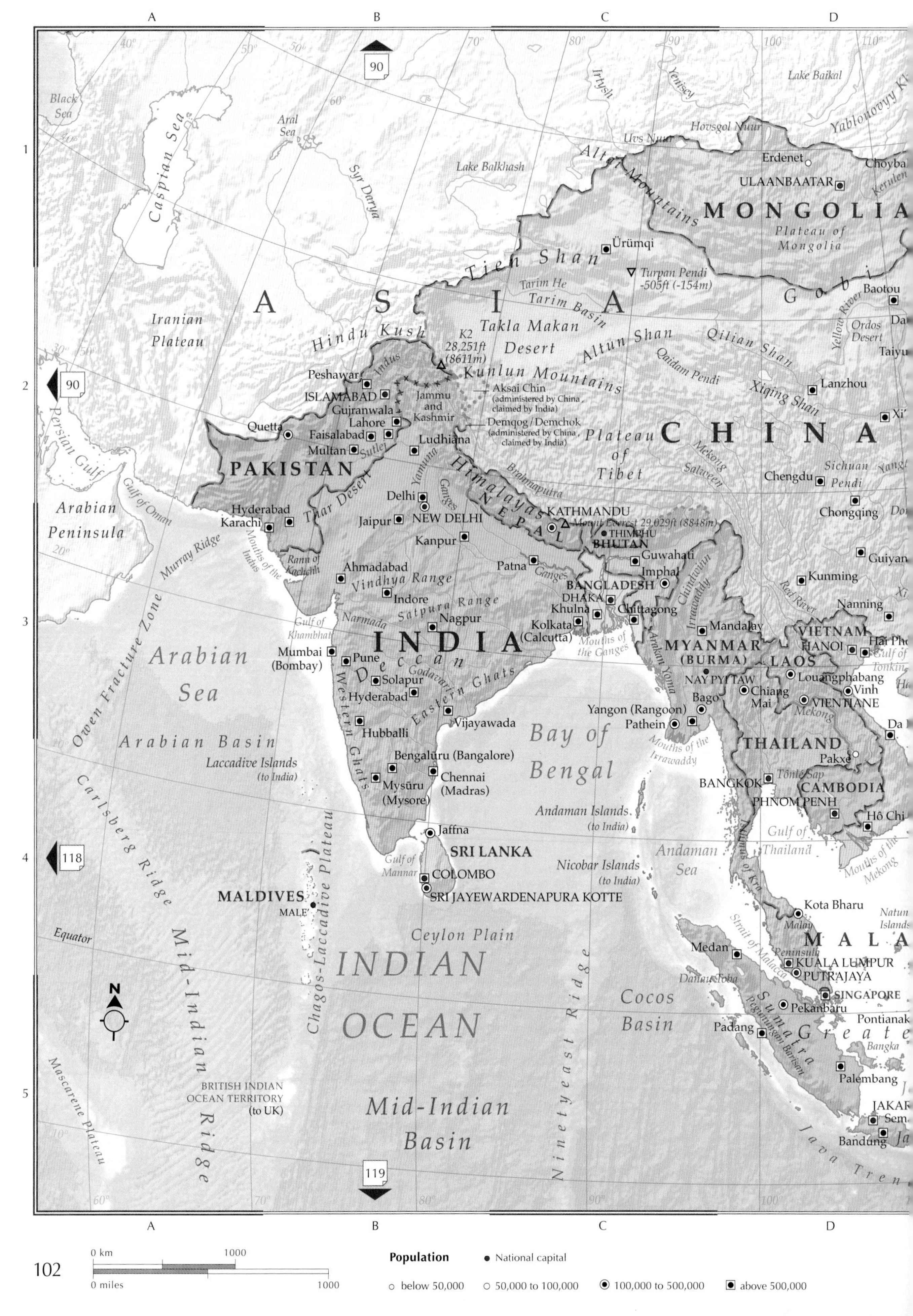

Black Sea

Caspian Sea

Aral Sea

Syr Darya

Lake Balkhash

Lake Baikal

Uvs Nuur

Hovsgol Nuur

Altai Mountains

Yablonovvy Kh

Kerulen

Erdenet

Choyba

ULAANBAATAR

MONGOLIA

Plateau of Mongolia

Gobi

Baotou

Da

Ordos Desert

Taiyu

Irtysh

Yenisey

Ürümqi

Tien Shan

Tarim He

Tarim Basin

▽ Turpan Pendi
-505ft (-154m)

Xiqing Shan

Qilian Shan

Lanzhou

Xi'

A S I A

Iranian Plateau

Hindu Kush

K2 28,251ft (8611m) △

Takla Makan Desert

Kunlun Mountains

Aksai Chin (administered by China, claimed by India)

Demqog / Demchok (administered by China, claimed by India)

Altun Shan

Qaidam Pendi

Plateau of Tibet

Mekong

Salween

C H I N A

Chengdu

Sichuan Pendi

Yangt

Peshawar

ISLAMABAD

Gujranwala

Lahore

Quetta

Faisalabad

Multan

Indus

Jammu and Kashmir

Ludhiana

Sutlej

Himalayas

Brahmaputra

Chongqing

Do

Persian Gulf

90

Gulf of Oman

Arabian Peninsula

PAKISTAN

Thar Desert

Yamuna

Delhi

Jaipur

NEW DELHI

Kanpur

Ganges

NEPAL

KATHMANDU

△ Mount Everest 29,029ft (8848m)

THIMPHU

BHUTAN

Guwahati

Imphal

Chindwin

Guiyan

Kunming

Nanning

Xi

Murray Ridge

Hyderabad

Karachi

Mouths of the Indus

Rann of Kachchh

Ahmadabad

Vindhya Range

Indore

Narmada

Satpura Range

Nagpur

Patna

Ganges

BANGLADESH

DHAKA

Khulna

Kolkata (Calcutta)

Chittagong

Mouths of the Ganges

Mandalay

MYANMAR (BURMA)

Irrawaddy

Arakan Yoma

VIETNAM

HANOI

Hai Pho

Gulf of Tonkin

LAOS

Owen Fracture Zone

Gulf of Khambhat

Mumbai (Bombay)

Pune

Deccan

Solapur

Hyderabad

Godavari

Western Ghats

Eastern Ghats

I N D I A

Vijayawada

Hubballi

NAY PYI TAW

Bago

Louangphabang

Chiang Mai

Vinh

VIENTIANE

Da

Mekong

Arabian Sea

Carlsberg Ridge

Arabian Basin

Laccadive Islands (to India)

Bengalūru (Bangalore)

Mysūru (Mysore)

Chennai (Madras)

Bay of Bengal

Yangon (Rangoon)

Pathein

Mouths of the Irrawaddy

THAILAND

Pakxe

Tônlé Sap

BANGKOK

CAMBODIA

PHNOM PENH

Hô Chi

118

Jaffna

SRI LANKA

COLOMBO

SRI JAYEWARDENAPURA KOTTE

Gulf of Mannar

Andaman Islands (to India)

Nicobar Islands (to India)

Andaman Sea

Gulf of Thailand

Mouths of the Mekong

Isthmus of Kra

Kota Bharu

Natun Islands

Malay Peninsula

M A L A

MALDIVES

MALE

Chagos-Laccadive Plateau

Equator

Ceylon Plain

Medan

Danau Toba

KUALA LUMPUR

PUTRAJAYA

SINGAPORE

Pekanbaru

Strait of Malacca

Mid-Indian Ridge

INDIAN OCEAN

Mascarene Plateau

119

BRITISH INDIAN OCEAN TERRITORY (to UK)

Mid-Indian Basin

Ninetyeast Ridge

Cocos Basin

Padang

Pagunungan Barisan

Gre a te

Pontianak

Bangka

JAKAR

Palembang

Sumatra

Bandung

Ja

Java Tren

N

0 km   1000

0 miles   1000

Population   ● National capital

○ below 50,000   ◎ 50,000 to 100,000   ◉ 100,000 to 500,000   ■ above 500,000

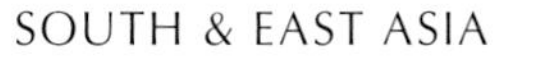

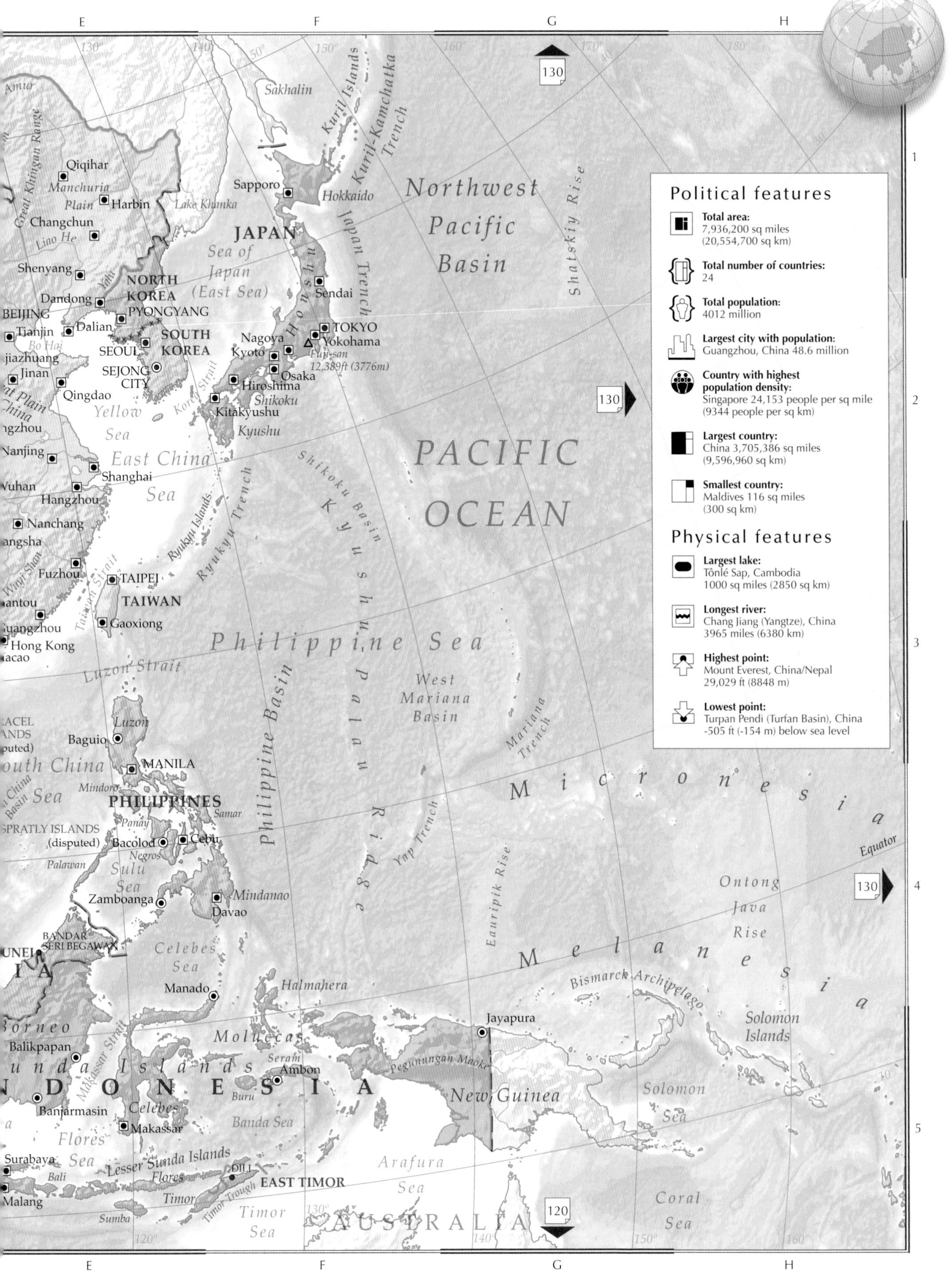

Political features

Total area:
7,936,200 sq miles
(20,554,700 sq km)

Total number of countries:
24

Total population:
4012 million

Largest city with population:
Guangzhou, China 48.6 million

Country with highest population density:
Singapore 24,153 people per sq mile
(9344 people per sq km)

Largest country:
China 3,705,386 sq miles
(9,596,960 sq km)

Smallest country:
Maldives 116 sq miles
(300 sq km)

Physical features

Largest lake:
Tônlé Sap, Cambodia
1000 sq miles (2850 sq km)

Longest river:
Chang Jiang (Yangtze), China
3965 miles (6380 km)

Highest point:
Mount Everest, China/Nepal
29,029 ft (8848 m)

Lowest point:
Turpan Pendi (Turfan Basin), China
-505 ft (-154 m) below sea level

Western China & Mongolia

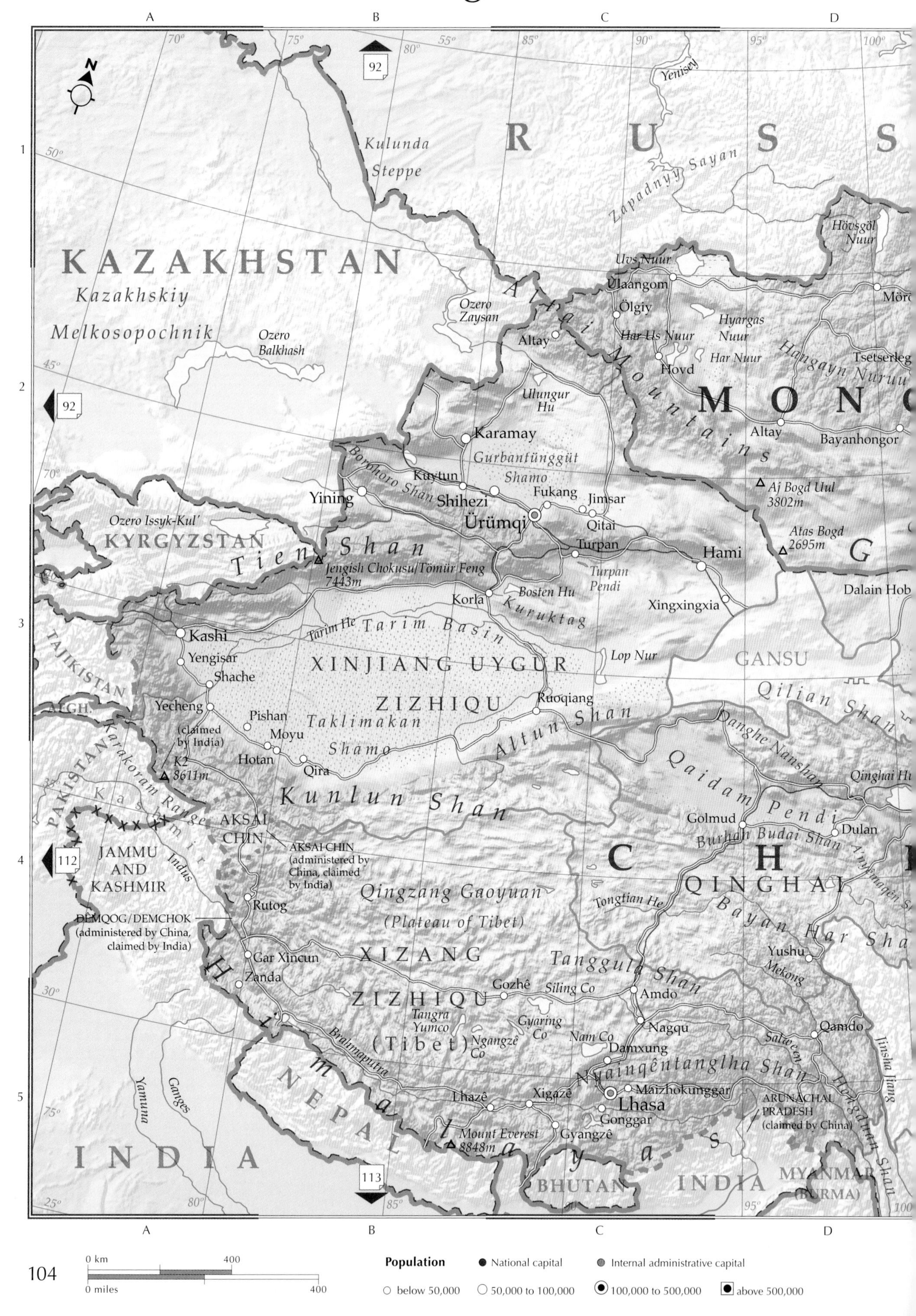

KAZAKHSTAN

Kazakhskiy

Melkosopochnik

Ozero Balkhash

R U S S

Kulunda Steppe

Zapadnyy Sayan

Yenisey

Hövsgöl Nuur

Ozero Zaysan

Uvs Nuur

Ulaangom

Ölgiy

Altay

Hyargas Nuur

Har Us Nuur

Hovd

Har Nuur

Mör

M O N G

Tsetserleg

Hangayn Nuruu

Ulungur Hu

Karamay

Gurbantünggüt Shamo

Altay

Bayanhongor

Aj Bogd Uul 3802m

Borohoro Shan

Kuytun

Shihezi

Fukang

Jimsar

Yining

Ürümqi

Qitai

Turpan

Hami

Atas Bogd 2695m

G

KYRGYZSTAN

Ozero Issyk-Kul'

Tien Shan

Jengish Chokusu/Tömür Feng 7443m

Korla

Bosten Hu

Turpan Pendi

Xingxingxia

Dalain Hob

Tarim He

Kuruktag

GANSU

Tarim Basin

TAJIKISTAN

AFGH.

Kashi

Yengisar

Shache

XINJIANG UYGUR

ZIZHIQU

Lop Nur

Ruoqiang

Qilian Shan

Yecheng

(claimed by India)

Pishan

Moyu

Taklimakan Shamo

Altun Shan

Danghe Nanshan

Qaidam Pendi

Qinghai Hu

Karakoram Range

K2 8611m

Hotan

Qira

Kunlun Shan

Burhan Budai Shan

Golmud

Dulan

QINGHAI

Kashmir

PAKISTAN

X X X X X

AKSAI CHIN

AKSAI CHIN (administered by China, claimed by India)

C H

H I

Indus

JAMMU AND KASHMIR

Rutog

Qingzang Gaoyuan (Plateau of Tibet)

Tongtian He

Bayan Har Sha

Yushu

Mekong

DÊMQOG/DEMCHOK (administered by China, claimed by India)

Gar Xincun

Zanda

XIZANG

Tanggula Shan

Amdo

Qamdo

Brahmaputra

ZIZHIQU

(Tibet)

Gozhê

Siling Co

Gyaring Co

Nam Co

Nagqu

Salween

Jinsha Jiang

Tangra Yumco

Ngangzê Co

Damxung

Nyainqêntanglha Shan

H I M A

Yamuna

Ganges

NEPAL

Lhazê

Xigazê

Maizhokunggar

Lhasa

Gonggar

ARUNACHAL PRADESH (claimed by China)

Mount Everest 8848m

Gyangzê

l a y a s

INDIA

BHUTAN

INDIA

MYANMAR (BURMA)

0 km   400

0 miles   400

Population   ● National capital   ● Internal administrative capital

○ below 50,000   ○ 50,000 to 100,000   ◉ 100,000 to 500,000   ◼ above 500,000

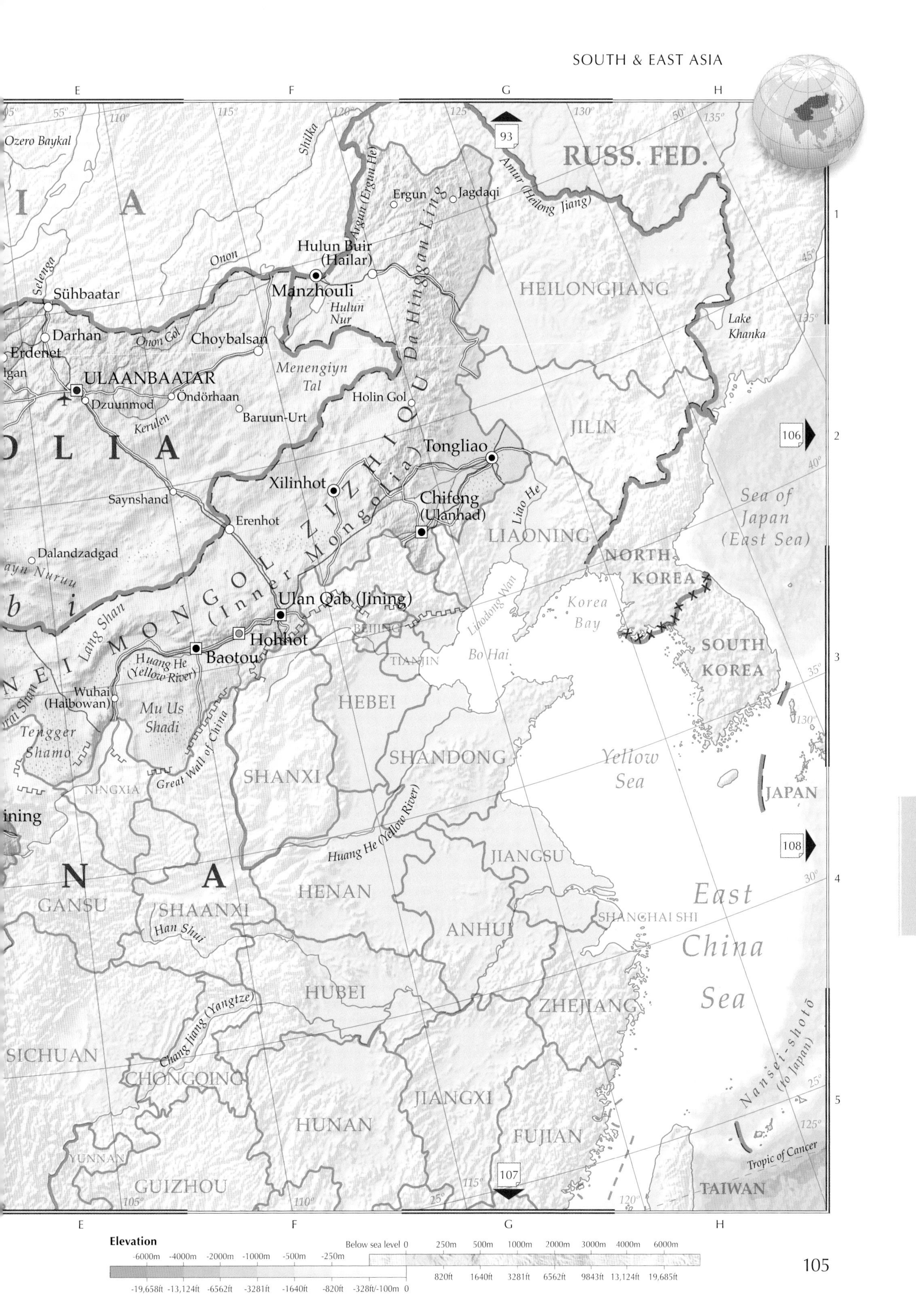

E   F   G   H

RUSS. FED.

93

Ozero Baykal

Shilka

Arguin (Ergun He)

Ergun   Jagdaqi

Amur (Heilong Jiang)

HEILONGJIANG

Onon

Hulun Buir
(Hailar)

Manzhouli

Hulun
Nur

Lake
Khanka

135°

106

Sühbaatar

Selenga

Onon Gol   Choybalsan

Darhan

Erdenet

lgan

Kerulen

Menengiyn
Tal

ULAANBAATAR

Dzuunmod   Öndörhaan

Baruun-Urt

Holin Gol

JILIN

Da Hinggan Ling

Sea of
Japan
(East Sea)

40°

Xilinhot

Tongliao

Saynshand

Erenhot

Chifeng
(Ulanhad)

Liao He

LIAONING

NORTH
KOREA

Korea
Bay

Dalandzadgad

ayn Nuruu

b   i

NEI MONGOL ZIZHOU
(Inner Mongolia)

Ulan Qab (Jining)

Lang Shan

rai Shan

Huang He
(Yellow River)

Hohhot

Baotou

BEIJING

TIANJIN

Liaodong Wan

Bo Hai

SOUTH
KOREA

35°

130°

Wuhai
(Haibowan)

Mu Us
Shadi

Tengger
Shamo

NINGXIA

Great Wall of China

HEBEI

SHANDONG

Yellow
Sea

JAPAN

108

ining

SHANXI

Huang He (Yellow River)

JIANGSU

30°

N   A

GANSU

SHAANXI

Han Shui

HENAN

ANHUI

SHANGHAI SHI

East
China
Sea

SICHUAN

HUBEI

Chang Jiang (Yangtze)

ZHEJIANG

CHONGQING

JIANGXI

HUNAN

FUJIAN

Nansei-shotō
(to Japan)

25°

125°

YUNNAN

GUIZHOU

107

Tropic of Cancer

TAIWAN

105°   110°   25°   115°   120°

E   F   G   H

Elevation

| -6000m | -4000m | -2000m | -1000m | -500m | -250m | Below sea level 0 | 250m | 500m | 1000m | 2000m | 3000m | 4000m | 6000m |

| -19,658ft | -13,124ft | -6562ft | -3281ft | -1640ft | -820ft | -328ft/-100m 0 | 820ft | 1640ft | 3281ft | 6562ft | 9843ft | 13,124ft | 19,685ft |

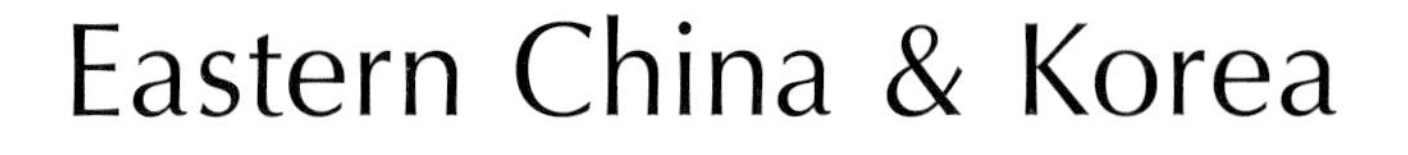

Eastern China & Korea

0 km   400
0 miles   400

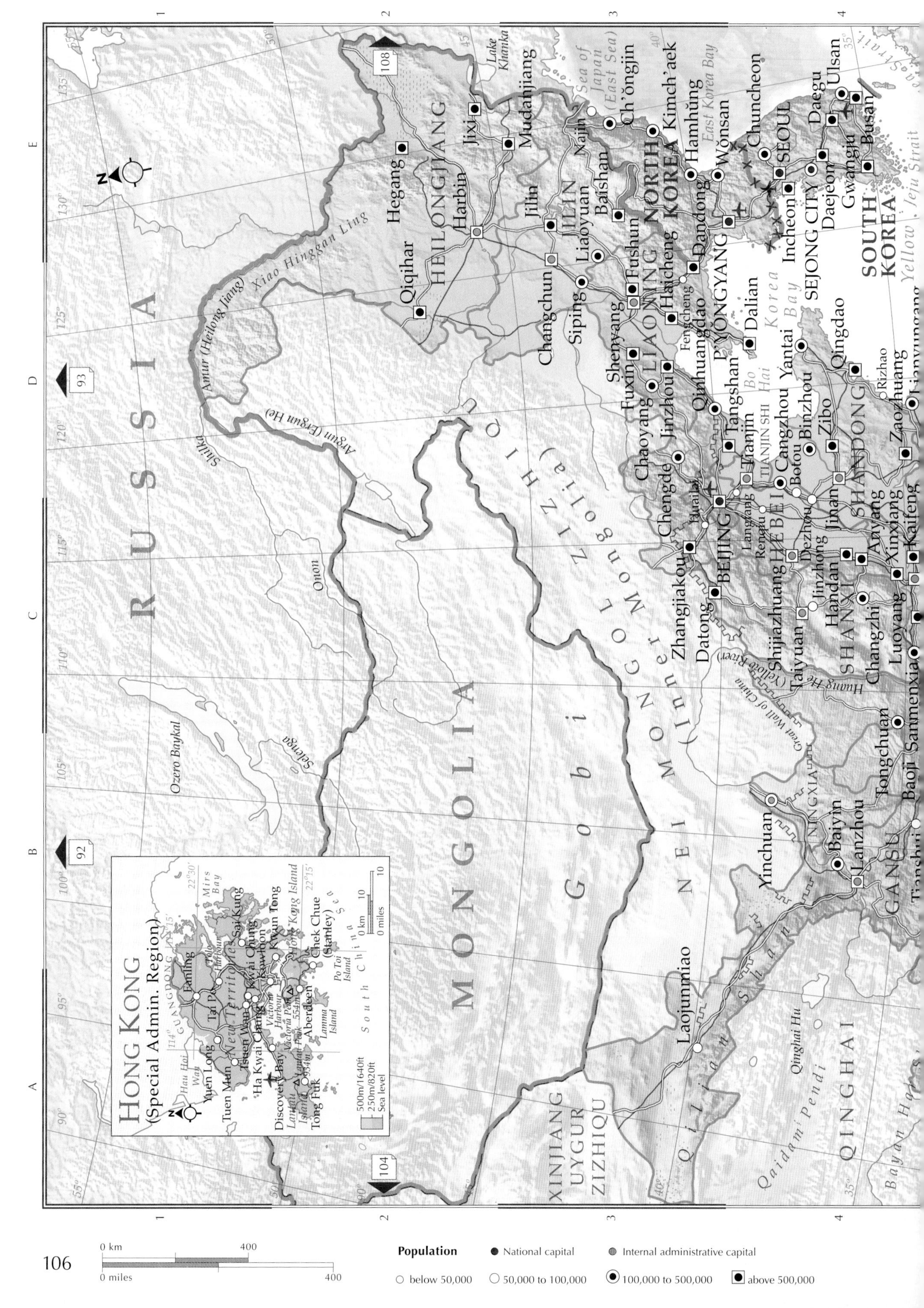

Population
- ○ National capital
- ○ Internal administrative capital
- ○ below 50,000
- ○ 50,000 to 100,000
- ● 100,000 to 500,000
- ■ above 500,000

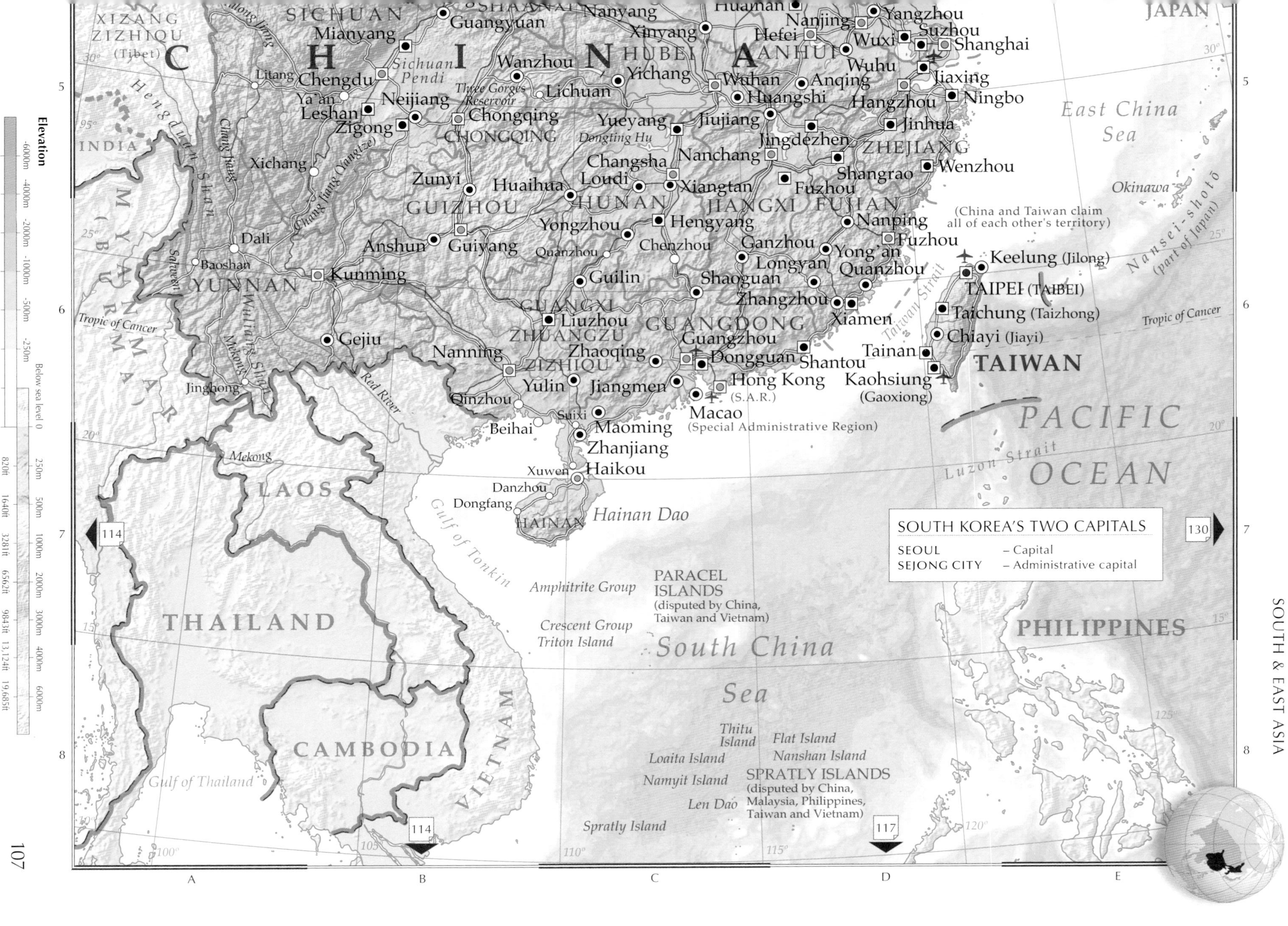
CHINA
JAPAN
TAIWAN
PHILIPPINES
THAILAND
CAMBODIA
LAOS
VIETNAM
MYANMAR (BURMA)
INDIA
PACIFIC OCEAN
East China Sea
South China Sea
Gulf of Tonkin
Gulf of Thailand
Luzon Strait
Taiwan Strait
Tropic of Cancer
Nansei-shoto (part of Japan)
Okinawa
(China and Taiwan claim all of each other's territory)
SOUTH KOREA'S TWO CAPITALS
SEOUL – Capital
SEJONG CITY – Administrative capital
SICHUAN
YUNNAN
GUIZHOU
CHONGQING
HUNAN
HUBEI
ANHUI
JIANGXI
FUJIAN
ZHEJIANG
GUANGDONG
GUANGXI ZHUANGZU ZIZHIQU
HAINAN
XIZANG ZIZHIQU (Tibet)
Shanghai
Suzhou
Yangzhou
Wuxi
Nanjing
Hefei
Jiaxing
Ningbo
Hangzhou
Jinhua
Wenzhou
Wuhu
Anqing
Huangshi
Wuhan
Xinyang
Nanyang
Yichang
Jingdezhen
Shangrao
Fuzhou
Nanping
Yong'an
Quanzhou
Xiamen
Zhangzhou
Shantou
Longyan
Ganzhou
Nanchang
Jiujiang
Xiangtan
Changsha
Loudi
Hengyang
Chenzhou
Shaoguan
Guangzhou
Dongguan
Hong Kong (S.A.R.)
Macao (Special Administrative Region)
Jiangmen
Maoming
Zhanjiang
Haikou
Zhaoqing
Liuzhou
Guilin
Yongzhou
Yueyang
Dongting Hu
Lichuan
Wanzhou
Chongqing
Three Gorges Reservoir
Guangyuan
Mianyang
Chengdu
Sichuan Pendi
Neijiang
Zigong
Leshan
Ya'an
Litang
Xichang
Kunming
Gejiu
Anshun
Guiyang
Zunyi
Huaihua
Nanning
Yulin
Qinzhou
Beihai
Suixi
Xuwen
Danzhou
Dongfang
Hainan Dao
Keelung (Jilong)
TAIPEI (TAIBEI)
Taichung (Taizhong)
Chiayi (Jiayi)
Tainan
Kaohsiung (Gaoxiong)
Dali
Baoshan
Jinghong
Chang Jiang
Chang Jiang (Yangtze)
Red River
Mekong
Salween
Hengduan Shan
Wuliang Shan
PARACEL ISLANDS (disputed by China, Taiwan and Vietnam)
Amphitrite Group
Crescent Group
Triton Island
SPRATLY ISLANDS (disputed by China, Malaysia, Philippines, Taiwan and Vietnam)
Thitu Island
Flat Island
Nanshan Island
Loaita Island
Namyit Island
Len Dao
Spratly Island
Elevation
Below sea level 0
250m 500m 1000m 2000m 3000m 4000m 6000m
820ft 1640ft 3281ft 6562ft 9843ft 13,124ft 19,685ft
-6000m -4000m -2000m -1000m -500m -250m
-19,658ft -13,124ft -6562ft -3281ft -1640ft -820ft -328ft/-100m 0

114
117
130

Japan

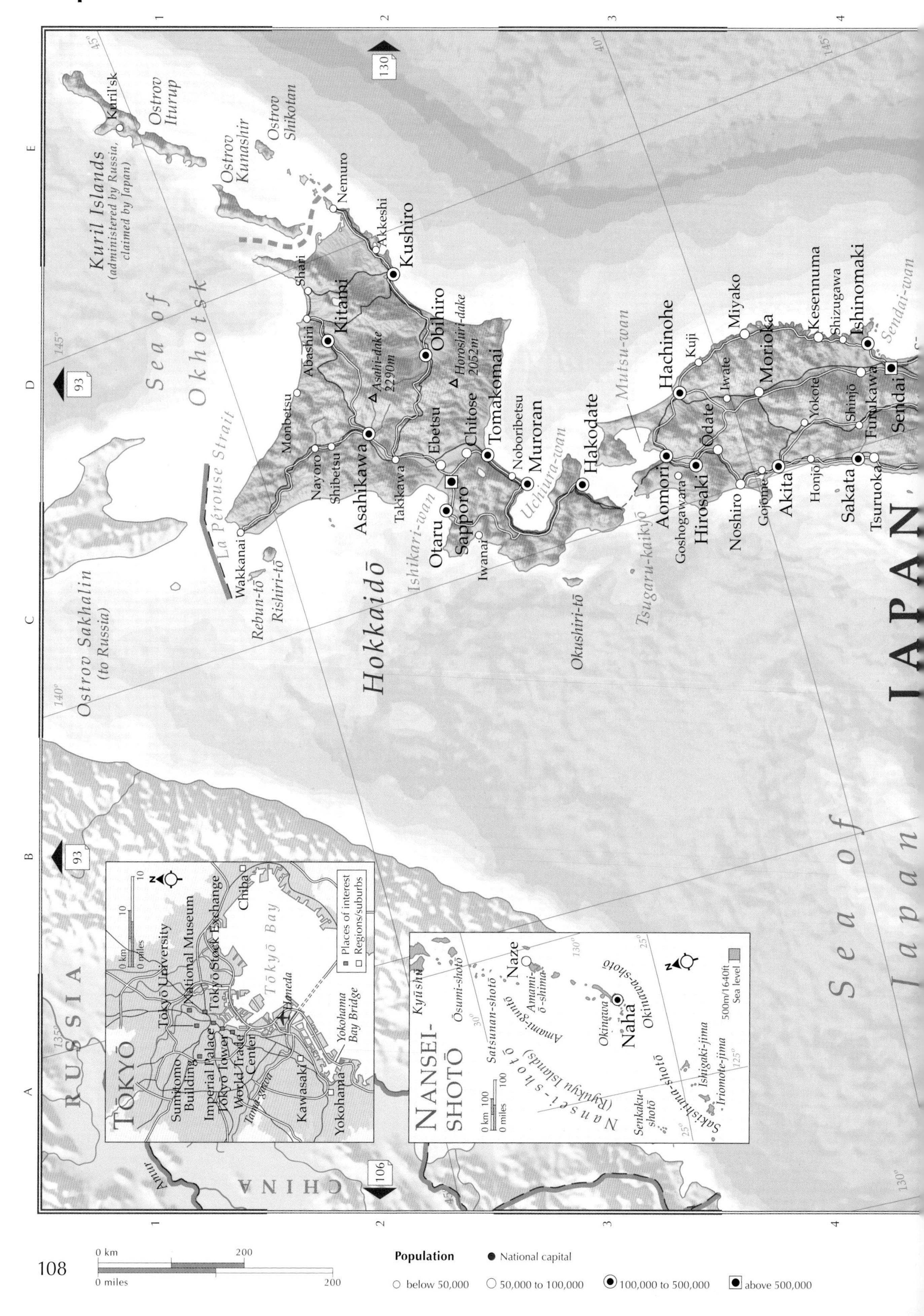

45°

130°

E

D

145°

93

C

140°

B

93

A

135°

Kuril'sk
Ostrov Iturup

Ostrov Kunashir
Ostrov Shikotan

Kuril Islands
(administered by Russia,
claimed by Japan)

Nemuro

Akkeshi

Kushiro

Sea of Okhotsk

Shari

Abashiri

Kitami

Monbetsu

△ Asahi-dake
2290m

Obihiro

△ Horoshiri-dake
2052m

Nayoro

Shibetsu

Asahikawa

Takikawa

Ebetsu

Chitose

Tomakomai

Noboribetsu

Muroran

Hakodate

Uchiura-wan

La Pérouse Strait

Wakkanai

Rebun-tō

Rishiri-tō

Otaru

Sapporo

Iwanai

Ishikari-wan

Hokkaidō

Okushiri-tō

Tsugaru-kaikyō

Ostrov Sakhalin
(to Russia)

Mutsu-wan

Hachinohe

Kuji

Iwate

Miyako

Morioka

Kesennuma

Shizugawa

Ishinomaki

Sendai-wan

Aomori

Goshogawara

Hirosaki

Ōdate

Noshiro

Gojōme

Yokote

Shinjō

Furukawa

Sendai

Akita

Honjō

Sakata

Tsuruoka

Sea of Japan

Japan

JAPAN

RUSSIA

130°

CHINA

106

TŌKYŌ

Chiba

Tōkyō University

National Museum

Tōkyō Stock Exchange

Sumitomo Building

Imperial Palace

Tōkyō Tower

World Trade Center

Kawasaki

Yokohama

Tōkyō Bay

Haneda

Yokohama Bay Bridge

0 km   10
0 miles   10

■ Places of interest
□ Regions/suburbs

NANSEI-SHOTŌ

Kyūshū

Ōsumi-shotō

Satsunan-shotō

Naze

Amami-guntō

Amami-ō-shima

Nansei-shotō (Ryūkyū Islands)

Okinawa

Naha

Okinawa-shotō

Sakishima-shotō

Ishigaki-jima

Iriomote-jima

Senkaku-shotō

130°

25°

30°

25°

125°

0 km   100
0 miles   100

500m/1640ft
Sea level

0 km   200

0 miles   200

Population   ● National capital

○ below 50,000   ○ 50,000 to 100,000   ◉ 100,000 to 500,000   ■ above 500,000

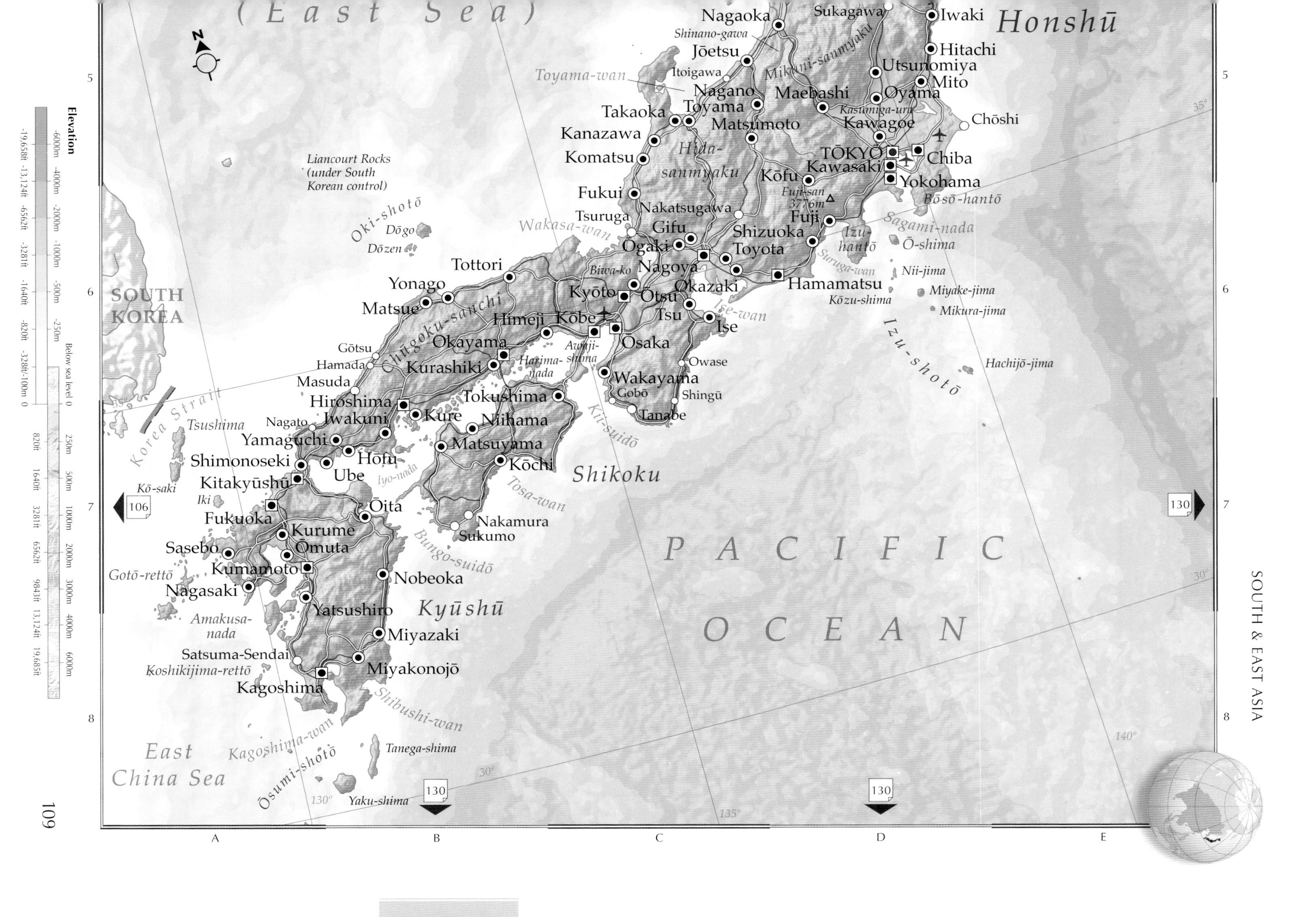

Honshū

Iwaki

Hitachi

Utsunomiya
Mito

Chōshi

Sukagawa

Oyama
Maebashi

Chiba
Yokohama

Kawagoe

TOKYO
Kawasaki

Bōsō-hantō

Sagami-nada

Ō-shima

Nii-jima

Miyake-jima

Mikura-jima

Hachijō-jima

Nagaoka

Jōetsu

Nagano

Matsumoto

Kōfu

Fuji

Fujisan
△ 3776m

Shizuoka

Hamamatsu

Izu-hantō

Izu-shotō

Suruga-wan

Kōzu-shima

P A C I F I C

Toyama

Takaoka

Kanazawa

Komatsu

Fukui

Matsumoto

Gifu

Nakatsugawa

Ogaki

Toyota

Nagoya

Ōtsu
Okazaki

Tsu

Ise

Owase

Shingū

Ise-wan

O C E A N

Shinano-gawa

Itoigawa

*Hida-
sanmyaku*

Tsuruga

Biwa-ko

Kyoto
Kōbe

Osaka

*Awaji-
shima*

Wakayama

Gobō

Tanabe

Kii-suidō

Shikoku

Tottori

Yonago

Matsue

Himeji

Okayama

Kurashiki

Tokushima

Niihama

Matsuyama

Kōchi

Nakamura

Sukumo

Tosa-wan

Wakasa-wan

Chūgoku-sanchi

*Hina-
nada*

Kure

Mihara

Bungo-suidō

Kyūshū

Nobeoka

Miyazaki

Miyakonojō

Shibushi-wan

Tanega-shima

Liancourt Rocks
(under South
Korean control)

Oki-shotō

Dōgo

Dōzen

Gōtsu

Hamada

Masuda

Hiroshima
Iwakuni

Hōfu

Ube

Ōita

Suō-nada

Katsushiro

Yaku-shima

(E a s t   S e a)

Toyama-wan

Nagato

Yamaguchi

Shimonoseki

Kitakyūshū

Fukuoka

Kurume

Ōmuta

Kumamoto

Satsuma-Sendai

Kagoshima

*Amakusa-
nada*

Koshikijima-rettō

Kagoshima-wan

Ōsumi-shotō

SOUTH
KOREA

Tsushima

Kō-saki

Iki

Sasebo

Nagasaki

Gotō-rettō

Koshikijima-rettō

Korea Strait

N

*East
China Sea*

130

130

130

130

106

35°

30°

140°

135°

30°

130°

130°

5

6

7

8

5

6

7

8

E

D

C

B

A

Elevation

Below sea level 0   250m   500m   1000m   2000m   3000m   4000m   6000m

-6000m   -4000m   -2000m   -1000m   -500m   -250m

-19,658ft   -13,124ft   -6562ft   -3281ft   -1640ft   -820ft   -328ft/-100m 0

820ft   1640ft   3281ft   6562ft   9843ft   13,124ft   19,685ft

109

South India & Sri Lanka

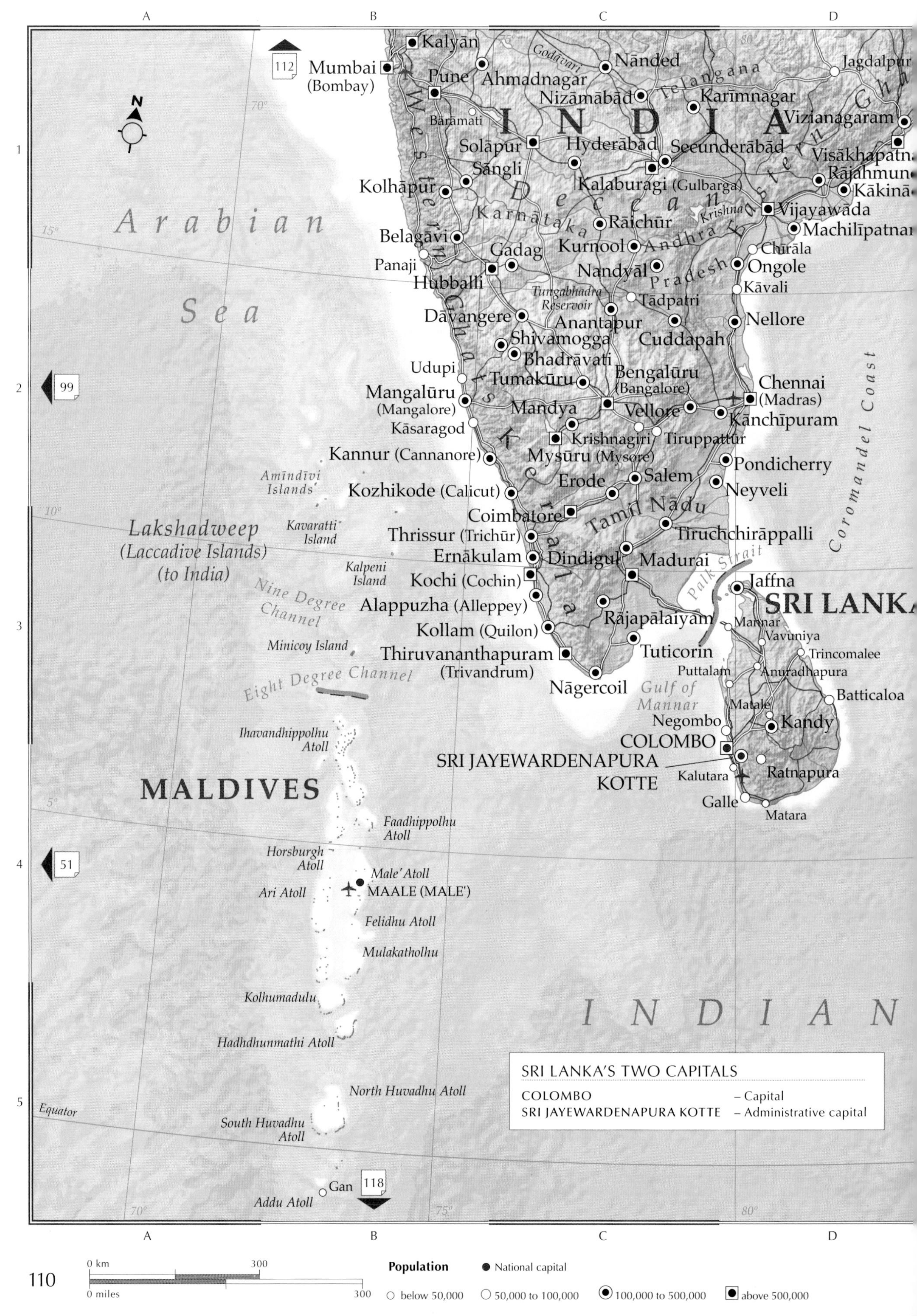

Kalyān
Mumbai (Bombay)
112
Pune
Ahmadnagar
Nānded
Jagdalpur
Bārāmati
Nizāmābād
Karīmnagar
Telangana
Solāpur
Hyderābād
Seeunderābād
Vizianagaram
Sāngli
Visākhapatn
Kolhāpur
Kalaburagi (Gulbarga)
Rājahmun
Karnataka
Kākinā
Belagāvi
Rāichūr
Krishna
Vijayawāda
Deccan
Machilīpatnar
Panaji
Gadag
Kurnool
Andhra
Chīrāla
Hubballi
Nandyāl
Pradesh
Ongole
Tungabhadra Reservoir
Tādpatri
Kāvali
Dāvangere
Anantapur
Nellore
Shivamogga
Cuddapah
Bhadrāvati
Bengalūru (Bangalore)
Udupi
Tumakūru
Chennai (Madras)
Mangalūru (Mangalore)
Mandya
Vellore
Kanchīpuram
Kāsaragod
Krishnagiri
Tiruppattūr
Kannur (Cannanore)
Mysūru (Mysore)
Pondicherry
Kozhikode (Calicut)
Erode
Salem
Neyveli
Coimbatore
Tamil Nādu
Tiruchchirāppalli
Thrissur (Trichūr)
Ernākulam
Dindigul
Madurai
Jaffna
Kochi (Cochin)
SRI LANKA
Alappuzha (Alleppey)
Rājapālaiyam
Mannar
Vavuniya
Kollam (Quilon)
Tuticorin
Trincomalee
Thiruvananthapuram (Trivandrum)
Puttalam
Anuradhapura
Nāgercoil
Gulf of Mannar
Batticaloa
Matale
Negombo
Kandy
COLOMBO
SRI JAYEWARDENAPURA KOTTE
Kalutara
Ratnapura
Galle
Matara

Arabian Sea

Lakshadweep (Laccadive Islands) (to India)
Amīndīvi Islands
Kavaratti Island
Kalpeni Island
Nine Degree Channel
Minicoy Island
Eight Degree Channel
Ihavandhippolhu Atoll

MALDIVES
Faadhippolhu Atoll
Horsburgh Atoll
Ari Atoll
Male' Atoll
MAALE (MALE')
Felidhu Atoll
Mulakatholhu
Kolhumadulu
Hadhdhunmathi Atoll
North Huvadhu Atoll
Equator
South Huvadhu Atoll
Addu Atoll
Gan
118

INDIAN

Coromandel Coast
Palk Strait

99
51

112
70°
N
15°
70°
10°
5°
Equator

SRI LANKA'S TWO CAPITALS

COLOMBO — Capital
SRI JAYEWARDENAPURA KOTTE — Administrative capital

0 km   300
0 miles   300

Population   ● National capital

○ below 50,000   ○ 50,000 to 100,000   ◉ 100,000 to 500,000   ◼ above 500,000

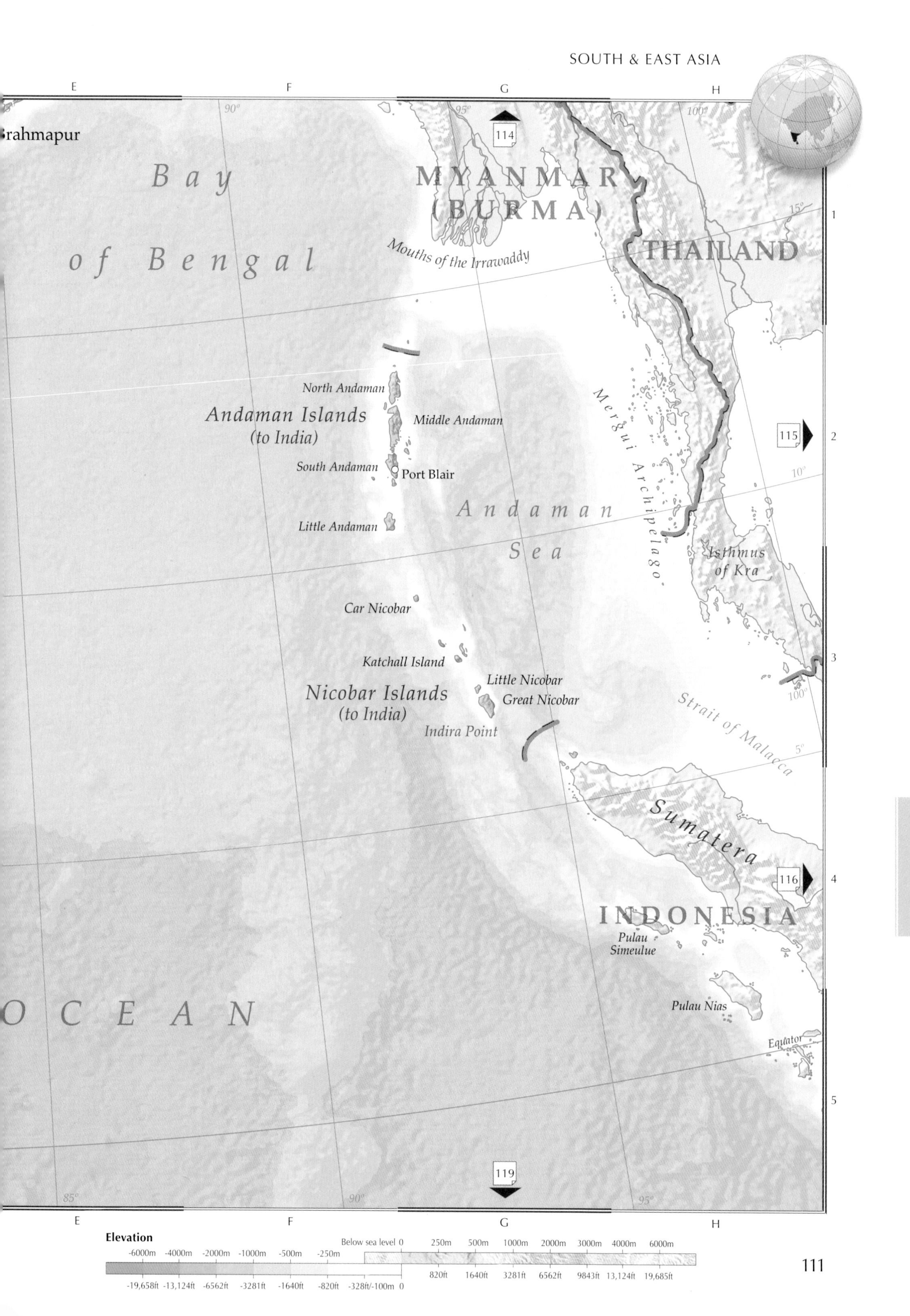

rahmapur

Bay

of Bengal

MYANMAR
(BURMA)

THAILAND

Mouths of the Irrawaddy

114

115

North Andaman

Andaman Islands
(to India)

Middle Andaman

South Andaman   Port Blair

Mergui Archipelago

Andaman

Sea

Little Andaman

*Isthmus
of Kra*

Car Nicobar

Katchall Island

Nicobar Islands
(to India)

Little Nicobar

Great Nicobar

Strait of Malacca

Indira Point

116

Sumatera

INDONESIA

*Pulau
Simeulue*

OCEAN

Pulau Nias

Equator

119

Elevation

-6000m	-4000m	-2000m	-1000m	-500m	-250m	Below sea level 0	250m	500m	1000m	2000m	3000m	4000m	6000m
-19,658ft	-13,124ft	-6562ft	-3281ft	-1640ft	-820ft	-328ft/-100m 0		820ft	1640ft	3281ft	6562ft	9843ft 13,124ft 19,685ft	

Northern India, Pakistan & Bangladesh

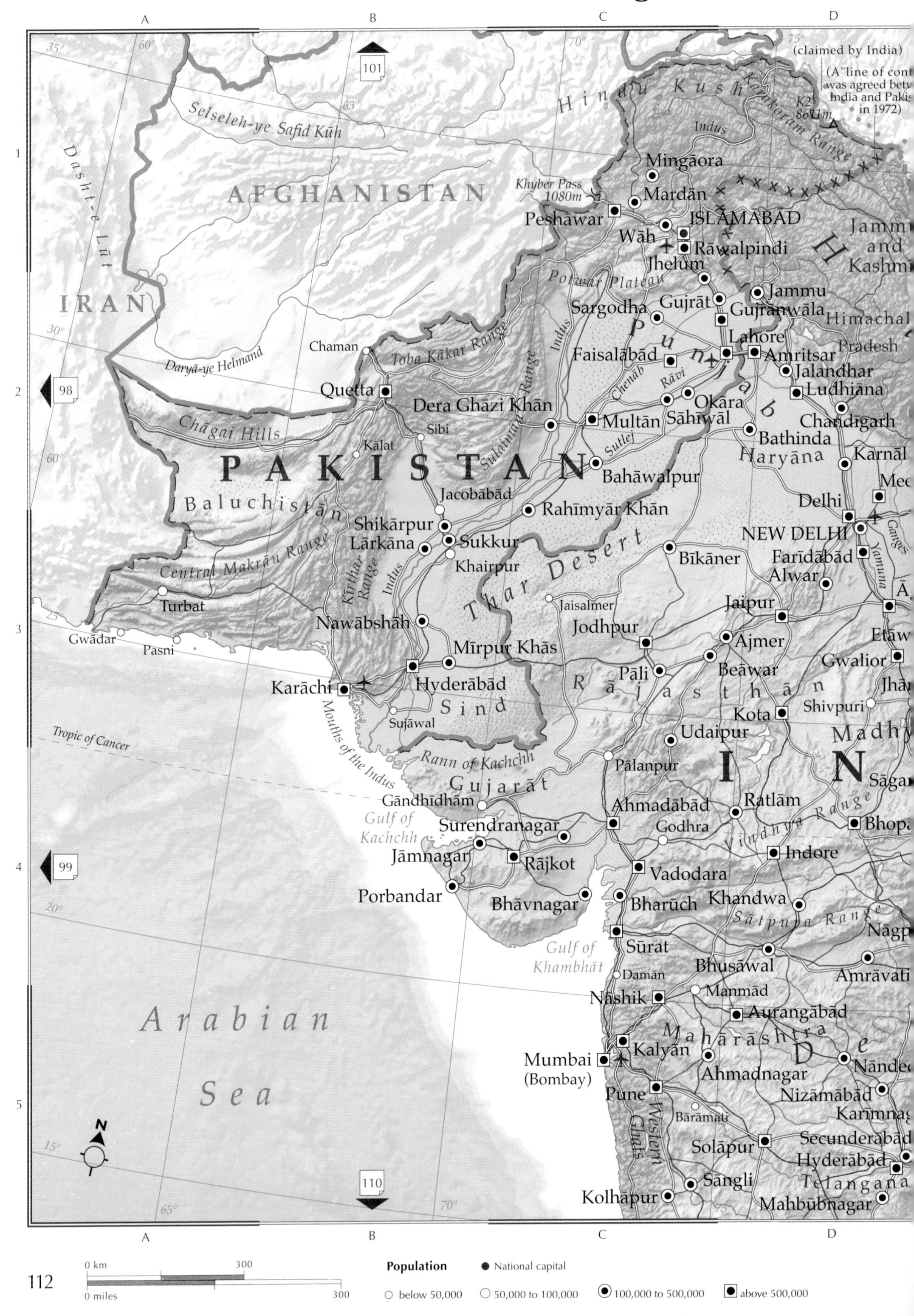

(claimed by India)

(A "line of cont
was agreed betv
India and Pakis
in 1972)

AFGHANISTAN

Selseleh-ye Safid Kūh

Dasht-e Lūt

Hindu Kush

Karakoram Range

K2
8611m

Indus

Mingāora

Khyber Pass
1080m

Mardān

Peshāwar

Wāh

ISLĀMĀBĀD

Rāwalpindi

Jamm
and
Kashm

Jhelum

IRAN

Jammu

Sargodha

Gujrāt

Gujrānwāla

Himachal
Pradesh

Chaman

Toba Kākar Range

Porwar Plateau

Indus

Lahore

Chenāb

Faisalābād

Amritsar

Jalandhar

Daryā-ye Helmand

Range

Rāvi

Ludhiāna

Quetta

Dera Ghāzi Khān

Multān

Okāra

Sāhīwāl

Chandīgarh

Chāgai Hills

Sibi

P A K I S T A N

Sutlej

Bathinda

Haryāna

Karnāl

Kalat

Bahāwalpur

Mee

Baluchistān

Jacobābād

Rahīmyār Khān

Delhi

Central Makrān Range

Shikārpur

Lārkāna

Sukkur

Bīkāner

NEW DELHI

Farīdābād

Ā

Kirthar Range

Khairpur

Thar Desert

Alwar

Turbat

Indus

Jaisalmer

Jaipur

Etaw

Gwādar

Pasni

Nawābshāh

Jodhpur

Ajmer

Gwalior

Jha

Mīrpur Khās

R

Pāli

Beāwar

Shivpuri

Karāchi

Hyderābād

Sind

ā

j

a

s

t

h

ā

n

Kota

Tropic of Cancer

Sujāwal

Rann of Kachchh

Udaipur

Madhy

Mouths of the Indus

Pālanpur

I

N

Gāndhīdhām

Gujarāt

Ahmadābād

Ratlām

Sāga

Gulf of
Kachchh

Surendranagar

Godhra

Bhopa

Vindhya Range

Jāmnagar

Rājkot

Indore

Vadodara

Khandwa

Porbandar

Bhāvnagar

Bharūch

Sātpura Range

Nāgp

Sūrat

Bhusāwal

Amrāvati

Gulf of
Khambhāt

Daman

Manmād

Nāshik

Aurangābād

A r a b i a n

Kalyān

M a h ā r ā s h t r a

D

e

S e a

Mumbai
(Bombay)

Ahmadnagar

Nānded

Pune

Nizāmābād

Western Ghāts

Bārāmati

Karimnag

Solāpur

Secunderābād

Hyderābād

Sangli

Telangāna

Kolhāpur

Mahbūbnagar

0 km   300

0 miles   300

Population   ● National capital

○ below 50,000   ○ 50,000 to 100,000   ◉ 100,000 to 500,000   ■ above 500,000

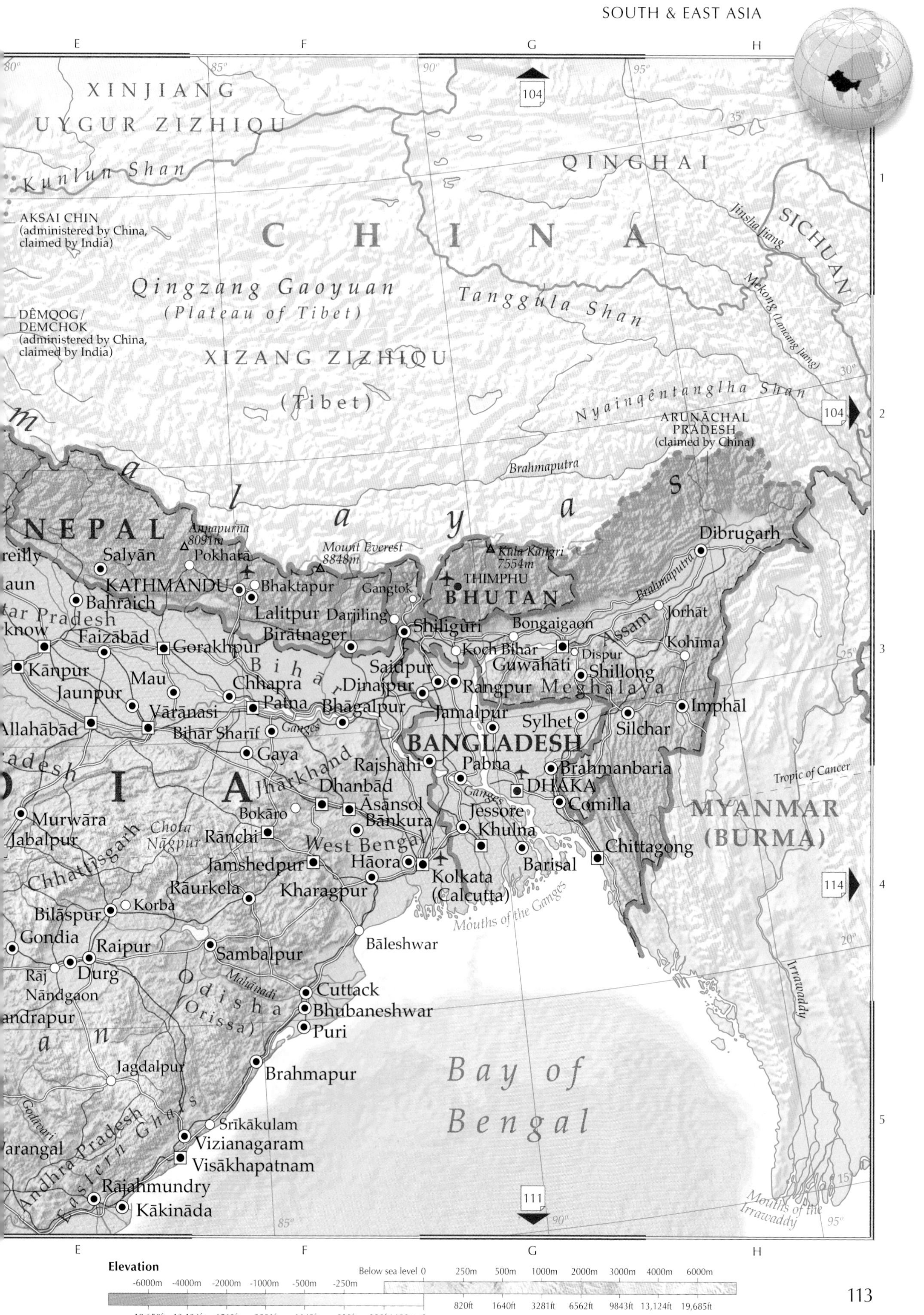

80°   85°   90°   95°

104

XINJIANG
UYGUR ZIZHIQU
Kunlun Shan

QINGHAI

35°

AKSAI CHIN
(administered by China,
claimed by India)

C H I N A

Qingzang Gaoyuan
(Plateau of Tibet)

Tanggula Shan

Jinsha Jiang

SICHUAN

1

DÊMQOG /
DEMCHOK
(administered by China,
claimed by India)

XIZANG ZIZHIQU
(Tibet)

Mekong (Lancang Jiang)

30°

104

2

Nyainqêntanglha Shan

ARUNÁCHAL
PRADESH
(claimed by China)

m   Brahmaputra

a   s

l   a   y

Annapurna
8091m   Mount Everest
8848m   Kula Kangri
7554m

Dibrugarh

NEPAL   Salyān   Pokhara

Brahmaputra

reilly   KATHMANDU   Bhaktapur   Gangtok   THIMPHU

aun   Bahraich   Lalitpur Darjiling   Shiliguri   BHUTAN   Bongaigaon   Jorhat

ar Pradesh   Biratnager   Koch Bihar   Assam   Kohima

know   Faizābād   Gorakhpur   Saidpur   Guwahāti   Dispur   25°   3

Kānpur   Mau   B i h a   Dinajpur   Rangpur   Meghālaya   Shillong   Imphāl

Jaunpur   Chhapra   Bhagalpur   Jamalpur   Sylhet

Vāranasi   Patna   Silchar

Allahābād   Bihar Sharif   Ganges   BANGLADESH

I   Bihar Sharif   Gaya   Rajshahi   Pabna   Brahmanbaria

D   Jharkhand   Dhanbād   Ganges   DHAKA   Comilla   Tropic of Cancer

Murwāra   Bokāro   Āsānsol   Jessore   MYANMAR

Jabalpur   Chota Nagpur   Rānchi   Bankura   Khulna   (BURMA)

Bilāspur   Korba   Jamshedpur   Hāora   Barisal   Chittagong

Gondia   Rāurkela   Kharagpur   Kolkata
(Calcutta)   114   4

Raipur   Sambalpur   Bāleshwar   Mouths of the Ganges

Rāj   Durg   20°

Nāndgaon   Mahānadi   Cuttack

andrapur   Odisha
(Orissa)   Bhubaneshwar

a   n   Puri

Jagdalpur   Brahmapur   B a y o f

B e n g a l   15°

arangal   Srīkākulam   Irrawaddy

Vizianagaram

Eastern Ghats   Visākhapatnam   111

Godavari   Rājahmundry   Mouths of the
Irrawaddy   5

Kākināda   85°   90°   95°

E   F   G   H

Elevation

-6000m -4000m -2000m -1000m -500m -250m   Below sea level 0   250m   500m 1000m 2000m 3000m 4000m 6000m

-19,658ft -13,124ft -6562ft -3281ft -1640ft -820ft -328ft/-100m 0   820ft   1640ft   3281ft   6562ft 9843ft 13,124ft 19,685ft

113

Mainland Southeast Asia

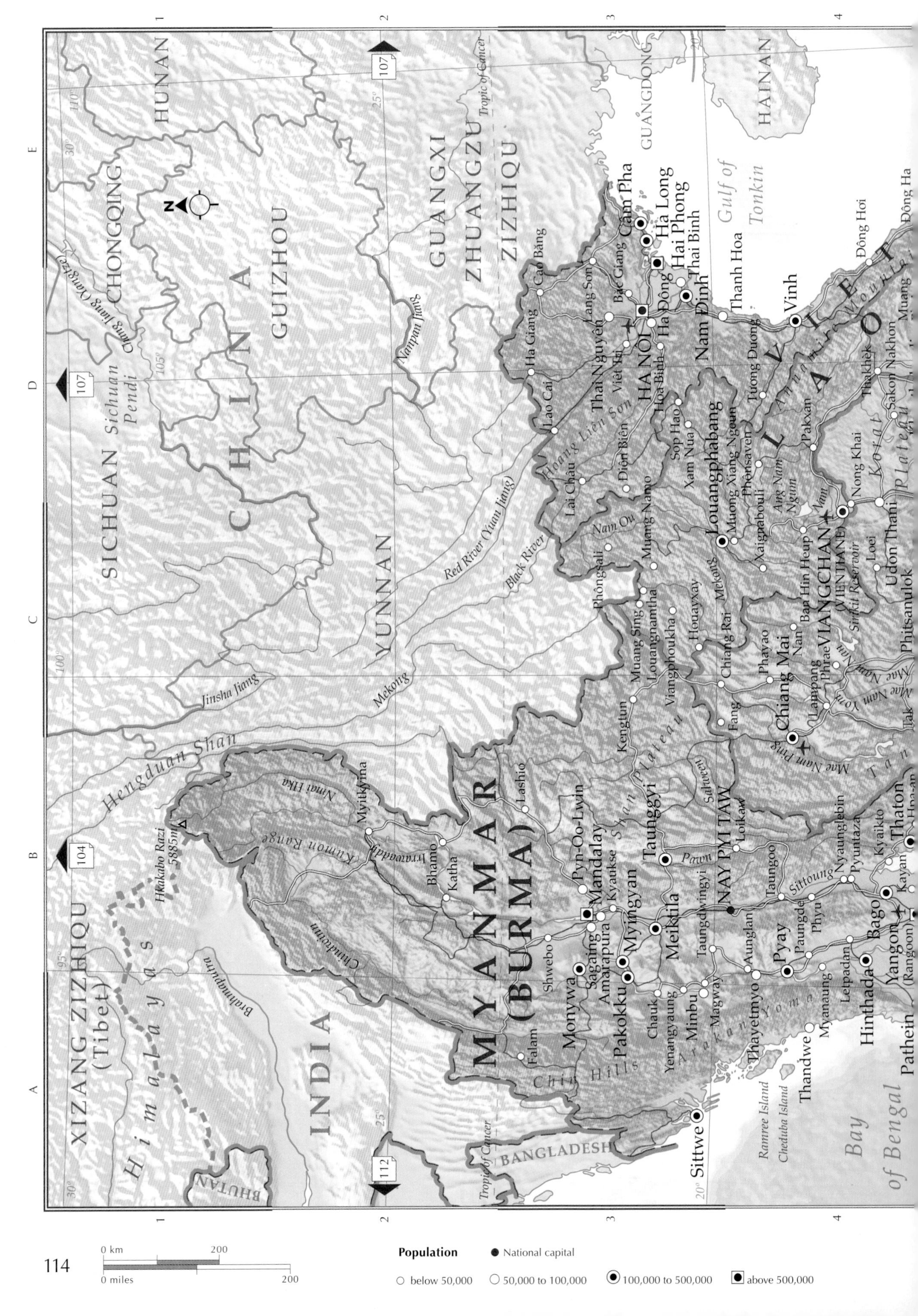

0 km   200
0 miles   200

Population

● National capital

○ below 50,000   ○ 50,000 to 100,000   ◉ 100,000 to 500,000   ■ above 500,000

HUNAN

XIZANG ZIZHIQU (Tibet)

Himalayas

CHONGQING

Sichuan Pendi

SICHUAN

C H I N A

GUIZHOU

GUANGXI ZHUANGZU ZIZHIQU

GUANGDONG

HAINAN

YUNNAN

Tropic of Cancer

Gulf of Tonkin

Jinsha Jiang

Mekong

Hengduan Shan

BHUTAN

INDIA

Brahmaputra

Chindwin

Kumon Range

Nmai Hka

Irrawaddy

M Y A N M A R (B U R M A)

Chin Hills

Arakan Yoma

Pegu Yoma

Salween

Shan Plateau

BANGLADESH

Bay of Bengal

Ramree Island
Cheduba Island

Sittwe

Falam

Monywa
Shwebo
Sagaing
Mandalay
Amarapura
Pakokku
Myingyan
Chauk
Yenangyaung
Minbu
Magway
Myanaung
Thayetmyo
Pyay
Paungde
Phyu
Letpadan
Hinthada
Pathein
Thandwe
Hkakabo Razi 5585m △
Myitkyina
Bhamo
Katha
Lashio
Pyin-Oo-Lwin
Kyaukse
Meiktila
Taungdwingyi
Pawn
Taungoo
NAY PYI TAW
Aunglan
Sittoung
Nyaunglebin
Pyuntaza
Kyaikto
Bago
Yangon (Rangoon)
Kayan
Thaton
Taunggyi
Loikaw
Kengtun
Muang Sing

Red River (Yuan Jiang)
Black River
Nam Ou
Nampan Jiang
Chang Jiang (Yangtze)

Cao Bằng
Hà Giang
Lào Cai
Lai Châu
Điện Biên
Hoàng Liên Sơn
Lạng Sơn
Thai Nguyên
Việt Trì
Bắc Giang
HANOI
Hòa Bình
Hải Dương
Cẩm Pha
Hạ Long
Hải Phòng
Thái Bình
Bắc Ninh
Nam Định
Ninh Bình
Thanh Hoa
Vinh
Đồng Hới
Đông Ha

V I E T N A M

L A O S

Phôngsali
Louangnamtha
Muang Xai
Sop Hao
Xam Nua
Muang Namo
Louangphabang
Muong Xiang Ngeun
Pôngsavan
Tuong Duong
Louangphabang
Houayxay
Viangphoukha
Xaignabouli
Anu Nam Ngum
Nam Ngum Reservoir
Pakxan
VIANGCHAN (VIENTIANE)
Nong Khai
Udon Thani
Loei
Sakon Nakhon
Nakhon

T H A I L A N D

Chiang Rai
Fang
Chiang Mai
Phayao
Nan
Lampang
Phrae
Mae Nam Ping
Mae Nam Nan
Mae Nam Yom
Korat Plateau
Uttaradit
Phitsanulok
Tak
Sukhothai
Sirikit Reservoir
Khorat

95°   100°   105°   110°
30°   25°   20°
Tropic of Cancer

N

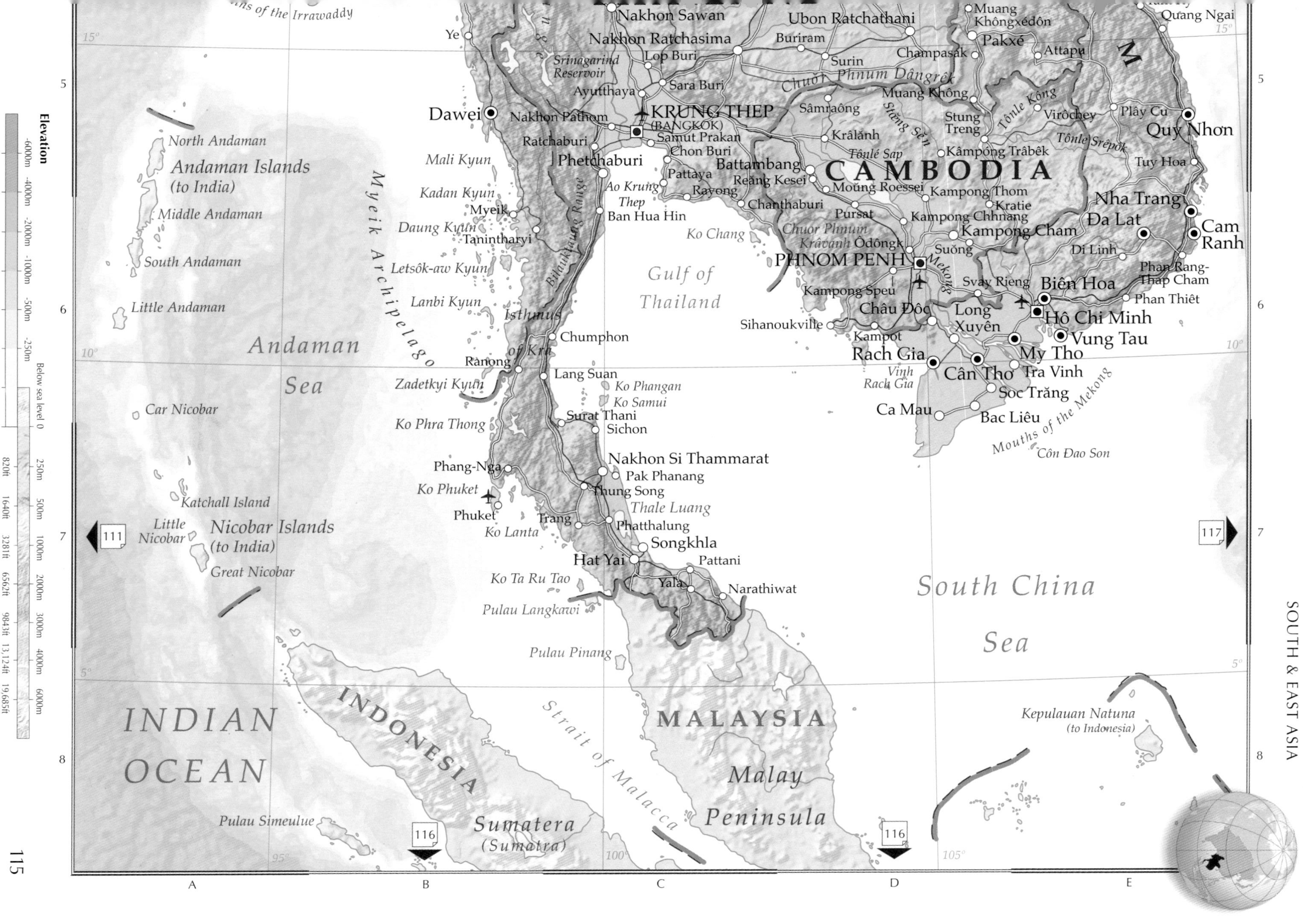

Quang Ngai

Quy Nhon
Play Cu
Tuy Hoa
Cam Ranh
Nha Trang
Da Lat
Di Linh
Phan Rang-
Thap Cham
Phan Thiet

M

Muang
Không xédôn
Pakxé
Attapu
Viròchey
Tônlé Kông
Stung
Treng
Kâmpóng
Trabêk
Kampong
Cham
Krâtie
Biên Hoa
Hô Chi Minh
Vung Tau

Muang Không
Kampong Thom
Suông
Svay Riêng
My Tho
Tra Vinh

CAMBODIA

Stoêng Sên

Kâmpóng Chhnang

Long
Xuyen
Can Tho
Soc Trang

Bac Liêu

Mouths of the Mekong

Côn Dao Son

Champasak
Surin
Muang Không
Kralánh
Moung Roessei
Pursat

Châu Dôc
Rach Gia

Ca Mau

South China

Sea

Ubon Ratchathani
Buriram
Sisaket
Chŏng
Phnum Dângrêk
Chuor Phnum
Krâcanh Odongk
Kampong Speu
Kampot
Vinh
Rach Gia

Kepulauan Natuna
(to Indonesia)

Nakhon Sawan
Nakhon Ratchasima
Lop Buri
Sara Buri
Battambang
Reäng Kesei
Chanthaburi
PHNOM PENH
Mekong
Pursat
Sihanoukville

Ko Chang

KRUNG THEP
(BANGKOK)
Samut Prakan
Pattaya
Rayong

Gulf of

Thailand

MALAYSIA

Malay Peninsula

Ayutthaya
Chon Buri
Ao Krung
Thep
Ban Hua Hin

Nakhon Si Thammarat
Pak Phanang
Chung Song
Thale Luang
Phatthalung
Songkhla
Pattani
Narathiwat

Nakhon Pathom
Ratchaburi
Phetchaburi

Chumphon
Lang Suan
Surat Thani
Sichon

Yala
Hat Yai

Strait of Malacca

Srinagarind
Reservoir

Bilauktaung Range

Kra
Isthmus of Kra

Ranong

Pak Phra Thong

Phang-Nga
Ko Phuket
Phuket
Ko Lanta

Trang

Ko Ta Ru Tao
Pulau Langkawi

Pulau Pinang

SUMATRA

INDONESIA

ns of the Irrawaddy

Ye
Dawei

Mali Kyun
Kadan Kyun
Myeik
Taninthayi
Daung Kyun
Letsôk-aw Kyun
Lambi Kyun

Zadetkyi Kyun

Myeik Archipelago

Andaman Islands
(to India)

North Andaman
Middle Andaman
South Andaman
Little Andaman

Andaman

Sea

Car Nicobar
Katchall Island

Little Nicobar
Great Nicobar

Nicobar Islands
(to India)

Pulau Simeulue

INDIAN

OCEAN

117

116

116

111

Elevation

-6000m	-4000m	-2000m	-1000m	-500m	-250m	
-19,658ft	-13,124ft	-6562ft	-3281ft	-1640ft	-820ft	-328ft/-100m 0

Below sea level 0   250m   500m   1000m   2000m   3000m   4000m   6000m

820ft   1640ft   3281ft   6562ft   9843ft   13,124ft   19,685ft

Maritime Southeast Asia

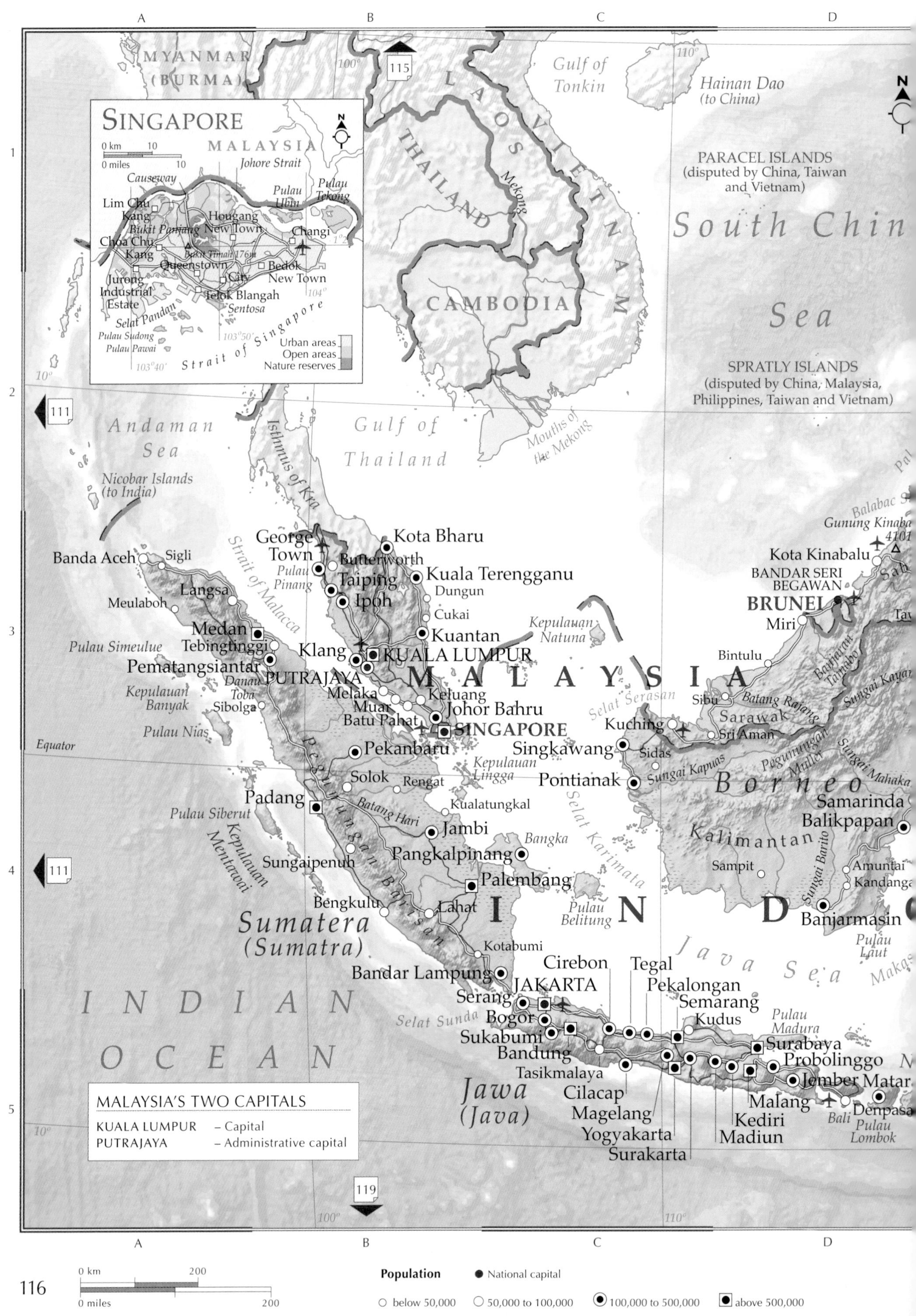

SINGAPORE

0 km 10
0 miles 10

MALAYSIA

Johore Strait
Causeway
Pulau Ubin
Pulau Tekong
Lim Chu Kang
Hougang New Town
Bukit Panjang New Town
Choa Chu Kang
Bukit Timah 176m
Changi
Queenstown
Bedok New Town
Jurong Industrial Estate
Telok Blangah
Sentosa
Selat Pandan
Pulau Sudong
Pulau Pawai
Strait of Singapore

Urban areas
Open areas
Nature reserves

MYANMAR (BURMA)

LAOS
VIETNAM
THAILAND
Mekong
CAMBODIA
Gulf of Tonkin
Hainan Dao (to China)

PARACEL ISLANDS (disputed by China, Taiwan and Vietnam)

South China Sea

SPRATLY ISLANDS (disputed by China, Malaysia, Philippines, Taiwan and Vietnam)

Andaman Sea
Nicobar Islands (to India)
Gulf of Thailand
Mouths of the Mekong

Isthmus of Kra

Banda Aceh
Sigli
George Town
Butterworth
Kota Bharu
Kuala Terengganu
Dungun
Cukai
Taiping
Ipoh
Pulau Pinang
Langsa
Meulaboh
Strait of Malacca
Kuantan
Kepulauan Natuna
Medan
Tebingtinggi
Klang
KUALA LUMPUR
Pematangsiantar
PUTRAJAYA
Pulau Simeulue
Danau Toba
Melaka
Keluang
Kepulauan Banyak
Sibolga
Muar
Batu Pahat
Johor Bahru
SINGAPORE
Pulau Nias
Pekanbaru
Singkawang
Kuching
Selat Serasan
Sibu
Batang Rajang
Sri Aman
Sidas
Bintulu
Kota Kinabalu
BANDAR SERI BEGAWAN
BRUNEI
Miri
Gunung Kinabalu 4101
Sabah

Equator
Solok
Rengat
Pontianak
Sungai Kapus
Kepulauan Lingga
Borneo
Padang
Pulau Siberut
Batang Hari
Kualatungkal
Sungai Kapuas
Samarinda
Balikpapan
Kepulauan Mentawai
Jambi
Bangka
Kalimantan
Pangkalpinang
Selat Karimata
Sampit
Amuntai
Kandanga
Sungaipenuh
Sungai Barito
Bengkulu
Lahat
Palembang
Pulau Belitung
Banjarmasin
Pulau Laut

Sumatera (Sumatra)
Kotabumi
Java Sea
Makas
Bandar Lampung
Cirebon
Tegal
Pekalongan
Semarang
Surabaya
INDIAN
Serang
JAKARTA
Kudus
Pulau Madura
Probolinggo
OCEAN
Selat Sunda
Bogor
Sukabumi
Bandung
Tasikmalaya
Cilacap
Jawa (Java)
Magelang
Yogyakarta
Surakarta
Kediri
Madiun
Malang
Jember
Matar
Bali
Pulau Lombok
Denpasar

MALAYSIA'S TWO CAPITALS
KUALA LUMPUR – Capital
PUTRAJAYA – Administrative capital

0 km 200
0 miles 200

Population ● National capital

○ below 50,000 ○ 50,000 to 100,000 ◉ 100,000 to 500,000 ◼ above 500,000

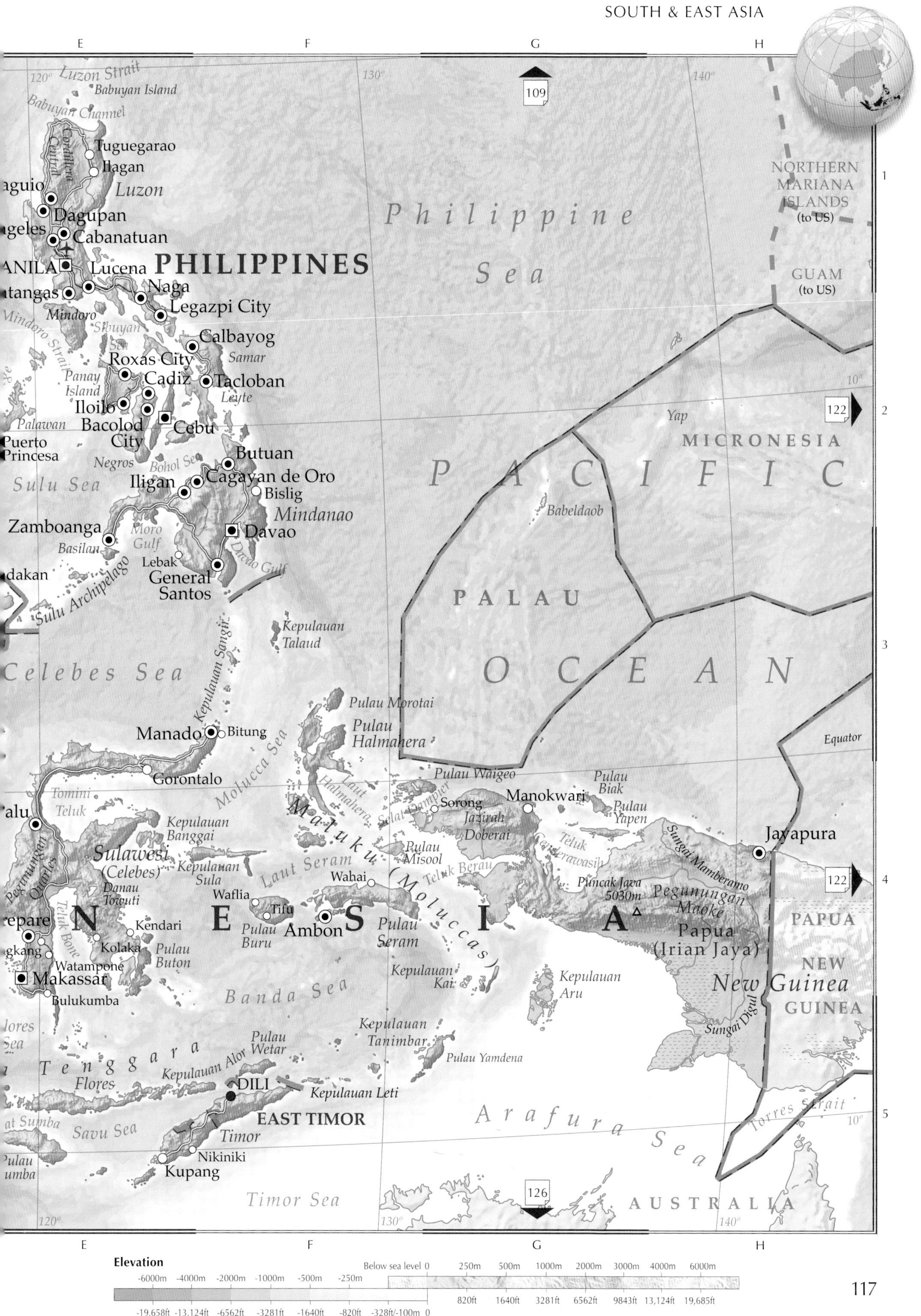

Luzon Strait
120°
Babuyan Island
Babuyan Channel
Tuguegarao
Ilagan
Cordillera
Central
Luzon
aguio
Dagupan
geles
Cabanatuan
ANILA
Lucena
PHILIPPINES
tangas
Naga
Mindoro
Legazpi City
Mindoro Strait
Sibuyan
Calbayog
Roxas City
Sibuyan Sea
Samar
Cadiz
Tacloban
Panay Island
Iloilo
Leyte
Bacolod City
Cebu
Palawan
Negros
Bohol Sea
Butuan
Puerto Princesa
Iligan
Cagayan de Oro
Bislig
Sulu Sea
Zamboanga
Mindanao
Basilan
Moro Gulf
Davao
dakan
Lebak
Davao Gulf
General Santos
Sulu Archipelago
Celebes Sea
Kepulauan Talaud

Philippine Sea

130°

Yap

MICRONESIA

P A C I F I C

Babeldaob

PALAU

O C E A N

Pulau Morotai
Manado
Bitung
Gorontalo
Pulau Halmahera
Kepulauan Sangir
140°
NORTHERN MARIANA ISLANDS (to US)
GUAM (to US)
Equator

Pulau Waigeo
Sorong
Jazirah Doberai
Manokwari
Pulau Biak
Pulau Yapen
Teluk Cenderawasih
Javapura
alu
Tomini Teluk
Kepulauan Banggai
Sulawesi (Celebes)
Kepulauan Sula
Danau Towuti
Waflia
Tifu
Pulau Buru
Wahai
Ambon
Pulau Seram
Laut Seram
Maluku (Moluccas)
Teluk Berau
Pulau Misool
Sungai Mamberamo
Puncak Jaya 5030m
Pegunungan Maoke
Papua (Irian Jaya)
PAPUA NEW GUINEA
epare
N E S I A
Kendari
Kolaka
Pulau Buton
Pulau
Watampone
Makassar
Bulukumba
Banda Sea
Kepulauan Kai
Kepulauan Aru
New Guinea
GUINEA
Sungai Digul
Torres Strait
ores Sea
Tenggara
Flores
Kepulauan Alor
Pulau Wetar
Pulau Yamdena
Kepulauan Tanimbar
Kepulauan Leti
A r a f u r a   S e a
at Sumba
Savu Sea
DILI
EAST TIMOR
Timor
Nikiniki
Kupang
Timor Sea
120°
130°
140°
A U S T R A L I A

Elevation

-6000m	-4000m	-2000m	-1000m	-500m	-250m	Below sea level 0	250m	500m	1000m	2000m	3000m	4000m	6000m

-19,658ft -13,124ft -6562ft -3281ft -1640ft -820ft -328ft/-100m 0
820ft 1640ft 3281ft 6562ft 9843ft 13,124ft 19,685ft

The Indian Ocean

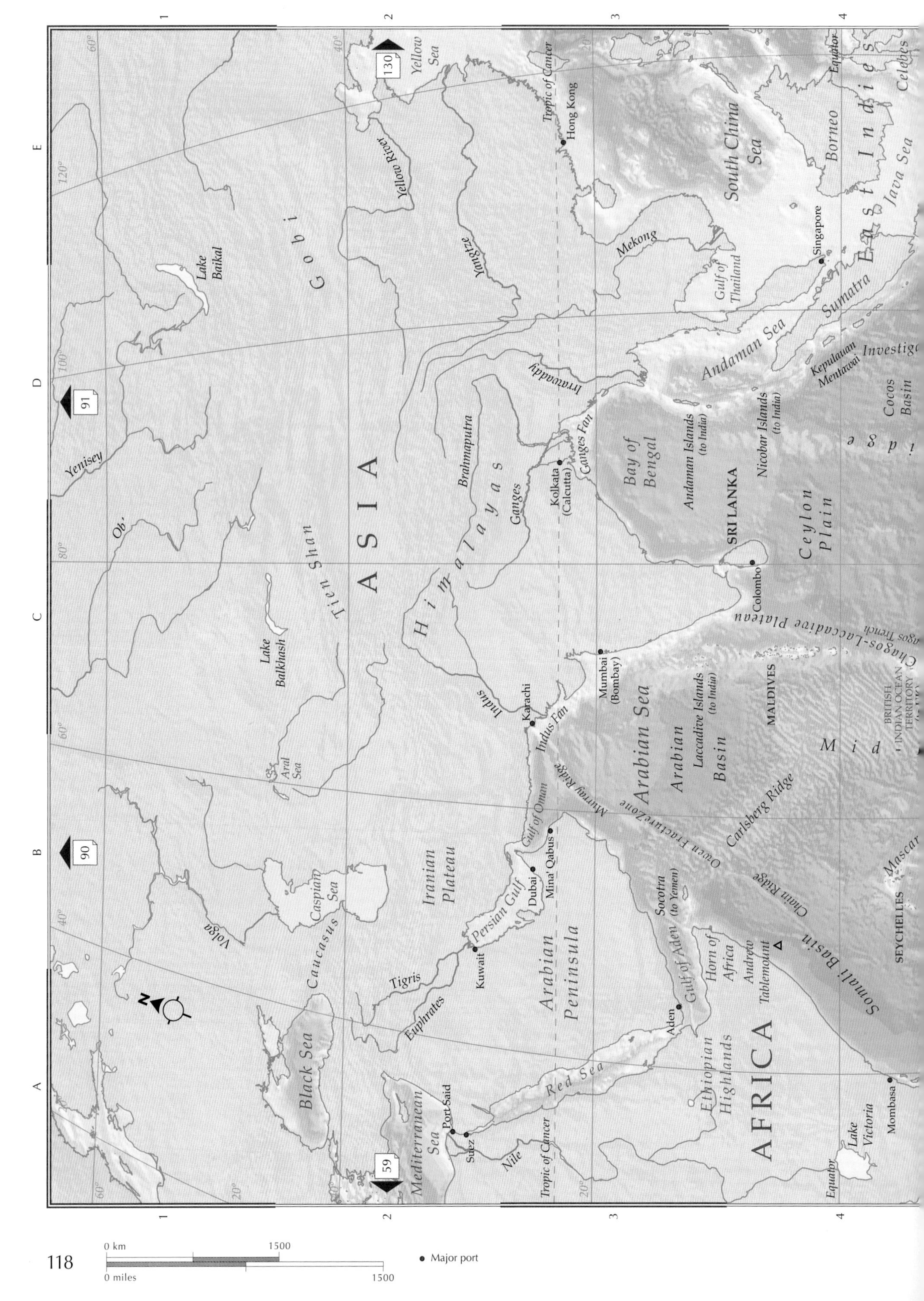

Yellow Sea

130

Tropic of Cancer

Hong Kong

South China Sea

Borneo

Equator

Celebes Sea

East Indies

Java Sea

Yellow River

Yangtze

Mekong

Gulf of Thailand

Sumatra

Singapore

Gobi

Lake Baikal

120°

91

Andaman Sea

Kepulauan Mentawai

Investig

100°

Irrawaddy

Ganges Fan

Bay of Bengal

Andaman Islands (to India)

Nicobar Islands (to India)

Cocos Basin

p i

Yenisey

Brahmaputra

Ganges

Kolkata (Calcutta)

SRI LANKA

Ceylon Plain

Ob'

80°

Tien Shan

A S I A

Himalayas

Colombo

Chagos-Laccadive Plateau

argos Trench

Lake Balkhash

Mumbai (Bombay)

Laccadive Islands (to India)

MALDIVES

BRITISH INDIAN OCEAN TERRITORY (to UK)

Mid

Karachi

Indus

Indus Fan

Arabian Sea

Arabian Basin

Aral Sea

60°

Murray Ridge

Gulf of Oman

Owen Fracturezone

Carlsberg Ridge

Chain Ridge

Mascar

Volga

Caspian Sea

Iranian Plateau

Persian Gulf

Dubai

Mina' Qabus

Somali Basin

SEYCHELLES

40°

Caucasus

Kuwait

Arabian Peninsula

Socotra

Horn of Africa

Andreto Tablemount

Black Sea

Tigris

Euphrates

Gulf of Aden (to Yemen)

Aden

Ethiopian Highlands

AFRICA

Red Sea

Lake Victoria

Mombasa

Mediterranean Sea

Port Said

Suez

Nile

Tropic of Cancer

Equabr

N

90

59

20°

0 km   1500

0 miles   1500

● Major port

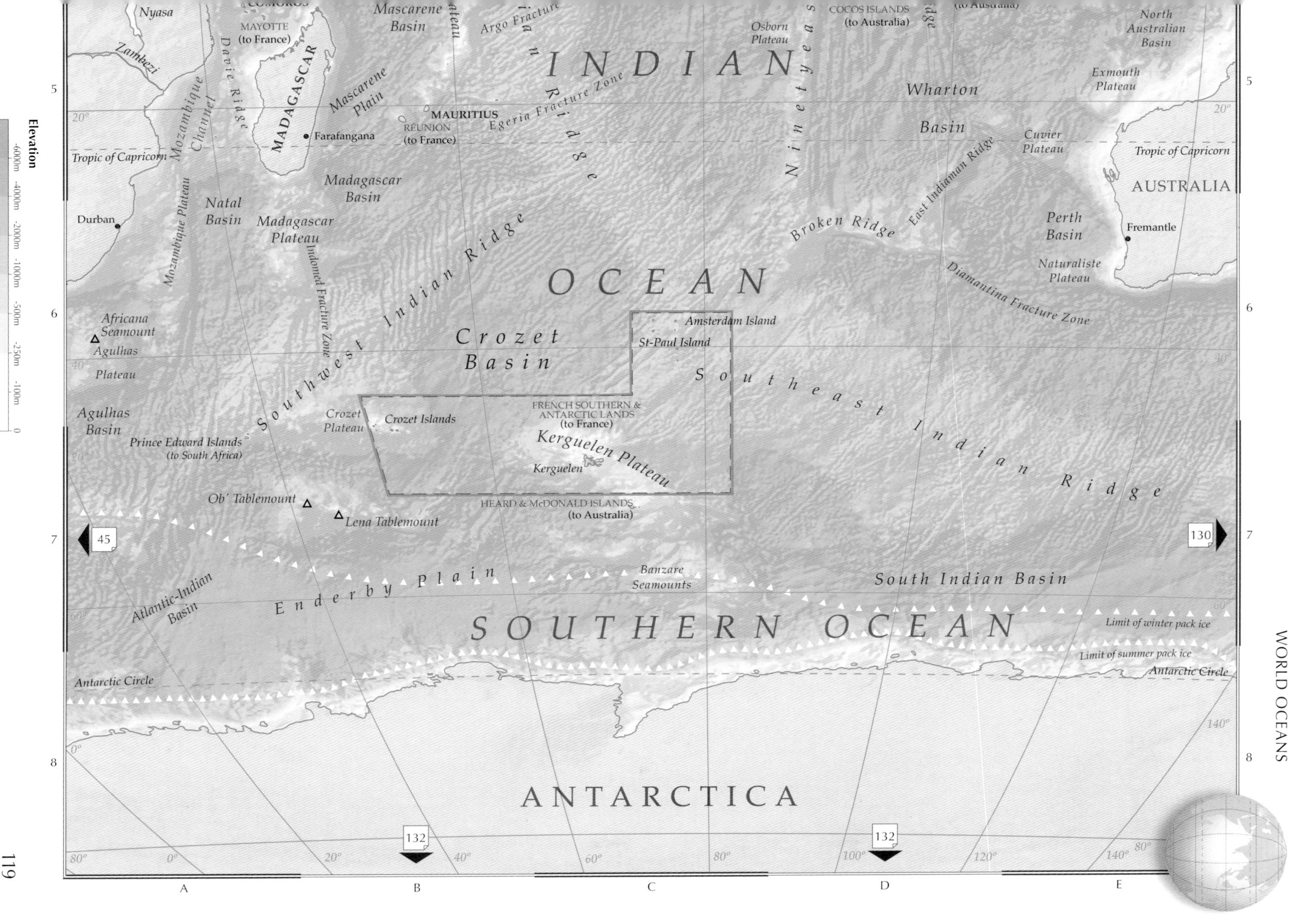

Elevation

-6000m	-4000m	-2000m	-1000m	-500m	-250m	-100m	0
-19,658ft	-13,124ft	-6562ft	-3281ft	-1640ft	-820ft	-328ft/-100m	0

Australasia & Oceania

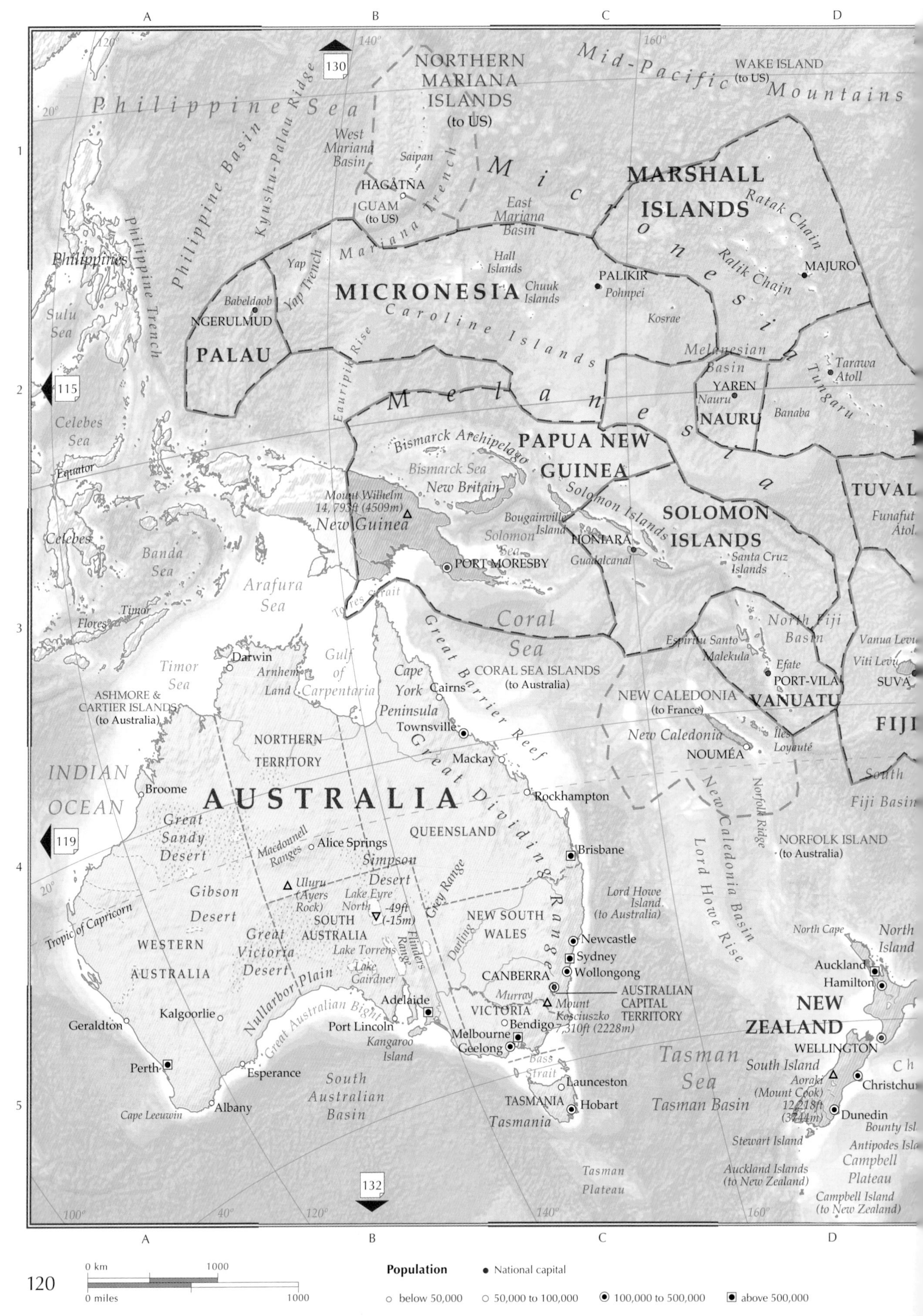

Philippine Sea

Philippine Basin

Kyushu-Palau Ridge

Philippine Trench

Philippines

Sulu Sea

Celebes Sea

Equator

Celebes

Timor

Flores

Banda Sea

Timor Sea

Arafura Sea

INDIAN OCEAN

NORTHERN MARIANA ISLANDS (to US)

West Mariana Basin

Saipan

Mariana Trench

HAGÁTÑA
GUAM (to US)

Yap

Yap Trench

Babeldaob

NGERULMUD

PALAU

East Mariana Basin

Eauripik Rise

MICRONESIA

Caroline Islands

Hall Islands

Chuuk Islands

PALIKIR • *Pohnpei*

Kosrae

Mic *ron* *e* *si* *a*

MARSHALL ISLANDS

Ratak Chain

Ralik Chain

MAJURO

Melanesian Basin

Tarawa Atoll

YAREN
Nauru **NAURU**
Banaba

Tungaru

TUVAL

Funafut Atol.

Mela *n* *e* *s* *i* *a*

Bismarck Archipelago

Bismarck Sea

New Britain

Mount Wilhelm 14,793ft (4509m) △

New Guinea

PAPUA NEW GUINEA

Solomon Islands

SOLOMON ISLANDS

Bougainville Island

Solomon Sea

HONIARA
Guadalcanal

Santa Cruz Islands

North Fiji Basin

Espiritu Santo
Malekula

Efate **PORT-VILA**

VANUATU

Vanua Levu

Viti Levu

SUVA

FIJI

South Fiji Basin

Torres Strait

PORT MORESBY

Coral Sea

CORAL SEA ISLANDS (to Australia)

NEW CALEDONIA (to France)

New Caledonia

NOUMÉA

Îles Loyauté

New Caledonia Ridge

Norfolk Ridge

NORFOLK ISLAND (to Australia)

Lord Howe Basin

Lord Howe Rise

Darwin

Arnhem Land

Gulf of Carpentaria

Cape York Peninsula

Cairns

Great Barrier Reef

Townsville

Mackay

Rockhampton

ASHMORE & CARTIER ISLANDS (to Australia)

NORTHERN TERRITORY

AUSTRALIA

Broome

Great Sandy Desert

Macdonnell Ranges • *Alice Springs*

Simpson Desert

QUEENSLAND

Brisbane

Lord Howe Island (to Australia)

Grey Range

Gibson Desert

Uluru (Ayers Rock) △

Lake Eyre North -49ft (-15m) ▽

SOUTH AUSTRALIA

Newcastle
Sydney
Wollongong

NEW SOUTH WALES

Darling

Murray

Tropic of Capricorn

WESTERN AUSTRALIA

Great Victoria Desert

Lake Torrens

Lake Gairdner

Flinders Range

CANBERRA
AUSTRALIAN CAPITAL TERRITORY

△ *Mount Kosciuszko 7,310ft (2228m)*

Geraldton

Kalgoorlie

Nullarbor Plain

Adelaide

Great Australian Bight

Port Lincoln

Kangaroo Island

VICTORIA

Bendigo

Melbourne
Geelong

Bass Strait

Perth

Esperance

South Australian Basin

Cape Leeuwin

Albany

Launceston

TASMANIA
Hobart

Tasmania

Tasman Sea

Tasman Basin

North Cape

North Island

Hamilton

Auckland

NEW ZEALAND

WELLINGTON

South Island

Aoraki (Mount Cook) 12,218ft (3744m) △

Christchu

Dunedin

Ch

Bounty Isl.

Antipodes Islan

Stewart Island

Campbell Plateau

Auckland Islands (to New Zealand)

Campbell Island (to New Zealand)

Tasman Plateau

Population   • National capital

0 km — 1000
0 miles — 1000

○ below 50,000   ○ 50,000 to 100,000   ◉ 100,000 to 500,000   ◼ above 500,000

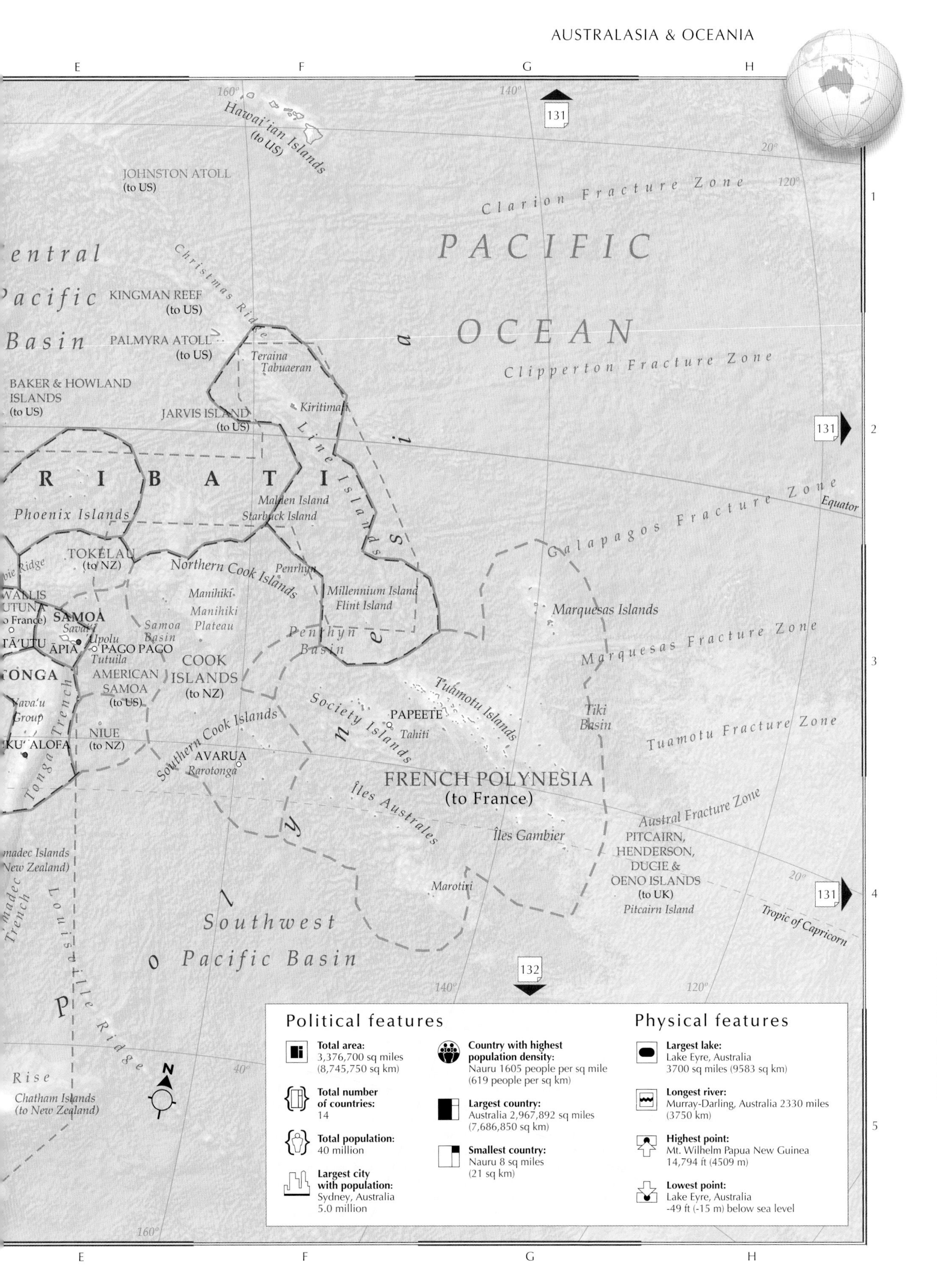

E   F   G   H

160°

Hawai'ian Islands
(to US)

131

20°

JOHNSTON ATOLL
(to US)

120°

Clarion Fracture Zone

1

PACIFIC

entral

Pacific

Christmas Ridge

KINGMAN REEF
(to US)

OCEAN

PALMYRA ATOLL
(to US)

Clipperton Fracture Zone

Basin

Teraina
Tabuaeran

BAKER & HOWLAND
ISLANDS
(to US)

Kiritimati

131

2

JARVIS ISLAND
(to US)

Line Islands

Equator

R I B A T I

Galapagos Fracture Zone

Phoenix Islands

Malden Island
Starbuck Island

Penrhyn

TOKELAU
(to NZ)

Northern Cook Islands

Marquesas Islands

WALLIS
UTUNA
o France)

Manihiki

Millennium Island
Flint Island

Marquesas Fracture Zone

SAMOA
Savai'i

Manihiki
Plateau

3

TĀ'UTU
Upolu
APIA   *Tutuila*

Samoa
Basin

Penrhyn
Basin

Tiki
Basin

PAGO PAGO

TONGA

COOK
ISLANDS
(to NZ)

Tuamotu Islands

Nava'u
Group

AMERICAN
SAMOA
(to US)

Society Islands

PAPEETE

Tuamotu Fracture Zone

KU'ALOFA

NIUE
(to NZ)

Tahiti

Southern Cook Islands

AVARUA
Rarotonga

FRENCH POLYNESIA
(to France)

Îles Australes

Austral Fracture Zone

20°

Íles Gambier

PITCAIRN,
HENDERSON,
DUCIE &
OENO ISLANDS
(to UK)

131

4

madec Islands
New Zealand)

Marotiri

Pitcairn Island

Tropic of Capricorn

Southwest

Pacific Basin

132

140°

120°

Rise

Chatham Islands
(to New Zealand)

N

Political features

Total area:
3,376,700 sq miles
(8,745,750 sq km)

**Total number
of countries:**
14

Total population:
40 million

**Largest city
with population:**
Sydney, Australia
5.0 million

**Country with highest
population density:**
Nauru 1605 people per sq mile
(619 people per sq km)

Largest country:
Australia 2,967,892 sq miles
(7,686,850 sq km)

Smallest country:
Nauru 8 sq miles
(21 sq km)

Physical features

Largest lake:
Lake Eyre, Australia
3700 sq miles (9583 sq km)

Longest river:
Murray-Darling, Australia 2330 miles
(3750 km)

Highest point:
Mt. Wilhelm Papua New Guinea
14,794 ft (4509 m)

Lowest point:
Lake Eyre, Australia
-49 ft (-15 m) below sea level

5

E   F   G   H

The Southwest Pacific

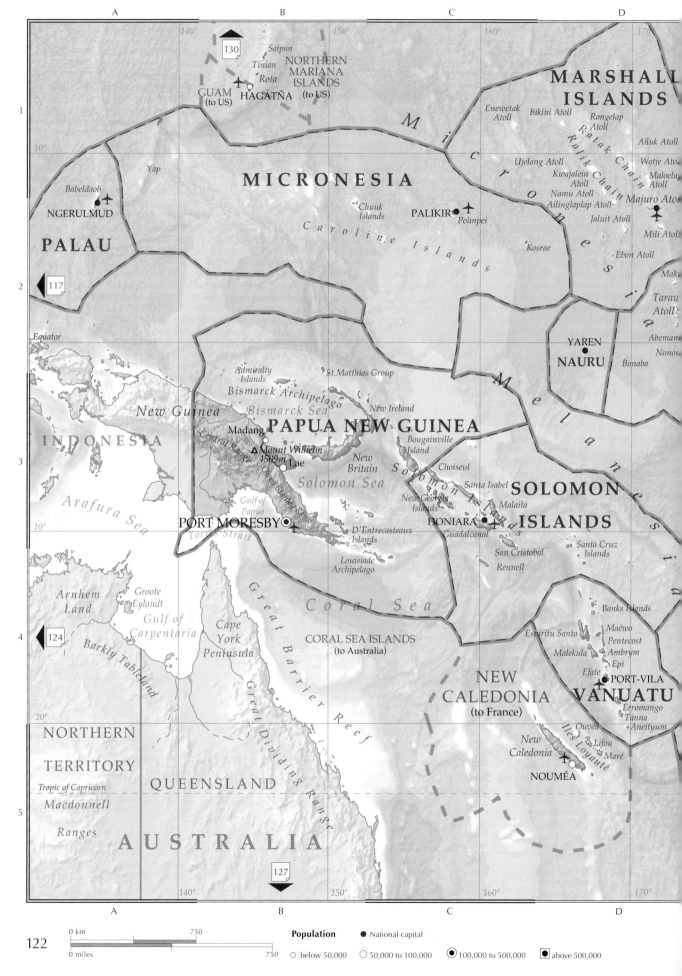

A

B

C

D

130

117

Saipan
Tinian
Rota

NORTHERN
MARIANA
ISLANDS
(to US)

GUAM
(to US)

HAGÅTÑA

MARSHALL
ISLANDS

Enewetak
Atoll

Bikini Atoll

Rongelap
Atoll

Ailuk Atoll

1

10°

Yap

MICRONESIA

Ujelang Atoll

Kwajalein
Atoll

Namu Atoll

Ailinglaplap Atoll

Wotje Ato.

Maloela
Atoll

Majuro Ato

Jaluit Atoll

Mili Atol.

Babeldaob

NGERULMUD

Chuuk
Islands

PALIKIR

Pohnpei

Caroline Islands

Kosrae

Ebon Atoll

Mak

2

PALAU

Tarau
Atoll

Equator

Abemam

YAREN

NAURU

Nonou

Banaba

Admiralty
Islands

St.Matthias Group

New Guinea

Bismarck Archipelago

Bismarck Sea

New Ireland

INDONESIA

Madang

PAPUA NEW GUINEA

Central Range

Mount Wilhelm
4509m

Lae

New
Britain

Bougainville
Island

Choiseul

Santa Isabel

SOLOMON

Queen Stanley Range

Solomon Sea

New Georgia
Islands

Malaita

3

10°

Arafura Sea

PORT MORESBY

Gulf of
Papua

Torres Strait

D'Entrecasteaux
Islands

HONIARA

Guadalcanal

ISLANDS

San Cristobal

Santa Cruz
Islands

Louisiade
Archipelago

Rennell

124

Arnhem
Land

Groote
Eylandt

Gulf of
Carpentaria

Cape
York
Peninsula

Coral Sea

Banks Islands

Espíritu Santo

Maéwo
Pentecost

CORAL SEA ISLANDS
(to Australia)

Malekula

Ambrym
Epi

4

Barkly Tableland

Great Barrier Reef

NEW
CALEDONIA
(to France)

Efate

VANUATU

PORT-VILA

Erromango
Tanna

Ouvéa

Aneityum

20°

Great Dividing Range

New
Caledonia

Îles Loyauté

Lifou

Maré

NORTHERN

TERRITORY

Tropic of Capricorn

QUEENSLAND

NOUMÉA

Macdonnell

5

Ranges

AUSTRALIA

127

A

B

C

D

140°

150°

160°

170°

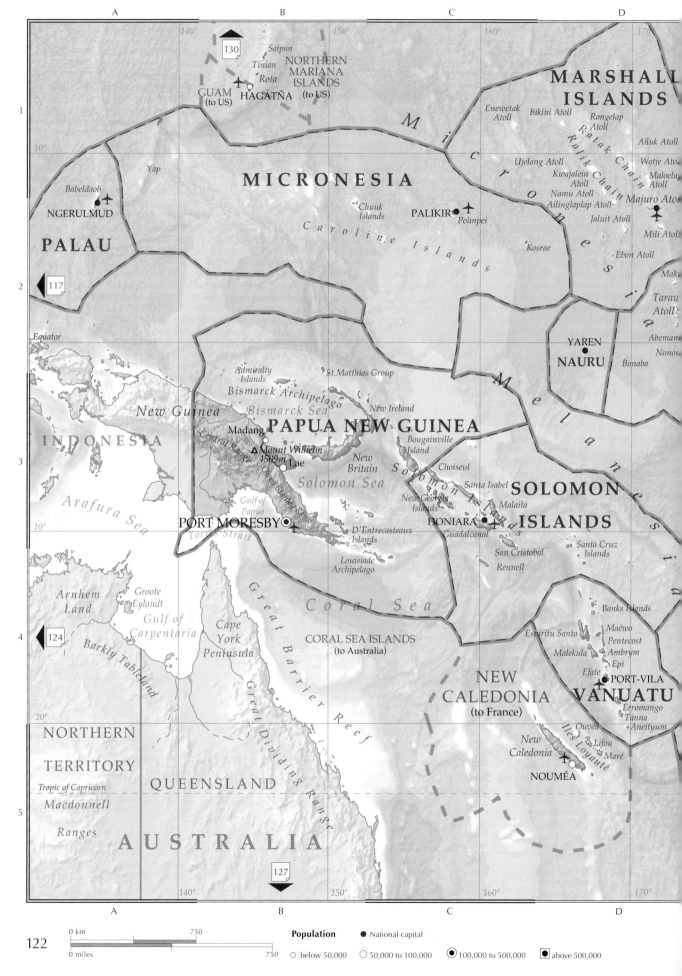

0 km   750

0 miles   750

Population   ● National capital

○ below 50,000   ○ 50,000 to 100,000   ◉ 100,000 to 500,000   ◼ above 500,000

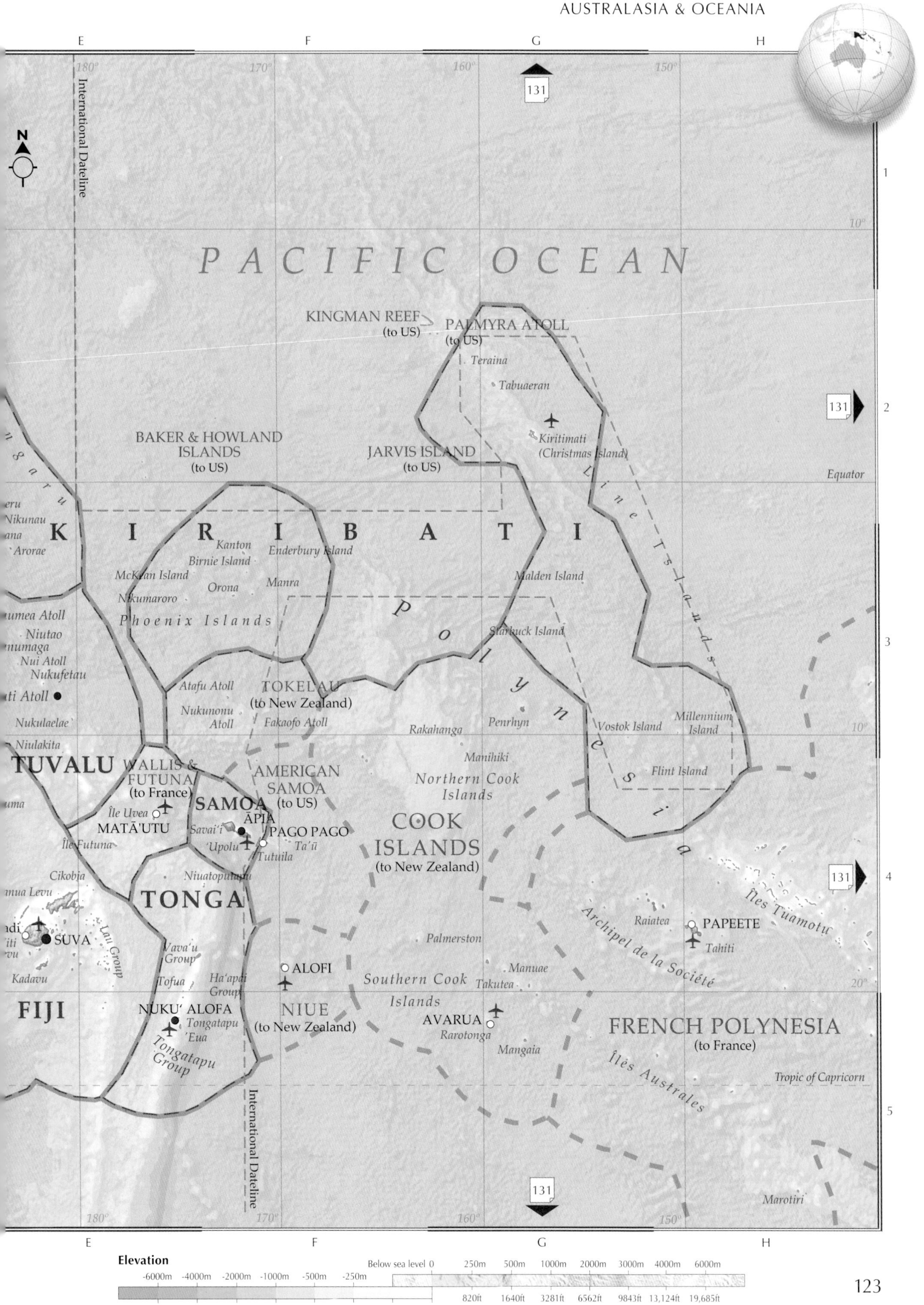

E   180°   F   170°   G   160°   150°   H

131

N

International Dateline

1

10°

PACIFIC OCEAN

131   2

Equator

KINGMAN REEF
(to US)

PALMYRA ATOLL
(to US)

Teraina

Tabuaeran

Kiritimati
(Christmas Island)

BAKER & HOWLAND
ISLANDS
(to US)

JARVIS ISLAND
(to US)

K I R I B A T I

Kanton

Birnie Island

Enderbury Island

McKean Island

Orona

Manra

Malden Island

Nikumaroro

P h o e n i x   I s l a n d s

Starbuck Island

P

o

l

y

n

e

s

i

a

Vostok Island

*Millennium
Island*

3

10°

Niutao

Nui Atoll

Nukufetau

TOKELAU
(to New Zealand)

Atafu Atoll

*Nukunonu
Atoll*

Fakaofo Atoll

Rakahanga

Penrhyn

Flint Island

Nukulaelae

Niulakita

TUVALU

WALLIS &
FUTUNA
(to France)

Île Uvea

MATĀ'UTU

SAMOA

AMERICAN
SAMOA
(to US)

ĀPIA

PAGO PAGO

Manihiki

*Northern Cook
Islands*

COOK
ISLANDS
(to New Zealand)

131   4

Savai'i

Upolu

Ta'ū

Tutuila

Île Futuna

Cikobia

anua Levu

Niuatoputapu

TONGA

*Va'ava'u
Group*

Tofua

Raiatea

PAPEETE

Tahiti

Archipel de la Société

Îles Tuamotu

SUVA

Kadavu

Lau Group

*Ha'apai
Group*

ALOFI

Palmerston

Manuae

Takutea

*Southern Cook
Islands*

20°

FIJI

NUKU'ALOFA

Tongatapu

'Eua

NIUE
(to New Zealand)

AVARUA

Rarotonga

Mangaia

FRENCH POLYNESIA
(to France)

*Tongatapu
Group*

International Dateline

Tropic of Capricorn

Îles Australes

5

131

Marotiri

E   180°   F   170°   G   160°   150°   H

Elevation

| -6000m | -4000m | -2000m | -1000m | -500m | -250m | Below sea level 0 | 250m | 500m | 1000m | 2000m | 3000m | 4000m | 6000m |

| -19,658ft | -13,124ft | -6562ft | -3281ft | -1640ft | -820ft | -328ft/-100m | 0 | 820ft | 1640ft | 3281ft | 6562ft | 9843ft | 13,124ft | 19,685ft |

Western Australia

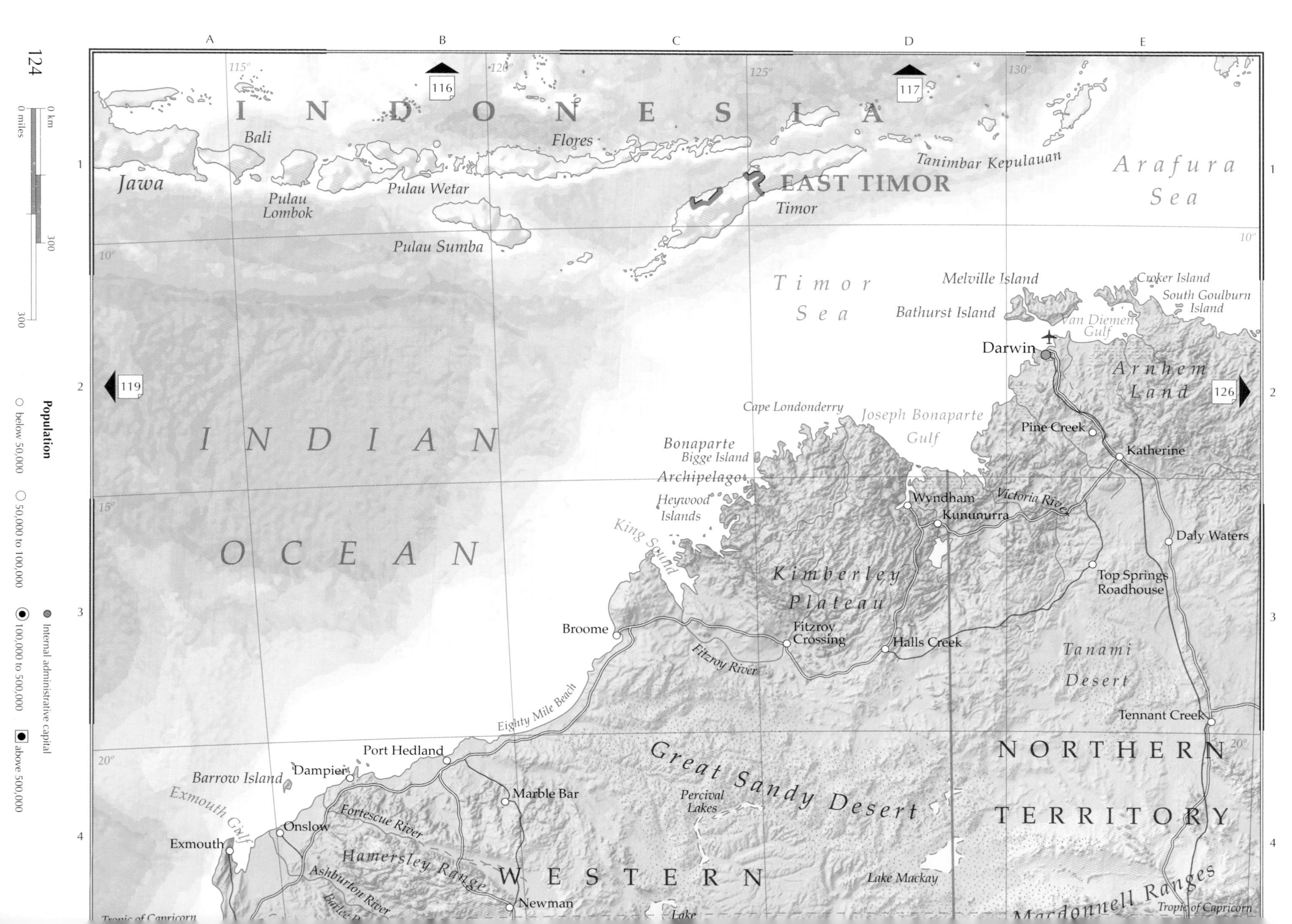

Arafura Sea

Croker Island
South Goulburn Island

Arnhem Land

Katherine

Daly Waters

NORTHERN

TERRITORY

Top Springs Roadhouse

Tennant Creek

Tanami Desert

Macdonnell Ranges

Tropic of Capricorn

Van Diemen Gulf

Pine Creek

Darwin

Melville Island

Bathurst Island

Victoria River

Kununurra

Wyndham

Joseph Bonaparte Gulf

Timor Sea

Timor

EAST TIMOR

INDONESIA

Tanimbar Kepulauan

117

Halls Creek

Kimberley Plateau

Great Sandy Desert

WESTERN

Lake Mackay

Lake

Cape Londonderry

Fitzroy Crossing

Fitzroy River

Perceval Lakes

Bonaparte Archipelago
Bigge Island
Heywood Islands

King Sound

125°

130°

Flores

Pulau Wetar

Pulau Sumba

116

Broome

Eighty Mile Beach

Marble Bar

Newman

Port Hedland

Fortescue River

Hamersley Range

Ashburton River

Battle R.

Bali

Pulau Lombok

Jawa

INDIAN

OCEAN

115°

120°

Dampier

Onslow

Barrow Island

Exmouth Gulf

Exmouth

119

126

Tropic of Capricorn

124

0 km   300

0 miles   300

Population

○ below 50,000   ○ 50,000 to 100,000   ⊙ 100,000 to 500,000   ◉ above 500,000

● Internal administrative capital

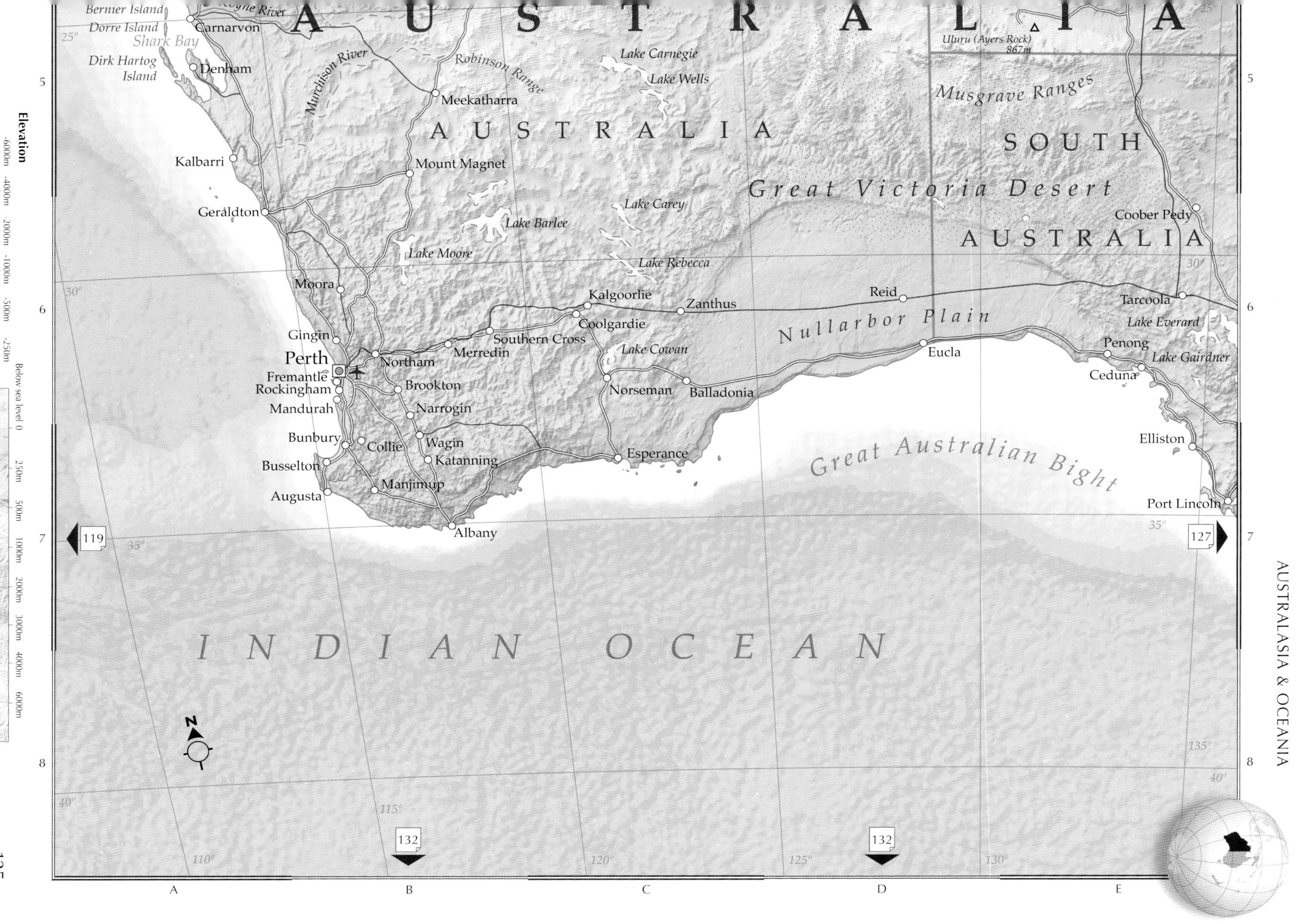

Berrier Island
Dorre Island
Dirk Hartog Island
Shark Bay

AUSTRALIA
Uluru (Ayers Rock) 867m
Musgrave Ranges
SOUTH
AUSTRALIA
Great Victoria Desert

Coober Pedy
Tarcoola
Lake Everard
Penong
Lake Gairdner
Ceduna
Lake Gairdner
Elliston
Port Lincoln

Nullarbor Plain
Reid
Eucla

Great Australian Bight

INDIAN OCEAN

Murchison River
Gascoyne River
Carnarvon
Denham

Robinson Range
Meekatharra
Lake Carnegie
Lake Wells
Lake Carey
Lake Rebecca
Zanthus
Balladonia

Kalbarri
Geraldton
Lake Moore
Lake Barlee
Mount Magnet
Lake Cowan
Kalgoorlie
Coolgardie
Southern Cross
Merredin
Norseman
Esperance

Moora
Gingin
Perth
Fremantle
Rockingham
Mandurah
Bunbury
Busselton
Augusta
Northam
Brookton
Narrogin
Wagin
Collie
Katanning
Manjimup
Albany

AUSTRALIA

Elevation

-6000m	-4000m	-2000m	-1000m	-500m	-250m	Below sea level 0	250m	500m	1000m	2000m	3000m	4000m	6000m
-19,658ft	-13,124ft	-6562ft	-3281ft	-1640ft	-820ft	-328ft/-100m 0	820ft	1640ft	3281ft	6562ft	9843ft	13,124ft	19,685ft

Eastern Australia

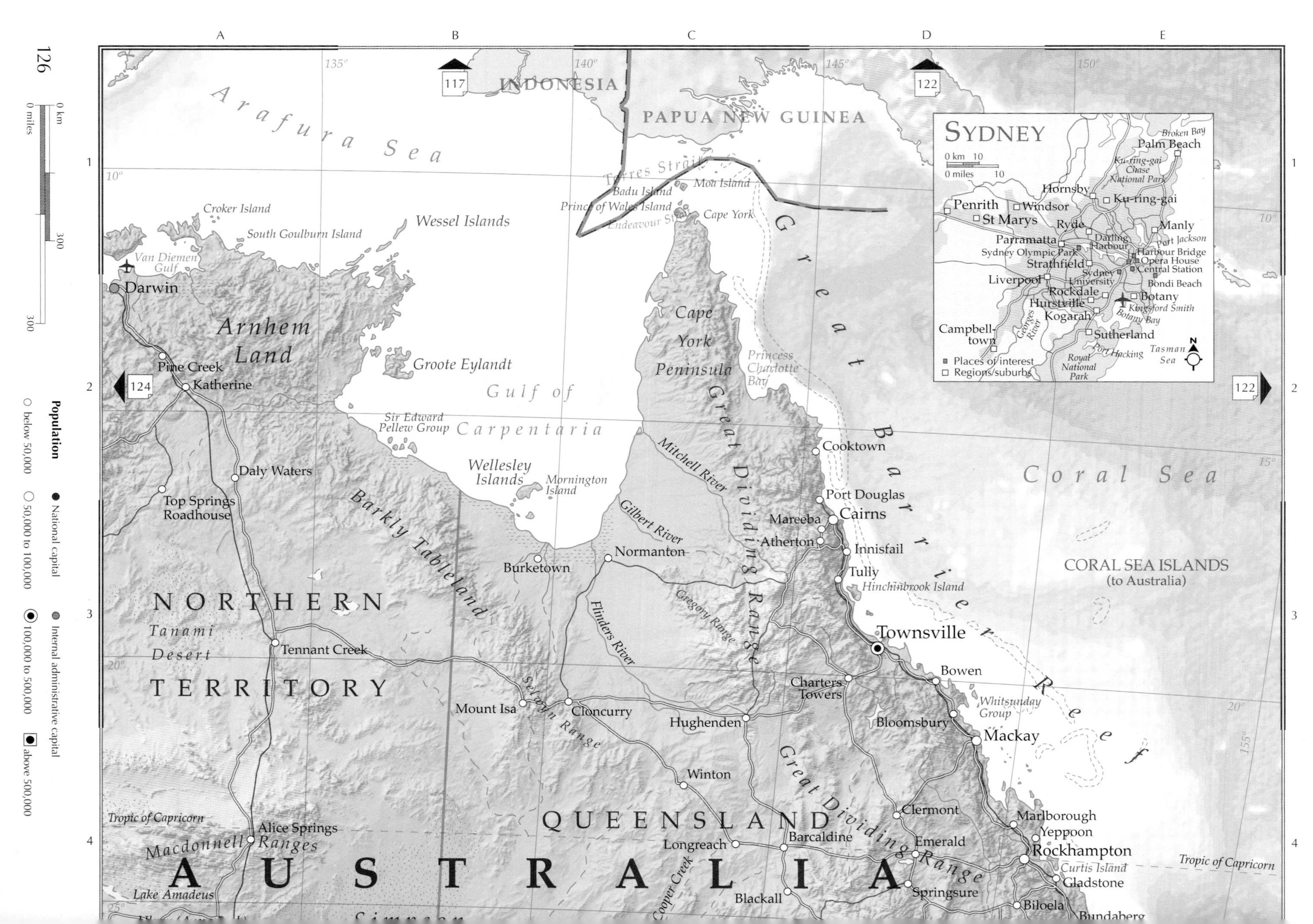

SYDNEY

Broken Bay
Palm Beach
Ku-ring-gai Chase National Park
Manly
Ku-ring-gai
Port Jackson
Harbour Bridge
Opera House
Central Station
Bondi Beach
Hornsby
Windsor
Ryde
Darling Harbour
Parramatta
Sydney Olympic Park
Botany
Sydney
Strathfield
University
Rockdale
Kingsford Smith
Botany Bay
St Marys
Liverpool
Hurstville
Kogarah
Sutherland
Penrith
Port Hacking
Georges River
Campbell town
Royal National Park

Places of interest
Regions/suburbs

0 km 10
0 miles 10

PAPUA NEW GUINEA

INDONESIA

Arafura Sea

Coral Sea

CORAL SEA ISLANDS
(to Australia)

Great Barrier Reef

Tropic of Capricorn

Van Diemen Gulf
Darwin
Croker Island
South Goulburn Island
Arnhem Land
Wessel Islands
Groote Eylandt
Sir Edward Pellew Group
Wellesley Islands
Mornington Island
Gulf of Carpentaria

Pine Creek
Katherine
Daly Waters
Top Springs Roadhouse
Tennant Creek
Tanami Desert
Alice Springs
Macdonnell Ranges
Lake Amadeus
Tropic of Capricorn

NORTHERN TERRITORY

Torres Strait
Prince of Wales Island
Badu Island
Moa Island
Endeavour Strait
Cape York
Cape York Peninsula
Princess Charlotte Bay

Cooktown
Port Douglas
Cairns
Mareeba
Atherton
Innisfail
Tully
Hinchinbrook Island

Mitchell River
Great Dividing Range
Gilbert River
Normanton
Gregory Range
Flinders River
Burketown
Selwyn Range
Mount Isa
Cloncurry
Barkly Tableland
Winton
Longreach
Cooper Creek
Blackall
Barcaldine

Townsville
Bowen
Whitsunday Group
Mackay
Bloomsbury
Charters Towers
Hughenden
Clermont
Emerald
Springsure
Marlborough
Yeppoon
Rockhampton
Curtis Island
Gladstone
Biloela
Bundaberg
Great Dividing Range

QUEENSLAND

AUSTRALIA

Population

● National capital
● Internal administrative capital
○ below 50,000
○ 50,000 to 100,000
◉ 100,000 to 500,000
▣ above 500,000

0 km 300
0 miles 300

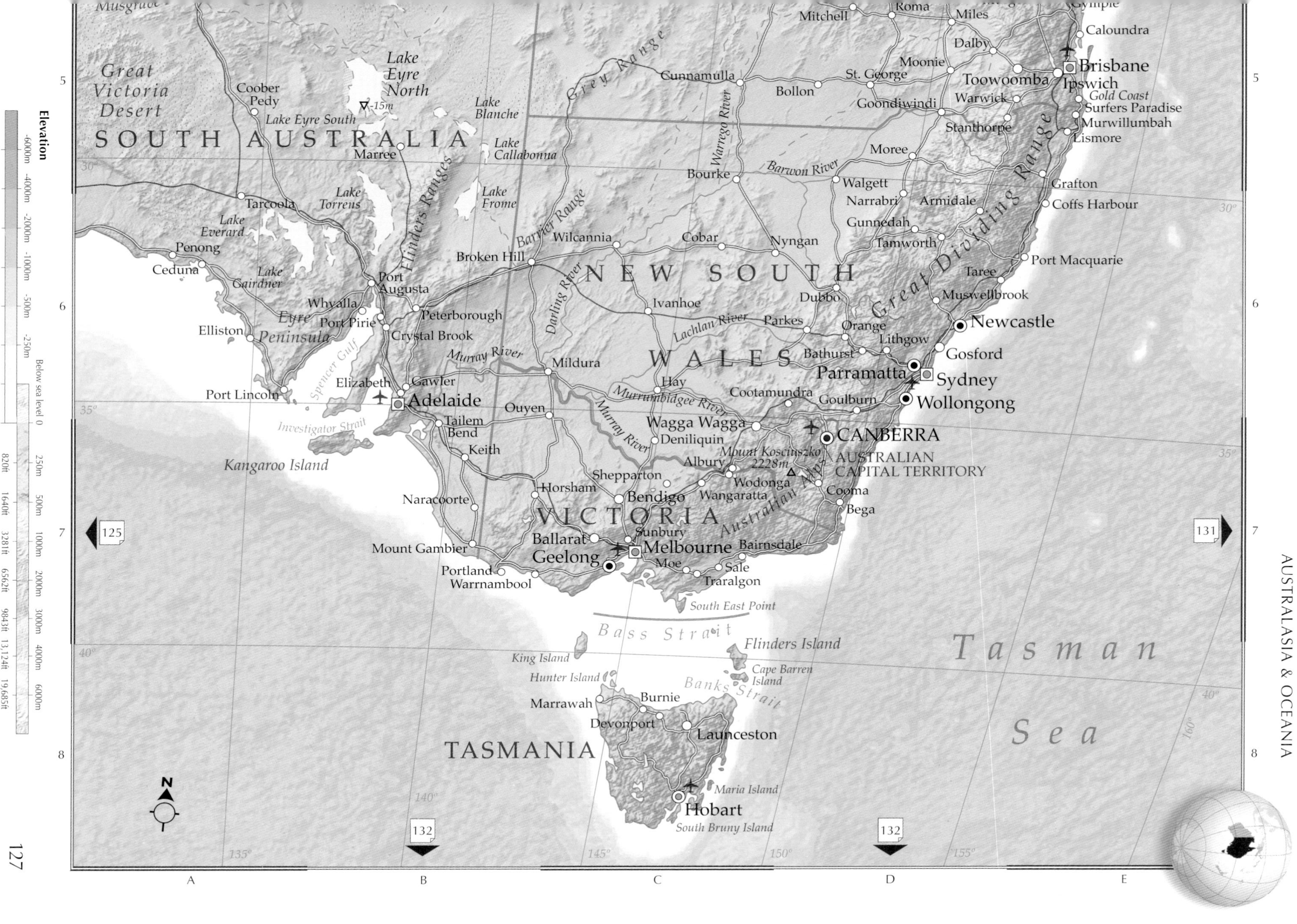

T a s m a n   S e a

132

132

125

AUSTRALASIA & OCEANIA

Caloundra
Brisbane
Ipswich
Gympie
Toowoomba
Warwick
Gold Coast
Surfers Paradise
Murwillumbah
Lismore
Miles
Dalby
Stanthorpe
Coffs Harbour
Roma
Moonie
Goondiwindi
Grafton
Mitchell
St. George
Port Macquarie
Bollon
Moree
Taree
Narrabri
Musswellbrook
Newcastle
Gunnedah
Tamworth
Gosford
Armidale
Walgett
Sydney
Wollongong
Barwon River
Nyngan
Dubbo
Lithgow
Parramatta
Orange
Bathurst
CANBERRA
AUSTRALIAN
CAPITAL TERRITORY
Cunnamulla
Warrego River
Bourke
Parkes
Cootamundra
Goulburn
Cooma
Bega
Cobar
Lachlan River
Mount Kosciuszko
2228m
Wodonga
Wangaratta
Bairnsdale
Ivanhoe
Hay
Wagga Wagga
Albury
Sale
Traralgon
South East Point
Flinders Island
Cape Barren
Island
Murrumbidgee River
Deniliquin
Bendigo
Moe
Melbourne
Banks Strait
Launceston
Maria Island
Hobart
Broken Hill
Barrier Range
Darling River
Mildura
Shepparton
Sunbury
Geelong
South Bruny Island
Lake
Frome
Lake
Callabonna
Horsham
Ballarat
Burnie
Devonport
Lake
Blanche
Peterborough
Ouyen
Bass Strait
Marrawah
Lake
Eyre
North
▽-15m
Marree
Port
Augusta
Crystal Brook
Tailem
Bend
Keith
Naracoorte
King Island
Hunter Island
TASMANIA
Lake
Torrens
Port Pirie
Peterborough
Elizabeth
Adelaide
Gawler
Mount Gambier
Portland
Warrnambool
Lake Eyre South
Whyalla
Cowell
Coober
Pedy
Flinders Ranges
Spencer Gulf
Investigator Strait
Tarcoola
Lake
Gairdner
Eyre
Peninsula
Port Lincoln
Kangaroo Island
Ceduna
Penong
Elliston
Great
Victoria
Desert
Lake
Everard

SOUTH AUSTRALIA

NEW SOUTH WALES

Great Dividing Range

VICTORIA

Elevation

| -6000m | -4000m | -2000m | -1000m | -500m | -250m | Below sea level 0 | 250m | 500m | 1000m | 2000m | 3000m | 4000m | 6000m |

-19,658ft -13,124ft -6562ft -3281ft -1640ft -820ft -328ft/-100m 0   820ft 1640ft 3281ft 6562ft 9843ft 13,124ft 19,685ft

New Zealand

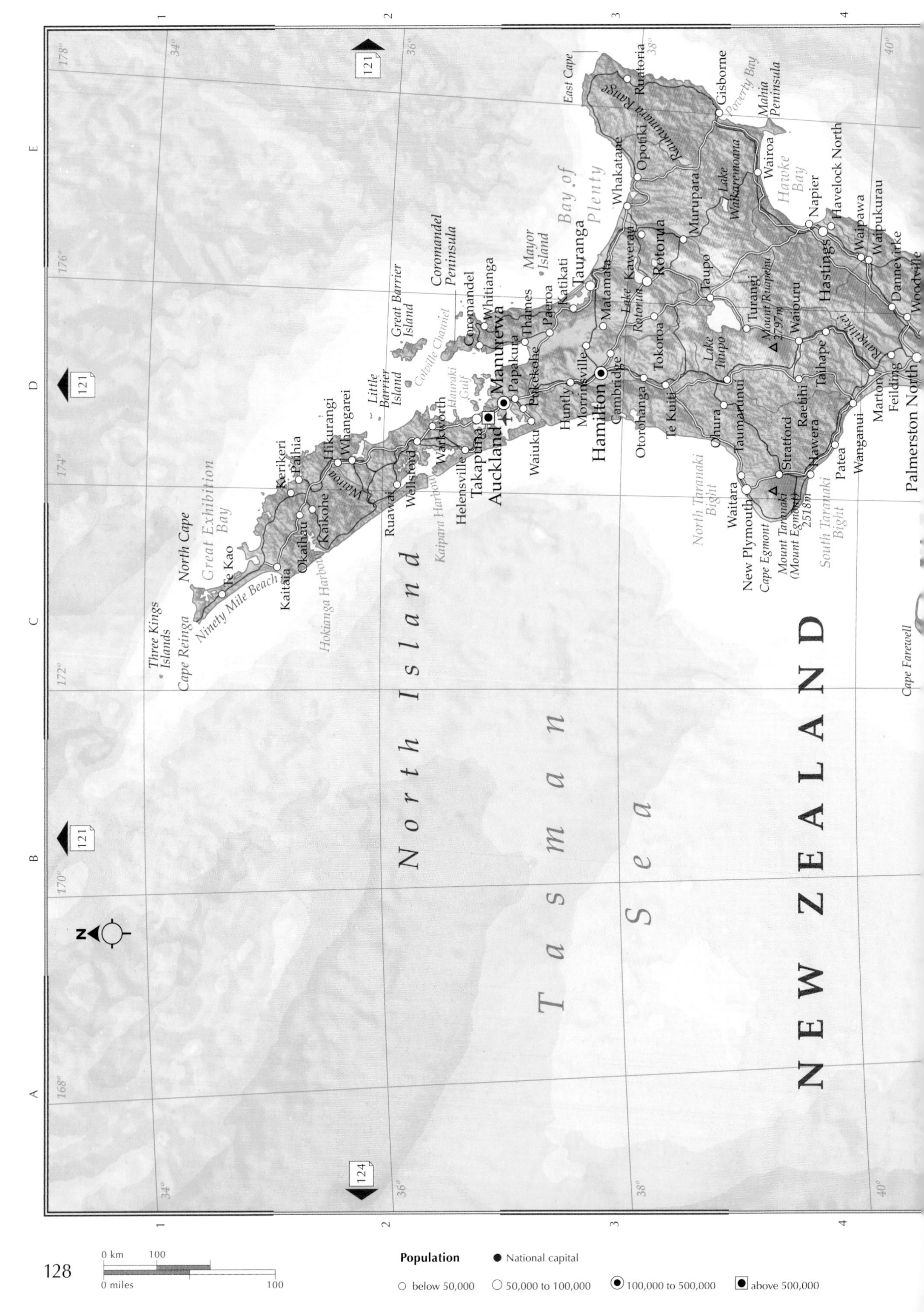

0 km   100

0 miles   100

Population   ● National capital

○ below 50,000   ○ 50,000 to 100,000   ◉ 100,000 to 500,000   ■ above 500,000

Three Kings Islands
Cape Reinga
North Cape
Te Kao
Ninety Mile Beach
Kaitaia
Okaihau
Kaikohe
Kerikeri
Pahia
Hikurangi
Whangarei
Warkworth
Wellsford
Helensville
Ruawai
Kaipara Harbour
Hokianga Harbour
Great Exhibition Bay
Wairoa
Takapuna
Auckland
Manurewa
Papakura
Waiuku
Waiuku
Pukekohe
Huntly
Morrinsville
Cambridge
Hamilton
Otorohanga
Te Kuiti
Ohura
Taumarunui
Waitara
New Plymouth
Cape Egmont
Mount Taranaki
(Mount Egmont)
2518m
Stratford
Hawera
Patea
Wanganui
Marton
Feilding
Palmerston North
Woodville
Danevirke
Waipukurau
Waipawa
Hastings
Havelock North
Napier
Hawke Bay
Wairoa
Mahia Peninsula
Gisborne
Poverty Bay
East Cape
Ruatoria
Raukumara Range
Opotiki
Whakatane
Murupara
Lake Waikaremoana
Taupo
Turangi
Mount Ruapehu
2797m
Waiouru
Raetihi
Tahape
Rangitikei
Rotorua
Lake Rotorua
Kawerau
Lake Rotoiti
Tokoroa
Lake Taupo
Matamata
Tauranga
Bay of Plenty
Katikati
Paeroa
Thames
Coromandel
Whitianga
Mayor Island
Coromandel Peninsula
Great Barrier Island
Little Barrier Island
Hauraki Gulf
Coville Channel

North Island

Tasman Sea

NEW ZEALAND

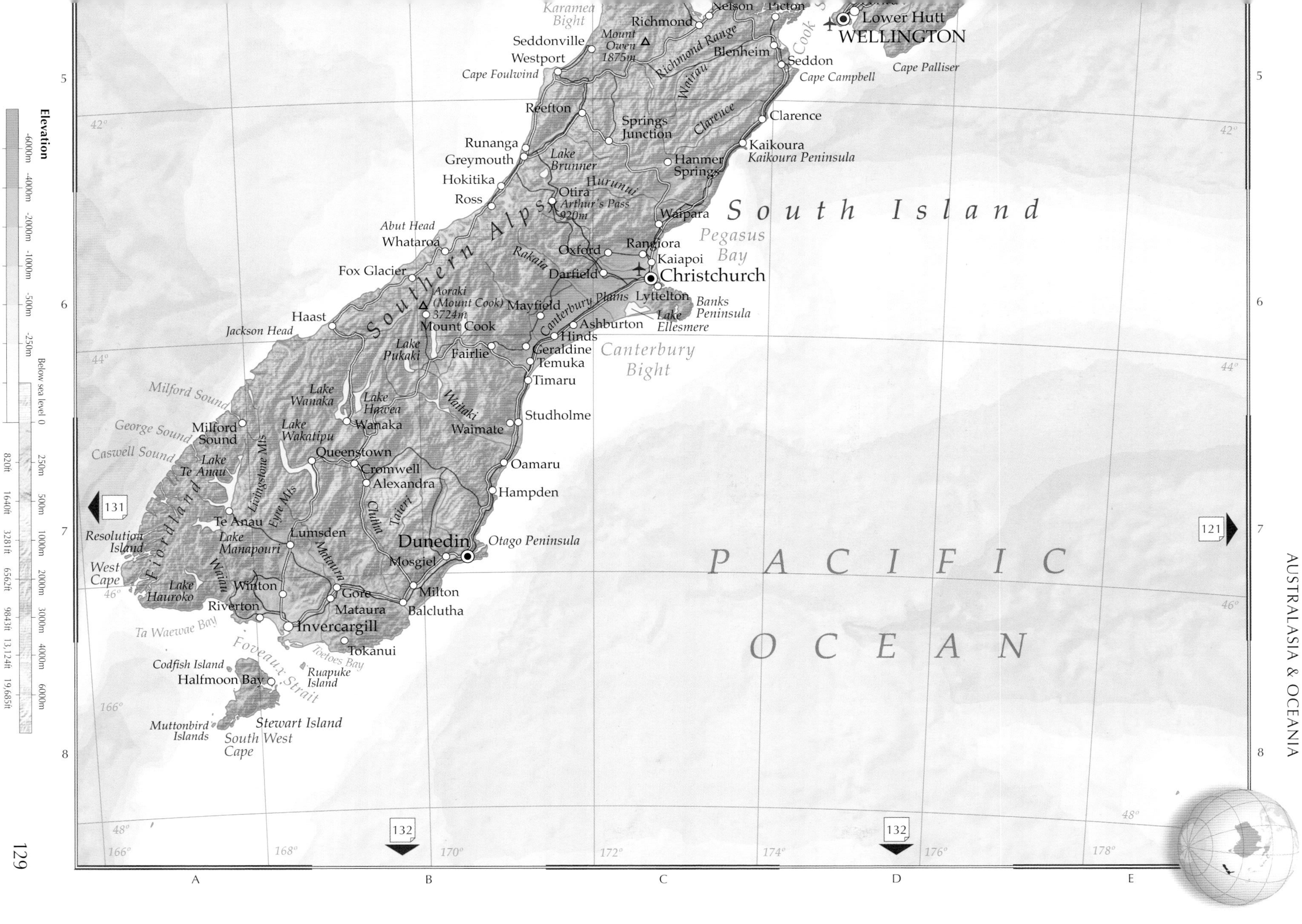

South Island

PACIFIC OCEAN

Lower Hutt
WELLINGTON
Cape Palliser
Cape Campbell
Picton
Seddon
Clarence
Kaikoura
Kaikoura Peninsula
Blenheim
Nelson
Richmond
Mount Owen 1875m
Richmond Range
Clarence
Waiau
Hanmer Springs
Springs Junction
Waipara
Rangiora
Kaiapoi
Christchurch
Lyttelton
Banks Peninsula
Pegasus Bay
Reefton
Lake Brunner
Otira
Arthur's Pass 920m
Hurunui
Oxford
Darfield
Ashburton
Hinds
Lake Ellesmere
Canterbury Bight
Seddonville
Westport
Cape Foulwind
Karamea Bight
Runanga
Greymouth
Hokitika
Ross
Rakaia
Mayfield
Waimakariri Plains
Geraldine
Temuka
Timaru
Studholme
Oamaru
Hampden
Abut Head
Whataroa
Fox Glacier
Aoraki (Mount Cook) 3724m
Mount Cook
Fairlie
Waitaki
Waimate
Jackson Head
Haast
Lake Pukaki
Lake Ohau
Lake Hawea
Wanaka
Cromwell
Alexandra
Clutha
Lumsden
Matauri
Gore
Mataura
Invercargill
Tokanui
Southern Alps
Lake Wanaka
Lake Wakatipu
Queenstown
Eyre Mts
Taieri
Dunedin
Mosgiel
Milton
Balclutha
Otago Peninsula
Milford Sound
George Sound
Caswell Sound
Milford
Sound
Lake Te Anau
Te Anau
Livingstone Mts
Lake Manapouri
Waiau
Winton
Riverton
Lake Hauroko
Fiordland
Resolution Island
West Cape
Codfish Island
Halfmoon Bay
Te Waewae Bay
Foveaux Strait
Muttonbird Islands
Ruapuke Island
Stewart Island
South West Cape

Elevation

| | | | | | | | | | | |
| -6000m | -4000m | -2000m | -1000m | -500m | -250m | | | | | |

Below sea level 0   250m   500m   1000m   2000m   3000m   4000m   6000m

820ft   1640ft   3281ft   6562ft   9843ft   13,124ft   19,685ft

-19,658ft   -13,124ft   -6562ft   -3281ft   -1640ft   -820ft   -328ft/-100m   0

The Pacific Ocean

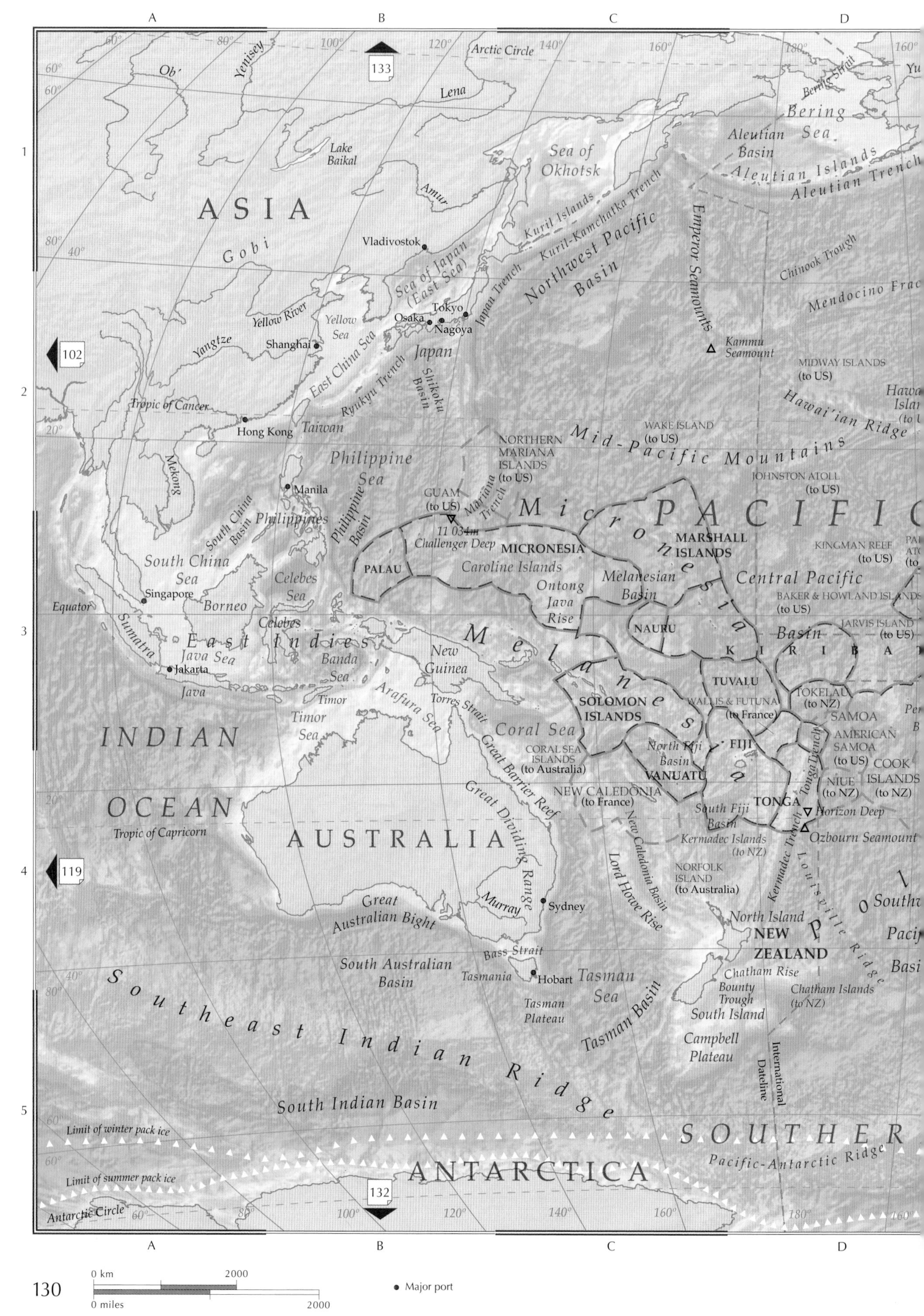

A B C D

133

102

ASIA
Ob'
Yenisey
Lena
Arctic Circle
Lake Baikal
Gobi
Amur
Sea of Okhotsk
Bering Sea
Aleutian Basin
Aleutian Islands
Aleutian Trench
Yu
Bering Strait
Vladivostok
Kuril Islands
Kuril-Kamchatka Trench
Northwest Pacific Basin
Emperor Seamounts
Chinook Trough
Mendocino Frac.
Yellow River
Sea of Japan (East Sea)
Tokyo
Osaka
Nagoya
Japan
Japan Trench
Shikoku Basin
Kammu Seamount
MIDWAY ISLANDS (to US)
Hawa
Yellow Sea
Shanghai
Yangtze
Tropic of Cancer
East China Sea
Ryukyu Trench
Hawaiian Ridge
Isla
(to
Hong Kong
Taiwan
20°
Philippine Sea
NORTHERN MARIANA ISLANDS (to US)
WAKE ISLAND (to US)
Mid-Pacific Mountains
Hawaiian Islan
(to
Mekong
Philippine Basin
GUAM (to US)
Mariana Trench
Micronesia
JOHNSTON ATOLL (to US)
PACIFIC
Manila
Philippines
11 034m Challenger Deep
MICRONESIA
MARSHALL ISLANDS
KINGMAN REEF (to US)
PA
ATO
(to
South China Basin
PALAU
Caroline Islands
Melanesian Basin
Central Pacific
BAKER & HOWLAND ISLANDS (to US)
Singapore
South China Sea
Celebes Sea
Ontong Java Rise
NAURU
Basin
JARVIS ISLAND (to US)
Equator
Borneo
Celebes
Melanesia
KIRIBA
Sumatra
East Indies
New Guinea
TUVALU
TOKELAU (to NZ)
Jakarta
Java Sea
Banda Sea
SOLOMON ISLANDS
WALLIS & FUTUNA (to France)
SAMOA
AMERICAN SAMOA (to US)
Per
Java
Timor
Arafura Sea
Torres Strait
North Fiji Basin
FIJI
COOK
INDIAN
Timor Sea
Coral Sea
CORAL SEA ISLANDS (to Australia)
VANUATU
NIUE (to NZ)
ISLANDS (to NZ)
Great Barrier Reef
NEW CALEDONIA (to France)
New Caledonia Basin
TONGA
Horizon Deep
Tonga Trench
OCEAN
Tropic of Capricorn
AUSTRALIA
South Fiji Basin
Kermadec Islands (to NZ)
Ozbourn Seamount
Kermadec Trench
119
Great Dividing Range
Lord Howe Rise
NORFOLK ISLAND (to Australia)
North Island
Louisville Ridge
P
O
Southe
Paci
Great Australian Bight
Murray
Sydney
NEW ZEALAND
Basi
South Australian Basin
Bass Strait
Tasmania
Hobart
Tasman Sea
Tasman Basin
Chatham Rise
Bounty Trough
Chatham Islands (to NZ)
Tasman Plateau
South Island
Campbell Plateau
International Dateline
Southeast Indian Ridge
South Indian Basin
SOUTHER
Limit of winter pack ice
Pacific-Antarctic Ridge
Limit of summer pack ice
ANTARCTICA
132
Antarctic Circle

A B C D

130

0 km   2000
0 miles   2000

● Major port

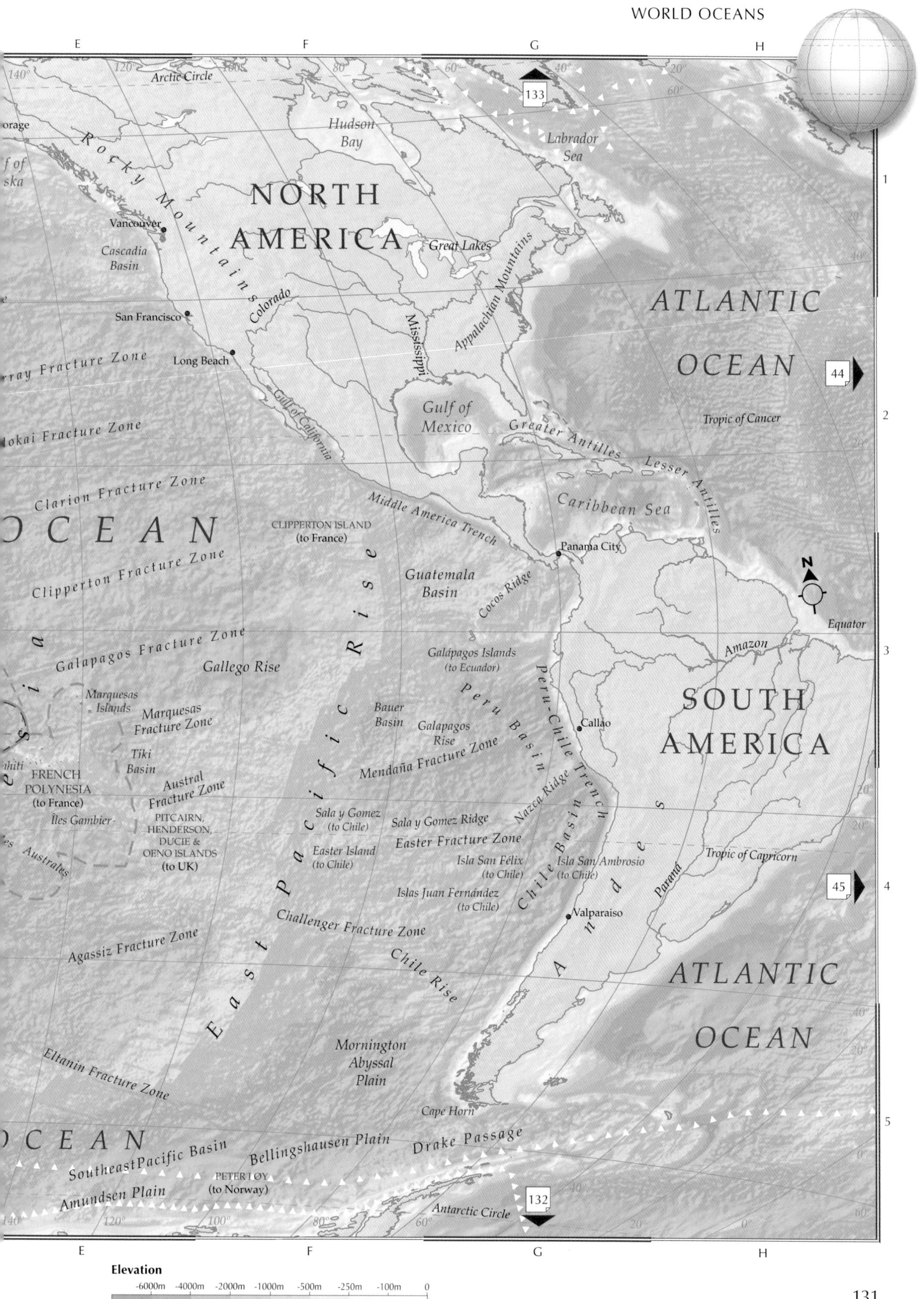

E · F · G · H

140° 120° 100° 80° 60° 40° 20° 0°

Arctic Circle

orage

f of ska

NORTH AMERICA

Hudson Bay

Vancouver

Rocky Mountains

Cascadia Basin

Labrador Sea

ATLANTIC OCEAN

San Francisco

Great Lakes

Colorado

Appalachian Mountains

44

rray Fracture Zone

Long Beach

Mississippi

Gulf of California

Gulf of Mexico

Greater Antilles

Tropic of Cancer

Lesser Antilles

okai Fracture Zone

Clarion Fracture Zone

OCEAN

Middle America Trench

CLIPPERTON ISLAND (to France)

Caribbean Sea

Clipperton Fracture Zone

Guatemala Basin

Panama City

N

Galapagos Fracture Zone

Cocos Ridge

Gallego Rise

Galápagos Islands (to Ecuador)

Equator

Amazon

Marquesas Islands

Marquesas Fracture Zone

Bauer Basin

Galapagos Rise

Peru Basin

SOUTH AMERICA

Callao

Tiki Basin

Mendaña Fracture Zone

FRENCH POLYNESIA (to France)

Austral Fracture Zone

Peru-Chile Trench

Nazca Ridge

Îles Gambier

PITCAIRN, HENDERSON, DUCIE & OENO ISLANDS (to UK)

Sala y Gomez (to Chile)

Sala y Gomez Ridge

Easter Fracture Zone

Easter Island (to Chile)

Isla San Félix (to Chile)

Isla San Ambrosio (to Chile)

Tropic of Capricorn

es Australes

East Pacific Rise

Islas Juan Fernández (to Chile)

Chile Basin

Paraná

45

Andes

Valparaiso

Agassiz Fracture Zone

Challenger Fracture Zone

Chile Rise

ATLANTIC

Mornington Abyssal Plain

Eltanin Fracture Zone

OCEAN

Cape Horn

Drake Passage

5

OCEAN

Southeast Pacific Basin

Bellingshausen Plain

PETER I ØY (to Norway)

Amundsen Plain

Antarctic Circle

140° 120° 100° 80° 60° 20° 0°

E · F · G · H

Elevation

-6000m	-4000m	-2000m	-1000m	-500m	-250m	-100m	0
-19,658ft	-13,124ft	-6562ft	-3281ft	-1640ft	-820ft	-328ft/-100m	0

Antarctica

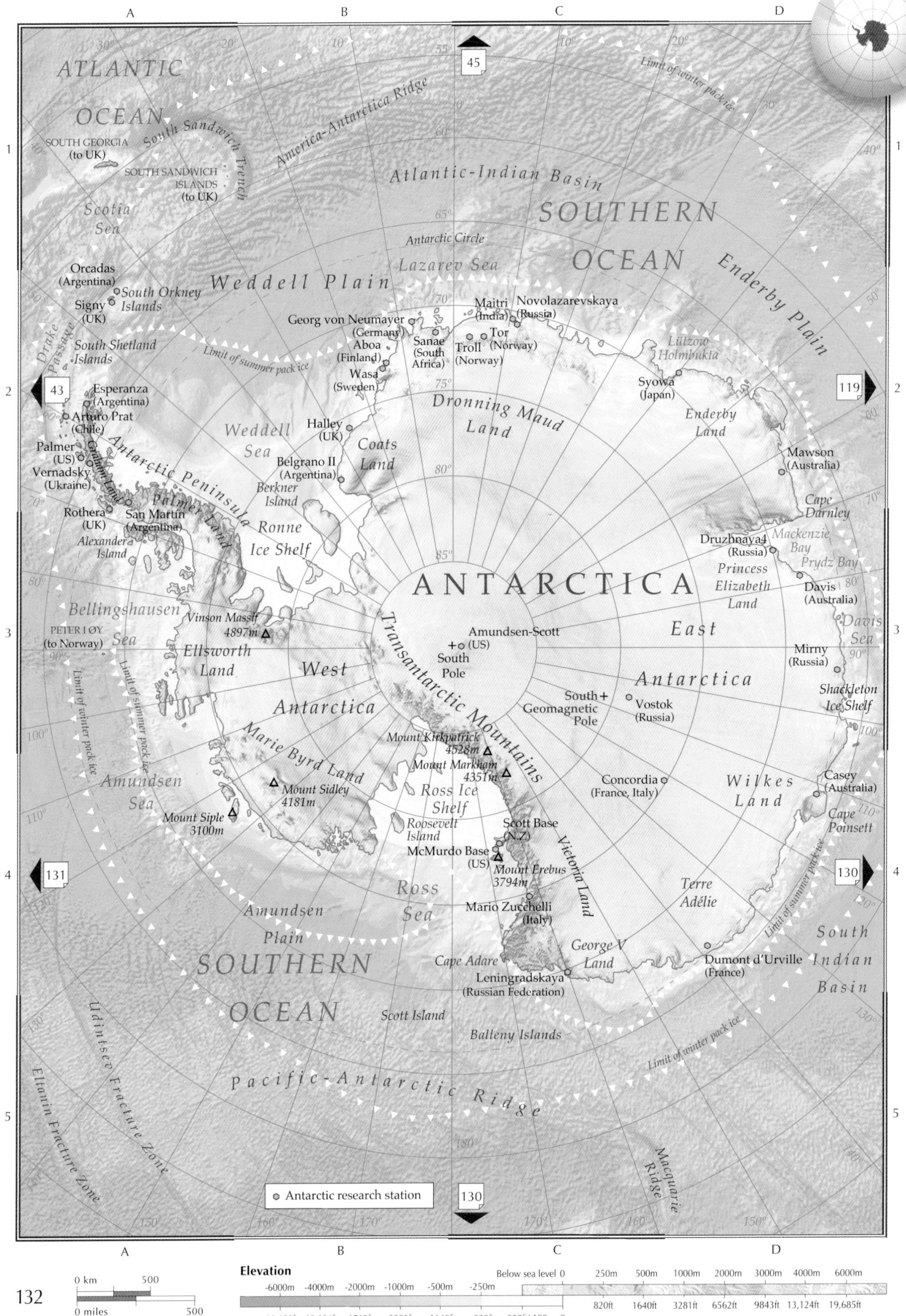

45

ATLANTIC

OCEAN

SOUTH GEORGIA
(to UK)

SOUTH SANDWICH
ISLANDS
(to UK)

South Sandwich Trench

America-Antarctica Ridge

Atlantic-Indian Basin

Scotia
Sea

Antarctic Circle

Lazarev Sea

Weddell Plain

SOUTHERN

OCEAN

Enderby Plain

Limit of winter pack ice

Orcadas
(Argentina)

South Orkney
Islands

Signy
(UK)

South Shetland
Islands

Drake Passage

Limit of summer pack ice

Georg von Neumayer
(Germany)

Aboa
(Finland)

Wasa
(Sweden)

Maitri
(India)

Novolazarevskaya
(Russia)

Sanae
(South
Africa)

Tor
(Norway)

Troll
(Norway)

Syowa
(Japan)

Enderby
Land

43

119

Esperanza
(Argentina)

Arturo Prat
(Chile)

Palmer
(US)

Vernadsky
(Ukraine)

Antarctic Peninsula

Graham Land

Weddell
Sea

Halley
(UK)

Belgrano II
(Argentina)

Berkner
Island

Dronning Maud
Land

Coats
Land

Mawson
(Australia)

Cape
Darnley

Rothera
(UK)

San Martin
(Argentina)

Palmer
Land

Ronne
Ice Shelf

Druzhnaya4
(Russia)

Mackenzie
Bay

Prydz Bay

Alexander
Island

Princess
Elizabeth
Land

Davis
(Australia)

ANTARCTICA

Bellingshausen
Sea

PETER I ØY
(to Norway)

Vinson Massif
4897m

Ellsworth
Land

West

Antarctica

Transantarctic Mountains

Amundsen-Scott
(US)

South
Pole

East

Antarctica

Davis
Sea

Mirny
(Russia)

Limit of winter pack ice

Limit of summer pack ice

South
Geomagnetic
Pole

Vostok
(Russia)

Shackleton
Ice Shelf

Amundsen
Sea

Marie Byrd Land

Mount Sidley
4181m

Mount Siple
3100m

Mount Kirkpatrick
4528m

Mount Markham
4351m

Ross Ice
Shelf

Roosevelt
Island

Scott Base
(N.Z.)

McMurdo Base
(US)

Mount Erebus
3794m

Concordia
(France, Italy)

Wilkes
Land

Casey
(Australia)

Cape
Poinsett

131

130

Amundsen

Plain

SOUTHERN

OCEAN

Ross
Sea

Cape Adare

Leningradskaya
(Russian Federation)

Mario Zucchelli
(Italy)

Victoria Land

George V
Land

Terre
Adélie

Dumont d'Urville
(France)

South
Indian
Basin

Scott Island

Balleny Islands

Udintsev Fracture Zone

Eltanin Fracture Zone

Macquarie
Ridge

Limit of winter pack ice

Pacific-Antarctic Ridge

● Antarctic research station

130

0 km   500

0 miles   500

Elevation

Below sea level 0   250m   500m   1000m   2000m   3000m   4000m   6000m

-6000m   -4000m   -2000m   -1000m   -500m   -250m

-328ft/-100m

820ft   1640ft   3281ft   6562ft   9843ft   13,124ft   19,685ft

-19,658ft   -13,124ft   -6562ft   -3281ft   -1640ft   -820ft

Arctic Ocean

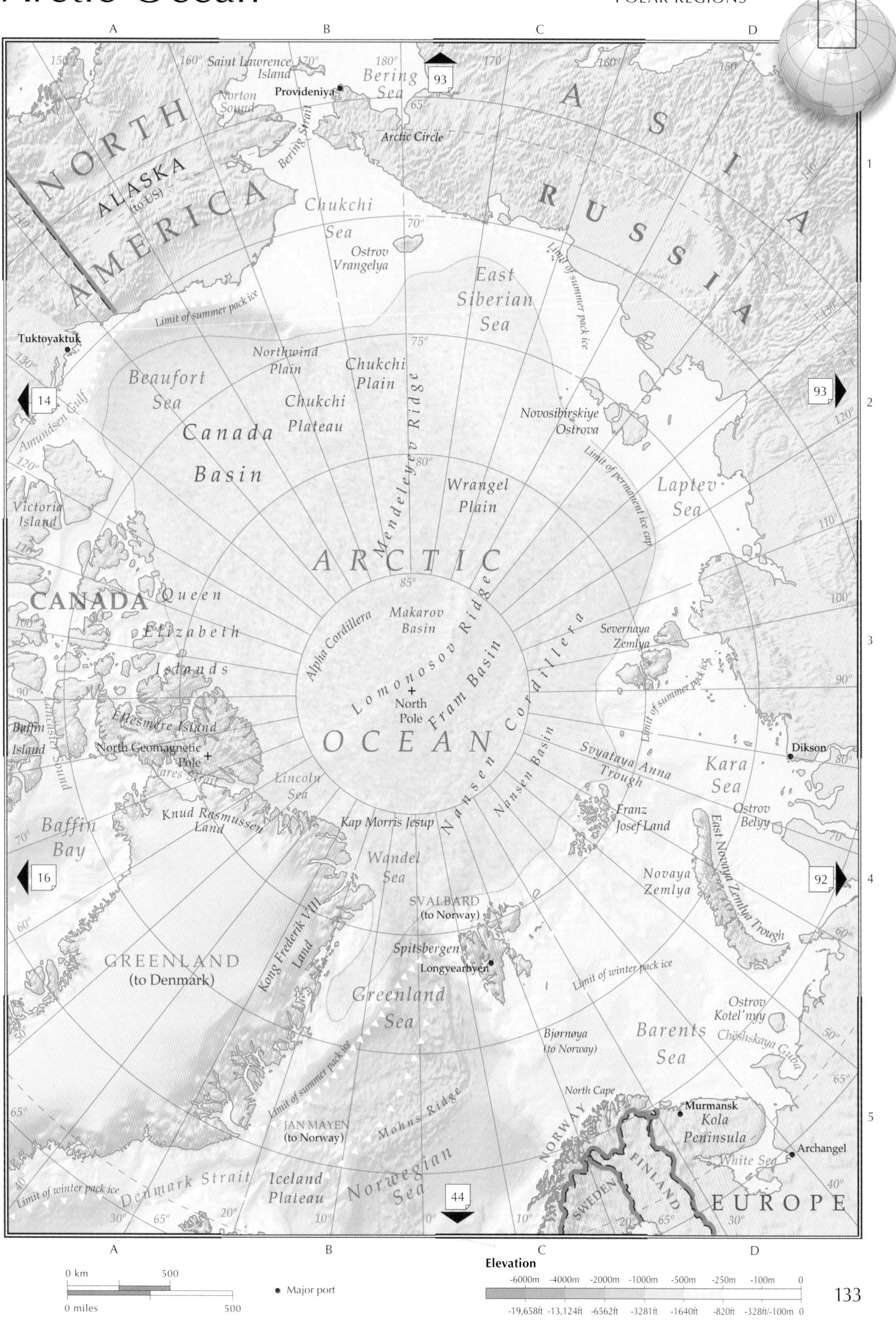

A
B
C
D

1

NORTH
ALASKA
(to US)
AMERICA
150°
160°
Saint Lawrence
Island
Norton
Sound
Provideniya
Bering Strait
170°
180°
Bering
Sea
170°
65°
Arctic Circle
160°
150°
93
A
S
I
A
RUSSIA
160°

Chukchi
Sea
70°
Ostrov
Vrangelya
East
Siberian
Sea
Limit of summer pack ice

Limit of summer pack ice
14
2
Tuktoyaktuk
130°
120°
Beaufort
Sea
Canada
Basin
Northwind
Plain
Chukchi
Plateau
Chukchi
Plain
75°
Mendeleyev Ridge
80°
Wrangel
Plain
Novosibirskiye
Ostrova
Limit of permanent ice cap
Laptev
Sea
110°
93
120°
2

3
Victoria
Island
CANADA
170°
120°
100°
90°
Queen
Elizabeth
Islands
Alpha Cordillera
Makarov
Basin
85°
Lomonosov Ridge
North
Pole
Fram Basin
ARCTIC
Nansen Cordillera
Severnaya
Zemlya
Svyataya Anna
Trough
Nansen Basin
90°
100°
Kara
Sea
Dikson
80°
3

4
Baffin
Island
Lancaster Sound
90°
70°
Baffin
Bay
Ellesmere Island
North Geomagnetic
Pole
Nares Strait
Knud Rasmussen
Land
Lincoln
Sea
Kap Morris Jesup
O C E A N
Wandel
Sea
SVALBARD
(to Norway)
Franz
Josef Land
Novaya
Zemlya
East Novaya Zemlya Trough
Ostrov
Belyy
70°
80°
16
4
60°
GREENLAND
(to Denmark)
Kong Frederik VIII
Land
50°
Spitsbergen
Longyearbyen
Greenland
Sea
Bjørnøya
(to Norway)
Limit of winter pack ice
60°
Ostrov
Kotel'nyy
Chëshskaya Guba
Barents
Sea
50°
92
4

5
65°
60°
Limit of winter pack ice
30°
65°
Denmark Strait
JAN MAYEN
(to Norway)
Limit of summer pack ice
Iceland
Plateau
20°
10°
Mohns Ridge
Norwegian
Sea
10°
44
North Cape
NORWAY
SWEDEN
FINLAND
20°
65°
Murmansk
Kola
Peninsula
White Sea
Archangel
40°
EUROPE
30°
5

A
B
C
D

0 km   500
0 miles   500
● Major port

Elevation
-6000m -4000m -2000m -1000m -500m -250m -100m 0
-19,658ft -13,124ft -6562ft -3281ft -1640ft -820ft -328ft/-100m 0

133

Country Profiles

This Factfile is intended as a guide to a world that is continually changing as political fashions and personalities come and go. Nevertheless, all the material in these factfiles has been researched from the most up-to-date and authoritative sources to give an incisive portrait of the geographical, political, and social characteristics that make each country so unique.

There are currently 196 independent countries in the world - more than at any previous time - and over 50 dependencies. Antarctica is the only land area on Earth that is not officially part of, and does not belong to, any single country.

AFGHANISTAN
Central Asia

Page 100 D4

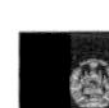

Landlocked in Central Asia, Afghanistan has suffered decades of conflict. The Islamist taliban, ousted by a US-led offensive in 2001, continue to resist subsequent elected governments.

Official name Islamic Republic of Afghanistan
Formation 1919 / 1919
Capital Kabul
Population 35.5 million / 141 people per sq mile (54 people per sq km)
Total area 250,000 sq miles (647,500 sq km)
Languages Pashtu*, Tajik, Dari*, Farsi, Uzbek, Turkmen
Religions Sunni Muslim 80%, Shi'a Muslim 19%, Other 1%
Ethnic mix Pashtun 38%, Tajik 25%, Hazara 19%, Uzbek and Turkmen 15%, Other 3%
Government Nonparty system
Currency Afghani = 100 puls
Literacy rate 32%
Calorie consumption 2090 kilocalories

ALBANIA
Southeast Europe

Page 79 C6

Lying at the southeastern end of the Adriatic Sea, Albania – or the "land of the eagles" – underwent upheavals after 1991 to emerge from its communist-period isolation.

Official name Republic of Albania
Formation 1912 / 1921
Capital Tirana
Population 2.9 million / 274 people per sq mile (106 people per sq km)
Total area 11,100 sq miles (28,748 sq km)
Languages Albanian*, Greek
Religions Muslim (mainly Sunni) 68%, Roman Catholic 12%, Albanian Orthodox 8%, Nonreligious 6%, Other 6%
Ethnic mix Albanian 98%, Greek 1%, Other 1%
Government Parliamentary system
Currency Lek = 100 qindarka (qintars)
Literacy rate 97%
Calorie consumption 3193 kilocalories

ALGERIA
North Africa

Page 48 C3

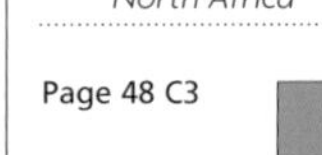

Lying mostly in the Sahara, this former French colony was riven by civil war after Islamists were denied electoral victory in 1992. Fighting has subsided but Islamic extremists remain a threat.

Official name People's Democratic Republic of Algeria
Formation 1962 / 1962
Capital Algiers
Population 41.3 million / 45 people per sq mile (17 people per sq km)
Total area 919,590 sq miles (2,381,740 sq km)
Languages Arabic*, Tamazight* (Kabyle, Shawia, Tamashek), French
Religions Sunni Muslim 99%, Christian & Jewish 1%
Ethnic mix Arab 75%, Berber 24%, European & Jewish 1%
Government Presidential system
Currency Algerian dinar = 100 centimes
Literacy rate 75%
Calorie consumption 3296 kilocalories

ANDORRA
Southwest Europe

Page 69 B6

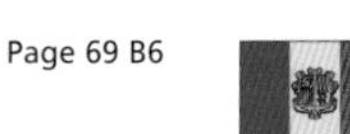

A tiny landlocked principality, Andorra lies between France and Spain, high in the eastern Pyrenees. Its economy, based on tourism, also features low tax and duty-free shopping.

Official name Principality of Andorra
Formation 1278 / 1278
Capital Andorra la Vella
Population 77,000 / 428 people per sq mile (166 people per sq km)
Total area 181 sq miles (468 sq km)
Languages Spanish, Catalan*, French, Portuguese
Religions Roman Catholic 94%, Other 6%
Ethnic mix Spanish 46%, Andorran 28%, Other 18%, French 8%
Government Parliamentary system
Currency Euro = 100 cents
Literacy rate 99%
Calorie consumption Not available

ANGOLA
Southern Africa

Page 56 B2

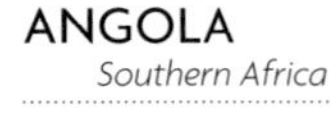

An oil- and diamond-rich former Portuguese colony in Southwest Africa, Angola is badly scarred from the 1975–2002 civil war. The removal of thousands of land mines continues.

Official name Republic of Angola
Formation 1975 / 1975
Capital Luanda
Population 29.8 million / 62 people per sq mile (24 people per sq km)
Total area 481,351 sq miles (1,246,700 sq km)
Languages Portuguese*, Umbundu, Kimbundu, Kikongo
Religions Roman Catholic 40%, Protestant 38%, Nonreligious 12%, Other (including animist) 10%
Ethnic mix Ovimbundu 37%, Ambundu 25%, Other 25%, Bakongo 13%
Government Presidential system
Currency Readjusted kwanza = 100 lwei
Literacy rate 66%
Calorie consumption 2473 kilocalories

ANTIGUA & BARBUDA
West Indies

Page 33 H3

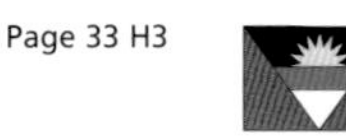

Lying on the Atlantic edge of the Leeward Islands, Antigua was in turn a Spanish, French, and British colony. Tourism is key, but Barbuda's beaches have suffered hurricane damage.

Official name Antigua and Barbuda
Formation 1981 / 1981
Capital St. John's
Population 100,000 / 588 people per sq mile (227 people per sq km)
Total area 170 sq miles (442 sq km)
Languages English*, English patois
Religions Other Christian 49%, Other 19%, Anglican 19%, Seventh-day Adventist 13%
Ethnic mix Black African 87%, Mixed race 5%, Hispanic 3%, Other 3%, White 2%
Government Parliamentary system
Currency East Caribbean dollar = 100 cents
Literacy rate 99%
Calorie consumption 2417 kilocalories

ARGENTINA
South America

Page 43 B5

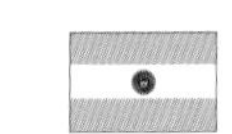

From semiarid lowlands, through fertile grasslands, to the glacial southern tip of South America, Argentina has enjoyed democratic rule since 1983 but struggled with high foreign debts.

Official name Argentine Republic
Formation 1816 / 1816
Capital Buenos Aires
Population 44.3 million / 42 people per sq mile (16 people per sq km)
Total area 1,068,296 sq miles (2,766,890 sq km)
Languages Spanish*, Italian, Amerindian languages
Religions Roman Catholic 71%, Protestant 15%, Nonreligious 11%, Other 3%
Ethnic mix Indo-European 97%, Mestizo 2%, Amerindian 1%
Government Presidential system
Currency Argentine peso = 100 centavos
Literacy rate 98%
Calorie consumption 3229 kilocalories

ARMENIA
Southwest Asia

Page 95 F3

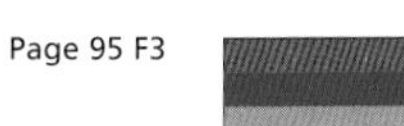

The smallest of the ex-Soviet republics, landlocked Armenia lies in the Lesser Caucasus mountains. It was the first country to adopt Christianity as the state religion, in the 4th century AD.

Official name Republic of Armenia
Formation 1991 / 1991
Capital Yerevan
Population 2.9 million / 252 people per sq mile (97 people per sq km)
Total area 11,506 sq miles (29,800 sq km)
Languages Armenian*, Azeri, Russian
Religions Orthodox Christian 89%, Other 8%, Nonreligious 2%, Armenian Catholic Church 1%
Ethnic mix Armenian 98%, Other 1%, Yezidi 1%
Government Parliamentary system
Currency Dram = 100 luma
Literacy rate 99%
Calorie consumption 2928 kilocalories

AUSTRALIA
Australasia & Oceania

Page 125 B5

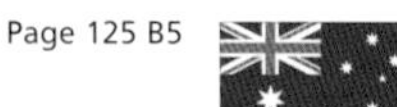

An island continent between the Indian and Pacific oceans, Australia was settled by Europeans from 1788, but recent immigrants are mostly Asian. Minerals underpin the economy.

Official name Commonwealth of Australia
Formation 1901 / 1901
Capital Canberra
Population 24.5 million / 8 people per sq mile (3 people per sq km)
Total area 2,967,893 sq miles (7,686,850 sq km)
Languages English*, Italian, Cantonese, Greek, Arabic, Vietnamese, Aboriginal languages
Religions Roman Catholic 28%, Nonreligious 24%, Other Christian 20%, Anglican 19%, Other 9%
Ethnic mix British 34%, Australian 27%, Other 18%, Irish 8%, Italian 4%, German 3%, Chinese 3%
Government Parliamentary system
Currency Australian dollar = 100 cents
Literacy rate 99%
Calorie consumption 3276 kilocalories

AUSTRIA
Central Europe

Page 73 D7

Nestled in Central Europe, Austria was created after the Austro-Hungarian Empire was defeated in World War I. Absorbed into Hitler's Germany in 1938, it re–emerged in 1955.

Official name Republic of Austria
Formation 1918 / 1919
Capital Vienna
Population 8.7 million / 272 people per sq mile (105 people per sq km)
Total area 32,378 sq miles (83,858 sq km)
Languages German*, Croatian, Slovenian, Hungarian (Magyar)
Religions Roman Catholic 75%, Nonreligious 12%, Other Christian 8%, Muslim 4%, Other 1%
Ethnic mix Austrian 93%, Croat, Slovene, and Hungarian 6%, Other 1%
Government Parliamentary system
Currency Euro = 100 cents
Literacy rate 99%
Calorie consumption 3768 kilocalories

AZERBAIJAN
Southwest Asia

Page 95 G2

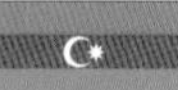

On the west coast of the Caspian Sea, oil-rich Azerbaijan regained its independence from the USSR in 1991. A territorial dispute with Armenia remains unresolved.

Official name Republic of Azerbaijan
Formation 1991 / 1991
Capital Baku
Population 9.8 million / 293 people per sq mile (113 people per sq km)
Total area 33,436 sq miles (86,600 sq km)
Languages Azeri*, Russian
Religions Shi'a Muslim 68%, Sunni Muslim 26%, Russian Orthodox 3%, Armenian Apostolic Church (Orthodox) 2%, Other 1%
Ethnic mix Azeri 91%, Other 3%, Armenian 2%, Russian 2%, Lazs 2%
Government Presidential system
Currency New manat = 100 gopik
Literacy rate 99%
Calorie consumption 3118 kilocalories

BAHAMAS, THE
West Indies

Page 32 C1

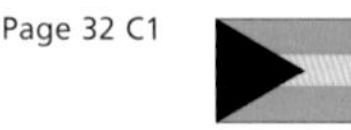

Located in the western Atlantic, off the Florida coast, the Bahamas comprise some 700 islands and 2400 cays; only 30 are inhabited. Financial services and shipping support the economy.

Official name Commonwealth of The Bahamas
Formation 1973 / 1973
Capital Nassau
Population 400,000 / 103 people per sq mile (40 people per sq km)
Total area 5382 sq miles (13,940 sq km)
Languages English*, English Creole, French Creole
Religions Baptist 36%, Other 20%, Anglican 14%, Roman Catholic 12%, Pentecostal 9%, Seventh-day Adventist 5%, Methodist 4%
Ethnic mix Black African 85%, European 12%, Asian and Hispanic 3%
Government Parliamentary system
Currency Bahamian dollar = 100 cents
Literacy rate 96%
Calorie consumption 2670 kilocalories

BAHRAIN
Southwest Asia

Page 98 C4

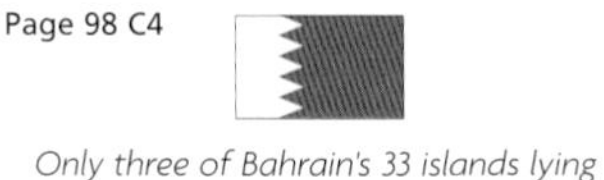

Only three of Bahrain's 33 islands lying between the Qatar peninsula and Saudi Arabian are inhabited. The first Gulf emirate to export oil, reserves are expected to last another 10 to 15 years.

Official name Kingdom of Bahrain
Formation 1971 / 1971
Capital Manama
Population 1.5 million / 5495 people per sq mile (2125 people per sq km)
Total area 239 sq miles (620 sq km)
Languages Arabic*
Religions Muslim (mainly Shi'a) 70%, Other 30%
Ethnic mix Bahraini 46%, Asian 46%, Other Arab 5%, Other 3%
Government Monarchical / parliamentary system
Currency Bahraini dinar = 1000 fils
Literacy rate 95%
Calorie consumption Not available

BANGLADESH
South Asia

Page 113 G3

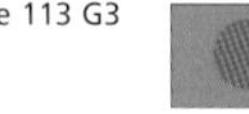

Low-lying Bangladesh on the Bay of Bengal suffers annual monsoon flooding. It seceded from Pakistan in 1971. Political instability and corruption are ongoing problems.

Official name People's Republic of Bangladesh
Formation 1971 / 1971
Capital Dhaka
Population 165 million / 3186 people per sq mile (1230 people per sq km)
Total area 55,598 sq miles (144,000 sq km)
Languages Bengali*, Urdu, Chakma, Marma (Magh), Garo, Khasi, Santhali, Tripuri, Mro
Religions Muslim (mainly Sunni) 90%, Hindu 9%, Other 1%
Ethnic mix Bengali 98%, Other 2%
Government Parliamentary system
Currency Taka = 100 poisha
Literacy rate 73%
Calorie consumption 2450 kilocalories

BARBADOS
West Indies

Page 33 H4

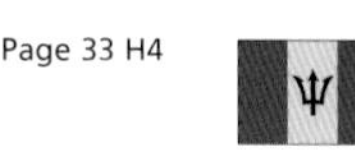

The most easterly of the Windward Islands, Barbados was under British rule from the 1620s. A sugar exporter in the 18th century, it now relies on tourism and financial services.

Official name Barbados
Formation 1966 / 1966
Capital Bridgetown
Population 300,000 / 1807 people per sq mile (698 people per sq km)
Total area 166 sq miles (430 sq km)
Languages Bajan (Barbadian English), English*
Religions Anglican 24%, Nonreligious 21%, Other 21%, Pentecostal 20%, Seventh-day Adventist 6%, Methodist 4%, Roman Catholic 4%
Ethnic mix Black African 93%, Mixed race 3%, White 3%, Other 1%
Government Parliamentary system
Currency Barbados dollar = 100 cents
Literacy rate 99%
Calorie consumption 2937 kilocalories

BELARUS
Eastern Europe

Page 85 B6

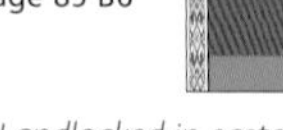

Landlocked in eastern Europe, forested Belarus, which means "White Russia," was reluctant to become independent of the USSR in 1991, and has been slow to reform its economy since.

Official name Republic of Belarus
Formation 1991 / 1991
Capital Minsk
Population 9.5 million / 119 people per sq mile (46 people per sq km)
Total area 80,154 sq miles (207,600 sq km)
Languages Belarussian*, Russian*
Religions Orthodox Christian 73%, Roman Catholic 12%, Other 12%, Nonreligious 3%
Ethnic mix Belarussian 86%, Russian 8%, Polish 3%, Other 2%, Ukrainian 1%
Government Presidential system
Currency Belarussian rouble = 100 kopeks
Literacy rate 99%
Calorie consumption 3250 kilocalories

BELGIUM
Northwest Europe

Page 65 B6

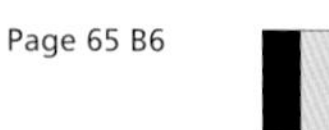

Located in Northwest Europe, Belgium has forests in the south and canals in the flat north. Its history and politics are marked by the division between its Flemish and Walloon communities.

Official name Kingdom of Belgium
Formation 1830 / 1919
Capital Brussels
Population 11.4 million / 900 people per sq mile (347 people per sq km)
Total area 11,780 sq miles (30,510 sq km)
Languages Dutch*, French*, German*
Religions Roman Catholic 88%, Other 10%, Muslim 2%
Ethnic mix Fleming 58%, Walloon 33%, Other 6%, Italian 2%, Moroccan 1%
Government Parliamentary system
Currency Euro = 100 cents
Literacy rate 99%
Calorie consumption 3733 kilocalories

BELIZE
Central America

Page 30 B1

The last Central American country to gain independence, this former British colony lies on the eastern shore of the Yucatan Peninsula. Offshore is the world's second-largest barrier reef.

Official name Belize
Formation 1981 / 1981
Capital Belmopan
Population 400,000 / 45 people per sq mile (18 people per sq km)
Total area 8867 sq miles (22,966 sq km)
Languages English Creole, Spanish, English*, Mayan, Garifuna (Carib)
Religions Roman Catholic 40%, Other Christian 34%, Nonreligious 16%, Other 10%
Ethnic mix Mestizo 49%, Creole 24%, Maya 10%, Other 7%, Garifuna 6%, Asian Indian 4%
Government Parliamentary system
Currency Belizean dollar = 100 cents
Literacy rate 75%
Calorie consumption 2751 kilocalories

BENIN
West Africa

Page 53 F4

Stretching north from the West African coast, this ex-French colony suffered military rule after independence but in recent decades has been a leading example of African democratization.

Official name Republic of Benin
Formation 1960 / 1960
Capital Porto-Novo; Cotonou
Population 11.2 million / 262 people per sq mile (101 people per sq km)
Total area 43,483 sq miles (112,620 sq km)
Languages Fon, Bariba, Yoruba, Adja, Houeda, Somba, French*
Religions Indigenous beliefs and Voodoo 50%, Christian 30%, Muslim 20%
Ethnic mix Fon 41%, Other 21%, Adja 16%, Yoruba 12%, Bariba 10%
Government Presidential system
Currency CFA franc = 100 centimes
Literacy rate 33%
Calorie consumption 2619 kilocalories

BHUTAN
South Asia

Page 113 G3

This landlocked Buddhist kingdom, perched in the eastern Himalayas between India and China, is carefully protecting its cultural identity from modernization and the outside world.

Official name Kingdom of Bhutan
Formation 1656 / 1865
Capital Thimphu
Population 800,000 / 44 people per sq mile (17 people per sq km)
Total area 18,147 sq miles (47,000 sq km)
Languages Dzongkha*, Nepali, Assamese
Religions Mahayana Buddhist 75%, Hindu 25%
Ethnic mix Drukpa 50%, Nepalese 35%, Other 15%
Government Monarchical / parliamentary system
Currency Ngultrum = 100 chetrum
Literacy rate 57%
Calorie consumption Not available

BOLIVIA
South America

Page 39 F3

Bolivia lies landlocked high in central South America. Mineral riches once made it the region's wealthiest state, but wars, coups, and poor governance have reduced it to the poorest.

Official name Plurinational State of Bolivia
Formation 1825 / 1938
Capital La Paz (administrative); Sucre (judicial)
Population 11.1 million / 27 people per sq mile (10 people per sq km)
Total area 424,162 sq miles (1,098,580 sq km)
Languages Aymara*, Quechua*, Spanish*
Religions Roman Catholic 77%, Protestant 16%, Nonreligious 4%, Other 3%
Ethnic mix Quechua 37%, Aymara 32%, Mixed race 13%, European 10%, Other 8%
Government Presidential system
Currency Boliviano = 100 centavos
Literacy rate 92%
Calorie consumption 2256 kilocalories

BOSNIA & HERZEGOVINA
Southeast Europe

Page 78 B3

In the mountainous western Balkans this state, born out of the bitter conflicts of Yugoslavia's collapse, has two key concerns: balancing ethnic rivalries, and integrating with Europe.

Official name Bosnia and Herzegovina
Formation 1992 / 1992
Capital Sarajevo
Population 3.5 million / 177 people per sq mile (68 people per sq km)
Total area 19,741 sq miles (51,129 sq km)
Languages Bosnian*, Serbian*, Croatian*
Religions Muslim (mainly Sunni) 53%, Orthodox Christian 35%, Roman Catholic 8%, Nonreligious 3%, Other 1%
Ethnic mix Bosniak 48%, Serb 34%, Croat 16%, Other 2%
Government Parliamentary system
Currency Marka = 100 pfeninga
Literacy rate 97%
Calorie consumption 3154 kilocalories

BOTSWANA
Southern Africa

Page 56 C3

Botswana, once the British protectorate of Bechuanaland, lies landlocked in Southern Africa. Diamonds provide it with a relatively prosperous economy, but the rate of HIV infection is high.

Official name Republic of Botswana
Formation 1966 / 1966
Capital Gaborone
Population 2.3 million / 11 people per sq mile (4 people per sq km)
Total area 231,803 sq miles (600,370 sq km)
Languages Setswana, English*, Shona, San, Khoikhoi, isiNdebele
Religions Christian (mainly Protestant) 80%, Nonreligious 15%, Traditional beliefs 4%, Other (including Muslim) 1%
Ethnic mix Tswana 79%, Kalanga 11%, Other 10%
Government Presidential system
Currency Pula = 100 thebe
Literacy rate 87%
Calorie consumption 2326 kilocalories

BRAZIL
South America

Page 40 C2

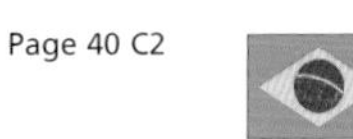

Brazil covers more than half of South America and is the site of the world's largest rain forest. It has immense natural resources and produces a third of the world's coffee.

Official name Federative Republic of Brazil
Formation 1822 / 1828
Capital Brasilia
Population 209 million / 64 people per sq mile (25 people per sq km)
Total area 3,286,470 sq miles (8,511,965 sq km)
Languages Portuguese*, German, Italian, Spanish, Polish, Japanese, Amerindian languages
Religions Roman Catholic 61%, Protestant 26%, Nonreligious 8%, Other 5%
Ethnic mix White 48%, Mixed race 43%, Black 8%, Other 1%
Government Presidential system
Currency Real = 100 centavos
Literacy rate 92%
Calorie consumption 3263 kilocalories

BRUNEI
Southeast Asia

Page 116 D3

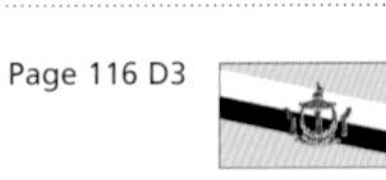

On the northwest coast of the island of Borneo, Brunei is surrounded and divided in two by the Malaysian state of Sarawak. Oil and gas revenues have brought a high standard of living.

Official name Brunei Darussalam
Formation 1984 / 1984
Capital Bandar Seri Begawan
Population 400,000 / 197 people per sq mile (76 people per sq km)
Total area 2228 sq miles (5770 sq km)
Languages Malay*, English, Chinese
Religions Muslim (mainly Sunni) 79%, Christian 9%, Buddhist 8%, Other 4%
Ethnic mix Malay 66%, Other 21%, Chinese 10%, Indigenous 3%
Government Monarchy
Currency Brunei dollar = 100 cents
Literacy rate 95%
Calorie consumption 2985 kilocalories

BULGARIA
Southeast Europe

Page 82 C2

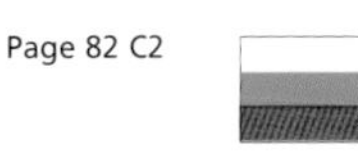

Bulgaria is located on the western shore of the Black Sea. After the fall of its communist regime in 1990, economic and political reform were slow, but EU membership was achieved in 2007.

Official name Republic of Bulgaria
Formation 1908 / 1947
Capital Sofia
Population 7.1 million / 166 people per sq mile (64 people per sq km)
Total area 42,822 sq miles (110,910 sq km)
Languages Bulgarian*, Turkish, Romani
Religions Orthodox Christian 75%, Muslim 15%, Nonreligious 5%, Other 3%, Protestant 1%, Roman Catholic 1%
Ethnic mix Bulgarian 85%, Turkish 9%, Roma 5%, Other 1%
Government Parliamentary system
Currency Lev = 100 stotinki
Literacy rate 98%
Calorie consumption 2829 kilocalories

BURKINA FASO
West Africa

Page 53 E4

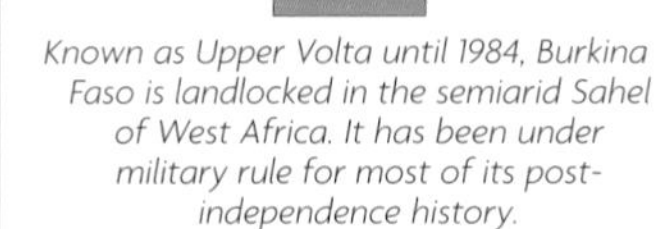

Known as Upper Volta until 1984, Burkina Faso is landlocked in the semiarid Sahel of West Africa. It has been under military rule for most of its post-independence history.

Official name Burkina Faso
Formation 1960 / 1960
Capital Ouagadougou
Population 19.2 million / 182 people per sq mile (70 people per sq km)
Total area 105,869 sq miles (274,200 sq km)
Languages Mossi, Fulani, French*, Tuareg, Dyula, Songhai
Religions Muslim 61%, Roman Catholic 19%, Traditional beliefs 15%, Protestant 4%, Other 1%
Ethnic mix Mossi 48%, Other 21%, Peul 10%, Lobi 7%, Bobo 7%, Mandé 7%
Government Presidential system
Currency CFA franc = 100 centimes
Literacy rate 35%
Calorie consumption 2720 kilocalories

BURUNDI
Central Africa

Page 51 B7

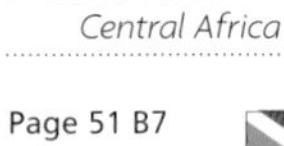

Small, landlocked Burundi lies just south of the Equator, on the Nile–Congo watershed. A decade of brutal conflict between Hutu and Tutsi from 1993 led to power-sharing in governance.

Official name Republic of Burundi
Formation 1962 / 1962
Capital Bujumbura
Population 10.9 million / 1101 people per sq mile (425 people per sq km)
Total area 10,745 sq miles (27,830 sq km)
Languages Kirundi*, French*, Kiswahili
Religions Roman Catholic 65%, Protestant 23%, Other 7%, Muslim 3%, Seventh-day Adventist 2%
Ethnic mix Hutu 85%, Tutsi 14%, Twa 1%
Government Presidential system
Currency Burundian franc = 100 centimes
Literacy rate 6.2%
Calorie consumption 1604 kilocalories

CAMBODIA
Southeast Asia

Page 115 D5

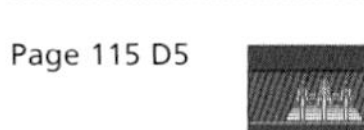

This ancient Southeast Asian kingdom suffered the brutal totalitarian Khmer Rouge regime in the 1970s and then a decade of Vietnamese puppet rule. Free elections were only held in 1993.

Official name Kingdom of Cambodia
Formation 1953 / 1953
Capital Phnom Penh
Population 16 million / 235 people per sq mile (91 people per sq km)
Total area 69,900 sq miles (181,040 sq km)
Languages Khmer*, French, Chinese, Vietnamese, Cham
Religions Buddhist 97%, Muslim 2%, Other (mostly Christian) 1%
Ethnic mix Khmer 90%, Vietnamese 5%, Other 4%, Chinese 1%
Government Parliamentary system
Currency Riel = 100 sen
Literacy rate 74%
Calorie consumption 2477 kilocalories

CAMEROON
Central Africa

Page 54 A4

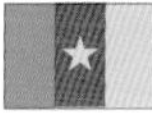

A former trading hub on the central West African coast, Cameroon was effectively a one-party state for 30 years. Elections since 1992 have brought no change in leadership.

Official name Republic of Cameroon
Formation 1960 / 1961
Capital Yaoundé
Population 24.1 million / 134 people per sq mile (52 people per sq km)
Total area 183,567 sq miles (475,400 sq km)
Languages Bamileke, Fang, Fulani, French*, English*
Religions Roman Catholic 35%, Traditional beliefs 25%, Muslim 22%, Protestant 18%
Ethnic mix Cameroon highlanders 31%, Other 21%, Equatorial Bantu 19%, Kirdi 11%, Fulani 10%, Northwestern Bantu 8%
Government Presidential system
Currency CFA franc = 100 centimes
Literacy rate 71%
Calorie consumption 2671 kilocalories

CANADA
North America

Page 15 E4

The world's second-largest country spans six time zones, extends north from its US border into the Arctic, and is rich in natural resources. Separatism is strong in French-speaking Québec.

Official name Canada
Formation 1867 / 1949
Capital Ottawa
Population 36.6 million / 10 people per sq mile (4 people per sq km)
Total area 3,855,171 sq miles (9,984,670 sq km)
Languages English*, French*, Chinese, Italian, German, Ukrainian, Portuguese, Inuktitut, Cree
Religions Roman Catholic 39%, Other Christian 28%, Nonreligious 24%, Other 6%, Muslim 3%
Ethnic mix European descent 80%, Asian 15%, First Nations, Métis, and Inuit 5%
Government Parliamentary system
Currency Canadian dollar = 100 cents
Literacy rate 99%
Calorie consumption 3494 kilocalories

CAPE VERDE
Atlantic Ocean

Page 52 A2

The mostly volcanic islands that make up Cape Verde lie off Africa's west coast. A Portuguese colony until 1975, it has been a stable democracy since its first multiparty elections in 1991.

Official name Republic of Cabo Verde
Formation 1975 / 1975
Capital Praia
Population 500,000 / 321 people per sq mile (124 people per sq km)
Total area 1557 sq miles (4033 sq km)
Languages Portuguese Creole, Portuguese*
Religions Roman Catholic 97%, Other 2%, Protestant (Church of the Nazarene) 1%
Ethnic mix Mestiço 71%, African 28%, European 1%
Government Presidential / parliamentary system
Currency Escudo = 100 centavos
Literacy rate 87%
Calorie consumption 2609 kilocalories

CENTRAL AFRICAN REPUBLIC
Central Africa

Page 54 C4

A landlocked plateau dividing the Chad and Congo river basins, the CAR has been plagued by rebellions since military rule ended in 1993. The arid north is sparsely populated.

Official name Central African Republic
Formation 1960 / 1960
Capital Bangui
Population 4.7 million / 20 people per sq mile (8 people per sq km)
Total area 240,534 sq miles (622,984 sq km)
Languages Sango, Banda, Gbaya, French*
Religions Traditional beliefs 35%, Roman Catholic 25%, Protestant 25%, Muslim 15%
Ethnic mix Baya 33%, Banda 27%, Other 17%, Mandjia 13%, Sara 10%
Government Presidential system
Currency CFA franc = 100 centimes
Literacy rate 37%
Calorie consumption 1879 kilocalories

CHAD
Central Africa

Page 54 C3

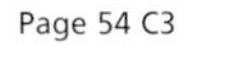

Landlocked in north Central Africa, Chad has been torn by intermittent periods of civil war since it gained independence from France in 1960. It became a net oil exporter in 2003.

Official name Republic of Chad
Formation 1960 / 1960
Capital N'Djaména
Population 14.9 million / 31 people per sq mile (12 people per sq km)
Total area 495,752 sq miles (1,284,000 sq km)
Languages French*, Sara, Arabic*, Maba
Religions Muslim 51%, Christian 35%, Animist 7%, Traditional beliefs 7%
Ethnic mix Other 30%, Sara 28%, Mayo-Kebbi 12%, Arab 12%, Ouaddai 9%, Kanem-Bornou 9%
Government Presidential system
Currency CFA franc = 100 centimes
Literacy rate 22%
Calorie consumption 2110 kilocalories

CHILE
South America

Page 42 B3

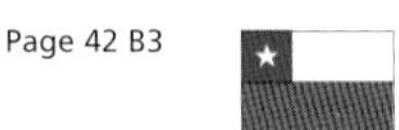

Extending in a ribbon down the Pacific coast of South America, Chile restored democracy in 1989 after a referendum rejected its military dictator. It is the world's largest copper producer.

Official name Republic of Chile
Formation 1818 / 1883
Capital Santiago
Population 18.1 million / 63 people per sq mile (24 people per sq km)
Total area 292,258 sq miles (756,950 sq km)
Languages Spanish*, Amerindian languages
Religions Roman Catholic 64%, Protestant 17%, Nonreligious 16%, Other 3%
Ethnic mix Mestizo and European 95%, Mapuche 4%, Other Amerindian 1%
Government Presidential system
Currency Chilean peso = 100 centavos
Literacy rate 96%
Calorie consumption 2979 kilocalories

CHINA
East Asia

Page 104 C4

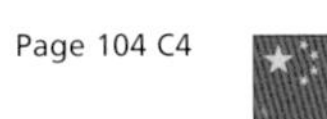

This vast East Asian country, home to a fifth of the global population, became a communist state in 1949. It has now emerged as one of the world's major political and economic powers.

Official name People's Republic of China
Formation 960 / 1999
Capital Beijing
Population 1.41 billion / 391 people per sq mile (151 people per sq km)
Total area 3,705,386 sq miles (9,596,960 sq km)
Languages Mandarin*, Wu, Cantonese, Hsiang, Min, Hakka, Kan
Religions Nonreligious or traditional beliefs 73%, Buddhist 16%, Other 7%, Christian 3%, Muslim 1%
Ethnic mix Han 92%, Other 4%, Zhuang 1%, Hui 1%, Manchu 1%, Uighur 1%
Government One-party state
Currency Renminbi (or yuan) = 10 jiao = 100 fen
Literacy rate 95%
Calorie consumption 3108 kilocalories

COLOMBIA
South America

Page 36 B3

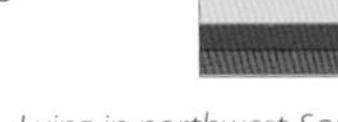

Lying in northwest South America, Colombia is noted for coffee, gold, emeralds, and narcotics trafficking. A 52-year civil war that displaced seven million people ended in 2016.

Official name Republic of Colombia
Formation 1819 / 1903
Capital Bogotá
Population 49.1 million / 122 people per sq mile (47 people per sq km)
Total area 439,733 sq miles (1,138,910 sq km)
Languages Spanish*, Amerindian languages
Religions Roman Catholic 79%, Protestant 13%, Nonreligious 6%, Other 2%
Ethnic mix Mestizo 58%, White 20%, European–African 14%, African 4%, African–Amerindian 3%, Amerindian 1%
Government Presidential system
Currency Colombian peso = 100 centavos
Literacy rate 94%
Calorie consumption 2804 kilocalories

COMOROS
Indian Ocean

Page 57 F2

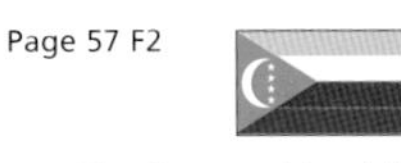

The Comoros islands lie between Mozambique and Madagascar. There have been many coups and secession attempts by the smaller islands since independence from France in 1975.

Official name Union of the Comoros
Formation 1975 / 1975
Capital Moroni
Population 800,000 / 929 people per sq mile (359 people per sq km)
Total area 838 sq miles (2170 sq km)
Languages Arabic*, Comoran*, French*
Religions Muslim (mainly Sunni) 98%, Other 1%, Roman Catholic 1%
Ethnic mix Comoran 97%, Other 3%
Government Presidential system
Currency Comoros franc = 100 centimes
Literacy rate 49%
Calorie consumption 2139 kilocalories

CONGO
Central Africa

Page 55 B5

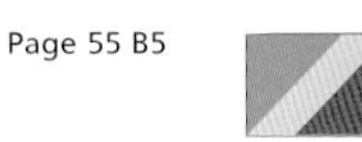

Astride the Equator in Central Africa, this former French colony emerged from 26 years of Marxist-Leninist rule in 1990, though the Marxist-era dictator seized power again in 1997.

Official name Republic of the Congo
Formation 1960 / 1960
Capital Brazzaville
Population 5.3 million / 40 people per sq mile (16 people per sq km)
Total area 132,046 sq miles (342,000 sq km)
Languages Kongo, Teke, Lingala, French*
Religions Traditional beliefs 50%, Roman Catholic 35%, Protestant 13%, Muslim 2%
Ethnic mix Bakongo 51%, Teke 17%, Other 16%, Mbochi 11%, Mbédé 5%
Government Presidential system
Currency CFA franc = 100 centimes
Literacy rate 79%
Calorie consumption 2208 kilocalories

CONGO, DEM. REP.
Central Africa

Page 55 C6

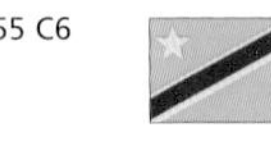

Straddling the Equator in east Central Africa, mineral-rich Dem. Rep. Congo is Africa's second-largest country. The former Belgian colony has endured years of corrupt rule and conflict.

Official name Democratic Republic of the Congo
Formation 1960 / 1960
Capital Kinshasa
Population 81.3 million / 93 people per sq mile (36 people per sq km)
Total area 905,563 sq miles (2,345,410 sq km)
Languages Kiswahili, Tshiluba, Kikongo, Lingala, French*
Religions Roman Catholic 50%, Protestant 20%, Traditional beliefs and other 20%, Muslim 10%
Ethnic mix Other 55%, Mongo, Luba, Kongo, and Mangbetu-Azande 45%
Government Presidential system
Currency Congolese franc = 100 centimes
Literacy rate 77%
Calorie consumption 1585 kilocalories

COSTA RICA
Central America

Page 31 E4

Costa Rica is the most stable country in Central America. It abolished its army in 1948 and its neutrality in foreign affairs is long-standing, but it has very strong ties with the US.

Official name Republic of Costa Rica
Formation 1838 / 1838
Capital San José
Population 4.9 million / 249 people per sq mile (96 people per sq km)
Total area 19,730 sq miles (51,100 sq km)
Languages Spanish*, English Creole, Bribri, Cabecar
Religions Roman Catholic 62%, Protestant 25%, Nonreligious 9%, Other 4%
Ethnic mix Mestizo and European 96%, Amerindian 3%, Black 1%
Government Presidential system
Currency Costa Rican colón = 100 céntimos
Literacy rate 97%
Calorie consumption 2848 kilocalories

CROATIA
Southeast Europe

Page 78 B2

Post-independence fighting afflicted this former Yugoslav republic until 1995. It is now capitalizing on its location on the eastern Adriatic coast and joined the EU in 2013.

Official name Republic of Croatia
Formation 1991 / 1991
Capital Zagreb
Population 4.2 million / 192 people per sq mile (74 people per sq km)
Total area 21,831 sq miles (56,542 sq km)
Languages Croatian*
Religions Roman Catholic 84%, Nonreligious 7%, Orthodox Christian 4%, Other 3%, Muslim 2%
Ethnic mix Croat 92%, Serb 4%, Other 3%, Bosniak 1%
Government Parliamentary system
Currency Kuna = 100 lipa
Literacy rate 99%
Calorie consumption 3059 kilocalories

CUBA
West Indies

Page 32 C2

Cuba is the largest island in the Caribbean and the only communist country in the Americas. It was led by Fidel Castro for almost 40 years until he handed over to his brother in 2008.

Official name Republic of Cuba
Formation 1902 / 1902
Capital Havana
Population 11.5 million / 269 people per sq mile (104 people per sq km)
Total area 42,803 sq miles (110,860 sq km)
Languages Spanish
Religions Nonreligious 49%, Roman Catholic 40%, Atheist 6%, Other 4%, Protestant 1%
Ethnic mix White 65%, Mulatto (mixed race) 25%, Black 10%
Government One-party state
Currency Cuban peso = 100 centavos
Literacy rate 99%
Calorie consumption 3409 kilocalories

CYPRUS
Southeast Europe

Page 80 C5

Cyprus lies south of Turkey in the eastern Mediterranean. Since 1974, it has been partitioned between the Turkish-occupied north and the Greek south (which joined the EU in 2004).

Official name Republic of Cyprus
Formation 1960 / 1960
Capital Nicosia
Population 1.2 million / 336 people per sq mile (130 people per sq km)
Total area 3571 sq miles (9250 sq km)
Languages Greek*, Turkish*
Religions Orthodox Christian 78%, Muslim 18%, Other 4%
Ethnic mix Greek 81%, Turkish 11%, Other 8%
Government Presidential system
Currency Euro = 100 cents (In TRNC, Turkish lira = 100 kurus)
Literacy rate 99%
Calorie consumption 2649 kilocalories

CZECH REPUBLIC (CZECHIA)
Central Europe

Page 77 A5

Landlocked in Central Europe, and formerly part of communist Czechoslovakia, it peacefully dissolved its federal union with Slovakia in 1993, and joined the EU in 2004.

Official name Czech Republic
Formation 1993 / 1993
Capital Prague
Population 10.6 million / 348 people per sq mile (134 people per sq km)
Total area 30,450 sq miles (78,866 sq km)
Languages Czech*, Slovak, Hungarian (Magyar)
Religions Nonreligious 72%, Roman Catholic 21%, Other 6%, Orthodox Christian 1%
Ethnic mix Czech 86%, Moravian 7%, Other 5%, Slovak 2%
Government Parliamentary system
Currency Czech koruna = 100 haleru
Literacy rate 99%
Calorie consumption 3256 kilocalories

DENMARK
Northern Europe

Page 63 A7

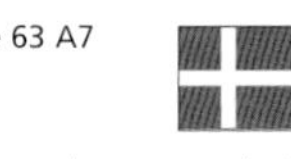

Denmark occupies the low-lying Jutland peninsula and over 400 islands. In the 1930s it set up one of the first welfare systems. Greenland and the Faroe Islands are self-governing territories.

Official name Kingdom of Denmark
Formation 950 / 1944
Capital Copenhagen
Population 5.7 million / 348 people per sq mile (135 people per sq km)
Total area 16,639 sq miles (43,094 sq km)
Languages Danish*
Religions Evangelical Lutheran 95%, Roman Catholic 3%, Muslim 2%
Ethnic mix Danish 96%, Other (including Scandinavian and Turkish) 3%, Faroese and Inuit 1%
Government Parliamentary system
Currency Danish krone = 100 øre
Literacy rate 99%
Calorie consumption 3367 kilocalories

DJIBOUTI
East Africa

Page 50 D4

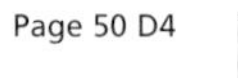

Once known as French Somaliland, this city state with a desert hinterland lies on the coast of the Horn of Africa. Its economy relies on its Red Sea port, a vital trade link for landlocked Ethiopia.

Official name Republic of Djibouti
Formation 1977 / 1977
Capital Djibouti
Population 1 million / 112 people per sq mile (43 people per sq km)
Total area 8494 sq miles (22,000 sq km)
Languages Somali, Afar, French*, Arabic*
Religions Muslim (mainly Sunni) 94%, Christian 6%
Ethnic mix Issa 60%, Afar 35%, Other 5%
Government Presidential system
Currency Djibouti franc = 100 centimes
Literacy rate 70%
Calorie consumption 2607 kilocalories

DOMINICA
West Indies

Page 33 H4

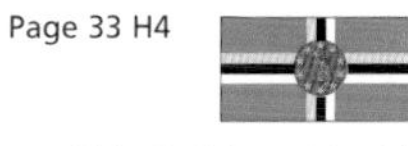

This Caribbean island, known for its lush flora and fauna, resisted European colonization until the 18th century, when it came first under French and then British rule.

Official name Commonwealth of Dominica
Formation 1978 / 1978
Capital Roseau
Population 74,000 / 254 people per sq mile (98 people per sq km)
Total area 291 sq miles (754 sq km)
Languages French Creole, English*
Religions Roman Catholic 62%, Protestant 30%, Nonreligious 6%, Other 2%
Ethnic mix Black 87%, Mixed race 9%, Carib 3%, Other 1%
Government Parliamentary system
Currency East Caribbean dollar = 100 cents
Literacy rate 88%
Calorie consumption 2931 kilocalories

DOMINICAN REPUBLIC
West Indies

Page 33 E2

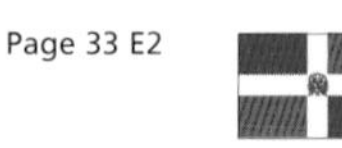

Occupying the eastern two-thirds of the island of Hispaniola, the Dominican Republic is the Caribbean's top tourist destination and largest economy. Ties with the US are strong.

Official name Dominican Republic
Formation 1865 / 1865
Capital Santo Domingo
Population 10.8 million / 578 people per sq mile (223 people per sq km)
Total area 18,679 sq miles (48,380 sq km)
Languages Spanish*, French Creole
Religions Roman Catholic 57%, Protestant 23%, Nonreligious 18%, Other 2%
Ethnic mix Mixed race 73%, European 16%, Black African 11%
Government Presidential system
Currency Dominican Republic peso = 100 centavos
Literacy rate 92%
Calorie consumption 2614 kilocalories

EAST TIMOR
Southeast Asia

Page 116 F5

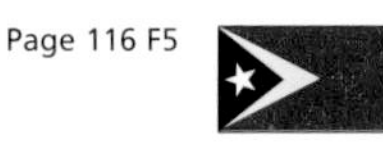

This former Portuguese colony on the island of Timor in the East Indies was invaded by Indonesia in 1975. In 1999 it voted for independence, achieved in 2002 after a turbulent transition.

Official name Democratic Republic of Timor-Leste
Formation 2002 / 2002
Capital Dili
Population 1.3 million / 230 people per sq mile 89 people per sq km)
Total area 5756 sq miles (14,874 sq km)
Languages Tetum* (Portuguese/Austronesian), Bahasa Indonesia, Portuguese*
Religions Roman Catholic 96%, Protestant 2%, Other 2%
Ethnic mix Papuan groups approx. 85%, Indonesian groups approx. 13%, Chinese 2%
Government Parliamentary system
Currency US dollar = 100 cents
Literacy rate 58%
Calorie consumption 2131 kilocalories

ECUADOR
South America

Page 38 A2

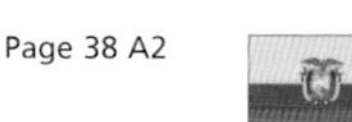

Once part of the Inca heartland on the northwest coast of South America, Ecuador is the world's leading banana exporter. Its territory includes the wildlife-rich Galapagos Islands.

Official name Republic of Ecuador
Formation 1830 / 1942
Capital Quito
Population 16.6 million / 155 people per sq mile (60 people per sq km)
Total area 109,483 sq miles (283,560 sq km)
Languages Spanish*, Quechua, other Amerindian languages
Religions Roman Catholic 79%, Protestant 13%, Nonreligious 5%, Other 3%
Ethnic mix Mestizo 79%, Black African 7%, Amerindian 7%, White 6%, Other 1%
Government Presidential system
Currency US dollar = 100 cents
Literacy rate 94%
Calorie consumption 2344 kilocalories

EGYPT
North Africa

Page 50 B2

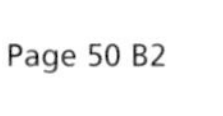

Egypt lies in Africa's northeast corner; the fertile Nile valley divides desert lands. Nearly 50 years of de facto military rule was interrupted in 2011 by the "Arab Spring" popular uprising.

Official name Arab Republic of Egypt
Formation 1936 / 1982
Capital Cairo
Population 97.6 million / 254 people per sq mile (98 people per sq km)
Total area 386,660 sq miles (1,001,450 sq km)
Languages Arabic*, French, English, Berber
Religions Muslim (mainly Sunni) 90%, Coptic Christian and other 9%, Other Christian 1%
Ethnic mix Egyptian 99%, Other 1%
Government Presidential system
Currency Egyptian pound = 100 piastres
Literacy rate 75%
Calorie consumption 3522 kilocalories

EL SALVADOR
Central America

Page 30 B3

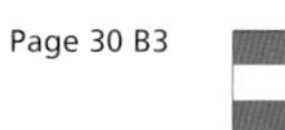

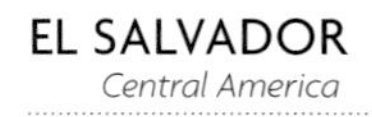

El Salvador is Central America's smallest country. Since a 12-year war between the US-backed army and left-wing guerrillas ended in 1992, crime and gang violence have been key issues.

Official name Republic of El Salvador
Formation 1841 / 1841
Capital San Salvador
Population 6.4 million / 800 people per sq mile (309 people per sq km)
Total area 8124 sq miles (21,040 sq km)
Languages Spanish*
Religions Roman Catholic 50%, Protestant 36%, Nonreligious 12%, Other 2
Ethnic mix Mestizo 86%, White 13%, Other and Amerindian 1%
Government Presidential system
Currency Salvadorean colón = 100 centavos; US dollar = 100 cents
Literacy rate 88%
Calorie consumption 2577 kilocalories

EQUATORIAL GUINEA
Central Africa

Page 55 A5

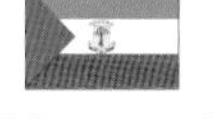

Equatorial Guinea comprises the Rio Muni mainland in west Central Africa and five islands. Free elections were first held in 1988, but the former ruling party still dominates.

Official name Republic of Equatorial Guinea
Formation 1968 / 1968
Capital Malabo
Population 1.3 million / 120 people per sq mile (46 people per sq km)
Total area 10,830 sq miles (28,051 sq km)
Languages Spanish*, Fang, Bubi, French*
Religions Roman Catholic 90%, Other 10%
Ethnic mix Fang 85%, Other 11%, Bubi 4%
Government Presidential system
Currency CFA franc = 100 centimes
Literacy rate 94%
Calorie consumption Not available

ERITREA
East Africa

Page 50 C4

Lying on the shores of the Red Sea, this former Italian colony was annexed by Ethiopia in 1952. It successfully seceded in 1993, following a 30-year war for independence.

Official name State of Eritrea
Formation 1993 / 2002
Capital Asmara
Population 5.1 million / 112 people per sq mile (43 people per sq km)
Total area 46,842 sq miles (121,320 sq km)
Languages Tigrinya*, English*, Tigre, Afar, Arabic*, Saho, Bilen, Kunama, Nara, Hadareb
Religions Christian 50%, Muslim 48%, Other 2%
Ethnic mix Tigray 50%, Tigre 31%, Other 9%, Saho 5%, Afar 5%
Government Presidential / parliamentary system
Currency Nakfa = 100 cents
Literacy rate 70%
Calorie consumption 1640 kilocalories

ESTONIA
Northeast Europe

Page 84 D2

The smallest, richest, most developed Baltic state, Estonia has emphasized advanced IT and integration with Europe since renouncing the Soviet model. It joined the EU in 2004.

Official name Republic of Estonia
Formation 1991 / 1991
Capital Tallinn
Population 1.3 million / 75 people per sq mile (29 people per sq km)
Total area 17,462 sq miles (45,226 sq km)
Languages Estonian*, Russian
Religions Nonreligious 45%, Orthodox Christian 25%, Lutheran 20%, Other 10%
Ethnic mix Estonian 70%, Russian 25%, Other 2%, Ukrainian 2%, Belarussian 1%
Government Parliamentary system
Currency Euro = 100 cents
Literacy rate 99%
Calorie consumption 3253 kilocalories

ETHIOPIA
East Africa

Page 51 C5

Ethiopia, the only African country to escape colonization, was a Marxist regime in 1974–1991. Now landlocked in the Horn of Africa, it has suffered economic, civil, and natural crises.

Official name Federal Democratic Republic of Ethiopia
Formation 1896 / 2002
Capital Addis Ababa
Population 105 million / 245 people per sq mile (95 people per sq km)
Total area 435,184 sq miles (1,127,127 sq km)
Languages Amharic*, Tigrinya, Galla, Sidamo, Somali, English, Arabic
Religions Christian 62%, Muslim 34%, Other 4%
Ethnic mix Oromo 34%, Amhara 27%, Other 23%, Somali 6%, Tigray 6%, Sidama 4%
Government Parliamentary system
Currency Birr = 100 cents
Literacy rate 39%
Calorie consumption 2131 kilocalories

FRANCE
Western Europe

Page 68 B4

Straddling Western Europe from the English Channel to the Mediterranean Sea, France was Europe's first modern republic. It is now one of the world's leading industrial powers.

Official name French Republic
Formation 987 / 1919
Capital Paris
Population 65 million / 306 people per sq mile (118 people per sq km)
Total area 211,208 sq miles (547,030 sq km)
Languages French*, Provençal, German, Breton, Catalan, Basque
Religions Christian 51%, Nonreligious 40%, Muslim 6%, Other 2%, Jewish 1%
Ethnic mix French 86%, North African 5%, Black 5%, German (Alsace) 2%, Breton 1%, Other 1%
Government Presidential / parliamentary system
Currency Euro = 100 cents
Literacy rate 99%
Calorie consumption 3482 kilocalories

GEORGIA
Southwest Asia

Page 95 F2

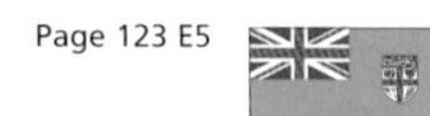

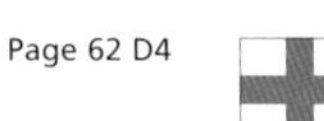

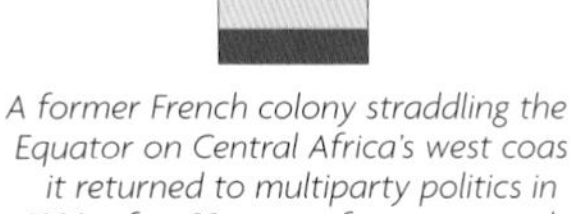

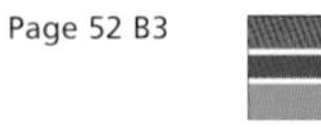

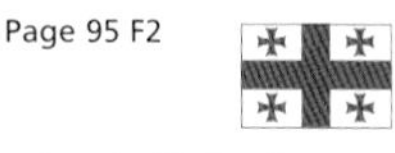

Located in the Caucasus on the Black Sea's eastern shore, Georgia is noted for its wine. Conflict broke out after the breakup of the USSR; the northern provinces have de facto autonomy.

Official name Georgia
Formation 1991 / 1991
Capital Tbilisi
Population 3.9 million / 145 people per sq mile (56 people per sq km)
Total area 26,911 sq miles (69,700 sq km)
Languages Georgian*, Russian, Azeri, Armenian, Mingrelian, Ossetian, Abkhazian (* in Abkhazia)
Religions Orthodox Christian 89%, Muslim 9%, Roman Catholic 1%, Other 1%
Ethnic mix Georgian 87%, Azeri 6%, Armenian 4%, Other 2%, Russian 1%
Government Presidential / Parliamentary system
Currency Lari = 100 tetri
Literacy rate 99%
Calorie consumption 2905 kilocalories

GREECE
Southeast Europe

Page 83 A5

The southernmost Balkan nation has a mountainous mainland and over 2000 islands, engendering its seafaring tradition. High state debt has led to recent unpopular austerity measures.

Official name Hellenic Republic
Formation 1829 / 1947
Capital Athens
Population 11.2 million / 222 people per sq mile (86 people per sq km)
Total area 50,942 sq miles (131,940 sq km)
Languages Greek*, Turkish, Macedonian, Albanian
Religions Orthodox Christian 90%, Nonreligious 4%, Other 4%, Muslim 2%
Ethnic mix Greek 98%, Other 2%
Government Parliamentary system
Currency Euro = 100 cents
Literacy rate 97%
Calorie consumption 3400 kilocalories

FIJI
Australasia & Oceania

Page 123 E5

Fiji is a volcanic archipelago of 882 islands in the southern Pacific Ocean. Tensions between ethnic Fijians and Indo-Fijians have provoked several coups. Sugar is the main export.

Official name Republic of Fiji
Formation 1970 / 1970
Capital Suva
Population 900,000 / 128 people per sq mile (49 people per sq km)
Total area 7054 sq miles (18,270 sq km)
Languages Fijian, English*, Hindi, Urdu, Tamil, Telugu
Religions Methodist 35%, Hindu 28%, Other Christian 21%, Roman Catholic 9%, Muslim 6%, Other and nonreligious 1%
Ethnic mix Melanesian 57%, Indian 38%, Other 5%
Government Parliamentary system
Currency Fiji dollar = 100 cents
Literacy rate 94%
Calorie consumption 2943 kilocalories

GABON
Central Africa

Page 55 A5

A former French colony straddling the Equator on Central Africa's west coast, it returned to multiparty politics in 1990, after 22 years of one-party rule. The economy relies on oil revenue.

Official name Gabonese Republic
Formation 1960 / 1960
Capital Libreville
Population 2 million / 20 people per sq mile (8 people per sq km)
Total area 103,346 sq miles (267,667 sq km)
Languages Fang, French*, Punu, Sira, Nzebi, Mpongwe
Religions Christian (mainly Roman Catholic) 55%, Traditional beliefs 40%, Other 4%, Muslim 1%
Ethnic mix Fang 26%, Shira-punu 24%, Other 16%, Foreign residents 15%, Nzabi-duma 11%, Mbédé-Teke 8%
Government Presidential system
Currency CFA franc = 100 centimes
Literacy rate 82%
Calorie consumption 2830 kilocalories

GERMANY
Northern Europe

Page 72 B4

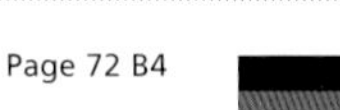

Germany is Europe's major economic power and a leading influence in the EU. Divided after World War II, its democratic west and communist east were re-unified in 1990.

Official name Federal Republic of Germany
Formation 1871 / 1990
Capital Berlin
Population 82.1 million / 608 people per sq mile (235 people per sq km)
Total area 137,846 sq miles (357,021 sq km)
Languages German*, Turkish
Religions Nonreligious 36%, Roman Catholic 29%, Protestant 26%, Muslim 5%, Other 4%
Ethnic mix German 81%, Other European 10%, Other 4%, Turkish 3%, Polish 2%
Government Parliamentary system
Currency Euro = 100 cents
Literacy rate 99%
Calorie consumption 3499 kilocalories

GRENADA
West Indies

Page 33 G5

The most southerly Windward gained worldwide notoriety in 1983, when the US invaded to sever its growing links with Cuba. It is the world's second-biggest nutmeg producer.

Official name Grenada
Formation 1974 / 1974
Capital St. George's
Population 100,000 / 763 people per sq mile (294 people per sq km)
Total area 131 sq miles (340 sq km)
Languages English*, English Creole
Religions Roman Catholic 68%, Anglican 17%, Other 15%
Ethnic mix Black African 82%, Mulatto (mixed race) 13%, East Indian 3%, Other 2%
Government Parliamentary system
Currency East Caribbean dollar = 100 cents
Literacy rate 96%
Calorie consumption 2447 kilocalories

FINLAND
Northern Europe

Page 62 D4

A low-lying country of forests and lakes, Finland joins Scandinavia to Russia. Its language is related to only two others in Europe. Finnish women were the first in Europe to get the vote, in 1906.

Official name Republic of Finland
Formation 1917 / 1947
Capital Helsinki
Population 5.5 million / 47 people per sq mile (18 people per sq km)
Total area 130,127 sq miles (337,030 sq km)
Languages Finnish*, Swedish*, Sámi
Religions Evangelical Lutheran 78%, Nonreligious 19%, Other 2%, Orthodox Christian 1%
Ethnic mix Finnish 93%, Other (including Sámi) 7%
Government Parliamentary system
Currency Euro = 100 cents
Literacy rate 99%
Calorie consumption 3368 kilocalories

GAMBIA, THE
West Africa

Page 52 B3

A narrow state along the Gambia River on Africa's west coast and surrounded by Senegal, Gambia was renowned for its stability until a coup in 1994; the coup leader then ruled until 2016.

Official name Republic of The Gambia
Formation 1965 / 1965
Capital Banjul
Population 2.1 million / 544 people per sq mile (210 people per sq km)
Total area 4363 sq miles (11,300 sq km)
Languages Mandinka, Fulani, Wolof, Jola, Soninke, English*
Religions Sunni Muslim 90%, Christian 8%, Traditional beliefs 2%
Ethnic mix Mandinka 42%, Fulani 18%, Wolof 16%, Jola 10%, Serahuli 9%, Other 5%
Government Presidential system
Currency Dalasi = 100 butut
Literacy rate 42%
Calorie consumption 2628 kilocalories

GHANA
West Africa

Page 53 E5

Once known as the Gold Coast, Ghana was the first colony in West Africa to gain independence. In recent decades multiparty democracy has been consolidated despite economic issues.

Official name Republic of Ghana
Formation 1957 / 1957
Capital Accra
Population 28.8 million / 324 people per sq mile (125 people per sq km)
Total area 92,100 sq miles (238,540 sq km)
Languages Twi, Fanti, Ewe, Ga, Adangbe, Gurma, Dagomba (Dagbani), English*
Religions Christian 71%, Muslim 18%, Traditional beliefs 5%, Nonreligious 5%, Other 1%
Ethnic mix Akan 47%, Gurma 17%, Ga-Dangme 14%, Other 9%, Ewe 7%, Guan 6%
Government Presidential system
Currency Cedi = 100 pesewas
Literacy rate 72%
Calorie consumption 3016 kilocalories

GUATEMALA
Central America

Page 30 A2

Once the heart of the Mayan civilization, the largest and most populous state on the Central American isthmus suffered years of civil war and military rule, but democracy is now flourishing.

Official name Republic of Guatemala
Formation 1838 / 1838
Capital Guatemala City
Population 16.9 million / 404 people per sq mile (156 people per sq km)
Total area 42,042 sq miles (108,890 sq km)
Languages Quiché, Mam, Cakchiquel, Kekchí, Spanish*
Religions Roman Catholic 50%, Protestant 41%, Nonreligious 6%, Other 3%
Ethnic mix Amerindian 60%, Mestizo 30%, Other 10%
Government Presidential system
Currency Quetzal = 100 centavos
Literacy rate 81%
Calorie consumption 2419 kilocalories

GUINEA
West Africa

Page 52 C4

A former French colony on Africa's west coast, Guinea chose a Marxist path, then came under army rule. Polls in 2010 brought fresh hope, though the recent Ebola epidemic was a setback.

Official name Republic of Guinea
Formation 1958 / 1958
Capital Conakry
Population 12.7 million / 134 people per sq mile (52 people per sq km)
Total area 94,925 sq miles (245,857 sq km)
Languages Pulaar, Malinké, Soussou, French*
Religions Muslim 89%, Christian 7%, Nonreligious 2%, Traditional beliefs and other 2%
Ethnic mix Peul 40%, Malinké 30%, Soussou 20%, Other 10%
Government Presidential system
Currency Guinea franc = 100 centimes
Literacy rate 32%
Calorie consumption 2566 kilocalories

GUINEA-BISSAU
West Africa

Page 52 B4

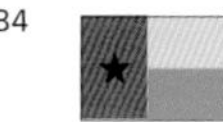

Known as Portuguese Guinea in colonial times, Guinea-Bissau is situated on Africa's west coast. One of the world's poorest countries, it has now become a transit point for cocaine trafficking.

Official name Republic of Guinea-Bissau
Formation 1974 / 1974
Capital Bissau
Population 1.9 million / 175 people per sq mile (68 people per sq km)
Total area 13,946 sq miles (36,120 sq km)
Languages Portuguese Creole, Balante, Fulani, Malinké, Portuguese*
Religions Muslim 54%, Christian 26%, Traditional beliefs 18%, Nonreligious 2%
Ethnic mix Balante 30%, Fulani 20%, Other 16%, Mandyako 14%, Mandinka 13%, Papel 7%
Government Presidential system
Currency CFA franc = 100 centimes
Literacy rate 46%
Calorie consumption 2292 kilocalories

GUYANA
South America

Page 37 F3

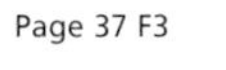

A land of rain forest, mountains, coastal plains, and savanna, Guyana is South America's only English-speaking state. It became a republic in 1970, four years after independence from Britain.

Official name Cooperative Republic of Guyana
Formation 1966 / 1966
Capital Georgetown
Population 800,000 / 11 people per sq mile (4 people per sq km)
Total area 83,000 sq miles (214,970 sq km)
Languages English Creole, Hindi, Tamil, Amerindian languages, English*
Religions Christian 57%, Hindu 28%, Muslim 10%, Other 5%
Ethnic mix East Indian 43%, Black African 30%, Mixed race 17%, Amerindian 9%, Other 1%
Government Presidential system
Currency Guyanese dollar = 100 cents
Literacy rate 86%
Calorie consumption 2764 kilocalories

HAITI
West Indies

Page 32 D3

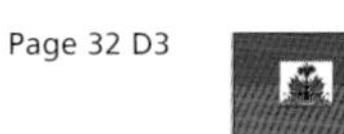

The western third of the Caribbean island of Hispaniola, Haiti became the world's first black republic in 1804. Natural disasters and periodic anarchy perpetuate its endemic poverty.

Official name Republic of Haiti
Formation 1804 / 1884
Capital Port-au-Prince
Population 11 million / 1034 people per sq mile (399 people per sq km)
Total area 10,714 sq miles (27,750 sq km)
Languages French Creole*, French*
Religions Roman Catholic 55%, Protestant 28%, Other (including Voodoo) 16%, Nonreligious 1%
Ethnic mix Black African 95%, Mulatto (mixed race) and European 5%
Government Presidential system
Currency Gourde = 100 centimes
Literacy rate 49%
Calorie consumption 2091 kilocalories

HONDURAS
Central America

Page 30 C2

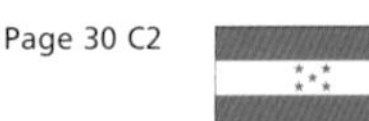

Straddling the Central American isthmus, Honduras returned to civilian rule in 1984, after a succession of military regimes. Crime is high and it has one of the world's worst murder rates.

Official name Republic of Honduras
Formation 1838 / 1838
Capital Tegucigalpa
Population 9.3 million / 215 people per sq mile (83 people per sq km)
Total area 43,278 sq miles (112,090 sq km)
Languages Spanish*, Garifuna (Carib), English Creole
Religions Roman Catholic 46%, Protestant 41%, Nonreligious 10%, Other 3%
Ethnic mix Mestizo 90%, Black African 5%, Amerindian 4%, White 1%
Government Presidential system
Currency Lempira = 100 centavos
Literacy rate 89%
Calorie consumption 2641 kilocalories

HUNGARY
Central Europe

Page 77 C6

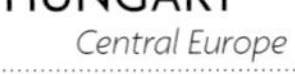

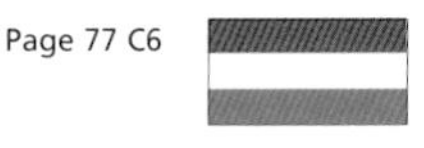

Hungary is bordered by seven states in Central Europe. After the fall of communism in 1989, it introduced political and economic reforms and joined the EU in 2004.

Official name Hungary
Formation 1918 / 1947
Capital Budapest
Population 9.7 million / 272 people per sq mile (105 people per sq km)
Total area 35,919 sq miles (93,030 sq km)
Languages Hungarian* (Magyar)
Religions Roman Catholic 52%, Nonreligious 21%, Presbyterian 13%, Other (mostly Protestant) 10%
Ethnic mix Magyar 92%, Roma 3%, Other 3%, German 2%
Government Parliamentary system
Currency Forint = 100 fillér
Literacy rate 99%
Calorie consumption 3037 kilocalories

ICELAND
Northwest Europe

Page 61 E4

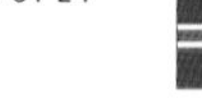

This northerly island outpost of Europe, sitting on the mid-Atlantic ridge, has stunning, sparsely inhabited volcanic terrain. Its economy crashed heavily in the 2008 global credit crunch.

Official name Republic of Iceland
Formation 1944 / 1944
Capital Reykjavik
Population 300,000 / 8 people per sq mile (3 people per sq km)
Total area 39,768 sq miles (103,000 sq km)
Languages Icelandic*
Religions Evangelical Lutheran 84%, Other (mostly Christian) 10%, Nonreligious 3%, Roman Catholic 3%
Ethnic mix Icelandic 94%, Other 5%, Danish 1%
Government Parliamentary system
Currency Icelandic króna = 100 aurar
Literacy rate 99%
Calorie consumption 3380 kilocalories

INDIA
South Asia

Page 112 D4

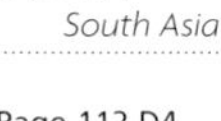

The Indian subcontinent, divided from the rest of Asia by the Himalayas, was once the jewel of the British empire. India is the world's largest democracy and second most populous country.

Official name Republic of India
Formation 1947 / 1947
Capital New Delhi
Population 1.34 billion / 1167 people per sq mile (450 people per sq km)
Total area 1,269,338 sq miles (3,287,590 sq km)
Languages Hindi*, English*, Urdu, Bengali, Marathi, Telugu, Tamil, Bihari, Gujarati, Kanarese
Religions Hindu 81%, Muslim 13%, Sikh 2%, Christian 2%, Buddhist 1%, Other 1%
Ethnic mix Indo-Aryan 72%, Dravidian 25%, Mongoloid and other 3%
Government Parliamentary system
Currency Indian rupee = 100 paise
Literacy rate 69%
Calorie consumption 2459 kilocalories

INDONESIA
Southeast Asia

Page 116 C4

The world's largest archipelago spans over 3100 miles (5000 km), from the Indian to the Pacific Ocean. Formerly the Dutch East Indies, it produces palm oil, rubber, spices, and natural gas.

Official name Republic of Indonesia
Formation 1949 / 1999
Capital Jakarta
Population 264 million / 381 people per sq mile (147 people per sq km)
Total area 741,096 sq miles (1,919,440 sq km)
Languages Javanese, Sundanese, Madurese, Bahasa Indonesia*, Dutch
Religions Sunni Muslim 87%, Protestant 7%, Roman Catholic 3%, Hindu 2%, Buddhist 1%
Ethnic mix Javanese 40%, Other 27%, Sundanese 16%, Coastal Malays 14%, Madurese 3%
Government Presidential system
Currency Rupiah = 100 sen
Literacy rate 95%
Calorie consumption 2777 kilocalories

IRAN
Southwest Asia

Page 98 C3

After the 1979 Islamist revolution led by Ayatollah Khomeini deposed the shah, this Middle Eastern country became the world's largest theocracy. It has large oil and natural gas reserves.

Official name Islamic Republic of Iran
Formation 1502 / 1990
Capital Tehran
Population 81.2 million / 129 people per sq mile (50 people per sq km)
Total area 636,293 sq miles (1,648,000 sq km)
Languages Farsi*, Azeri, Luri, Gilaki, Mazanderani, Kurdish, Turkmen, Arabic, Baluchi
Religions Shi'a Muslim 90%, Sunni Muslim 9%, Other 1%
Ethnic mix Persian 51%, Azari 24%, Other 10%, Lur and Bakhtiari 8%, Kurdish 7%
Government Islamic theocracy
Currency Iranian rial = 100 dinars
Literacy rate 85%
Calorie consumption 3094 kilocalories

IRAQ
Southwest Asia

Page 98 B3

Oil-rich Iraq is situated in the central Middle East. A US-led invasion in 2003 toppled Saddam Hussein's regime, but sectarian violence since then has caused political and social turmoil.

Official name Republic of Iraq
Formation 1932 / 1990
Capital Baghdad
Population 38.3 million / 227 people per sq mile (88 people per sq km)
Total area 168,753 sq miles (437,072 sq km)
Languages Arabic*, Kurdish*, Turkic languages, Armenian, Assyrian
Religions Shi'a Muslim 60%, Sunni Muslim 35%, Other (including Christian) 5%
Ethnic mix Arab 80%, Kurdish 15%, Turkmen 3%, Other 2%
Government Parliamentary system
Currency New Iraqi dinar = 1000 fils
Literacy rate 44%
Calorie consumption 2545 kilocalories

IRELAND
Northwest Europe

Page 67 A6

British rule ended in 1922 for 80% of the island of Ireland, which became the Irish Republic in 1949. The economy is now recovering after suffering heavily in the 2008 global financial crisis.

Official name Ireland
Formation 1922 / 1922
Capital Dublin
Population 4.8 million / 180 people per sq mile (70 people per sq km)
Total area 27,135 sq miles (70,280 sq km)
Languages English*, Irish*
Religions Roman Catholic 86%, Other Christian 6%, Nonreligious 6%, Muslim 1%, Other 1%
Ethnic mix Irish 86%, Other White 9%, Asian 2%, Other 2%, Black 1%
Government Parliamentary system
Currency Euro = 100 cents
Literacy rate 99%
Calorie consumption 3600 kilocalories

ISRAEL
Southwest Asia

Page 97 A7

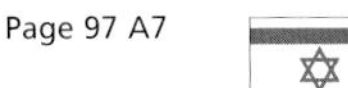

In 1948 this Jewish state was carved out of Palestine on the east coast of the Mediterranean. It has gained land from its Arab neighbors, and the status of the Palestinians remains unresolved.

Official name State of Israel
Formation 1948 / 1994
Capital Jerusalem (not internationally recognized)
Population 8.3 million / 1057 people per sq mile (408 people per sq km)
Total area 8019 sq miles (20,770 sq km)
Languages Hebrew*, Arabic*, Yiddish, German, Russian, Polish, Romanian, Persian
Religions Jewish 81%, Muslim (mainly Sunni) 14%, Druze 2%, Christian 2%, Other and nonreligious 1%
Ethnic mix Jewish 81%, Arab 18%, Other 1%
Government Parliamentary system
Currency Shekel = 100 agorot
Literacy rate 98%
Calorie consumption 3610 kilocalories

ITALY
Southern Europe

Page 74 B3

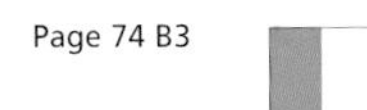

A boot-shaped peninsula jutting into the Mediterranean, Italy is a world leader in product design, fashion, and textiles. Divisions exist between the industrial north and poorer south.

Official name Italian Republic
Formation 1861 / 1947
Capital Rome
Population 59.4 million / 523 people per sq mile (202 people per sq km)
Total area 116,305 sq miles (301,230 sq km)
Languages Italian*, German, French, Rhaeto-Romanic, Sardinian
Religions Roman Catholic 90%, Nonreligious 6%, Muslim 2%, Other Christian 2%
Ethnic mix Italian 92%, Other European 5%, Other 2%, North African (mainly Moroccan) 1%
Government Parliamentary system
Currency Euro = 100 cents
Literacy rate 99%
Calorie consumption 3579 kilocalories

IVORY COAST (CÔTE D'IVOIRE)
West Africa

Page 52 D4

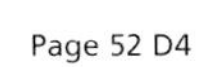

One of the larger countries on the West African coast, this ex-French colony is the world's biggest cocoa producer. Coups and recent conflicts have destroyed its reputation for stability.

Official name Republic of Côte d'Ivoire
Formation 1960 / 1960
Capital Yamoussoukro
Population 24.3 million / 198 people per sq mile (76 people per sq km)
Total area 124,502 sq miles (322,460 sq km)
Languages Akan, French*, Krou, Voltaique
Religions Muslim 43%, Nonreligious or traditional beliefs 23%, Roman Catholic 17%, Evangelical 12%, Other Christian 4%, Other 1%
Ethnic mix Akan 42%, Voltaique 18%, Mandé du Nord 17%, Krou 11%, Mandé du Sud 10%, Other 2%
Government Presidential system
Currency CFA franc = 100 centimes
Literacy rate 44%
Calorie consumption 2799 kilocalories

JAMAICA
West Indies

Page 32 C3

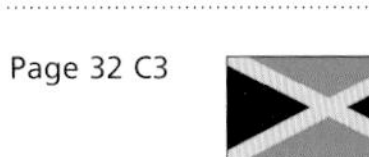

Colonized by Spain and then Britain, Jamaica was the first Caribbean island to gain independence in the postwar era. Jamaican popular music culture developed reggae, ska, and dancehall.

Official name Jamaica
Formation 1962 / 1962
Capital Kingston
Population 2.9 million / 694 people per sq mile (268 people per sq km)
Total area 4243 sq miles (10,990 sq km)
Languages English Creole, English*
Religions Church of God 26%, Nonreligious 22%, Other Christian 21%, Seventh-day Adventist 12%, Pentecostal 11%, Other 8%
Ethnic mix Black African 92%, Mulatto (mixed race) 6%, East Indian 1%, Other 1%
Government Parliamentary system
Currency Jamaican dollar = 100 cents
Literacy rate 88%
Calorie consumption 2746 kilocalories

JAPAN
East Asia

Page 108 C4

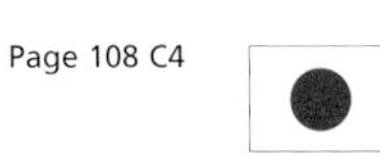

Japan has four main islands and over 3000 smaller ones. It rebuilt after defeat in World War II to become one of the world's biggest economies. It retains its emperor as head of state.

Official name Japan
Formation 1590 / 1972
Capital Tokyo
Population 128 million / 877 people per sq mile (339 people per sq km)
Total area 145,882 sq miles (377,835 sq km)
Languages Japanese*, Korean, Chinese
Religions Buddhist 50%, Nonreligious 23%, Shinto 16%, Christian 10%, Muslim 1%
Ethnic mix Japanese 99%, Other (mainly Korean) 1%
Government Parliamentary system
Currency Yen = 100 sen
Literacy rate 99%
Calorie consumption 2726 kilocalories

JORDAN
Southwest Asia

Page 97 B6

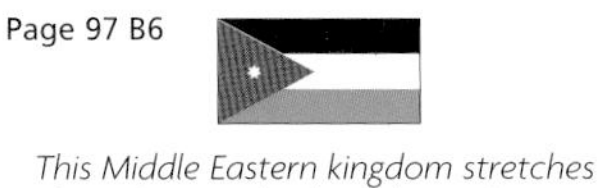

This Middle Eastern kingdom stretches from the east bank of the Jordan River into largely uninhabited desert. Calls for greater democratization have engendered some reforms.

Official name Hashemite Kingdom of Jordan
Formation 1946 / 1967
Capital Amman
Population 9.7 million / 283 people per sq mile (109 people per sq km)
Total area 35,637 sq miles (92,300 sq km)
Languages Arabic*
Religions Sunni Muslim 92%, Christian 6%, Other 2%
Ethnic mix Arab 98%, Circassian 1%, Armenian 1%
Government Monarchy
Currency Jordanian dinar = 1000 fils
Literacy rate 98%
Calorie consumption 3100 kilocalories

KAZAKHSTAN
Central Asia

Page 92 B4

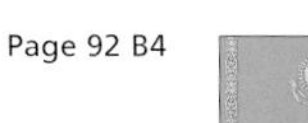

Second-largest of the former Soviet republics, mineral-rich Kazakhstan is Central Asia's major economic power. The former communist leader remains in charge, facing little opposition.

Official name Republic of Kazakhstan
Formation 1991 / 1991
Capital Astana
Population 18.2 million / 17 people per sq mile (7 people per sq km)
Total area 1,049,150 sq miles (2,717,300 sq km)
Languages Kazakh*, Russian, Ukrainian, German, Uzbek, Tatar, Uighur
Religions Muslim (mainly Sunni) 71%, Christian (mainly Orthodox) 26%, Nonreligious 3%
Ethnic mix Kazakh 63%, Russian 24%, Other 6%, Uzbek 3%, Ukrainian 2%, Uighur 1%, Tatar 1%
Government Presidential system
Currency Tenge = 100 tiyn
Literacy rate 99%
Calorie consumption 3264 kilocalories

JAPAN
East Asia

KENYA
East Africa

Page 51 C6

Straddling the Equator on Africa's east coast, Kenya has known both stable periods and internal strife since independence in 1963. Corruption is now a key political issue.

Official name Republic of Kenya
Formation 1963 / 1963
Capital Nairobi
Population 49.7 million / 227 people per sq mile (88 people per sq km)
Total area 224,961 sq miles (582,650 sq km)
Languages Kiswahili*, English*, Kikuyu, Luo, Kalenjin, Kamba
Religions Other Christian 60%, Roman Catholic 23%, Muslim 11%, Other 4%, Nonreligious 2%
Ethnic mix Other 35%, Kikuyu 17%, Luhya 14%, Kalenjin 13%, Luo 11%, Kamba 10%
Government Presidential system
Currency Kenya shilling = 100 cents
Literacy rate 79%
Calorie consumption 2206 kilocalories

KIRIBATI
Australasia & Oceania

Page 123 F3

Part of the British colony of the Gilbert and Ellice Islands until independence in 1979, Kiribati comprises 33 islands in the mid-Pacific Ocean. Phosphate deposits on Banaba ran out in 1980.

Official name Republic of Kiribati
Formation 1979 / 1979
Capital Tarawa Atoll
Population 100,000 / 365 people per sq mile (141 people per sq km)
Total area 277 sq miles (717 sq km)
Languages English*, Kiribati
Religions Roman Catholic 56%, Kiribati Protestant Church 34%, Mormon 5%, Baha'i 2%, Seventh-day Adventist 2%, Other 1%
Ethnic mix Micronesian 99%, Other 1%
Government Presidential system
Currency Australian dollar = 100 cents
Literacy rate 99%
Calorie consumption 3040 kilocalories

KOSOVO (not fully recognized)
Southeast Europe

Page 79 D5

NATO intervention in 1999 ended ethnic cleansing by the Serbs of Kosovo's majority Albanian population, and nine years later the region unilaterally declared independence from Serbia.

Official name Republic of Kosovo
Formation 2008 / 2008
Capital Prishtine/Priština
Population 1.9 million / 451 people per sq mile (174 people per sq km)
Total area 4212 sq miles (10,908 sq km)
Languages Albanian*, Serbian*, Bosniak, Gorani, Roma, Turkish
Religions Muslim 92%, Orthodox Christian 4%, Roman Catholic 4%
Ethnic mix Albanian 92%, Serb 4%, Bosniak and Gorani 2%, Roma 1%, Turkish 1%
Government Parliamentary system
Currency Euro = 100 cents
Literacy rate 92%
Calorie consumption Not available

KUWAIT
Southwest Asia

Page 98 C4

Kuwait, on the Persian Gulf, was a British protectorate from 1914 to 1961. Oil-rich since the 1950s, it was annexed briefly in 1990 by Iraq but US-led intervention restored the ruling amir.

Official name State of Kuwait
Formation 1961 / 1961
Capital Kuwait City
Population 4.1 million / 596 people per sq mile (230 people per sq km)
Total area 6880 sq miles (17,820 sq km)
Languages Arabic*, English
Religions Sunni Muslim 45%, Shi'a Muslim 40%, Christian, Hindu, and other 15%
Ethnic mix Asian 39%, Kuwaiti 37%, Other Arab 21%, African 2%, Other 1%
Government Monarchy
Currency Kuwaiti dinar = 1000 fils
Literacy rate 96%
Calorie consumption 3501 kilocalories

KYRGYZSTAN
Central Asia

Page 101 F2

This mountainous, landlocked state in Central Asia is the most rural of the ex-Soviet republics. Popular protests ousted the long-term president in 2005 and his successor in 2010.

Official name Kyrgyz Republic
Formation 1991 / 1991
Capital Bishkek
Population 6 million / 78 people per sq mile (30 people per sq km)
Total area 76,641 sq miles (198,500 sq km)
Languages Kyrgyz*, Russian*, Uzbek, Tatar, Ukrainian
Religions Muslim (mainly Sunni) 70%, Orthodox Christian 30%
Ethnic mix Kyrgyz 71%, Uzbek 14%, Russian 8%, Other 4%, Dungan 1%, Uighur 1%, Tajik 1%
Government Presidential / parliamentary system
Currency Som = 100 tyiyn
Literacy rate 99%
Calorie consumption 2817 kilocalories

LAOS
Southeast Asia

Page 114 D4

Landlocked Laos suffered a long civil war after French rule ended, and was badly bombed by US forces engaged in Vietnam. It has been under communist rule since 1975.

Official name Lao People's Democratic Republic
Formation 1953 / 1953
Capital Viangchan (Vientiane)
Population 6.9 million / 77 people per sq mile (30 people per sq km)
Total area 91,428 sq miles (236,800 sq km)
Languages Lao*, Mon-Khmer, Yao, Vietnamese, Chinese, French
Religions Buddhist 67%, Other 31%, Christian 2%
Ethnic mix Lao Loum 66%, Lao Theung 30%, Other 2%, Lao Soung 2%
Government One-party state
Currency Kip = 100 at
Literacy rate 58%
Calorie consumption 2451 kilocalories

LATVIA
Northeast Europe

Page 84 C3

Situated on the low-lying eastern shores of the Baltic Sea, Latvia, like its Baltic neighbors, regained its independence at the collapse of the USSR in 1991. It retains a large Russian population.

Official name Republic of Latvia
Formation 1991 / 1991
Capital Riga
Population 1.9 million / 76 people per sq mile (29 people per sq km)
Total area 24,938 sq miles (64,589 sq km)
Languages Latvian*, Russian
Religions Orthodox Christian 31%, Roman Catholic 23%, Nonreligious 21%, Lutheran 19%, Other 6%
Ethnic mix Latvian 62%, Russian 27%, Belarussian 3%, Other 3%, Polish 2%, Ukrainian 2%, Lithuanian 1%
Government Parliamentary system
Currency Euro = 100 cents
Literacy rate 99%
Calorie consumption 3174 kilocalories

LEBANON
Southwest Asia

Page 96 A4

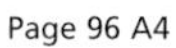

Lebanon is dwarfed by its two powerful neighbors, Syria and Israel. Muslims and Christians fought a 14-year civil war until agreeing to share power in 1989, however, instability continues.

Official name Lebanese Republic
Formation 1941 / 1941
Capital Beirut
Population 6.1 million / 1544 people per sq mile (596 people per sq km)
Total area 4015 sq miles (10,400 sq km)
Languages Arabic*, French, Armenian, Assyrian
Religions Muslim 60%, Christian 39%, Other 1%
Ethnic mix Arab 95%, Armenian 4%, Other 1%
Government Parliamentary system
Currency Lebanese pound = 100 piastres
Literacy rate 91%
Calorie consumption 3066 kilocalories

LESOTHO
Southern Africa

Page 56 D5

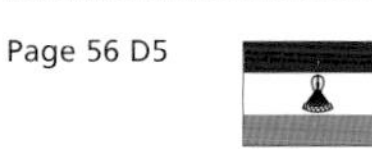

Lesotho lies within South Africa, on whom it is economically dependent. Elections in 1993 ended military rule, but South Africa has had to intervene in politics since. AIDS is a problem.

Official name Kingdom of Lesotho
Formation 1966 / 1966
Capital Maseru
Population 2.2 million / 188 people per sq mile (72 people per sq km)
Total area 11,720 sq miles (30,355 sq km)
Languages English*, Sesotho*, isiZulu
Religions Christian 90%, Traditional beliefs 10%
Ethnic mix Sotho 99%, European and Asian 1%
Government Parliamentary system
Currency Loti = 100 lisente; South African rand = 100 cents
Literacy rate 77%
Calorie consumption 2529 kilocalories

LIBERIA
West Africa

Page 52 C5

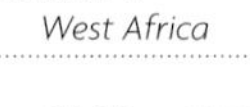

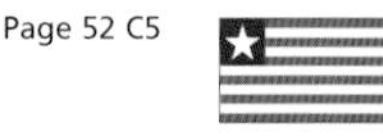

Facing the Atlantic Ocean, Liberia is Africa's oldest republic, founded in 1847 by freed US slaves. Recovery from the 1990s' civil war has been set back by the recent Ebola epidemic.

Official name Republic of Liberia
Formation 1847 / 1847
Capital Monrovia
Population 4.7 million / 126 people per sq mile (49 people per sq km)
Total area 43,000 sq miles (111,370 sq km)
Languages Kpelle, Vai, Bassa, Kru, Grebo, Kissi, Gola, Lorma, English*
Religions Christian 86%, Muslim 12%, Nonreligious 1%, Traditional beliefs and other 1%
Ethnic mix Indigenous tribes (12 groups) 50%, Kpellé 20%, Bassa 14%, Gio 8%, Krou 6%, Other 2%
Government Presidential system
Currency Liberian dollar = 100 cents
Literacy rate 43%
Calorie consumption 2204 kilocalories

LIBYA
North Africa

Page 49 F3

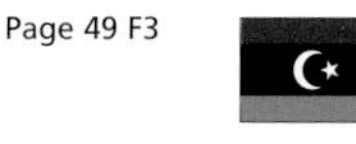

On the Mediterranean coast, Libya was ruled from 1969 by the idiosyncratic Col. Gaddafi. The 2011 "Arab Spring" turned to civil war, toppling his regime, but leaving Libya in anarchy.

Official name Libya
Formation 1951 / 1951
Capital Tripoli
Population 6.4 million / 9 people per sq mile (4 people per sq km)
Total area 679,358 sq miles (1,759,540 sq km)
Languages Arabic*, Tuareg
Religions Muslim (mainly Sunni) 97%, Other 3%
Ethnic mix Arab and Berber 97%, Other 3%
Government Transitional regime
Currency Libyan dinar = 1000 dirhams
Literacy rate 90%
Calorie consumption 3211 kilocalories

LIECHTENSTEIN
Central Europe

Page 73 B7

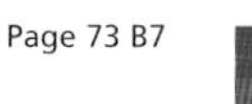

Tucked in the Alps between Switzerland and Austria, Liechtenstein became an independent principality of the Holy Roman Empire in 1719. Switzerland handles its foreign affairs and defense.

Official name Principality of Liechtenstein
Formation 1719 / 1719
Capital Vaduz
Population 38,000 / 613 people per sq mile (238 people per sq km)
Total area 62 sq miles (160 sq km)
Languages German*, Alemannish dialect, Italian
Religions Roman Catholic 78%, Protestant 9%, Muslim 6%, Nonreligious 5%, Orthodox Christian 1%, Other 1%
Ethnic mix Liechtensteiner 66%, Other 12%, Swiss 10%, Austrian 6%, German 3%, Italian 3%
Government Parliamentary system
Currency Swiss franc = 100 rappen/centimes
Literacy rate 99%
Calorie consumption Not available

LITHUANIA
Northeast Europe

Page 84 B4

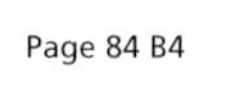

A flat land of lakes, moors, and bogs, Lithuania is the largest of the three Baltic states. It has historical ties to Poland and was the first former Soviet republic to declare independence.

Official name Republic of Lithuania
Formation 1991 / 1991
Capital Vilnius
Population 2.9 million / 115 people per sq mile (44 people per sq km)
Total area 25,174 sq miles (65,200 sq km)
Languages Lithuanian*, Russian
Religions Roman Catholic 75%, Christian 14%, Nonreligious 6%, Orthodox Christian 3%, Other 2%
Ethnic mix Lithuanian 85%, Polish 7%, Russian 6%, Belarussian 1%, Other 1%
Government Parliamentary system
Currency Euro = 100 cents
Literacy rate 99%
Calorie consumption 3417 kilocalories

LUXEMBOURG
Northwest Europe

Page 65 D8

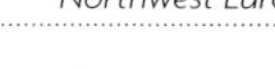

Part of the forested Ardennes plateau in Northwest Europe, Luxembourg is Europe's last independent duchy and one of its richest states. It is a banking center and hosts EU institutions.

Official name Grand Duchy of Luxembourg
Formation 1867 / 1867
Capital Luxembourg
Population 600,000 / 601 people per sq mile (232 people per sq km)
Total area 998 sq miles (2586 sq km)
Languages Luxembourgish*, German*, French*
Religions Roman Catholic 97%, Protestant, Orthodox Christian, and Jewish 3%
Ethnic mix Luxembourger 62%, Foreign residents 38%
Government Parliamentary system
Currency Euro = 100 cents
Literacy rate 99%
Calorie consumption 3539 kilocalories

MACEDONIA
Southeast Europe

Page 79 D6

This ex-Yugoslav state is landlocked in the southern Balkans. Its EU candidacy is held back over Greek fears that its name implies a claim to its own northern province of Macedonia.

Official name Republic of Macedonia
Formation 1991 / 1991
Capital Skopje
Population 2.1 million / 212 people per sq mile (82 people per sq km)
Total area 9781 sq miles (25,333 sq km)
Languages Macedonian*, Albanian*, Turkish, Romani, Serbian
Religions Orthodox Christian 65%, Muslim 33%, Other 2%
Ethnic mix Macedonian 64%, Albanian 25%, Turkish 4%, Roma 3%, Other 2%, Serb 2%
Government Presidential / parliamentary system
Currency Macedonian denar = 100 deni
Literacy rate 98%
Calorie consumption 2949 kilocalories

MADAGASCAR
Indian Ocean

Page 57 F4

Off Africa's southeast coast, this former French colony is the world's fourth-largest island. Free elections in 1993 ended 18 years of socialism, but power struggles have blighted politics since.

Official name Republic of Madagascar
Formation 1960 / 1960
Capital Antananarivo
Population 25.6 million / 114 people per sq mile (44 people per sq km)
Total area 226,656 sq miles (587,040 sq km)
Languages Malagasy*, French*, English*
Religions Traditional beliefs 52%, Christian (mainly Roman Catholic) 41%, Other 7%
Ethnic mix Other Malay 46%, Merina 26%, Betsimisaraka 15%, Betsileo 12%, Other 1%
Government Presidential / parliamentary system
Currency Ariary = 5 iraimbilanja
Literacy rate 72%
Calorie consumption 2052 kilocalories

MALAWI
Southern Africa

Page 57 E1

This landlocked former British colony lies along the Great Rift Valley and Lake Nyasa, Africa's third-largest lake. Multiparty elections in 1994 ended three decades of single-party rule.

Official name Republic of Malawi
Formation 1964 / 1964
Capital Lilongwe
Population 18.6 million / 512 people per sq mile (198 people per sq km)
Total area 45,745 sq miles (118,480 sq km)
Languages Chewa, Lomwe, Yao, Ngoni, English*
Religions Christian (mainly Protestant) 83%, Muslim 13%, Nonreligious 2%, Other 2%
Ethnic mix Bantu 99%, Other 1%
Government Presidential system
Currency Malawi kwacha = 100 tambala
Literacy rate 62%
Calorie consumption 2367 kilocalories

MALAYSIA
Southeast Asia

Page 116 B3

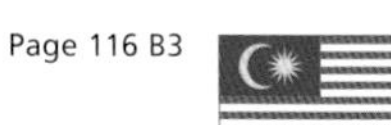

Three separate territories, Peninsular Malaysia, and Sarawak and Sabah on Borneo, make up Malaysia. Relations between indigenous Malays and the Chinese minority dominate politics.

Official name Malaysia
Formation 1963 / 1965
Capital Kuala Lumpur; Putrajaya (administrative)
Population 31.6 million / 249 people per sq mile (96 people per sq km)
Total area 127,316 sq miles (329,750 sq km)
Languages Bahasa Malaysia*, Malay, Chinese, Tamil, English
Religions Muslim (mainly Sunni) 62%, Buddhist 20%, Christian 9%, Hindu 6%, Other 3%
Ethnic mix Malay 50%, Chinese 22%, Indigenous tribes 12%, Other 9%, Indian 7%
Government Parliamentary system
Currency Ringgit = 100 sen
Literacy rate 93%
Calorie consumption 2916 kilocalories

MALDIVES
Indian Ocean

Page 110 A4

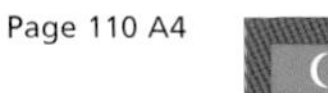

Of this group of over 1000 small low-lying coral islands in the Indian Ocean, only 200 are inhabited. A few families dominate politics and have reversed the electoral upsets of 2008 and 2009.

Official name Republic of Maldives
Formation 1965 / 1965
Capital Maale (Male')
Population 400,000 / 3448 people per sq mile (1333 people per sq km)
Total area 116 sq miles (300 sq km)
Languages Dhivehi* (Maldivian), Sinhala, Tamil, Arabic
Religions Sunni Muslim 94%, Hindu 3%, Christian 2%, Buddhist 1%
Ethnic mix Arab–Sinhalese–Malay 100%
Government Presidential system
Currency Rufiyaa = 100 laari
Literacy rate 99%
Calorie consumption 2732 kilocalories

MALI
West Africa

Page 53 E2

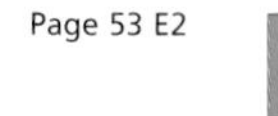

Mali's power as a trans-Saharan trading empire peaked 700 years ago. Modern Mali, a one-party state until 1992, called on former colonial power France to suppress Islamist rebels since 2013.

Official name Republic of Mali
Formation 1960 / 1960
Capital Bamako
Population 18.5 million / 39 people per sq mile (15 people per sq km)
Total area 478,764 sq miles (1,240,000 sq km)
Languages Bambara, Fulani, Senufo, Soninke, French*
Religions Muslim (mainly Sunni) 90%, Traditional beliefs 6%, Christian 4%
Ethnic mix Bambara 52%, Other 14%, Fulani 11%, Saracolé 7%, Soninka 7%, Tuareg 5%, Mianka 4%
Government Presidential system
Currency CFA franc = 100 centimes
Literacy rate 33%
Calorie consumption 2890 kilocalories

MALTA
Southern Europe

Page 80 A5

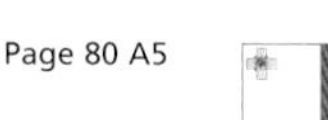

The Maltese archipelago lies off Sicily. Only Malta, Kemmuna, and Gozo are inhabited. Its mid-Mediterranean location has made it a gateway for illegal migration from Africa to Europe.

Official name Republic of Malta
Formation 1964 / 1964
Capital Valletta
Population 400,000 / 3226 people per sq mile (1250 people per sq km)
Total area 122 sq miles (316 sq km)
Languages Maltese*, English*
Religions Roman Catholic 98%, Other and nonreligious 2%
Ethnic mix Maltese 96%, Other 4%
Government Parliamentary system
Currency Euro = 100 cents
Literacy rate 93%
Calorie consumption 3378 kilocalories

MARSHALL ISLANDS
Australasia & Oceania

Page 122 D1

This group of 34 atolls was under US rule as part of the UN Trust Territory of the Pacific Islands until 1986. The economy depends on US aid and rent for the US missile base on Kwajalein.

Official name Republic of the Marshall Islands
Formation 1986 / 1986
Capital Majuro Atoll
Population 53,000 / 757 people per sq mile (292 people per sq km)
Total area 70 sq miles (181 sq km)
Languages Marshallese*, English*, Japanese, German
Religions Protestant 81%, Other 11%, Roman Catholic 8%
Ethnic mix Micronesian 90%, Other 10%
Government Presidential system
Currency US dollar = 100 cents
Literacy rate 98%
Calorie consumption Not available

MAURITANIA
West Africa

Page 52 C2

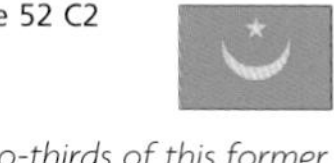

Two-thirds of this former French colony is desert. The Maures oppress the black minority. Despite multiparty polls since 1991, military leaders have held power until 2005 and since 2008.

Official name Islamic Republic of Mauritania
Formation 1960 / 1960
Capital Nouakchott
Population 4.4 million / 11 people per sq mile (4 people per sq km)
Total area 397,953 sq miles (1,030,700 sq km)
Languages Arabic*, Hassaniyah Arabic, Wolof, French
Religions Sunni Muslim 100%
Ethnic mix Maure 81%, Wolof 7%, Tukolor 5%, Other 4%, Soninka 3%
Government Presidential system
Currency Ouguiya = 5 khoums
Literacy rate 46%
Calorie consumption 2876 kilocalories

MAURITIUS
Indian Ocean

Page 57 H3

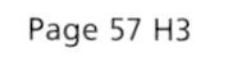

East of Madagascar in the Indian Ocean, Mauritius became a republic 24 years after independence from Britain. Its diversified economy includes tourism, financial services, and outsourcing.

Official name Republic of Mauritius
Formation 1968 / 1968
Capital Port Louis
Population 1.3 million / 1811 people per sq mile (699 people per sq km)
Total area 718 sq miles (1860 sq km)
Languages French Creole, Hindi, Urdu, Tamil, Chinese, English*, French
Religions Hindu 48%, Roman Catholic 26%, Muslim 17%, Other Christian 7%, Other 2%
Ethnic mix Indo-Mauritian 68%, Creole 27%, Sino-Mauritian 3%, Franco-Mauritian 2%
Government Parliamentary system
Currency Mauritian rupee = 100 cents
Literacy rate 93%
Calorie consumption 3065 kilocalories

MEXICO
North America

Page 28 D3

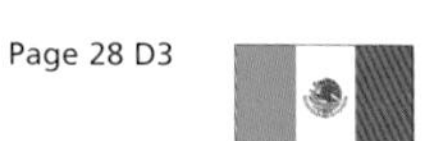

Located between the US and the Central American states, Mexico was a Spanish colony for 300 years. Sprawling Mexico City is built on the site of the Aztec capital, Tenochtitlán.

Official name United Mexican States
Formation 1836 / 1848
Capital Mexico City
Population 129 million / 175 people per sq mile (68 people per sq km)
Total area 761,602 sq miles (1,972,550 sq km)
Languages Spanish*, Nahuatl, Mayan, Zapotec, Mixtec, Otomi, Totonac, Tzotzil, Tzeltal
Religions Roman Catholic 81%, Protestant 9%, Nonreligious 7%, Other 3%
Ethnic mix Mestizo 60%, Amerindian 30%, European 9%, Other 1%
Government Presidential system
Currency Mexican peso = 100 centavos
Literacy rate 94%
Calorie consumption 3072 kilocalories

MICRONESIA
Australasia & Oceania

Page 122 B1

The Federated States of Micronesia, situated in the western Pacific, comprises 607 islands and atolls grouped into four main island states. The economy relies on US aid.

Official name Federated States of Micronesia
Formation 1986 / 1986
Capital Palikir (Pohnpei Island)
Population 100,000 / 369 people per sq mile (142 people per sq km)
Total area 271 sq miles (702 sq km)
Languages Trukese, Pohnpeian, Kosraean, Yapese, English*
Religions Roman Catholic 53%, Protestant 43%, Other 3%, Nonreligious 1%
Ethnic mix Chuukese 49%, Pohnpeian 24%, Other 14%, Kosraean 6%, Yapese 5%, Asian 2%
Government Nonparty system
Currency US dollar = 100 cents
Literacy rate 81%
Calorie consumption Not available

MOLDOVA
Southeast Europe

Page 86 D3

The smallest and most densely populated of the ex-Soviet republics, Moldova has strong linguistic and cultural ties with Romania to the west. It exports tobacco, wine, and fruit.

Official name Republic of Moldova
Formation 1991 / 1991
Capital Chisinau
Population 4.1 million / 315 people per sq mile (122 people per sq km)
Total area 13,067 sq miles (33,843 sq km)
Languages Moldovan*, Ukrainian, Russian
Religions Orthodox Christian 92%, Other 6%, Nonreligious 2%
Ethnic mix Moldovan 76%, Ukrainian 9%, Russian 6%, Gagauz 4%, Romanian 2%, Bulgarian 2%, Other 1%
Government Parliamentary system
Currency Moldovan leu = 100 bani
Literacy rate 99%
Calorie consumption 2714 kilocalories

MONACO
Southern Europe

Page 69 E6

The destiny of this tiny enclave on France's Côte d'Azur was changed in 1863 when its prince opened a casino. A jet-set image and thriving service sector define its modern identity.

Official name Principality of Monaco
Formation 1861 / 1861
Capital Monaco
Population 38,000 / 50,667 people per sq mile (19,487 people per sq km)
Total area 0.75 sq miles (1.95 sq km)
Languages French*, Italian, Monégasque, English
Religions Roman Catholic 89%, Protestant 6%, Other 5%
Ethnic mix French 47%, Other 21%, Italian 16%, Monégasque 16%
Government Monarchical / parliamentary system
Currency Euro = 100 cents
Literacy rate 99%
Calorie consumption Not available

MONGOLIA
East Asia

Page 104 D2

Vast Mongolia is sparsely populated and mostly desert. Under the sway of its giant neighbors, Russia and China, it was communist from independence from China in 1924 until 1990.

Official name Mongolia
Formation 1924 / 1924
Capital Ulaanbaatar
Population 3.1 million / 5 people per sq mile (2 people per sq km)
Total area 604,247 sq miles (1,565,000 sq km)
Languages Khalkha Mongolian*, Kazakh, Chinese, Russian
Religions Tibetan Buddhist 53%, Nonreligious 38%, Muslim 3%, Shamanist 3%, Christian 2%, Other 1%
Ethnic mix Khalkh 82%, Other 9%, Kazakh 4%, Dorvod 3%, Bayad 2%
Government Presidential / parliamentary system
Currency Tugrik (tögrög) = 100 möngö
Literacy rate 98%
Calorie consumption 2510 kilocalories

MONTENEGRO
Southeast Europe

Page 79 C5

Part of the former Yugoslavia, the tiny republic of Montenegro broke away from Serbia in 2006. Its attractive coast and mountains are a big tourist draw. It hopes to join the EU soon.

Official name Montenegro
Formation 2006 / 2006
Capital Podgorica
Population 600,000 / 113 people per sq mile (43 people per sq km)
Total area 5332 sq miles (13,812 sq km)
Languages Montenegrin*, Serbian, Albanian, Bosniak, Croatian
Religions Orthodox Christian 74%, Muslim 20%, Roman Catholic 4%, Nonreligious 1%, Other 1%
Ethnic mix Montenegrin 43%, Serb 32%, Other 12%, Bosniak 8%, Albanian 5%
Government Parliamentary system
Currency Euro = 100 cents
Literacy rate 98%
Calorie consumption 3491 kilocalories

MOROCCO
North Africa

Page 48 C2

A former French colony in northwest Africa, Morocco has occupied the disputed territory of Western Sahara since 1975. The king has handed more power to parliament since 2011.

Official name Kingdom of Morocco
Formation 1956 / 1969
Capital Rabat
Population 35.7 million / 207 people per sq mile (80 people per sq km)
Total area 172,316 sq miles (446,300 sq km)
Languages Arabic*, Tamazight* (Berber), French, Spanish
Religions Muslim (mainly Sunni) 99%, Other (mostly Christian) 1%
Ethnic mix Arab 70%, Berber 29%, European 1%
Government Monarchical / parliamentary system
Currency Moroccan dirham = 100 centimes
Literacy rate 69%
Calorie consumption 3403 kilocalories

MOZAMBIQUE
Southern Africa

Page 57 E3

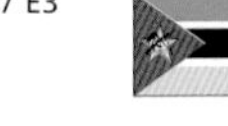

Mozambique, on the southeast African coast, frequently suffers both floods and droughts. It was torn by civil war from 1977 to 1992 as the Marxist state fought South African-backed rebels.

Official name Republic of Mozambique
Formation 1975 / 1975
Capital Maputo
Population 29.7 million / 98 people per sq mile (38 people per sq km)
Total area 309,494 sq miles (801,590 sq km)
Languages Makua, Xitsonga, Sena, Lomwe, Portuguese*
Religions Roman Catholic 28%, Nonreligious 19%, Muslim 18%, Traditional beliefs 16%, Other 19%
Ethnic mix Makua Lomwe 47%, Tsonga 23%, Malawi 12%, Shona 11%, Yao 4%, Other 3%
Government Presidential system
Currency New metical = 100 centavos
Literacy rate 51%
Calorie consumption 2283 kilocalories

MYANMAR (BURMA)
Southeast Asia

Page 114 A3

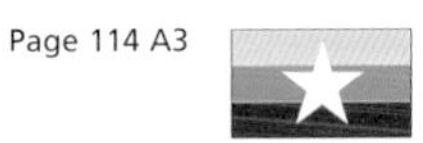

Myanmar, on the eastern shores of the Bay of Bengal and the Andaman Sea, suffered years of ethnic conflict and repressive military rule following independence from Britain in 1948.

Official name Republic of the Union of Myanmar
Formation 1948 / 1948
Capital Nay Pyi Taw
Population 53.4 million / 210 people per sq mile (81 people per sq km)
Total area 261,969 sq miles (678,500 sq km)
Languages Burmese* (Myanmar), Shan, Karen, Rakhine, Chin, Yangbye, Kachin, Mon
Religions Buddhist 90%, Christian 6%, Muslim 2%, Animist 1%, Other 1%
Ethnic mix Burman (Bamah) 68%, Other 12%, Shan 9%, Karen 7%, Rakhine 4%
Government Presidential system
Currency Kyat = 100 pyas
Literacy rate 76%
Calorie consumption 2571 kilocalories

NAMIBIA
Southern Africa

Page 56 B3

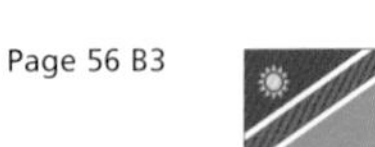

On Africa's southwest coast, this mineral-rich ex-German colony was governed by South Africa from 1915 to 1990. The white minority controls the economy, a legacy of apartheid.

Official name Republic of Namibia
Formation 1990 / 1994
Capital Windhoek
Population 2.5 million / 8 people per sq mile (3 people per sq km)
Total area 318,694 sq miles (825,418 sq km)
Languages Ovambo, Kavango, English*, Bergdama, German, Afrikaans
Religions Christian 90%, Traditional beliefs 10%
Ethnic mix Ovambo 50%, Other tribes 22%, Kavango 9%, Herero 7%, Damara 7%, Other 5%
Government Presidential system
Currency Namibian dollar = 100 cents; South African rand = 100 cents
Literacy rate 88%
Calorie consumption 2171 kilocalories

NAURU
Australasia & Oceania

Page 122 D3

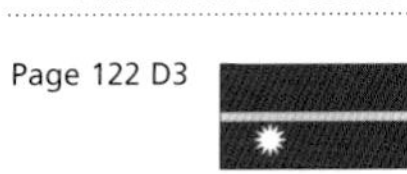

The world's smallest republic, 2480 miles (4000 km) northeast of Australia, grew rich from its phosphate deposits, but these have almost run out and poor investment has caused financial crisis.

Official name Republic of Nauru
Formation 1968 / 1968
Capital None (Yaren de facto capital)
Population 13,000 / 1605 people per sq mile (619 people per sq km)
Total area 8.1 sq miles (21 sq km)
Languages Nauruan*, Kiribati, Chinese, Tuvaluan, English
Religions Nauruan Congregational Church 60%, Roman Catholic 35%, Other 5%
Ethnic mix Nauruan 93%, Chinese 5%, Other Pacific islanders 1%, European 1%
Government Nonparty system
Currency Australian dollar = 100 cents
Literacy rate 95%
Calorie consumption Not available

NEPAL
South Asia

Page 113 E3

Nestled in the Himalayas, Nepal had an absolute monarch until 1990. Unstable coalitions typify politics. Abolition of the monarchy was a condition for ending the Maoist rebellion in 2008.

Official name Federal Democratic Republic of Nepal
Formation 1769 / 1769
Capital Kathmandu
Population 29.3 million / 555 people per sq mile (214 people per sq km)
Total area 54,363 sq miles (140,800 sq km)
Languages Nepali*, Maithili, Bhojpuri
Religions Hindu 82%, Buddhist 9%, Other (including Christian) 5%, Muslim 4%
Ethnic mix Other 52%, Chhetri 17%, Hill Brahman 12%, Magar 7%, Tharu 7%, Tamang 6%
Government Parliamentary system
Currency Nepalese rupee = 100 paisa
Literacy rate 60%
Calorie consumption 2673 kilocalories

NETHERLANDS
Northwest Europe

Page 64 C3

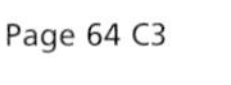

Astride the delta of four major rivers in northwest Europe, the Netherlands was ruled by Spain until 1648. It has a long trading tradition, and Rotterdam remains Europe's largest port.

Official name Kingdom of the Netherlands
Formation 1648 / 1839
Capital Amsterdam; The Hague (administrative)
Population 17 million / 1298 people per sq mile (501 people per sq km)
Total area 16,033 sq miles (41,526 sq km)
Languages Dutch*, Frisian
Religions Roman Catholic 36%, Other 34%, Protestant 27%, Muslim 3%
Ethnic mix Dutch 82%, Other 12%, Surinamese 2%, Turkish 2%, Moroccan 2%
Government Parliamentary system
Currency Euro = 100 cents
Literacy rate 99%
Calorie consumption 3228 kilocalories

NEW ZEALAND
Australasia & Oceania

Page 128 A4

This former British colony, on the Pacific Rim, has a volcanic, more populous North Island and a mountainous South Island. It was the first country to give women the vote, in 1893.

Official name New Zealand
Formation 1947 / 1947
Capital Wellington
Population 4.7 million / 45 people per sq mile (17 people per sq km)
Total area 103,737 sq miles (268,680 sq km)
Languages English*, Maori*
Religions Nonreligious 36%, Other Christian 16%, Anglican 15%, Roman Catholic 14%, Presbyterian 11%, Other 8%
Ethnic mix European 60%, Other 19%, Maori 14%, Chinese 4%, Samoan 3%
Government Parliamentary system
Currency New Zealand dollar = 100 cents
Literacy rate 99%
Calorie consumption 3137 kilocalories

NICARAGUA
Central America

Page 30 D3

Nicaragua, at the heart of Central America, plans to build a canal to rival Panama. Left-wing Sandinistas threw out a brutal dictator in 1978, then faced conflict with US-backed Contras.

Official name Republic of Nicaragua
Formation 1838 / 1838
Capital Managua
Population 6.2 million / 135 people per sq mile (52 people per sq km)
Total area 49,998 sq miles (129,494 sq km)
Languages Spanish*, English Creole, Miskito
Religions Roman Catholic 50%, Protestant 40%, Nonreligious 7%, Other 3%
Ethnic mix Mestizo 69%, White 17%, Black 9%, Amerindian 5%
Government Presidential system
Currency Córdoba oro = 100 centavos
Literacy rate 78%
Calorie consumption 2638 kilocalories

NIGER
West Africa

Page 53 G3

Landlocked Niger is linked to the sea by the River Niger. This ex-French colony has suffered coups, military rule, civil unrest, and severe droughts. It is one of the poorest countries in the world.

Official name Republic of Niger
Formation 1960 / 1960
Capital Niamey
Population 21.5 million / 44 people per sq mile (17 people per sq km)
Total area 489,188 sq miles (1,267,000 sq km)
Languages Hausa, Djerma, Fulani, Tuareg, Teda, French*
Religions Muslim 99%, Other (including Christian) 1%
Ethnic mix Hausa 55%, Djerma and Songhai 21%, Tuareg 9%, Peul 9%, Kanuri 5%, Other 1%
Government Presidential system
Currency CFA franc = 100 centimes
Literacy rate 16%
Calorie consumption 2547 kilocalories

NIGERIA
West Africa

Page 53 G4

Nigeria has Africa's largest population, whose religious and ethnic rivalries have brought down both civilian and military regimes in the past. Islamic extremists are one current challenge.

Official name Federal Republic of Nigeria
Formation 1960 / 1961
Capital Abuja
Population 191 million / 543 people per sq mile (210 people per sq km)
Total area 356,667 sq miles (923,768 sq km)
Languages Hausa, English*, Yoruba, Ibo
Religions Muslim 50%, Christian 40%, Traditional beliefs 10%
Ethnic mix Other 29%, Hausa 21%, Yoruba 21%, Ibo 18%, Fulani 11%
Government Presidential system
Currency Naira = 100 kobo
Literacy rate 51%
Calorie consumption 2700 kilocalories

NORTH KOREA
East Asia

Page 106 E3

The maverick communist state in Korea's northern half has been isolated from the outside world since 1948. Its shattered state-run economy leaves people short of food and power.

Official name Democratic People's Republic of Korea
Formation 1948 / 1953
Capital Pyongyang
Population 25.5 million / 549 people per sq mile (212 people per sq km)
Total area 46,540 sq miles (120,540 sq km)
Languages Korean*
Religions Atheist 100%
Ethnic mix Korean 100%
Government One-party state
Currency North Korean won = 100 chon
Literacy rate 99%
Calorie consumption 2094 kilocalories

NORWAY
Northern Europe

Page 63 A5

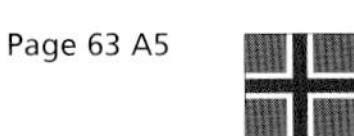

Lying on the rugged western coast of Scandinavia, most people live in southern, coastal areas. Oil and gas wealth has brought one of the world's best standards of living.

Official name Kingdom of Norway
Formation 1905 / 1905
Capital Oslo
Population 5.3 million / 45 people per sq mile (17 people per sq km)
Total area 125,181 sq miles (324,220 sq km)
Languages Norwegian* (Bokmål "book language" and Nynorsk "new Norsk"), Sámi
Religions Evangelical Lutheran 88%, Other and nonreligious 8%, Muslim 2%, Roman Catholic 1%, Pentecostal 1%
Ethnic mix Norwegian 93%, Other 6%, Sámi 1%
Government Parliamentary system
Currency Norwegian krone = 100 øre
Literacy rate 99%
Calorie consumption 3485 kilocalories

OMAN
Southwest Asia

Page 99 D6

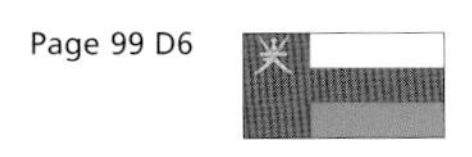

Situated on the eastern corner of the Arabian Peninsula, Oman is the least developed of the Gulf states, despite modest oil exports. The current sultan has been in power since 1970.

Official name Sultanate of Oman
Formation 1951 / 1951
Capital Muscat
Population 4.6 million / 56 people per sq mile (22 people per sq km)
Total area 82,031 sq miles (212,460 sq km)
Languages Arabic*, Baluchi, Farsi, Hindi, Punjabi
Religions Ibadi Muslim 75%, Other Muslim and Hindu 25%
Ethnic mix Arab 88%, Baluchi 4%, Indian and Pakistani 3%, Persian 3%, African 2%
Government Monarchy
Currency Omani rial = 1000 baisa
Literacy rate 93%
Calorie consumption 3143 kilocalories

PAKISTAN
South Asia

Page 112 B2

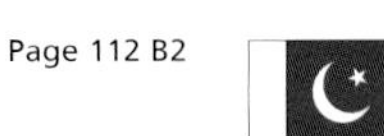

Once part of British India, Pakistan was created in 1947 as a Muslim state. Today, this nuclear-armed country is struggling to deal with complex domestic and international tensions.

Official name Islamic Republic of Pakistan
Formation 1947 / 1971
Capital Islamabad
Population 197 million / 662 people per sq mile (256 people per sq km)
Total area 310,401 sq miles (803,940 sq km)
Languages Punjabi, Sindhi, Pashtu, Urdu*, Baluchi, Brahui
Religions Sunni Muslim 77%, Shi'a Muslim 20%, Hindu 2%, Christian 1%
Ethnic mix Punjabi 56%, Pathan (Pashtun) 15%, Sindhi 14%, Mohajir 7%, Other 4%, Baluchi 4%
Government Parliamentary system
Currency Pakistani rupee = 100 paisa
Literacy rate 57%
Calorie consumption 2440 kilocalories

PALAU
Australasia & Oceania

Page 122 A2

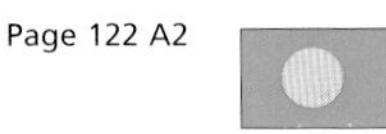

This archipelago of over 200 islands, only ten of which are inhabited, lies in the western Pacific Ocean. Until 1994 it was under US administration. The economy relies on US aid and tourism.

Official name Republic of Palau
Formation 1994 / 1994
Capital Ngerulmud
Population 21,000 / 107 people per sq mile (41 people per sq km)
Total area 177 sq miles (458 sq km)
Languages Palauan*, English*, Japanese, Angaur, Tobi, Sonsorolese
Religions Roman Catholic 49%, Protestant 33%, Modekngei 9%, Other 8%, Nonreligious 1%
Ethnic mix Palauan 73%, Filipino 16%, Other Asian 7%, Other Micronesian 3%, Other 1%
Government Nonparty system
Currency US dollar = 100 cents
Literacy rate 97%
Calorie consumption Not available

PANAMA
Central America

Page 31 F5

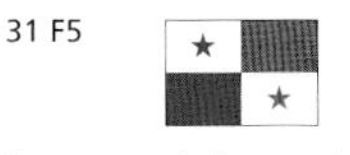

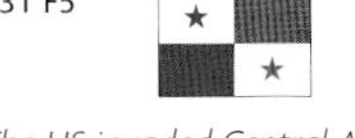

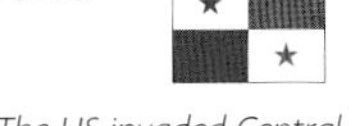

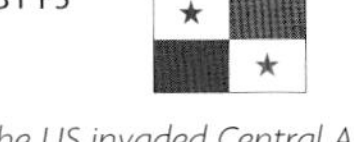

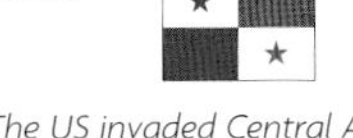

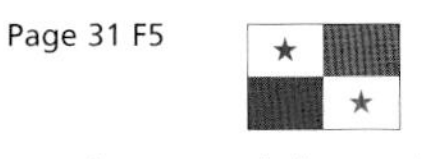

The US invaded Central America's southernmost country in 1989 to oust its dictator. The Panama Canal is a vital shortcut for shipping between the Atlantic and Pacific oceans.

Official name Republic of Panama
Formation 1903 / 1903
Capital Panama City
Population 4.1 million / 140 people per sq mile (54 people per sq km)
Total area 30,193 sq miles (78,200 sq km)
Languages English Creole, Spanish*, Amerindian languages, Chibchan languages
Religions Roman Catholic 70%, Protestant 19%, Nonreligious 7%, Other 4%
Ethnic mix Mestizo 70%, Black 14%, White 10%, Amerindian 6%
Government Presidential system
Currency Balboa = 100 centésimos; US dollar
Literacy rate 94%
Calorie consumption 2733 kilocalories

PAPUA NEW GUINEA
Australasia & Oceania

Page 122 B3

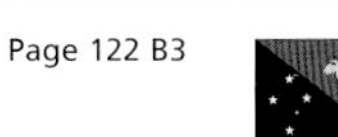

The world's most linguistically diverse country, mineral-rich PNG occupies the east of the island of New Guinea and several other island groups. It was administered by Australia before 1975.

Official name Independent State of Papua New Guinea
Formation 1975 / 1975
Capital Port Moresby
Population 8.3 million / 47 people per sq mile (18 people per sq km)
Total area 178,703 sq miles (462,840 sq km)
Languages Pidgin English, Papuan, English*, Motu, 800 (est.) native languages
Religions Protestant 60%, Roman Catholic 37%, Other 3%
Ethnic mix Melanesian and mixed race 100%
Government Parliamentary system
Currency Kina = 100 toea
Literacy rate 63%
Calorie consumption 2193 kilocalories

PARAGUAY
South America

Page 42 D2

South America's longest dictatorship held power in landlocked Paraguay from 1954 to 1989. Now under democratic rule, the country's economy is still largely agricultural.

Official name Republic of Paraguay
Formation 1811 / 1938
Capital Asunción
Population 6.8 million / 44 people per sq mile (17 people per sq km)
Total area 157,046 sq miles (406,750 sq km)
Languages Guaraní*, Spanish*, German
Religions Roman Catholic 89%, Protestant (including Mennonite) 7%, Other 3%, Nonreligious 1%
Ethnic mix Mestizo 91%, Other 7%, Amerindian 2%
Government Presidential system
Currency Guaraní = 100 céntimos
Literacy rate 95%
Calorie consumption 2589 kilocalories

PERU
South America

Page 38 C3

On the Pacific coast of South America, Peru was once the heart of the Inca empire, before the Spanish conquest in the 16th century. It elected its first Amerindian president in 2001.

Official name Republic of Peru
Formation 1824 / 1941
Capital Lima
Population 32.2 million / 65 people per sq mile (25 people per sq km)
Total area 496,223 sq miles (1,285,200 sq km)
Languages Spanish*, Quechua*, Aymara
Religions Roman Catholic 76%, Protestant 17%, Nonreligious 4%, Other 3%
Ethnic mix Amerindian 45%, Mestizo 37%, White 15%, Other 3%
Government Presidential system
Currency New sol = 100 céntimos
Literacy rate 94%
Calorie consumption 2700 kilocalories

PHILIPPINES
Southeast Asia

Page 117 E1

This 7107-island archipelago between the South China Sea and the Pacific is prone to earthquakes and volcanic activity. A 21-year dictatorship ended in 1986; politics since has been volatile.

Official name Republic of the Philippines
Formation 1946 / 1946
Capital Manila
Population 105 million / 911 people per sq mile (352 people per sq km)
Total area 115,830 sq miles (300,000 sq km)
Languages Filipino*, English*, Tagalog, Cebuano, Ilocano, Hiligaynon, many other local languages
Religions Roman Catholic 81%, Other Christian 11%, Muslim 5%, Other 3%
Ethnic mix Other 34%, Tagalog 28%, Cebuano 13%, Ilocano 9%, Hiligaynon 8%, Bisaya 8%
Government Presidential system
Currency Philippine peso = 100 centavos
Literacy rate 96%
Calorie consumption 2570 kilocalories

POLAND
Northern Europe

Page 76 B3

Poland's low-lying plains extend from the Baltic Sea into the heart of Europe. It has undergone massive political and economic change since the fall of communism. It joined the EU in 2004.

Official name Republic of Poland
Formation 1918 / 1945
Capital Warsaw
Population 38.2 million / 325 people per sq mile (125 people per sq km)
Total area 120,728 sq miles (312,685 sq km)
Languages Polish*
Religions Roman Catholic 87%, Nonreligious 7%, Other 5%, Orthodox Christian 1%
Ethnic mix Polish 97%, Silesian 2%, Other 1%
Government Parliamentary system
Currency Zloty = 100 groszy
Literacy rate 99%
Calorie consumption 3451 kilocalories

PORTUGAL
Southwest Europe

Page 70 B3

Portugal, on the Iberian Peninsula, is the westernmost country in mainland Europe. Isolated under 44 years of dictatorship until 1974, it modernized fast after joining the EU in 1986.

Official name Portuguese Republic
Formation 1139 / 1640
Capital Lisbon
Population 10.3 million / 290 people per sq mile (112 people per sq km)
Total area 35,672 sq miles (92,391 sq km)
Languages Portuguese*
Religions Roman Catholic 88%, Nonreligious 7%, Other Christian 4%, Other 1%
Ethnic mix Portuguese 98%, African and other 2%
Government Parliamentary system
Currency Euro = 100 cents
Literacy rate 94%
Calorie consumption 3477 kilocalories

QATAR
Southwest Asia

Page 98 C4

Projecting north from the Arabian Peninsula into the Persian Gulf, Qatar is mostly flat, semiarid desert. Massive reserves of oil and gas have made it one of the world's wealthiest states.

Official name State of Qatar
Formation 1971 / 1971
Capital Doha
Population 2.6 million / 612 people per sq mile (236 people per sq km)
Total area 4416 sq miles (11,437 sq km)
Languages Arabic*
Religions Muslim (mainly Sunni) 78%, Other 14%, Christian 8%
Ethnic mix Qatari 20%, Other Arab 20%, Indian 20%, Nepalese 13%, Filipino 10%, Other 10%, Pakistani 7%
Government Monarchy
Currency Qatar riyal = 100 dirhams
Literacy rate 98%
Calorie consumption Not available

ROMANIA
Southeast Europe

Page 86 B4

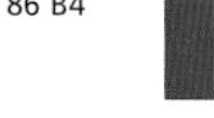

Romania lies on the western shores of the Black Sea. Its communist regime was overthrown in 1989 and, despite a slow transition to a free-market economy, it joined the EU in 2007.

Official name Romania
Formation 1878 / 1947
Capital Bucharest
Population 19.7 million / 222 people per sq mile (86 people per sq km)
Total area 91,699 sq miles (237,500 sq km)
Languages Romanian*, Hungarian (Magyar), Romani, German
Religions Orthodox Christian 86%, Other 8%, Roman Catholic 5%, Nonreligious 1%
Ethnic mix Romanian 89%, Magyar 7%, Roma 3%, Other 1%
Government Presidential / Parliamentary system
Currency New Romanian leu = 100 bani
Literacy rate 99%
Calorie consumption 3358 kilocalories

RUSSIA
Europe / Asia

Page 92 D4

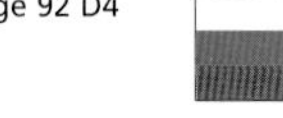

The world's largest country, with vast mineral and energy reserves, Russia dominated the former USSR and is still a major power. It has over 150 ethnic groups, many with their own territory.

Official name Russian Federation
Formation 1480 / 1991
Capital Moscow
Population 144 million / 22 people per sq mile (8 people per sq km)
Total area 6,592,735 sq miles (17,075,200 sq km)
Languages Russian*, Tatar, Ukrainian, Chavash, various other national languages
Religions Orthodox Christian 71%, Nonreligious 15%, Muslim 11%, Other Christian 2%, Other 1%
Ethnic mix Russian 81%, Other 11%, Tatar 4%, Ukrainian 1%, Bashkir 1%, Chavash 1%, Chechen 1%
Government Presidential / parliamentary system
Currency Russian rouble = 100 kopeks
Literacy rate 99%
Calorie consumption 3361 kilocalories

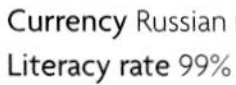

RWANDA
Central Africa

Page 51 B6

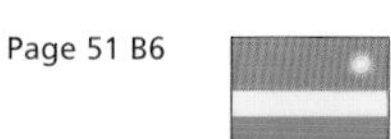

Rwanda lies just south of the Equator in Central Africa. Ethnic violence flared into genocide in 1994, when almost a million died. The main victims, the Tutsi, dominate government now.

Official name Republic of Rwanda
Formation 1962 / 1962
Capital Kigali
Population 12.2 million / 1266 people per sq mile (489 people per sq km)
Total area 10,169 sq miles (26,338 sq km)
Languages Kinyarwanda*, French*, Kiswahili, English*
Religions Roman Catholic 44%, Protestant 38%, Seventh-day Adventist 12%; Other 4%, Muslim 2%
Ethnic mix Hutu 85%, Tutsi 14%, Other (including Twa) 1%
Government Presidential system
Currency Rwanda franc = 100 centimes
Literacy rate 68%
Calorie consumption 2228 kilocalories

ST KITTS AND NEVIS
West Indies

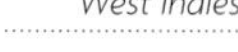

Page 33 G3

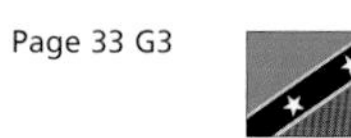

Saint Kitts and Nevis are part of the Caribbean Leeward Islands. A former British colony, the country is a popular tourist destination. Less-developed Nevis is famed for its hot springs.

Official name Federation of Saint Christopher and Nevis
Formation 1983 / 1983
Capital Basseterre
Population 55,000 / 545 people per sq mile (211 people per sq km)
Total area 101 sq miles (261 sq km)
Languages English*, English Creole
Religions Anglican 33%, Methodist 29%, Other 22%, Moravian 9%, Roman Catholic 7%
Ethnic mix Black 95%, Mixed race 3%, White 1%, Other and Amerindian 1%
Government Parliamentary system
Currency East Caribbean dollar = 100 cents
Literacy rate 98%
Calorie consumption 2492 kilocalories

ST LUCIA
West Indies

Page 33 G4

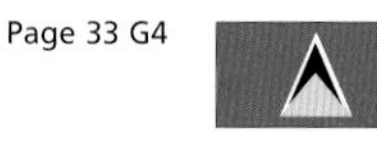

One of the most beautiful Caribbean Windward Islands, Saint Lucia retains both French and British influences from its colonial history. Tourism and fruit production dominate the economy.

Official name Saint Lucia
Formation 1979 / 1979
Capital Castries
Population 178,000 / 745 people per sq mile (287 people per sq km)
Total area 239 sq miles (620 sq km)
Languages English*, French Creole
Religions Roman Catholic 68%, Seventh-day Adventist 9%, Other Christian 9%, Pentecostal 6%, Nonreligious 5%, Rastafarian 2%, Other 1%
Ethnic mix Black 84%, Mulatto (mixed race) 12%, Asian 3%, Other 1%
Government Parliamentary system
Currency East Caribbean dollar = 100 cents
Literacy rate 95%
Calorie consumption 2595 kilocalories

ST VINCENT & THE GRENADINES
West Indies

Page 33 G4

Formerly ruled by Britain, these volcanic islands form part of the Caribbean Windward Islands. The economy relies on tourism and bananas, and it is the world's largest arrowroot producer.

Official name Saint Vincent and the Grenadines
Formation 1979 / 1979
Capital Kingstown
Population 110,000 / 733 people per sq mile (282 people per sq km)
Total area 150 sq miles (389 sq km)
Languages English*, English Creole
Religions Other Christian 48%, Anglican 18%, Pentecostal 18%, Nonreligious 9%, Other 7%
Ethnic mix Black 73%, Mulatto (mixed race) 20%, Carib 4%, Asian 2%, Other 1%
Government Parliamentary system
Currency East Caribbean dollar = 100 cents
Literacy rate 88%
Calorie consumption 2968 kilocalories

SAMOA
Australasia & Oceania

Page 123 F4

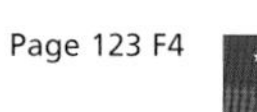

Samoa, in the southern Pacific, was ruled by New Zealand before 1962. Four of the nine islands are inhabited. The traditional Samoan way of life is communal and conservative.

Official name Independent State of Samoa
Formation 1962 / 1962
Capital Apia
Population 200,000 / 183 people per sq mile (71 people per sq km)
Total area 1104 sq miles (2860 sq km)
Languages Samoan*, English*
Religions Other Christian 78%, Roman Catholic 20%, Other 2%
Ethnic mix Polynesian 91%, Euronesian 7%, Other 2%
Government Parliamentary system
Currency Tala = 100 sene
Literacy rate 99%
Calorie consumption 2960 kilocalories

SAN MARINO
Southern Europe

Page 74 C3

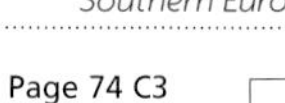

Perched on the slopes of Monte Titano in the Italian Appennino, San Marino has been a city-state since the 4th century AD, and was recognized as independent by the pope in 1631.

Official name Republic of San Marino
Formation 1631 / 1631
Capital San Marino
Population 34,000 / 1417 people per sq mile (557 people per sq km)
Total area 23.6 sq miles (61 sq km)
Languages Italian*
Religions Roman Catholic 93%, Other and nonreligious 7%
Ethnic mix Sammarinese 88%, Italian 10%, Other 2%
Government Parliamentary system
Currency Euro = 100 cents
Literacy rate 99%
Calorie consumption Not available

SAO TOME & PRINCIPE
West Africa

Page 55 A5

This ex-Portuguese colony off Africa's west coast has two main islands and smaller islets. Multiparty democracy, adopted in 1990, ended 15 years of Marxism. Cocoa is the main export.

Official name Democratic Republic of São Tomé and Príncipe
Formation 1975 / 1975
Capital São Tomé
Population 200,000 / 539 people per sq mile (208 people per sq km)
Total area 386 sq miles (1001 sq km)
Languages Portuguese Creole, Portuguese*
Religions Roman Catholic 56%, Nonreligious 21%, Other Christian 15%, Other 8%
Ethnic mix Black 90%, Portuguese and Creole 10%
Government Presidential / Parliamentary system
Currency Dobra = 100 céntimos
Literacy rate 90%
Calorie consumption 2400 kilocalories

SAUDI ARABIA
Southwest Asia

Page 99 B5

The desert kingdom of Saudi Arabia, rich in oil and gas, covers an area the size of Western Europe. It includes Islam's holiest cities, Medina and Mecca. Women's rights are restricted.

Official name Kingdom of Saudi Arabia
Formation 1932 / 1932
Capital Riyadh
Population 32.9 million / 40 people per sq mile (16 people per sq km)
Total area 756,981 sq miles (1,960,582 sq km)
Languages Arabic*
Religions Sunni Muslim 85%, Shi'a Muslim 15%
Ethnic mix Arab 72%, Foreign residents (mostly south and southeast Asian) 20%, Afro-Asian 8%
Government Monarchy
Currency Saudi riyal = 100 halalat
Literacy rate 94%
Calorie consumption 3255 kilocalories

SENEGAL
West Africa

Page 52 B3

This ex-French colony was ruled by one party for 40 years after independence, despite the adoption of multipartyism in 1981. Its capital, Dakar, stands on the westernmost cape of Africa.

Official name Republic of Senegal
Formation 1960 / 1960
Capital Dakar
Population 15.9 million / 214 people per sq mile (83 people per sq km)
Total area 75,749 sq miles (196,190 sq km)
Languages Wolof, Pulaar, Serer, Diola, Mandinka, Malinké, Soninké, French*
Religions Sunni Muslim 95%, Christian (mainly Roman Catholic) 4%, Traditional beliefs 1%
Ethnic mix Wolof 43%, Serer 15%, Peul 14%, Other 14%, Toucouleur 9%, Diola 5%
Government Presidential system
Currency CFA franc = 100 centimes
Literacy rate 43%
Calorie consumption 2456 kilocalories

SERBIA
Southeast Europe

Page 78 D4

The former Yugoslavia began breaking up in 1991, and Serbia has found itself the sole successor republic. It refuses to acknowledge the 2008 secession of Albanian-dominated Kosovo.

Official name Republic of Serbia
Formation 2006 / 2008
Capital Belgrade
Population 8.8 million / 294 people per sq mile (114 people per sq km)
Total area 29,905 sq miles (77,453 sq km)
Languages Serbian*, Hungarian (Magyar)
Religions Orthodox Christian 88%, Roman Catholic 4%, Nonreligious 4%, Muslim 2%, Other 2%
Ethnic mix Serb 87%, Magyar 4%, Other 3%, Roma 2%, Bosniak 2%, Croat 1%, Slovak 1%
Government Parliamentary system
Currency Serbian dinar = 100 para
Literacy rate 99%
Calorie consumption 2728 kilocalories

SEYCHELLES
Indian Ocean

Page 57 G1

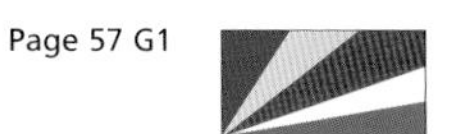

This ex-British colony spans 115 islands in the Indian Ocean. Multiparty polls in 1993 ended 14 years of one-party rule. Unique flora includes the world's largest seed, the coco-de-mer.

Official name Republic of Seychelles
Formation 1976 / 1976
Capital Victoria
Population 100,000 / 962 people per sq mile (370 people per sq km)
Total area 176 sq miles (455 sq km)
Languages French Creole*, English*, French*
Religions Roman Catholic 84%, Anglican 6%, Other Christian 5%, Hindu 2%, Other and nonreligious 2%, Muslim 1%
Ethnic mix Creole 89%, Indian 5%, Other 4%, Chinese 2%
Government Presidential system
Currency Seychelles rupee = 100 cents
Literacy rate 92%
Calorie consumption 2426 kilocalories

SIERRA LEONE
West Africa

Page 52 C4

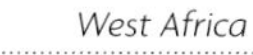

Founded in 1787 as a British colony for freed slaves, Sierra Leone gained independence in 1961. Recovery from civil war in the 1990s was set back by West Africa's recent Ebola epidemic.

Official name Republic of Sierra Leone
Formation 1961 / 1961
Capital Freetown
Population 7.6 million / 275 people per sq mile (106 people per sq km)
Total area 27,698 sq miles (71,740 sq km)
Languages Mende, Temne, Krio, English*
Religions Muslim 60%, Christian 30%, Traditional beliefs 10%
Ethnic mix Mende 35%, Temne 32%, Other 21%, Limba 8%, Kuranko 4%
Government Presidential system
Currency Leone = 100 cents
Literacy rate 32%
Calorie consumption 2404 kilocalories

SINGAPORE
Southeast Asia

Page 116 A1

A city state linked to the southern tip of the Malay Peninsula by a causeway, Singapore is one of Asia's major commercial centers. Politics has been dominated for decades by one party.

Official name Republic of Singapore
Formation 1965 / 1965
Capital Singapore
Population 5.7 million / 24,153 people per sq mile (9344 people per sq km)
Total area 250 sq miles (648 sq km)
Languages Mandarin*, Malay*, Tamil*, English*
Religions Christian 31%, Buddhist 28%, Nonreligious 14%, Muslim 13%, Taoist 9%, Hindu 4%, Other 1%
Ethnic mix Chinese 74%, Malay 14%, Indian 9%, Other 3%
Government Parliamentary system
Currency Singapore dollar = 100 cents
Literacy rate 97%
Calorie consumption Not available

SLOVAKIA
Central Europe

Page 77 C6

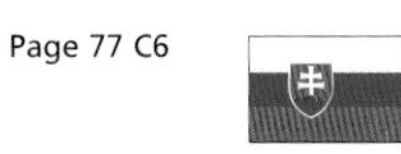

After 900 years of Hungarian control, Slovakia was the less-developed half of communist Czechoslovakia in the 20th century. It became independent in 1993 and joined the EU in 2004.

Official name Slovak Republic
Formation 1993 / 1993
Capital Bratislava
Population 5.4 million / 285 people per sq mile (110 people per sq km)
Total area 18,859 sq miles (48,845 sq km)
Languages Slovak*, Hungarian (Magyar), Czech
Religions Roman Catholic 69%, Nonreligious 15%, Other Christian 11%, Greek Catholic (Uniate) 4%, Other 1%
Ethnic mix Slovak 87%, Magyar 9%, Roma 2%, Other 1%, Czech 1%
Government Parliamentary system
Currency Euro = 100 cents
Literacy rate 99%
Calorie consumption 2944 kilocalories

SLOVENIA
Central Europe

Page 73 D8

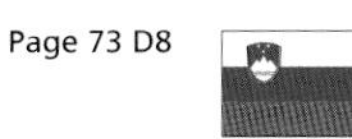

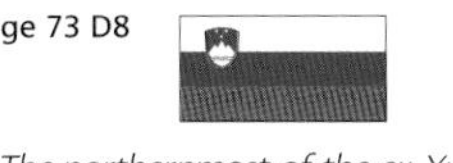

The northernmost of the ex-Yugoslav republics was the first to break away, with little violence, in 1991. It always had the closest links with Western Europe, and joined the EU in 2004.

Official name Republic of Slovenia
Formation 1991 / 1991
Capital Ljubljana
Population 2.1 million / 269 people per sq mile (104 people per sq km)
Total area 7820 sq miles (20,253 sq km)
Languages Slovenian*
Religions Roman Catholic 75%, Nonreligious 18%, Muslim 3%, Orthodox Christian 3%, Other (mostly Protestant) 1%
Ethnic mix Slovene 92%, Other 3%, Serb 2%, Croat 2%, Bosniak 1%
Government Parliamentary system
Currency Euro = 100 cents
Literacy rate 99%
Calorie consumption 3168 kilocalories

SOLOMON ISLANDS
Australasia & Oceania

Page 122 C3

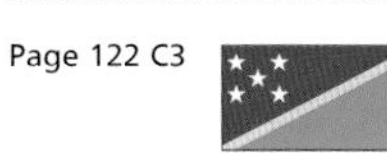

This archipelago of around 1000 islands scattered in the southwest Pacific was formerly ruled by Britain. Most people live on six main islands. Ethnic conflict from 1998 led to devolved governance.

Official name Solomon Islands
Formation 1978 / 1978
Capital Honiara
Population 600,000 / 56 people per sq mile (21 people per sq km)
Total area 10,985 sq miles (28,450 sq km)
Languages English*, Pidgin English, Melanesian Pidgin, around 120 native languages
Religions Church of Melanesia (Anglican) 34%, Roman Catholic 19%, South Seas Evangelical Church 17%, Methodist 11%, Other 19%
Ethnic mix Melanesian 93%, Polynesian 4%, Other 3%
Government Parliamentary system
Currency Solomon Islands dollar = 100 cents
Literacy rate 77%
Calorie consumption 2391 kilocalories

SOMALIA
East Africa

Page 51 E5

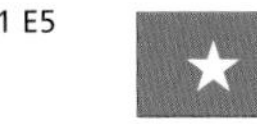

Italian and British Somaliland were united to create this semiarid state on the Horn of Africa. Anarchy since 1991 has caused mass hunger, a refugee crisis, and ineffective central authority.

Official name Federal Republic of Somalia
Formation 1960 / 1960
Capital Mogadishu
Population 14.7 million / 61 people per sq mile (23 people per sq km)
Total area 246,199 sq miles (637,657 sq km)
Languages Somali*, Arabic*, English, Italian
Religions Sunni Muslim 99%, Christian 1%
Ethnic mix Somali 85%, Other 15%
Government Nonparty system
Currency Somali shilin = 100 senti
Literacy rate 24%
Calorie consumption 1696 kilocalories

SOUTH AFRICA
Southern Africa

Page 56 C4

Mineral-rich South Africa was settled by the Dutch and the British. Multiracial polls in 1994 ended decades of white minority rule and apartheid. AIDS, poverty, and crime are problems.

Official name Republic of South Africa
Formation 1934 / 1994
Capital Pretoria (Tshwane); Cape Town; Bloemfontein
Population 56.7 million / 120 people per sq mile (46 people per sq km)
Total area 471,008 sq miles (1,219,912 sq km)
Languages English*, isiZulu*, isiXhosa*, Afrikaans*, Sepedi*, Setswana*, Sesotho*, Xitsonga*, siSwati*, Tshivenda*, isiNdebele*
Religions Christian 81%, Nonreligious 15%, Muslim 2%, Hindu 1%, Other 1%
Ethnic mix Black 80%, White 9%, Colored 9%, Asian 2%
Government Presidential system
Currency Rand = 100 cents
Literacy rate 94%
Calorie consumption 3022 kilocalories

SOUTH KOREA
East Asia

Page 106 E4

Allied with the US, the southern half of the Korean peninsula was separated from the communist North in 1948. It is the world's leading shipbuilder and a major force in high-tech industries.

Official name Republic of Korea
Formation 1948 / 1953
Capital Seoul; Sejong City (administrative)
Population 51 million / 1338 people per sq mile (517 people per sq km)
Total area 38,023 sq miles (98,480 sq km)
Languages Korean*
Religions Nonreligious 47%, Mahayana Buddhist 23%, Other Christian 18%, Roman Catholic 11%, Other 1%
Ethnic mix Korean 100%
Government Presidential system
Currency South Korean won = 100 chon
Literacy rate 99%
Calorie consumption 3334 kilocalories

SOUTH SUDAN
East Africa

Page 51 B5

Landlocked and little developed, this mostly Christian region seceded from the mainly Muslim north of Sudan in 2011 after years of civil war. Oil production is the economic mainstay.

Official name Republic of South Sudan
Formation 2011 / 2011
Capital Juba
Population 12.6 million / 51 people per sq mile (20 people per sq km)
Total area 248,777 sq miles (644,329 sq km)
Languages Arabic, Dinka, Nuer, Zande, Bari, Shilluk, Lotuko, English*
Religions Over 50% Christian/traditional beliefs
Ethnic mix Dinka 40%, Nuer 15%, Shilluk/Anwak 10%, Azande 10%, Arab 10%, Bari 10%, Other 5%
Government Transitional regime
Currency South Sudan Pound = 100 piastres
Literacy rate 37%
Calorie consumption Not available

SPAIN
Southwest Europe

Page 70 D2

At the gateway to the Mediterranean, Spain became a world power once united in 1492. A vigorous regionalism now exists, with separatist movements in the Basque Country and Catalonia.

Official name Kingdom of Spain
Formation 1492 / 1713
Capital Madrid
Population 46.4 million / 241 people per sq mile (93 people per sq km)
Total area 194,896 sq miles (504,782 sq km)
Languages Spanish*, Catalan*, Galician*, Basque*
Religions Roman Catholic 71%, Nonreligious 26%, Other 3%
Ethnic mix Castilian Spanish 72%, Catalan 17%, Galician 6%, Basque 2%, Other 2%, Roma 1%
Government Parliamentary system
Currency Euro = 100 cents
Literacy rate 98%
Calorie consumption 3174 kilocalories

SRI LANKA
South Asia

Page 110 D3

A former British colony, the island republic of Sri Lanka is separated from India by the narrow Palk Strait. A brutal 26-year civil war between the Sinhalese and Tamils ended in 2009.

Official name Democratic Socialist Republic of Sri Lanka
Formation 1948 / 1948
Capital Colombo; Sri Jayewardenapura Kotte
Population 20.9 million / 836 people per sq mile (323 people per sq km)
Total area 25,332 sq miles (65,610 sq km)
Languages Sinhala*, Tamil*, Sinhala-Tamil, English
Religions Buddhist 70%, Hindu 13%, Muslim 10%, Christian (mainly Roman Catholic) 7%
Ethnic mix Sinhalese 75%, Tamil 15%, Moor 9%, Other 1%
Government Presidential / parliamentary system
Currency Sri Lanka rupee = 100 cents
Literacy rate 91%
Calorie consumption 2539 kilocalories

SUDAN
East Africa

Page 50 B4

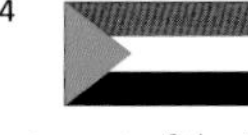

On the west coast of the Red Sea, Sudan has been ruled by a military Islamic regime since a coup in 1989. In 2011, it lost its southern third (and most of its oil reserves) after years of civil war.

Official name Republic of the Sudan
Formation 1956 / 2011
Capital Khartoum
Population 40.5 million / 56 people per sq mile (22 people per sq km)
Total area 718,722 sq miles (1,861,481 sq km)
Languages Arabic*, Nubian, Beja, Fur
Religions Almost 100% Muslim (mainly Sunni)
Ethnic mix Arab 60%, Other 18%, Nubian 10%, Beja 8%, Fur 3%, Zaghawa 1%
Government Presidential system
Currency New Sudanese pound = 100 piastres
Literacy rate 73%
Calorie consumption 2336 kilocalories

SURINAME
South America

Page 37 G3

This former Dutch colony on the north coast of South America has some of the world's richest bauxite reserves. The military ruler in the 1980s was elected president from 2010.

Official name Republic of Suriname
Formation 1975 / 1975
Capital Paramaribo
Population 600,000 / 10 people per sq mile (4 people per sq km)
Total area 63,039 sq miles (163,270 sq km)
Languages Sranan (creole), Dutch*, Javanese, Sarnami Hindi, Saramaccan, Chinese, Carib
Religions Christian 50%, Hindu 23%, Muslim 14%, Other 13%
Ethnic mix East Indian 27%, Creole 18%, Black 15%, Javanese 15%, Mixed race 13%, Other 12%
Government Presidential / parliamentary system
Currency Surinamese dollar = 100 cents
Literacy rate 93%
Calorie consumption 2753 kilocalories

SWAZILAND
Southern Africa

Page 56 D4

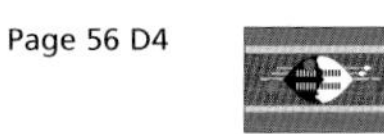

This tiny kingdom, ruled by Britain until 1968, depends economically on its neighbor South Africa. Its absolute monarch has banned political parties. It has the world's highest rate of HIV.

Official name Kingdom of Swaziland
Formation 1968 / 1968
Capital Mbabane; Lobamba
Population 1.4 million / 211 people per sq mile (81 people per sq km)
Total area 6704 sq miles (17,363 sq km)
Languages English*, siSwati*, isiZulu, Xitsonga
Religions Traditional beliefs 40%, Other 30%, Roman Catholic 20%, Muslim 10%
Ethnic mix Swazi 97%, Other 3%
Government Monarchy
Currency Lilangeni = 100 cents
Literacy rate 83%
Calorie consumption 2329 kilocalories

SWEDEN
Northern Europe

Page 62 B4

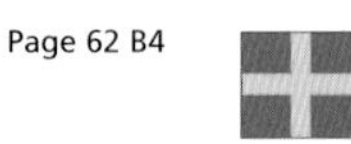

Densely forested Sweden is the largest and most populous Scandinavian country and stretches into the Arctic Circle. Its strong industrial base helps to fund its extensive welfare system.

Official name Kingdom of Sweden
Formation 1523 / 1921
Capital Stockholm
Population 9.9 million / 62 people per sq mile (24 people per sq km)
Total area 173,731 sq miles (449,964 sq km)
Languages Swedish*, Finnish, Sámi
Religions Evangelical Lutheran 75%, Other 13%, Muslim 5%, Other Protestant 5%, Roman Catholic 2%
Ethnic mix Swedish 86%, Foreign-born or first-generation immigrant 12%, Finnish & Sámi 2%
Government Parliamentary system
Currency Swedish krona = 100 öre
Literacy rate 99%
Calorie consumption 3179 kilocalories

SWITZERLAND
Central Europe

Page 73 A7

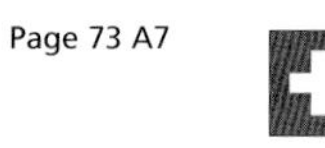

One of the world's richest countries, with a long tradition of neutrality, this mountainous nation lies at the center of Europe geographically, but outside it politically, having not joined the EU.

Official name Swiss Confederation
Formation 1291 / 1857
Capital Bern
Population 8.5 million / 554 people per sq mile (214 people per sq km)
Total area 15,942 sq miles (41,290 sq km)
Languages German*, Swiss-German, French*, Italian*, Romansch*
Religions Roman Catholic 39%, Other Christian 34%, Nonreligious 21%, Muslim 5%, Other 1%
Ethnic mix German 64%, French 20%, Other 9.5%, Italian 6%, Romansch 0.5%
Government Parliamentary system
Currency Swiss franc = 100 rappen/centimes
Literacy rate 99%
Calorie consumption 3391 kilocalories

SYRIA
Southwest Asia

Page 96 B3

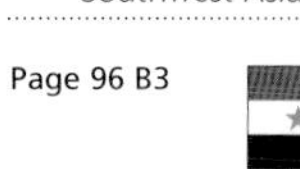

Syria's borders were drawn in 1941 at the end of French rule. Suppression of pro-democracy protests in 2011 erupted into civil war; Islamists controlled the Euphrates Valley in 2014-2017.

Official name Syrian Arab Republic
Formation 1941 / 1967
Capital Damascus
Population 18.3 million / 258 people per sq mile (99 people per sq km)
Total area 71,498 sq miles (184,180 sq km)
Languages Arabic*, French, Kurdish, Armenian, Circassian, Turkic languages, Assyrian, Aramaic
Religions Sunni Muslim 74%, Alawi 12%, Christian 10%, Druze 3%, Other 1%
Ethnic mix Arab 90%, Kurdish 9%, Armenian, Turkmen, and Circassian 1%
Government Presidential system
Currency Syrian pound = 100 piastres
Literacy rate 85%
Calorie consumption 3106 kilocalories

TAIWAN
East Asia

Page 107 D6

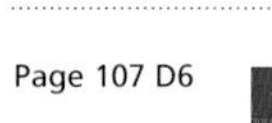

China's nationalist government fled to Taiwan in 1949 when ousted by the communists. China regards the island, 80 miles (130 km) southeast of the mainland, as a renegade province.

Official name Republic of China (ROC)
Formation 1949 / 1949
Capital Taipei
Population 23.5 million / 1887 people per sq mile (728 people per sq km)
Total area 13,892 sq miles (35,980 sq km)
Languages Amoy Chinese, Mandarin Chinese*, Hakka Chinese
Religions Buddhist, Confucianist, and Taoist 93%, Christian 5%, Other 2%
Ethnic mix Han (pre-20th-century migration) 84%, Han (20th-century migration) 14%, Aboriginal 2%
Government Presidential system
Currency Taiwan dollar = 100 cents
Literacy rate 98%
Calorie consumption 2997 kilocalories

TAJIKISTAN
Central Asia

Page 101 F3

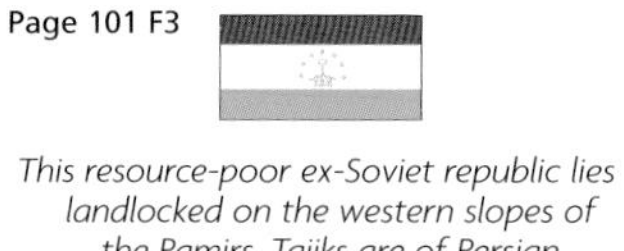

This resource-poor ex-Soviet republic lies landlocked on the western slopes of the Pamirs. Tajiks are of Persian (Iranian) origin rather than Turkic like their Central Asian neighbors.

Official name Republic of Tajikistan
Formation 1991 / 1991
Capital Dushanbe
Population 8.9 million / 161 people per sq mile (62 people per sq km)
Total area 55,251 sq miles (143,100 sq km)
Languages Tajik*, Uzbek, Russian
Religions Sunni Muslim 95%, Shi'a Muslim 3%, Other 2%
Ethnic mix Tajik 84%, Uzbek 12%, Other 2%, Kyrgyz 1%, Russian 1%
Government Presidential system
Currency Somoni = 100 diram
Literacy rate 99%
Calorie consumption 2201 kilocalories

TANZANIA
East Africa

Page 51 B7

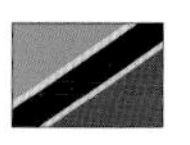

This East African state was formed in 1964 by the union of Tanganyika and Zanzibar. A third of its area is game reserve or national park, including Africa's highest peak, Mt. Kilimanjaro.

Official name United Republic of Tanzania
Formation 1964 / 1964
Capital Dodoma
Population 57.3 million / 167 people per sq mile (65 people per sq km)
Total area 364,898 sq miles (945,087 sq km)
Languages Kiswahili*, Sukuma, Chagga, Nyamwezi, Hehe, Makonde, Yao, Sandawe, English*
Religions Christian 63%, Muslim 35%, Other 2%
Ethnic mix Native African (over 120 tribes) 99%, European, Asian, and Arab 1%
Government Presidential system
Currency Tanzanian shilling = 100 cents
Literacy rate 78%
Calorie consumption 2208 kilocalories

THAILAND
Southeast Asia

Page 115 C5

Thailand lies at the heart of the Indochinese Peninsula. Formerly Siam, it has been an independent kingdom for most of its history. The military has frequently intervened in politics.

Official name Kingdom of Thailand
Formation 1238 / 1907
Capital Bangkok
Population 69 million / 350 people per sq mile (135 people per sq km)
Total area 198,455 sq miles (514,000 sq km)
Languages Thai*, Chinese, Malay, Khmer, Mon, Karen, Miao
Religions Buddhist 94%, Muslim 5%, Other (including Christian) 1%
Ethnic mix Thai 83%, Chinese 12%, Malay 3%, Khmer and Other 2%
Government Transitional regime
Currency Baht = 100 satang
Literacy rate 93%
Calorie consumption 2784 kilocalories

TOGO
West Africa

Page 53 F4

Togo lies sandwiched between Ghana and Benin in West Africa. Its long-term military leader, and then his son and successor, have won every election held there since 1993.

Official name Togolese Republic
Formation 1960 / 1960
Capital Lomé
Population 7.8 million / 371 people per sq mile (143 people per sq km)
Total area 21,924 sq miles (56,785 sq km)
Languages Ewe, Kabye, Gurma, French*
Religions Christian 47%, Traditional beliefs 33%, Muslim 14%, Other 6%
Ethnic mix Ewe 46%, Other African 41%, Kabye 12%, European 1%
Government Presidential system
Currency CFA franc = 100 centimes
Literacy rate 64%
Calorie consumption 2454 kilocalories

TONGA
Australasia & Oceania

Page 123 E4

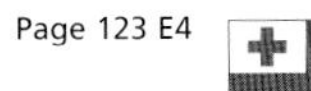

Northeast of New Zealand, Tonga is a 170-island archipelago, 45 of which are inhabited. Politics is effectively controlled by the king, though limited democratic reforms are taking place.

Official name Kingdom of Tonga
Formation 1970 / 1970
Capital Nuku'alofa
Population 100,000/ 360 people per sq mile (139 people per sq km)
Total area 289 sq miles (748 sq km)
Languages English*, Tongan*
Religions Free Wesleyan 38%, Church of Jesus Christ of Latter-day Saints 17%, Roman Catholic 16%, Other Christian 16%, Free Church of Tonga 12%, Other 1%
Ethnic mix Tongan 98%, Other 2%
Government Monarchy
Currency Pa'anga (Tongan dollar) = 100 seniti
Literacy rate 99%
Calorie consumption Not available

TRINIDAD AND TOBAGO
West Indies

Page 33 H5

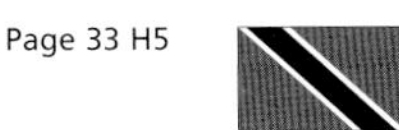

This former British colony is the most southerly of the Windward Islands, just 9 miles (15 km) off the coast of Venezuela. Politics is mainly polarized by race. Oil and gas are exported.

Official name Republic of Trinidad and Tobago
Formation 1962 / 1962
Capital Port of Spain
Population 1.4 million / 707 people per sq mile (273 people per sq km)
Total area 1980 sq miles (5128 sq km)
Languages English Creole, English*, Hindi, French, Spanish
Religions Protestant 38%, Roman Catholic 24%, Hindu 20%, Other 12%, Muslim 6%
Ethnic mix East Indian 38%, Black 36%, Mixed race 24%, White and Chinese 1%, Other 1%
Government Parliamentary system
Currency Trinidad and Tobago dollar = 100 cents
Literacy rate 99%
Calorie consumption 3052 kilocalories

TUNISIA
North Africa

Page 49 E2

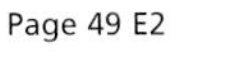

North Africa's smallest country, one of the more liberal yet stable Arab states, had only two post-independence rulers until the "Arab Spring" of 2011. Moderate Islamists then won power.

Official name Republic of Tunisia
Formation 1956 / 1956
Capital Tunis
Population 11.5 million / 192 people per sq mile (74 people per sq km)
Total area 63,169 sq miles (163,610 sq km)
Languages Arabic*, French
Religions Muslim (mainly Sunni) 98%, Christian 1%, Jewish 1%
Ethnic mix Arab and Berber 98%, European 1%, Jewish 1%
Government Presidential / parliamentary system
Currency Tunisian dinar = 1000 millimes
Literacy rate 79%
Calorie consumption 3349 kilocalories

TURKEY
Asia / Europe

Page 94 B3

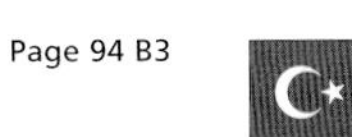

With land in Europe and Asia, Turkey guards the entrance to the Black Sea. The secular/Islamic divide is key to its politics. It is the only Muslim member of NATO, and hopes to join the EU.

Official name Republic of Turkey
Formation 1923 / 1939
Capital Ankara
Population 80.7 million / 272 people per sq mile (105 people per sq km)
Total area 301,382 sq miles (780,580 sq km)
Languages Turkish*, Kurdish, Arabic, Circassian, Armenian, Greek, Georgian, Ladino
Religions Muslim (mainly Sunni) 99%, Other 1%
Ethnic mix Turkish 70%, Kurdish 20%, Other 8%, Arab 2%
Government Parliamentary system
Currency Turkish lira = 100 kurus
Literacy rate 96%
Calorie consumption 3706 kilocalories

TURKMENISTAN
Central Asia

Page 100 B2

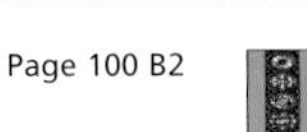

Stretching from the Caspian Sea into Central Asia's deserts, this ex-Soviet state exploits vast gas reserves. The pre-independence president built a personality cult and ruled until 2007.

Official name Turkmenistan
Formation 1991 / 1991
Capital Ashgabat
Population 5.8 million / 31 people per sq mile (12 people per sq km)
Total area 188,455 sq miles (488,100 sq km)
Languages Turkmen*, Uzbek, Russian, Kazakh, Tatar
Religions Sunni Muslim 89%, Orthodox Christian 9%, Other 2%
Ethnic mix Turkmen 85%, Other 6%, Uzbek 5%, Russian 4%
Government Presidential system
Currency New manat = 100 tenge
Literacy rate 99%
Calorie consumption 2840 kilocalories

TUVALU
Australasia & Oceania

Page 123 E3

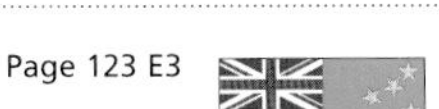

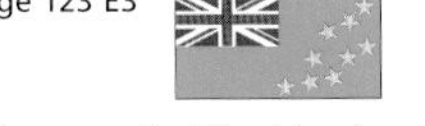

Known as the Ellice Islands under British rule, Tuvalu is a chain of nine atolls in the Central Pacific. It has the world's smallest GNI, but made substantial earnings leasing its ".tv" internet suffix.

Official name Tuvalu
Formation 1978 / 1978
Capital Funafuti Atoll
Population 11,000 / 1100 people per sq mile (423 people per sq km)
Total area 10 sq miles (26 sq km)
Languages Tuvaluan, Kiribati, English*
Religions Church of Tuvalu 91%, Other (mostly Protestant) 5%, Seventh-day Adventist 2%, Baha'i 2%
Ethnic mix Polynesian 96%, Micronesian 4%
Government Nonparty system
Currency Australian dollar = 100 cents; Tuvaluan dollar = 100 cents
Literacy rate 95%
Calorie consumption Not available

UGANDA
East Africa

Page 51 B6

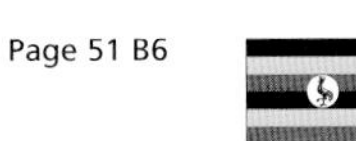

Landlocked Uganda faced ethnic strife under 1970s' dictator Idi Amin. From 1986, reconciliation was aided by two decades of "no-party" democracy, but insurgency continued in the north.

Official name Republic of Uganda
Formation 1962 / 1962
Capital Kampala
Population 42.9 million / 557 people per sq mile (215 people per sq km)
Total area 91,135 sq miles (236,040 sq km)
Languages Luganda, Nkole, Chiga, Lango, Acholi, Teso, Lugbara, English*
Religions Roman Catholic 42%, Protestant 42%, Muslim (mainly Sunni) 12%, Other 4%
Ethnic mix Other 50%, Baganda 17%, Banyakole 10%, Basoga 9%, Bakiga 7%, Iteso 7%
Government Presidential system
Currency Uganda shilling = 100 cents
Literacy rate 70%
Calorie consumption 2130 kilocalories

UKRAINE
Eastern Europe

Page 86 C2

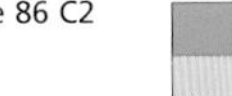

Bordered by seven states, fertile Ukraine was the "breadbasket" of the USSR. Its political divide between pro-Russian sentiment and assertive nationalism exploded into civil war in 2014.

Official name Ukraine
Formation 1991 / 1991
Capital Kiev
Population 44.2 million / 190 people per sq mile (73 people per sq km)
Total area 223,089 sq miles (603,700 sq km)
Languages Ukrainian*, Russian, Tatar
Religions Orthodox Christian 78%, Roman Catholic 10%, Nonreligious 7%, Other 5%
Ethnic mix Ukrainian 78%, Russian 17%, Other 4%, Belarussian 1%
Government Presidential / Parliamentary system
Currency Hryvna = 100 kopiykas
Literacy rate 99%
Calorie consumption 3138 kilocalories

UNITED ARAB EMIRATES
Southwest Asia

Page 99 D5

Bordering the Persian Gulf on the north of the Arabian Peninsula, the United Arab Emirates is a federation of seven states. Wealth once relied on pearls, but oil and gas are now exported.

Official name United Arab Emirates
Formation 1971 / 1972
Capital Abu Dhabi
Population 9.4 million / 291 people per sq mile (112 people per sq km)
Total area 32,000 sq miles (82,880 sq km)
Languages Arabic*, Farsi, Indian and Pakistani languages, English
Religions Muslim (mainly Sunni) 96%, Christian, Hindu, and other 4%
Ethnic mix Asian 60%, Emirian 25%, Other Arab 12%, European 3%
Government Monarchy
Currency UAE dirham = 100 fils
Literacy rate 90%
Calorie consumption 3280 kilocalories

UNITED KINGDOM
Northwest Europe

Page 67 C5

Lying across the English Channel from France, the UK comprises England, Wales, Scotland, and Northern Ireland. Its prominent role in world affairs is a legacy of its once-vast empire.

Official name United Kingdom of Great Britain and Northern Ireland
Formation 1707 / 1922
Capital London
Population 66.2 million / 710 people per sq mile (274 people per sq km)
Total area 94,525 sq miles (244,820 sq km)
Languages English*, Welsh (* in Wales), Gaelic, Irish
Religions Christian 64%, Nonreligious 28%, Muslim 5%, Other 2%, Hindu 1%
Ethnic mix White 87%, Indian and Pakistani 4%, Other 3%, Black 3%, Other Asian 2%, Bengali 1%
Government Parliamentary system
Currency Pound sterling = 100 pence
Literacy rate 99%
Calorie consumption 3424 kilocalories

UNITED STATES
North America

Page 13 B5

Stretching across the most temperate part of North America, and with many natural resources, the USA is the sole truly global superpower and has the world's largest economy.

Official name United States of America
Formation 1776 / 1959
Capital Washington D.C.
Population 324 million / 92 people per sq mile (35 people per sq km)
Total area 3,717,792 sq miles (9,626,091 sq km)
Languages English*, Spanish, Chinese, French, Polish, German, Tagalog, Vietnamese, Italian, Korean, Russian
Religions Protestant 47%, Nonreligious 23%, Roman Catholic 21%, Other 6%, Jewish 2%, Muslim 1%
Ethnic mix White 60%, Hispanic 17%, Black American/African 14%, Asian 6%, Other 3%
Government Presidential system
Currency US dollar = 100 cents
Literacy rate 99%
Calorie consumption 3682 kilocalories

URUGUAY
South America

Page 42 D4

Uruguay, in southeastern South America, has much rich low-lying pasture land and is a major wool exporter. Military rule from 1973 to 1985 has given way to democracy.

Official name Oriental Republic of Uruguay
Formation 1828 / 1828
Capital Montevideo
Population 3.5 million / 52 people per sq mile (20 people per sq km)
Total area 68,039 sq miles (176,220 sq km)
Languages Spanish*
Religions Roman Catholic 42%, Nonreligious 37%, Protestant 15%, Other 6%
Ethnic mix White 88%, Black 7%, Mestizo 5%, Other 1%
Government Presidential system
Currency Uruguayan peso = 100 centésimos
Literacy rate 98%
Calorie consumption 3050 kilocalories

UZBEKISTAN
Central Asia

Page 100 D2

The most populous of the Central Asian republics lies on the ancient Silk Road. Today, its main exports are cotton, oil, gas, and gold. Its pre-independence ruler held power until he died in 2016.

Official name Republic of Uzbekistan
Formation 1991 / 1991
Capital Tashkent
Population 31.9 million / 185 people per sq mile (71 people per sq km)
Total area 172,741 sq miles (447,400 sq km)
Languages Uzbek*, Russian, Tajik, Kazakh
Religions Sunni Muslim 88%, Orthodox Christian 9%, Other 3%
Ethnic mix Uzbek 80%, Other 6%, Russian 6%, Tajik 5%, Kazakh 3%
Government Presidential system
Currency Som = 100 tiyin
Literacy rate 99%
Calorie consumption 2760 kilocalories

VANUATU
Australasia & Oceania

Page 122 D4

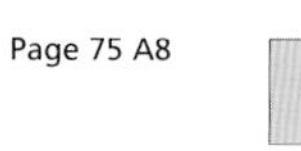

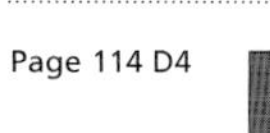

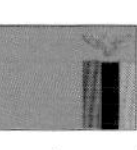

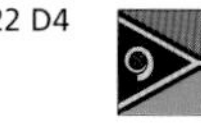

This South Pacific archipelago of 82 islands and islets boasts the world's highest per capita density of languages. Until independence, it was under joint Anglo-French rule.

Official name Republic of Vanuatu
Formation 1980 / 1980
Capital Port Vila
Population 300,000 / 64 people per sq mile (25 people per sq km)
Total area 4710 sq miles (12,200 sq km)
Languages Bislama* (Melanesian pidgin), English*, French*, other indigenous languages
Religions Other 33%, Presbyterian 28%, Anglican 15%, Seventh-day Adventist 12%, Roman Catholic 12%
Ethnic mix ni-Vanuatu 99%, Other 1%
Government Parliamentary system
Currency Vatu = 100 centimes
Literacy rate 83%
Calorie consumption 2836 kilocalories

VATICAN CITY
Southern Europe

Page 75 A8

The Vatican City, seat of the Roman Catholic Church, is a walled enclave in Rome. It is the world's smallest country. Its head, the pope, is elected for life by a college of cardinals.

Official name Vatican City
Formation 1929 / 1929
Capital Vatican City
Population 1000 / 5882 people per sq mile (2273 people per sq km)
Total area 0.17 sq miles (0.44 sq km)
Languages Italian*, Latin*
Religions Roman Catholic 100%
Ethnic mix Most resident lay persons are Italian
Government Papal state
Currency Euro = 100 cents
Literacy rate 99%
Calorie consumption Not available

VENEZUELA
South America

Page 36 D2

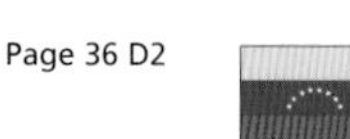

Located on the Caribbean coast of South America, Venezuela has the continent's most urbanized society, and some of the largest known oil deposits outside the Middle East.

Official name Bolivarian Republic of Venezuela
Formation 1830 / 1830
Capital Caracas
Population 32 million / 94 people per sq mile (36 people per sq km)
Total area 352,143 sq miles (912,050 sq km)
Languages Spanish*, Amerindian languages
Religions Roman Catholic 73%, Protestant 17%, Nonreligious 7%, Other 3%
Ethnic mix Mestizo 69%, White 20%, Black 9%, Amerindian 2%
Government Presidential system
Currency Bolívar fuerte = 100 céntimos
Literacy rate 97%
Calorie consumption 2631 kilocalories

VIETNAM
Southeast Asia

Page 114 D4

The eastern strip of the Indochinese Peninsula, Vietnam was partitioned in 1954, and only reunited after the communist north's victory in the devastating 1962–75 Vietnam War.

Official name Socialist Republic of Vietnam
Formation 1976 / 1976
Capital Hanoi
Population 95.5 million / 760 people per sq mile (294 people per sq km)
Total area 127,243 sq miles (329,560 sq km)
Languages Vietnamese*, Chinese, Thai, Khmer, Muong, Nung, Miao, Yao, Jarai
Religions Nonreligious 81%, Buddhist 9%, Roman Catholic 7%, Hoa Hao 1%, Cao Dai 1%, Other 1%
Ethnic mix Vietnamese 86%, Other 8%, Tay 2%, Thai 2%, Muong 1%, Khome 1%
Government One-party state
Currency Dông = 10 hao = 100 xu
Literacy rate 94%
Calorie consumption 2745 kilocalories

YEMEN
Southwest Asia

Page 99 C7

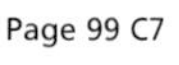

The Arab world's only Marxist regime and a military-run republic united in 1990 to form Yemen, stretching across southern Arabia. Tribal insurgency and militant Islamism have caused conflict.

Official name Republic of Yemen
Formation 1990 / 1990
Capital Saana
Population 28.3 million / 130 people per sq mile (50 people per sq km)
Total area 203,849 sq miles (527,970 sq km)
Languages Arabic*
Religions Sunni Muslim 55%, Shi'a Muslim 42%, Christian, Hindu, and Jewish 3%
Ethnic mix Arab 99%, Afro-Arab, Indian, Somali, and European 1%
Government Transitional regime
Currency Yemeni rial = 100 fils
Literacy rate 66%
Calorie consumption 2223 kilocalories

ZAMBIA
Southern Africa

Page 56 C2

Landlocked in southern Africa, copper-rich Zambia (once known as Northern Rhodesia) has seen its politics dogged by corruption both before and after the end of single-party rule in 1991.

Official name Republic of Zambia
Formation 1964 / 1964
Capital Lusaka
Population 17.1 million / 60 people per sq mile (23 people per sq km)
Total area 290,584 sq miles (752,614 sq km)
Languages Bemba, Tonga, Nyanja, Lozi, Lala-Bisa, Nsenga, English*
Religions Protestant 75%, Roman Catholic 20%, Other (including Muslim) 3%, Nonreligious 2%
Ethnic mix Bemba 34%, Other African 26%, Tonga 16%, Nyanja 14%, Lozi 9%, European 1%
Government Presidential system
Currency New Zambian kwacha = 100 ngwee
Literacy rate 83%
Calorie consumption 1930 kilocalories

ZIMBABWE
Southern Africa

Page 56 D3

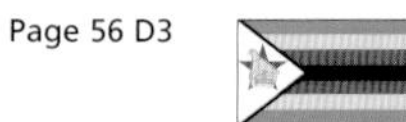

Full independence from Britain in 1980 ended 15 years of troubled white-minority rule. Poor governance, violent land redistribution, and severe drought have destroyed the economy.

Official name Republic of Zimbabwe
Formation 1980 / 1980
Capital Harare
Population 16.5 million / 111 people per sq mile (43 people per sq km)
Total area 150,803 sq miles (390,580 sq km)
Languages Shona, isiNdebele, English*
Religions Syncretic 50%, Christian 25%, Traditional beliefs 24%, Other (including Muslim) 1%
Ethnic mix Shona 71%, Ndebele 16%, Other African 11%, White 1%, Asian 1%,
Government Presidential system
Currency Zimbabwe dollar suspended in 2009; US dollar, SA rand, & 7 other currencies legal tender
Literacy rate 89%
Calorie consumption 2110 kilocalories

Overseas Territories and Dependencies

Despite the rapid process of decolonization since the end of the Second World War, around 10 million people in more than 50 territories around the world continue to live under the protection of a parent state.

AUSTRALIA

ASHMORE & CARTIER ISLANDS
Indian Ocean
Claimed 1931
Capital not applicable
Area 2 sq miles (5 sq km)
Population None

CHRISTMAS ISLAND
Indian Ocean
Claimed 1958
Capital The Settlement
Area 52 sq miles (135 sq km)
Population 2205

COCOS ISLANDS
Indian Ocean
Claimed 1955
Capital West Island
Area 5.5 sq miles (14 sq km)
Population 596

CORAL SEA ISLANDS
Southwest Pacific
Claimed 1969
Capital None
Area Less than 1.2 sq miles (3 sq km)
Population below 10 (scientists)

HEARD & McDONALD ISLANDS
Indian Ocean
Claimed 1947
Capital not applicable
Area 161 sq miles (417 sq km)
Population None

NORFOLK ISLAND
Southwest Pacific
Claimed 1774
Capital Kingston
Area 13.3 sq miles (34 sq km)
Population 1748

DENMARK

FAROE ISLANDS
North Atlantic
Claimed 1380
Capital Tórshavn
Area 540 sq miles (1399 sq km)
Population 49,120

GREENLAND
North Atlantic
Claimed 1380
Capital Nuuk
Area 840,000 sq miles (2,175,516 sq km)
Population 56,190

FRANCE

CLIPPERTON ISLAND
East Pacific
Claimed 1935
Capital not applicable
Area 2.7 sq miles (7 sq km)
Population None

FRENCH GUIANA
South America
Claimed 1817
Capital Cayenne
Area 35,135 sq miles (90,996 sq km)
Population 276,000

FRENCH POLYNESIA
South Pacific
Claimed 1843
Capital Papeete
Area 1,608 sq miles (4165 sq km)
Population 280,210

GUADELOUPE
West Indies
Claimed 1635
Capital Basse-Terre
Area 629 sq miles (1628 sq km)
Population 400,187

MARTINIQUE
West Indies
Claimed 1635
Capital Fort-de-France
Area 425 sq miles (1100 sq km)
Population 396,000

MAYOTTE
Indian Ocean
Claimed 1843
Capital Mamoudzou
Area 144 sq miles (374 sq km)
Population 212,645

NEW CALEDONIA
Southwest Pacific
Claimed 1853
Capital Nouméa
Area 7,374 sq miles (19,103 sq km)
Population 278,000

RÉUNION
Indian Ocean
Claimed 1638
Capital Saint-Denis
Area 970 sq miles (2512 sq km)
Population 867,000

ST. PIERRE & MIQUELON
North America
Claimed 1604
Capital Saint-Pierre
Area 93 sq miles (242 sq km)
Population 5533

WALLIS & FUTUNA
South Pacific
Claimed 1842
Capital Matá'Utu
Area 106 sq miles (274 sq km)
Population 15,714

NETHERLANDS

ARUBA
West Indies
Claimed 1643
Capital Oranjestad
Area 75 sq miles (194 sq km)
Population 104,810

BONAIRE
West Indies
Claimed 1816
Capital Kralendijk
Area 113 sq miles (294 sq km)
Population 19,400

CURAÇAO
West Indies
Claimed 1815
Capital Willemstad
Area 171 sq miles (444 sq km)
Population 160,000

SABA
West Indies
Claimed 1816
Capital The Bottom
Area 5 sq miles (13 sq km)
Population 1846

SINT-EUSTATIUS
West Indies
Claimed 1784
Capital Oranjestad
Area 8 sq miles (21 sq km)
Population 3200

SINT-MAARTEN
West Indies
Claimed 1648
Capital Phillipsburg
Area 13 sq miles (34 sq km)
Population 40,010

NEW ZEALAND

COOK ISLANDS
South Pacific
Claimed 1901
Capital Avarua
Area 91 sq miles (235 sq km)
Population 11,700

NIUE
South Pacific
Claimed 1901
Capital Alofi
Area 102 sq miles (264 sq km)
Population 1618

TOKELAU
South Pacific
Claimed 1926
Capital not applicable
Area 4 sq miles (10 sq km)
Population 1285

NORWAY

BOUVET ISLAND
South Atlantic
Claimed 1928
Capital not applicable
Area 22 sq miles (58 sq km)
Population None

JAN MAYEN
North Atlantic
Claimed 1929
Capital not applicable
Area 147 sq miles (381 sq km)
Population 18 (scientists)

PETER I ISLAND
Antarctica
Claimed 1931
Capital not applicable
Area 69 sq miles (180 sq km)
Population None

SVALBARD
Arctic Ocean
Claimed 1920
Capital Longyearbyen
Area 24,289 sq miles (62,906 sq km)
Population 2752

UNITED KINGDOM

ANGUILLA
West Indies
Claimed 1650
Capital The Valley
Area 37 sq miles (96 sq km)
Population 17,087

ASCENSION ISLAND
South Atlantic
Claimed 1673
Capital Georgetown
Area 34 sq miles (88 sq km)
Population 806

BERMUDA
North Atlantic
Claimed 1612
Capital Hamilton
Area 20 sq miles (53 sq km)
Population 65,330

BRITISH INDIAN OCEAN TERRITORY
Indian Ocean
Claimed 1814
Capital Diego Garcia
Area 23 sq miles (60 sq km)
Population 4200

BRITISH VIRGIN ISLANDS
West Indies
Claimed 1672
Capital Road Town
Area 59 sq miles (153 sq km)
Population 30,660

CAYMAN ISLANDS
West Indies
Claimed 1670
Capital George Town
Area 100 sq miles (259 sq km)
Population 60,770

FALKLAND ISLANDS
South Atlantic
Claimed 1832
Capital Stanley
Area 4699 sq miles (12,173 sq km)
Population 3198

GIBRALTAR
Southwest Europe
Claimed 1713
Capital Gibraltar
Area 2.5 sq miles (6.5 sq km)
Population 34,410

GUERNSEY
Northwest Europe
Claimed 1066
Capital St Peter Port
Area 25 sq miles (65 sq km)
Population 66,502

ISLE OF MAN
Northwest Europe
Claimed 1765
Capital Douglas
Area 221 sq miles (572 sq km)
Population 83,740

JERSEY
Northwest Europe
Claimed 1066
Capital St. Helier
Area 45 sq miles (116 sq km)
Population 98,840

MONTSERRAT
West Indies
Claimed 1632
Capital Brades (de facto); Plymouth (de jure)
Area 40 sq miles (102 sq km)
Population 5215

PITCAIRN GROUP OF ISLANDS
South Pacific
Claimed 1887
Capital Adamstown
Area 18 sq miles (47 sq km)
Population 54

ST. HELENA
South Atlantic
Claimed 1673
Capital Jamestown
Area 47 sq miles (122 sq km)
Population 4800

SOUTH GEORGIA &
  THE SOUTH SANDWICH ISLANDS
South Atlantic
Capital not applicable
Claimed 1775
Area 1387 sq miles (3592 sq km)
Population None

TRISTAN DA CUNHA
South Atlantic
Claimed 1612
Capital Edinburgh
Area 38 sq miles (98 sq km)
Population 293

TURKS & CAICOS ISLANDS
West Indies
Claimed 1766
Capital Cockburn Town
Area 166 sq miles (430 sq km)
Population 34,900

UNITED STATES OF AMERICA

AMERICAN SAMOA
South Pacific
Claimed 1900
Capital Pago Pago
Area 75 sq miles (195 sq km)
Population 55,600

BAKER & HOWLAND ISLANDS
Central Pacific
Claimed 1856
Capital not applicable
Area 0.54 sq miles (1.4 sq km)
Population None

GUAM
West Pacific
Claimed 1898
Capital Hagåtña
Area 212 sq miles (549 sq km)
Population 162,900

JARVIS ISLAND
Central Pacific
Claimed 1856
Capital not applicable
Area 1.7 sq miles (4.5 sq km)
Population None

NORTHERN MARIANA ISLANDS
West Pacific
Claimed 1947
Capital Saipan
Area 177 sq miles (457 sq km)
Population 55,020

PALMYRA ATOLL
Central Pacific
Claimed 1898
Capital not applicable
Area 5 sq miles (12 sq km)
Population None

PUERTO RICO
West Indies
Claimed 1898
Capital San Juan
Area 3515 sq miles (9104 sq km)
Population 3.7 million

VIRGIN ISLANDS
West Indies
Claimed 1917
Capital Charlotte Amalie
Area 137 sq miles (355 sq km)
Population 102,950

WAKE ISLAND
Central Pacific
Claimed 1898
Capital not applicable
Area 2.5 sq miles (6.5 sq km)
Population 150 (US air base)

Geographical comparisons

Largest countries

Russia	6,592,735 sq miles	(17,075,200 sq km)
Canada	3,855,171 sq miles	(9,984,670 sq km)
USA	3,717,792 sq miles	(9,626,091 sq km)
China	3,705,386 sq miles	(9,596,960 sq km)
Brazil	3,286,470 sq miles	(8,511,965 sq km)
Australia	2,967,893 sq miles	(7,686,850 sq km)
India	1,269,338 sq miles	(3,287,590 sq km)
Argentina	1,068,296 sq miles	(2,766,890 sq km)
Kazakhstan	1,049,150 sq miles	(2,717,300 sq km)
Algeria	919,590 sq miles	(2,381,740 sq km)

Smallest countries

Vatican City	0.17 sq miles	(0.44 sq km)
Monaco	0.75 sq miles	(1.95 sq km)
Nauru	8 sq miles	(21 sq km)
Tuvalu	10 sq miles	(26 sq km)
San Marino	24 sq miles	(61 sq km)
Liechtenstein	62 sq miles	(160 sq km)
Marshall Islands	70 sq miles	(181 sq km)
St. Kitts & Nevis	101 sq miles	(261 sq km)
Maldives	116 sq miles	(300 sq km)
Malta	122 sq miles	(316 sq km)

Largest islands

Greenland	840,000 sq miles	(2,175,600 sq km)
New Guinea	312,000 sq miles	(808,000 sq km)
Borneo	292,222 sq miles	(757,050 sq km)
Madagascar	226,656 sq miles	(587,040 sq km)
Sumatra	202,300 sq miles	(524,000 sq km)
Baffin Island	183,800 sq miles	(476,000 sq km)
Honshu	88,800 sq miles	(230,000 sq km)
Britain	88,700 sq miles	(229,800 sq km)
Victoria Island	81,900 sq miles	(212,000 sq km)
Ellesmere Island	75,700 sq miles	(196,000 sq km)

Richest countries   (GNI per capita, in US$)

Monaco	186,950
Liechtenstein	115,530
Norway	93,530
Switzerland	84,550
Qatar	83,990
Luxembourg	77,480
Denmark	60,270
Australia	60,050
Sweden	57,900
USA	55,990

Poorest countries   (GNI per capita, in US$)

Burundi	260
Central African Republic	320
Malawi	350
Liberia	380
Niger	390
Congo, Dem. Republic	410
Madagascar	420
The Gambia	470
Guinea	470
North Korea	500

Most populous countries

China	1.41 billion
India	1.34 billion
USA	324 million
Indonesia	264 million
Brazil	209 million
Pakistan	197 million

Most populous countries *continued*

Nigeria	174 million
Bangladesh	157 million
Russia	143 million
Japan	127 million

Least populous countries

Vatican City	1000
Tuvalu	11,000
Nauru	13,000
Palau	21,000
San Marino	34,000
Monaco	38,000
Liechtenstein	38,000
Marshall Islands	53,000
St. Kitts & Nevis	55,000
Dominica	74,000

Most densely populated countries

Monaco	50,667 people per sq mile (19,487 per sq km)
Singapore	24,153 people per sq mile (9344 per sq km)
Vatican City	5882 people per sq mile (2273 per sq km)
Bahrain	5495 people per sq mile (2125 per sq km)
Maldives	3448 people per sq mile (1333 per sq km)
Malta	3226 people per sq mile (1250 per sq km)
Bangladesh	3186 people per sq mile (1230 per sq km)
Taiwan	1887 people per sq mile (728 per sq km)
Mauritius	1811 people per sq mile (699 per sq km)
Barbados	1807 people per sq mile (698 per sq km)

Most sparsely populated countries

Mongolia	5 people per sq mile (2 per sq km)
Namibia	8 people per sq mile (3 per sq km)
Australia	8 people per sq mile (3 per sq km)
Iceland	8 people per sq mile (3 per sq km)
Libya	9 people per sq mile (4 per sq km)
Suriname	10 people per sq mile (4 per sq km)
Canada	10 people per sq mile (4 per sq km)
Mauritania	11 people per sq mile (4 per sq km)
Botswana	11 people per sq mile (4 per sq km)
Guyana	11 people per sq mile (4 per sq km)

Most widely spoken languages

1. Chinese (Mandarin)	6. Portuguese
2. Spanish	7. Bengali
3. English	8. Russian
4. Hindi	9. Japanese
5. Arabic	10. Javanese

Largest conurbations

Guangzhou (China)	48,600,000
Tokyo (Japan)	39,800,000
Shanghai (China)	31,100,000
Jakarta (Indonesia)	28,900,000
Delhi (India)	27,200,000
Karachi (Pakistan)	25,100,000
Seoul (South Korea)	24,800,000
Manila (Philippines)	24,100,000
Mumbai (India)	23,600,000
Mexico City (Mexico)	22,300,000
New York (USA)	22,200,000
São Paulo (Brazil)	21,900,000
Beijing (China)	20,700,000
Dhaka (Bangladesh)	17,900,000
Osaka (Japan)	17,800,000
Los Angeles (USA)	17,700,000
Lagos (Nigeria)	17,600,000
Bangkok (Thailand)	17,400,000

Largest conurbations *continued*

Cairo (Egypt)..17,100,000
Moscow (Russia)...17,100,000
Kolkata (India)...16,200,000
Buenos Aires (Argentina)................................16,000,000
Istanbul (Turkey)...14,600,000
London (UK)..14,500,000
Tehran (Iran)...14,000,000

Longest rivers

Nile (Northeast Africa).............................4160 miles.................(6695 km)
Amazon (South America).........................4049 miles.................(6516 km)
Yangtze (China)..3915 miles.................(6299 km)
Mississippi/Missouri (USA)......................3710 miles.................(5969 km)
Ob'-Irtysh (Russia)...................................3461 miles.................(5570 km)
Yellow River (China)................................3395 miles.................(5464 km)
Congo (Central Africa).............................2900 miles.................(4667 km)
Mekong (Southeast Asia).........................2749 miles.................(4425 km)
Lena (Russia)..2734 miles.................(4400 km)
Mackenzie (Canada).................................2640 miles.................(4250 km)
Yenisey (Russia).......................................2541 miles.................(4090 km)

Highest mountains   (Height above sea level)

Everest...29,029 ft.....................(8848 m)
K2..28,251 ft.....................(8611 m)
Kanchenjunga I...........................28,210 ft.....................(8598 m)
Makalu I......................................27,767 ft.....................(8463 m)
Cho Oyu......................................26,907 ft.....................(8201 m)
Dhaulagiri I.................................26,796 ft.....................(8167 m)
Manaslu I.....................................26,783 ft.....................(8163 m)
Nanga Parbat I............................26,661 ft.....................(8126 m)
Annapurna I.................................26,547 ft.....................(8091 m)
Gasherbrum I...............................26,471 ft.....................(8068 m)

Largest bodies of inland water   (Area & depth)

Caspian Sea.............143,243 sq miles (371,000 sq km).....3215 ft (980 m)
Lake Superior..............32,151 sq miles (83,270 sq km).....1289 ft (393 m)
Lake Victoria ....26,560 sq miles (68,880 sq km)......328 ft (100 m)
Lake Huron ....23,436 sq miles (60,700 sq km)......751 ft (229 m)
Lake Michigan.............22,402 sq miles (58,020 sq km)......922 ft (281 m)
Lake Tanganyika..........12,703 sq miles (32,900 sq km)...4700 ft (1435 m)
Great Bear Lake ....12,274 sq miles (31,790 sq km).....1047 ft (319 m)
Lake Baikal ....11,776 sq miles (30,500 sq km)...5712 ft (1741 m)
Great Slave Lake ....10,981 sq miles (28,440 sq km).......459 ft (140 m)
Lake Erie.......................9915 sq miles (25,680 sq km).........197 ft (60 m)

Deepest ocean features

Challenger Deep, Mariana Trench (Pacific)...............36,201 ft (11,034 m)
Vityaz III Depth, Tonga Trench (Pacific)....................35,704 ft (10,882 m)
Vityaz Depth, Kurile-Kamchatka Trench (Pacific) ......34,588 ft (10,542 m)
Cape Johnson Deep, Philippine Trench (Pacific) ....34,441 ft (10,497 m)
Kermadec Trench (Pacific) ....32,964 ft (10,047 m)
Ramapo Deep, Japan Trench (Pacific)........................32,758 ft (9984 m)
Milwaukee Deep, Puerto Rico Trench (Atlantic) ....30,185 ft (9200 m)
Argo Deep, Torres Trench (Pacific) ....30,070 ft (9165 m)
Meteor Depth, South Sandwich Trench (Atlantic) ....30,000 ft (9144 m)
Planet Deep, New Britain Trench (Pacific)...................29,988 ft (9140 m)

Greatest waterfalls   (Mean flow of water)

Boyoma (Congo, Dem. Rep.)................600,400 cu. ft/sec (17,000 cu.m/sec)
Khone (Laos/Cambodia).......................410,000 cu. ft/sec (11,600 cu.m/sec)
Niagara (USA/Canada)...........................195,000 cu. ft/sec (5500 cu.m/sec)
Grande (Uruguay)...................................160,000 cu. ft/sec (4500 cu.m/sec)
Paulo Afonso (Brazil)..............................100,000 cu. ft/sec (2800 cu.m/sec)
Urubupunga (Brazil)..................................97,000 cu. ft/sec (2750 cu.m/sec)
Iguaçu (Argentina/Brazil)..........................62,000 cu. ft/sec (1700 cu.m/sec)
Maribondo (Brazil) ....53,000 cu. ft/sec (1500 cu.m/sec)
Victoria (Zimbabwe) ....39,000 cu. ft/sec (1100 cu.m/sec)

Greatest waterfalls *continued*

Kabalega (Uganda)..................................42,000 cu. ft/sec (1200 cu.m/sec)
Churchill (Canada) ....35,000 cu. ft/sec (1000 cu.m/sec)
Cauvery (India)..33,000 cu. ft/sec (900 cu.m/sec)

Highest waterfalls

Angel (Venezuela)..............................3212 ft.....................(979 m)
Tugela (South Africa)..........................3110 ft.....................(948 m)
Utigard (Norway) ....2625 ft.....................(800 m)
Mongefossen (Norway)......................2539 ft.....................(774 m)
Mtarazi (Zimbabwe)...........................2500 ft.....................(762 m)
Yosemite (USA)...................................2425 ft.....................(739 m)
Ostre Mardola Foss (Norway) ....2156 ft.....................(657 m)
Tyssestrengane (Norway)...................2119 ft.....................(646 m)
*Cuquenan (Venezuela)......................2001 ft.....................(610 m)
Sutherland (New Zealand)..................1903 ft.....................(580 m)
*Kjellfossen (Norway)..........................1841 ft.....................(561 m)
indicates that the total height is a single leap

Largest deserts

Sahara ....3,450,000 sq miles (9,065,000 sq km)
Gobi ....500,000 sq miles (1,295,000 sq km)
Ar Rub al Khali ....289,600 sq miles (750,000 sq km)
Great Victorian ....249,800 sq miles (647,000 sq km)
Sonoran ....120,000 sq miles (311,000 sq km)
Kalahari ....120,000 sq miles (310,800 sq km)
Garagum ....115,800 sq miles (300,000 sq km)
Takla Makan ....100,400 sq miles (260,000 sq km)
Namib.................................52,100 sq miles (135,000 sq km)
Thar ....33,670 sq miles (130,000 sq km)

NB – Most of Antarctica is a polar desert, with only 2 inches (50 mm) of precipitation annually

Hottest inhabited places

Djibouti (Djibouti) ....86.0°F......................(30.0°C)
Timbouctou (Mali)......................84.7°F......................(29.3°C)
Tirunelveli (India)........................84.7°F......................(29.3°C)
Tuticorin (India)..........................84.7°F......................(29.3°C)
Nellore (India)............................84.5°F......................(29.2°C)
Santa Marta (Colombia)..............84.5°F......................(29.2°C)
Aden (Yemen) ....84.0°F......................(29.0°C)
Madurai (India)...........................84.0°F......................(29.0°C)
Niamey (Niger) ....84.0°F......................(29.0°C)

Driest inhabited places

Aswān (Egypt) ....0.02 in......................(0.5 mm)
Luxor (Egypt)................................0.03 in......................(0.7 mm)
Arica (Chile).................................0.04 in......................(1.1 mm)
Ica (Peru)......................................0.10 in......................(2.3 mm)
Antofagasta (Chile).......................0.20 in......................(4.9 mm)
El Minya (Egypt)...........................0.20 in......................(5.1 mm)
Asyūt (Egypt)................................0.20 in......................(5.2 mm)
Callao (Peru).................................0.50 in......................(12.0 mm)
Trujillo (Peru).................................0.55 in......................(14.0 mm)
El Faiyûm (Egypt)..........................0.80 in......................(19.0 mm)

Wettest inhabited places

Buenaventura (Colombia) ....265 in......................(6743 mm)
Monrovia (Liberia)..................................202 in......................(5131 mm)
Pago Pago (American Samoa).................196 in......................(4990 mm)
Moulmein (Myanmar)............................191 in......................(4852 mm)
Lae (Papua New Guinea)........................183 in......................(4645 mm)
Baguio (Luzon I., Philippines).................180 in......................(4573 mm)
Sylhet (Bangladesh)................................176 in......................(4457 mm)
Padang (Sumatra, Indonesia)..................166 in......................(4225 mm)
Bogor (Java, Indonesia)...........................166 in......................(4225 mm)
Conakry (Guinea)....................................171 in......................(4341 mm)

Aa – Almere

A

Aa *see* Gauja
Aachen *72 A4 Dut.* Aken, *Fr.* Aix-la-Chapelle; *anc.* Aquae Grani, Aquisgranum. Nordrhein-Westfalen, W Germany
Aaiún *see* Laâyoune
Aalborg *63 B7 var.* Ålborg, Ålborg-Nørresundby; *anc.* Alburgum. Nordjylland, N Denmark
Aalen *73 B6* Baden-Württemberg, S Germany
Aalsmeer *64 C3* Noord-Holland, C Netherlands
Aalst *65 B6* Oost-Vlaanderen, C Belgium
Aalten *64 E4* Gelderland, E Netherlands
Aalter *65 B5* Oost-Vlaanderen, NW Belgium
Aanaarjärvi *see* Inarijärvi
Äänekoski *63 D5* Keski-Suomi, W Finland
Aar *see* Aare
Aare *73 A7 var.* Aar. *river* W Switzerland
Aarhus *63 B7 var.* Århus. C Denmark
Aarlen *see* Arlon
Aat *see* Ath
Aba *55 E5* Orientale, NE Dem. Rep. Congo
Aba *53 G5* Abia, S Nigeria
Ābā as Su'ūd *see* Najrān
Abaco Island *see* Great Abaco, N The Bahamas
Ābādān *98 C4* Khūzestān, SW Iran
Abadan *100 C3 prev.* Bezmein, Büzmeýin, *Rus.* Byuzmeyin. Ahal Welaýaty, C Turkmenistan
Abai *see* Blue Nile
Abakan *92 D4* Respublika Khakasiya, S Russia
Abancay *38 D4* Apurímac, SE Peru
Abariringa *see* Kanton
Abashiri *108 D2 var.* Abasiri. Hokkaidō, NE Japan
Abasiri *see* Abashiri
Ābay Wenz *see* Blue Nile
Ābbaia *see* Ābaya Hāyk'
Abbatis Villa *see* Abbeville
Abbazia *see* Opatija
Abbeville *68 C2 anc.* Abbatis Villa. Somme, N France
'Abd al 'Azīz, Jabal *96 D2 mountain range* NE Syria
Abéché *54 C3 var.* Abécher, Abeshr. Ouaddaï, SE Chad
Abécher *see* Abéché
Abela *see* Ávila
Abellinum *see* Avellino
Abemama *122 D2 var.* Apamama; *prev.* Roger Simpson Island. *atoll* Tungaru, W Kiribati
Abengourou *53 E5* E Ivory Coast
Aberbrothock *see* Arbroath
Abercorn *see* Mbala
Aberdeen *66 D3 anc.* Devana. NE Scotland, United Kingdom
Aberdeen *23 E2* South Dakota, N USA
Aberdeen *24 B2* Washington, NW USA
Abergwaun *see* Fishguard
Abertawe *see* Swansea
Aberystwyth *67 C6* W Wales, United Kingdom
Abeshr *see* Abéché
Abhā *99 B6* 'Asīr, SW Saudi Arabia
Abidavichy *85 D7 Rus.* Obidovichi. Mahilyowskaya Voblasts', E Belarus
Abidjan *53 E5* S Ivory Coast
Abilene *27 F3* Texas, SW USA
Abingdon *see* Pinta, Isla
Abkhazia *see* Ap'khazet'i
Åbo *see* Turku
Aboa *132 B2 Finnish research station* Antarctica
Aboisso *53 E5* SE Ivory Coast
Abo, Massif d' *54 B1 mountain range* NW Chad
Abomey *53 F5* S Benin
Abou-Déïa *54 C3* Salamat, SE Chad
Aboudouhour *see* Abū aḍ Ḍuhūr
Abou Kémal *see* Abū Kamāl
Abrantes *70 B3 var.* Abrántes. Santarém, C Portugal
Abrashlare *see* Brezovo
Abrolhos Bank *34 E4 undersea bank* W Atlantic Ocean
Abrova *85 B6 Rus.* Obrovo. Brestskaya Voblasts', SW Belarus
Abrud *86 B4 Ger.* Gross-Schlatten, *Hung.* Abrudbánya. Alba, SW Romania
Abrudbánya *see* Abrud
Abruzzese, Appennino *74 C4 mountain range* C Italy
Absaroka Range *22 B2 mountain range* Montana/Wyoming, NW USA
Abū aḍ Ḍuhūr *96 B3 Fr.* Aboudouhour. Idlib, NW Syria
Abu Dhabi *see* Abū Ẓabī
Abu Hamed *50 C3* River Nile, N Sudan
Abū Ḥardān *96 E3 var.* Hajîne. Dayr az Zawr, E Syria
Abuja *53 G4 country capital* (Nigeria) Federal Capital District, C Nigeria
Abū Kamāl *96 E3 Fr.* Abou Kémal. Dayr az Zawr, E Syria
Abula *see* Ávila
Abunã, Rio *40 C2 var.* Río Abuná. *river* Bolivia/Brazil
Abut Head *129 B6 headland* South Island, New Zealand
Abuye Meda *50 D4 mountain* C Ethiopia
Abū Ẓabī *99 C5 var.* Abū Ẓabī, *Eng.* Abu Dhabi. *country capital* (United Arab Emirates) Abū Ẓaby, C United Arab Emirates
Abū Ẓabī *see* Abū Ẓabī
Abyaḍ, Al Baḥr al *see* White Nile
Abyla *see* Ávila
Abyssinia *see* Ethiopia
Acalayong *55 A5* SW Equatorial Guinea
Acaponeta *28 D4* Nayarit, C Mexico
Acapulco *29 E5 var.* Acapulco de Juárez. Guerrero, S Mexico
Acapulco de Juárez *see* Acapulco
Acarai Mountains *37 F4 Sp.* Serra Acaraí. *mountain range* Brazil/Guyana
Acaraí, Serra *see* Acarai Mountains
Acarigua *36 D2* Portuguesa, N Venezuela
Accra *53 E5 country capital* (Ghana) SE Ghana
Achacachi *39 E4* La Paz, W Bolivia
Ach'ara *95 F2 var.* Ajaria. *autonomous republic* SW Georgia
Acklins Island *32 C2 island* SE The Bahamas
Aconcagua, Cerro *42 B4 mountain* W Argentina
Açores/Açores, Arquipélago dos/Açores, Ilhas dos *see* Azores
A Coruña *70 B1 Cast.* La Coruña, *Eng.* Corunna; *anc.* Caronium. Galicia, NW Spain
Acre *40 C2 off.* Estado do Acre. *state/region* W Brazil
Açu *see* Assu
Acunum Acusio *see* Montélimar
Ada *78 D3* Vojvodina, N Serbia
Ada *27 G2* Oklahoma, C USA
Ada Bazar *see* Adapazarı
Adalia *see* Antalya
Adalia, Gulf of *see* Antalya Körfezi
Adama *see* Nazrēt
'Adan *99 B7 Eng.* Aden. SW Yemen
Adana *94 D4 var.* Seyhan. Adana, S Turkey
Adâncata *see* Horlivka
Adapazarı *94 B2 prev.* Ada Bazar. Sakarya, NW Turkey
Adare, Cape *132 B4 cape* Antarctica
Ad Dahna *98 C4 desert* E Saudi Arabia
Ad Dakhla *48 A4 var.* Dakhla. SW Western Sahara
Ad Dalanj *see* Dilling
Ad Damar *see* Ed Damer
Ad Damazīn *see* Ed Damazin
Ad Dāmir *see* Ed Damer
Ad Dammān *98 C4 var.* Dammām. Ash Sharqīyah, NE Saudi Arabia
Ad Dāmūr *see* Damoûr
Ad Dawḥah *98 C4 Eng.* Doha. *country capital* (Qatar) C Qatar
Aḍ Ḍiffah *see* Libyan Plateau
Addis Ababa *see* Ādīs Ābeba
Addoo Atoll *see* Addu Atoll
Addu Atoll *110 A5 var.* Addoo Atoll, Seenu Atoll. *atoll* S Maldives
Adelaide *127 B6 state capital* South Australia
Adelsberg *see* Postojna
Aden *see* 'Adan
Aden, Gulf of *99 C7 gulf* SW Arabian Sea
Adige *74 C2 Ger.* Etsch. *river* N Italy
Adirondack Mountains *19 F2 mountain range* New York, NE USA
Ādīs Ābeba *51 C5 Eng.* Addis Ababa. *country capital* (Ethiopia) Ādīs Ābeba, C Ethiopia
Adıyaman *95 E4* Adıyaman, SE Turkey
Adjud *86 C4* Vrancea, E Romania
Admiralty Islands *122 B3 island group* N Papua New Guinea
Adra *71 E5* Andalucía, S Spain
Adrar *48 D3* C Algeria
Adrian *18 C3* Michigan, N USA
Adrianople/Adrianopolis *see* Edirne
Adriatico, Mare *see* Adriatic Sea
Adriatic Sea *81 E2 Alb.* Deti Adriatik, *It.* Mare Adriatico, *SCr.* Jadransko More, *Slvn.* Jadransko Morje. *sea* N Mediterranean Sea
Adriatik, Deti *see* Adriatic Sea
Adycha *93 F2 river* NE Russia
Aegean Sea *83 C5 Gk.* Aigaíon Pelagos, Aigaío Pélagos, *Turk.* Ege Denizi. *sea* NE Mediterranean Sea
Aegviidu *84 D2 Ger.* Charlottenhof. Harjumaa, NW Estonia
Aegyptus *see* Egypt
Aelana *see* Al 'Aqabah
Aelok *see* Ailuk Atoll
Aelōnlaplap *see* Ailinglaplap Atoll
Aemona *see* Ljubljana
Aeolian Islands *75 C6 var.* Isole Lipari, *Eng.* Aeolian Islands, Lipari Islands. *island group* S Italy
Aeolian Islands *see* Eolie, Isole
Æsernia *see* Isernia
Afar Depression *see* Danakil Desert
Afars et des Issas, Territoire Français des *see* Djibouti
Afghānestān, Dowlat-e Eslāmī-ye *see* Afghanistan
Afghanistan *100 C4 off.* Islamic Republic of Afghanistan, *Per.* Dowlat-e Eslāmī-ye Afghānestān; *prev.* Republic of Afghanistan. *country* C Asia
Afmadow *51 D6* Jubbada Hoose, S Somalia
Africa *46 continent*
Africa, Horn of *46 E4 physical region* Ethiopia/Somalia
Africana Seamount *119 A6 seamount* SW Indian Ocean
'Afrīn *96 B2* Ḥalab, N Syria
Afyon *94 B3 prev.* Afyonkarahisar. Afyon, W Turkey
Agadès *see* Agadez
Agadez *53 G3 prev.* Agadès. Agadez, C Niger
Agadir *48 B3* SW Morocco
Agana/Agaña *see* Hagåtña
Āgaro *51 C5* Oromīya, C Ethiopia
Agassiz Fracture Zone *121 G5 fracture zone* S Pacific Ocean
Agatha *see* Agde
Agathónisi *83 D6 island* Dodekánisa, Greece, Aegean Sea
Agde *69 C6 anc.* Agatha. Hérault, S France
Agedabia *see* Ajdābiyā
Agen *69 B5 anc.* Aginnum. Lot-et-Garonne, SW France
Agendicum *see* Sens
Aghri Dagh *see* Büyükağrı Dağı
Agiá *82 B4 var.* Ayiá. Thessalía, C Greece
Agialoúsa *see* Yenierenköy
Agía Marína *83 E6* Léros, Dodekánisa, Greece, Aegean Sea
Aginnum *see* Agen
Ágios Efstrátios *82 D4 var.* Áyios Evstrátios, Hagios Evstrátios. *island* E Greece
Ágios Nikólaos *83 D8 var.* Áyios Nikólaos. Kríti, Greece, E Mediterranean Sea
Āgra *112 D3* Uttar Pradesh, N India
Agra and Oudh, United Provinces of *see* Uttar Pradesh
Agram *see* Zagreb
Ağrı *95 F3 var.* Karaköse; *prev.* Karakılısse. Ağrı, NE Turkey
Agri Dagi *see* Büyükağrı Dağı
Agrigento *75 C7 Gk.* Akragas; *prev.* Girgenti. Sicilia, Italy, C Mediterranean Sea
Agriovótano *83 C5* Évvoia, C Greece
Agropoli *75 D5* Campania, S Italy
Aguachica *36 B2* Cesar, N Colombia
Aguadulce *31 F5* Coclé, S Panama
Agua Prieta *28 B1* Sonora, NW Mexico
Aguascalientes *28 D4* Aguascalientes, C Mexico
Aguaytía *38 C3* Ucayali, C Peru
Aguilas *71 E4* Murcia, SE Spain
Aguililla *28 D4* Michoacán, SW Mexico
Agulhas Basin *47 D8 undersea basin* SW Indian Ocean
Agulhas, Cape *56 C5 headland* SW South Africa
Agulhas Plateau *45 D6 undersea plateau* SW Indian Ocean
Ahaggar *53 F2 high plateau region* SE Algeria
Ahlen *72 B4* Nordrhein-Westfalen, W Germany
Ahmadābād *112 C4 var.* Ahmedabad. Gujarāt, W India
Ahmadnagar *112 C5 var.* Ahmednagar. Mahārāshtra, W India
Ahmedabad *see* Ahmadābād
Ahmednagar *see* Ahmadnagar
Ahuachapán *30 B3* Ahuachapán, W El Salvador
Ahvāz *98 C3 var.* Ahwāz; *prev.* Nāsiri. Khūzestān, SW Iran
Ahvenanmaa *see* Åland
Ahwāz *see* Ahvāz
Aigaíon Pelagos/Aigaío Pélagos *see* Aegean Sea
Aígina *83 C6 var.* Aíyina, Egina. Aígina, C Greece
Aígio *83 B5 var.* Egio; *prev.* Aíyion. Dytikí Ellás, S Greece
Aiken *21 E2* South Carolina, SE USA
Ailinglaplap Atoll *122 D2 var.* Aelōnlaplap. *atoll* Ralik Chain, S Marshall Islands
Ailuk Atoll *122 D1 var.* Aelok. *atoll* Ratak Chain, NE Marshall Islands
Ainaži *84 D3 Est.* Heinaste, *Ger.* Hainasch. Limbaži, N Latvia
'Aïn Ben Tili *52 D1* Tiris Zemmour, N Mauritania
Aintab *see* Gaziantep
Aïoun el Atrous/Aïoun el Atroûss *see* 'Ayoûn el 'Atroûs
Aiquile *39 F4* Cochabamba, C Bolivia
Aïr *see* Aïr, Massif de l'
Air du Azbine *see* Aïr, Massif de l'
Aïr, Massif de l' *53 G2 var.* Aïr, Air du Azbine, Asben. *mountain range* NC Niger
Aiud *86 B4 Ger.* Strassburg, *Hung.* Nagyenyed; *prev.* Engeten. Alba, SW Romania
Aix *see* Aix-en-Provence
Aix-en-Provence *69 D6 var.* Aix; *anc.* Aquae Sextiae. Bouches-du-Rhône, SE France
Aix-la-Chapelle *see* Aachen
Aíyina *see* Aígina
Aíyion *see* Aígio
Aizkraukle *84 C4* Aizkraukle, S Latvia
Ajaccio *69 E7* Corse, France, C Mediterranean Sea
Ajaria *see* Ach'ara
Ajastan *see* Armenia
Aj Bogd Uul *104 D2 mountain* SW Mongolia
Ajdābiyā *49 G2 var.* Agedabia, Ajdābiyah. NE Libya
Ajdābiyah *see* Ajdābiyā
Ajjinena *see* El Geneina
Ajmer *112 D3 var.* Ajmere. Rājasthān, N India
Ajo *26 A3* Arizona, SW USA
Akaba *see* Al 'Aqabah
Akamagaseki *see* Shimonoseki
Akasha *50 B3* Northern, N Sudan
Akchâr *52 C2 desert* W Mauritania
Aken *see* Aachen
Akermanceaster *see* Bath
Akhaltsikhe *95 F2 prev.* Akhalts'ikhe. SW Georgia
Akhalts'ikhe *see* Akhaltsikhe
Akhisar *94 A3* Manisa, W Turkey
Akhmīm *50 B2 var.* Akhmîm; *anc.* Panopolis. C Egypt
Akhtubinsk *89 C7* Astrakhanskaya Oblast', SW Russia
Akhtyrka *see* Okhtyrka
Akimiski Island *16 C3 island* Nunavut, C Canada
Akinovka *87 F4* Zaporiz'ka Oblast', S Ukraine
Akita *108 D4* Akita, Honshū, C Japan
Akjoujt *52 C2 prev.* Fort-Repoux. Inchiri, W Mauritania
Akkeshi *108 E2* Hokkaidō, NE Japan
Aklavik *14 D3* Northwest Territories, NW Canada
Akmola *see* Astana
Akmolinsk *see* Astana
Aknavásár *see* Târgu Ocna
Akpatok Island *17 E1 island* Nunavut, E Canada
Akragas *see* Agrigento
Akron *18 D4* Ohio, N USA
Akrotiri *80 C5 var.* Akrotírion. *UK air base* S Cyprus
Akrotírion *see* Akrotiri
Akrotírion Aksai Chin *102 B2 Chin.* Aksayqin. *disputed region* China/India
Aksaray *94 C4* Aksaray, C Turkey
Aksayqin *see* Aksai Chin
Akşehir *94 B4* Konya, W Turkey
Aktash *see* Oqtosh
Aktau *92 A4 Kaz.* Aqtaū; *prev.* Shevchenko. Mangistau, W Kazakhstan
Aktjubinsk/Aktyubinsk *see* Aktobe
Aktobe *92 B4 Kaz.* Aqtöbe; *prev.* Aktjubinsk, Aktyubinsk. Aktyubinsk, NW Kazakhstan
Aktsyabrski *85 C7 Rus.* Oktyabr'skiy; *prev.* Karpilovka. Homyel'skaya Voblasts', SE Belarus
Aktyubinsk *see* Aktobe
Akula *55 C5* Equateur, NW Dem. Rep. Congo
Akureyri *61 E4* Nordhurland Eystra, N Iceland
Akyab *see* Sittwe
Alabama *20 C2 off.* State of Alabama, *also known as* Camellia State, Heart of Dixie, The Cotton State, Yellowhammer State. *state* S USA
Alabama River *20 C3 river* Alabama, S USA
Alaca *94 C3* Çorum, N Turkey
Alacant *see* Alicante
Alagoas *41 G2 off.* Estado de Alagoas. *state/region* E Brazil
Alais *see* Alès
Alajuela *31 E4* Alajuela, C Costa Rica
Alakanuk *14 C2* Alaska, USA
Al 'Alamayn *50 B1 var.* El 'Alamein. N Egypt
Al 'Amārah *98 C3 var.* Amara. Maysān, E Iraq
Alamo *25 D6* Nevada, W USA
Alamogordo *26 D3* New Mexico, SW USA
Alamosa *22 C5* Colorado, C USA
Åland *63 C6 var.* Aland Islands, *Fin.* Ahvenanmaa. *island group* SW Finland
Aland Islands *see* Åland
Aland Sea *see* Ålands Hav
Ålands Hav *63 C6 var.* Aland Sea. *strait* Baltic Sea/Gulf of Bothnia
Alanya *94 C4* Antalya, S Turkey
Alappuzha *110 C3 var.* Alleppey. Kerala, SW India
Al 'Aqabah *97 B8 var.* Akaba, Aqaba, 'Aqaba; *anc.* Aelana, Elath. Al 'Aqabah, SW Jordan
Al 'Arabīyah as Su'ūdīyah *see* Saudi Arabia
Alasca, Golfo de *see* Alaska, Gulf of
Alaşehir *94 A4* Manisa, W Turkey
Al 'Ashārah *96 E3 var.* Ashara. Dayr az Zawr, E Syria
Alaska *14 C3 off.* State of Alaska, *also known as* Land of the Midnight Sun, The Last Frontier, Seward's Folly; *prev.* Russian America. *state* NW USA
Alaska, Gulf of *14 C4 var.* Golfo de Alasca. *gulf* Canada/USA
Alaska Peninsula *14 C3 peninsula* Alaska, USA
Alaska Range *12 B2 mountain range* Alaska, USA
Al-Asnam *see* Chlef
Alattio *see* Alta
Al Awaynāt *see* Al 'Uwaynāt
Alaykel'/Alay-Kuu *see* Këk-Art
Al 'Aynā *97 B7* Al Karak, W Jordan
Alazeya *93 G2 river* NE Russia
Al Bāb *96 B2* Ḥalab, N Syria
Albacete *71 E3* Castilla-La Mancha, C Spain
Al Baghdādī *98 B3 var.* Khān al Baghdādī. Al Anbār, SW Iraq
Al Bāha *see* Al Bāḥah
Al Bāḥah *99 B5 var.* Al Bāha. Al BāḤah, SW Saudi Arabia
Al Baḥrayn *see* Bahrain
Alba Iulia *86 B4 Ger.* Weissenburg, *Hung.* Gyulafehérvár; *prev.* Bălgrad, Karlsburg, Károly-Fehérvár. Alba, W Romania
Albania *79 C7 off.* Republic of Albania, *Alb.* Republika e Shqipqërisë, Shqipëria; *prev.* People's Socialist Republic of Albania. *country* SE Europe
Albania *see* Aubagne
Albany *125 B7* Western Australia
Albany *20 D3* Georgia, SE USA
Albany *19 F3 state capital* New York, NE USA
Albany *24 B3* Oregon, NW USA
Albany *16 C3 river* Ontario, S Canada
Alba Regia *see* Székesfehérvár
Al Bāridah *96 C4 var.* Bāridah. Ḥimṣ, C Syria
Al Baṣrah *98 C3 Eng.* Basra, *hist.* Busra, Bussora. Al Baṣrah, SE Iraq
Al Batrūn *see* Batroûn
Al Bawītī *50 B2 var.* Bawîṭi C Egypt
Al Bayḍā' *49 G2 var.* Beida. NE Libya
Albemarle Island *see* Isabela, Isla
Albemarle Sound *21 G1 inlet* W Atlantic Ocean
Albergaria-a-Velha *70 B2* Aveiro, N Portugal
Albert *68 C3* Somme, N France
Alberta *15 E4 province* SW Canada
Albert Edward Nyanza *see* Edward, Lake
Albert, Lake *51 B6 var.* Albert Nyanza, Lac Mobutu Sese Seko. *lake* Uganda/Dem. Rep. Congo
Albert Lea *23 F3* Minnesota, N USA
Albert Nyanza *see* Albert, Lake
Albertville *see* Kalemie
Albi *69 C6 anc.* Albiga. Tarn, S France
Albiga *see* Albi
Ålborg *see* Aalborg
Ålborg-Nørresundby *see* Aalborg
Alborz, Reshteh-ye Kūhhā-ye *98 C2 Eng.* Elburz Mountains. *mountain range* N Iran
Albuquerque *26 D2* New Mexico, SW USA
Al Burayqah *see* Marsá al Burayqah
Alburgum *see* Aalborg
Albury *127 C7* New South Wales, SE Australia
Alcácer do Sal *70 B4* Setúbal, W Portugal
Alcalá de Henares *71 E3 Ar.* Alkal'a; *anc.* Complutum. Madrid, C Spain
Alcamo *75 C7* Sicilia, Italy, C Mediterranean Sea
Alcañiz *71 F2* Aragón, NE Spain
Alcántara, Embalse de *70 C3 reservoir* W Spain
Alcaudete *70 D4* Andalucía, S Spain
Alcázar *see* Ksar-el-Kebir
Alcazarquivir *see* Ksar-el-Kebir
Alcoi *see* Alcoy
Alcoy *71 F4 Cat.* Alcoi. País Valenciano, E Spain
Aldabra Group *57 G2 island group* SW Seychelles
Aldan *93 F3 river* NE Russia
al Dar al Baida *see* Rabat
Alderney *68 A2 island* Channel Islands
Aleg *52 C3* Brakna, SW Mauritania
Aleksandriya *see* Oleksandriya
Aleksandropol' *see* Gyumri
Aleksandrovka *see* Oleksandrivka
Aleksandrovsk *see* Zaporizhzhya
Aleksin *89 B5* Tul'skaya Oblast', W Russia
Aleksinac *78 E4* Serbia, SE Serbia
Alençon *68 B3* Orne, N France
Alenquer *41 E2* Pará, NE Brazil
Alep/Aleppo *see* Ḥalab
Alert *15 F1* Ellesmere Island, Nunavut, N Canada
Alès *69 C6 prev.* Alais. Gard, S France
Aleşd *86 B3 Hung.* Élesd. Bihor, SW Romania
Alessandria *74 B2 Fr.* Alexandrie. Piemonte, N Italy
Ålesund *63 A5* Møre og Romsdal, S Norway
Aleutian Basin *91 G3 undersea basin* Bering Sea
Aleutian Islands *14 A3 island group* Alaska, USA
Aleutian Range *12 A2 mountain range* Alaska, USA
Aleutian Trench *91 H3 trench* S Bering Sea
Alexander Archipelago *14 D4 island group* Alaska, USA
Alexander City *20 D2* Alabama, S USA
Alexander Island *132 A3 island* Antarctica
Alexander Range *see* Kirghiz Range
Alexandra *129 B7* Otago, South Island, New Zealand
Alexándreia *82 B4 var.* Alexándria. Kentrikí Makedonía, N Greece
Alexandretta *see* İskenderun
Alexandretta, Gulf of *see* İskenderun Körfezi
Alexandria *50 B1 Ar.* Al Iskandarīyah. N Egypt
Alexandria *86 C5* Teleorman, S Romania
Alexandria *20 B3* Louisiana, S USA
Alexandria *23 F2* Minnesota, N USA
Alexándria *see* Alexándreia
Alexandrie *see* Alessandria
Alexandroúpoli *82 D3 var.* Alexandroúpolis, *Turk.* Dedeagaç, Dedeagach. Anatolikí Makedonía kai Thráki, NE Greece
Alexandroúpolis *see* Alexandroúpoli
Al Fāshir *see* El Fasher
Alfatar *82 E1* Silistra, NE Bulgaria
Alfeiós *83 B6 prev.* Alfiós; *anc.* Alpheius, Alpheus. *river* S Greece
Alfiós *see* Alfeiós
Alföld *see* Great Hungarian Plain
Al-Furāt *see* Euphrates
Alga *92 B4 Kaz.* Algha. Aktyubinsk, NW Kazakhstan
Algarve *70 B4 cultural region* S Portugal
Algeciras *70 C5* Andalucía, SW Spain
Algemesí *71 F3* País Valenciano, E Spain
Al-Genain *see* El Geneina
Alger *49 E1 var.* Algiers, El Djazaïr, Al Jazair. *country capital* (Algeria) N Algeria
Algeria *48 C3 off.* Democratic and Popular Republic of Algeria. *country* N Africa
Algeria, Democratic and Popular Republic of *see* Algeria
Algerian Basin *58 C5 var.* Balearic Plain. *undersea basin* W Mediterranean Sea
Algha *see* Alga
Al Ghābah *99 E5 var.* Ghaba. C Oman
Alghero *75 A5* Sardegna, Italy, C Mediterranean Sea
Al Ghurdaqah *50 C2 var.* Hurghada, Ghurdaqah. E Egypt
Algiers *see* Alger
Al Golea *see* El Goléa
Algona *23 F3* Iowa, C USA
Al Hajar al Gharbi *99 D5 mountain range* N Oman
Al Hamad *see* Syrian Desert
Al Ḥasakah *96 D2 var.* Al Hasijah, El Haseke, *Fr.* Hassetché. Al Ḥasakah, NE Syria
Al Hasijah *see* Al Ḥasakah
Al Ḥillah *98 B3 var.* Hilla. Bābil, C Iraq
Al Ḥisā *97 B7* Aṭ Ṭafīlah, W Jordan
Al Ḥudaydah *99 B6 Eng.* Hodeida. W Yemen
Al Hufūf *98 C4 var.* Hofuf. Ash Sharqīyah, NE Saudi Arabia
Aliákmon *see* Aliákmonas
Aliákmonas *82 B4 prev.* Aliákmon; *anc.* Haliacmon. *river* N Greece
Alíartos *83 C5* Stereá Ellás, C Greece
Alicante *71 F4 Cat.* Alacant, *Lat.* Lucentum. País Valenciano, SE Spain
Alice *27 G5* Texas, SW USA
Alice Springs *126 A4* Northern Territory, C Australia
Alifu Atoll *see* Ari Atoll
Aligandí *31 G4* Kuna Yala, NE Panama
Aliki *see* Alykí
Alima *55 B6 river* C Congo
Al Imārāt al 'Arabīyahal Muttaḥidah *see* United Arab Emirates
Alindao *54 C4* Basse-Kotto, S Central African Republic
Aliquippa *18 D4* Pennsylvania, NE USA
Al Iskandarīyah *see* Alexandria
Al Ismā'īlīya *50 B1 var.* Ismailia, Ismâ'ilîya. N Egypt
Alistráti *82 C3* Kentrikí Makedonía, NE Greece
Alivéri *83 C5 var.* Alivérion. Évvoia, C Greece
Alivérion *see* Alivéri
Al Jabal al Akhḍar *49 G2 mountain range* NE Libya
Al Jafr *97 B7* Ma'ān, S Jordan
Al Jaghbūb *49 H3* NE Libya
Al Jahrā' *98 C4 var.* Al Jahrah, Jahra. C Kuwait
Al Jahrah *see* Al Jahrā'
Al Jamāhīrīyah al 'Arabīyah al Lībīyah ash Sha'bīyah al Ishtirākīy *see* Libya
Al Jawf *98 B4 off.* Jauf. Al Jawf, NW Saudi Arabia
Al Jawlān *see* Golan Heights
Al Jazair *see* Alger
Al Jazirah *96 E2 physical region* Iraq/Syria
Al Jīzah *see* Gîza
Al Junaynah *see* El Geneina
Alkal'a *see* Alcalá de Henares
Al Karak *97 B7 var.* El Kerak, Karak, Kerak; *anc.* Kir Moab, Kir of Moab. Al Karak, W Jordan
Al-Kasr al-Kebir *see* Ksar-el-Kebir
Al Khalīl *see* Hebron
Al Khārijah *50 B2 var.* El Khârga. C Egypt
Al Khums *49 F2 var.* Homs, Khoms, Khums. NW Libya
Alkmaar *64 C2* Noord-Holland, NW Netherlands
Al Kufrah *49 H4* SE Libya
Al Kūt *98 C3 var.* Kūt al 'Amārah, Kut al Imara. Wāsiṭ, E Iraq
Al-Kuwait *see* Al Kuwayt
Al Kuwayt *98 C4 var.* Al-Kuwait, *Eng.* Kuwait, Kuwait City; *prev.* Qurein. *country capital* (Kuwait) E Kuwait
Al Lādhiqīyah *96 A3 Eng.* Latakia, *Fr.* Lattaquié; *anc.* Laodicea, Laodicea ad Mare. Al Lādhiqīyah, W Syria
Allahābād *113 E3* Uttar Pradesh, N India
Allanmyo *see* Aunglan
Allegheny Plateau *19 E3 mountain range* New York/Pennsylvania, NE USA
Allenstein *see* Olsztyn
Allentown *19 F4* Pennsylvania, NE USA
Alleppey *see* Alappuzha
Alliance *22 D3* Nebraska, C USA
Al Līth *99 B5* Makkah, SW Saudi Arabia
Al Lubnān *see* Lebanon
Alma-Ata *see* Almaty
Almada *70 B4* Setúbal, W Portugal
Al Madīnah *99 A5 Eng.* Medina. Al Madīnah, W Saudi Arabia
Al Mafraq *97 B6 var.* Mafraq. Al Mafraq, N Jordan
Al Mahdīyah *see* Mahdia
Al Mahrah *99 C6 mountain range* E Yemen
Al Majma'ah *98 B4* Ar Riyāḍ, C Saudi Arabia
Al Mālikīyah *96 E1 var.* Malkiye. Al Ḥasakah, N Syria
Almalyk *see* Olmaliq
Al Mamlakah *see* Morocco
Al Mamlaka al Urdunīya al Hashemīyah *see* Jordan
Al Manāmah *98 C4 Eng.* Manama. *country capital* (Bahrain) N Bahrain
Al Manāṣif *96 E3 mountain range* E Syria
Almansa *71 F4* Castilla-La Mancha, C Spain
Al-Mariyya *see* Almería
Al Marj *49 G2 var.* Barka, *It.* Barce. NE Libya
Almaty *92 C5 var.* Alma-Ata. Almaty, SE Kazakhstan
Al Mawṣil *98 B2 Eng.* Mosul. Nīnawá, N Iraq
Al Mayādīn *96 D3 var.* Mayadin, *Fr.* Meyadine. Dayr az Zawr, E Syria
Al Mazra' *see* Al Mazra'ah
Al Mazra'ah *97 B6 var.* Al Mazra', Mazra'a. Al Karak, W Jordan
Almelo *64 E3* Overijssel, E Netherlands
Almendra, Embalse de *70 C2 reservoir* Castilla-León, NW Spain
Almendralejo *70 C4* Extremadura, W Spain
Almere *64 C3 var.* Almere-stad. Flevoland, C Netherlands

Arizona – Balearic Plain

Arizona 26 A2 off. State of Arizona, also known as Copper State, Grand Canyon State. state SW USA
Arkansas 20 A1 off. State of Arkansas, also known as The Land of Opportunity. state S USA
Arkansas City 23 F5 Kansas, C USA
Arkansas River 27 G1 river C USA
Arkhangel'sk 92 B2 Eng. Archangel. Arkhangel'skaya Oblast', NW Russia
Arkoí 83 E6 island Dodekánisa, Greece, Aegean Sea
Arles 69 D6 var. Arles-sur-Rhône; anc. Arelas, Arelate. Bouches-du-Rhône, SE France
Arles-sur-Rhône see Arles
Arlington 27 G2 Texas, SW USA
Arlington 19 E4 Virginia, NE USA
Arlon 65 D8 Dut. Aarlen, Ger. Arel, Lat. Orolaunum. Luxembourg, SE Belgium
Armagh 67 B5 Ir. Ard Mhacha. S Northern Ireland, United Kingdom
Armagnac 69 B6 cultural region S France
Armenia 36 B3 Quindío, W Colombia
Armenia 95 F3 off. Republic of Armenia, var. Ajastan, Arm. Hayastani Hanrapetut'yun; prev. Armenian Soviet Socialist Republic. country SW Asia
Armenian Soviet Socialist Republic see Armenia
Armenia, Republic of see Armenia
Armidale 127 D6 New South Wales, SE Australia
Armstrong 16 B3 Ontario, S Canada
Armyans'k 87 F4 Rus. Armyansk. Respublika Krym, S Ukraine
Arnaía 82 C4 Cont. Arnea. Kentrikí Makedonía, N Greece
Arnaud 60 A3 river Québec, E Canada
Arnea see Arnaía
Arnedo 71 E2 La Rioja, N Spain
Arnhem 64 D4 Gelderland, SE Netherlands
Arnhem Land 126 A2 physical region Northern Territory, N Australia
Arno 74 B3 river C Italy
Arnold 23 G4 Missouri, C USA
Arnswalde see Choszczno
Aroe Islands see Aru, Kepulauan
Arorae 123 E3 atoll Tungaru, W Kiribati
Arrabona see Győr
Ar Rahad see Er Rahad
Ar Ramādī 98 B3 var. Ramadi, Rumadiya. Al Anbār, SW Iraq
Ar Rāmī 96 C4 Ḥimṣ, C Syria
Ar Ramthā 97 B5 var. Ramtha. Irbid, N Jordan
Arran, Isle of 66 C4 island SW Scotland, United Kingdom
Ar Raqqah 96 C2 var. Rakka; anc. Nicephorium. Ar Raqqah, N Syria
Arras 68 C2 anc. Nemetocenna. Pas-de-Calais, N France
Ar Rawḍatayn 98 C4 var. Raudhatain. N Kuwait
Arretium see Arezzo
Arriaca see Guadalajara
Arriaga 29 G5 Chiapas, SE Mexico
Ar Riyāḍ 99 C5 Eng. Riyadh. country capital (Saudi Arabia) Ar Riyāḍ, C Saudi Arabia
Ar Rub 'al Khali 99 C6 Eng. Empty Quarter, Great Sandy Desert. desert SW Asia
Ar Rustāq 99 E5 var. Rostak, Rustaq. N Oman
Ar Ruṭbah 98 B3 var. Rutba. Al Anbār, SW Iraq
Árta 83 A5 anc. Ambracia. Ípeiros, W Greece
Artashat 95 F3 S Armenia
Artemisa 32 B2 La Habana, W Cuba
Artesia 26 D3 New Mexico, SW USA
Arthur's Pass 129 C6 pass South Island, New Zealand
Artigas 42 D3 prev. San Eugenio, San Eugenio del Cuareim. Artigas, N Uruguay
Art'ik 95 F2 W Armenia
Artois 68 C2 cultural region N France
Artsiz see Artsyz
Artsyz 86 D4 Rus. Artsiz. Odes'ka Oblast', SW Ukraine
Arturo Prat 132 A2 Chilean research station South Shetland Islands, Antarctica
Artvin 95 F2 Artvin, NE Turkey
Arua 51 B6 NW Uganda
Aruba 36 C1 var. Oruba. Dutch self-governing territory S West Indies
Aru Islands see Aru, Kepulauan
Aru, Kepulauan 117 G4 Eng. Aru Islands; prev. Aroe Islands. island group E Indonesia
Arunāchal Pradesh 113 G3 prev. North East Frontier Agency, North East Frontier Agency of Assam. cultural region NE India
Arusha 51 C7 Arusha, N Tanzania
Arviat 15 G4 prev. Eskimo Point. Nunavut, C Canada
Arvidsjaur 62 C4 Norrbotten, N Sweden
Arys' 92 B5 Kaz. Arys. Yuzhnyy Kazakhstan, S Kazakhstan
Arys see Arys'
Asadābād 101 F4 var. Asaḏābād; prev. Chaghasarāy. Konar, E Afghanistan
Asaḏābād see Asadābād
Asad, Buḥayrat al 96 C2 Eng. Lake Assad. lake N Syria
Asahi-dake 108 D2 mountain Hokkaidō, N Japan
Asahikawa 108 D2 Hokkaidō, N Japan
Asamankese 53 E5 SE Ghana
Āsānsol 113 F4 West Bengal, NE India
Asben see Aïr, Massif de l'
Ascension Fracture Zone 47 A5 tectonic Feature C Atlantic Ocean
Ascension Island see St Helena, Ascension and Tristan da Cunha
Ascoli Piceno 74 C4 anc. Asculum Picenum. Marche, C Italy
Asculum Picenum see Ascoli Piceno
'Aseb 50 D4 var. Assab, Amh. Āseb. SE Eritrea
Assen 64 E2 Drenthe, NE Netherlands
Aşgabat 100 C3 prev. Ashgabat, Ashkhabad, Poltoratsk. country capital (Turkmenistan) Ahal Welaýaty, C Turkmenistan
Ashara see Al 'Ashārah
Ashburton 129 C6 Canterbury, South Island, New Zealand
Ashburton River 124 A4 river Western Australia
Ashdod 97 A6 anc. Azotos, Lat. Azotus. Central, W Israel
Asheville 21 E1 North Carolina, SE USA
Ashgabat see Aşgabat
Ashkelon 97 A6 prev. Ashqelon. Southern, C Israel
Ashkhabad see Aşgabat
Ashland 24 B4 Oregon, NW USA
Ashland 18 B1 Wisconsin, N USA
Ashmore and Cartier Islands 120 A3 Australian external territory E Indian Ocean
Ashmyany 85 C5 Rus. Oshmyany. Hrodzyenskaya Voblasts', W Belarus
Ashqelon see Ashkelon
Ash Shadādah 96 D2 var. Ash Shaddādah, Jisr ash Shadadi, Shaddādī, Shedadi, Tell Shedadi. Al Ḥasakah, NE Syria
Ash Shaddādah see Ash Shadādah
Ash Sharah 97 B7 var. Esh Sharā. mountain range W Jordan
Ash Shāriqah 98 D4 Eng. Sharjah. Ash Shāriqah, NE United Arab Emirates
Ash Shawbak 97 B7 Ma'ān, W Jordan
Ash Shiḥr 99 C6 SE Yemen
Asia 90 continent
Asinara 74 A4 island W Italy
Asipovichy 85 D6 Rus. Osipovichi. Mahilyowskaya Voblasts', C Belarus
Aşkale 95 E3 Erzurum, NE Turkey
Askersund 63 C6 Örebro, C Sweden
Asmara see Asmera
Asmera 50 C4 var. Asmara. country capital (Eritrea) C Eritrea
Aspadana see Eşfahān
Asphaltites, Lacus see Dead Sea
Aspinwall see Colón
Assab see 'Aseb
As Sabkhah 96 D2 var. Sabkha. Ar Raqqah, NE Syria
Assad, Lake see Asad, Buḥayrat al
Aş Şafāwī 97 C6 Al Mafraq, N Jordan
Aş Şaḥrā' ash Sharqīyah see Sahara el Sharqîya
As Salamīyah see Salamīyah
'Assal, Lac 46 E4 lake C Djibouti
As Salṭ 97 B6 var. Salt. Al Balqā', NW Jordan
Assamaka see Assamakka
Assamakka 53 F2 var. Assamaka. Agadez, NW Niger
As Samāwah 98 B3 var. Samawa. Al Muthannā, S Iraq
Assenede 65 B5 Oost-Vlaanderen, NW Belgium
Assiout see Asyūṭ
Assiut see Asyūṭ
Assling see Jesenice
Assouan see Aswān
Assu 41 G2 var. Açu. Rio Grande do Norte, E Brazil
Assuan see Aswān
As Sukhnah 96 C3 var. Sukhne, Fr. Soukhné. Ḥimṣ, C Syria
As Sulaymānīyah 98 C3 var. Sulaimaniya, Kurd. Slēmānī. As Sulaymānīyah, NE Iraq
As Sulayyil 99 B5 Ar Riyāḍ, S Saudi Arabia
Aş Şuwār 96 D2 var. Şuwār. Dayr az Zawr, E Syria
As Suwaydā' 97 B5 var. El Suweida, Es Suweida, Suweida, Fr. Soueida. As Suwaydā', SW Syria
As Suways see Suez
Asta Colonia see Asti
Astacus see Izmit
Astana 92 C4 prev. Akmola, Akmolinsk, Tselinograd, Aqmola. country capital (Kazakhstan) Akmola, N Kazakhstan
Asta Pompeia see Asti
Astarabad see Gorgān
Asterābād see Gorgān
Asti 74 A2 anc. Asta Colonia, Asta Pompeia, Hasta Colonia, Hasta Pompeia. Piemonte, NW Italy
Astigi see Ecija
Astipálaia see Astypálaia
Astorga 70 C1 anc. Asturica Augusta. Castilla-León, N Spain
Astrabad see Gorgān
Astrakhan' 89 C7 Astrakhanskaya Oblast', SW Russia
Asturias 70 C1 autonomous community NW Spain
Asturias see Oviedo
Asturica Augusta see Astorga
Astypálaia 83 D7 var. Astipálaia, It. Stampalia. island Kykládes, Greece, Aegean Sea
Asunción 42 D2 country capital (Paraguay) Central, S Paraguay
Aswān 50 B2 var. Assouan, Assuan, Aswân; anc. Syene. SE Egypt
Aswân see Aswān
Asyūṭ 50 B2 var. Assiout, Assiut, Asyût, Siut; anc. Lycopolis. C Egypt
Asyût see Asyūṭ
Atacama Desert 42 B2 Eng. Atacama Desert. desert N Chile
Atacama Desert see Atacama, Desierto de
Atafu Atoll 123 E3 island NW Tokelau
Atamyrat 100 D3 prev. Kerki. Lebap Welaýaty, E Turkmenistan
Aṭâr 52 C2 Adrar, W Mauritania
Atas Bogd 104 D3 mountain SW Mongolia
Atascadero 25 B7 California, W USA
Atatürk Barajī 95 E4 reservoir S Turkey
Atbara 50 C3 var. 'Aṭbārah. River Nile, NE Sudan
'Aṭbārah/'Aṭbarah, Nahr see Atbara
Atbasar 92 C4 Akmola, N Kazakhstan
Atchison 23 F4 Kansas, C USA
Aternum see Pescara
Ath 65 B6 var. Aat. Hainaut, SW Belgium
Athabasca 15 E5 Alberta, SW Canada
Athabasca 15 E5 var. Athabaska. river Alberta, SW Canada
Athabasca, Lake 15 F4 lake Alberta/Saskatchewan, SW Canada
Athabaska see Athabasca
Athenae see Athína
Athens 21 E2 Georgia, SE USA
Athens 18 D4 Ohio, N USA
Athens 27 G3 Texas, SW USA
Athens see Athína
Atherton 126 D3 Queensland, NE Australia
Athína 83 C6 Eng. Athens, prev. Athínai; anc. Athenae. country capital (Greece) Attikí, C Greece
Athínai see Athína
Athlone 67 B5 Ir. Baile Átha Luain. C Ireland
Ath Thawrah see Madīnat ath Thawrah
Ati 54 C3 Batha, C Chad
Atikokan 16 B4 Ontario, S Canada
Atka 93 G3 Magadanskaya Oblast', E Russia
Atka 14 A3 Atka Island, Alaska, USA
Atlanta 20 D2 state capital Georgia, SE USA
Atlanta 27 H2 Texas, SW USA
Atlantic City 19 F4 New Jersey, NE USA
Atlantic-Indian Basin 45 D7 undersea basin SW Indian Ocean
Atlantic-Indian Ridge 47 B8 undersea ridge SW Indian Ocean
Atlantic Ocean 44 B4 ocean
Atlas Mountains 48 C2 mountain range NW Africa
Atlasovo 93 H3 Kamchatskaya Oblast', E Russia
Atlas Saharien 48 D2 var. Saharan Atlas. mountain range Algeria/Morocco
Atlas, Tell see Atlas Tellien
Atlas Tellien 80 C3 Eng. Tell Atlas. mountain range N Algeria
Atlin 14 D4 British Columbia, W Canada
Aṭ Ṭafīlah 97 B7 var. Et Tafila, Tafila. Aṭ Ṭafīlah, W Jordan
Aṭ Ṭā'if 99 B5 Makkah, W Saudi Arabia
Attaleia/Attalia see Antalya
At Tall al Abyaḍ 96 C2 var. Tall al Abyaḍ, Tell Abyad, Fr. Tell Abiad. Ar Raqqah, N Syria
Aṭ Ṭanf 96 D4 Ḥimṣ, S Syria
Attapu 115 E5 var. Samakhixai, Attopeu. Attapu, S Laos
Attawapiskat 16 C3 Ontario, C Canada
Attawapiskat 16 C3 river Ontario, S Canada
At Tibnī 96 D2 var. Tibnī. Dayr az Zawr, NE Syria
Attopeu see Attapu
Attu Island 14 A2 island Aleutian Islands, Alaska, USA
Atyrau 92 B4 prev. Gur'yev. Atyrau, W Kazakhstan
Aubagne 69 D6 anc. Albania. Bouches-du-Rhône, SE France
Aubange 65 D8 Luxembourg, SE Belgium
Aubervilliers 68 E1 Seine-St-Denis, Île-de-France, N France Europe
Auburn 24 B2 Washington, NW USA
Auch 69 B6 Lat. Augusta Auscorum, Elimberrum. Gers, S France
Auckland 128 D2 Auckland, North Island, New Zealand
Auckland Islands 120 C5 island group S New Zealand
Audern see Audru
Audincourt 68 E4 Doubs, E France
Audru 84 D2 Ger. Audern. Pärnumaa, SW Estonia
Augathella 127 D5 Queensland, E Australia
Augsbourg see Augsburg
Augsburg 73 C6 Fr. Augsbourg; anc. Augusta Vindelicorum. Bayern, S Germany
Augusta 125 A7 Western Australia
Augusta 21 E2 Georgia, SE USA
Augusta 19 G2 state capital Maine, NE USA
Augusta see London
Augusta Auscorum see Auch
Augusta Emerita see Mérida
Augusta Praetoria see Aosta
Augusta Trajana see Stara Zagora
Augusta Treverorum see Trier
Augusta Vangionum see Worms
Augusta Vindelicorum see Augsburg
Augustobona Tricassium see Troyes
Augustodurum see Bayeux
Augustoritum Lemovicensium see Limoges
Augustów 76 E2 Rus. Avgustov. Podlaskie, NE Poland
Aulie Ata/Auliye-Ata see Taraz
Aunglan 114 B4 var. Allanmyo, Myaydo. Magway, C Myanmar (Burma)
Auob 56 B4 var. Oup. river Namibia/South Africa
Aurangābād 112 D5 Mahārāshtra, C India
Auray 68 A3 Morbihan, NW France
Aurelia Aquensis see Baden-Baden
Aurelianum see Orléans
Aurès, Massif de l' 80 C4 mountain range NE Algeria
Aurillac 69 C5 Cantal, C France
Aurium see Ourense
Aurora 37 F2 NW Guyana
Aurora 22 D4 Colorado, C USA
Aurora 18 B3 Illinois, N USA
Aurora 23 G5 Missouri, C USA
Aurora see Maéwo, Vanuatu
Aus 56 B4 Karas, SW Namibia
Ausa see Vic
Aussig see Ústí nad Labem
Austin 23 G3 Minnesota, N USA
Austin 27 G3 state capital Texas, SW USA
Australes, Archipel des see Australes, Îles
Australes et Antarctiques Françaises, Terres see French Southern and Antarctic Lands
Australes, Îles 121 F4 var. Archipel des Australes, Îles Tubuai, Tubuai Islands, Eng. Austral Islands. island group SW French Polynesia
Austral Fracture Zone 121 H4 tectonic feature S Pacific Ocean
Australia 120 A4 off. Commonwealth of Australia. country
Australia, Commonwealth of see Australia
Australian Alps 127 C7 mountain range SE Australia
Australian Capital Territory 127 D7 prev. Federal Capital Territory. territory SE Australia
Australie, Bassin Nord de l' see North Australian Basin
Austral Islands see Australes, Îles
Austrava see Ostrov
Austria 73 D7 off. Republic of Austria, Ger. Österreich. country C Europe
Austria, Republic of see Austria
Autesiodorum see Auxerre
Autissiodorum see Auxerre
Autricum see Chartres
Auvergne 69 C5 cultural region C France
Auxerre 68 C4 anc. Autesiodorum, Autissiodorum. Yonne, C France
Avaricum see Bourges
Avarua 123 G5 dependent territory capital (Cook Islands) Rarotonga, S Cook Islands
Avasfelsőfalu see Negrești-Oaș
Ávdira 82 C3 Anatolikí Makedonía kai Thráki, NE Greece
Aveiro 70 B2 anc. Talabriga. Aveiro, W Portugal
Avela see Ávila
Avellino 75 D5 anc. Abellinum. Campania, S Italy
Avenio see Avignon
Avesta 63 C6 Dalarna, C Sweden
Aveyron 69 C6 river S France
Avezzano 74 C4 Abruzzo, C Italy
Avgustov see Augustów
Aviemore 66 C3 N Scotland, United Kingdom
Avignon 69 D6 anc. Avenio. Vaucluse, SE France
Ávila 70 D3 var. Avila; anc. Abela, Abula, Abyla, Avela. Castilla-León, C Spain
Avilés 70 C1 Asturias, NW Spain
Avranches 68 B3 Manche, N France
Avveel see Ivalo, Finland
Avvil see Ivalo
Awaji-shima 109 C6 island SW Japan
Āwash 51 D5 Āfar, NE Ethiopia
Awbārī 49 F3 SW Libya
Ax see Dax
Axel 65 B5 Zeeland, SW Netherlands
Axel Heiberg Island 15 E1 var. Axel Heiburg. island Nunavut, N Canada
Axel Heiburg see Axel Heiberg Island
Axiós see Vardar
Ayacucho 38 D4 Ayacucho, S Peru
Ayagoz 92 C5 var. Ayaguz, Kaz. Ayakoz. river E Kazakhstan
Ayamonte 70 C4 Andalucía, S Spain
Ayaviri 39 E4 Puno, S Peru
Aydarko'l Ko'li 101 E2 Rus. Ozero Aydarkul'. lake C Uzbekistan
Aydarkul', Ozero see Aydarko'l Ko'li
Aydın 94 A4 var. Aïdin; anc. Tralles Aydin. Aydın, SW Turkey
Ayers Rock see Uluru
Ayeyarwady see Irrawaddy
Ayiá see Agiá
Áyios Evstrátios see Ágios Efstrátios
Áyios Nikólaos see Ágios Nikólaos
Ayorou 53 E3 Tillabéri, W Niger
'Ayoûn el 'Atroûs 52 D3 var. Aïoun el Atrous, Aïoun el Atroûss. Hodh el Gharbi, SE Mauritania
Ayr 66 C4 W Scotland, United Kingdom
Ayteke Bi 92 B4 Kaz. Zhangaqazaly; prev. Novokazalinsk. Kzylorda, SW Kazakhstan
Aytos 82 E2 Burgas, E Bulgaria
Ayutthaya 115 C5 var. Phra Nakhon Si Ayutthaya. Phra Nakhon Si Ayutthaya, C Thailand
Ayvalık 94 A3 Balıkesir, W Turkey
Azahar, Costa del 71 F3 coastal region E Spain
Azaouâd 53 E3 desert C Mali
Azärbaycan/Azärbaycan Respublikasī see Azerbaijan
A'zāz 96 B2 Ḥalab, NW Syria
Azerbaijan 95 G2 off. Azerbaijani Republic, Az. Azärbaycan, Azärbaycan Respublikası; prev. Azerbaijan SSR. country SE Asia
Azerbaijani Republic see Azerbaijan
Azerbaijan SSR see Azerbaijan
Azimabad see Patna
Azizie see Telish
Azogues 38 B2 Cañar, S Ecuador
Azores 70 A4 var. Açores, Ilhas dos Açores, Port. Arquipélago dos Açores. island group Portugal, NE Atlantic Ocean
Azores-Biscay Rise 58 A3 undersea rise E Atlantic Ocean
Azotos/Azotus see Ashdod
Azoum, Bahr 54 C3 seasonal river SE Chad
Azov, Sea of 81 H1 Rus. Azovskoye More, Ukr. Azovs'ke More. sea NE Black Sea
Azovs'ke More/Azovskoye More see Azov, Sea of
Azraq, Wāḥat al 97 C6 oasis N Jordan
Aztec 26 C1 New Mexico, SW USA
Azuaga 70 C4 Extremadura, W Spain
Azuero, Península de 31 F5 peninsula S Panama
Azul 43 D5 Buenos Aires, E Argentina
Azur, Côte d' 69 E6 coastal region SE France
'Azza see Gaza
Az Zaqāzīq 50 B1 var. Zagazig. N Egypt
Az Zarqā' 97 B6 NW Jordan
Az Zāwiyah 49 F2 var. Zawia. NW Libya
Az Zilfī 98 B4 Ar Riyāḍ, N Saudi Arabia

B

Baalbek 96 B4 var. Ba'labakk; anc. Heliopolis. E Lebanon
Baardheere 51 D6 var. Bardere, It. Bardera. Gedo, SW Somalia
Baarle-Hertog 65 C5 Antwerpen, N Belgium
Baarn 64 C3 Utrecht, C Netherlands
Babadag 86 D5 Tulcea, SE Romania
Babahoyo 38 B2 prev. Bodegas. Los Ríos, C Ecuador
Bābā, Kūh-e 101 E4 mountain range C Afghanistan
Babayevo 88 B4 Vologodskaya Oblast', NW Russia
Babeldaob 122 A1 var. Babeldaop, Babelthuap. island N Palau
Babeldaop see Babeldaob
Bab el Mandeb 99 B7 strait Gulf of Aden/Red Sea
Babelthuap see Babeldaob
Babian Jiang see Black River
Babruysk 85 D7 Rus. Bobruysk. Mahilyowskaya Voblasts', E Belarus
Babuyan Channel 117 E1 channel N Philippines
Babuyan Islands 117 E1 island group N Philippines
Bacabal 41 F2 Maranhão, E Brazil
Bacău 86 C4 Hung. Bákó. Bacău, NE Romania
Băc Bô, Vinh see Tonkin, Gulf of
Băc Giang 114 D3 Ha Băc, N Vietnam
Bacheykava 85 D5 Rus. Bocheykovo. Vitsyebskaya Voblasts', N Belarus
Back 15 F3 river Nunavut, N Canada
Bačka Palanka 78 D3 prev. Palanka. Serbia, NW Serbia
Bačka Topola 78 D3 Hung. Topolya; prev. Hung. Bácstopolya. Vojvodina, N Serbia
Bac Liêu 115 D6 var. Vinh Loi. Minh Hai, S Vietnam
Bacolod 103 E4 off. Bacolod City. Negros, C Philippines
Bacolod City see Bacolod
Bácsszenttamás see Srbobran
Bácstopolya see Bačka Topola
Bactra see Balkh
Badajoz 70 C4 anc. Pax Augusta. Extremadura, W Spain
Baden-Baden 73 B6 anc. Aurelia Aquensis. Baden-Württemberg, SW Germany
Bad Freienwalde 72 D3 Brandenburg, NE Germany
Badger State see Wisconsin
Bad Hersfeld 72 B4 Hessen, C Germany
Bad Homburg see Bad Homburg vor der Höhe
Bad Homburg vor der Höhe 73 B5 var. Bad Homburg. Hessen, W Germany
Bad Ischl 73 D7 Oberösterreich, N Austria
Bad Krozingen 73 A6 Baden-Württemberg, SW Germany
Badlands 22 D2 physical region North Dakota/South Dakota, N USA
Badu Island 126 C1 island Queensland, NE Australia
Bad Vöslau 73 E6 Niederösterreich, NE Austria
Baeterrae/Baeterrae Septimanorum see Béziers
Baetic Cordillera/Baetic Mountains see Béticos, Sistemas
Bafatá 52 C4 C Guinea-Bissau
Baffin Bay 15 G2 bay Canada/Greenland
Baffin Island 15 G2 island Nunavut, NE Canada
Bafing 52 C3 river W Africa
Bafoussam 54 A4 Ouest, W Cameroon
Bafra 94 D2 Samsun, N Turkey
Bāft 98 D4 Kermān, S Iran
Bagaces 30 D4 Guanacaste, NW Costa Rica
Bagdad see Baghdād
Bagé 41 E5 Rio Grande do Sul, S Brazil
Baghdād 98 B3 var. Bagdad, country capital (Iraq) Baghdād, C Iraq
Baghlān 101 E3 Baghlān, NE Afghanistan
Bago 114 B4 var. Pegu. Bago, SW Myanmar (Burma)
Bagoé 52 D4 river Ivory Coast/Mali
Bagrationovsk 84 A4 Ger. Preussisch Eylau. Kaliningradskaya Oblast', W Russia
Bagrax Hu see Bosten Hu
Baguio 117 E1 off. Baguio City. Luzon, N Philippines
Baguio City see Baguio
Bagzane, Monts 53 F3 mountain N Niger
Bahama Islands see Bahamas, The
Bahamas, The 32 C2 off. Commonwealth of The Bahamas. country N West Indies
Bahamas 13 D6 var. Bahama Islands. island group N West Indies
Bahamas, Commonwealth of The see Bahamas, The
Baharly 100 C3 var. Bäherden, Rus. Bakharden; prev. Bakherden. Ahal Welaýaty, C Turkmenistan
Bahāwalpur 112 C2 Punjab, E Pakistan
Bäherden see Baharly
Bahia 41 F3 off. Estado da Bahia. state/region E Brazil
Bahía Blanca 43 C5 Buenos Aires, E Argentina
Bahia, Estado da see Bahia
Bahir Dar 50 C4 var. Bahr Dar, Bahrdar Giyorgis. Āmara, N Ethiopia
Bahraich 113 E3 Uttar Pradesh, N India
Bahrain 98 C4 off. State of Bahrain, Dawlat al Bahrayn, Ar. Al Baḥrayn, prev. Bahrein; anc. Tylos, Tyros. country SW Asia
Bahrain, State of see Bahrain
Bahrayn, Dawlat al see Bahrain
Bahr Dar/Bahrdar Giyorgis see Bahir Dar
Bahrein see Bahrain
Bahr el, Azraq see Blue Nile
Bahr Tabariya, Sea of see Tiberias, Lake
Bahushewsk 85 E6 Rus. Bogushëvsk. Vitsyebskaya Voblasts', NE Belarus
Baia Mare 86 B3 Ger. Frauenbach, Hung. Nagybánya; prev. Neustadt. Maramureș, NW Romania
Baia Sprie 86 B3 Ger. Mittelstadt, Hung. Felsőbánya. Maramureș, NW Romania
Baïbokoum 54 B4 Logone-Oriental, SW Chad
Baidoa see Baydhabo
Baie-Comeau 17 E3 Québec, SE Canada
Baikal, Lake 93 E4 Eng. Lake Baikal. lake S Russia
Baikal, Lake see Baykal, Ozero
Baile Átha Cliath see Dublin
Baile Átha Luain see Athlone
Bailén 70 D4 Andalucía, S Spain
Baile na Mainistreach see Newtownabbey
Băilești 86 B5 Dolj, SW Romania
Ba Illi 54 B3 Chari-Baguirmi, SW Chad
Bainbridge 20 D3 Georgia, SE USA
Bā'ir see Bāyir
Baireuth see Bayreuth
Bairnsdale 127 C7 Victoria, SE Australia
Baishan 107 E3 prev. Hunjiang. Jilin, NE China
Baiyin 106 B4 Gansu, C China
Baja 77 C7 Bács-Kiskun, S Hungary
Baja California 26 A4 Eng. Lower California. peninsula NW Mexico
Baja California Norte 28 B2 state NW Mexico
Bajo Boquete see Boquete
Bajram Curri 79 D5 Kukës, N Albania
Bakala 54 C4 Ouaka, C Central African Republic
Bakan see Shimonoseki
Baker 24 C3 Oregon, NW USA
Baker and Howland Islands 123 E2 US unincorporated territory W Polynesia
Baker Lake 15 F3 Nunavut, N Canada
Bakersfield 25 C7 California, W USA
Bakharden see Baharly
Bakhchisaray see Bakhchysaray
Bakhchysaray 87 F5 Rus. Bakhchisaray. Respublika Krym, S Ukraine
Bakherden see Baharly
Bakhmach 87 F1 Chernihivs'ka Oblast', N Ukraine
Bākhtarān see Kermānshāh
Bakı 95 H2 Eng. Baku. country capital (Azerbaijan) E Azerbaijan
Bákó see Bacău
Bakony 77 C7 Eng. Bakony Mountains, Ger. Bakonywald. mountain range W Hungary
Bakony Mountains/Bakonywald see Bakony
Baku see Bakı
Bakwanga see Mbuji-Mayi
Balabac Island 107 C8 island W Philippines
Balabac, Selat see Balabac Strait
Balabac Strait 116 D2 var. Selat Balabac. strait Malaysia/Philippines
Ba'labakk see Baalbek
Balaguer 71 F2 Cataluña, NE Spain
Balakovo 89 C6 Saratovskaya Oblast', W Russia
Bālā Morghāb see Bālā Murghāb
Bālā Murghāb 100 D4 prev. Bālā Morghāb. Laghmān, NW Afghanistan
Balashov 89 B6 Saratovskaya Oblast', W Russia
Balasore see Bāleshwar
Balaton, Lake 77 C7 var. Lake Balaton, Ger. Plattensee. lake W Hungary
Balaton, Lake see Balaton
Balbina, Represa 40 D1 reservoir NW Brazil
Balboa 31 G4 Panamá, C Panama
Balcarce 43 D5 Buenos Aires, E Argentina
Balclutha 129 B7 Otago, South Island, New Zealand
Baldy Mountain 22 C1 mountain Montana, NW USA
Bâle see Basel
Balearic Plain see Algerian Basin

Baleares Major – Bergeyk

Baleares Major see Mallorca
Balearic Islands 71 G3 Eng. Balearic Islands. island group Spain, W Mediterranean Sea
Balearic Islands see Baleares, Islas
Balearis Minor see Menorca
Baleine, Rivière à la 17 E2 river Québec, E Canada
Balen 65 C5 Antwerpen, N Belgium
Bāleshwar 113 F4 prev. Balasore. Odisha, E India
Bālgrad see Alba Iulia
Bali 116 D5 island C Indonesia
Balıkesir 94 A3 Balıkesir, W Turkey
Balīkh, Nahr 96 C2 river N Syria
Balikpapan 116 D4 Borneo, C Indonesia
Balkanabat 100 B2 Rus. Nebitdag. Balkan Welaýaty, W Turkmenistan
Balkan Mountains 82 C2 Bul./SCr. Stara Planina. mountain range Bulgaria/Serbia
Balkh 101 E3 anc. Bactra. Balkh, N Afghanistan
Balkhash 92 C5 Kaz. Balqash. Karagandy, SE Kazakhstan
Balkhash, Lake see Balkhash, Ozero
Balkhash, Ozero 92 C5 Eng. Lake Balkhash, Kaz. Balqash. lake SE Kazakhstan
Balladonia 125 C6 Western Australia
Ballarat 127 C7 Victoria, SE Australia
Balleny Islands 132 B5 island group Antarctica
Ballinger 27 F3 Texas, SW USA
Balochistān see Baluchistān
Balqash see Balkhash/Balkhash, Ozero
Balş 86 B5 Olt, S Romania
Balsas 41 F2 Maranhão, E Brazil
Balsas, Río 29 E5 var. Río Mexcala. river S Mexico
Bal'shavik 85 D7 Rus. Bol'shevik. Homyel'skaya Voblasts', SE Belarus
Balta 86 D3 Odes'ka Oblast', SW Ukraine
Bălţi 86 D3 Rus. Bel'tsy. N Moldova
Baltic Port see Paldiski
Baltic Sea 63 C7 Ger. Ostee, Rus. Baltiskoye More. sea N Europe
Baltimore 19 F4 Maryland, NE USA
Baltischport/Baltiski see Paldiski
Baltiskoye More see Baltic Sea
Baltkrievija see Belarus
Baluchistān 112 B3 var. Balochistān, Beluchistan. province SW Pakistan
Balvi 84 D4 Balvi, NE Latvia
Balykchy 101 G2 Kir. Ysyk-Köl; prev. Issyk-Kul', Rybach'ye. Issyk-Kul'skaya Oblast', NE Kyrgyzstan
Balzers 72 E2 S Liechtenstein
Bam 98 E4 Kermān, SE Iran
Bamako 52 D4 country capital (Mali) Capital District, SW Mali
Bambari 54 C4 Ouaka, C Central African Republic
Bamberg 73 C5 Bayern, SE Germany
Bamenda 54 A4 Nord-Ouest, W Cameroon
Banaba 122 D2 var. Ocean Island. island Tungaru, W Kiribati
Banaras see Vārānasi
Bandaaceh 116 A3 var. Banda Atjeh; prev. Koetaradja, Kutaradja, Kutaraja. Sumatera, W Indonesia
Banda Atjeh see Bandaaceh
Banda, Laut see Banda Sea
Bandama 52 D5 var. Bandama Fleuve. river S Ivory Coast
Bandama Fleuve see Bandama
Bandar 'Abbās see Bandar-e 'Abbās
Bandarbeyla 51 E5 var. Bender Beila, Bender Beyla. Bari, NE Somalia
Bandar-e 'Abbās 98 D4 var. Bandar 'Abbās; prev. Gombroon. Hormozgān, S Iran
Bandar-e Būshehr 98 C4 var. Būshehr, Eng. Bushire. Būshehr, S Iran
Bandar-e Kangān 98 D4 var. Kangān. Būshehr, S Iran
Bandar-e Khamīr 98 D4 Hormozgān, S Iran
Bandar-e Langeh see Bandar-e Lengeh
Bandar-e Lengeh 98 D4 var. Bandar-e Langeh, Lingeh. Hormozgān, S Iran
Bandar Kassim see Boosaaso
Bandar Lampung 116 C4 var. Bandarlampung, Tanjungkarang-Telukbetung; prev. Tandjoengkarang, Tanjungkarang, Teloekbetoeng, Telukbetung. Sumatera, W Indonesia
Bandarlampung see Bandar Lampung
Bandar Maharani see Muar
Bandar Masulipatnam see Machilīpatnam
Bandar Penggaram see Batu Pahat
Bandar Seri Begawan 116 D3 prev. Brunei Town. country capital (Brunei) N Brunei
Banda Sea 117 F5 var. Laut Banda. sea E Indonesia
Bandiagara 53 E3 Mopti, C Mali
Bandırma 94 A3 var. Penderma. Balıkesir, NW Turkey
Bandjarmasin see Banjarmasin
Bandoeng see Bandung
Bandundu 55 C6 prev. Banningville. Bandundu, W Dem. Rep. Congo
Bandung 116 C5 prev. Bandoeng. Jawa, C Indonesia
Bangalore see Bengalūru
Bangassou 54 D4 Mbomou, SE Central African Republic
Banggai, Kepulauan 117 E4 island group C Indonesia
Banghāzī 49 G2 Eng. Bengazi, Benghazi, It. Bengasi. NE Libya
Bangka, Pulau 116 C4 island W Indonesia
Bangkok see Ao Krung Thep
Bangkok, Bight of see Krung Thep, Ao
Bangladesh 113 G3 off. People's Republic of Bangladesh; prev. East Pakistan. country S Asia
Bangladesh, People's Republic of see Bangladesh
Bangor 67 C6 NW Wales, United Kingdom
Bangor 67 B5 Ir. Beannchar. E Northern Ireland, United Kingdom
Bangor 19 G2 Maine, NE USA
Bang Pla Soi see Chon Buri
Bangui 55 B5 country capital (Central African Republic) Ombella-Mpoko, SW Central African Republic
Bangweulu, Lake 51 B8 var. Lake Bengweulu. lake N Zambia
Ban Hat Yai see Hat Yai
Ban Hin Heup 114 C4 Viangchan, C Laos
Ban Houayxay/Ban Houei Sai see Houayxay
Ban Hua Hin 115 C6 var. Hua Hin. Prachuap Khiri Khan, SW Thailand
Bani 52 D3 river S Mali
Banias see Bāniyās
Banijska Palanka see Glina
Banī Suwayf 50 B2 var. Beni Suef. N Egypt
Bāniyās 96 B3 var. Banias, Baniyas, Paneas. Tarṭūs, W Syria
Banjak, Kepulauan see Banyak, Kepulauan
Banja Luka 78 B3 Republika Srpska, NW Bosnia and Herzegovina
Banjarmasin 116 D4 prev. Bandjarmasin. Borneo, C Indonesia
Banjul 52 B3 prev. Bathurst. country capital (The Gambia) W The Gambia
Banks, Îles see Banks Islands
Banks Island 15 E2 island Northwest Territories, NW Canada
Banks Islands 122 D4 Fr. Îles Banks. island group N Vanuatu
Banks Lake 24 B1 reservoir Washington, NW USA
Banks Peninsula 129 C6 peninsula South Island, New Zealand
Banks Strait 127 C8 strait SW Tasman Sea
Bānkura 113 F4 West Bengal, NE India
Ban Mak Khaeng see Udon Thani
Banmo see Bhamo
Banningville see Bandundu
Bañolas see Banyoles
Ban Pak Phanang see Pak Phanang
Ban Sichon see Sichon
Banská Bystrica 77 C6 Ger. Neusohl, Hung. Besztercebánya. Banskobystricky Kraj, C Slovakia
Bantry Bay 67 A7 Ir. Bá Bheanntraí. bay SW Ireland
Banya 82 E2 Burgas, E Bulgaria
Banyak, Kepulauan 116 A3 prev. Kepulauan Banjak. island group NW Indonesia
Banyo 54 B4 Adamaoua, NW Cameroon
Banyoles 71 G2 var. Bañolas. Cataluña, NE Spain
Banzare Seamounts 119 C7 seamount range S Indian Ocean
Banzart see Bizerte
Baoji 106 B4 var. Pao-chi, Paoki. Shaanxi, C China
Baoro 54 B4 Nana-Mambéré, W Central African Republic
Baoshan 106 A6 var. Pao-shan. Yunnan, SW China
Baotou 105 F3 var. Pao-t'ou, Paotow. Nei Mongol Zizhiqu, N China
Ba'qūbah 98 B3 var. Qubba. Diyālá, C Iraq
Baquerizo Moreno see Puerto Baquerizo Moreno
Bar 79 C5 It. Antivari. S Montenegro
Baraawe 51 D6 It. Brava. Shabeellaha Hoose, S Somalia
Bārāmati 112 C5 Mahārāshtra, W India
Baranavichy 85 B6 Pol. Baranowicze, Rus. Baranovichi. Brestskaya Voblasts', SW Belarus
Baranovichi/Baranowicze see Baranavichy
Barbados 33 G1 country SE West Indies
Barbastro 71 F2 Aragón, NE Spain
Barbate de Franco 70 C5 Andalucía, S Spain
Barbuda 33 G3 island N Antigua and Barbuda
Barcaldine 126 C4 Queensland, E Australia
Barcarozsnyó see Râşnov
Barcău see Berettyó
Barce see Al Marj
Barcelona 71 G2 anc. Barcino, Barcinona. Cataluña, E Spain
Barcelona 37 E2 Anzoátegui, NE Venezuela
Barcino/Barcinona see Barcelona
Barcoo see Cooper Creek
Barcs 77 C7 Somogy, SW Hungary
Bardaï 54 C1 Borkou-Ennedi-Tibesti, N Chad
Bardejov 77 D5 Ger. Bartfeld, Hung. Bártfa. Presovský Kraj, E Slovakia
Bardera/Bardere see Baardheere
Barduli see Barletta
Bareilly 113 E3 var. Bareli. Uttar Pradesh, N India
Bareli see Bareilly
Barendrecht 64 C4 Zuid-Holland, SW Netherlands
Barentin 68 C3 Seine-Maritime, N France
Barentsburg 61 G2 Spitsbergen, W Svalbard
Barentsevo More/Barents Havet see Barents Sea
Barentsøya 61 G2 island E Svalbard
Barents Sea 88 C2 Nor. Barents Havet, Rus. Barentsevo More. sea Arctic Ocean
Bar Harbor 19 H2 Mount Desert Island, Maine, NE USA
Bari 75 E5 var. Bari delle Puglie; anc. Barium. Puglia, SE Italy
Bāridah see Al Bāridah
Bari delle Puglie see Bari
Barikot see Barīkowṭ
Barīkowṭ 101 F4 var. Barikot. Konar, NE Afghanistan
Barillas 30 A2 var. Santa Cruz Barillas. Huehuetenango, NW Guatemala
Barinas 36 C2 Barinas, W Venezuela
Barisal 113 G4 Barisal, S Bangladesh
Barisan, Pegunungan 116 B4 mountain range Sumatera, W Indonesia
Barito, Sungai 116 D4 river Borneo, C Indonesia
Barium see Bari
Barka see Al Marj
Barkly Tableland 126 B3 plateau Northern Territory/Queensland, N Australia
Bârlad 86 D4 prev. Birlad. Vaslui, E Romania
Barlavento, Ilhas de 52 A2 var. Windward Islands. island group N Cape Verde
Bar-le-Duc 68 D3 var. Bar-sur-Ornain. Meuse, NE France
Barlee, Lake 125 B6 lake Western Australia
Barlee Range 124 A4 mountain range Western Australia
Barletta 75 D5 anc. Barduli. Puglia, SE Italy
Barlinek 76 B3 Ger. Berlinchen. Zachodnio-pomorskie, NW Poland
Barmen-Elberfeld see Wuppertal
Barmouth 67 C6 NW Wales, United Kingdom
Barnaul 92 D4 Altayskiy Kray, C Russia
Barnet 67 A7 United Kingdom
Barnstaple 67 C7 SW England, United Kingdom
Baroda see Vadodara
Baroghil Pass 101 F3 var. Kowtal-e Barowghīl. pass Afghanistan/Pakistan
Baron'ki 85 E7 Rus. Boron'ki. Mahilyowskaya Voblasts', E Belarus
Barowghīl, Kowtal-e see Baroghil Pass
Barquisimeto 36 C2 Lara, NW Venezuela
Barra 66 B3 island NW Scotland, United Kingdom
Barra de Río Grande 31 E3 Región Autónoma Atlántico Sur, E Nicaragua
Barranca 38 C3 Lima, W Peru
Barrancabermeja 36 B2 Santander, N Colombia
Barranquilla 36 B1 Atlántico, N Colombia
Barreiro 70 B4 Setúbal, W Portugal
Barrier Range 127 C6 hill range New South Wales, SE Australia
Barrow 14 D2 Alaska, USA
Barrow 67 B6 Ir. An Bhearú. river SE Ireland
Barrow-in-Furness 67 C5 NW England, United Kingdom
Barrow Island 124 A4 island Western Australia
Barstow 25 C7 California, W USA
Bar-sur-Ornain see Bar-le-Duc
Bartang 101 F3 river SE Tajikistan
Bartenstein see Bartoszyce
Bártfa/Bartfeld see Bardejov
Bartica 37 F3 N Guyana
Bartın 94 C2 Bartın, NW Turkey
Bartlesville 27 G1 Oklahoma, C USA
Bartoszyce 76 D2 Ger. Bartenstein. Warmińsko-mazurskie, NE Poland
Baruun-Urt 105 F2 Sühbaatar, E Mongolia
Barú, Volcán 31 E5 var. Volcán de Chiriquí. volcano W Panama
Barwon River 127 D5 river New South Wales, SE Australia
Barysaw 85 D6 Rus. Borisov. Minskaya Voblasts', NE Belarus
Basarabeasca 86 D4 Rus. Bessarabka. SE Moldova
Basel 73 A7 Eng. Basle, Fr. Bâle. Basel-Stadt, NW Switzerland
Basilan 117 E3 island Sulu Archipelago, SW Philippines
Basle see Basel
Basra see Al Baṣrah
Bassano del Grappa 74 C2 Veneto, NE Italy
Bassein see Pathein
Basseterre 33 G4 country capital (Saint Kitts and Nevis) Saint Kitts, Saint Kitts and Nevis
Basse-Terre 33 G3 dependent territory capital (Guadeloupe) Basse Terre, SW Guadeloupe
Basse Terre 33 G4 island W Guadeloupe
Bassikounou 52 D3 Hodh ech Chargui, SE Mauritania
Bass, Îlots de see Marotiri
Bass Strait 127 C7 strait SE Australia
Bassum 72 B3 Niedersachsen, NW Germany
Bastia 69 E7 Corse, France, C Mediterranean Sea
Bastogne 65 D7 Luxembourg, SE Belgium
Bastrop 20 B2 Louisiana, S USA
Bastyn' 85 B7 Rus. Bostyn'. Brestskaya Voblasts', SW Belarus
Basuo see Dongfang
Basutoland see Lesotho
Bata 55 A5 NW Equatorial Guinea
Batae Coritanorum see Leicester
Batajnica 78 D3 Vojvodina, N Serbia
Batangas 117 E2 off. Batangas City. Luzon, N Philippines
Batangas City see Batangas
Batavia see Jakarta
Bătdâmbâng see Battambang
Batéké, Plateaux 55 B6 plateau S Congo
Bath 67 D7 hist. Akermanceaster; anc. Aquae Calidae, Aquae Solis. SW England, United Kingdom
Bathinda 112 D2 Punjab, NW India
Bathsheba 33 G1 E Barbados
Bathurst 127 D6 New South Wales, SE Australia
Bathurst 17 F4 New Brunswick, SE Canada
Bathurst see Banjul
Bathurst Island 124 D2 island Northern Territory, N Australia
Bathurst Island 15 F2 island Parry Islands, Nunavut, N Canada
Batin, Wadi al 98 C4 dry watercourse SW Asia
Batman 95 E4 var. Iluh. Batman, SE Turkey
Batna 49 E2 NE Algeria
Baton Rouge 20 B3 state capital Louisiana, S USA
Batroûn 96 A4 var. Al Batrūn. N Lebanon
Battambang 115 C5 Khmer. Bătdâmbâng. NW Cambodia
Batticaloa 110 D3 Eastern Province, E Sri Lanka
Battipaglia 75 D5 Campania, S Italy
Battle Born State see Nevada
Batumi 95 F2 prev. Bat'umi. W Georgia
Bat'umi see Batumi
Batu Pahat 116 B3 prev. Bandar Penggaram. Johor, Peninsular Malaysia
Bauchi 53 G4 Bauchi, NE Nigeria
Bauer Basin 131 F3 undersea basin E Pacific Ocean
Bauska 84 C3 Ger. Bauske. Bauska, S Latvia
Bauske see Bauska
Bautzen 72 D4 Lus. Budyšin. Sachsen, E Germany
Bauzanum see Bolzano
Bavaria see Bayern
Bavarian Alps 73 C7 Ger. Bayrische Alpen. mountain range Austria/Germany
Bavière see Bayern
Bavispe, Río 28 C2 river NW Mexico
Bawīṭī see Al Bawīṭī
Bawku 53 E4 N Ghana
Bayamo 32 C3 Granma, E Cuba
Bayan Har Shan 104 D4 var. Bayan Khar. mountain range C China
Bayanhongor 104 D2 Bayanhongor, C Mongolia
Bayan Khar see Bayan Har Shan
Bayano, Lago 31 G4 lake E Panama
Bay City 18 C3 Michigan, N USA
Bay City 27 G4 Texas, SW USA
Baydhabo 51 D6 var. Baydhowa, Isha Baydhabo, It. Baidoa. Bay, SW Somalia
Baydhowa see Baydhabo
Bayern 73 C6 Eng. Bavaria, Fr. Bavière. state SE Germany
Bayeux 68 B3 anc. Augustodurum. Calvados, N France
Bāyir 97 C7 var. Bā'ir. Ma'ān, S Jordan
Bay Islands 30 C1 Eng. Bay Islands. island group N Honduras
Bay Islands see Bahía, Islas de la
Baymak 89 D6 Respublika Bashkortostan, W Russia
Bayonne 69 A6 anc. Lapurdum. Pyrénées-Atlantiques, SW France
Bayou State see Mississippi
Bayram-Ali see Baýramaly
Baýramaly 100 D3 var. Bayramaly; prev. Bayram-Ali. Mary Welaýaty, S Turkmenistan
Bayreuth 73 C5 var. Baireuth. Bayern, SE Germany
Bayrische Alpen see Bavarian Alps
Bayrūt see Beyrouth
Bay State see Massachusetts
Baysun see Boysun
Bayt Laḥm see Bethlehem
Baytown 27 H4 Texas, SW USA
Baza 71 E4 Andalucía, S Spain
Bazargic see Dobrich
Bazin see Pezinok
Beagle Channel 43 C8 channel Argentina/Chile
Béal Feirste see Belfast
Beannchar see Bangor, Northern Ireland, UK
Bear Island see Bjørnøya
Bear Lake 24 E4 lake Idaho/Utah, NW USA
Beas de Segura 71 E4 Andalucía, S Spain
Beata, Isla 33 E3 island SW Dominican Republic
Beatrice 23 F4 Nebraska, C USA
Beaufort Sea 14 D2 sea Arctic Ocean
Beaufort-Wes see Beaufort West
Beaufort West 56 C5 Afr. Beaufort-Wes. Western Cape, SW South Africa
Beaumont 27 H3 Texas, SW USA
Beaune 68 D4 Côte d'Or, C France
Beauvais 68 C3 anc. Bellovacum, Caesaromagus. Oise, N France
Beaver Island 18 C2 island Michigan, N USA
Beaver Lake 27 H1 reservoir Arkansas, C USA
Beaver River 27 F1 river Oklahoma, C USA
Beaver State see Oregon
Beāwar 112 C3 Rājasthān, N India
Bečej 78 D3 Ger. Altbetsche, Hung. Óbecse, Rácz-Becse; prev. Magyar-Becse, Stari Bečej. Vojvodina, N Serbia
Béchar 48 D2 prev. Colomb-Béchar. W Algeria
Beckley 18 D5 West Virginia, NE USA
Bécs see Wien
Bedford 67 D6 E England, United Kingdom
Bedum 64 E1 Groningen, NE Netherlands
Beehive State see Utah
Be'er Menuha 97 B7 prev. Be'ér Menuḥa. Southern, S Israel
Be'ér Menuḥa see Be'er Menuha
Beernem 65 A5 West-Vlaanderen, NW Belgium
Beersheba see Be'er Sheva
Be'er Sheva 97 A7 var. Beersheba, Ar. Bir es Saba; prev. Be'ér Sheva'. Southern, S Israel
Be'ér Sheva' see Be'er Sheva
Beesel 65 D5 Limburg, SE Netherlands
Beeville 27 G4 Texas, SW USA
Bega 127 D7 New South Wales, SE Australia
Begoml' see Byahoml'
Begovat see Bekobod
Behagle see Laï
Behar see Bihār
Beibu Wan see Tonkin, Gulf of
Beida see Al Bayḍā'
Beihai 106 B6 Guangxi Zhuangzu Zizhiqu, S China
Beijing 106 C3 var. Pei-ching, Peking; prev. Pei-p'ing. country capital (China) Beijing Shi, E China
Beilen 64 E2 Drenthe, NE Netherlands
Beira 57 E3 Sofala, C Mozambique
Beirut see Beyrouth
Beit Lekhem see Bethlehem
Beiuş 86 B3 Hung. Belényes. Bihor, NW Romania
Beja 70 B4 anc. Pax Julia. Beja, SE Portugal
Béjar 70 C3 Castilla-León, N Spain
Bejraburi see Phetchaburi
Bekabad see Bekobod
Békás see Bicaz
Bek-Budi see Qarshi
Békéscsaba 77 D7 Rom. Bichiş-Ciaba. Békés, SE Hungary
Bekobod 101 E2 Rus. Bekabad; prev. Begovat. Toshkent Viloyati, E Uzbekistan
Bela Crkva 78 E3 Ger. Weisskirchen, Hung. Fehértemplom. Vojvodina, W Serbia
Belagāvi 110 B1 prev. Belgaum. Karnātaka, W India
Belarus 85 B6 off. Republic of Belarus, var. Belorussia, Latv. Baltkrievija; prev. Belorussian SSR, Rus. Belorusskaya SSR. country E Europe
Belarus, Republic of see Belarus
Belau see Palau
Belaya Tserkov' see Bila Tserkva
Bełchatów 76 C4 var. Belchatow. Łódzski, C Poland
Belchatow see Bełchatów
Belcher, Îles see Belcher Islands
Belcher Islands 16 C2 Fr. Îles Belcher. island group Nunavut, SE Canada
Beledweyne 51 D5 var. Belet Huen, It. Belet Uen. Hiiraan, C Somalia
Belém 41 F1 var. Pará. state capital Pará, N Brazil
Belén 30 D4 Rivas, SW Nicaragua
Belen 26 D2 New Mexico, SW USA
Belényes see Beiuş
Belet Huen/Belet Uen see Beledweyne
Belfast 67 B5 Ir. Béal Feirste. national capital E Northern Ireland, United Kingdom
Belfield 22 D2 North Dakota, N USA
Belfort 68 E4 Territoire-de-Belfort, E France
Belgard see Białogard
Belgaum see Belagāvi
Belgian Congo see Congo (Democratic Republic of)
België/Belgique see Belgium
Belgium 65 B6 off. Kingdom of Belgium, Dut. België, Fr. Belgique. country NW Europe
Belgium, Kingdom of see Belgium
Belgorod 89 A6 Belgorodskaya Oblast', W Russia
Belgrano II 132 A2 Argentinian research station Antarctica
Belice see Belize/Belize City
Beligrad see Berat
Beli Manastir 78 C3 Hung. Pélmonostor; prev. Monostor. Osijek-Baranja, NE Croatia
Bélinga 55 B5 Ogooué-Ivindo, NE Gabon
Belitung, Pulau 116 C4 island W Indonesia
Belize 30 B1 Sp. Belice; prev. British Honduras, Colony of Belize. country Central America
Belize 30 B1 river Belize/Guatemala
Belize see Belize City
Belize City 30 C1 var. Belize, Sp. Belice. Belize, NE Belize
Belize, Colony of see Belize
Beljak see Villach
Belkofski 14 B3 Alaska, USA
Belle Île 68 A4 island NW France
Belle Isle, Strait of 17 G3 strait Newfoundland and Labrador, E Canada
Bellenz see Bellinzona
Belleville 18 B4 Illinois, N USA
Bellevue 23 F4 Iowa, C USA
Bellevue 24 B2 Washington, NW USA
Bellingham 24 B1 Washington, NW USA
Belling Hausen Mulde see Southeast Pacific Basin
Bellingshausen Abyssal Plain see Bellingshausen Plain
Bellingshausen Plain 131 F5 var. Bellingshausen Abyssal Plain. abyssal plain SE Pacific Ocean
Bellingshausen Sea 132 A3 sea Antarctica
Bellinzona 73 B8 Ger. Bellenz. Ticino, S Switzerland
Bello 36 B2 Antioquia, W Colombia
Bello Horizonte see Belo Horizonte
Bellovacum see Beauvais
Bellville 56 B5 Western Cape, SW South Africa
Belmopan 30 C1 country capital (Belize) Cayo, C Belize
Belogradchik 82 B1 Vidin, NW Bulgaria
Belo Horizonte 41 F4 prev. Bello Horizonte. state capital Minas Gerais, SE Brazil
Belomorsk 88 B3 Respublika Kareliya, NW Russia
Beloretsk 89 D6 Respublika Bashkortostan, W Russia
Belorussia/Belorussian SSR see Belarus
Belorusskaya Gryada see Byelaruskaya Hrada
Belorusskaya SSR see Belarus
Beloshchel'ye see Nar'yan-Mar
Belostok see Białystok
Belovár see Bjelovar
Beloye More 88 C3 Eng. White Sea. sea NW Russia
Belozërsk 88 B4 Vologodskaya Oblast', NW Russia
Belton 27 G3 Texas, SW USA
Bel'tsy see Bălţi
Beluchistan see Baluchistān
Belukha, Gora 92 D5 mountain Kazakhstan/Russia
Belynichi see Byalynichy
Belyy, Ostrov 92 D2 island N Russia
Bemaraha 57 F3 var. Plateau du Bemaraha. mountain range W Madagascar
Bemaraha, Plateau du see Bemaraha
Bemidji 23 F1 Minnesota, N USA
Bemmel 64 D4 Gelderland, SE Netherlands
Benaco see Garda, Lago di
Benares see Vārānasi
Benavente 70 D2 Castilla-León, N Spain
Bend 24 B3 Oregon, NW USA
Bender see Tighina
Bender Beila/Bender Beyla see Bandarbeyla
Bender Cassim/Bender Qaasim see Boosaaso
Bendern 72 E1 NW Liechtenstein Europe
Bendery see Tighina
Bendigo 127 C7 Victoria, SE Australia
Beneschau see Benešov
Beneški Zaliv see Venice, Gulf of
Benešov 77 B5 Ger. Beneschau. Středočeský Kraj, W Czech Republic (Czechia)
Benevento 75 D5 anc. Beneventum, Malventum. Campania, S Italy
Beneventum see Benevento
Bengal, Bay of 102 C4 bay N Indian Ocean
Bengalūru 110 C2 prev. Bangalore. state capital Karnātaka, S India
Bengasi see Banghāzī
Bengbu 106 D5 var. Peng-pu. Anhui, E China
Benghazi see Banghāzī
Bengkulu 116 B4 prev. Bengkoeloe, Benkoelen, Benkulen. Sumatera, W Indonesia
Benguela 56 A2 var. Benguella. Benguela, W Angola
Benguella see Benguela
Bengweulu, Lake see Bangweulu, Lake
Ben Hope 66 B2 mountain N Scotland, United Kingdom
Beni 55 E5 Nord-Kivu, NE Dem. Rep. Congo
Benidorm 71 F4 País Valenciano, SE Spain
Beni-Mellal 48 C2 C Morocco
Benin 53 F4 off. Republic of Benin; prev. Dahomey. country W Africa
Benin, Bight of 53 F5 gulf W Africa
Benin City 53 F5 Edo, SW Nigeria
Benin, Republic of see Benin
Beni, Río 39 E3 river N Bolivia
Beni Suef see Banī Suwayf
Ben Nevis 66 C3 mountain N Scotland, United Kingdom
Bénoué see Benue
Benson 26 B3 Arizona, SW USA
Bent Jbail 97 A5 var. Bint Jubayl. S Lebanon
Benton 20 B1 Arkansas, C USA
Benue 54 B4 Fr. Bénoué. river Cameroon/Nigeria
Beograd 78 D3 Eng. Belgrade. Serbia, N Serbia
Berane 79 D5 prev. Ivangrad. E Montenegro
Berat 79 C6 var. Berati, SCr. Beligrad. Berat, C Albania
Berātāu see Berettyó
Berati see Berat
Berau, Teluk 117 G4 var. MacCluer Gulf. bay Papua, E Indonesia
Berbera 50 D4 Sahil, NW Somalia
Berbérati 55 B5 Mambéré-Kadéï, SW Central African Republic
Berck-Plage 68 C2 Pas-de-Calais, N France
Berdichev see Berdychiv
Berdyans'k 87 G4 Rus. Berdyansk; prev. Osipenko. Zaporiz'ka Oblast', SE Ukraine
Berdychiv 86 D2 Rus. Berdichev. Zhytomyrs'ka Oblast', N Ukraine
Beregovo/Beregszász see Berehove
Berehove 86 B3 Cz. Berehovo, Hung. Beregszász, Rus. Beregovo. Zakarpats'ka Oblast', W Ukraine
Berehovo see Berehove
Bereket 100 B2 prev. Rus. Gazandzhyk, Kazandzhik, Turkm. Gazanjyk. Balkan Welaýaty, W Turkmenistan
Beretău see Berettyó
Berettyó 77 D7 Rom. Barcău; prev. Berătău, Beretău. river Hungary/Romania
Berettyóújfalu 77 D6 Hajdú-Bihar, E Hungary
Berezhany 86 C2 Pol. Brzeżany. Ternopil's'ka Oblast', W Ukraine
Berezina see Byerezino
Berezniki 89 D5 Permskaya Oblast', NW Russia
Berga 71 G2 Cataluña, NE Spain
Bergamo 74 B2 anc. Bergomum. Lombardia, N Italy
Bergara 71 E1 País Vasco, N Spain
Bergen 72 D2 Mecklenburg-Vorpommern, NE Germany
Bergen 64 C2 Noord-Holland, NW Netherlands
Bergen 63 A5 Hordaland, S Norway
Bergen see Mons
Bergerac 69 B5 Dordogne, SW France
Bergeyk 65 C5 Noord-Brabant, S Netherlands

Bergomum – Bramaputra

Bergomum see Bergamo
Bergse Maas 64 D4 river S Netherlands
Beringen 65 C5 Limburg, NE Belgium
Beringov Proliv see Bering Strait
Bering Sea 14 A2 sea N Pacific Ocean
Bering Strait 14 C2 Rus. Beringov Proliv. strait Bering Sea/Chukchi Sea
Berja 71 E5 Andalucía, S Spain
Berkeley 25 B6 California, W USA
Berkner Island 132 A2 island Antarctica
Berkovitsa 82 C2 Montana, NW Bulgaria
Berlin 72 D3 country capital (Germany) Berlin, NE Germany
Berlin 19 G2 New Hampshire, NE USA
Berlinchen see Barlinek
Bermejo, Río 42 C2 river N Argentina
Bermeo 71 E1 País Vasco, N Spain
Bermuda 13 D6 var. Bermuda Islands, Bermudas; prev. Somers Islands. UK crown colony NW Atlantic Ocean
Bermuda Islands see Bermuda
Bermuda Rise 13 E6 undersea rise C Sargasso Sea
Bermudas see Bermuda
Bern 73 A7 Fr. Berne. country capital (Switzerland) Bern, W Switzerland
Bernau 72 D3 Brandenburg, NE Germany
Bernburg 72 C4 Sachsen-Anhalt, C Germany
Berne see Bern
Berner Alpen 73 A7 var. Berner Oberland, Eng. Bernese Oberland. mountain range SW Switzerland
Berner Oberland/Bernese Oberland see Berner Alpen
Bernier Island 125 A5 island Western Australia
Beroea see Ḩalab
Berry 68 C4 cultural region C France
Berry Islands 32 C1 island group N The Bahamas
Bertoua 55 B5 Est, E Cameroon
Beru 123 E2 var. Peru. atoll Tungaru, W Kiribati
Berwick-upon-Tweed 66 D4 N England, United Kingdom
Berytus see Beyrouth
Besançon 68 D4 anc. Besontium, Vesontio. Doubs, E France
Beskra see Biskra
Besontium see Besançon
Bessarabka see Basarabeasca
Beszterce see Bistriţa
Besztercebánya see Banská Bystrica
Betafo 57 G3 Antananarivo, C Madagascar
Betanzos 70 B1 Galicia, NW Spain
Bethlehem 56 D4 Free State, C South Africa
Bethlehem 97 B6 var. Beit Lekhem, Ar. Bayt Laḩm, Heb. Bet Leḥem. C West Bank
Béticos, Sistemas 70 D4 var. Sistema Penibético, Eng. Baetic Cordillera, Baetic Mountains. mountain range S Spain
Bet Leḥem see Bethlehem
Bétou 55 C5 Likouala, N Congo
Bette, Picco 49 G4 var. Bikkū Bītti, It. Picco Bette. mountain S Libya
Bette, Picco see Bette, Picco
Beulah 18 C2 Michigan, N USA
Beuthen see Bytom
Beveren 65 B5 Oost-Vlaanderen, N Belgium
Beverley 67 D5 E England, United Kingdom
Bexley 67 B8 Bexley, SE England, United Kingdom
Beyla 52 D4 SE Guinea
Beyrouth 96 A4 var. Bayrūt, Eng. Beirut; anc. Berytus. country capital (Lebanon) W Lebanon
Beyşehir 94 B4 Konya, SW Turkey
Beyşehir Gölü 94 B4 lake C Turkey
Béziers 69 C6 anc. Baeterrae, Baeterrae Septimanorum, Julia Beterrae. Hérault, S France
Bezmein see Abadan
Bezwada see Vijayawāda
Bhadrāvati 110 C2 Karnātaka, SW India
Bhāgalpur 113 F3 Bihār, NE India
Bhaktapur 113 F3 Central, C Nepal
Bhamo 114 B2 var. Banmo. Kachin State, N Myanmar (Burma)
Bhārat see India
Bharūch 112 C4 Gujarāt, W India
Bhaunagar see Bhāvnagar
Bhāvnagar 112 C4 prev. Bhaunagar. Gujarāt, W India
Bheanntraí, Bá see Bantry Bay
Bhopāl 112 D4 state capital Madhya Pradesh, C India
Bhubaneshwar 113 F5 prev. Bhubaneswar, Bhuvaneshwar. state capital Odisha, E India
Bhubaneswar see Bhubaneshwar
Bhuket see Phuket
Bhusaval see Bhusāwal
Bhusāwal 112 D4 prev. Bhusaval. Mahārāshtra, C India
Bhutan 113 G3 off. Kingdom of Bhutan, var. Druk-yul. country S Asia
Bhutan, Kingdom of see Bhutan
Bhuvaneshwar see Bhubaneshwar
Biak, Pulau 117 G4 island E Indonesia
Biała Podlaska 76 E3 Lubelskie, E Poland
Białogard 76 B2 Ger. Belgard. Zachodnio-pomorskie, NW Poland
Białystok 76 E3 Rus. Belostok, Bielostok. Podlaskie, NE Poland
Bianco, Monte see Blanc, Mont
Biarritz 69 A6 Pyrénées-Atlantiques, SW France
Bicaz 86 C3 Hung. Békás. Neamţ, NE Romania
Bichiş-Ciaba see Békéscsaba
Biddeford 19 G2 Maine, NE USA
Bideford 67 C7 SW England, United Kingdom
Biel 73 A7 Fr. Bienne. Bern, W Switzerland
Bielefeld 72 B4 Nordrhein-Westfalen, NW Germany
Bielitz/Bielitz-Biala see Bielsko-Biała
Bielostok see Białystok
Bielsko-Biała 77 C5 Ger. Bielitz, Bielitz-Biala. Śląskie, S Poland
Bielsk Podlaski 76 E3 Białystok, E Poland
Bien Bien see Điện Biên
Biên Đông see South China Sea
Biên Hoa 115 E6 Đông Nai, S Vietnam
Bienne see Biel
Bienville, Lac 16 D2 lake Québec, C Canada
Bié, Planalto do 56 B2 var. Bié Plateau. plateau C Angola
Bié Plateau see Bié, Planalto do
Big Cypress Swamp 21 E5 wetland Florida, SE USA
Bigge Island 124 C2 island Western Australia
Bighorn Mountains 22 C2 mountain range Wyoming, C USA
Bighorn River 22 C2 river Montana/Wyoming, NW USA
Bignona 52 B3 SW Senegal
Bigorra see Tarbes
Bigosovo see Bihosava
Big Sioux River 23 E2 river Iowa/South Dakota, N USA
Big Spring 27 E3 Texas, SW USA
Bihać 78 B3 Federacija Bosna I Hercegovina, NW Bosnia and Herzegovina
Bihār 113 F3 prev. Behar. cultural region N India
Bihār see Bihār Sharīf
Biharamulo 51 B7 Kagera, NW Tanzania
Bihār Sharīf 113 F3 var. Bihār. Bihār, N India
Bihosava 85 D5 Rus. Bigosovo. Vitsyebskaya Voblasts', NW Belarus
Bijeljina 78 C3 Republika Srpska, NE Bosnia and Herzegovina
Bijelo Polje 79 D5 E Montenegro
Bikāner 112 C3 Rājasthān, NW India
Bikin 93 G4 Khabarovskiy Kray, SE Russia
Bikini Atoll 122 C1 var. Pikinni. atoll Ralik Chain, NW Marshall Islands
Bikkū Bītti see Bette, Picco
Bilāspur 113 E4 Chhattīsgarh, C India
Biläsuvar 95 H3 Rus. Bilyasuvar; prev. Pushkino. SE Azerbaijan
Bila Tserkva 87 E2 Rus. Belaya Tserkov'. Kyyivs'ka Oblast', N Ukraine
Bilauktaung Range 115 C6 var. Thanintari Taungdan. mountain range Myanmar (Burma)/ Thailand
Bilbao 71 E1 Basq. Bilbo. País Vasco, N Spain
Bilbo see Bilbao
Bilecik 94 B3 Bilecik, NW Turkey
Billings 22 C2 Montana, NW USA
Bilma, Grand Erg de 53 H3 desert NE Niger
Biloela 126 D4 Queensland, E Australia
Biloxi 20 C3 Mississippi, S USA
Biltine 54 C3 Biltine, E Chad
Bilwi see Puerto Cabezas
Bilyasuvar see Biläsuvar
Bilzen 65 D6 Limburg, NE Belgium
Bimini Islands 32 C1 island group W The Bahamas
Binche 65 B7 Hainaut, S Belgium
Bindloe Island see Marchena, Isla
Bin Ghalfān, Jazā'ir see Ḩalāniyāt, Juzur al
Binghamton 19 F3 New York, NE USA
Bingöl 95 E3 Bingöl, E Turkey
Bint Jubayl see Bent Jbail
Bintulu 116 D3 Sarawak, East Malaysia
Binzhou 106 D4 Shandong, E China
Bío Bío, Río 43 B5 river C Chile
Bioco, Isla de 55 A5 var. Bioko, Eng. Fernando Po, Sp. Fernando Póo; prev. Macías Nguema Biyogo. island NW Equatorial Guinea
Bioko see Bioco, Isla de
Birāk 49 F3 var. Brak. C Libya
Birao 54 D3 Vakaga, NE Central African Republic
Birātnagar 113 F3 Eastern, SE Nepal
Bir es Saba see Be'er Sheva
Birjand 98 E3 Khorāsān-e Janūbī, E Iran
Birkenfeld 73 A5 Rheinland-Pfalz, SW Germany
Birkenhead 67 C5 NW England, United Kingdom
Birlad see Bârlad
Birmingham 67 C6 C England, United Kingdom
Birmingham 20 C2 Alabama, S USA
Bir Moghrein see Bîr Mogreïn
Bîr Mogreïn 52 C1 var. Bir Moghrein; prev. Fort-Trinquet. Tiris Zemmour, N Mauritania
Birnie Island 123 E3 atoll Phoenix Islands, C Kiribati
Birnin Konni 53 F3 var. Birni-Nkonni. Tahoua, SW Niger
Birni-Nkonni see Birnin Konni
Birobidzhan 93 G4 Yevreyskaya Avtonomnaya Oblast', SE Russia
Birsen see Biržai
Birsk 89 D5 Respublika Bashkortostan, W Russia
Biržai 84 C4 Ger. Birsen. Panevėžys, NE Lithuania
Birżebbuġa 80 B5 SE Malta
Bisanthe see Tekirdağ
Bisbee 26 B3 Arizona, SW USA
Biscaia, Baía de see Biscay, Bay of
Biscay, Bay of 58 B4 Sp. Golfo de Vizcaya, Port. Baía de Biscaia. bay France/Spain
Biscay Plain 58 B3 abyssal plain SE Bay of Biscay
Bischofsburg see Biskupiec
Bishah, Wadi 99 B5 dry watercourse C Saudi Arabia
Bishkek 101 G2 var. Pishpek; prev. Frunze. country capital (Kyrgyzstan) Chuyskaya Oblast', N Kyrgyzstan
Bishop's Lynn see King's Lynn
Bishrī, Jabal 96 D3 mountain range E Syria
Biskara see Biskra
Biskra 49 E2 var. Beskra, Biskara. NE Algeria
Biskupiec 76 D2 Ger. Bischofsburg. Warmińsko-Mazurskie, NE Poland
Bislig 117 F2 Mindanao, S Philippines
Bismarck 23 E2 state capital North Dakota, N USA
Bismarck Archipelago 122 B3 island group NE Papua New Guinea
Bismarck Sea 122 B3 sea W Pacific Ocean
Bisnulok see Phitsanulok
Bissau 52 B4 country capital (Guinea-Bissau) W Guinea-Bissau
Bistriţa 86 B3 Ger. Bistritz, Hung. Beszterce; prev. Nösen. Bistriţa-Năsăud, N Romania
Bistritz see Bistriţa
Bitam 55 B5 Woleu-Ntem, N Gabon
Bitburg 73 A5 Rheinland-Pfalz, SW Germany
Bitlis 95 F3 Bitlis, SE Turkey
Bitoeng see Bitung
Bitola 79 D6 Turk. Monastir; prev. Bitolj. S Macedonia
Bitolj see Bitola
Bitonto 75 D5 anc. Butuntum. Puglia, SE Italy
Bitterroot Range 24 D2 mountain range Idaho/ Montana, NW USA
Bitung 117 F3 prev. Bitoeng. Sulawesi, C Indonesia
Biu 53 H4 Borno, E Nigeria
Biwa-ko 109 C6 lake Honshū, SW Japan
Bizerta see Bizerte
Bizerte 49 E1 Ar. Banzart, Eng. Bizerta. N Tunisia
Bjelovar 78 B2 Hung. Belovár. Bjelovar-Bilogora, N Croatia
Bjeshkët e Namuna see North Albanian Alps
Björneborg see Pori
Bjørnøya 61 F3 Eng. Bear Island. island N Norway
Blackall 126 C4 Queensland, E Australia
Black Drin 79 D6 Alb. Lumi i Drinit të Zi, SCr. Crni Drim. river Albania/Macedonia
Blackfoot 24 E4 Idaho, NW USA
Black Forest 73 B6 Eng. Black Forest. mountain range SW Germany
Black Forest see Schwarzwald
Black Hills 22 D3 mountain range South Dakota/ Wyoming, N USA
Blackpool 67 C5 NW England, United Kingdom
Black Range 26 C2 mountain range New Mexico, SW USA
Black River 32 A5 W Jamaica
Black River 114 C3 Chin. Babian Jiang, Lixian Jiang, Fr. Rivière Noire, Vtn. Sông Đa. river China/Vietnam
Black Rock Desert 25 C5 desert Nevada, W USA
Black Sand Desert see Garagum
Black Sea 94 B1 var. Euxine Sea, Bul. Cherno More, Rom. Marea Neagră, Rus. Chernoye More, Turk. Karadeniz, Ukr. Chorne More. sea Asia/Europe
Black Sea Lowland 87 E4 Ukr. Prychornomor'ska Nyzovyna. depression SE Europe
Black Volta 53 E4 var. Borongo, Mouhoun, Moun Hou, Fr. Volta Noire. river W Africa
Blackwater 67 A6 Ir. An Abhainn Mhór. river S Ireland
Blackwater State see Nebraska
Blagoevgrad 82 C3 prev. Gorna Dzhumaya. Blagoevgrad, W Bulgaria
Blagoveshchensk 93 G4 Amurskaya Oblast', SE Russia
Blahovishchens'ke 87 E3 Rus. Ulyanovka. Kirovohrads'ka Oblast', C Ukraine
Blake Plateau 13 D6 var. Blake Terrace. undersea plateau W Atlantic Ocean
Blake Terrace see Blake Plateau
Blanca, Bahía 43 C5 bay E Argentina
Blanca, Costa 71 F4 physical region SE Spain
Blanche, Lake 127 B5 lake South Australia
Blanc, Mont 69 D5 It. Monte Bianco. mountain France/Italy
Blanco, Cape 24 A4 headland Oregon, NW USA
Blanes 71 G2 Cataluña, NE Spain
Blankenberge 65 A5 West-Vlaanderen, NW Belgium
Blankenheim 73 A5 Nordrhein-Westfalen, W Germany
Blanquilla, Isla 37 E1 var. La Blanquilla. island N Venezuela
Blanquilla, La see Blanquilla, Isla
Blantyre 57 E2 var. Blantyre-Limbe. Southern, S Malawi
Blantyre-Limbe see Blantyre
Blaricum 64 C3 Noord-Holland, C Netherlands
Blatnitsa see Durankulak
Blenheim 129 C5 Marlborough, South Island, New Zealand
Blesae see Blois
Blida 48 D2 var. El Boulaida, El Boulaïda. N Algeria
Bloemfontein 56 C4 var. Mangaung. country capital (South Africa-judicial capital) Free State, C South Africa
Blois 68 C4 anc. Blesae. Loir-et-Cher, C France
Bloomfield 26 C1 New Mexico, SW USA
Bloomington 18 B4 Illinois, N USA
Bloomington 18 C4 Indiana, N USA
Bloomington 23 F2 Minnesota, N USA
Bloomsbury 126 D3 Queensland, NE Australia
Bluefield 18 D5 West Virginia, NE USA
Bluefields 31 E3 Región Autónoma Atlántico Sur, SE Nicaragua
Bluegrass State see Kentucky
Blue Hen State see Delaware
Blue Law State see Connecticut
Blue Mountain Peak 32 B5 mountain E Jamaica
Blue Mountains 24 C3 mountain range Oregon/ Washington, NW USA
Blue Nile 50 C4 var. Abai, Bahr el, Azraq, Amh. Ābay Wenz, Ar. An Nīl al Azraq. river Ethiopia/Sudan
Blumenau 41 E5 Santa Catarina, S Brazil
Blythe 25 D8 California, W USA
Blytheville 20 C1 Arkansas, C USA
Bo 52 C4 S Sierra Leone
Boaco 30 D3 Boaco, S Nicaragua
Boa Vista 40 D1 state capital Roraima, NW Brazil
Boa Vista 52 A3 island Ilhas de Barlavento, E Cape Verde
Bobaomby, Tanjona 57 G2 Fr. Cap d'Ambre. headland N Madagascar
Bobigny 68 E1 Seine-St-Denis, N France
Bobo-Dioulasso 52 D4 SW Burkina
Bobrinets see Bobrynets'
Bobruysk see Babruysk
Bobrynets' 87 E3 Rus. Bobrinets. Kirovohrads'ka Oblast', C Ukraine
Boca Raton 21 F5 Florida, SE USA
Bocay 30 D2 Jinotega, N Nicaragua
Bocheykovo see Bacheykava
Bocholt 72 A4 Nordrhein-Westfalen, W Germany
Bochum 72 A4 Nordrhein-Westfalen, W Germany
Bocşa 86 A4 Ger. Bokschen, Hung. Boksánbánya. Caraş-Severin, SW Romania
Bodaybo 93 F4 Irkutskaya Oblast', E Russia
Bodegas see Babahoyo
Boden 62 D4 Norrbotten, N Sweden
Bodensee see Constance, Lake, C Europe
Bodmin 67 C7 SW England, United Kingdom
Bodø 62 C3 Nordland, C Norway
Bodrum 94 A4 Muğla, SW Turkey
Boeloekoemba see Bulukumba
Boende 55 C5 Equateur, C Dem. Rep. Congo
Boeroe see Buru, Pulau
Boetoeng see Buton, Pulau
Bogale 114 B4 Ayeyarwady, SW Myanmar (Burma)
Bogalusa 20 B3 Louisiana, S USA
Bogatynia 76 B4 Ger. Reichenau. Dolnośląskie, SW Poland
Boğazlıyan 94 D3 Yozgat, C Turkey
Bogendorf see Łuków
Bogor 116 C5 Dut. Buitenzorg. Jawa, C Indonesia
Bogotá 36 B3 prev. Santa Fe, Santa Fe de Bogotá. country capital (Colombia) Cundinamarca, C Colombia
Bogushëvsk see Bahushewsk
Boguslav see Bohuslav
Bo Hai 106 D4 var. Gulf of Chihli. gulf NE China
Bohemia 77 A5 Cz. Čechy, Ger. Böhmen. W Czech Republic (Czechia)
Bohemian Forest 73 C5 Cz. Ceský Les, Šumava, Ger. Böhmerwald. mountain range C Europe
Böhmen see Bohemia
Böhmerwald see Bohemian Forest
Böhmisch-Krumau see Český Krumlov
Bohol Sea 117 E2 var. Mindanao Sea. sea S Philippines
Bohuslav 87 E2 Rus. Boguslav. Kyyivs'ka Oblast', N Ukraine
Boise 24 D3 var. Boise City. state capital Idaho, NW USA
Boise City 27 E1 Oklahoma, C USA
Boise City see Boise
Bois, Lac des see Woods, Lake of the
Bois-le-Duc see 's-Hertogenbosch
Boizenburg 72 C3 Mecklenburg-Vorpommern, N Germany
Bojador see Boujdour
Bojnūrd 98 D2 var. Bujnurd. Khorāsān-e Shemālī, N Iran
Bokāro 113 F4 Jhārkhand, N India
Boké 52 C4 W Guinea
Bokhara see Buxoro
Boksánbánya/Bokschen see Bocşa
Bol 54 B3 Lac, W Chad
Bolgatanga 53 E4 N Ghana
Bolgrad see Bolhrad
Bolhrad 86 D4 Rus. Bolgrad. Odes'ka Oblast', SW Ukraine
Bolívar, Pico 36 C2 mountain W Venezuela
Bolivia 39 F3 off. Republic of Bolivia. country W South America
Bolivia, Republic of see Bolivia
Bollène 69 D6 Vaucluse, SE France
Bollnäs 63 C5 Gävleborg, C Sweden
Bollon 127 D5 Queensland, C Australia
Bologna 74 C3 Emilia-Romagna, N Italy
Bol'shevik see Bal'shavik
Bol'shevik, Ostrov 93 E2 island Severnaya Zemlya, N Russia
Bol'shezemel'skaya Tundra 88 E3 physical region NW Russia
Bol'shoy Lyakhovskiy, Ostrov 93 F2 island NE Russia
Bolton 67 D5 prev. Bolton-le-Moors. NW England, United Kingdom
Bolton-le-Moors see Bolton
Bolu 94 B3 Bolu, NW Turkey
Bolungarvík 61 E4 Vestfirdhir, NW Iceland
Bolyarovo 82 D3 prev. Pashkeni. Yambol, E Bulgaria
Bolzano 74 C1 Ger. Bozen; anc. Bauzanum. Trentino-Alto Adige, N Italy
Boma 55 B6 Bas-Congo, W Dem. Rep. Congo
Bombay see Mumbai
Bomu 54 D4 var. Mbomou, Mbomu, M'Bomu. river Central African Republic/Dem. Rep. Congo
Bonaire 33 F5 Dutch special municipality S West Indies
Bonanza 30 D2 Región Autónoma Atlántico Norte, NE Nicaragua
Bonaparte Archipelago 124 C2 island group Western Australia
Bon, Cap 80 D3 headland N Tunisia
Bonda 55 B6 Ogooué-Lolo, C Gabon
Bondoukou 53 E4 E Ivory Coast
Bône see Annaba, Algeria
Bone see Watampone, Indonesia
Bone, Teluk 117 E4 bay Sulawesi, C Indonesia
Bongaigaon 113 G3 Assam, NE India
Bongo, Massif des 54 D4 var. Chaîne des Mongos. mountain range NE Central African Republic
Bongor 54 B3 Mayo-Kébbi, SW Chad
Bonifacio 69 E7 Corse, France, C Mediterranean Sea
Bonifacio, Bocche de/Bonifacio, Bouches de see Bonifacio, Strait of
Bonifacio, Strait of 74 A4 Fr. Bouches de Bonifacio, It. Bocche di Bonifacio. strait C Mediterranean Sea
Bonn 73 A5 Nordrhein-Westfalen, W Germany
Bononia see Vidin, Bulgaria
Bononia see Boulogne-sur-Mer, France
Boosaaso 50 E4 var. Bandar Kassim, Bender Qaasim, Bosaso, It. Bender Cassim. Bari, N Somalia
Boothia Felix see Boothia Peninsula
Boothia, Gulf of 15 F2 gulf Nunavut, NE Canada
Boothia Peninsula 15 F2 prev. Boothia Felix. peninsula Nunavut, NE Canada
Boppard 73 A5 Rheinland-Pfalz, W Germany
Boquete 31 E5 var. Bajo Boquete. Chiriquí, W Panama
Boquillas 28 D2 var. Boquillas del Carmen. Coahuila, NE Mexico
Boquillas del Carmen see Boquillas
Bor 78 E4 Serbia, E Serbia
Bor 51 B5 Jonglei, C South Sudan
Borås 63 B7 Västra Götaland, S Sweden
Borbetomagus see Worms
Borborema, Planalto da 34 E3 plateau NE Brazil
Bordeaux 69 B5 anc. Burdigala. Gironde, SW France
Bordj Omar Driss 49 E3 E Algeria
Borgå see Porvoo
Børgefjelt 62 C4 mountain range C Norway
Borger 64 E2 Drenthe, NE Netherlands
Borger 27 E1 Texas, SW USA
Borgholm 63 C7 Kalmar, S Sweden
Borgo Maggiore 74 E1 NW San Marino
Borislav see Boryslav
Borisoglebsk 89 B6 Voronezhskaya Oblast', W Russia
Borisov see Barysaw
Borlänge 63 C6 Dalarna, C Sweden
Borne 64 E3 Overijssel, E Netherlands
Borneo 116 C4 island Brunei/Indonesia/Malaysia
Bornholm 63 B8 island E Denmark
Borohoro Shan 104 B2 mountain range NW China
Borongo see Black Volta
Boron'ki see Baron'ki
Borosjenő see Ineu
Borovan 82 C2 Vratsa, NW Bulgaria
Borovichi 88 B4 Novgorodskaya Oblast', W Russia
Borovo 78 C3 Vukovar-Srijem, NE Croatia
Borriana 71 F3 var. Burriana. País Valenciano, E Spain
Borşa 86 C3 Hung. Borsa. Maramureş, N Romania
Boryslav 86 B2 Pol. Borysław, Rus. Borislav. L'vivs'ka Oblast', NW Ukraine
Borysław see Boryslav
Bosanska Dubica 78 B3 var. Kozarska Dubica. Republika Srpska, NW Bosnia and Herzegovina
Bosanska Gradiška 78 B3 var. Gradiška. Republika Srpska, N Bosnia and Herzegovina
Bosanski Novi 78 B3 var. Novi Grad. Republika Srpska, NW Bosnia and Herzegovina
Bosanski Šamac 78 C3 var. Šamac. Republika Srpska, N Bosnia and Herzegovina
Bosaso see Boosaaso
Bösing see Pezinok
Boskovice 77 B5 Ger. Boskowitz. Jihomoravský Kraj, SE Czech Republic (Czechia)
Boskowitz see Boskovice
Bosna 78 C4 river N Bosnia and Herzegovina
Bosnia and Herzegovina 78 B3 off. Republic of Bosnia and Herzegovina. country SE Europe
Bosnia and Herzegovina, Republic of see Bosnia and Herzegovina
Boso-hanto 109 D6 peninsula Honshū, S Japan
Bosphorus/Bosporus see İstanbul Boğazı
Bosporus Cimmerius see Kerch Strait
Bosporus Thracius see İstanbul Boğazı
Bossangoa 54 C4 Ouham, C Central African Republic
Bossembélé 54 C4 Ombella-Mpoko, C Central African Republic
Bossier City 20 A2 Louisiana, S USA
Bosten Hu 104 C3 var. Bagrax Hu. lake NW China
Boston 67 E6 prev. St.Botolph's Town. E England, United Kingdom
Boston 19 G3 state capital Massachusetts, NE USA
Boston Mountains 20 B1 mountain range Arkansas, C USA
Bostyn' see Bastyn'
Botany 126 E2 New South Wales, E Australia
Botany Bay 126 E2 inlet New South Wales, SE Australia
Boteti 56 C3 var. Botletle. river N Botswana
Bothnia, Gulf of 63 D5 Fin. Pohjanlahti, Swe. Bottniska Viken. gulf N Baltic Sea
Botletle see Boteti
Botoşani 86 C3 Hung. Botosány. Botoşani, NE Romania
Botosány see Botoşani
Botou 106 C4 prev. Bozhen. Hebei, E China
Botrange 65 D6 mountain E Belgium
Botswana 56 C3 off. Republic of Botswana. country S Africa
Botswana, Republic of see Botswana
Bottniska Viken see Bothnia, Gulf of
Bouar 54 B4 Nana-Mambéré, W Central African Republic
Bou Craa 48 B3 var. Bu Craa. NW Western Sahara
Bougainville Island 120 B3 island NE Papua New Guinea
Bougaroun, Cap 80 C3 headland NE Algeria
Bougouni 52 D4 Sikasso, SW Mali
Boujdour 48 A3 var. Bojador. W Western Sahara
Boulder 22 C4 Colorado, C USA
Boulder 22 B2 Montana, NW USA
Boulogne see Boulogne-sur-Mer
Boulogne-Billancourt 68 D1 Île-de-France, N France Europe
Boulogne-sur-Mer 68 C2 var. Boulogne; anc. Bononia, Gesoriacum, Gessoriacum. Pas-de-Calais, N France
Boûmdeïd 52 C3 var. Boumdeït. Assaba, S Mauritania
Boumdeït see Boûmdeïd
Boundiali 52 D4 N Ivory Coast
Bountiful 22 B4 Utah, W USA
Bounty Basin see Bounty Trough
Bounty Islands 120 D5 island group S New Zealand
Bounty Trough 130 C5 var. Bounty Basin. trough S Pacific Ocean
Bourbonnais 68 C4 cultural region C France
Bourbon Vendée see la Roche-sur-Yon
Bourg see Bourg-en-Bresse
Bourgas see Burgas
Bourge-en-Bresse see Bourg-en-Bresse
Bourg-en-Bresse 69 D5 var. Bourg, Bourge-en-Bresse. Ain, E France
Bourges 68 C4 anc. Avaricum. Cher, C France
Bourgogne 68 C4 Eng. Burgundy. cultural region E France
Bourke 127 C5 New South Wales, SE Australia
Bournemouth 67 D7 S England, United Kingdom
Boutilimit 52 C3 Trarza, SW Mauritania
Bouvet Island 45 D7 Norwegian dependency S Atlantic Ocean
Bowen 126 D3 Queensland, NE Australia
Bowling Green 18 B5 Kentucky, S USA
Bowling Green 18 C3 Ohio, N USA
Boxmeer 64 D4 Noord-Brabant, SE Netherlands
Boyarka 87 E2 Kyyivs'ka Oblast', N Ukraine
Boychinovtsi 82 C2 prev. Lekhchevo. Montana, NW Bulgaria
Boysun 101 E3 Rus. Baysun. Surkhondaryo Viloyati, S Uzbekistan
Bozeman 22 B2 Montana, NW USA
Bozen see Bolzano
Bozhen see Botou
Bozüyük 94 B3 Bilecik, NW Turkey
Brač 78 B4 var. Brach, It. Brazza; anc. Brattia. island S Croatia
Bracara Augusta see Braga
Brach see Brač
Brades 33 G3 de facto dependent territory capital, de jure capital, Plymouth, destroyed by volcano in 1995 (Montserrat) SW Montserrat
Bradford 67 D5 N England, United Kingdom
Brady 27 F3 Texas, SW USA
Braga 70 B2 anc. Bracara Augusta. Braga, NW Portugal
Bragança 70 C2 Eng. Braganza; anc. Julio Briga. Bragança, NE Portugal
Braganza see Bragança
Brahestad see Raahe
Brahmanbaria 113 G4 Chittagong, E Bangladesh
Brahmapur 113 F5 Odisha, E India
Brahmaputra 113 H3 var. Padma, Tsangpo, Ben. Jamuna, Chin. Yarlung Zangbo Jiang, Ind. Bramaputra, Dihang, Siang. river S Asia
Brăila 86 D4 Brăila, E Romania
Braine-le-Comte 65 B6 Hainaut, SW Belgium
Brainerd 23 F2 Minnesota, N USA
Brak see Birāk
Bramaputra see Brahmaputra

Brampton – Caprivizipfel

Brampton *16 D5* Ontario, S Canada
Branco, Rio *34 C3* river N Brazil
Brandberg *56 A3* mountain NW Namibia
Brandenburg *72 C3* var. Brandenburg an der Havel. Brandenburg, NE Germany
Brandenburg an der Havel see Brandenburg
Brandon *15 F5* Manitoba, S Canada
Braniewo *76 D2* Ger. Braunsberg. Warmińsko-mazurskie, N Poland
Brasil see Brazil
Brasília *41 F3* country capital (Brazil) Distrito Federal, C Brazil
Brasil, República Federativa do see Brazil
Brașov *86 C4* Ger. Kronstadt, Hung. Brassó; prev. Orașul Stalin. Brașov, C Romania
Brassó see Brașov
Bratislava *77 C6* Ger. Pressburg, Hung. Pozsony. country capital (Slovakia) Bratislavský Kraj, W Slovakia
Bratsk *93 E4* Irkutskaya Oblast', C Russia
Brattia see Brač
Braunsberg see Braniewo
Braunschweig *72 C4* Eng./Fr. Brunswick. Niedersachsen, N Germany
Brava see Baraawe
Brava, Costa *71 H2* coastal region NE Spain
Bravo del Norte, Río/Bravo, Río see Grande, Rio
Bravo, Río *28 C1* river Mexico/USA North America
Brawley *25 D8* California, W USA
Brazil *40 C2* off. Federative Republic of Brazil, Port. República Federativa do Brasil, Sp. Brasil; prev. United States of Brazil. country South America
Brazil Basin *45 C5* var. Brazilian Basin, Brazil'skaya Kotlovina. undersea basin W Atlantic Ocean
Brazil, Federative Republic of see Brazil
Brazilian Basin see Brazil Basin
Brazilian Highlands see Central, Planalto
Brazil'skaya Kotlovina see Brazil Basin
Brazil, United States of see Brazil
Brazos River *27 G3* river Texas, SW USA
Brazza see Brač
Brazzaville *55 B6* country capital (Congo) Capital District, S Congo
Brčko *78 C3* Republika Srpska, NE Bosnia and Herzegovina
Brecht *65 C5* Antwerpen, N Belgium
Brecon Beacons *67 C6* mountain range S Wales, United Kingdom
Breda *64 C4* Noord-Brabant, S Netherlands
Bree *65 D5* Limburg, NE Belgium
Bregalnica *79 E6* river E Macedonia
Bregenz *73 B7* anc. Brigantium. Vorarlberg, W Austria
Bregovo *82 B1* Vidin, NW Bulgaria
Bremen *72 B3* Fr. Brême. Bremen, NW Germany
Bremerhaven *72 B3* Bremen, NW Germany
Bremerton *24 B2* Washington, NW USA
Brenham *27 G3* Texas, SW USA
Brenner, Col du/Brennero, Passo del see Brenner Pass
Brenner Pass *74 C1* var. Brenner Sattel, Fr. Col du Brenner, Ger. Brennerpass, It. Passo del Brennero. pass Austria/Italy
Brennerpass see Brenner Pass
Brenner Sattel see Brenner Pass
Brescia *74 B2* anc. Brixia. Lombardia, N Italy
Breslau see Wrocław
Bressanone *74 C1* Ger. Brixen. Trentino-Alto Adige, N Italy
Brest *85 A6* Pol. Brześć nad Bugiem, Rus. Brest-Litovsk; prev. Brześć Litewski. Brestskaya Voblasts', SW Belarus
Brest *68 A3* Finistère, NW France
Brest-Litovsk see Brest
Bretagne *68 A3* Eng. Brittany, Lat. Britannia Minor. cultural region NW France
Brewster, Kap see Kangikajik
Brewton *20 C3* Alabama, S USA
Brezhnev see Naberezhnyye Chelny
Brezovo *82 D2* prev. Abrashlare. Plovdiv, C Bulgaria
Bria *54 D4* Haute-Kotto, C Central African Republic
Briançon *69 D5* anc. Brigantio. Hautes-Alpes, SE France
Bricgstow see Bristol
Bridgeport *19 F3* Connecticut, NE USA
Bridgetown *33 G2* country capital (Barbados) SW Barbados
Bridlington *67 D5* E England, United Kingdom
Bridport *67 D7* S England, United Kingdom
Brieg see Brzeg
Brig *73 A7* Fr. Brigue, It. Briga. Valais, SW Switzerland
Briga see Brig
Brigantio see Briançon
Brigantium see Bregenz
Brigham City *22 B3* Utah, W USA
Brighton *67 E7* SE England, United Kingdom
Brighton *22 D4* Colorado, C USA
Brigue see Brig
Brindisi *75 E5* anc. Brundisium, Brundusium. Puglia, SE Italy
Briovera see St-Lô
Brisbane *127 E5* state capital Queensland, E Australia
Bristol *67 D7* anc. Bricgstow. SW England, United Kingdom
Bristol *19 F3* Connecticut, NE USA
Bristol *18 D5* Tennessee, S USA
Bristol Bay *14 B3* bay Alaska, USA
Bristol Channel *67 C7* inlet England/Wales, United Kingdom
Britain *58 C3* var. Great Britain. island United Kingdom
Britannia Minor see Bretagne
British Columbia *14 D4* Fr. Colombie-Britannique. province SW Canada
British Guiana see Guyana
British Honduras see Belize
British Indian Ocean Territory *119 B5* UK dependent territory C Indian Ocean
British Isles *67* island group NW Europe
British North Borneo see Sabah
British Solomon Islands Protectorate see Solomon Islands
British Virgin Islands *33 F3* var. Virgin Islands. UK dependent territory E West Indies
Brittany see Bretagne
Briva Curretia see Brive-la-Gaillarde
Briva Isarae see Pontoise
Brive see Brive-la-Gaillarde
Brive-la-Gaillarde *69 C5* prev. Brive; anc. Briva Curretia. Corrèze, C France
Brixen see Bressanone
Brixia see Brescia
Brno *77 B5* Ger. Brünn. Jihomoravský Kraj, SE Czech Republic (Czechia)
Brocēni *84 B3* Saldus, SW Latvia
Brod/Bród see Slavonski Brod
Brodeur Peninsula *15 F2* peninsula Baffin Island, Nunavut, NE Canada
Brod na Savi see Slavonski Brod
Brodnica *76 C3* Ger. Buddenbrock. Kujawski-pomorskie, C Poland
Broek-in-Waterland *64 C3* Noord-Holland, C Netherlands
Broken Arrow *27 G1* Oklahoma, C USA
Broken Bay *126 E1* bay New South Wales, SE Australia
Broken Hill *127 B6* New South Wales, SE Australia
Broken Ridge *119 D6* undersea plateau S Indian Ocean
Bromberg see Bydgoszcz
Bromley *67 B8* United Kingdom
Brookhaven *20 B3* Mississippi, S USA
Brookings *23 F3* South Dakota, N USA
Brooks Range *14 D2* mountain range Alaska, USA
Brookton *125 B6* Western Australia
Broome *124 B3* Western Australia
Broomfield *22 D4* Colorado, C USA
Broucsella see Brussel/Bruxelles
Brovary *87 E2* Kyyivs'ka Oblast', N Ukraine
Brownfield *27 E2* Texas, SW USA
Brownsville *27 G5* Texas, SW USA
Brownwood *27 F3* Texas, SW USA
Brozha *85 D7* Mahilyowskaya Voblasts', E Belarus
Bruges see Brugge
Brugge *65 A5* Fr. Bruges. West-Vlaanderen, NW Belgium
Brummen *64 D3* Gelderland, E Netherlands
Brundisium/Brundusium see Brindisi
Brunei *116 D3* off. Brunei Darussalam, Mal. Negara Brunei Darussalam. country SE Asia
Brunei Darussalam see Brunei
Brunei Town see Bandar Seri Begawan
Brünn see Brno
Brunner, Lake *129 C5* lake South Island, New Zealand
Brunswick *21 E3* Georgia, SE USA
Brunswick see Braunschweig
Brusa see Bursa
Brus Laguna *30 D2* Gracias a Dios, E Honduras
Brussa see Bursa
Brussel *65 C6* var. Brussels, Fr. Bruxelles, Ger. Brüssel; anc. Broucsella. country capital (Belgium) Brussels, C Belgium
Brüssel/Brussels see Brussel/Bruxelles
Brüx see Most
Bruxelles see Brussel
Bryan *27 G3* Texas, SW USA
Bryansk *89 A5* Bryanskaya Oblast', W Russia
Brzeg *76 C4* Ger. Brieg; anc. Civitas Altae Ripae. Opolskie, S Poland
Brześć Litewski/Brześć nad Bugiem see Brest
Brzeżany see Berezhany
Bucaramanga *36 B2* Santander, N Colombia
Buchanan *52 C5* prev. Grand Bassa. SW Liberia
Buchanan, Lake *27 F3* reservoir Texas, SW USA
Bucharest see București
Buckeye State see Ohio
Bu Craa see Bou Craa
București *86 C5* Eng. Bucharest, Ger. Bukarest, prev. Altenburg; anc. Cetatea Damboviței. country capital (Romania) București, S Romania
Buda-Kashalyova *85 D7* Rus. Buda-Koshelëvo. Homyel'skaya Voblasts', SE Belarus
Buda-Koshelëvo see Buda-Kashalyova
Budapest *77 C6* off. Budapest Főváros, SCr. Budimpešta. country capital (Hungary) Pest, N Hungary
Budapest Főváros see Budapest
Budaun *112 D3* Uttar Pradesh, N India
Buddenbrock see Brodnica
Budimpešta see Budapest
Budweis see České Budějovice
Budyšin see Bautzen
Buena Park *24 E2* California, W USA North America
Buenaventura *36 A3* Valle del Cauca, W Colombia
Buena Vista *39 G4* Santa Cruz, C Bolivia
Buena Vista *71 H5* S Gibraltar Europe
Buenos Aires *42 D4* hist. Santa Maria del Buen Aire. country capital (Argentina) Buenos Aires, E Argentina
Buenos Aires *31 E5* Puntarenas, SE Costa Rica
Buenos Aires, Lago *43 B6* var. Lago General Carrera. lake Argentina/Chile
Buffalo *19 E3* New York, NE USA
Buffalo Narrows *15 F4* Saskatchewan, C Canada
Buff Bay *32 B5* E Jamaica
Buftea *86 C5* Ilfov, S Romania
Bug *59 E3* Bel. Zakhodni Buh, Eng. Western Bug, Rus. Zapadnyy Bug, Ukr. Zakhidnyy Buh. river E Europe
Buga *36 B3* Valle del Cauca, W Colombia
Bughotu see Santa Isabel
Buguruslan *89 D6* Orenburgskaya Oblast', W Russia
Buitenzorg see Bogor
Bujalance *70 D4* Andalucía, S Spain
Bujanovac *79 E5* SE Serbia
Bujnurd see Bojnūrd
Bujumbura *51 B7* prev. Usumbura. country capital (Burundi) W Burundi
Bukarest see București
Bukavu *55 E6* prev. Costermansville. Sud-Kivu, E Dem. Rep. Congo
Bukhara see Buxoro
Bukoba *51 B6* Kagera, NW Tanzania
Bülach *73 B7* Zürich, NW Switzerland
Bulawayo *56 D3* Matabeleland North, SW Zimbabwe
Bulgan *105 E2* Bulgan, N Mongolia
Bulgaria *82 C2* off. Republic of Bulgaria, Bul. Bŭlgariya; prev. People's Republic of Bulgaria. country SE Europe
Bulgaria, People's Republic of see Bulgaria
Bulgaria, Republic of see Bulgaria
Bŭlgariya see Bulgaria
Bullion State see Missouri
Bull Shoals Lake *20 B1* reservoir Arkansas/Missouri, C USA
Bulukumba *117 E4* prev. Boeloekoemba. Sulawesi, C Indonesia
Bumba *55 D5* Equateur, N Dem. Rep. Congo
Bunbury *125 A7* Western Australia
Bundaberg *126 E4* Queensland, E Australia
Bungo-suido *109 B7* strait SW Japan
Bunia *55 E5* Orientale, NE Dem. Rep. Congo
Bünyan *94 D3* Kayseri, C Turkey
Buraida see Buraydah
Buraydah *98 B4* var. Buraida. Al Qaşim, N Saudi Arabia
Burdigala see Bordeaux
Burdur *94 B4* var. Buldur. Burdur, SW Turkey
Burdur Gölü *94 B4* salt lake SW Turkey
Burē *50 C4* Āmara, N Ethiopia
Burgas *82 E2* var. Bourgas. Burgas, E Bulgaria
Burgaski Zaliv *82 E2* gulf E Bulgaria
Burgos *70 D2* Castilla-León, N Spain
Burgundy see Bourgogne
Burhan Budai Shan *104 D4* mountain range C China
Buriram *115 D5* var. Buri Ram, Puriramya. Buri Ram, E Thailand
Buri Ram see Buriram
Burjassot *71 F3* País Valenciano, E Spain
Burkburnett *27 F2* Texas, SW USA
Burketown *126 B3* Queensland, NE Australia
Burkina see Burkina Faso
Burkina Faso *53 E4* off. Burkina Faso; var. Burkina; prev. Upper Volta. country W Africa
Burley *24 D4* Idaho, NW USA
Burlington *23 G4* Iowa, C USA
Burlington *19 F2* Vermont, NE USA
Burma see Myanmar
Burnie *127 C8* Tasmania, SE Australia
Burns *24 C3* Oregon, NW USA
Burnside *15 F3* river Nunavut, NW Canada
Burnsville *23 F2* Minnesota, N USA
Burrel *79 D6* var. Burreli. Dibër, C Albania
Burreli see Burrel
Burriana see Borriana
Bursa *94 B3* var. Brussa, prev. Brusa; anc. Prusa. Bursa, NW Turkey
Būr Sa'īd *50 B1* var. Port Said. N Egypt
Burtnieks *84 C3* var. Burtnieks Ezers. lake N Latvia
Burtnieks Ezers see Burtnieks
Burundi *51 B7* off. Republic of Burundi; prev. Kingdom of Burundi, Urundi. country C Africa
Burundi, Kingdom of see Burundi
Burundi, Republic of see Burundi
Buru, Pulau *117 F4* prev. Boeroe. island E Indonesia
Busan *107 E4* off. Busan Gwang-yeoksi, prev. Pusan, Jap. Fusan. SE South Korea
Busan Gwang-yeoksi see Busan
Buşayrah *96 D3* Dayr az Zawr, E Syria
Būshehr/Bushire see Bandar-e Būshehr
Busra see Al Başrah, Iraq
Busselton *125 A7* Western Australia
Bussora see Al Başrah
Buta *55 D5* Orientale, N Dem. Rep. Congo
Butembo *55 E5* Nord-Kivu, NE Dem. Rep. Congo
Butler *19 E4* Pennsylvania, NE USA
Buton, Pulau *117 E4* var. Pulau Butung; prev. Boetoeng. island C Indonesia
Bütow see Bytów
Butte *22 B2* Montana, NW USA
Butterworth *116 B3* Pinang, Peninsular Malaysia
Button Islands *17 E1* island group Nunavut, NE Canada
Butuan *117 F2* off. Butuan City. Mindanao, S Philippines
Butuan City see Butuan
Butung, Pulau see Buton, Pulau
Butuntum see Bitonto
Buulobarde *51 D5* var. Buulo Berde. Hiiraan, C Somalia
Buulo Berde see Buulobarde
Buur Gaabo *51 D6* Jubbada Hoose, S Somalia
Buxoro *100 D2* var. Bokhara, Rus. Bukhara. Buxoro Viloyati, C Uzbekistan
Buynaksk *89 B8* Respublika Dagestan, SW Russia
Büyükağrı Dağı *95 F3* var. Aghri Dagh, Agri Dagi, Koh I Noh, Masis, Eng. Great Ararat, Mount Ararat. mountain E Turkey
Büyükmenderes Nehri *94 A4* river SW Turkey
Buzău *86 C4* Buzău, SE Romania
Büzmeýin see Abadan
Buzuluk *89 D6* Orenburgskaya Oblast', W Russia
Byahoml' *85 D5* Rus. Begoml'. Vitsyebskaya Voblasts', N Belarus
Byalynichy *85 D6* Rus. Belynichi. Mahilyowskaya Voblasts', E Belarus
Byan Tumen see Choybalsan
Byarezina *85 D6* prev. Byerezino, Rus. Berezina. river C Belarus
Bydgoszcz *76 C3* Ger. Bromberg. Kujawski-pomorskie, C Poland
Byelaruskaya Hrada *85 B6* Rus. Belorusskaya Gryada. ridge N Belarus
Byerezino see Byarezina
Byron Island see Nikunau
Bystrovka see Kemin
Bytča *77 C5* Žilinský Kraj, N Slovakia
Bytom *77 C5* Ger. Beuthen. Śląskie, S Poland
Bytów *76 C2* Ger. Bütow. Pomorskie, N Poland
Byuzmeyin see Abadan
Byval'ki *85 D8* Homyel'skaya Voblasts', SE Belarus
Byzantium see İstanbul

C

Caála *56 B2* var. Kaala, Robert Williams, Port. Vila Robert Williams. Huambo, C Angola
Caazapá *42 D3* Caazapá, S Paraguay
Caballo Reservoir *26 C3* reservoir New Mexico, SW USA
Cabanaquinta *70 D1* var. Cabañaquinta. Asturias, N Spain
Cabañaquinta see Cabanaquinta
Cabanatuan *117 E1* off. Cabanatuan City. Luzon, N Philippines
Cabanatuan City see Cabanatuan
Cabillonum see Chalon-sur-Saône
Cabimas *36 C1* Zulia, NW Venezuela
Cabinda *56 A1* var. Kabinda. Cabinda, NW Angola
Cabinda *56 A1* var. Kabinda. province NW Angola
Cabo Verde, Republic of see Cape Verde
Cahora Bassa, Albufeira de *56 D2* var. Lake Cabora Bassa. reservoir NW Mozambique
Cabora Bassa, Lake see Cahora Bassa, Albufeira de
Caborca *28 B1* Sonora, NW Mexico
Cabot Strait *17 G4* strait E Canada
Cabo Verde, Ilhas do see Cape Verde
Cabras, Ilha das *54 E2* island S Sao Tome and Principe, Africa, E Atlantic Ocean
Cabrera, Illa de *71 G3* island E Spain
Cáceres *70 C3* Ar. Qazris. Extremadura, W Spain
Cachimbo, Serra do *41 E2* mountain range C Brazil
Caconda *56 B2* Huíla, C Angola
Čadca *77 C5* Hung. Csaca. Žilinský Kraj, N Slovakia
Cadillac *18 C2* Michigan, N USA
Cadiz *117 E2* off. Cadiz City. Negros, C Philippines
Cádiz *70 C5* anc. Gades, Gadier, Gadir, Gadire. Andalucía, SW Spain
Cadiz City see Cadiz
Cádiz, Golfo de *70 B5* Eng. Gulf of Cadiz. gulf Portugal/Spain
Cadiz, Gulf of see Cádiz, Golfo de
Cadurcum see Cahors
Caen *68 B3* Calvados, N France
Caene/Caenepolis see Qinā
Caerdydd see Cardiff
Caer Glou see Gloucester
Caer Gybi see Holyhead
Caerleon see Chester
Caer Luel see Carlisle
Caesaraugusta see Zaragoza
Caesarea Mazaca see Kayseri
Caesarobriga see Talavera de la Reina
Caesarodunum see Tours
Caesaromagus see Beauvais
Caesena see Cesena
Cafayate *42 C2* Salta, N Argentina
Cagayan de Oro *117 E2* off. Cagayan de Oro City. Mindanao, S Philippines
Cagayan de Oro City see Cagayan de Oro
Cagliari *75 A6* anc. Caralis. Sardegna, Italy, C Mediterranean Sea
Caguas *33 F3* E Puerto Rico
Cahors *69 C5* anc. Cadurcum. Lot, S France
Cahul *86 D4* Rus. Kagul. S Moldova
Caicos Passage *32 D2* strait The Bahamas/Turks and Caicos Islands
Caiffa see Hefa
Cailungo *74 E1* N San Marino
Caiphas see Hefa
Cairns *126 D3* Queensland, NE Australia
Cairo *50 B2* var. El Qâhira, Ar. Al Qāhirah. country capital (Egypt) N Egypt
Cairo *18 B5* Illinois, N USA
Caisleán an Bharraigh see Castlebar
Cajamarca *38 B3* prev. Caxamarca. Cajamarca, NW Peru
Čakovec *78 B2* Ger. Csakathurn, Hung. Csáktornya; prev. Ger. Tschakathurn. Medimurje, N Croatia
Calabar *53 G5* Cross River, S Nigeria
Calabozo *36 D2* Guárico, C Venezuela
Calafat *86 B5* Dolj, SW Romania
Calafate see El Calafate
Calahorra *71 E2* La Rioja, N Spain
Calais *68 C2* Pas-de-Calais, N France
Calais *19 H2* Maine, NE USA
Calais, Pas de see Dover, Strait of
Calama *42 B2* Antofagasta, N Chile
Călăras see Călărași
Călărași *86 D3* var. Călăras, Rus. Kalarash. C Moldova
Călărași *86 C5* Călărași, SE Romania
Calatayud *71 E2* Aragón, NE Spain
Calbayog *117 E2* off. Calbayog City. Samar, C Philippines
Calbayog City see Calbayog
Calcutta see Kolkata
Caldas da Rainha *70 B3* Leiria, W Portugal
Caldera *42 B3* Atacama, N Chile
Caldwell *24 C3* Idaho, NW USA
Caledonia *30 C1* Corozal, N Belize
Caleta Olivia *43 B6* Santa Cruz, SE Argentina
Calgary *15 E5* Alberta, SW Canada
Cali *36 B3* Valle del Cauca, W Colombia
Calicut see Kozhikode
California *25 B7* off. State of California, also known as El Dorado, The Golden State. state W USA
California, Golfo de *28 B2* Eng. Gulf of California; prev. Sea of Cortez. gulf W Mexico
California, Gulf of see California, Golfo de
Călimănești *86 B4* Vâlcea, SW Romania
Calisia see Kalisz
Callabonna, Lake *127 B5* lake South Australia
Callao *38 C4* Callao, W Peru
Callatis see Mangalia
Callosa de Segura *71 F4* País Valenciano, E Spain
Calmar see Kalmar
Caloundra *127 E5* Queensland, E Australia
Caltanissetta *75 C7* Sicilia, Italy, C Mediterranean Sea
Caluula *50 E4* Bari, NE Somalia
Calvinia *56 C5* W South Africa
Camabatela *56 B1* Cuanza Norte, NW Angola
Camacupa *56 B2* var. General Machado, Port. Vila General Machado. Bié, C Angola
Camagüey *32 C2* prev. Puerto Príncipe. Camagüey, C Cuba
Camagüey, Archipiélago de *32 C2* island group C Cuba
Camana *39 E4* var. Camaná. Arequipa, SW Peru
Camargue *69 D6* physical region SE France
Ca Mau *115 D6* var. Quan Long. Minh Hai, S Vietnam
Cambay, Gulf of see Khambhāt, Gulf of
Camberia see Chambéry
Cambodia *115 D5* off. Kingdom of Cambodia, var. Democratic Kampuchea, Roat Kampuchea, Cam. Kampuchea; prev. People's Democratic Republic of Kampuchea. country SE Asia
Cambodia, Kingdom of see Cambodia
Cambrai *68 C2* Flem. Kambryk, prev. Cambray; anc. Cameracum. Nord, N France
Cambray see Cambrai
Cambrian Mountains *67 C6* mountain range C Wales, United Kingdom
Cambridge *32 A4* W Jamaica
Cambridge *128 D3* Waikato, North Island, New Zealand
Cambridge *67 E6* Lat. Cantabrigia. E England, United Kingdom
Cambridge *19 F4* Maryland, NE USA
Cambridge *18 D4* Ohio, NE USA
Cambridge Bay *15 F3* var. Ikaluktutiak. Victoria Island, Nunavut, NW Canada
Camden *20 B2* Arkansas, C USA
Camellia State see Alabama
Cameracum see Cambrai
Cameroon *54 A4* off. Republic of Cameroon, Fr. Cameroun. country W Africa
Cameroon, Republic of see Cameroon
Cameroun see Cameroon
Camocim *41 F2* Ceará, E Brazil
Camopi *37 H3* E French Guiana
Campamento *30 C2* Olancho, C Honduras
Campania *75 D5* Eng. Champagne. region S Italy
Campbell, Cape *129 D5* headland South Island, New Zealand
Campbell Island *120 D5* island S New Zealand
Campbell Plateau *120 D5* undersea plateau SW Pacific Ocean
Campbell River *14 D5* Vancouver Island, British Columbia, SW Canada
Campeche *29 G4* Campeche, SE Mexico
Campeche, Bahía de *29 F4* Eng. Bay of Campeche. bay E Mexico
Campeche, Bay of see Campeche, Bahía de
Cẩm Pha *114 E3* Quang Ninh, N Vietnam
Câmpina *86 C4* prev. Cîmpina. Prahova, SE Romania
Campina Grande *41 G2* Paraíba, E Brazil
Campinas *41 F4* São Paulo, S Brazil
Campobasso *75 D5* Molise, C Italy
Campo Criptana see Campo de Criptana
Campo de Criptana *71 E3* var. Campo Criptana. Castilla-La Mancha, C Spain
Campo dos Goytacazes *41 F4* var. Campos. Rio de Janeiro, SE Brazil
Campo Grande *41 E4* state capital Mato Grosso do Sul, SW Brazil
Campos see Campo dos Goytacazes
Câmpulung *86 B4* prev. Câmpulung-Mușcel, Cîmpulung. Argeș, S Romania
Câmpulung-Mușcel see Câmpulung
Campus Stellae see Santiago de Compostela
Cam Ranh *115 E6* Khanh Hoa, S Vietnam
Canada *12 B4* country N North America
Canada Basin *12 C2* undersea basin Arctic Ocean
Canadian River *27 E2* river SW USA
Çanakkale *94 A3* var. Dardanelli; prev. Chanak, Kale Sultanie. Çanakkale, W Turkey
Cananea *28 B1* Sonora, NW Mexico
Canarreos, Archipiélago de los *32 B2* island group W Cuba
Canary Islands *48 A2* Eng. Canary Islands. island group Spain, NE Atlantic Ocean
Canary Islands see Canarias, Islas
Cañas *30 D4* Guanacaste, NW Costa Rica
Canaveral, Cape *21 E4* headland Florida, SE USA
Canavieiras *41 G3* Bahia, E Brazil
Canberra *120 C4* country capital (Australia) Australian Capital Territory, SE Australia
Cancún *29 H3* Quintana Roo, SE Mexico
Candia see Irákleio
Canea see Chaniá
Cangzhou *106 D4* Hebei, E China
Caniapiscau *17 E2* river Québec, E Canada
Caniapiscau, Réservoir de *16 D3* reservoir Québec, C Canada
Canik Dağları *94 D2* mountain range N Turkey
Canillo *69 A7* Canillo, C Andorra Europe
Çankırı *94 C3* var. Chankiri; anc. Gangra, Germanicopolis. Çankın, N Turkey
Cannanore see Kannur
Cannes *69 D6* Alpes-Maritimes, SE France
Canoas *41 E5* Rio Grande do Sul, S Brazil
Canon City *22 C5* Colorado, C USA
Cantabria *70 D1* autonomous community N Spain
Cantábrica, Cordillera *70 C1* mountain range N Spain
Cantabrigia see Cambridge
Cantaura *37 E2* Anzoátegui, NE Venezuela
Canterbury *67 E7* hist. Cantwaraburh; anc. Durovernum, Lat. Cantuaria. SE England, United Kingdom
Canterbury Bight *129 C6* bight South Island, New Zealand
Canterbury Plains *129 C6* plain South Island, New Zealand
Cần Thơ *115 E6* Cần Thơ, S Vietnam
Canton *20 B2* Mississippi, S USA
Canton *18 D4* Ohio, N USA
Canton see Guangzhou
Canton Island see Kanton
Cantuaria/Cantwaraburh see Canterbury
Canyon *27 E2* Texas, SW USA
Cao Băng *114 D3* var. Caobang. Cao Băng, N Vietnam
Caobang see Cao Băng
Cap-Breton, Île du see Cape Breton Island
Cape Barren Island *127 C8* island Furneaux Group, Tasmania, SE Australia
Cape Basin *47 B7* undersea basin S Atlantic Ocean
Cape Breton Island *17 G4* Fr. Île du Cap-Breton. island Nova Scotia, SE Canada
Cape Charles *19 F5* Virginia, NE USA
Cape Coast *53 E5* prev. Cape Coast Castle. S Ghana
Cape Coast Castle see Cape Coast
Cape Girardeau *23 H5* Missouri, C USA
Capelle aan den IJssel *64 C4* Zuid-Holland, SW Netherlands
Cape Palmas see Harper
Cape Saint Jacques see Vung Tau
Cape Town *56 B5* var. Ekapa, Afr. Kaapstad, Kapstad. country capital (South Africa-legislative capital) Western Cape, SW South Africa
Cape Verde *52 A2* off. Republic of Cabo Verde, Port. Cabo Verde, Ilhas do Cabo Verde. country E Atlantic Ocean
Cape Verde Basin *44 C4* undersea basin E Atlantic Ocean
Cape Verde Plain *44 C4* abyssal plain E Atlantic Ocean
Cape York Peninsula *126 C2* peninsula Queensland, N Australia
Cap-Haïtien *32 D3* var. Le Cap. N Haiti
Capira *31 G5* Panamá, C Panama
Capitán Pablo Lagerenza *42 D1* var. Mayor Pablo Lagerenza. Chaco, N Paraguay
Capodistria see Koper
Capri *75 C5* island S Italy
Caprivi Concession see Caprivi Strip
Caprivi Strip *56 C3* Ger. Caprivizipfel; prev. Caprivi Concession. cultural region NE Namibia
Caprivizipfel see Caprivi Strip

Cap Saint-Jacques – Ch'ing Hai

Cap Saint-Jacques see Vung Tau
Caquetá, Río 36 C5 var. Rio Japurá, Yapurá. river Brazil/Colombia
Caquetá, Río see Japurá, Rio
CAR see Central African Republic
Caracal 86 B5 Olt, S Romania
Caracaraí 40 D1 Rondônia, W Brazil
Caracas 36 D1 country capital (Venezuela) Distrito Federal, N Venezuela
Caralis see Cagliari
Caratasca, Laguna de 31 E2 lagoon NE Honduras
Carballiño see O Carballiño
Carbón, Laguna del 43 B7 physical feature SE Argentina
Carbondale 18 B5 Illinois, N USA
Carbonia 75 A6 var. Carbonia Centro. Sardegna, Italy, C Mediterranean Sea
Carbonia Centro see Carbonia
Carcaso see Carcassonne
Carcassonne 69 C6 anc. Carcaso. Aude, S France
Cardamomes, Chaîne des see Krâvanh, Chuŏr Phnum
Cardamom Mountains see Krâvanh, Chuŏr Phnum
Cárdenas 32 B2 Matanzas, W Cuba
Cardiff 67 C7 Wel. Caerdydd. national capital S Wales, United Kingdom
Cardigan Bay 67 C6 bay W Wales, United Kingdom
Carei 86 B3 Ger. Gross-Karol, Karol, Hung. Nagykároly; prev. Careii-Mari. Satu Mare, NW Romania
Careii-Mari see Carei
Carey, Lake 125 B6 lake Western Australia
Cariaco 37 E1 Sucre, NE Venezuela
Caribbean Sea 32 C4 sea W Atlantic Ocean
Caribrod see Dimitrovgrad
Carlisle 66 C4 anc. Caer Luel, Luguvallium, Luguvallum. NW England, United Kingdom
Carlow 67 B6 Ir. Ceatharlach. SE Ireland
Carlsbad 26 D3 New Mexico, SW USA
Carlsbad see Karlovy Vary
Carlsberg Ridge 118 B4 undersea ridge S Arabian Sea
Carlsruhe see Karlsruhe
Carmana/Carmania see Kermān
Carmarthen 67 C6 SW Wales, United Kingdom
Carmaux 69 C6 Tarn, S France
Carmel 18 C4 Indiana, N USA
Carmelita 30 B1 Petén, N Guatemala
Carmen 29 G4 var. Ciudad del Carmen. Campeche, SE Mexico
Carmona 70 C4 Andalucía, S Spain
Carmona see Uíge
Carnaro see Kvarner
Carnarvon 125 A5 Western Australia
Carnegie, Lake 125 B5 salt lake Western Australia
Car Nicobar 111 F3 island Nicobar Islands, India, NE Indian Ocean
Caroço, Ilha 54 E1 island N Sao Tome and Principe, Africa, E Atlantic Ocean
Carolina 41 F2 Maranhão, E Brazil
Caroline Island see Millennium Island
Caroline Islands 122 B2 island group C Micronesia
Carolopois see Châlons-en-Champagne
Caroní, Río 37 E3 river E Venezuela
Caronium see A Coruña
Carora 36 C1 Lara, N Venezuela
Carpathian Mountains 59 E4 var. Carpathians, Cz./Pol. Karpaty, Ger. Karpaten. mountain range E Europe
Carpathians see Carpathian Mountains
Carpathos/Carpathus see Kárpathos
Carpații Meridionalii 86 B4 var. Alpi Transilvaniei, Carpații Sudici, Eng. South Carpathians, Transylvanian Alps, Ger. Südkarpaten, Transsylvanische Alpen, Hung. Déli-Kárpátok, Erdélyi-Havasok. mountain range C Romania
Carpații Sudici see Carpații Meridionalii
Carpentaria, Gulf of 126 B2 gulf N Australia
Carpi 74 C2 Emilia-Romagna, N Italy
Carrara 74 B3 Toscana, C Italy
Carson City 25 C5 state capital Nevada, W USA
Carson Sink 25 C5 salt flat Nevada, W USA
Carstensz, Puntjak see Jaya, Puncak
Cartagena 36 B1 var. Cartagena de los Indes. Bolívar, NW Colombia
Cartagena 71 F4 anc. Carthago Nova. Murcia, SE Spain
Cartagena de los Indes see Cartagena
Cartago 31 E4 Cartago, C Costa Rica
Carthage 23 F5 Missouri, C USA
Carthago Nova see Cartagena
Cartwright 17 F2 Newfoundland and Labrador, E Canada
Carúpano 37 E1 Sucre, NE Venezuela
Carusbur see Cherbourg
Caruthersville 23 H5 Missouri, C USA
Cary 21 F1 North Carolina, SE USA
Casablanca 48 C2 Ar. Dar-el-Beida. NW Morocco
Casa Grande 26 B2 Arizona, SW USA
Cascade Range 24 B3 mountain range Oregon/Washington, NW USA
Cascadia Basin 12 A4 undersea basin NE Pacific Ocean
Cascais 70 B4 Lisboa, C Portugal
Caserta 75 D5 Campania, S Italy
Casey 132 D4 Australian research station Antarctica
Čáslav 77 B5 Ger. Tschaslau. Střední Čechy, C Czech Republic (Czechia)
Casper 22 C3 Wyoming, C USA
Caspian Depression 89 B7 Kaz. Kaspiy Mangy Oypaty, Rus. Prikaspiyskaya Nizmennost'. depression Kazakhstan/Russia
Caspian Sea 92 A4 Az. Xäzär Dänizi, Kaz. Kaspiy Tengizi, Per. Baḩr-e Khazar, Daryā-ye Khazar, Rus. Kaspiyskoye More. inland sea Asia/Europe
Cassai see Kasai
Cassel see Kassel
Castamoni see Kastamonu
Casteggio 74 B2 Lombardia, N Italy
Castelló de la Plana see Castellón de la Plana
Castellón see Castellón de la Plana
Castellón de la Plana 71 F3 var. Castellón, Cat. Castelló de la Plana. País Valenciano, E Spain
Castelnaudary 69 C6 Aude, S France
Castelo Branco 70 C3 Castelo Branco, C Portugal
Castelsarrasin 69 B6 Tarn-et-Garonne, S France
Castelvetrano 75 C7 Sicilia, Italy, C Mediterranean Sea
Castilla-La Mancha 71 E3 autonomous community NE Spain
Castilla-León 70 C2 var. Castillia y Leon. autonomous community NW Spain
Castillia y Leon see Castilla-León
Castlebar 67 A5 Ir. Caisleán an Bharraigh. W Ireland
Castleford 67 D5 N England, United Kingdom
Castle Harbour 20 B5 inlet Bermuda, NW Atlantic Ocean
Castra Regina see Regensburg
Castricum 64 C3 Noord-Holland, W Netherlands
Castries 33 F1 country capital (Saint Lucia) N Saint Lucia
Castro 43 B6 Los Lagos, W Chile
Castrovillari 75 D6 Calabria, SW Italy
Castuera 70 D4 Extremadura, W Spain
Caswell Sound 129 A7 sound South Island, New Zealand
Catacamas 30 D2 Olancho, C Honduras
Catacaos 38 B3 Piura, NW Peru
Catalan Bay 71 H4 bay E Gibraltar, Mediterranean Sea
Cataluña 71 G2 N Spain
Catamarca see San Fernando del Valle de Catamarca
Catania 75 D7 Sicilia, Italy, C Mediterranean Sea
Catanzaro 75 D6 Calabria, SW Italy
Catarroja 71 F3 País Valenciano, E Spain
Cat Island 32 C1 island C The Bahamas
Catskill Mountains 19 F3 mountain range New York, NE USA
Cattaro see Kotor
Cauca, Río 36 B2 river N Colombia
Caucasia 36 B2 Antioquia, NW Colombia
Caucasus 59 G4 Rus. Kavkaz. mountain range Georgia/Russia
Caura, Río 37 E3 river C Venezuela
Cavaia see Kavajë
Cavalla 52 D5 var. Cavally, Cavally Fleuve. river Ivory Coast/Liberia
Cavally/Cavally Fleuve see Cavalla
Caviana de Fora, Ilha 41 E1 var. Ilha Caviana. island N Brazil
Caviana, Ilha see Caviana de Fora, Ilha
Cawnpore see Kānpur
Caxamarca see Cajamarca
Caxito 56 B1 Bengo, NW Angola
Cayenne 37 H3 dependent territory/arrondissement capital (French Guiana) NE French Guiana
Cayes 32 D3 var. Les Cayes. SW Haiti
Cayman Brac 32 B3 island E Cayman Islands
Cayman Islands 32 B3 UK dependent territory W West Indies
Cayo see San Ignacio
Cay Sal 32 B2 islet SW The Bahamas
Cazin 78 B3 Federacija Bosna I Hercegovina, NW Bosnia and Herzegovina
Cazorla 71 E4 Andalucía, S Spain
Ceadâr-Lunga see Ciadîr-Lunga
Ceará 41 F2 off. Estado do Ceará. state/region C Brazil
Ceará see Fortaleza
Ceara Abyssal Plain see Ceará Plain
Ceará, Estado do see Ceará
Ceará Plain 34 E3 var. Ceara Abyssal Plain. abyssal plain W Atlantic Ocean
Ceatharlach see Carlow
Cébaco, Isla 31 F5 island SW Panama
Cebu 117 E2 off. Cebu City. Cebu, C Philippines
Cebu City see Cebu
Čechy see Bohemia
Cecina 74 B3 Toscana, C Italy
Cedar City 22 A5 Utah, W USA
Cedar Falls 23 G3 Iowa, C USA
Cedar Lake 16 A2 lake Manitoba, C Canada
Cedar Rapids 23 G3 Iowa, C USA
Cedros, Isla 28 A2 island W Mexico
Ceduna 127 A6 South Australia
Cefalù 75 C7 anc. Cephaloedium. Sicilia, Italy, C Mediterranean Sea
Celebes see Sulawesi
Celebes Sea 117 E3 Ind. Laut Sulawesi. sea Indonesia/Philippines
Celje 73 E7 Ger. Cilli. C Slovenia
Celldömölk 77 C6 Vas, W Hungary
Celle 72 B3 var. Zelle. Niedersachsen, N Germany
Celovec see Klagenfurt
Celtic Sea 67 B7 Ir. An Mhuir Cheilteach. sea SW British Isles
Celtic Shelf 58 B3 continental shelf E Atlantic Ocean
Cenderawasih, Teluk 117 G4 var. Teluk Irian, Teluk Sarera. bay W Pacific Ocean
Cenon 69 B5 Gironde, SW France
Centennial State see Colorado
Centrafricaine, République see Central African Republic
Central African Republic 54 C4 var. République Centrafricaine, abbrev. CAR; prev. Ubangi-Shari, Oubangui-Chari, Territoire de l'Oubangui-Chari. country C Africa
Central, Cordillera 36 B3 mountain range W Colombia
Cordillera Central 33 E3 mountain range C Dominican Republic
Cordillera Central 31 F5 mountain range C Panama
Central, Cordillera 117 E1 mountain range Luzon, N Philippines
Central Group see Inner Islands
Centralia 24 B2 Washington, NW USA
Central Indian Ridge see Mid-Indian Ridge
Central Makran Range 112 A3 mountain range W Pakistan
Central Pacific Basin 120 D1 undersea basin C Pacific Ocean
Central, Planalto 41 F3 var. Brazilian Highlands. mountain range E Brazil
Central Provinces and Berar see Madhya Pradesh
Central Range 122 B3 mountain range NW Papua New Guinea
Central Russian Upland see Srednerusskaya Vozvyshennost'
Central Siberian Plateau 92 D3 var. Central Siberian Uplands, Eng. Central Siberian Plateau. mountain range N Russia
Central Siberian Plateau/Central Siberian Uplands see Srednesibirskoye Ploskogor'ye
Central, Sistema 70 D3 mountain range C Spain
Central Valley 25 B6 valley California, W USA
Centum Cellae see Civitavecchia
Ceos see Tziá
Cephaloedium see Cefalù
Ceram see Seram, Pulau
Ceram Sea see Laut Seram
Cerasus see Giresun
Cereté 36 B2 Córdoba, NW Colombia
Cergy-Pontoise see Pontoise
Cerignola 75 D5 Puglia, SE Italy
Çerkeş 94 C2 Çankin, N Turkey
Cernăuţi see Chernivtsi
Cernay 68 E4 Haut-Rhin, NE France
Cerro de Pasco 38 C3 Pasco, C Peru
Cervera 71 F2 Cataluña, NE Spain
Cervino, Monte see Matterhorn
Cesena 74 C3 anc. Caesena. Emilia-Romagna, N Italy
Cēsis 84 D3 Ger. Wenden. Cēsis, C Latvia
Česká Republika see Czech Republic (Czechia)
České Budějovice 77 B5 Ger. Budweis. Jihočeský Kraj, S Czech Republic (Czechia)
Český Krumlov 77 A5 var. Böhmisch-Krumau, Ger. Krummau. Jihočeský Kraj, S Czech Republic (Czechia)
Ceský Les see Bohemian Forest
Cetatea Damboviței see București
Cetinje 79 C5 It. Cettigne. S Montenegro
Cette see Sète
Cettigne see Cetinje
Ceuta 48 C2 enclave Spain, N Africa
Cévennes 69 C6 mountain range S France
Ceyhan 94 D4 Adana, S Turkey
Ceylanpınar 95 E4 Şanlıurfa, SE Turkey
Ceylon see Sri Lanka
Ceylon Plain 102 B4 abyssal plain N Indian Ocean
Ceyre to the Caribs see Marie-Galante
Chachapoyas 38 B2 Amazonas, NW Peru
Chachevichy 85 D6 Rus. Chechevichi. Mahilyowskaya Voblasts', E Belarus
Chaco see Gran Chaco
Chad 54 C3 off. Republic of Chad, Fr. Tchad. country C Africa
Chad, Lake 54 B3 Fr. Lac Tchad. lake C Africa
Chad, Republic of see Chad
Chadron 22 D3 Nebraska, C USA
Chadyr-Lunga see Ciadîr-Lunga
Chagai Hills 112 A2 var. Chāh Gay. mountain range Afghanistan/Pakistan
Chaghasarāy see Asadābād
Chagos-Laccadive Plateau 102 B4 undersea plateau N Indian Ocean
Chagos Trench 119 C5 trench N Indian Ocean
Chāh Gay see Chāgai Hills
Chaillu, Massif du 55 B6 mountain range C Gabon
Chain Ridge 118 B4 undersea ridge W Indian Ocean
Chajul 30 B2 Quiché, W Guatemala
Chakhānsūr 100 D5 Nīmrūz, SW Afghanistan
Chala 38 D4 Arequipa, SW Peru
Chalatenango 30 C3 Chalatenango, N El Salvador
Chalcidice see Chalkidikí
Chalcis see Chalkída
Chalki 83 E7 island Dodekánisa, Greece, Aegean Sea
Chalkída 83 C5 var. Halkida, prev. Khalkís; anc. Chalcis. Evvoia, E Greece
Chalkidikí 82 C4 var. Khalkidhikí; anc. Chalcidice. peninsula NE Greece
Challans 68 B4 Vendée, NW France
Challapata 39 F4 Oruro, SW Bolivia
Challenger Deep 130 B3 trench W Pacific Ocean
Challenger Fracture Zone 131 F4 tectonic feature SE Pacific Ocean
Châlons-en-Champagne 68 D3 prev. Châlons-sur-Marne, hist. Arcae Remorum; anc. Carolopois. Marne, NE France
Châlons-sur-Marne see Châlons-en-Champagne
Chalon-sur-Saône 68 D4 anc. Cabillonum. Saône-et-Loire, C France
Cha Mai see Thung Song
Chaman 112 B2 Baluchistān, SW Pakistan
Chambéry 69 D5 anc. Camberia. Savoie, E France
Champagne 68 D3 cultural region N France
Champagne see Campania
Champaign 18 B4 Illinois, N USA
Champasak 115 D5 Champasak, S Laos
Champlain, Lake 19 F2 lake Canada/USA
Champotón 29 G4 Campeche, SE Mexico
Chanak see Çanakkale
Chañaral 42 B3 Atacama, N Chile
Chan-chiang/Chanchiang see Zhanjiang
Chandeleur Islands 20 C3 island group Louisiana, S USA
Chandīgarh 112 D2 state capital Punjab, N India
Chandrapur 113 E5 Mahārāshtra, C India
Changan see Xi'an, Shaanxi, C China
Changane 57 E3 river S Mozambique
Changchun 106 D3 var. Ch'angch'un, Ch'ang-ch'un; prev. Hsinking. province capital Jilin, NE China
Ch'angch'un/Ch'ang-ch'un see Changchun
Chang Jiang 106 A5 Eng. Yangtze; var. Yangtze Kiang. river SW China
Changkiakow see Zhangjiakou
Chang, Ko 115 C6 island S Thailand
Changsha 106 C5 var. Ch'angsha, Ch'ang-sha. province capital Hunan, S China
Ch'angsha/Ch'ang-sha see Changsha
Changzhi 106 C4 Shanxi, C China
Chaniá 83 C7 var. Hania, Khaniá, Eng. Canea; anc. Cydonia. Kríti, Greece, E Mediterranean Sea
Chañi, Nevado de 42 B2 mountain NW Argentina
Chankiri see Çankırı
Channel Islands 67 C8 Fr. Iles Normandes. island group S English Channel
Channel Islands 25 B8 island group California, W USA
Channel-Port aux Basques 17 G4 Newfoundland and Labrador, SE Canada
Channel, The see English Channel
Channel Tunnel 68 C2 tunnel France/United Kingdom
Chantabun/Chantaburi see Chanthaburi
Chantada 70 C1 Galicia, NW Spain
Chanthaburi 115 C6 var. Chantabun, Chantaburi. Chantaburi, S Thailand
Chanute 23 F5 Kansas, C USA
Chaouèn see Chefchaouen
Chaoyang 106 D3 Liaoning, NE China
Chapala, Lago de 28 D4 lake C Mexico
Chapan, Gora 100 B3 mountain C Turkmenistan
Chapayevsk 89 C6 Samarskaya Oblast', W Russia
Chaplynka 87 F4 Khersons'ka Oblast', S Ukraine
Chapra see Chhapra
Charcot Seamounts 58 B3 seamount range E Atlantic Ocean
Chardzhev see Türkmenabat
Chardzhou/Chardzhui see Türkmenabat
Charente 69 B5 cultural region W France
Charente 69 B5 river W France
Chari 54 B3 var. Shari. river Central African Republic/Chad
Chārīkār 101 E4 Parvān, NE Afghanistan
Charity 37 F2 NW Guyana
Chärjew see Türkmenabat
Charkhlik/Charkhliq see Ruoqiang
Charleroi 65 C7 Hainaut, S Belgium
Charlesbourg 17 E4 Québec, SE Canada
Charles de Gaulle 68 E1 (Paris) Seine-et-Marne, N France
Charles Island 16 D1 island Nunavut, NE Canada
Charles Island see Santa María, Isla
Charleston 21 F2 South Carolina, SE USA
Charleston 18 D5 state capital West Virginia, NE USA
Charleville 127 D5 Queensland, E Australia
Charleville-Mézières 68 D3 Ardennes, N France
Charlie-Gibbs Fracture Zone 44 C2 tectonic feature N Atlantic Ocean
Charlotte 21 E1 North Carolina, SE USA
Charlotte Amalie 33 F3 prev. Saint Thomas. dependent territory capital (Virgin Islands (US)) Saint Thomas, N Virgin Islands (US)
Charlotte Harbor 21 E5 inlet Florida, SE USA
Charlottenhof see Aegviidu
Charlottesville 19 E5 Virginia, NE USA
Charlottetown 17 F4 province capital Prince Edward Island, Prince Edward Island, SE Canada
Charlotte Town see Roseau, Dominica
Charsk see Shar
Charters Towers 126 D3 Queensland, NE Australia
Chartres 68 C3 anc. Autricum, Civitas Carnutum. Eure-et-Loir, C France
Chashniki 85 D5 Vitsyebskaya Voblasts', N Belarus
Châteaubriant 68 B4 Loire-Atlantique, NW France
Châteaudun 68 C3 Eure-et-Loir, C France
Châteauroux 68 C4 prev. Indreville. Indre, C France
Château-Thierry 68 C3 Aisne, N France
Châtelet 65 C7 Hainaut, S Belgium
Châtelherault see Châtellerault
Châtellerault 68 B4 var. Châtelherault. Vienne, W France
Chatham Island see San Cristóbal, Isla
Chatham Island Rise see Chatham Rise
Chatham Islands 121 E5 island group New Zealand, SW Pacific Ocean
Chatham Rise 120 D5 var. Chatham Island Rise. undersea rise S Pacific Ocean
Chatkal Range 101 F2 Rus. Chatkal'skiy Khrebet. mountain range Kyrgyzstan/Uzbekistan
Chatkal'skiy Khrebet see Chatkal Range
Chāttagām see Chittagong
Chattahoochee River 20 D3 river SE USA
Chattanooga 20 D1 Tennessee, S USA
Chatyr-Tash 101 G2 Narynskaya Oblast', C Kyrgyzstan
Châu Đôc 115 D6 var. Chauphu, Chau Phu. An Giang, S Vietnam
Chauk 114 A3 Magway, W Myanmar (Burma)
Chaumont 68 D4 prev. Chaumont-en-Bassigny. Haute-Marne, N France
Chaumont-en-Bassigny see Chaumont
Chau Phu see Châu Đôc
Chausy see Chavusy
Chaves 70 C2 anc. Aquae Flaviae. Vila Real, N Portugal
Chávez, Isla see Santa Cruz, Isla
Chavusy 85 E6 Rus. Chausy. Mahilyowskaya Voblasts', E Belarus
Chaykovskiy 89 D5 Permskaya Oblast', NW Russia
Cheb 77 A5 Ger. Eger. Karlovarský Kraj, W Czech Republic (Czechia)
Cheboksary 89 C5 Chuvashskaya Respublika, W Russia
Cheboygan 18 C2 Michigan, N USA
Chechaouèn see Chefchaouen
Chech, Erg 52 D1 desert Algeria/Mali
Chechevichi see Chachevichy
Che-chiang see Zhejiang
Cheduba Island 114 A4 island W Myanmar (Burma)
Chefchaouen 48 C2 var. Chaouèn, Chechaouèn, Sp. Xauen. N Morocco
Chefoo see Yantai
Cheju-do see Jeju-do
Cheju Strait see Jeju Strait
Chekiang see Zhejiang
Cheleken see Hazar
Chelkar see Shalkar
Chełm 76 E4 Rus. Kholm. Lubelskie, SE Poland
Chełmno 76 C3 Ger. Culm, Kulm. Kujawski-pomorskie, C Poland
Chełmża 76 C3 Ger. Culmsee, Kulmsee. Kujawski-pomorskie, C Poland
Cheltenham 67 D6 C England, United Kingdom
Chelyabinsk 92 C3 Chelyabinskaya Oblast', C Russia
Chemnitz 72 D4 prev. Karl-Marx-Stadt. Sachsen, E Germany
Chemulpo see Incheon
Chenāb 112 C2 river India/Pakistan
Chengchiatun see Liaoyuan
Ch'eng-chou/Chengchow see Zhengzhou
Chengde 106 D3 var. Jehol. Hebei, E China
Chengdu 106 B5 var. Chengtu, Ch'eng-tu. province capital Sichuan, C China
Chenghsien see Zhengzhou
Chengtu/Ch'eng-tu see Chengdu
Chennai 110 D2 prev. Madras. state capital Tamil Nādu, S India
Chenstokhov see Częstochowa
Chen Xian/Chenxian/Chen Xiang see Chenzhou
Chenzhou 106 C6 var. Chenxian, Chen Xian, Chen Xiang. Hunan, S China
Chepelare 82 C3 Smolyan, S Bulgaria
Chepén 38 B3 La Libertad, C Peru
Cher 68 C4 river C France
Cherbourg 68 B3 anc. Carusbur. Manche, N France
Cherepovets 88 B4 Vologodskaya Oblast', NW Russia
Chergui, Chott ech 48 D2 salt lake NW Algeria
Cherikov see Cherykaw
Cherkassy see Cherkasy
Cherkasy 87 E2 Rus. Cherkassy. Cherkas'ka Oblast', C Ukraine
Cherkessk 89 B7 Karachayevo-Cherkesskaya Respublika, SW Russia
Chernigov see Chernihiv
Chernihiv 87 E1 Rus. Chernigov. Chernihivs'ka Oblast', NE Ukraine
Chernivtsi 86 C3 Ger. Czernowitz, Rom. Cernăuţi, Rus. Chernovtsy. Chernivets'ka Oblast', W Ukraine
Cherno More see Black Sea
Chernomorskoye see Chornomors'ke
Chernovtsy see Chernivtsi
Chernoye More see Black Sea
Chernyakhovsk 84 A4 Ger. Insterburg. Kaliningradskaya Oblast', W Russia
Cherry Hill 19 F4 New Jersey, NE USA
Cherski Range see Cherskogo, Khrebet
Cherskiy 93 G2 Respublika Sakha (Yakutiya), NE Russia
Cherskogo, Khrebet 93 F2 var. Cherski Range. mountain range NE Russia
Cherso see Cres
Cherven' see Chervyen'
Chervonograd see Chervonohrad
Chervonohrad 86 C2 Rus. Chervonograd. L'vivs'ka Oblast', NW Ukraine
Chervyen' 85 D6 Rus. Cherven'. Minskaya Voblasts', C Belarus
Cherykaw 85 E7 Rus. Cherikov. Mahilyowskaya Voblasts', E Belarus
Chesapeake Bay 19 F5 inlet NE USA
Chesha Bay see Chëshskaya Guba
Chëshskaya Guba 133 D5 var. Archangel Bay, Chesha Bay, Dvina Bay. bay NW Russia
Chester 67 C6 Wel. Caerleon, hist. Legaceaster, Lat. Deva, Devana Castra. C England, United Kingdom
Chetumal 29 H4 var. Payo Obispo. Quintana Roo, SE Mexico
Cheviot Hills 66 D4 hill range England/Scotland, United Kingdom
Cheyenne 22 D4 state capital Wyoming, C USA
Cheyenne River 22 D3 river South Dakota/Wyoming, N USA
Chezdi-Oșorheiu see Târgu Secuiesc
Chhapra 113 F3 prev. Chapra. Bihār, N India
Chhattīsgarh 113 E4 cultural region E India
Chiai see Chiayi
Chia-i see Chiayi
Chiang-hsi see Jiangxi
Chiang Mai 114 B4 var. Chiangmai, Chiengmai, Kiangmai. Chiang Mai, NW Thailand
Chiangmai see Chiang Mai
Chiang Rai 114 C3 var. Chianpai, Chienrai, Muang Chiang Rai. Chiang Rai, NW Thailand
Chiang-su see Jiangsu
Chianning/Chian-ning see Nanjing
Chianpai see Chiang Rai
Chianti 74 C3 cultural region C Italy
Chiapa see Chiapa de Corzo
Chiapa de Corzo 29 G5 var. Chiapa. Chiapas, SE Mexico
Chiayi 106 D6 var. Chiai, Chia-i, Kiayi, Jiayi, Jap. Kagi. C Taiwan
Chiba 108 B1 var. Tiba. Chiba, Honshū, S Japan
Chibougamau 16 D3 Québec, SE Canada
Chicago 18 B3 Illinois, N USA
Ch'i-ch'i-ha-erh see Qiqihar
Chickasha 27 G2 Oklahoma, C USA
Chiclayo 38 B3 Lambayeque, NW Peru
Chico 25 B5 California, W USA
Chico, Río 43 B7 river SE Argentina
Chico, Río 43 B6 river S Argentina
Chicoutimi 17 E4 Québec, SE Canada
Chiengmai see Chiang Mai
Chienrai see Chiang Rai
Chiesanuova 74 D2 SW San Marino
Chieti 74 D4 var. Teate. Abruzzo, C Italy
Chifeng 105 G2 var. Ulanhad. Nei Mongol Zizhiqu, N China
Chigirin see Chyhyryn
Chih-fu see Yantai
Chihli see Hebei
Chihli, Gulf of see Bo Hai
Chihuahua 28 C2 Chihuahua, NW Mexico
Childress 27 F2 Texas, SW USA
Chile 42 B3 off. Republic of Chile. country SW South America
Chile Basin 35 A5 undersea basin E Pacific Ocean
Chile Chico 43 B6 Aisén, W Chile
Chile, Republic of see Chile
Chile Rise 35 A7 undersea rise SE Pacific Ocean
Chilia-Nouă see Kiliya
Chililabombwe 56 D2 Copperbelt, C Zambia
Chi-lin see Jilin
Chillán 43 B5 Bío Bío, C Chile
Chillicothe 18 D4 Ohio, N USA
Chill Mhantáin, Sléibhte see Wicklow Mountains
Chiloé, Isla de 43 A6 var. Isla Grande de Chiloé. island W Chile
Chilpancingo 29 E5 var. Chilpancingo de los Bravos. Guerrero, S Mexico
Chilpancingo de los Bravos see Chilpancingo
Chilung see Keelung
Chimán 31 G5 Panamá, E Panama
Chimbay see Chimboy
Chimborazo 38 A1 volcano C Ecuador
Chimbote 38 C3 Ancash, W Peru
Chimboy 100 D1 Rus. Chimbay. Qoraqalpog'iston Respublikasi, NW Uzbekistan
Chimkent see Shymkent
Chimoio 57 E3 Manica, C Mozambique
China 102 C2 off. People's Republic of China, Chin. Chung-hua Jen-min Kung-ho-kuo, Zhonghua Renmin Gongheguo; prev. Chinese Empire. country E Asia
Chi-nan/Chinan see Jinan
Chinandega 30 C3 Chinandega, NW Nicaragua
China, People's Republic of see China
China, Republic of see Taiwan
Chincha Alta 38 D4 Ica, SW Peru
Chin-chiang see Quanzhou
Chin-chou/Chinchow see Jinzhou
Chindwin see Chindwinn
Chindwinn 114 B2 var. Chindwin. river N Myanmar (Burma)
Chinese Empire see China
Ch'ing Hai see Qinghai Hu, China

Chinghai – Cranbrook

Chinghai see Qinghai
Chingola 56 D2 Copperbelt, C Zambia
Ching-Tao/Ch'ing-tao see Qingdao
Chinguețți 52 C2 var. Chinguetti. Adrar, C Mauritania
Chin Hills 114 A3 mountain range W Myanmar (Burma)
Chinhsien see Jinzhou
Chinnereth see Tiberias, Lake
Chinook Trough 91 H4 trough N Pacific Ocean
Chioggia 74 C2 anc. Fossa Claudia. Veneto, NE Italy
Chíos 83 D5 var. Hios, Khíos, It. Scio, Turk. Sakiz-Adasi. Chíos, E Greece
Chíos 83 D5 var. Khíos. island E Greece
Chipata 56 D2 prev. Fort Jameson. Eastern, E Zambia
Chiquián 38 C3 Ancash, W Peru
Chiquimula 30 B2 Chiquimula, SE Guatemala
Chīrāla 110 D1 Andhra Pradesh, E India
Chirchik see Chirchiq
Chirchiq 101 E2 Rus. Chirchik. Toshkent Viloyati, E Uzbekistan
Chiriqui Gulf 31 E5 Eng. Chiriqui Gulf. gulf SW Panama
Chiriqui Gulf see Chiriquí, Golfo de
Chiriquí, Laguna de 31 E5 lagoon NW Panama
Chiriquí, Volcán de see Barú, Volcán
Chirripó, Cerro see Chirripó Grande, Cerro
Chirripó Grande, Cerro 30 D4 var. Cerro Chirripó. mountain SE Costa Rica
Chisec 30 B2 Alta Verapaz, C Guatemala
Chisholm 23 F1 Minnesota, N USA
Chisimaio/Chisimayu see Kismaayo
Chişinău 86 D4 Rus. Kishinev. country capital (Moldova) C Moldova
Chita 93 F4 Chitinskaya Oblast', S Russia
Chitangwiza see Chitungwiza
Chitato 56 C1 Lunda Norte, NE Angola
Chitina 14 D3 Alaska, USA
Chitose 108 D2 var. Titose. Hokkaidō, NE Japan
Chitré 31 F5 Herrera, S Panama
Chittagong 113 G4 Ben. Chāttagām. Chittagong, SE Bangladesh
Chitungwiza 56 D3 prev. Chitangwiza. Mashonaland East, NE Zimbabwe
Chkalov see Orenburg
Chlef 48 D2 var. Ech Cheliff, Ech Chleff; prev. Al-Asnam, El Asnam, Orléansville. NW Algeria
Chocolate Mountains 25 D8 mountain range California, W USA
Chodorów see Khodoriv
Chodzież 76 C3 Wielkopolskie, C Poland
Choele Choel 43 C5 Río Negro, C Argentina
Choiseul 122 C3 var. Lauru. island NW Solomon Islands
Chojnice 76 C2 Ger. Konitz. Pomorskie, N Poland
Ch'ok'ē 50 C4 var. Choke Mountains. mountain range NW Ethiopia
Choke Mountains see Ch'ok'ē
Cholet 68 B4 Maine-et-Loire, NW France
Choluteca 30 C3 Choluteca, S Honduras
Choluteca, Río 30 C3 river SW Honduras
Choma 56 D2 Southern, S Zambia
Chomutov 76 A4 Ger. Komotau. Ústecký Kraj, NW Czech Republic (Czechia)
Chona 91 E2 river C Russia
Chon Buri 115 C5 prev. Bang Pla Soi. Chon Buri, S Thailand
Chone 38 A1 Manabí, W Ecuador
Ch'ŏngjin 107 E3 NE North Korea
Chongqing 107 B5 var. Ch'ung-ching, Ch'ung-ch'ing, Chungking, Pahsien, Tchongking, Yuzhou. Chongqing, C China
Chongqing 107 B5 province C China
Chonnacht see Connaught
Chonos, Archipiélago de los 43 A6 island group S Chile
Chóra 83 D7 Kykládes, Greece, Aegean Sea
Chóra Sfakíon 83 C8 var. Sfákia. Kríti, Greece, E Mediterranean Sea
Chorne More see Black Sea
Chornomors'k 87 E4 Rus. Illichivs'k. Odes'ka Oblast', SW Ukraine
Chornomors'ke 87 E4 Rus. Chernomorskoye. Respublika Krym, S Ukraine
Chorokh/Chorokhi see Çoruh Nehri
Chortkiv 86 C2 Rus. Chortkov. Ternopil's'ka Oblast', W Ukraine
Chortkov see Chortkiv
Chorzów 77 C5 Ger. Königshütte; prev. Królewska Huta. Śląskie, S Poland
Chośebuz see Cottbus
Chōsen-kaikyō see Korea Strait
Chōshi 109 D5 var. Tyôsi. Chiba, Honshū, S Japan
Chosŏn-minjujuŭi-inmin-kanghwaguk see North Korea
Choszczno 76 B3 Ger. Arnswalde. Zachodnio-pomorskie, NW Poland
Chota Nagpur 113 E4 plateau N India
Choûm 52 C2 Adrar, C Mauritania
Choybalsan 105 F2 prev. Byan Tumen. Dornod, E Mongolia
Christchurch 129 C6 Canterbury, South Island, New Zealand
Christiana 32 B5 C Jamaica
Christiania see Oslo
Christiansand see Kristiansand
Christianshåb see Qasigiannguit
Christiansund see Kristiansund
Christmas Island 119 D5 Australian external territory E Indian Ocean
Christmas Island see Kiritimati
Christmas Ridge 121 E1 undersea ridge C Pacific Ocean
Chuan see Sichuan
Ch'uan-chou see Quanzhou
Chubek see Moskva
Chubut, Río 43 B6 river SE Argentina
Ch'u-chiang see Shaoguan
Chudskoye Ozero see Peipus, Lake
Chugoku-sanchi 109 B6 mountain range Honshū, SW Japan
Chuí see Chuy
Chukai see Cukai
Chukchi Plain 133 B2 abyssal plain Arctic Ocean
Chukchi Plateau 12 C2 undersea plateau Arctic Ocean
Chukchi Sea 12 B2 Rus. Chukotskoye More. sea Arctic Ocean
Chukotskoye More see Chukchi Sea
Chula Vista 25 C8 California, W USA
Chulucanas 38 B2 Piura, NW Peru
Chulym 92 D4 river C Russia
Chumphon 115 C6 var. Jumporn. Chumphon, SW Thailand
Chuncheon 107 E4 prev. Ch'unch'ŏn, Jap. Shunsen. N South Korea
Ch'unch'ŏn see Chuncheon
Ch'ung-ch'ing/Ch'ung-ching see Chongqing
Chung-hua Jen-min Kung-ho-kuo see China
Chungking see Chongqing
Chunya 93 E3 river C Russia
Chuquicamata 42 B2 Antofagasta, N Chile
Chuquisaca see Sucre
Chur 73 B7 Fr. Coire, It. Coira, Rmsch. Cuera, Quera; anc. Curia Rhaetorum. Graubünden, E Switzerland
Churchill 15 G4 Manitoba, C Canada
Churchill 16 B2 river Manitoba/Saskatchewan, C Canada
Churchill 17 F2 river Newfoundland and Labrador, E Canada
Chuska Mountains 26 C1 mountain range Arizona/New Mexico, SW USA
Chusovoy 89 D5 Permskaya Oblast', NW Russia
Chust see Khust
Chuuk Islands 122 B2 var. Hogoley Islands; prev. Truk Islands. island group Caroline Islands, C Micronesia
Chuy 42 E4 var. Chuí. Rocha, E Uruguay
Chyhyryn 87 E2 Rus. Chigirin. Cherkas'ka Oblast', N Ukraine
Chystyakove 87 H3 Rus. Torez. Donets'ka Oblast', SE Ukraine
Ciadir-Lunga 86 D4 var. Ceadâr-Lunga, Rus. Chadyr-Lunga. S Moldova
Cide 94 C2 Kastamonu, N Turkey
Ciechanów 76 D3 prev. Zichenau. Mazowieckie, C Poland
Ciego de Ávila 32 C2 Ciego de Ávila, C Cuba
Ciénaga 36 B1 Magdalena, N Colombia
Cienfuegos 32 B2 Cienfuegos, C Cuba
Cieza 71 E4 Murcia, SE Spain
Cihanbeyli 94 C3 Konya, C Turkey
Cikobia 123 E4 prev. Thikombia. island N Fiji
Cilacap 116 C5 prev. Tjilatjap. Jawa, C Indonesia
Cill Airne see Killarney
Cill Chainnigh see Kilkenny
Cilli see Celje
Cill Mhantáin see Wicklow
Cimpina see Câmpina
Cimpulung see Câmpulung
Cina Selatan, Laut see South China Sea
Cincinnati 18 C4 Ohio, N USA
Ciney 65 C7 Namur, SE Belgium
Cinto, Monte 69 E7 mountain Corse, France, C Mediterranean Sea
Cintra see Sintra
Cipolletti 43 B5 Río Negro, C Argentina
Cirebon 116 C4 prev. Tjirebon. Jawa, S Indonesia
Cirkvenica see Crikvenica
Cirò Marina 75 E6 Calabria, S Italy
Cirquenizza see Crikvenica
Cisnădie 86 B4 Ger. Heltau, Hung. Nagydisznód. Sibiu, SW Romania
Citharista see la Ciotat
Citlaltépetl see Orizaba, Volcán Pico de
Citrus Heights 25 B5 California, W USA
Ciudad Acuña see Villa Acuña
Ciudad Bolívar 37 E2 prev. Angostura. Bolívar, E Venezuela
Ciudad Camargo 28 D2 Chihuahua, N Mexico
Ciudad Cortés see Cortés
Ciudad Darío 30 D3 var. Dario. Matagalpa, W Nicaragua
Ciudad de Dolores Hidalgo see Dolores Hidalgo
Ciudad de Guatemala 30 B2 Eng. Guatemala City; prev. Santiago de los Caballeros. country capital (Guatemala) Guatemala, C Guatemala
Ciudad del Carmen see Carmen
Ciudad del Este 42 E2 prev. Ciudad Presidente Stroessner, Presidente Stroessner, Puerto Presidente Stroessner. Alto Paraná, SE Paraguay
Ciudad Delicias see Delicias
Ciudad de México see México
Ciudad de Panama see Panamá
Ciudad Guayana 37 E2 prev. San Tomé de Guayana, Santo Tomé de Guayana. Bolívar, NE Venezuela
Ciudad Guzmán 28 D4 Jalisco, SW Mexico
Ciudad Hidalgo 29 G5 Chiapas, SE Mexico
Ciudad Juárez 28 C1 Chihuahua, N Mexico
Ciudad Lerdo 28 D3 Durango, C Mexico
Ciudad Madero 29 E3 var. Villa Cecilia. Tamaulipas, C Mexico
Ciudad Mante 29 E3 Tamaulipas, C Mexico
Ciudad Miguel Alemán 29 E2 Tamaulipas, C Mexico
Ciudad Obregón 28 B2 Sonora, NW Mexico
Ciudad Ojeda 36 C1 Zulia, NW Venezuela
Ciudad Porfirio Díaz see Piedras Negras
Ciudad Presidente Stroessner see Ciudad del Este
Ciudad Quesada see Quesada
Ciudad Real 70 D3 Castilla-La Mancha, C Spain
Ciudad-Rodrigo 70 C3 Castilla-León, N Spain
Ciudad Trujillo see Santo Domingo
Ciudad Valles 29 E3 San Luis Potosí, C Mexico
Ciudad Victoria 29 E3 Tamaulipas, C Mexico
Ciutadella 71 H3 var. Ciutadella de Menorca. Menorca, Spain, W Mediterranean Sea
Ciutadella Ciutadella de Menorca see Ciutadella
Civitanova Marche 74 D3 Marche, C Italy
Civitas Altae Ripae see Brzeg
Civitas Carnutum see Chartres
Civitas Eburovicum see Évreux
Civitavecchia 74 C4 anc. Centum Cellae, Trajani Portus. Lazio, C Italy
Claremore 27 G1 Oklahoma, C USA
Clarence 129 C5 Canterbury, South Island, New Zealand
Clarence 129 C5 river South Island, New Zealand
Clarence Town 32 D2 Long Island, C The Bahamas
Clarinda 23 F4 Iowa, C USA
Clarion Fracture Zone 131 E2 tectonic feature NE Pacific Ocean
Clarión, Isla 28 A5 island W Mexico
Clark Fork 22 A1 river Idaho/Montana, NW USA
Clark Hill Lake 21 E2 var. J.Storm Thurmond Reservoir. reservoir Georgia/South Carolina, SE USA
Clarksburg 18 D4 West Virginia, NE USA
Clarksdale 20 B2 Mississippi, S USA
Clarksville 20 C1 Tennessee, S USA
Clausentum see Southampton
Clayton 27 E1 New Mexico, SW USA
Clearwater 21 E4 Florida, SE USA
Clearwater Mountains 24 D2 mountain range Idaho, NW USA
Cleburne 27 G3 Texas, SW USA
Clermont 126 D4 Queensland, E Australia
Clermont-Ferrand 69 C5 Puy-de-Dôme, C France
Cleveland 18 D3 Ohio, N USA
Cleveland 20 D1 Tennessee, S USA
Clifton 26 C2 Arizona, SW USA
Clinton 20 B2 Mississippi, S USA
Clinton 27 F1 Oklahoma, C USA
Clipperton Fracture Zone 131 E3 tectonic feature E Pacific Ocean
Clipperton Island 13 A7 French dependency of French Polynesia E Pacific Ocean
Cloncurry 126 B3 Queensland, C Australia
Clonmel 67 B6 Ir. Cluain Meala. S Ireland
Cloppenburg 72 B3 Niedersachsen, NW Germany
Cloquet 23 G2 Minnesota, N USA
Cloud Peak 22 C3 mountain Wyoming, C USA
Clovis 27 E2 New Mexico, SW USA
Cluain Meala see Clonmel
Cluj see Cluj-Napoca
Cluj-Napoca 86 B3 Ger. Klausenburg, Hung. Kolozsvár; prev. Cluj. Cluj, NW Romania
Clutha 129 B7 river South Island, New Zealand
Clyde 66 C4 river W Scotland, United Kingdom
Coari 40 D2 Amazonas, N Brazil
Coast Mountains 14 D4 Fr. Chaîne Côtière. mountain range Canada/USA
Coast Ranges 24 A4 mountain range W USA
Coats Island 15 G3 island Nunavut, NE Canada
Coats Land 132 B2 physical region Antarctica
Coatzacoalcos 29 G4 var. Quetzalcoalco; prev. Puerto México. Veracruz-Llave, E Mexico
Cobán 30 B2 Alta Verapaz, C Guatemala
Cobar 127 C6 New South Wales, SE Australia
Cobija 39 E3 Pando, NW Bolivia
Coblence/Coblenz see Koblenz
Coburg 73 C5 Bayern, SE Germany
Coca see Puerto Francisco de Orellana
Cocanada see Kākināda
Cochabamba 39 F4 hist. Oropeza. Cochabamba, C Bolivia
Cochin see Kochi
Cochinos, Bahía de 32 B2 Eng. Bay of Pigs. bay SE Cuba
Cochrane 16 C4 Ontario, S Canada
Cochrane 43 B7 Aisén, S Chile
Cocibolca see Nicaragua, Lago de
Cockade State see Maryland
Cockburn Town 33 E2 San Salvador, E The Bahamas
Cockpit Country, The 32 A4 physical region W Jamaica
Cocobeach 55 A5 Estuaire, NW Gabon
Coconino Plateau 26 B1 plain Arizona, SW USA
Coco, Río 31 E2 var. Río Wanki, Segoviao Wangki. river Honduras/Nicaragua
Cocos Basin 102 C5 undersea basin E Indian Ocean
Cocos Island Ridge see Cocos Ridge
Cocos Islands 119 D5 island group E Indian Ocean
Cocos Ridge 13 C8 var. Cocos Island Ridge. undersea ridge E Pacific Ocean
Cod, Cape 19 G3 headland Massachusetts, NE USA
Codfish Island 129 A8 island SW New Zealand
Codlea 86 C4 Ger. Zeiden, Hung. Feketehalom. Braşov, C Romania
Cody 22 C2 Wyoming, C USA
Coeur d'Alene 24 C2 Idaho, NW USA
Coevorden 64 E2 Drenthe, NE Netherlands
Coffs Harbour 127 E6 New South Wales, SE Australia
Cognac 69 B5 anc. Compniacum. Charente, W France
Cohalm see Rupea
Coiba, Isla de 31 E5 island SW Panama
Coihaique 43 B6 var. Coyhaique. Aisén, S Chile
Coimbatore 110 C3 Tamil Nādu, S India
Coimbra 70 B3 anc. Conimbria, Conimbriga. Coimbra, W Portugal
Coín 70 D5 Andalucía, S Spain
Coira/Coire see Chur
Coirib, Loch see Corrib, Lough
Colby 23 E4 Kansas, C USA
Colchester 67 E6 Connecticut, NE USA
Coleman 27 F3 Texas, SW USA
Coleraine 66 B4 Ir. Cúil Raithin. N Northern Ireland, United Kingdom
Colesberg 56 C5 Northern Cape, C South Africa
Colima 28 D4 Colima, S Mexico
Coll 66 B3 island W Scotland, United Kingdom
College Station 27 G3 Texas, SW USA
Collie 125 A7 Western Australia
Collipo see Leiria
Colmar 68 E4 Ger. Kolmar. Haut-Rhin, NE France
Cöln see Köln
Cologne see Köln
Colomb-Béchar see Béchar
Colombia 36 B3 off. Republic of Colombia. country N South America
Colombian Basin 34 A1 undersea basin SW Caribbean Sea
Colombia, Republic of see Colombia
Colombie-Britannique see British Columbia
Colombo 110 C4 country capital (Sri Lanka) Western Province, W Sri Lanka
Colón 31 G4 prev. Aspinwall. Colón, C Panama
Colón, Archipiélago de see Galápagos Islands
Colón Ridge 13 B8 undersea ridge E Pacific Ocean
Colorado 22 C4 off. State of Colorado, also known as Centennial State, Silver State. state C USA
Colorado City 27 F3 Texas, SW USA
Colorado Plateau 26 B1 plateau W USA
Colorado, Río 43 C5 river E Argentina
Colorado, Río see Colorado River
Colorado River 13 B5 var. Río Colorado. river Mexico/USA
Colorado River 27 G4 river Texas, SW USA
Colorado Springs 22 D5 Colorado, C USA
Columbia 19 E4 Maryland, NE USA
Columbia 23 G4 Missouri, C USA
Columbia 21 E2 state capital South Carolina, SE USA
Columbia 20 C1 Tennessee, S USA
Columbia River 24 B3 river Canada/USA
Columbia Plateau 24 C3 plateau Idaho/Oregon, NW USA
Columbus 20 D2 Georgia, SE USA
Columbus 18 C4 Indiana, N USA
Columbus 20 C2 Mississippi, S USA
Columbus 23 F4 Nebraska, C USA
Columbus 18 D4 state capital Ohio, N USA
Colville Channel 128 D2 channel North Island, New Zealand
Colville River 14 D2 river Alaska, USA
Comacchio 74 C3 var. Commachio; anc. Comactium. Emilia-Romagna, N Italy
Comactium see Comacchio
Comalcalco 29 G4 Tabasco, SE Mexico
Coma Pedrosa, Pic de 69 A7 mountain NW Andorra
Comarapa 39 F4 Santa Cruz, C Bolivia
Comayagua 30 C2 Comayagua, W Honduras
Comer See see Como, Lago di
Comilla 113 G4 Ben. Kumillā. Chittagong, E Bangladesh
Comino see Kemmuna
Comitán 29 G5 var. Comitán de Domínguez. Chiapas, SE Mexico
Comitán de Domínguez see Comitán
Commachio see Comacchio
Commissioner's Point 20 A5 headland W Bermuda
Communism Peak 101 F3 prev. Qullai Kommunizm. mountain E Tajikistan
Como 74 B2 anc. Comum. Lombardia, N Italy
Comodoro Rivadavia 43 B6 Chubut, SE Argentina
Como, Lake 74 B2 var. Lario, Eng. Lake Como, Ger. Comer See. lake N Italy
Como, Lake see Como, Lago di
Comores, République Fédérale Islamique des see Comoros
Comoros 57 F2 off. Federal Islamic Republic of the Comoros, Fr. République Fédérale Islamique des Comores. country W Indian Ocean
Comoros, Federal Islamic Republic of the see Comoros
Compiègne 68 C3 Oise, N France
Complutum see Alcalá de Henares
Compniacum see Cognac
Compostella see Santiago de Compostela
Comrat 86 D4 Rus. Komrat. S Moldova
Comum see Como
Conakry 52 C4 country capital (Guinea) SW Guinea
Conca see Cuenca
Concarneau 68 A3 Finistère, NW France
Concepción 39 G3 Santa Cruz, E Bolivia
Concepción 43 B5 Bío Bío, C Chile
Concepción 42 D2 var. Villa Concepción. Concepción, C Paraguay
Concepción see La Concepción
Concepción de la Vega see La Vega
Conchos, Río 26 D4 river NW Mexico
Conchos, Río 28 D2 river C Mexico
Concord 19 G3 state capital New Hampshire, NE USA
Concordia 42 D4 Entre Ríos, E Argentina
Concordia 23 E4 Kansas, C USA
Concordia 132 C4 French/Italian research station Antarctica
Côn Đao see Côn Đao Son
Côn Đao Son 115 E7 var. Côn Đao, Con Son. island S Vietnam
Condate see Rennes, Ille-et-Vilaine, France
Condate see St-Claude, Jura, France
Condega 30 D3 Estelí, NW Nicaragua
Condivincum see Nantes
Confluentes see Koblenz
Công Hoa Xa Hôi Chu Nghia Viêt Nam see Vietnam
Congo 55 D5 off. Republic of the Congo, Fr. Moyen-Congo; prev. Middle Congo. country C Africa
Congo 55 C6 off. Democratic Republic of Congo; prev. Zaire, Belgian Congo, Congo (Kinshasa). country C Africa
Congo 55 C6 var. Kongo, Fr. Zaire. river C Africa
Congo Basin 55 C6 drainage basin W Dem. Rep. Congo
Congo/Congo (Kinshasa) see Congo (Democratic Republic of)
Coni see Cuneo
Conimbria/Conimbriga see Coimbra
Conjeeveram see Kānchīpuram
Connacht see Connaught
Connaught 67 A5 var. Connacht, Ir. Chonnacht, Cúige. province W Ireland
Connecticut 19 F3 off. State of Connecticut, also known as Blue Law State, Constitution State, Land of Steady Habits, Nutmeg State. state NE USA
Connecticut 19 G3 river Canada/USA
Conroe 27 G3 Texas, SW USA
Consentia see Cosenza
Consolación del Sur 32 A2 Pinar del Río, W Cuba
Con Son see Côn Đao Son
Constance see Konstanz
Constance, Lake 73 B7 Ger. Bodensee. lake C Europe
Constanța 86 D5 var. Küstendje, Eng. Constanza, Ger. Konstanza, Turk. Küstence. Constanța, SE Romania
Constantia see Coutances
Constantia see Konstanz
Constantine 49 E2 var. Qacentina, Ar. Qoussantîna. NE Algeria
Constantinople see İstanbul
Constantiola see Oltenița
Constanz see Konstanz
Constanza see Constanța
Constitution State see Connecticut
Coo see Kos
Coober Pedy 127 A5 South Australia
Cookeville 20 D1 Tennessee, S USA
Cook Islands 123 F4 self-governing territory in free association with New Zealand S Pacific Ocean
Cook, Mount see Aoraki
Cook Strait 129 D5 var. Raukawa. strait New Zealand
Cooktown 126 D2 Queensland, NE Australia
Coolgardie 125 B6 Western Australia
Cooma 127 D7 New South Wales, SE Australia
Coomassie see Kumasi
Coon Rapids 23 F2 Minnesota, N USA
Cooper Creek 126 C4 var. Barcoo, Cooper's Creek. seasonal river Queensland/South Australia, Australia
Cooper's Creek see Cooper Creek
Coos Bay 24 A3 Oregon, NW USA
Cootamundra 127 D6 New South Wales, SE Australia
Copacabana 39 E4 La Paz, W Bolivia
Copenhagen see København
Copiapó 42 B3 Atacama, N Chile
Copperas Cove 27 G3 Texas, SW USA
Coppermine see Kugluktuk
Copper State see Arizona
Coquilhatville see Mbandaka
Coquimbo 42 B3 Coquimbo, N Chile
Corabia 86 B5 Olt, S Romania
Coral Harbour 15 G3 var. Salliq. Southampton Island, Nunavut, NE Canada
Coral Sea 120 B3 sea SW Pacific Ocean
Coral Sea Islands 122 B4 Australian external territory SW Pacific Ocean
Corantijn Rivier see Courantyne River
Corcovado, Golfo 43 B6 gulf S Chile
Corcyra Nigra see Korčula
Cordele 20 D3 Georgia, SE USA
Córdoba 42 C3 Córdoba, C Argentina
Córdoba 29 F4 Veracruz-Llave, E Mexico
Córdoba 70 D4 var. Cordoba, Eng. Cordova; anc. Corduba. Andalucía, SW Spain
Cordova 14 C3 Alaska, USA
Cordova/Cordoba see Córdoba
Corduba see Córdoba
Corentyne River see Courantyne River
Corfu see Kérkyra
Coria 70 C3 Extremadura, W Spain
Corinth 20 C1 Mississippi, S USA
Corinth see Kórinthos
Corinth, Gulf of/Corinthiacus Sinus see Korinthiakós Kólpos
Corinthus see Kórinthos
Corinto 30 C3 Chinandega, NW Nicaragua
Cork 67 A6 Ir. Corcaigh. S Ireland
Çorlu 94 A2 Tekirdağ, NW Turkey
Corner Brook 17 G3 Newfoundland, Newfoundland and Labrador, E Canada
Cornhusker State see Nebraska
Corn Islands 31 E3 var. Corn Islands. island group SE Nicaragua
Corn Islands see Maíz, Islas del
Cornwallis Island 15 F2 Island Nunavut, N Canada
Coro 36 C1 prev. Santa Ana de Coro. Falcón, NW Venezuela
Corocoro 39 F4 La Paz, W Bolivia
Coromandel 128 D2 Waikato, North Island, New Zealand
Coromandel Coast 110 D2 coast E India
Coromandel Peninsula 128 D2 peninsula North Island, New Zealand
Coronado, Bahía de 30 D5 bay S Costa Rica
Coronel Dorrego 43 C5 Buenos Aires, E Argentina
Coronel Oviedo 42 D2 Caaguazú, SE Paraguay
Corozal 30 C1 Corozal, N Belize
Corpus Christi 27 G4 Texas, SW USA
Corrales 26 D2 New Mexico, SW USA
Corrib, Lough 67 A5 Ir. Loch Coirib. lake W Ireland
Corrientes 42 D3 Corrientes, NE Argentina
Corriza see Korçë
Corsica 69 E7 Eng. Corsica. island France, C Mediterranean Sea
Corsica see Corse
Corsicana 27 G3 Texas, SW USA
Cortegana 70 C4 Andalucía, S Spain
Cortés 31 E5 var. Ciudad Cortés. Puntarenas, SE Costa Rica
Cortez, Sea of see California, Golfo de
Cortina d'Ampezzo 74 C1 Veneto, NE Italy
Coruche 70 B3 Santarém, C Portugal
Çoruh Nehri 95 E3 Geor. Chorokh, Rus. Chorokhi. river Georgia/Turkey
Çorum 94 D3 var. Chorum. Çorum, N Turkey
Corunna see A Coruña
Corvallis 24 B3 Oregon, NW USA
Corvo 70 A5 var. Ilha do Corvo. island Azores, Portugal, NE Atlantic Ocean
Corvo, Ilha do see Corvo
Cos see Kos
Cosenza 75 D6 anc. Consentia. Calabria, SW Italy
Cosne-Cours-sur-Loire 68 C4 Nièvre, Bourgogne, C France Europe
Costa Mesa 24 D2 California, W USA North America
Costa Rica 31 E4 off. Republic of Costa Rica. country Central America
Costa Rica, Republic of see Costa Rica
Costermansville see Bukavu
Cotagaita 39 F5 Potosí, S Bolivia
Côte d'Ivoire see Ivory Coast
Côte d'Ivoire, République de la see Ivory Coast
Côte d'Or 68 D4 cultural region C France
Côte Française des Somalis see Djibouti
Côtière, Chaîne see Coast Mountains
Cotonou 53 F5 var. Kotonu. country capital (Benin - seat of government) S Benin
Cotrone see Crotone
Cotswold Hills 67 D6 var. Cotswolds. hill range S England, United Kingdom
Cotswolds see Cotswold Hills
Cottbus 72 D4 Lus. Chośebuz; prev. Kottbus. Brandenburg, E Germany
Cotton State, The see Alabama
Cotyora see Ordu
Couentrey see Coventry
Council Bluffs 23 F4 Iowa, C USA
Courantyne River 37 G4 var. Corantijn Rivier, Corentyne River. river Guyana/Suriname
Courland Lagoon 84 A4 Ger. Kurisches Haff, Rus. Kurskiy Zaliv. lagoon Lithuania/Russia
Courtrai see Kortrijk
Coutances 68 B3 anc. Constantia. Manche, N France
Couvin 65 C7 Namur, S Belgium
Coventry 67 D6 anc. Couentrey. C England, United Kingdom
Covilhã 70 C3 Castelo Branco, E Portugal
Cowan, Lake 125 B6 lake Western Australia
Coxen Hole see Roatán
Coxin Hole see Roatán
Coyhaique see Coihaique
Coyote State, The see South Dakota
Cozhê 104 C5 Xizang Zizhiqu, W China
Cozumel, Isla 29 H3 island SE Mexico
Cracovia/Cracow see Kraków
Cradock 56 C5 Eastern Cape, S South Africa
Craig 22 C4 Colorado, C USA
Craiova 86 B5 Dolj, SW Romania
Cranbrook 15 E5 British Columbia, SW Canada

Crane – Djéma

Crane see The Crane
Cranz see Zelenogradsk
Crawley 67 E7 SE England, United Kingdom
Cremona 74 B2 Lombardia, N Italy
Creole State see Louisiana
Cres 78 A3 It. Cherso; anc. Crexa. island W Croatia
Crescent City 24 A4 California, W USA
Crescent Group 106 C7 island group C Paracel Islands
Creston 23 F4 Iowa, C USA
Crestview 20 D3 Florida, SE USA
Crete see Kríti
Créteil 68 E2 Val-de-Marne, N France
Crete, Sea of/Creticum, Mare see Kritikó Pélagos
Creuse 68 B4 river C France
Crewe 67 D6 C England, United Kingdom
Crexa see Cres
Crikvenica 78 A3 It. Cirquenizza; prev. Cirkvenica, Crjkvenica. Primorje-Gorski Kotar, NW Croatia
Crimea see Kryms'kyy Pivostriv
Cristóbal 31 G4 Colón, C Panama
Cristóbal Colón, Pico 36 B1 mountain N Colombia
Cristur/Cristuru Săcuiesc see Cristuru Secuiesc
Cristuru Secuiesc 86 C4 prev. Cristur, Cristuru Săcuiesc, Sitaș Cristuru, Ger. Kreutz, Hung. Székelykeresztúr, Szitás-Keresztúr. Harghita, C Romania
Crjkvenica see Crikvenica
Crna Gora see Montenegro
Crna Reka 79 D6 river S Macedonia
Crni Drim see Black Drin
Croatia 78 B3 off. Republic of Croatia, Ger. Kroatien, SCr. Hrvatska. country SE Europe
Croatia, Republic of see Croatia
Crocodile see Limpopo
Croia see Krujë
Croker Island 124 E2 island Northern Territory, N Australia
Cromwell 129 B7 Otago, South Island, New Zealand
Crooked Island 32 D2 island SE The Bahamas
Crooked Island Passage 32 D2 channel SE The Bahamas
Crookston 23 F1 Minnesota, N USA
Crossen see Krosno Odrzańskie
Croton/Crotona see Crotone
Crotone 75 E6 var. Cotrone; anc. Croton, Crotona. Calabria, SW Italy
Croydon 67 A8 SE England, United Kingdom
Crozet Basin 119 B6 undersea basin S Indian Ocean
Crozet Islands 119 B7 island group French Southern and Antarctic Lands
Crozet Plateau 119 B7 var. Crozet Plateaus. undersea plateau SW Indian Ocean
Crozet Plateaus see Crozet Plateau
Crystal Brook 127 B6 South Australia
Csaca see Čadca
Csakathurn/Csáktornya see Čakovec
Csíkszereda see Miercurea-Ciuc
Csorna 77 C6 Győr-Moson-Sopron, NW Hungary
Csurgó 77 C7 Somogy, SW Hungary
Cuando 56 C2 var. Kwando. river S Africa
Cuango see Kwango
Cuanza 56 B1 var. Kwanza. river C Angola
Cuauhtémoc 28 C2 Chihuahua, N Mexico
Cuautla 29 E4 Morelos, S Mexico
Cuba 32 B2 off. Republic of Cuba. country W West Indies
Cubal 56 B2 Benguela, W Angola
Cubango 56 B2 var. Kuvango, Port. Vila Artur de Paiva, Vila da Ponte. Huíla, SW Angola
Cubango 56 B2 var. Kavango, Kavengo, Kubango, Okavango, Okavanggo. river S Africa
Cuba, Republic of see Cuba
Cúcuta 36 C2 var. San José de Cúcuta. Norte de Santander, N Colombia
Cuddapah 110 C2 Andhra Pradesh, S India
Cuenca 38 B2 Azuay, S Ecuador
Cuenca 71 E3 anc. Conca. Castilla-La Mancha, C Spain
Cuera see Chur
Cuernavaca 29 E4 Morelos, S Mexico
Cuiabá 41 E3 prev. Cuyabá. state capital Mato Grosso, SW Brazil
Cúige see Connaught
Cúige Laighean see Leinster
Cúige Mumhan see Munster
Cuijck 64 D4 Noord-Brabant, SE Netherlands
Cúil Raithin see Coleraine
Cuito 56 B2 var. Kwito. river SE Angola
Cukai 116 B3 var. Chukai, Kemaman. Terengganu, Peninsular Malaysia
Cularo see Grenoble
Culiacán 28 C3 var. Culiacán Rosales, Culiacán-Rosales. Sinaloa, C Mexico
Culiacán-Rosales/Culiacán Rosales see Culiacán
Cullera 71 F3 País Valenciano, E Spain
Cullman 20 C2 Alabama, S USA
Culm see Chełmno
Culmsee see Chełmża
Cumaná 37 E1 Sucre, NE Venezuela
Cumbal, Nevado de 36 A4 elevation S Colombia
Cumberland 19 E4 Maryland, NE USA
Cumberland Plateau 20 D1 plateau E USA
Cumberland Sound 15 H3 inlet Baffin Island, Nunavut, NE Canada
Cumpas 28 B2 Sonora, NW Mexico
Cuneo 74 A2 Fr. Coni. Piemonte, NW Italy
Cunnamulla 127 C5 Queensland, E Australia
Ćuprija 78 E4 Serbia, E Serbia
Curaçao 33 E5 Dutch self-governing territory S West Indies
Curia Rhaetorum see Chur
Curicó 42 B4 Maule, C Chile
Curieta see Krk
Curitiba 41 E4 prev. Curytiba. state capital Paraná, S Brazil
Curtbunar see Tervel
Curtea de Argeș 86 C4 var. Curtea-de-Arges. Argeș, S Romania
Curtea-de-Arges see Curtea de Argeș
Curtici 86 A4 Ger. Kurtitsch, Hung. Kürtös. Arad, W Romania
Curtis Island 126 E4 island Queensland, SE Australia
Curytiba see Curitiba
Curzola see Korčula
Cusco 39 E4 var. Cuzco. Cusco, C Peru
Cusset 69 C5 Allier, C France
Cutch, Gulf of see Kachchh, Gulf of
Cuttack 113 F4 Odisha, E India
Cuvier Plateau 119 E6 undersea plateau E Indian Ocean
Cuxhaven 72 B2 Niedersachsen, NW Germany
Cuyabá see Cuiabá
Cuyuni, Río see Cuyuni River
Cuyuni River 37 F3 var. Río Cuyuni. river Guyana/Venezuela
Cuzco see Cusco
Cyclades see Kykládes
Cydonia see Chaniá
Cymru see Wales
Cyprus 80 C4 off. Republic of Cyprus, Gk. Kypros, Turk. Kıbrıs, Kıbrıs Cumhuriyeti. country E Mediterranean Sea
Cyprus, Republic of see Cyprus
Cythnos see Kýthnos
Czech Republic 77 A5 var. Czechia Cz. Česká Republika. country C Europe
Czechia see Czech Republic
Czenstochau see Częstochowa
Czernowitz see Chernivtsi
Częstochowa 76 C4 Ger. Czenstochau, Tschenstochau, Rus. Chenstokhov. Śląskie, S Poland
Człuchów 76 C3 Ger. Schlochau. Pomorskie, NW Poland

D

Dabajuro 36 C1 Falcón, NW Venezuela
Dabeiba 36 B2 Antioquia, NW Colombia
Dąbrowa Tarnowska 77 D5 Małopolskie, S Poland
Dabryn' 85 C8 Rus. Dobryn'. Homyel'skaya Voblasts', SE Belarus
Dacca see Dhaka
Daegu 107 E4 off. Daegu Gwang-yeoksi, prev. Taegu, Jap. Taikyū. SE South Korea
Daegu Gwang-yeoksi see Daegu
Dagana 52 B3 N Senegal
Dagda 84 D4 Krāslava, SE Latvia
Dagden see Hiiumaa
Dagenham 67 B8 United Kingdom
Dağliq Quarabağ see Nagornyy-Karabakh
Dagö see Hiiumaa
Dagupan 117 E1 off. Dagupan City. Luzon, N Philippines
Dagupan City see Dagupan
Da Hinggan Ling 105 G1 Eng. Great Khingan Range. mountain range NE China
Dahm, Ramlat 99 B6 desert NW Yemen
Dahomey see Benin
Daihoku see Taibei
Daimiel 70 D3 Castilla-La Mancha, C Spain
Daimoniá 83 B7 Pelopónnisos, S Greece
Dainan see Tainan
Daingin, Bá an see Dingle Bay
Dairen see Dalian
Daejeon 107 E4 off. Daejeon Gwang-yeoksi, prev. Taejŏn, Jap. Taiden. C South Korea
Daejeon Gwang-yeoksi see Daejeon
Dakar 52 B3 country capital (Senegal) W Senegal
Dakhla see Ad Dakhla
Dakoro 53 G3 Maradi, S Niger
Đakovica see Gjakovë
Đakovo 78 C3 var. Djakovo, Hung. Diakovár. Osijek-Baranja, E Croatia
Dakshin see Deccan
Dalain Hob 104 D3 var. Ejin Qi. Nei Mongol Zizhiqu, N China
Dalai Nor see Hulun Nur
Dalaman 94 A4 Muğla, SW Turkey
Dalandzadgad 105 E3 Ömnögovĭ, S Mongolia
Đa Lat 115 E6 Lâm Đông, S Vietnam
Dalby 127 D5 Queensland, E Australia
Dale City 19 E4 Virginia, NE USA
Dalhart 27 E1 Texas, SW USA
Dali 106 A6 var. Xiaguan. Yunnan, SW China
Dalian 106 D4 var. Dairen, Dalien, Jay Dairen, Lüda, Ta-lien, Rus. Dalny. Liaoning, NE China
Dalien see Dalian
Dallas 27 G2 Texas, SW USA
Dalmacija 78 B4 Eng. Dalmatia, Ger. Dalmatien, It. Dalmazia. cultural region S Croatia
Dalmatia/Dalmatien/Dalmazia see Dalmacija
Dalny see Dalian
Dalton 20 D1 Georgia, SE USA
Dálvvadis see Jokkmokk
Daly Waters 126 A2 Northern Territory, N Australia
Damachava 85 A6 var. Damachova, Pol. Domaczewo, Rus. Domachëvo. Brestskaya Voblasts', SW Belarus
Damachova see Damachava
Damān 112 C4 Damān and Diu, W India
Damara 54 C4 Ombella-Mpoko, S Central African Republic
Damas see Dimashq
Damasco see Dimashq
Damascus see Dimashq
Damavand, Qolleh-ye 98 D3 mountain N Iran
Damietta see Dumyāṭ
Dammām see Ad Dammān
Damoûr 97 A5 var. Ad Dāmūr. W Lebanon
Dampier 124 A4 Western Australia
Dampier, Selat 117 F4 strait Papua, E Indonesia
Damqawt 99 D6 var. Damqut. E Yemen
Damqut see Damqawt
Damxung 104 C5 var. Gongtang. Xizang Zizhiqu, W China
Danakil Desert 50 D4 var. Afar Depression, Danakil Plain. desert E Africa
Danakil Plain see Danakil Desert
Danané 52 D5 W Ivory Coast
Đă Nẵng 115 E5 prev. Tourane. Quang Nam-Đa Nẵng, C Vietnam
Danborg see Daneborg
Dandong 106 D3 var. Tan-tung; prev. An-tung. Liaoning, NE China
Daneborg 61 E3 var. Danborg. Tunu, N Greenland
Dänew see Galkynyş
Dangara see Danghara
Dangerous Archipelago see Tuamotu, Îles
Danghara 101 E3 Rus. Dangara. SW Tajikistan
Danghe Nanshan 104 D3 mountain range W China
Dang Raek, Phanom/Dangrek, Chaîne des see Dângrêk, Chuŏr Phnum
Dângrêk, Chuŏr Phnum 115 D5 var. Phanom Dang Raek, Phanom Dong Rak, Fr. Chaîne des Dangrek. mountain range Cambodia/Thailand
Dangriga 30 C1 prev. Stann Creek. Stann Creek, E Belize
Danish West Indies see Virgin Islands (US)
Danlí 30 D2 El Paraíso, S Honduras
Danmark see Denmark
Danmarksstraedet see Denmark Strait
Dannenberg 72 C3 Niedersachsen, N Germany
Dannevirke 128 D4 Manawatu-Wanganui, North Island, New Zealand
Dantzig see Gdańsk
Danube 59 E4 Bul. Dunav, Cz. Dunaj, Ger. Donau, Hung. Duna, Rom. Dunărea. river C Europe
Danubian Plain see Dunavska Ravnina
Danum see Doncaster
Danville 19 E5 Virginia, NE USA
Danxian/Dan Xian see Danzhou
Danzhou 106 C7 prev. Danxian, Dan Xian, Nada. Hainan, S China
Danzig see Gdańsk
Danziger Bucht see Danzig, Gulf of
Danzig, Gulf of 76 C2 var. Gulf of Gdańsk, Ger. Danziger Bucht, Pol. Zakota Gdańska, Rus. Gdan'skaya Bukhta. gulf N Poland
Daqm see Duqm
Dar'ā 97 B5 var. Der'a, Fr. Déraa. Dar'ā, SW Syria
Darabani 86 C3 Botoșani, NW Romania
Daraut-Kurgan see Daroot-Korgon
Dardanelles 94 A2 Eng. Dardanelles. strait NW Turkey
Dardanelles see Çanakkale Boğazı
Dardanelli see Çanakkale
Dar-el-Beida see Casablanca
Dar es Salaam 51 C7 Dar es Salaam, E Tanzania
Darfield 129 C6 Canterbury, South Island, New Zealand
Darfur 50 A4 var. Darfur Massif. cultural region W Sudan
Darfur Massif see Darfur
Darhan 105 E2 Darhan Uul, N Mongolia
Darién, Golfo del see Darien, Gulf of
Darien, Gulf of 36 A2 Sp. Golfo del Darién. gulf S Caribbean Sea
Darien, Isthmus of see Panama, Istmo de
Darién, Serranía del 31 H5 mountain range Colombia/Panama
Dario see Ciudad Darío
Dariorigum see Vannes
Darjeeling see Dārjiling
Dārjiling 113 F3 prev. Darjeeling. West Bengal, NE India
Darling River 127 C6 river New South Wales, SE Australia
Darlington 67 D5 N England, United Kingdom
Darmstadt 73 B5 Hessen, SW Germany
Darnah 49 G2 var. Dérna. NE Libya
Darnley, Cape 132 D2 cape Antarctica
Daroca 71 E2 Aragón, NE Spain
Daroot-Korgon 101 F3 var. Daraut-Kurgan. Oshskaya Oblast', SW Kyrgyzstan
Dartford 67 B8 SE England, United Kingdom
Dartmoor 67 C7 moorland SW England, United Kingdom
Dartmouth 17 F4 Nova Scotia, SE Canada
Darvaza see Derweze, Turkmenistan
Darwin 124 D2 prev. Palmerston, Port Darwin. territory capital Northern Territory, N Australia
Darwin, Isla 38 A4 island Galápagos Islands, W Ecuador
Dashhowuz see Daşoguz
Dashkawka 85 D6 Rus. Dashkovka. Mahilyowskaya Voblasts', E Belarus
Dashkovka see Dashkawka
Daşoguz 100 C2 Rus. Dashkhovuz, Turkm. Dashhowuz; prev. Tashauz. Daşoguz Welaýaty, N Turkmenistan
Đa, Sông see Black River
Datong 106 C3 var. Tatung, Ta-t'ung. Shanxi, C China
Daugava see Western Dvina
Daugavpils 84 D4 Ger. Dünaburg; prev. Rus. Dvinsk. Daugvapils, SE Latvia
Daung Kyun 115 B6 island S Myanmar (Burma)
Dauphiné 69 D5 cultural region E France
Dāvangere 110 C2 Karnātaka, W India
Davao 117 F3 off. Davao City. Mindanao, S Philippines
Davao City see Davao
Davao Gulf 117 F3 gulf Mindanao, S Philippines
Davenport 23 G3 Iowa, C USA
David 31 E5 Chiriquí, W Panama
Davie Ridge 119 A5 undersea ridge W Indian Ocean
Davis 132 D3 Australian research station Antarctica
Davis Sea 132 D3 sea Antarctica
Davis Strait 60 B3 strait Baffin Bay/Labrador Sea
Dawei 115 B5 var. Tavoy, Htawei. Tanintharyi, S Myanmar (Burma)
Dawlat Qaṭar see Qatar
Dax 69 B6 var. Ax; anc. Aquae Augustae, Aquae Tarbelicae. Landes, SW France
Dayr az Zawr 96 D3 var. Deir ez Zor. Dayr az Zawr, E Syria
Dayton 18 C4 Ohio, N USA
Daytona Beach 21 E4 Florida, SE USA
De Aar 56 C5 Northern Cape, C South Africa
Dead Sea 97 B6 var. Bahret Lut, Lacus Asphaltites, Ar. Al Baḥr al Mayyit, Baḥrat Lūt, Heb. Yam HaMelaḥ. salt lake Israel/Jordan
Deán Funes 42 C3 Córdoba, C Argentina
Death Valley 25 C7 valley California, W USA
Deatnu 62 D2 Fin. Tenojoki, Nor. Tana. river Finland/Norway
Debar 79 D6 Ger. Dibra, Turk. Debre. W Macedonia
De Behagle see Laï
Dębica 77 D5 Podkarpackie, SE Poland
De Bildt see De Bilt
De Bilt 64 C3 var. De Bildt. Utrecht, C Netherlands
Dębno 76 B3 Zachodnio-pomorskie, NW Poland
Debre see Debar
Debrecen 77 D6 Ger. Debreczin, Rom. Debrețin; prev. Debreczen. Hajdú-Bihar, E Hungary
Debreczen/Debreczin see Debrecen
Debrețin see Debrecen
Decatur 20 C1 Alabama, S USA
Decatur 18 B4 Illinois, N USA
Deccan 112 D5 Hind. Dakshin. plateau C India
Děčín 76 B4 Ger. Tetschen. Ústecký Kraj, NW Czech Republic (Czechia)
Dedeagaç/Dedeagach see Alexandroúpoli
Dedemsvaart 64 E3 Overijssel, E Netherlands
Dee 66 C3 river NE Scotland, United Kingdom
Deering 14 C2 Alaska, USA
Deés see Dej
Deggendorf 73 D6 Bayern, SE Germany
Değirmenlik 80 C5 Gk. Kythréa. N Cyprus
Deh Bīd see Şafāshahr
Dehli see Delhi
Deh Shū see Dīshū
Deinze 65 B5 Oost-Vlaanderen, NW Belgium
Deir ez Zor see Dayr az Zawr
Deirgeirt, Loch see Derg, Lough
Dej 86 B3 Hung. Dés; prev. Deés. Cluj, NW Romania
De Jouwer see Joure
Dekéleia see Dhekelia
Dékoa 54 C4 Kémo, C Central African Republic
De Land 21 E4 Florida, SE USA
Delano 25 C7 California, W USA
Delārām see Dilārām
Delaware 18 D4 Ohio, N USA
Delaware 19 F4 off. State of Delaware, also known as Blue Hen State, Diamond State, First State. state NE USA
Delft 64 B4 Zuid-Holland, W Netherlands
Delfzijl 64 E1 Groningen, NE Netherlands
Delgo 50 B3 Northern, N Sudan
Delhi 112 D3 var. Dehli, Hind. Dilli, hist. Shahjahanabad. union territory capital Delhi, N India
Delicias 28 D2 var. Ciudad Delicias. Chihuahua, N Mexico
Déli-Kárpátok see Carpații Meridionalii
Delmenhorst 72 B3 Niedersachsen, NW Germany
Del Rio 27 F4 Texas, SW USA
Deltona 21 E4 Florida, SE USA
Demba 55 D6 Kasai-Occidental, C Dem. Rep. Congo
Dembia 54 D4 Mbomou, SE Central African Rep.
Demchok see Dêmqog
Demerara Plain 34 C2 abyssal plain W Atlantic Ocean
Deming 26 C3 New Mexico, SW USA
Demmin 72 C2 Mecklenburg-Vorpommern, NE Germany
Demopolis 20 C2 Alabama, S USA
Dêmqog 104 A5 var. Demchok. disputed region China/India
Denali 14 C3 var. Mount McKinley. mountain Alaska, USA
Denau see Denov
Dender 65 B6 Fr. Dendre. river W Belgium
Dendre see Dender
Denekamp 64 E3 Overijssel, E Netherlands
Den Haag see 's-Gravenhage
Denham 125 A5 Western Australia
Den Ham 64 E3 Overijssel, E Netherlands
Den Helder 64 C2 Noord-Holland, NW Netherlands
Dénia 71 F4 País Valenciano, E Spain
Deniliquin 127 C7 New South Wales, SE Australia
Denison 23 F3 Iowa, C USA
Denison 27 G2 Texas, SW USA
Denizli 94 B4 Denizli, SW Turkey
Denmark 63 A7 off. Kingdom of Denmark, Dan. Danmark; anc. Hafnia. country N Europe
Denmark, Kingdom of see Denmark
Denmark Strait 60 D4 var. Danmarksstraedet. strait Greenland/Iceland
Dennery 33 F1 E Saint Lucia
Denov 101 E3 Rus. Denau. Surkhondaryo Viloyati, S Uzbekistan
Denpasar 116 D5 prev. Paloe. Bali, C Indonesia
Denton 27 G2 Texas, SW USA
D'Entrecasteaux Islands 122 B3 island group SE Papua New Guinea
Denver 22 D4 state capital Colorado, C USA
Der'a/Derá/Déraa see Dar'ā
Dera Ghāzi Khān 112 C2 var. Dera Ghāzikhān. Punjab, C Pakistan
Dera Ghāzikhān see Dera Ghāzi Khān
Đeravica 79 D5 mountain S Serbia
Derbent 89 B8 Respublika Dagestan, SW Russia
Derby 67 D6 C England, United Kingdom
Dereli see Gónnoi
Dergachi see Derhachi
Derg, Lough 67 A6 Ir. Loch Deirgeirt. lake W Ireland
Derhachi 87 G2 Rus. Dergachi. Kharkivs'ka Oblast', E Ukraine
De Ridder 20 A3 Louisiana, S USA
Dérna see Darnah
Derry see Londonderry
Dertosa see Tortosa
Derventa 78 B3 Republika Srpska, N Bosnia and Herzegovina
Derweze 100 C2 Rus. Darvaza. Ahal Welaýaty, C Turkmenistan
Dés see Dej
Deschutes River 24 B3 river Oregon, NW USA
Desē 50 C4 var. Desse, It. Dessie. Āmara, N Ethiopia
Deseado, Río 43 B7 river S Argentina
Desertas, Ilhas 48 A2 island group Madeira, Portugal, NE Atlantic Ocean
Deshu see Dīshū
Des Moines 23 F3 state capital Iowa, C USA
Desna 87 E2 river Russia/Ukraine
Dessau 72 C4 Sachsen-Anhalt, E Germany
Desse see Desē
Dessie see Desē
Destêrro see Florianópolis
Detroit 18 D3 Michigan, N USA
Detroit Lakes 23 F2 Minnesota, N USA
Deurne 65 D5 Noord-Brabant, SE Netherlands
Deutschendorf see Poprad
Deutsch-Eylau see Iława
Deutsch Krone see Wałcz
Deutschland/Deutschland, Bundesrepublik see Germany
Deutsch-Südwestafrika see Namibia
Deva 86 B4 Ger. Diemrich, Hung. Déva. Hunedoara, W Romania
Déva see Deva
Deva see Chester
Devana see Aberdeen
Devana Castra see Chester
Đevđelija see Gevgelija
Deventer 64 D3 Overijssel, E Netherlands
Devils Lake 23 E1 North Dakota, N USA
Devoll see Devollit, Lumi i
Devollit, Lumi i 79 D6 var. Devoll. river SE Albania
Devon Island 15 F2 prev. North Devon Island. island Parry Islands, Nunavut, NE Canada
Devonport 127 C8 Tasmania, SE Australia
Devrek 94 C2 Zonguldak, N Turkey
Dexter 23 H5 Missouri, C USA
Deynau see Galkynyş
Dezfūl 98 C3 var. Dizful. Khūzestān, SW Iran
Dezhou 106 D4 Shandong, E China
Dhaka 113 G4 prev. Dacca. country capital (Bangladesh) Dhaka, C Bangladesh
Dhanbād 113 F4 Jhārkhand, NE India
Dhekelia 80 C5 Gk. Dekéleia. UK air base SE Cyprus
Dhidhimótikhon see Didymóteicho
Dhíkti Ori see Díkti
Dhodhekánisos see Dodekánisa
Dhomokós see Domokós
Dhráma see Dráma
Dhrepanon, Akrotírio see Drépano, Akrotírio
Dhún na nGall, Bá see Donegal Bay
Dhuusa Marreeb 51 E5 var. Dusa Marreb, It. Dusa Mareb. Galguduud, C Somalia
Diakovár see Đakovo
Diamantina, Chapada 41 F3 mountain range E Brazil
Diamantina Fracture Zone 119 E6 tectonic feature E Indian Ocean
Diamond State see Delaware
Diarbekr see Diyarbakır
Dibio see Dijon
Dibra see Debar
Dibrugarh 113 H3 Assam, NE India
Dickinson 22 D2 North Dakota, N USA
Dicle see Tigris
Didimotiho see Didymóteicho
Didymóteicho 82 D3 var. Dhidhimótikhon, Didimotiho. Anatolikí Makedonía kai Thráki, NE Greece
Diedenhofen see Thionville
Diekirch 65 D7 Diekirch, C Luxembourg
Diemrich see Deva
Điện Biên 114 D3 var. Bien Bien, Dien Bien Phu. Lai Châu, N Vietnam
Dien Bien Phu see Điện Biên
Diepenbeek 65 D6 Limburg, NE Belgium
Diepholz 72 B3 Niedersachsen, NW Germany
Dieppe 68 C2 Seine-Maritime, N France
Dieren 64 D4 Gelderland, E Netherlands
Differdange 65 D8 Luxembourg, SW Luxembourg
Digne 69 D6 var. Digne-les-Bains. Alpes-de-Haute-Provence, SE France
Digne-les-Bains see Digne
Digoel see Digul, Sungai
Digoin 68 C4 Saône-et-Loire, C France
Digul, Sungai 117 H5 prev. Digoel. river Papua, E Indonesia
Dihang see Brahmaputra
Dijlah see Tigris
Dijon 68 D4 anc. Dibio. Côte d'Or, C France
Dikhil 50 D4 SW Djibouti
Dikson 92 D2 Taymyrskiy (Dolgano-Nenetskiy) Avtonomnyy Okrug, N Russia
Díkti 83 D8 var. Dhíkti Ori. mountain range Kríti, Greece, E Mediterranean Sea
Dilārām 100 D5 prev. Delārām. Nīmrūz, SW Afghanistan
Dili 117 F5 var. Dilli, Dilly. country capital (East Timor) N East Timor
Dilia 53 G3 var. Dillia. river SE Niger
Di Linh 115 E6 Lâm Đông, S Vietnam
Dilli see Dili, East Timor
Dilli see Delhi, India
Dillia see Dilia
Dilling 50 B4 var. Ad Dalanj. Southern Kordofan, C Sudan
Dillon 22 B2 Montana, NW USA
Dilly see Dili
Dilolo 55 D7 Katanga, S Dem. Rep. Congo
Dimashq 97 B5 var. Ash Shām, Esh Sham, Eng. Damascus, Fr. Damas, It. Damasco. country capital (Syria) Dimashq, SW Syria
Dimitrovgrad 82 D3 Khaskovo, S Bulgaria
Dimitrovgrad 89 C6 prev. Caribrod. Serbia, SE Serbia
Dimitrovo see Pernik
Dimovo 82 B1 Vidin, NW Bulgaria
Dinajpur 113 F3 Rajshahi, NW Bangladesh
Dinan 68 B3 Côtes d'Armor, NW France
Dinant 65 C7 Namur, S Belgium
Dinar 94 B4 Afyon, SW Turkey
Dinara see Dinaric Alps
Dinaric Alps 78 C4 var. Dinara. mountain range Bosnia and Herzegovina/Croatia
Dindigul 110 C3 Tamil Nādu, SE India
Dingle Bay 67 A6 Ir. Bá an Daingin. bay SW Ireland
Dinguiraye 52 C4 N Guinea
Diourbel 52 B3 W Senegal
Dirē Dawa 51 D5 Dirē Dawa, E Ethiopia
Dirk Hartog Island 125 A5 island Western Australia
Dirschau see Tczew
Disappointment, Lake 124 C4 salt lake Western Australia
Discovery Bay 32 B4 Middlesex, Jamaica, Greater Antilles, C Jamaica Caribbean Sea
Dīshū 100 D5 prev. Deh Shū, var. Deshu. Helmand, S Afghanistan
Disko Bugt see Qeqertarsuup Tunua
Dispur 113 G3 state capital Assam, NE India
Divinópolis 41 F4 Minas Gerais, SE Brazil
Divo 52 D5 S Ivory Coast
Divodurum Mediomatricum see Metz
Diyarbakır 95 E4 var. Diarbekr; anc. Amida. Diyarbakır, SE Turkey
Dizful see Dezfūl
Djailolo see Halmahera, Pulau
Djajapura see Jayapura
Djakarta see Jakarta
Djakovo see Đakovo
Djambala 55 B6 Plateaux, C Congo
Djambi see Jambi
Djambi see Hari, Batang
Djanet 49 E4 prev. Fort Charlet. SE Algeria
Djéblé see Jablah
Djelfa 48 D2 var. El Djelfa. N Algeria
Djéma 54 D4 Haut-Mbomou, E Central African Republic

Djember – El Muglad

Djember *see* Jember
Djérablous *see* Jarābulus
Djerba 49 F2 var. Djerba, Jazīrat Jarbah. island E Tunisia
Djerba *see* Jerba, Île de
Djérem 54 B4 river C Cameroon
Djevdjelija *see* Gevgelija
Djibouti 50 D4 var. Jibuti. country capital (Djibouti) E Djibouti
Djibouti 50 D4 off. Republic of Djibouti, var. Jibuti; prev. French Somaliland, French Territory of the Afars and Issas, Fr. Côte Française des Somalis, Territoire Français des Afars et des Issas. country E Africa
Djibouti, Republic of *see* Djibouti
Djokjakarta *see* Yogyakarta
Djourab, Erg du 54 C2 desert N Chad
Djúpivogur 61 E5 Austurland, SE Iceland
Dmitriyevsk *see* Makiyivka
Dnepr *see* Dnieper
Dneprodzerzhinskoye Vodokhranilishche *see* Dniprodzerzhyns'ke Vodoskhovyshche
Dneprorudnoye *see* Dniprorudne
Dnestr *see* Dniester
Dnieper 59 F4 Bel. Dnyapro, Rus. Dnepr, Ukr. Dnipro. river E Europe
Dnieper Lowland 87 E2 Bel. Prydnyaprowskaya Nizina, Ukr. Prydniprovs'ka Nyzovyna. lowlands Belarus/Ukraine
Dniester 59 E4 Rom. Nistru, Rus. Dnestr, Ukr. Dnister; anc. Tyras. river Moldova/Ukraine
Dnipro *see* Dnieper
Dniprodzerzhyns'k *see* Kam"yans'ke
Dniprodzerzhyns'ke Vodoskhovyshche 87 F3 Rus. Dneprodzerzhinskoye Vodokhranilishche. reservoir C Ukraine
Dnipro 87 F3 Rus. Dnipropetrovs'k; prev. Yekaterinoslav. Dnipropetrovs'ka Oblast', E Ukraine
Dnipropetrovs'k *see* Dnipro
Dniprorudne 87 F3 Rus. Dneprorudnoye. Zaporiz'ka Oblast', SE Ukraine
Dnister *see* Dniester
Dnyapro *see* Dnieper
Doba 54 C4 Logone-Oriental, S Chad
Döbeln 72 D4 Sachsen, E Germany
Doberai Peninsula 117 G4 Dut. Vogelkop. peninsula Papua, E Indonesia
Doboj 78 C3 Republiks Srpska, N Bosnia and Herzegovina
Dobre Miasto 76 D2 Ger. Guttstadt. Warmińsko-mazurskie, NE Poland
Dobrich 82 E1 Rom. Bazargic; prev. Tolbukhin. Dobrich, NE Bulgaria
Dobrush 85 D7 Homyel'skaya Voblasts', SE Belarus
Dobryn' *see* Dabryn'
Dodecanese *see* Dodekánisa
Dodekánisa 83 D6 var. Nóties Sporádes, Eng. Dodecanese; prev. Dhodhekánisos, Dodekanisos. island group SE Greece
Dodekanisos *see* Dodekánisa
Dodge City 23 E5 Kansas, C USA
Dodoma 47 D5 country capital (Tanzania) Dodoma, C Tanzania
Dogana 74 E1 NE San Marino Europe
Dogo 109 B6 island Oki-shotō, SW Japan
Dogondoutchi 53 F3 Dosso, SW Niger
Dogrular *see* Pravda
Doğubayazıt 95 F3 Ağrı, E Turkey
Doğu Karadeniz Dağları 95 E3 var. Anadolu Dağları. mountain range NE Turkey
Doha *see* Ad Dawḩah
Doire *see* Londonderry
Dokdo *see* Liancourt Rocks
Dokkum 64 D1 Friesland, N Netherlands
Dokuchayevs'k 87 G3 var. Dokuchayevsk. Donets'ka Oblast', SE Ukraine
Dokuchayevsk *see* Dokuchayevs'k
Doldrums Fracture Zone 44 C4 fracture zone W Atlantic Ocean
Dôle 68 D4 Jura, E France
Dolina *see* Dolyna
Dolinskaya *see* Dolyns'ka
Dolisie 55 B6 prev. Loubomo. Niari, S Congo
Dolna Oryakhovitsa 82 D2 prev. Polikrayshte. Veliko Tŭrnovo, N Bulgaria
Dolni Chiflik 82 E2 prev. Rudnik. Varna, E Bulgaria
Dolomites 74 C1 var. Dolomiti, Eng. Dolomites. mountain range NE Italy
Dolomites/Dolomiti *see* Dolomitiche, Alpi
Dolores 42 D4 Buenos Aires, E Argentina
Dolores 30 B1 Petén, N Guatemala
Dolores 42 D4 Soriano, SW Uruguay
Dolores Hidalgo 29 E4 var. Ciudad de Dolores Hidalgo. Guanajuato, C Mexico
Dolyna 86 B2 Rus. Dolina. Ivano-Frankivs'ka Oblast', W Ukraine
Dolyns'ka 87 F3 Rus. Dolinskaya. Kirovohrads'ka Oblast', S Ukraine
Domachëvo/Domaczewo *see* Damachava
Dombås 63 B5 Oppland, S Norway
Domel Island *see* Letsôk-aw Kyun
Domesnes, Cape *see* Kolkasrags
Domeyko 42 B3 Atacama, N Chile
Dominica 33 H4 off. Commonwealth of Dominica. country E West Indies
Dominica Channel *see* Martinique Passage
Dominica, Commonwealth of *see* Dominica
Dominican Republic 33 E2 country C West Indies
Domokós 83 B5 var. Dhomokós. Stereá Elláda, C Greece
Don 89 B6 var. Duna, Tanais. river SW Russia
Donau *see* Danube
Donauwörth 73 C6 Bayern, S Germany
Don Benito 70 C3 Extremadura, W Spain
Doncaster 67 D5 anc. Danum. N England, United Kingdom
Dondo 56 B1 Cuanza Norte, NW Angola
Donegal 67 B5 Ir. Dún na nGall. Donegal, NW Ireland
Donegal Bay 67 A5 Ir. Bá Dhún na nGall. bay NW Ireland
Donets 87 G2 river Russia/Ukraine
Donets'k 87 G3 Rus. Donetsk; prev. Stalino. Donets'ka Oblast', E Ukraine
Dongfang 106 B7 var. Basuo. Hainan, S China
Dongguan 106 C6 Guangdong, S China
Đông Hà 114 E4 Quang Tri, C Vietnam
Dong Hai *see* East China Sea
Đông Hơi 114 D4 Quang Binh, C Vietnam
Dongliao *see* Liaoyuan
Dongola 50 B3 var. Donqola, Dunqulah. Northern, N Sudan
Dongou 55 C5 Likouala, NE Congo
Dong Rak, Phanom *see* Dângrêk, Chuŏr Phnum
Dongting Hu 106 C5 var. Tung-t'ing Hu. lake S China
Donostia 71 E1 País Vasco, N Spain *see also* San Sebastián
Donqola *see* Dongola
Doolow 51 D5 Sumalē, E Ethiopia
Doornik *see* Tournai
Door Peninsula 18 C2 peninsula Wisconsin, N USA
Dooxo Nugaaleed 51 E5 var. Nogal Valley. valley E Somalia
Dordogne 69 B5 cultural region SW France
Dordogne 69 B5 river W France
Dordrecht 64 C4 var. Dordt, Dort. Zuid-Holland, SW Netherlands
Dordt *see* Dordrecht
Dorohoi 86 C3 Botoşani, NE Romania
Dorotea 62 C4 Västerbotten, N Sweden
Dorpat *see* Tartu
Dorre Island 125 A5 island Western Australia
Dort *see* Dordrecht
Dortmund 72 A4 Nordrhein-Westfalen, W Germany
Dos Hermanas 70 C4 Andalucía, S Spain
Dospad Dagh *see* Rhodope Mountains
Dospat 82 C3 Smolyan, S Bulgaria
Dothan 20 D3 Alabama, S USA
Dotnuva 84 B4 Kaunas, C Lithuania
Douai 68 C2 prev. Douay; anc. Duacum. Nord, N France
Douala 55 A5 var. Duala. Littoral, W Cameroon
Douay *see* Douai
Douglas 67 C5 dependent territory capital (Isle of Man) E Isle of Man
Douglas 26 C3 Arizona, SW USA
Douglas 22 D3 Wyoming, C USA
Douma *see* Dūmā
Douro *see* Duero
Douvres *see* Dover
Dover 67 E7 Fr. Douvres, Lat. Dubris Portus. SE England, United Kingdom
Dover 19 F4 state capital Delaware, NE USA
Dover, Strait of 68 C2 var. Straits of Dover, Fr. Pas de Calais. strait England, United Kingdom/France
Dover, Straits of *see* Dover, Strait of
Dovrefjell 63 B5 plateau S Norway
Downpatrick 67 B5 Ir. Dún Pádraig. SE Northern Ireland, United Kingdom
Dozen 109 B6 island Oki-shotō, SW Japan
Drâa, Hammada du *see* Dra, Hamada du
Drač/Draç *see* Durrës
Drachten 64 D2 Friesland, N Netherlands
Drăgăşani 86 B5 Vâlcea, SW Romania
Dragoman 82 B2 Sofiya, W Bulgaria
Dra, Hamada du 48 C3 var. Hammada du Drâa, Haut Plateau du Dra. plateau W Algeria
Dra, Haut Plateau du *see* Dra, Hamada du
Drahichyn 85 B6 Pol. Drohiczyn Poleski, Rus. Drogichin. Brestskaya Voblasts', SW Belarus
Drakensberg 56 D5 mountain range Lesotho/South Africa
Drake Passage 35 B8 passage Atlantic Ocean/Pacific Ocean
Dralfa 82 D2 Tŭrgovishte, N Bulgaria
Dráma 82 C3 var. Dhráma. Anatolikí Makedonía kai Thráki, NE Greece
Dramburg *see* Drawsko Pomorskie
Drammen 63 B6 Buskerud, S Norway
Drau *see* Drava
Drava 78 C3 var. Drau, Eng. Drave, Hung. Dráva. river C Europe
Dráva/Drave *see* Drau/Drava
Drawsko Pomorskie 76 B3 Ger. Dramburg. Zachodnio-pomorskie, NW Poland
Drépano, Akrotírio 82 C4 var. Akrotírio Dhrepanon. headland N Greece
Drepanum *see* Trapani
Dresden 72 D4 Sachsen, E Germany
Drin *see* Drinit, Lumi i
Drina 78 C3 river Bosnia and Herzegovina/Serbia
Drinit, Lumi i 79 D5 var. Drin. river NW Albania
Drinit të Zi, Lumi i *see* Black Drin
Drissa *see* Drysa
Drobeta-Turnu Severin 86 B5 prev. Turnu Severin. Mehedinţi, SW Romania
Drogheda 67 B5 Ir. Droichead Átha. NE Ireland
Drogichin *see* Drahichyn
Drogobych *see* Drohobych
Drohiczyn Poleski *see* Drahichyn
Drohobych 86 B2 Pol. Drohobycz, Rus. Drogobych. L'vivs'ka Oblast', NW Ukraine
Drohobycz *see* Drohobych
Droichead Átha *see* Drogheda
Drôme 69 D5 cultural region E France
Dronning Maud Land 132 B2 physical region Antarctica
Drontheim *see* Trondheim
Drug *see* Durg
Druk-yul *see* Bhutan
Drummondville 17 E4 Québec, SE Canada
Druskienniki *see* Druskininkai
Druskininkai 85 B5 Pol. Druskienniki. Alytus, S Lithuania
Druzhnaya4 132 D3 Russian research station Antarctica
Dryden 16 B3 Ontario, C Canada
Drysa 85 D5 Rus. Drissa. river N Belarus
Duacum *see* Douai
Duala *see* Douala
Dubai *see* Dubayy
Dubăsari 86 D3 Rus. Dubossary. NE Moldova
Dubawnt 15 F4 river Nunavut, NW Canada
Dubayy 98 D4 Eng. Dubai. Dubayy, NE United Arab Emirates
Dubbo 127 D6 New South Wales, SE Australia
Dublin 67 B5 Ir. Baile Átha Cliath; anc. Eblana. country capital (Ireland) Dublin, E Ireland
Dublin 21 E2 Georgia, SE USA
Dubno 86 C2 Rivnens'ka Oblast', NW Ukraine
Dubossary *see* Dubăsari
Dubris Portus *see* Dover
Dubrovnik 79 B5 It. Ragusa. Dubrovnik-Neretva, SE Croatia
Dubuque 23 G3 Iowa, C USA
Dudelange 65 D8 var. Forge du Sud, Ger. Dudelingen. Luxembourg, S Luxembourg
Dudelingen *see* Dudelange
Duero 70 D2 Port. Douro. river Portugal/Spain
Duesseldorf *see* Düsseldorf
Duffel 65 C5 Antwerpen, C Belgium
Dugi Otok 78 A4 var. Isola Grossa, It. Isola Lunga. island W Croatia
Duinekerke *see* Dunkerque
Duisburg 72 A4 prev. Duisburg-Hamborn. Nordrhein-Westfalen, W Germany
Duisburg-Hamborn *see* Duisburg
Duiven 64 D4 Gelderland, E Netherlands
Duk Faiwil 51 B5 Jonglei, C South Sudan
Dulan 104 D4 var. Qagan Us. Qinghai, C China
Dulce, Golfo 31 E5 gulf S Costa Rica
Dulce, Golfo *see* Izabal, Lago de
Dülmen 72 A4 Nordrhein-Westfalen, W Germany
Dulovo 82 E1 Silistra, NE Bulgaria
Duluth 23 G2 Minnesota, N USA
Dūmā 97 B5 Fr. Douma. Dimashq, SW Syria
Dumas 27 E1 Texas, SW USA
Dumfries 66 C4 S Scotland, United Kingdom
Dumont d'Urville 132 C4 French research station Antarctica
Dumyāţ 50 B1 Eng. Damietta. N Egypt
Duna *see* Danube, C Europe
Düna *see* Western Dvina
Duna *see* Don
Dünaburg *see* Daugavpils
Dunaj *see* Wien, Austria
Dunaj *see* Danube, C Europe
Dunapentele *see* Dunaújváros
Dunărea *see* Danube
Dunaújváros 77 C7 prev. Dunapentele, Sztálinváros. Fejér, C Hungary
Dunav *see* Danube
Dunavska Ravnina 82 C2 Eng. Danubian Plain. lowlands N Bulgaria
Duncan 27 G2 Oklahoma, C USA
Dundalk 67 B5 Ir. Dún Dealgan. Louth, NE Ireland
Dún Dealgan *see* Dundalk
Dundee 56 D4 KwaZulu/Natal, E South Africa
Dundee 66 C4 E Scotland, United Kingdom
Dunedin 129 B7 Otago, South Island, New Zealand
Dunfermline 66 C4 C Scotland, United Kingdom
Dungu 55 E5 Orientale, NE Dem. Rep. Congo
Dungun 116 B3 var. Kuala Dungun. Terengganu, Peninsular Malaysia
Dunholme *see* Durham
Dunkerque 68 C2 Eng. Dunkirk, Flem. Duinekerke; prev. Dunquerque. Nord, N France
Dunkirk *see* Dunkerque
Dún Laoghaire 67 B6 Eng. Dunleary; prev. Kingstown. E Ireland
Dunleary *see* Dún Laoghaire
Dún Pádraig *see* Downpatrick
Dunquerque *see* Dunkerque
Dunqulah *see* Dongola
Dupnitsa 82 C2 prev. Marek, Stanke Dimitrov. Kyustendil, W Bulgaria
Duqm 99 E5 var. Daqm. E Oman
Durance 69 D6 river SE France
Durango 28 D3 var. Victoria de Durango. Durango, W Mexico
Durango 22 C5 Colorado, C USA
Durankulak 82 E1 Rom. Răcari; prev. Blatnitsa, Duranulac. Dobrich, NE Bulgaria
Durant 27 G2 Oklahoma, C USA
Duranulac *see* Durankulak
Durazzo *see* Durrës
Durban 56 D4 var. Port Natal. KwaZulu/Natal, E South Africa
Durbe 84 B3 Ger. Durben. Liepāja, W Latvia
Durben *see* Durbe
Durg 113 E4 prev. Drug. Chhattīsgarh, C India
Durham 67 D5 hist. Dunholme. N England, United Kingdom
Durham 21 F1 North Carolina, SE USA
Durocortorum *see* Reims
Durostorum *see* Silistra
Durovernum *see* Canterbury
Durrës 79 C6 var. Durrësi, Dursi, It. Durazzo, SCr. Drač, Turk. Draç. Durrës, W Albania
Durrësi *see* Durrës
Dursi *see* Durrës
Durūz, Jabal ad 97 C5 mountain SW Syria
D'Urville Island 128 C4 island C New Zealand
Dusa Mareb/Dusa Marreb *see* Dhuusa Marreeb
Dushanbe 101 E3 var. Dyushambe; prev. Stalinabad, Taj. Stalinobod. country capital (Tajikistan) W Tajikistan
Düsseldorf 72 A4 var. Duesseldorf. Nordrhein-Westfalen, W Germany
Dŭstí 101 E3 Rus. Dusti. SW Tajikistan
Dutch East Indies *see* Indonesia
Dutch Guiana *see* Suriname
Dutch Harbor 14 B3 Unalaska Island, Alaska, USA
Dutch New Guinea *see* Papua
Duzdab *see* Zāhedān
Dvina Bay *see* Chëshskaya Guba
Dvinsk *see* Daugavpils
Dyanev *see* Galkynyş
Dyersburg 20 C1 Tennessee, S USA
Dyushambe *see* Dushanbe
Dza Chu *see* Mekong
Dzaudzhikau *see* Vladikavkaz
Dzerzhinsk 89 C5 Nizhegorodskaya Oblast', W Russia
Dzerzhinskiy *see* Nar'yan-Mar
Dzhalal-Abad 101 F2 Kir. Jalal-Abad. Dzhalal-Abadskaya Oblast', W Kyrgyzstan
Dzhambul *see* Taraz
Dzhankoy 87 F4 Respublika Krym, S Ukraine
Dzharkurgan *see* Jarqo'rg'on
Dzhelandy 101 F3 SE Tajikistan
Dzhergalan 101 G2 Kir. Jyrgalan. Issyk-Kul'skaya Oblast', NE Kyrgyzstan
Dzhezkazgan *see* Zhezkazgan
Dzhizak *see* Jizzax
Dzhugdzhur, Khrebet 93 G3 mountain range E Russia
Dzhusaly *see* Zhosaly
Działdowo 76 D3 Warmińsko-Mazurskie, C Poland
Dzuunmod 105 E2 Töv, C Mongolia
Dzüün Soyonï Nuruu *see* Vostochnyy Sayan
Dzvina *see* Western Dvina

E

Eagle Pass 27 F4 Texas, SW USA
East Açores Fracture Zone *see* East Azores Fracture Zone
East Antarctica 132 C3 var. Greater Antarctica. physical region Antarctica
East Australian Basin *see* Tasman Basin
East Azores Fracture Zone 44 C3 var. East Açores Fracture Zone. tectonic feature E Atlantic Ocean
Eastbourne 67 E7 SE England, United Kingdom
East Cape 128 E3 headland North Island, New Zealand
East China Sea 103 E2 Chin. Dong Hai. sea W Pacific Ocean
Easter Fracture Zone 131 G4 tectonic feature E Pacific Ocean
Easter Island *see* Pascua, Isla de
Eastern Desert *see* Şaḩrā' ash Sharqīyah
Eastern Ghats 102 B3 mountain range SE India
Eastern Sayans *see* Vostochnyy Sayan
Eastern Sierra Madre *see* Madre Oriental, Sierra
East Falkland 43 D8 var. Isla Soledad. island E Falkland Islands
East Frisian Islands 72 A3 Eng. East Frisian Islands. island group NW Germany
East Frisian Islands *see* Ostfriesische Inseln
East Grand Forks 23 E1 Minnesota, N USA
East Indiaman Ridge 119 D6 undersea ridge E Indian Ocean
East Indies 130 A3 island group SE Asia
East Kilbride 66 C4 S Scotland, United Kingdom
East Korea Bay 107 E3 bay E North Korea
Eastleigh 67 D7 S England, United Kingdom
East London 56 D5 Afr. Oos-Londen; prev. Emonti, Port Rex. Eastern Cape, S South Africa
Eastmain 16 D3 river Québec, C Canada
East Mariana Basin 120 B1 undersea basin W Pacific Ocean
East Novaya Zemlya Trough 90 C1 var. Novaya Zemlya Trough. trough W Kara Sea
East Pacific Rise 131 F4 undersea rise E Pacific Ocean
East Pakistan *see* Bangladesh
East Saint Louis 18 B4 Illinois, N USA
East Scotia Basin 45 C7 undersea basin SE Scotia Sea
East Sea 108 A4 var. Sea of Japan, Rus. Yapanskoye More. Sea NW Pacific Ocean
East Siberian Sea *see* Vostochno-Sibirskoye More
East Timor 117 F5 var. Loro Sae; prev. Portuguese Timor, Timor Timur. country S Indonesia
Eau Claire 18 A2 Wisconsin, N USA
Eau Claire, Lac à L' *see* St. Clair, Lake
Eauripik Rise 120 B2 undersea rise W Pacific Ocean
Ebensee 73 D6 Oberösterreich, N Austria
Eberswalde-Finow 72 D3 Brandenburg, E Germany
Ebetsu 108 D2 var. Ebetu. Hokkaidō, NE Japan
Ebetu *see* Ebetsu
Eblana *see* Dublin
Ebolowa 55 A5 Sud, S Cameroon
Ebon Atoll 122 D2 var. Epoon. atoll Ralik Chain, S Marshall Islands
Ebora *see* Évora
Eboracum *see* York
Ebro 71 E2 river NE Spain
Eburacum *see* York
Ebusus *see* Eivissa
Ecbatana *see* Hamadān
Ech Cheliff/Ech Chleff *see* Chlef
Echo Bay 15 E3 Northwest Territories, NW Canada
Echt 65 D5 Limburg, SE Netherlands
Ecija 70 D4 anc. Astigi. Andalucía, SW Spain
Eckengraf *see* Viesīte
Ecuador 38 B1 off. Republic of Ecuador. country NW South America
Ecuador, Republic of *see* Ecuador
Ed Da'ein 50 A4 Southern Darfur, W Sudan
Ed Damazin 50 C4 var. Ad Damazīn. Blue Nile, E Sudan
Ed Damer 50 C3 var. Ad Dāmir, Ad Damar. River Nile, NE Sudan
Ed Debba 50 B3 Northern, N Sudan
Ede 64 D4 Gelderland, C Netherlands
Ede 53 F5 Osun, SW Nigeria
Edéa 55 A5 Littoral, SW Cameroon
Edessa *see* Şanlıurfa
Edfu *see* Idfū
Edgeøya 61 G2 island S Svalbard
Edgware 67 A7 Harrow, SE England, United Kingdom
Edinburg 27 G5 Texas, SW USA
Edinburgh 66 C4 national capital S Scotland, United Kingdom
Edingen *see* Enghien
Edirne 94 A2 Eng. Adrianople; anc. Adrianopolis, Hadrianopolis. Edirne, NW Turkey
Edmonds 24 B2 Washington, NW USA
Edmonton 15 E5 province capital Alberta, SW Canada
Edmundston 17 E4 New Brunswick, SE Canada
Edna 27 G4 Texas, SW USA
Edolo 74 B1 Lombardia, N Italy
Edremit 94 A3 Balıkesir, NW Turkey
Edward, Lake 55 E5 var. Albert Edward Nyanza, Edward Nyanza, Lac Idi Amin, Lake Rutanzige. lake Uganda/Dem. Rep. Congo
Edward Nyanza *see* Edward, Lake
Edwards Plateau 27 F3 plain Texas, SW USA
Edzo 31 E4 prev. Rae-Edzo. Northwest Territories, NW Canada
Eeklo 65 B5 var. Eekloo. Oost-Vlaanderen, NW Belgium
Eekloo *see* Eeklo
Eems *see* Ems
Eersel 65 C5 Noord-Brabant, S Netherlands
Eesti Vabariik *see* Estonia
Efate 122 D4 var. Efate, Fr. Vaté; prev. Sandwich Island. island C Vanuatu
Efate *see* Efate
Effingham 18 B4 Illinois, N USA
Eforie-Sud 86 D5 Constanţa, E Romania
Egadi Is. 75 B7 island group S Italy
Ege Denizi *see* Aegean Sea
Eger 77 D6 Ger. Erlau. Heves, NE Hungary
Eger *see* Cheb, Czech Republic (Czechia)
Egeria Fracture Zone 119 C5 tectonic feature W Indian Ocean
Éghezèe 65 C6 Namur, C Belgium
Egina *see* Aígina
Egio *see* Aígio
Egmont *see* Taranaki, Mount
Egmont, Cape 128 C4 headland North Island, New Zealand
Egoli *see* Johannesburg
Egypt 50 B2 off. Arab Republic of Egypt, Ar. Jumhūrīyah Mişr al 'Arabīyah, prev. United Arab Republic; anc. Aegyptus. country NE Africa
Eibar 71 E1 País Vasco, N Spain
Eibergen 64 E3 Gelderland, E Netherlands
Eidfjord 63 A5 Hordaland, S Norway
Eier-Berg *see* Suur Munamägi
Eifel 73 A5 plateau W Germany
Eiger 73 B7 mountain C Switzerland
Eigg 66 B3 island W Scotland, United Kingdom
Eight Degree Channel 110 B3 channel India/Maldives
Eighty Mile Beach 124 B4 beach Western Australia
Eijsden 65 D6 Limburg, SE Netherlands
Eilat *see* Elat
Eindhoven 65 D5 Noord-Brabant, S Netherlands
Eipel *see* Ipel'
Éire *see* Ireland
Éireann, Muir *see* Irish Sea
Eisenhüttenstadt 72 D4 Brandenburg, E Germany
Eisenmarkt *see* Hunedoara
Eisenstadt 73 E6 Burgenland, E Austria
Eisleben 72 C4 Sachsen-Anhalt, C Germany
Eivissa 71 G3 var. Iviza, Cast. Ibiza; anc. Ebusus. Ibiza, Spain, W Mediterranean Sea
Ejea de los Caballeros 71 E2 Aragón, NE Spain
Ejin Qi *see* Dalain Hob
Ekapa *see* Cape Town
Ekaterinodar *see* Krasnodar
Ekvyvatapskiy Khrebet 93 G1 mountain range NE Russia
El 'Alamein *see* Al 'Alamayn
El Asnam *see* Chlef
Elat 97 B8 var. Eilat, Elath. Southern, S Israel
Elat, Gulf of *see* Aqaba, Gulf of
Elath *see* Elat, Israel
Elath *see* Al 'Aqabah, Jordan
El'Atrun 50 B3 Northern Darfur, NW Sudan
Elâzığ 95 E3 var. Elâzig, Elâziz. Elâzığ, E Turkey
Elba 74 B4 island Archipelago Toscano, C Italy
Elbasan 79 D6 var. Elbasani. Elbasan, C Albania
Elbasani *see* Elbasan
Elbe 58 D3 Cz. Labe. river Czech Republic (Czechia)/Germany
Elbert, Mount 22 C4 mountain Colorado, C USA
Elbing *see* Elbląg
Elbląg 76 C2 var. Elblag, Ger. Elbing. Warmińsko-Mazurskie, NE Poland
El Boulaida/El Boulaïda *see* Blida
El'brus 89 A8 var. Gora El'brus. mountain SW Russia
El'brus, Gora *see* El'brus
El Burgo de Osma 71 F2 Castilla-León, C Spain
Elburz Mountains *see* Alborz, Reshteh-ye Kūhhā-ye
El Cajon 25 C8 California, W USA
El Calafate 43 B7 var. Calafate. Santa Cruz, S Argentina
El Callao 37 E2 Bolívar, E Venezuela
El Campo 27 G4 Texas, SW USA
El Carmen de Bolívar 36 B2 Bolívar, NW Colombia
El Cayo *see* San Ignacio
El Centro 25 D8 California, W USA
Elche 71 F4 Cat. Elx; anc. Ilici, Lat. Illicis. País Valenciano, E Spain
Elda 71 F4 País Valenciano, E Spain
El Djazaïr *see* Alger
El Djelfa *see* Djelfa
Eldorado 42 E3 Misiones, NE Argentina
El Dorado 28 C3 Sinaloa, C Mexico
El Dorado 20 B2 Arkansas, C USA
El Dorado 23 F5 Kansas, C USA
El Dorado 37 F2 Bolívar, E Venezuela
El Dorado *see* California
Eldoret 51 C6 Rift Valley, W Kenya
Elektrostal' 89 B5 Moskovskaya Oblast', W Russia
Elemi Triangle 51 B5 disputed region Kenya/Sudan
Elephant Butte Reservoir 26 C2 reservoir New Mexico, SW USA
Élesd *see* Aleşd
Eleuthera Island 32 C1 island N The Bahamas
El Fasher 50 A4 var. Al Fāshir. Northern Darfur, W Sudan
El Ferrol/El Ferrol del Caudillo *see* Ferrol
El Gedaref *see* Gedaref
El Geneina 50 A4 var. Ajjinena, Al-Genain, Al Junaynah. Western Darfur, W Sudan
Elgin 66 C3 NE Scotland, United Kingdom
Elgin 18 B3 Illinois, N USA
El Gîza *see* Gîza
El Goléa 48 D3 var. Al Golea. C Algeria
El Ḩank 52 D1 cliff N Mauritania
El Haseke *see* Al Ḩasakah
Elimberrum *see* Auch
Eliocroca *see* Lorca
Élisabethville *see* Lubumbashi
Elista 89 B7 Respublika Kalmykiya, SW Russia
Elizabeth 127 B6 South Australia
Elizabeth City 21 G1 North Carolina, SE USA
Elizabethtown 18 C5 Kentucky, S USA
El-Jadida 48 C2 prev. Mazagan. W Morocco
Ełk 76 E2 Ger. Lyck. Warmińsko-mazurskie, NE Poland
Elk City 27 F1 Oklahoma, C USA
El Khalil *see* Hebron
El Khârga *see* Al Khārijah
Elkhart 18 C3 Indiana, N USA
El Khartûm *see* Khartoum
Elk River 23 F2 Minnesota, N USA
El Kuneitra *see* Al Qunayţirah
Ellás *see* Greece
Ellef Ringnes Island 15 E1 island Nunavut, N Canada
Ellen, Mount 22 B5 mountain Utah, W USA
Ellensburg 24 B2 Washington, NW USA
Ellesmere Island 15 F1 island Queen Elizabeth Islands, Nunavut, N Canada
Ellesmere, Lake 129 C6 lake South Island, New Zealand
Ellice Islands *see* Tuvalu
Elliston 127 A6 South Australia
Ellsworth Land 132 A3 physical region Antarctica
El Mahbas 48 B3 var. Mahbés. SW Western Sahara
El Mina 96 B4 var. Al Mīnā'. N Lebanon
El Minya *see* Al Minyā
Elmira 19 E3 New York, NE USA
El Mreyyé 52 D2 desert E Mauritania
Elmshorn 72 B3 Schleswig-Holstein, N Germany
El Muglad 50 B4 Western Kordofan, C Sudan

El Obeid – Fortaleza

El Obeid 50 B4 var. Al Obayyid, Al Ubayyiḍ. Northern Kordofan, C Sudan
El Ouâdi see El Oued
El Oued 49 E2 var. Al Oued, El Ouâdi, El Wad. NE Algeria
Eloy 26 B2 Arizona, SW USA
El Paso 26 D3 Texas, SW USA
El Porvenir 31 G4 Kuna Yala, N Panama
El Progreso 30 C2 Yoro, NW Honduras
El Puerto de Santa María 70 C5 Andalucía, S Spain
El Qâhira see Cairo
El Quneitra see Al Qunayṭirah
El Quseir see Al Quṣayr
El Quweira see Al Quwayrah
El Rama 31 E3 Región Autónoma Atlántico Sur, SE Nicaragua
El Real 31 H5 var. El Real de Santa María. Darién, SE Panama
El Real de Santa María see El Real
El Reno 27 F1 Oklahoma, C USA
El Salvador 30 B3 off. Republica de El Salvador. country Central America
El Salvador, Republica de see El Salvador
Elsass see Alsace
El Sáuz 28 C2 Chihuahua, N Mexico
El Serrat 69 A7 N Andorra Europe
Elst 64 D4 Gelderland, E Netherlands
El Sueco 28 C2 Chihuahua, N Mexico
El Suweida see As Suwaydā'
El Suweis see Suez
Eltanin Fracture Zone 131 E5 tectonic feature SE Pacific Ocean
El Tigre 37 E2 Anzoátegui, NE Venezuela
Elvas 70 C4 Portalegre, C Portugal
El Vendrell 71 G2 Cataluña, NE Spain
El Vigía 36 C2 Mérida, NW Venezuela
El Wad see El Oued
Elwell, Lake 22 B1 reservoir Montana, NW USA
Elx see Elche
Ely 25 D5 Nevada, W USA
El Yopal see Yopal
Emajõgi 84 D3 Ger. Embach. river SE Estonia
Emāmrūd see Shāhrūd
Emāmshahr see Shāhrūd
Emba 92 B4 Kaz. Embi. Aktyubinsk, W Kazakhstan
Embach see Emajõgi
Embi see Emba
Emden 72 A3 Niedersachsen, NW Germany
Emerald 126 D4 Queensland, E Australia
Emerald Isle see Montserrat
Emesa see Ḥimṣ
Emmaste 84 C2 Hiiumaa, W Estonia
Emmeloord 64 D2 Flevoland, N Netherlands
Emmen 64 E2 Drenthe, NE Netherlands
Emmendingen 73 A6 Baden-Württemberg, SW Germany
Emona see Ljubljana
Emonti see East London
Emory Peak 27 E4 mountain Texas, SW USA
Empalme 28 B2 Sonora, NW Mexico
Emperor Seamounts 91 G3 seamount range NW Pacific Ocean
Empire State of the South see Georgia
Emporia 23 F5 Kansas, C USA
Empty Quarter see Ar Rub 'al Khālī
Ems 72 A3 Dut. Eems. river NW Germany
Enaretrāsk see Inarijärvi
Encamp 69 A8 Encamp, C Andorra Europe
Encarnación 42 D3 Itapúa, S Paraguay
Encinitas 25 C8 California, W USA
Encs 77 D6 Borsod-Abaúj-Zemplén, NE Hungary
Endeavour Strait 126 C1 strait Queensland, NE Australia
Enderbury Island 123 F3 atoll Phoenix Islands, C Kiribati
Enderby Land 132 C2 physical region Antarctica
Enderby Plain 132 D2 abyssal plain S Indian Ocean
Endersdorf see Jędrzejów
Enewetak Atoll 122 C1 var. Ānewetak, Eniwetok. atoll Ralik Chain, W Marshall Islands
Enfield 67 A7 United Kingdom
Engeten see Aiud
Enghien 65 B6 Dut. Edingen. Hainaut, SW Belgium
England 67 D5 Lat. Anglia. cultural region England, United Kingdom
Englewood 22 D4 Colorado, C USA
English Channel 67 D8 var. The Channel, Fr. la Manche. channel NW Europe
Engure 84 C3 Tukums, W Latvia
Engures Ezers 84 B3 lake NW Latvia
Enguri 95 F1 Rus. Inguri. river NW Georgia
Enid 27 F1 Oklahoma, C USA
Enikale Strait see Kerch Strait
Eniwetok see Enewetak Atoll
En Nâqoûra 97 A5 var. An Nāqūrah. SW Lebanon
En Nazira see Natzrat
Ennedi 54 D2 plateau E Chad
Ennis 67 A6 Ir. Inis. Clare, W Ireland
Ennis 27 G3 Texas, SW USA
Enniskillen 67 B5 var. Inniskilling, Ir. Inis Ceithleann. SW Northern Ireland, United Kingdom
Enns 73 D6 river C Austria
Enschede 64 E3 Overijssel, E Netherlands
Ensenada 28 A1 Baja California Norte, NW Mexico
Entebbe 51 B6 S Uganda
Entroncamento 70 B3 Santarém, C Portugal
Enugu 53 G5 Enugu, S Nigeria
Epanomí 82 B4 Kentrikí Makedonía, N Greece
Epéna 55 B5 Likouala, NE Congo
Eperies/Eperjes see Prešov
Epi 122 D4 var. Épi. island C Vanuatu
Épi see Epi
Épinal 68 D4 Vosges, NE France
Epiphania see Ḥamāh
Epoon see Ebon Atoll
Epsom 67 A8 United Kingdom
Equality State see Wyoming
Equatorial Guinea 55 A5 off. Equatorial Guinea, Republic of. country C Africa
Equatorial Guinea, Republic of see Equatorial Guinea
Erautini see Johannesburg
Erbil see Arbīl
Erciş 95 F3 Van, E Turkey
Erdély see Transylvania
Erdélyi-Havasok see Carpații Meridionalii
Erdenet 105 E2 Orhon, N Mongolia
Erdi 54 C2 plateau NE Chad
Erdi Ma 54 D2 desert NE Chad
Erebus, Mount 132 B4 volcano Ross Island, Antarctica
Ereğli 94 C4 Konya, S Turkey
Erenhot 105 F2 var. Erlian. Nei Mongol Zizhiqu, NE China
Erfurt 72 C4 Thüringen, C Germany
Ergene Çayı see Ergene Irmağī
Ergene Irmağī 94 A2 var. Ergene Çayı. river NW Turkey
Ergun 105 F1 var. Labudalin; prev. Ergun Youqi. Nei Mongol Zizhiqu, N China
Ergun He see Argun
Ergun Youqi see Ergun
Erie 18 D3 Pennsylvania, NE USA
Érié, Lac see Erie, Lake
Erie, Lake 18 D3 Fr. Lac Érié. lake Canada/USA
Eritrea 50 C4 off. State of Eritrea, Ērtra. country E Africa
Eritrea, State of see Eritrea
Erivan see Yerevan
Erlangen 73 C5 Bayern, S Germany
Erlau see Eger
Erlian see Erenhot
Ermelo 64 D3 Gelderland, C Netherlands
Ermióni 83 C6 Pelopónnisos, S Greece
Ermoúpoli 83 D6 var. Hermoupolis; prev. Ermoúpolis. Sýyros, Kykládes, Greece, Aegean Sea
Ermoúpolis see Ermoúpoli
Ernākulam 110 C3 Kerala, SW India
Erode 110 C2 Tamil Nādu, SE India
Erquelinnes 65 B7 Hainaut, S Belgium
Er-Rachidia 48 C2 var. Ksar al Soule. E Morocco
Er Rahad 50 B4 var. Ar Rahad. Northern Kordofan, C Sudan
Erromango 122 D4 island S Vanuatu
Ertis see Irtysh, C Asia
Ērtra see Eritrea
Erzerum see Erzurum
Erzgebirge 73 C5 Cz. Krušné Hory, Eng. Ore Mountains. mountain range Czech Republic (Czechia)/Germany
Erzincan 95 E3 var. Erzinjan. Erzincan, E Turkey
Erzinjan see Erzincan
Erzurum 95 E3 prev. Erzerum. Erzurum, NE Turkey
Esbjerg 63 A7 Ribe, W Denmark
Esbo see Espoo
Escaldes 69 A8 Escaldes Engordany, C Andorra Europe
Escanaba 18 C2 Michigan, N USA
Escaut see Scheldt
Esch-sur-Alzette 65 D8 Luxembourg, S Luxembourg
Esclaves, Grand Lac des see Great Slave Lake
Escondido 25 C8 California, W USA
Escuinapa 28 D3 var. Escuinapa de Hidalgo. Sinaloa, C Mexico
Escuinapa de Hidalgo see Escuinapa
Escuintla 30 B2 Escuintla, S Guatemala
Escuintla 29 G5 Chiapas, SE Mexico
Esenguly 100 B3 Rus. Gasan-Kuli. Balkan Welaýaty, W Turkmenistan
Eşfahān 98 C3 Eng. Isfahan; anc. Aspadana. Eşfahān, C Iran
Esh Sharā see Ash Sharāh
Esil see Ishim, Kazakhstan/Russia
Eskimo Point see Arviat
Eskişehir 94 B3 var. Eskishehr. Eskişehir, W Turkey
Eskishehr see Eskişehir
Eslāmābād 98 C3 var. Eslāmābād-e Gharb; prev. Harunabad, Shāhābād. Kermānshāhān, W Iran
Eslāmābād-e Gharb see Eslāmābād
Esmeraldas 38 A1 Esmeraldas, N Ecuador
Esna see Isnā
España see Spain
Espanola 26 D1 New Mexico, SW USA
Esperance 125 B7 Western Australia
Esperanza 28 B2 Sonora, NW Mexico
Esperanza 132 A2 Argentinian research station Antarctica
Espinal 36 B3 Tolima, C Colombia
Espinhaço, Serra do 34 D4 mountain range SE Brazil
Espírito Santo 41 F4 off. Estado do Espírito Santo. region E Brazil
Espírito Santo 41 F4 off. Estado do Espírito Santo. state E Brazil
Espírito Santo, Estado do see Espírito Santo
Espiritu Santo 122 C4 var. Santo. island W Vanuatu
Espoo 63 D6 Swe. Esbo. Uusimaa, S Finland
Esquel 43 B6 Chubut, SW Argentina
Essaouira 48 B2 prev. Mogador. W Morocco
Esseg see Osijek
Es Semara see Smara
Essen 65 C5 Antwerpen, N Belgium
Essen 72 A4 var. Essen an der Ruhr. Nordrhein-Westfalen, W Germany
Essen an der Ruhr see Essen
Essequibo River 37 F3 river C Guyana
Es Suweida see As Suwaydā'
Estacado, Llano 27 E2 plain New Mexico/Texas, SW USA
Estados, Isla de los 43 C8 prev. Eng. Staten Island. island S Argentina
Estância 41 G3 Sergipe, E Brazil
Estelí 30 D3 Estelí, NW Nicaragua
Estella 71 E1 Bas. Lizarra. Navarra, N Spain
Estepona 70 D5 Andalucía, S Spain
Estevan 15 F5 Saskatchewan, S Canada
Estland see Estonia
Estonia 84 D2 off. Republic of Estonia, Est. Eesti Vabariik, Ger. Estland, Latv. Igaunija; prev. Estonian SSR, Rus. Estonskaya SSR. country NE Europe
Estonian SSR see Estonia
Estonia, Republic of see Estonia
Estonskaya SSR see Estonia
Estrela, Serra da 70 C3 mountain range C Portugal
Estremadura see Extremadura
Estremoz 70 C4 Évora, S Portugal
Eszék see Osijek
Esztergom 77 C6 Ger. Gran; anc. Strigonium. Komárom-Esztergom, N Hungary
Étalle 65 D8 Luxembourg, SE Belgium
Etāwah 112 D3 Uttar Pradesh, N India
Ethiopia 51 C5 off. Federal Democratic Republic of Ethiopia; prev. Abyssinia, People's Democratic Republic of Ethiopia. country E Africa
Ethiopia, Federal Democratic Republic of see Ethiopia
Ethiopian Highlands 51 C5 var. Ethiopian Plateau. plateau N Ethiopia
Ethiopian Plateau see Ethiopian Highlands
Ethiopia, People's Democratic Republic of see Ethiopia
Etna, Monte 75 C7 Eng. Mount Etna. volcano Sicilia, Italy, C Mediterranean Sea
Etna, Mount see Etna, Monte
Etosha Pan 56 B3 salt lake N Namibia
Etoumbi 55 B5 Cuvette Ouest, NW Congo
Etsch see Adige
Et Tafila see Aṭ Ṭafilah
Ettelbrück 65 D8 Diekirch, C Luxembourg
'Eua 123 E5 prev. Middleburg Island. island Tongatapu Group, SE Tonga
Euboea 83 C5 Lat. Euboea. island C Greece
Euboea see Évvoia
Eucla 125 D6 Western Australia
Euclid 18 D3 Ohio, N USA
Eufaula Lake 27 G1 var. Eufaula Reservoir. reservoir Oklahoma, C USA
Eufaula Reservoir see Eufaula Lake
Eugene 24 B3 Oregon, NW USA
Eumolpias see Plovdiv
Eupen 65 D6 Liège, E Belgium
Euphrates 90 B4 Ar. Al-Furāt, Turk. Firat Nehri. river SW Asia
Eureka 25 A5 California, W USA
Eureka 22 A1 Montana, NW USA
Europa Point 71 H5 headland S Gibraltar
Europe 58 continent
Eutin 72 C2 Schleswig-Holstein, N Germany
Euxine Sea see Black Sea
Evansdale 23 G3 Iowa, C USA
Evanston 18 B3 Illinois, N USA
Evanston 22 B4 Wyoming, C USA
Evansville 18 B5 Indiana, N USA
Eveleth 23 G1 Minnesota, N USA
Everard, Lake 127 A6 salt lake South Australia
Everest, Mount 104 B5 Chin. Qomolangma Feng, Nep. Sagarmāthā. mountain China/Nepal
Everett 24 B2 Washington, NW USA
Everglades, The 21 F5 wetland Florida, SE USA
Evje 63 A6 Aust-Agder, S Norway
Evmolpia see Plovdiv
Évora 70 B4 anc. Ebora, Lat. Liberalitas Julia. Évora, C Portugal
Évreux 68 C3 anc. Civitas Eburovicum. Eure, N France
Évros see Maritsa
Évry 68 E2 Essonne, N France
Ewarton 32 B5 C Jamaica
Excelsior Springs 23 F4 Missouri, C USA
Exe 67 C7 river SW England, United Kingdom
Exeter 67 C7 anc. Isca Damnoniorum. SW England, United Kingdom
Exmoor 67 C7 moorland SW England, United Kingdom
Exmouth 124 A4 Western Australia
Exmouth 67 C7 SW England, United Kingdom
Exmouth Gulf 124 A4 gulf Western Australia
Exmouth Plateau 119 E5 undersea plateau E Indian Ocean
Extremadura 70 C3 var. Estremadura. autonomous community W Spain
Exuma Cays 32 C1 islets C The Bahamas
Exuma Sound 32 C1 sound C The Bahamas
Eyre Mountains 129 A7 mountain range South Island, New Zealand
Eyre North, Lake 127 A5 salt lake South Australia
Eyre Peninsula 127 A6 peninsula South Australia
Eyre South, Lake 127 A5 salt lake South Australia
Ezo see Hokkaidō

F

Faadhippolhu Atoll 110 B4 var. Fadiffolu, Lhaviyani Atoll. atoll N Maldives
Fabens 26 D3 Texas, SW USA
Fada 54 C2 Borkou-Ennedi-Tibesti, E Chad
Fada-Ngourma 53 E4 E Burkina
Fadiffolu see Faadhippolhu Atoll
Faenza 74 C3 anc. Faventia. Emilia-Romagna, N Italy
Faeroe Islands see Faero Islands
Færøerne see Faroe Islands
Faetano 74 E2 E San Marino
Făgăraş 86 C4 Ger. Fogarasch, Hung. Fogaras. Braşov, C Romania
Fagibina, Lake see Faguibine, Lac
Fagne 65 C7 hill range S Belgium
Faguibine, Lac 53 E3 var. Lake Fagibina. lake NW Mali
Fahlun see Falun
Fahraj 98 E4 Kermān, SE Iran
Faial 70 A5 var. Ilha do Faial. island Azores, Portugal, NE Atlantic Ocean
Faial, Ilha do see Faial
Faifo see Hôi An
Fairbanks 14 D3 Alaska, USA
Fairfield 25 B6 California, W USA
Fair Isle 66 D2 island NE Scotland, United Kingdom
Fairlie 129 B6 Canterbury, South Island, New Zealand
Fairmont 23 F3 Minnesota, N USA
Faisalābād 112 C2 prev. Lyallpur. Punjab, NE Pakistan
Faizābād 113 E3 Uttar Pradesh, N India
Faizabad/Faizābād see Feyzābād
Fakaofo Atoll 123 F3 island SE Tokelau
Falam 114 A3 Chin State, W Myanmar (Burma)
Falconara Marittima 74 C3 Marche, C Italy
Falkenau an der Eger see Sokolov
Falkland Islands 43 D7 var. Falklands, Islas Malvinas. UK dependent territory SW Atlantic Ocean
Falkland Plateau 35 D7 var. Argentine Rise. undersea feature SW Atlantic Ocean
Falklands see Falkland Islands
Falknov nad Ohří see Sokolov
Fallbrook 25 C8 California, W USA
Falmouth 32 A4 W Jamaica
Falmouth 67 C7 SW England, United Kingdom
Falster 63 B8 island SE Denmark
Fălticeni 86 C3 Hung. Falticsén. Suceava, NE Romania
Falticsén see Fălticeni
Falun 63 C6 var. Fahlun. Kopparberg, C Sweden
Famagusta see Gazimağusa
Famagusta Bay see Gazimağusa Körfezi
Famenne 65 C7 physical region SE Belgium
Fang 114 C3 Chiang Mai, NW Thailand
Fanning Island see Tabuaeran
Fanø 74 C3 island W Denmark
Farafangana 57 G4 Fianarantsoa, SE Madagascar
Farāh 100 D4 var. Farah, Fararud. Farāh, W Afghanistan
Farah Rud 100 D4 river W Afghanistan
Faranah 52 C4 Haute-Guinée, S Guinea
Fararud see Farāh
Farasan, Jaza'ir 99 A6 island group SW Saudi Arabia
Farewell, Cape 128 C4 headland South Island, New Zealand
Farewell, Cape see Nunap Isua
Fargo 23 F2 North Dakota, N USA
Farg'ona 101 F2 Rus. Fergana; prev. Novyy Margilan. Farg'ona Viloyati, E Uzbekistan
Faribault 23 F2 Minnesota, N USA
Farīdābād 112 D3 Haryāna, N India
Farkhor 101 E3 Rus. Parkhar. SW Tajikistan
Farmington 23 G5 Missouri, C USA
Farmington 26 C1 New Mexico, SW USA
Faro 70 B5 Faro, S Portugal
Faroe-Iceland Ridge 58 C1 undersea ridge NW Norwegian Sea
Faroe Islands 61 E5 var. Faero Islands, Dan. Færøerne, Faer. Føroyar. Self-governing territory of Denmark N Atlantic Ocean
Faroe-Shetland Trough 58 C2 trough NE Atlantic Ocean
Farquhar Group 57 G2 island group S Seychelles
Fars, Khalij-e see Persian Gulf
Farvel, Kap see Nunap Isua
Fastiv 87 E2 Rus. Fastov. Kyyivs'ka Oblast', NW Ukraine
Fastov see Fastiv
Fauske 62 C3 Nordland, C Norway
Faventia see Faenza
Faxa Bay see Faxaflói
Faxaflói 60 D5 Eng. Faxa Bay. bay W Iceland
Faya 54 C2 prev. Faya-Largeau, Largeau. Borkou-Ennedi-Tibesti, N Chad
Faya-Largeau see Faya
Fayetteville 20 A1 Arkansas, C USA
Fayetteville 21 F1 North Carolina, SE USA
Fdérick see Fdérik
Fdérik 52 C2 var. Fdérick, Fr. Fort Gouraud. Tiris Zemmour, NW Mauritania
Fear, Cape 21 F2 headland Bald Head Island, North Carolina, SE USA
Fécamp 68 B3 Seine-Maritime, N France
Fédala see Mohammedia
Federal Capital Territory see Australian Capital Territory
Fehérgyarmat 77 E6 Szabolcs-Szatmár-Bereg, E Hungary
Fehértemplom see Bela Crkva
Fehmarn 72 C2 island N Germany
Fehmarn Belt 72 C2 Dan. Femern Bælt, Ger. Fehmarnbelt. strait Denmark /Germany
Fehmarnbelt Fehmarn Belt/Femer Bælt
Feijó 40 C2 Acre, W Brazil
Feilding 128 D4 Manawatu-Wanganui, North Island, New Zealand
Feira see Feira de Santana
Feira de Santana 41 G3 var. Feira. Bahia, E Brazil
Feketehalom see Codlea
Felanitx 71 G3 Mallorca, Spain, W Mediterranean Sea
Felicitas Julia see Lisboa
Felidhu Atoll 110 B4 atoll C Maldives
Felipe Carrillo Puerto 29 H4 Quintana Roo, SE Mexico
Felixstowe 67 E6 E England, United Kingdom
Fellin see Viljandi
Felsőbánya see Baia Sprie
Felsőmuzslya see Mužlja
Femunden 63 B5 lake S Norway
Fénérive see Fenoarivo Atsinanana
Fengcheng 106 D3 var. Feng-cheng, Fenghwangcheng. Liaoning, NE China
Feng-cheng see Fengcheng
Fenghwangcheng see Fengcheng
Fengtien see Shenyang, China
Fengtien see Liaoning, China
Fenoarivo Atsinanana 57 G3 Fr. Fénérive. Toamasina, E Madagascar
Fens, The 67 E6 wetland E England, United Kingdom
Feodosiya 87 F5 var. Kefe, It. Kaffa; anc. Theodosia. Respublika Krym, S Ukraine
Ferdinand see Montana, Bulgaria
Ferdinandsberg see Oțelu Roșu
Féres 82 D3 Anatolikí Makedonía kai Thráki, NE Greece
Fergana see Farg'ona
Fergus Falls 23 F2 Minnesota, N USA
Ferizaj 79 D5 Serb. Uroševac. C Kosovo
Ferkessédougou 52 D4 N Ivory Coast
Fermo 74 C4 anc. Firmum Picenum. Marche, C Italy
Fernandina, Isla 38 A5 var. Narborough Island. island Galápagos Islands, Ecuador, E Pacific Ocean
Fernando de Noronha 41 H2 island E Brazil
Fernando Po/Fernando Póo see Bioco, Isla de
Ferrara 74 C2 anc. Forum Alieni. Emilia-Romagna, N Italy
Ferreñafe 38 B3 Lambayeque, W Peru
Ferro see Hierro
Ferrol 70 B1 var. El Ferrol; prev. El Ferrol del Caudillo. Galicia, NW Spain
Fertő see Neusiedler See
Ferwerd see Ferwert
Ferwert 64 D1 Dutch. Ferwerd. Friesland, N Netherlands
Fès 48 C2 Eng. Fez. N Morocco
Feteşti 86 D5 Ialomiţa, SE Romania
Fethiye 94 B4 Muğla, SW Turkey
Fetlar 66 D1 island NE Scotland, United Kingdom
Feuilles, Rivière aux 16 D2 river Québec, E Canada
Feyzābād 101 F3 var. Faizabad, Faizābād, Feyzābād, Fyzabad. Badakhshān, NE Afghanistan
Feyzābād see Feyzābād
Fez see Fès
Fianarantsoa 57 F3 Fianarantsoa, C Madagascar
Fianga 54 B4 Mayo-Kébbi, SW Chad
Fier 79 C6 var. Fieri. Fier, SW Albania
Fieri see Fier
Figeac 69 C5 Lot, S France
Figig see Figuig
Figueira da Foz 70 B3 Coimbra, W Portugal
Figueres 71 G2 Cataluña, E Spain
Figuig 48 D2 var. Figig. E Morocco
Fiji 123 E5 off. Republic of Fiji, Fij. Viti. country SW Pacific Ocean
Fiji, Republic of see Fiji
Filadelfia 30 D4 Guanacaste, W Costa Rica
Filiaşi 86 B5 Dolj, SW Romania
Filipstad 63 B6 Värmland, C Sweden
Finale Ligure 74 A3 Liguria, NW Italy
Finchley 67 A7 United Kingdom
Findlay 18 C4 Ohio, N USA
Finike 94 B4 Antalya, SW Turkey
Finland 62 D4 off. Republic of Finland, Fin. Suomen Tasavalta, Suomi. country N Europe
Finland, Gulf of 63 D6 Est. Soome Laht, Fin. Suomenlahti, Ger. Finnischer Meerbusen, Rus. Finskiy Zaliv, Swe. Finska Viken. gulf E Baltic Sea
Finland, Republic of see Finland
Finnischer Meerbusen see Finland, Gulf of
Finnmarksvidda 62 D2 physical region N Norway
Finska Viken/Finskiy Zaliv see Finland, Gulf of
Finsterwalde 72 D4 Brandenburg, E Germany
Fiordland 129 A7 physical region South Island, New Zealand
Fiorina 74 E1 NE San Marino
Firat Nehri see Euphrates
Firenze 74 C3 Eng. Florence; anc. Florentia. Toscana, C Italy
Firmum Picenum see Fermo
First State see Delaware
Fischbacher Alpen 73 E7 mountain range E Austria
Fischhausen see Primorsk
Fish 56 B4 var. Vis. river S Namibia
Fishguard 67 C6 Wel. Abergwaun. SW Wales, United Kingdom
Fisterra, Cabo 70 B1 headland NW Spain
Fitzroy Crossing 124 C3 Western Australia
Fitzroy River 124 C3 river Western Australia
Fiume see Rijeka
Flagstaff 26 B2 Arizona, SW USA
Flanders 65 A6 Dut. Vlaanderen, Fr. Flandre. cultural region Belgium/France
Flandre see Flanders
Flathead Lake 22 B1 lake Montana, NW USA
Flat Island 106 C8 island NE Spratly Islands
Flatts Village 20 B5 var. The Flatts Village. C Bermuda
Flensburg 72 B2 Schleswig-Holstein, N Germany
Flessingue see Vlissingen
Flickertail State see North Dakota
Flinders Island 127 C8 island Furneaux Group, Tasmania, SE Australia
Flinders Ranges 127 B6 mountain range South Australia
Flinders River 126 C3 river Queensland, NE Australia
Flin Flon 15 F5 Manitoba, C Canada
Flint 18 C3 Michigan, N USA
Flint Island 123 G4 island Line Islands, E Kiribati
Floreana, Isla see Santa María, Isla
Florence 20 C1 Alabama, S USA
Florence 21 F2 South Carolina, SE USA
Florence see Firenze
Florencia 36 B4 Caquetá, S Colombia
Florentia see Firenze
Flores 30 B1 Petén, N Guatemala
Flores 117 E5 island Nusa Tenggara, C Indonesia
Flores 70 A5 island Azores, Portugal, NE Atlantic Ocean
Flores, Laut see Flores Sea
Flores Sea 116 D5 Ind. Laut Flores. sea C Indonesia
Floriano 41 F2 Piauí, E Brazil
Florianópolis 41 F5 prev. Destêrro. state capital Santa Catarina, S Brazil
Florida 42 D4 Florida, S Uruguay
Florida 21 E4 off. State of Florida, also known as Peninsular State, Sunshine State. state SE USA
Florida Bay 21 E5 bay Florida, SE USA
Florida Keys 21 E5 island group Florida, SE USA
Florida, Straits of 32 B1 strait Atlantic Ocean/Gulf of Mexico
Flórina 82 B4 var. Phlórina. Dytikí Makedonía, N Greece
Florissant 23 G4 Missouri, C USA
Floúda, Akrotírio 83 D7 headland Astypálaia, Kykládes, Greece, Aegean Sea
Flushing see Vlissingen
Flylân see Vlieland
Foča 78 C4 var. Srbinje. SE Bosnia and Herzegovina
Focşani 86 C4 Vrancea, E Romania
Fogaras/Fogarasch see Făgăraş
Foggia 75 D5 Puglia, SE Italy
Fogo 52 A3 island Ilhas de Sotavento, SW Cape Verde
Foix 69 B6 Ariège, S France
Folégandros 83 C7 island Kykládes, Greece, Aegean Sea
Foleyet 16 C4 Ontario, S Canada
Foligno 74 C4 Umbria, C Italy
Folkestone 67 E7 SE England, United Kingdom
Fond du Lac 18 B2 Wisconsin, N USA
Fonseca, Golfo de see Fonseca, Gulf of
Fonseca, Gulf of 30 C3 Sp. Golfo de Fonseca. gulf C Central America
Fontainebleau 68 C3 Seine-et-Marne, N France
Fontenay-le-Comte 68 B4 Vendée, NW France
Fontvieille 69 B8 SW Monaco Europe
Fonyód 77 C7 Somogy, W Hungary
Foochow see Fuzhou
Foochow see Fuzhou
Forchheim 73 C5 Bayern, SE Germany
Forel, Mont 60 D4 mountain SE Greenland
Forfar 66 C3 E Scotland, United Kingdom
Forge du Sud see Dudelange
Forlì 74 C3 anc. Forum Livii. Emilia-Romagna, N Italy
Formentera 71 G4 anc. Ophiusa, Lat. Frumentum. island Islas Baleares, Spain, W Mediterranean Sea
Formosa 42 D2 Formosa, NE Argentina
Formosa/Formo'sa see Taiwan
Formosa, Serra 41 E3 mountain range C Brazil
Formosa Strait see Taiwan Strait
Føroyar see Faroe Islands
Forrest City 20 B1 Arkansas, C USA
Fort Albany 16 C3 Ontario, C Canada
Fortaleza 39 F2 Pando, N Bolivia
Fortaleza 41 G2 prev. Ceará. state capital Ceará, NE Brazil

Fort-Archambault – Giżycko

Fort-Archambault see Sarh
Fort-Bayard see Zhanjiang
Fort-Cappolani see Tidjikja
Fort Charlet see Djanet
Fort-Chimo see Kuujjuaq
Fort Collins 22 D4 Colorado, C USA
Fort-Crampel see Kaga Bandoro
Fort Davis 27 E3 Texas, SW USA
Fort-de-France 33 H4 prev. Fort-Royal. dependent territory capital (Martinique) W Martinique
Fort Dodge 23 F3 Iowa, C USA
Fortescue River 124 A4 river Western Australia
Fort-Foureau see Kousséri
Fort Frances 16 B4 Ontario, S Canada
Fort Good Hope 15 E3 var. Rádeyilikóé. Northwest Territories, NW Canada
Fort Gouraud see Fdérik
Forth 66 C4 river C Scotland, United Kingdom
Forth, Firth of 66 C4 estuary E Scotland, United Kingdom
Fortín General Eugenio Garay see General Eugenio A. Garay
Fort Jameson see Chipata
Fort-Lamy see Ndjamena
Fort Lauderdale 21 F5 Florida, SE USA
Fort Liard 15 E4 var. Liard. Northwest Territories, W Canada
Fort Madison 23 G4 Iowa, C USA
Fort McMurray 15 E4 Alberta, C Canada
Fort McPherson 14 D3 var. McPherson. Northwest Territories, NW Canada
Fort Morgan 22 D4 Colorado, C USA
Fort Myers 21 E5 Florida, SE USA
Fort Nelson 15 E4 British Columbia, W Canada
Fort Peck Lake 22 C1 reservoir Montana, NW USA
Fort Pierce 21 F4 Florida, SE USA
Fort Providence 15 E4 var. Providence. Northwest Territories, W Canada
Fort-Repoux see Akjoujt
Fort Rosebery see Mansa
Fort Rousset see Owando
Fort-Royal see Fort-de-France
Fort St. John 15 E4 British Columbia, W Canada
Fort Scott 23 F5 Kansas, C USA
Fort Severn 16 C2 Ontario, C Canada
Fort-Shevchenko 92 A4 Mangistau, W Kazakhstan
Fort-Sibut see Sibut
Fort Simpson 15 E4 var. Simpson. Northwest Territories, W Canada
Fort Smith 15 E4 Northwest Territories, W Canada
Fort Smith 20 B1 Arkansas, C USA
Fort Stockton 27 E3 Texas, SW USA
Fort-Trinquet see Bir Mogreïn
Fort Vermilion 15 E4 Alberta, W Canada
Fort Victoria see Masvingo
Fort Walton Beach 20 C3 Florida, SE USA
Fort Wayne 18 C4 Indiana, N USA
Fort William 66 C3 N Scotland, United Kingdom
Fort Worth 27 G2 Texas, SW USA
Fort Yukon 14 D3 Alaska, USA
Forum Alieni see Ferrara
Forum Livii see Forlì
Fossa Claudia see Chioggia
Fougamou 55 A6 Ngounié, C Gabon
Fougères 68 B3 Ille-et-Vilaine, NW France
Fou-hsin see Fuxin
Foulwind, Cape 129 B5 headland South Island, New Zealand
Foumban 54 A4 Ouest, NW Cameroon
Fou-shan see Fushun
Foveaux Strait 129 A8 strait S New Zealand
Foxe Basin 15 G3 sea Nunavut, N Canada
Fox Glacier 129 B6 West Coast, South Island, New Zealand
Fraga 71 F2 Aragón, NE Spain
Fram Basin 133 C3 var. Amundsen Basin. undersea basin Arctic Ocean
France 68 B4 off. French Republic, It./Sp. Francia; prev. Gaul, Gaule, Lat. Gallia. country W Europe
Franceville 55 B6 var. Massoukou, Masuku. Haut-Ogooué, E Gabon
Francfort see Frankfurt am Main
Franche-Comté 68 D4 cultural region E France
Francia see France
Francis Case, Lake 23 E3 reservoir South Dakota, N USA
Francisco Escárcega 29 G4 Campeche, SE Mexico
Francistown 56 D3 North East, NE Botswana
Franconian Jura see Fränkische Alb
Frankenalb see Fränkische Alb
Frankenstein/Frankenstein in Schlesien see Ząbkowice Śląskie
Frankfort 18 C5 state capital Kentucky, S USA
Frankfort on the Main see Frankfurt am Main
Frankfurt see Frankfurt am Main, Germany
Frankfurt see Słubice, Poland
Frankfurt am Main 73 B5 var. Frankfurt, Fr. Francfort; prev. Eng. Frankfort on the Main. Hessen, SW Germany
Frankfurt an der Oder 72 D3 Brandenburg, E Germany
Fränkische Alb 73 C6 var. Frankenalb, Eng. Franconian Jura. mountain range S Germany
Franklin 20 C1 Tennessee, S USA
Franklin D. Roosevelt Lake 24 C1 reservoir Washington, NW USA
Franz Josef Land 92 D1 Eng. Franz Josef Land. island group N Russia
Franz Josef Land see Frantsa-Iosifa, Zemlya
Fraserburgh 66 D3 NE Scotland, United Kingdom
Fraser Island 126 E4 var. Great Sandy Island. island Queensland, E Australia
Frauenbach see Baia Mare
Frauenburg see Saldus, Latvia
Fredericksburg 19 E5 Virginia, NE USA
Fredericton 17 F4 province capital New Brunswick, SE Canada
Frederikshåb see Paamiut
Fredrikshald see Halden
Fredrikstad 63 B6 Østfold, S Norway
Freeport 32 C1 Grand Bahama Island, N The Bahamas
Freeport 27 H4 Texas, SW USA
Free State see Maryland
Freetown 52 C4 country capital (Sierra Leone) W Sierra Leone
Freiburg see Freiburg im Breisgau, Germany
Freiburg im Breisgau 73 A6 var. Freiburg, Fr. Fribourg-en-Brisgau. Baden-Württemberg, SW Germany
Freiburg in Schlesien see Świebodzice
Fremantle 125 A6 Western Australia
Fremont 23 F4 Nebraska, C USA
French Guiana 37 H3 var. Guiana, Guyane. French overseas department N South America
French Guinea see Guinea
French Polynesia 121 F4 French overseas territory S Pacific Ocean
French Republic see France
French Somaliland see Djibouti
French Southern and Antarctic Lands 119 B7 Fr. Terres Australes et Antarctiques Françaises. French overseas territory S Indian Ocean
French Sudan see Mali
French Territory of the Afars and Issas see Djibouti
French Togoland see Togo
Fresnillo 28 D3 var. Fresnillo de González Echeverría. Zacatecas, C Mexico
Fresnillo de González Echeverría see Fresnillo
Fresno 25 C6 California, W USA
Frías 42 C3 Catamarca, N Argentina
Fribourg-en-Brisgau see Freiburg im Breisgau
Friedek-Mistek see Frýdek-Místek
Friedrichshafen 73 B7 Baden-Württemberg, S Germany
Friendly Islands see Tonga
Frisches Haff see Vistula Lagoon
Frobisher Bay 60 B3 inlet Baffin Island, Nunavut, NE Canada
Frobisher Bay see Iqaluit
Frohavet 62 B4 sound C Norway
Frome, Lake 127 B6 salt lake South Australia
Frontera 29 G4 Tabasco, SE Mexico
Frontignan 69 C6 Hérault, S France
Frostviken see Kvarnbergsvattnet
Frøya 62 A4 island W Norway
Frumentum see Formentera
Frunze see Bishkek
Frýdek-Místek 77 C5 Ger. Friedek-Mistek. Moravskoslezský Kraj, E Czech Republic (Czechia)
Fu-chien see Fujian
Fu-chou see Fuzhou
Fuengirola 70 D5 Andalucía, S Spain
Fuerte Olimpo 42 D2 var. Olimpo. Alto Paraguay, NE Paraguay
Fuerte, Río 26 C5 river C Mexico
Fuerteventura 48 B3 island Islas Canarias, Spain, NE Atlantic Ocean
Fuhkien see Fujian
Fu-hsin see Fuxin
Fuji 109 D6 var. Huzi. Shizuoka, Honshū, S Japan
Fujian 106 D6 var. Fu-chien, Fuhkien, Fukien, Min, Fujian Sheng. province SE China
Fujian Sheng see Fujian
Fuji, Mount/Fujiyama see Fuji-san
Fuji-san 109 C6 var. Fujiyama, Eng. Mount Fuji. mountain Honshū, SE Japan
Fukang 104 C2 Xinjiang Uygur Zizhiqu, W China
Fukien see Fujian
Fukui 109 C6 var. Hukui. Fukui, Honshū, SW Japan
Fukuoka 109 A7 var. Hukuoka, hist. Najima. Fukuoka, Kyūshū, SW Japan
Fukushima 108 D4 var. Hukusima. Fukushima, Honshū, C Japan
Fulda 73 B5 Hessen, C Germany
Funafuti Atoll 123 E3 atoll and capital (Tuvalu) C Tuvalu
Funchal 48 A2 Madeira, Portugal, NE Atlantic Ocean
Fundy, Bay of 17 F5 bay Canada/USA
Fünen see Fyn
Fünfkirchen see Pécs
Furnes see Veurne
Fürth 73 C5 Bayern, S Germany
Furukawa 108 D4 var. Hurukawa, Ōsaki. Miyagi, Honshū, C Japan
Fusan see Busan
Fushë Kosovë 79 D5 Serb. Kosovo Polje. C Kosovo
Fushun 106 D3 var. Fou-shan, Fu-shun. Liaoning, NE China
Fu-shun see Fushun
Fusin see Fuxin
Füssen 73 C7 Bayern, S Germany
Futog 78 D3 Vojvodina, NW Serbia
Futuna, Île 123 E4 island S Wallis and Futuna
Fuxin 106 D3 var. Fou-hsin, Fu-hsin, Fusin. Liaoning, NE China
Fuzhou 106 D6 var. Foochow, Fu-chou. province capital Fujian, SE China
Fyn 63 B8 Ger. Fünen. island C Denmark
FYR Macedonia/FYROM see Macedonia
Fyzabad see Feyzābād

G

Gaafu Alifu Atoll see North Huvadhu Atoll
Gaalkacyo 51 E5 var. Galka'yo, It. Galcaio. Mudug, C Somalia
Gabela 56 B2 Cuanza Sul, W Angola
Gaberones see Gaborone
Gabès 49 E2 var. Qābis. E Tunisia
Gabès, Golfe de 49 F2 Ar. Khalīj Qābis. gulf E Tunisia
Gabon 55 B6 off. Gabonese Republic. country C Africa
Gabonese Republic see Gabon
Gaborone 56 C4 prev. Gaberones. country capital (Botswana) South East, SE Botswana
Gabrovo 82 D2 Gabrovo, N Bulgaria
Gadag 110 C1 Karnātaka, W India
Gades/Gadier/Gadir/Gadire see Cádiz
Gadsden 20 D2 Alabama, S USA
Gaeta 75 C5 Lazio, C Italy
Gaeta, Golfo di 75 C5 var. Gulf of Gaeta. gulf C Italy
Gaeta, Gulf of see Gaeta, Golfo di
Gäfle see Gävle
Gafsa 49 E2 var. Qafşah. W Tunisia
Gagnoa 52 D5 C Ivory Coast
Gagra 95 E1 NW Georgia
Gaillac 69 C6 var. Gaillac-sur-Tarn. Tarn, S France
Gaillac-sur-Tarn see Gaillac
Gaillimh see Galway
Gaillimhe, Cuan na see Galway Bay
Gainesville 21 E3 Florida, SE USA
Gainesville 20 D2 Georgia, SE USA
Gainesville 27 G2 Texas, SW USA
Lake Gairdner 127 A6 salt lake South Australia
Gaizina Kalns see Gaiziņkalns
Gaiziņkalns 84 C3 var. Gaizina Kalns. mountain E Latvia
Galán, Cerro 42 B3 mountain NW Argentina
Galanta 77 C6 Hung. Galánta. Trnavský Kraj, W Slovakia
Galapagos Fracture Zone 131 E3 tectonic feature E Pacific Ocean
Galápagos Islands 131 F3 var. Islas de los Galápagos, Sp. Archipiélago de Colón, Eng. Galapagos Islands, Tortoise Islands. island group Ecuador, E Pacific Ocean
Galapagos Islands see Galápagos Islands
Galápagos, Islas de los see Galápagos Islands
Galapagos Rise 131 F3 undersea rise E Pacific Ocean
Galashiels 66 C4 SE Scotland, United Kingdom
Galați 86 D4 Ger. Galatz. Galați, E Romania
Galatz see Galați
Galcaio see Gaalkacyo
Galesburg 18 B3 Illinois, N USA
Galicia 70 B1 anc. Gallaecia. autonomous community NW Spain
Galicia Bank 58 B4 undersea bank E Atlantic Ocean
Galilee, Sea of see Tiberias, Lake
Galka'yo see Gaalkacyo
Galkynyş 100 D3 prev. Rus. Deynau, Dyanev, Turkm. Dänew. Lebap Welaýaty, NE Turkmenistan
Gallaecia see Galicia
Galle 110 D4 prev. Point de Galle. Southern Province, SW Sri Lanka
Gallego Rise 131 F3 undersea rise E Pacific Ocean
Gallegos see Río Gallegos
Gallia see France
Gallipoli 75 E6 Puglia, SE Italy
Gällivare 62 C3 Lapp. Váhtjer. Norrbotten, N Sweden
Gallup 26 C1 New Mexico, SW USA
Galtat-Zemmour 48 B3 C Western Sahara
Galveston 27 H4 Texas, SW USA
Galway 67 A5 Ir. Gaillimh. W Ireland
Galway Bay 67 A6 Ir. Cuan na Gaillimhe. bay W Ireland
Gámas see Kaamanen
Gambell 14 C2 Saint Lawrence Island, Alaska, USA
Gambia see Gambia, The
Gambia 52 C3 Fr. Gambie. river W Africa
Gambia, Republic of The see Gambia, The
Gambia, The 52 B3 off. Republic of The Gambia var. Gambia. country W Africa
Gambie see Gambia, The
Gambier, Îles 121 G4 island group E French Polynesia
Gamboma 55 B6 Plateaux, E Congo
Gamlakarleby see Kokkola
Gan 110 B5 Addu Atoll, C Maldives
Gan see Gansu, China
Gan see Jiangxi, China
Ganaane see Juba
Gäncä 95 G2 Rus. Gyandzha; prev. Kirovabad, Yelisavetpol. W Azerbaijan
Gand see Gent
Gandajika 55 D7 Kasai-Oriental, S Dem. Rep. Congo
Gander 17 G3 Newfoundland and Labrador, SE Canada
Gāndhīdhām 112 C4 Gujarāt, W India
Gandia 71 F3 prev. Gandía. País Valenciano, E Spain
Gandía see Gandia
Ganges 113 F3 Ben. Padma. river Bangladesh/India
Ganges Cone see Ganges Fan
Ganges Fan 118 D3 var. Ganges Cone. undersea fan N Bay of Bengal
Ganges, Mouths of the 113 G4 delta Bangladesh/India
Gangra see Çankırı
Gangtok 113 F3 state capital Sikkim, N India
Gansos, Lago dos see Goose Lake
Gansu 106 B4 var. Gan, Gansu Sheng, Kansu. province N China
Gansu Sheng see Gansu
Gantsevichi see Hantsavichy
Ganzhou 106 D6 Jiangxi, S China
Gao 53 E3 Gao, E Mali
Gaocheng see Litang
Gaoual 52 C4 N Guinea
Gaoxiong see Kaohsiung
Gap 69 D5 anc. Vapincum. Hautes-Alpes, SE France
Gaplaňgyr Platosy 100 C2 Rus. Plato Kaplangky. ridge Turkmenistan/Uzbekistan
Gar see Gar Xincun
Garabil Belentligi 100 D3 Rus. Vozvyshennost' Karabil'. mountain range S Turkmenistan
Garabogaz Aylagy 100 B2 Rus. Zaliv Kara-Bogaz-Gol. bay NW Turkmenistan
Garachiné 31 G5 Darién, SE Panama
Garagum 100 C3 var. Garagumy, Qara Qum, Eng. Black Sand Desert, Kara Kum; prev. Peski Karakumy. desert C Turkmenistan
Garagum Canal 100 D3 var. Kara Kum Canal, Rus. Karagumskiy Kanal, Karakumskiy Kanal. canal C Turkmenistan
Garagumy see Garagum
Gara Khitrino 82 D2 Shumen, NE Bulgaria
Gárassavon see Kaaresuvanto
Garda, Lago di 74 C2 var. Benaco, Eng. Lake Garda, Ger. Gardasee. lake NE Italy
Garda, Lake see Garda, Lago di
Gardasee see Garda, Lago di
Garden City 23 E5 Kansas, C USA
Garden State, The see New Jersey
Gardēz 101 E4 prev. Gardīz. E Afghanistan
Gardīz see Gardēz
Gardner Island see Nikumaroro
Garegegasnjárga see Karigasniemi
Gargždai 84 B3 Klaipėda, W Lithuania
Garissa 51 D6 Coast, E Kenya
Garland 27 G2 Texas, SW USA
Garoe see Garoowe
Garonne 69 B5 anc. Garumna. river S France
Garoowe 51 E5 var. Garoe. Nugaal, N Somalia
Garoua 54 B4 var. Garua. Nord, N Cameroon
Garrygala see Magtymguly
Garry Lake 15 F3 lake Nunavut, N Canada
Garsen 51 D6 Coast, S Kenya
Garua see Garoua
Garumna see Garonne
Garwolin 76 D4 Mazowieckie, E Poland
Gar Xincun 104 A4 prev. Gar. Xizang Zizhiqu, W China
Gary 18 B3 Indiana, N USA
Garzón 36 B4 Huila, S Colombia
Gasan-Kuli see Esenguly
Gascogne 69 B6 Eng. Gascony. cultural region S France
Gascony see Gascogne
Gascoyne River 125 A5 river Western Australia
Gaspé 17 F3 Québec, SE Canada
Gaspé, Péninsule de 17 E4 var. Péninsule de la Gaspésie. peninsula Québec, SE Canada
Gaspésie, Péninsule de la see Gaspé, Péninsule de
Gastonia 21 E1 North Carolina, SE USA
Gastoúni 83 B6 Dytikí Elláda, S Greece
Gatchina 88 B4 Leningradskaya Oblast', NW Russia
Gatineau 16 D4 Québec, SE Canada
Gatooma see Kadoma
Gatún, Lake 31 F4 reservoir C Panama
Gauhāti see Guwāhāti
Gauja 84 D3 Ger. Aa. river Estonia/Latvia
Gaul/Gaule see France
Gauteng see Johannesburg, South Africa
Gāvbandī 98 D4 Hormozgān, S Iran
Gávdos 83 C8 island SE Greece
Gavere 65 B6 Oost-Vlaanderen, NW Belgium
Gävle 63 C6 var. Gäfle; prev. Gefle. Gävleborg, C Sweden
Gawler 127 B6 South Australia
Gaya 113 F3 Bihār, N India
Gaya see Kyjov
Gayndah 127 E5 Queensland, E Australia
Gaysin see Haysyn
Gaza 97 A6 Ar. Ghazzah, Heb. 'Azza. NE Gaza Strip
Gaz-Achak see Gazojak
Gazandzhyk/Gazanjyk see Bereket
Gaza Strip 97 A7 Ar. Qiţā Ghazzah. disputed region SW Asia
Gaziantep 94 D4 var. Gazi Antep; prev. Aintab, Antep. Gaziantep, S Turkey
Gazi Antep see Gaziantep
Gazimağusa 80 D5 var. Famagusta, Gk. Ammóchostos. E Cyprus
Gazimağusa Körfezi 80 C5 var. Famagusta Bay, Gk. Kólpos Ammóchostos. bay E Cyprus
Gazli 100 D2 Buxoro Viloyati, C Uzbekistan
Gazojak 100 D2 Rus. Gaz-Achak. Lebap Welaýaty, NE Turkmenistan
Gbanga 52 D5 var. Gbarnga. N Liberia
Gbarnga see Gbanga
Gdańsk 76 C2 Fr. Dantzig, Ger. Danzig. Pomorskie, N Poland
Gdan'skaya Bukhta/Gdańsk, Gulf of see Danzig, Gulf of
Gdańska, Zakota see Danzig, Gulf of
Gdingen see Gdynia
Gdynia 76 C2 Ger. Gdingen. Pomorskie, N Poland
Gedaref 50 C4 var. Al Qaḑārif, El Gedaref. Gedaref, E Sudan
Gediz 94 B3 Kütahya, W Turkey
Gediz Nehri 94 A3 river W Turkey
Geel 65 C5 var. Gheel. Antwerpen, N Belgium
Geelong 127 C7 Victoria, SE Australia
Ge'e'mu see Golmud
Gefle see Gävle
Geilo 63 A5 Buskerud, S Norway
Gejiu 106 B6 var. Kochiu. Yunnan, S China
Gëkdepe see Gökdepe
Gela 75 C7 prev. Terranova di Sicilia. Sicilia, Italy, C Mediterranean Sea
Geldermalsen 64 C4 Gelderland, C Netherlands
Geleen 65 D6 Limburg, SE Netherlands
Gelib see Jilib
Gellinsor 51 E5 Mudug, C Somalia
Gembloux 65 C6 Namur, Belgium
Gemena 55 C5 Equateur, NW Dem. Rep. Congo
Gem of the Mountains see Idaho
Gemona del Friuli 74 D2 Friuli-Venezia Giulia, NE Italy
Gem State see Idaho
Genalë Wenz see Juba
Genck see Genk
General Alvear 42 B4 Mendoza, W Argentina
General Carrera, Lago see Buenos Aires, Lago
General Eugenio A. Garay 42 C1 var. Fortín General Eugenio Garay; prev. Yrendagüé. Nueva Asunción, NW Paraguay
General José F.Uriburu see Zárate
General Machado see Camacupa
General Santos 117 F3 off. General Santos City. Mindanao, S Philippines
General Santos City see General Santos
Gênes see Genova
Geneva see Genève
Geneva, Lake 73 A7 Fr. Lac de Genève, Lac Léman, le Léman, Ger. Genfer See. lake France/Switzerland
Genève 73 A7 Eng. Geneva, Ger. Genf, It. Ginevra. Genève, SW Switzerland
Genève, Lac de see Geneva, Lake
Genf see Genève
Genfer See see Geneva, Lake
Genichesk see Heniches'k
Genk 65 D6 var. Genck. Limburg, NE Belgium
Gennep 64 D4 Limburg, SE Netherlands
Genoa see Genova
Genoa, Gulf of 74 A3 Eng. Gulf of Genoa. gulf NW Italy
Genoa, Gulf of see Genova, Golfo di
Genova 80 D1 Eng. Genoa; anc. Genua, Fr. Gênes. Liguria, NW Italy
Genovesa, Isla 38 B5 var. Tower Island. island Galápagos Islands, Ecuador, E Pacific Ocean
Gent 65 B5 Eng. Ghent, Fr. Gand. Oost-Vlaanderen, NW Belgium
Genua see Genova
Geok-Tepe see Gökdepe
George 56 C5 Western Cape, S South Africa
George 60 A4 river Newfoundland and Labrador/Québec, E Canada
George, Lake 21 E3 lake Florida, SE USA
Georgenburg see Jurbarkas
Georges Bank 13 D5 undersea bank W Atlantic Ocean
George Sound 129 A7 sound South Island, New Zealand
Georges River 126 D2 river New South Wales, E Australia
Georgetown 37 F2 country capital (Guyana) N Guyana
George Town 32 C2 Great Exuma Island, C The Bahamas
George Town 32 B3 var. Georgetown. dependent territory capital (Cayman Islands) Grand Cayman, SW Cayman Islands
George Town 116 B3 var. Penang, Pinang. Pinang, Peninsular Malaysia
Georgetown 21 F2 South Carolina, SE USA
Georgetown see George Town
George V Land 132 C4 physical region Antarctica
Georgia 95 F2 off. Republic of Georgia, Geor. Sak'art'velo, Rus. Gruzinskaya SSR, Gruziya. country SW Asia
Georgia 20 D2 off. State of Georgia, also known as Empire State of the South, Peach State. state SE USA
Georgian Bay 18 D2 lake bay Ontario, S Canada
Georgia, Republic of see Georgia
Georgia, Strait of 24 A1 strait British Columbia, W Canada
Georgi Dimitrov see Kostenets
Georgiu-Dezh see Liski
Georg von Neumayer 132 A2 German research station Antarctica
Gera 72 C4 Thüringen, E Germany
Geráki 83 B6 Pelopónnisos, S Greece
Geraldine 129 B6 Canterbury, South Island, New Zealand
Geraldton 125 A6 Western Australia
Geral, Serra 35 D5 mountain range S Brazil
Gerede 94 C2 Bolu, N Turkey
Gereshk 100 D5 Helmand, SW Afghanistan
Gering 22 D3 Nebraska, C USA
German East Africa see Tanzania
Germanicopolis see Çankırı
Germanicum, Mare/German Ocean see North Sea
German Southwest Africa see Namibia
Germany 72 B4 off. Federal Republic of Germany, Bundesrepublik Deutschland, Ger. Deutschland. country N Europe
Germany, Federal Republic of see Germany
Geroliménas 83 B7 Pelopónnisos, S Greece
Gerona see Girona
Gerpinnes 65 C7 Hainaut, S Belgium
Gerunda see Girona
Gerze 94 D2 Sinop, N Turkey
Gesoriacum see Boulogne-sur-Mer
Gessoriacum see Boulogne-sur-Mer
Getafe 70 D3 Madrid, C Spain
Gevaş 95 F3 Van, SE Turkey
Gevgeli see Gevgelija
Gevgelija 79 E6 var. Đevđelija, Djevdjelija, Turk. Gevgeli. SE Macedonia
Ghaba see Al Ghābah
Ghana 53 E5 off. Republic of Ghana. country W Africa
Ghanzi 56 C3 var. Khanzi. Ghanzi, W Botswana
Gharandal 97 B7 Al 'Aqabah, SW Jordan
Gharbt, Jabal al see Liban, Jebel
Ghardaïa 48 D2 N Algeria
Gharvān see Gharyān
Gharyān 49 F2 var. Gharvān. NW Libya
Ghawdex see Gozo
Ghaznī 101 E4 var. Ghazni. Ghaznī, E Afghanistan
Ghazzah see Gaza
Gheel see Geel
Ghent see Gent
Gheorgheni 86 C4 prev. Gheorghieni, Sîn-Miclăuş, Ger. Niklasmarkt, Hung. Gyergyószentmiklós. Harghita, C Romania
Gheorghieni see Gheorgheni
Ghōrīān 100 D4 prev. Ghūrīān. Herāt, W Afghanistan
Ghūdara 101 F3 var. Gudara, Rus. Kudara. SE Tajikistan
Ghurdaqah see Al Ghurdaqah
Ghūrīān see Ghōrīān
Giamame see Jamaame
Giannitsá 82 B4 var. Yiannitsá. Kentrikí Makedonía, N Greece
Gibraltar 71 G4 UK dependent territory SW Europe
Gibraltar, Bay of 71 G5 bay Gibraltar/Spain Europe Mediterranean Sea Atlantic Ocean
Gibraltar, Détroit de/Gibraltar, Estrecho de see Gibraltar, Strait of
Gibraltar, Strait of 70 C5 Fr. Détroit de Gibraltar, Sp. Estrecho de Gibraltar. strait Atlantic Ocean/Mediterranean Sea
Gibson Desert 125 B5 desert Western Australia
Giedraičiai 85 C5 Utena, E Lithuania
Giessen 73 B5 Hessen, W Germany
Gifu 109 C6 var. Gihu. Gifu, Honshū, SW Japan
Giganta, Sierra de la 28 B3 mountain range NW Mexico
Gihu see Gifu
G'ijduvon 100 D2 Rus. Gizhduvon. Buxoro Viloyati, C Uzbekistan
Gijón 70 D1 var. Xixón. Asturias, NW Spain
Gila River 26 A2 river Arizona, SW USA
Gilbert Islands see Tungaru
Gilbert River 126 C3 river Queensland, NE Australia
Gilf Kebir Plateau see Haḑabat al Jilf al Kabīr
Gillette 22 D3 Wyoming, C USA
Gilolo see Halmahera, Pulau
Gilroy 25 B6 California, W USA
Gimie, Mount 33 F1 mountain C Saint Lucia
Gimma see Jīma
Ginevra see Genève
Gingin 125 A6 Western Australia
Giohar see Jawhar
Gipeswic see Ipswich
Girardot 36 B3 Cundinamarca, C Colombia
Giresun 95 E2 var. Kerasunt; anc. Cerasus, Pharnacia. Giresun, NE Turkey
Girgenti see Agrigento
Girin see Jilin
Girne 80 C5 Gk. Kerýneia, Kyrenia. N Cyprus
Giron see Kiruna
Girona 71 G2 var. Gerona; anc. Gerunda. Cataluña, NE Spain
Gisborne 128 E3 Gisborne, North Island, New Zealand
Gissar Range 101 E3 Rus. Gissarskiy Khrebet. mountain range Tajikistan/Uzbekistan
Gissarskiy Khrebet see Gissar Range
Githio see Gýtheio
Giulianova 74 D4 Abruzzi, C Italy
Giumri see Gyumri
Giurgiu 86 C5 Giurgiu, S Romania
Giza 50 B1 var. Al Jīzah, El Gîza, Gizeh. N Egypt
Gizhduvon see G'ijduvon
Giżycko 76 D2 Ger. Lötzen. Warmińsko-Mazurskie, NE Poland

Gjakovë – Haast

Gjakovë *79 D5 Serb.* Đakovica. W Kosovo
Gjilan *79 D5 Serb.* Gnjilane. E Kosovo
Gjinokastër *see* Gjirokastër
Gjirokastër *79 C7 var.* Gjirokastra; *prev.* Gjinokastër, *Gk.* Argyrokastron, *It.* Argirocastro. Gjirokastër, S Albania
Gjirokastra *see* Gjirokastër
Gjoa Haven *15 F3 var.* Uqsuqtuuq. King William Island, Nunavut, NW Canada
Gjøvik *63 B5* Oppland, S Norway
Glace Bay *17 G4* Cape Breton Island, Nova Scotia, SE Canada
Gladstone *126 E4* Queensland, E Australia
Glåma *63 B5 var.* Glommen. *river* S Norway
Glasgow *66 C4* S Scotland, United Kingdom
Glavinitsa *82 D1 prev.* Pravda, Dogrular. Silistra, NE Bulgaria
Glavn'a Morava *see* Velika Morava
Glazov *89 D5* Udmurtskaya Respublika, NW Russia
Gleiwitz *see* Gliwice
Glendale *26 B2* Arizona, SW USA
Glendive *22 D2* Montana, NW USA
Glens Falls *19 F3* New York, NE USA
Glevum *see* Gloucester
Glina *78 B3 var.* Banijska Palanka. Sisak-Moslavina, NE Croatia
Glittertind *63 A5 mountain* S Norway
Gliwice *77 C5 Ger.* Gleiwitz. Śląskie, S Poland
Globe *26 B2* Arizona, SW USA
Globino *see* Hlobyne
Glogau *see* Głogów
Głogów *76 B4 Ger.* Glogau, Glogow. Dolnośląskie, SW Poland
Glogow *see* Głogów
Glomma *see* Glåma
Glommen *see* Glåma
Gloucester *67 D6 hist.* Caer Glou, *Lat.* Glevum. C England, United Kingdom
Głowno *76 D4* Łódź, C Poland
Glubokoye *see* Hlybokaye
Glukhov *see* Hlukhiv
Gnesen *see* Gniezno
Gniezno *76 C3 Ger.* Gnesen. Weilkopolskie, C Poland
Gnjilane *see* Gjilan
Gobabis *56 B3* Omaheke, E Namibia
Gobi *104 D3 desert* China/Mongolia
Gobō *109 C6* Wakayama, Honshū, SW Japan
Godāvari *102 B3 var.* Godavari. *river* C India
Godavari *see* Godāvari
Godhavn *see* Qeqertarsuaq
Godhra *112 C4* Gujarāt, W India
Göding *see* Hodonín
Godoy Cruz *42 B4* Mendoza, W Argentina
Godthaab/Godthåb *see* Nuuk
Godwin Austen, Mount *see* K2
Goede Hoop, Kaap de *see* Good Hope, Cape of
Goeie Hoop, Kaap die *see* Good Hope, Cape of
Goeree *64 B4 island* SW Netherlands
Goes *65 B5* Zeeland, SW Netherlands
Goettingen *see* Göttingen
Gogebic Range *18 B1 hill range* Michigan/Wisconsin, N USA
Goiânia *41 E3 prev.* Goyania. *state capital* Goiás, C Brazil
Goiás *41 E3 off.* Estado de Goiás; *prev.* Goiaz, Goyaz. *state/region* C Brazil
Goiás, Estado de *see* Goiás
Goiaz *see* Goiás
Goidhoo Atoll *see* Horsburgh Atoll
Gojōme *108 D4* Akita, Honshū, NW Japan
Gökçeada *82 D4 var.* Imroz Adasī, *Gk.* Imbros. *island* NW Turkey
Gökdepe *100 C3 Rus.* Gëkdepe, Geok-Tepe. Ahal Welaýaty, C Turkmenistan
Göksun *94 D4* Kahramanmaraş, C Turkey
Gol *63 A5* Buskerud, S Norway
Golan Heights *97 B5 Ar.* Al Jawlān, *Heb.* HaGolan. *mountain range* SW Syria
Golaya Pristan *see* Hola Prystan'
Gołdap *76 E2 Ger.* Goldap. Warmińsko-Mazurskie, NE Poland
Gold Coast *127 E5 cultural region* Queensland, E Australia
Golden Bay *128 C4 bay* South Island, New Zealand
Golden State, The *see* California
Goldingen *see* Kuldīga
Goldsboro *21 F1* North Carolina, SE USA
Goleniów *76 B3 Ger.* Gollnow. Zachodnio-pomorskie, NW Poland
Gollnow *see* Goleniów
Golmo *see* Golmud
Golmud *104 D4 var.* Ge'e'mu, Golmo, *Chin.* Ko-erh-mu. Qinghai, C China
Golovanevsk *see* Holovanivs'k
Golub-Dobrzyń *76 C3* Kujawski-pomorskie, C Poland
Goma *55 E6* Nord-Kivu, NE Dem. Rep. Congo
Gombi *53 H4* Adamawa, E Nigeria
Gombroon *see* Bandar-e 'Abbās
Gomel' *see* Homyel'
Gomera *48 A3 island* Islas Canarias, Spain, NE Atlantic Ocean
Gómez Palacio *28 D3* Durango, C Mexico
Gonaïves *32 D3 var.* Les Gonaïves. N Haiti
Gonâve, Île de la *32 D3 island* C Haiti
Gondar *see* Gonder
Gonder *50 C4 var.* Gondar. Āmara, NW Ethiopia
Gondia *113 E4* Mahārāshtra, C India
Gonggar *104 C5 var.* Gyixong. Xizang Zizhiqu, W China
Gongola *53 G4 river* E Nigeria
Gongtang *see* Damxung
Gonni/Gónnos *see* Gónnoi
Gónnoi *82 B4 var.* Gonni, Gónnos; *prev.* Derelí. Thessalía, C Greece
Good Hope, Cape of *56 B5 Afr.* Kaap de Goede Hoop, Kaap die Goeie Hoop. *headland* SW South Africa
Goodland *22 D4* Kansas, C USA
Goondiwindi *127 D5* Queensland, E Australia
Goor *64 E3* Overijssel, E Netherlands
Goose Green *43 D7 var.* Prado del Ganso. East Falkland, Falkland Islands
Goose Lake *24 B4 var.* Lago dos Gansos. *lake* California/Oregon, W USA
Gopher State *see* Minnesota
Göppingen *73 B6* Baden-Württemberg, SW Germany
Góra Kalwaria *92 D4* Mazowieckie, C Poland
Gorakhpur *113 E3* Uttar Pradesh, N India
Gorany *see* Harany
Goražde *78 C4* Federacija Bosna I Hercegovina, SE Bosnia and Herzegovina
Gorbovichi *see* Harbavichy
Goré *54 C4* Logone-Oriental, S Chad
Gorē *51 C5* Oromīya, C Ethiopia
Gore *129 B7* Southland, South Island, New Zealand
Gorgān *98 D2 var.* Astarabad, Astrabad, Gurgan, *prev.* Asterābād; *anc.* Hyrcania. Golestān, N Iran
Gori *95 F2* C Georgia
Gorinchem *64 C4 var.* Gorkum. Zuid-Holland, C Netherlands
Goris *95 G3* SE Armenia
Gorki *see* Horki
Gor'kiy *see* Nizhniy Novgorod
Gorkum *see* Gorinchem
Görlitz *72 D4* Sachsen, E Germany
Görlitz *see* Zgorzelec
Gorlovka *see* Horlivka
Gorna Dzhumaya *see* Blagoevgrad
Gornja Mužlja *see* Mužlja
Gornji Milanovac *78 C4* Serbia, C Serbia
Gorodets *see* Haradzyets
Gorodishche *see* Horodyshche
Gorodnya *see* Horodnya
Gorodok *see* Haradok
Gorodok/Gorodok Yagellonski *see* Horodok
Gorontalo *117 E4* Sulawesi, C Indonesia
Gorontalo, Teluk *see* Tomini, Gulf of
Gorssel *64 D3* Gelderland, E Netherlands
Goryn *see* Horyn'
Gorzów Wielkopolski *76 B3 Ger.* Landsberg, Landsberg an der Warthe. Lubuskie, W Poland
Gosford *127 D6* New South Wales, SE Australia
Goshogawara *108 D3 var.* Gosyogawara. Aomori, Honshū, C Japan
Gospić *78 A3* Lika-Senj, C Croatia
Gostivar *79 D6* W Macedonia
Gosyogawara *see* Goshogawara
Göteborg *63 B7 Eng.* Gothenburg. Västra Götaland, S Sweden
Gotel Mountains *53 G5 mountain range* E Nigeria
Gotha *72 C4* Thüringen, C Germany
Gothenburg *see* Göteborg
Gotland *63 C7 island* SE Sweden
Goto-retto *109 A7 island group* SW Japan
Gotska Sandön *84 B1 island* SE Sweden
Gōtsu *109 B6 var.* Gôtu. Shimane, Honshū, SW Japan
Göttingen *72 B4 var.* Goettingen. Niedersachsen, C Germany
Gottschee *see* Kočevje
Gottwaldov *see* Zlín
Gôtu *see* Gōtsu
Gouda *64 C4* Zuid-Holland, C Netherlands
Gough Fracture Zone *45 C6 tectonic feature* S Atlantic Ocean
Gough Island *47 B8 island* Tristan da Cunha, S Atlantic Ocean
Gouin, Réservoir *16 D4 reservoir* Québec, SE Canada
Goulburn *127 D6* New South Wales, SE Australia
Goundam *53 E3* Tombouctou, NW Mali
Gouré *53 G3* Zinder, SE Niger
Goverla, Gora *see* Hoverla, Hora
Governador Valadares *41 F4* Minas Gerais, SE Brazil
Govi Altayn Nuruu *105 E3 mountain range* S Mongolia
Goya *42 D3* Corrientes, NE Argentina
Goyania *see* Goiânia
Goyaz *see* Goiás
Goz Beïda *54 C3* Ouaddaï, SE Chad
Gozo *75 C8 var.* Ghawdex. *island* N Malta
Graciosa *70 A5 var.* Ilha Graciosa. *island* Azores, Portugal, NE Atlantic Ocean
Graciosa, Ilha *see* Graciosa
Gradačac *78 C3* Federacija Bosna I Hercegovina, N Bosnia and Herzegovina
Gradaús, Serra dos *41 E3 mountain range* C Brazil
Gradiška *see* Bosanska Gradiška
Grafton *127 E5* New South Wales, SE Australia
Grafton *23 E1* North Dakota, N USA
Graham Land *132 A2 physical region* Antarctica
Grajewo *76 E3* Podlaskie, NE Poland
Grampian Mountains *66 C3 mountain range* C Scotland, United Kingdom
Gran *see* Esztergom, Hungary
Granada *30 D3* Granada, SW Nicaragua
Granada *70 D5* Andalucía, S Spain
Gran Canaria *48 A3 var.* Grand Canary. *island* Islas Canarias, Spain, NE Atlantic Ocean
Gran Chaco *42 D2 var.* Chaco. *lowland plain* South America
Grand Bahama Island *32 B1 island* N The Bahamas
Grand Banks of Newfoundland *12 E4 undersea basin* NW Atlantic Ocean
Grand Bassa *see* Buchanan
Grand Canary *see* Gran Canaria
Grand Canyon *26 A1 canyon* Arizona, SW USA
Grand Canyon State *see* Arizona
Grand Cayman *32 B3 island* SW Cayman Islands
Grand Duchy of Luxembourg *see* Luxembourg
Grande, Bahía *43 B7 bay* S Argentina
Grande-Comore *see* Ngazidja
Grande de Chiloé, Isla *see* Chiloé, Isla de
Grande Prairie *15 E4* Alberta, W Canada
Grand Erg Occidental *48 D3 desert* W Algeria
Grand Erg Oriental *49 E3 desert* Algeria/Tunisia
Rio Grande *29 E2 var.* Río Bravo, *Sp.* Río Bravo del Norte, Bravo del Norte. *river* Mexico/USA
Grande Terre *33 G3 island* E West Indies
Grand Falls *17 G3* Newfoundland, Newfoundland and Labrador, SE Canada
Grand Forks *23 E1* North Dakota, N USA
Grandichi *see* Hrandzichy
Grand Island *23 E4* Nebraska, C USA
Grand Junction *22 C4* Colorado, C USA
Grand Paradis *see* Gran Paradiso
Grand Rapids *18 C3* Michigan, N USA
Grand Rapids *23 F1* Minnesota, N USA
Grand-Saint-Bernard, Col du *see* Great Saint Bernard Pass
Grand-Santi *37 G3* W French Guiana
Granite State *see* New Hampshire
Gran Lago *see* Nicaragua, Lago de
Gran Malvina *see* West Falkland
Gran Paradiso *74 A2 Fr.* Grand Paradis. *mountain* NW Italy
Gran San Bernardo, Passo di *see* Great Saint Bernard Pass
Gran Santiago *see* Santiago
Grants *26 C2* New Mexico, SW USA
Grants Pass *24 B4* Oregon, NW USA
Granville *68 B3* Manche, N France
Gratianopolis *see* Grenoble
Gratz *see* Graz
Graudenz *see* Grudziądz
Graulhet *69 C6* Tarn, S France
Grave *64 D4* Noord-Brabant, SE Netherlands
Grayling *14 C2* Alaska, USA
Graz *73 E7 prev.* Gratz. Steiermark, SE Austria
Great Abaco *32 C1 var.* Abaco Island. *island* N The Bahamas
Great Alfold *see* Great Hungarian Plain
Great Ararat *see* Büyükağrı Dağı
Great Australian Bight *125 D7 bight* S Australia
Great Barrier Island *128 D2 island* N New Zealand
Great Barrier Reef *126 D2 reef* Queensland, NE Australia
Great Basin *25 C5 basin* W USA
Great Bear Lake *15 E3 Fr.* Grand Lac de l'Ours. *lake* Northwest Territories, NW Canada
Great Belt *63 B8 var.* Store Bælt, *Eng.* Great Belt, Storebelt. *channel* Baltic Sea/Kattegat
Great Belt *see* Storebælt
Great Bend *23 E5* Kansas, C USA
Great Britain *see* Britain
Great Dividing Range *126 D4 mountain range* NE Australia
Greater Antilles *32 D3 island group* West Indies
Greater Caucasus *95 G2 mountain range* Azerbaijan/Georgia/Russia Asia/Europe
Greater Sunda Islands *102 D5 var.* Sunda Islands. *island group* Indonesia
Great Exhibition Bay *128 C1 inlet* North Island, New Zealand
Great Exuma Island *32 C2 island* C The Bahamas
Great Falls *22 B1* Montana, NW USA
Great Grimsby *see* Grimsby
Great Hungarian Plain *77 C7 var.* Great Alfold, Plain of Hungary, *Hung.* Alföld. *plain* SE Europe
Great Inagua *32 D2 var.* Inagua Islands. *island* S The Bahamas
Great Indian Desert *see* Thar Desert
Great Khingan Range *see* Da Hinggan Ling
Great Lake *see* Tônlé Sap
Great Lakes *13 C5 lakes* Ontario, Canada/USA
Great Lakes State *see* Michigan
Great Meteor Seamount *see* Great Meteor Tablemount
Great Meteor Tablemount *44 B3 var.* Great Meteor Seamount. *seamount* E Atlantic Ocean
Great Nicobar *111 G3 island* Nicobar Islands, India, NE Indian Ocean
Great Plain of China *103 E2 plain* E China
Great Plains *23 E3 var.* High Plains. *plains* Canada/USA
Great Rift Valley *51 C5 var.* Rift Valley. *depression* Asia/Africa
Great Ruaha *51 C7 river* S Tanzania
Great Saint Bernard Pass *74 A1 Fr.* Col du Grand-Saint-Bernard, *It.* Passo del Gran San Bernardo. *pass* Italy/Switzerland
Great Salt Desert *see* Kavīr, Dasht-e
Great Salt Lake *22 A3 salt lake* Utah, W USA
Great Salt Lake Desert *22 A4 plain* Utah, W USA
Great Sand Sea *49 H3 desert* Egypt/Libya
Great Sandy Desert *124 C4 desert* Western Australia
Great Sandy Desert *see* Ar Rub 'al Khālī
Great Sandy Island *see* Fraser Island
Great Slave Lake *15 E4 Fr.* Grand Lac des Esclaves. *lake* Northwest Territories, NW Canada
Great Socialist People's Libyan Arab Jamahiriya *see* Libya
Great Sound *20 A5 sound* Bermuda, NW Atlantic Ocean
Great Victoria Desert *125 C5 desert* South Australia/Western Australia
Great Wall of China *106 C4 ancient monument* N China Asia
Great Yarmouth *67 E6 var.* Yarmouth. E England, United Kingdom
Grebenka *see* Hrebinka
Gredos, Sierra de *70 D3 mountain range* W Spain
Greece *83 A5 off.* Hellenic Republic, *Gk.* Ellás; *anc.* Hellas. *country* SE Europe
Greeley *22 D4* Colorado, C USA
Green Bay *18 B2* Wisconsin, N USA
Green Bay *18 B2 lake bay* Michigan/Wisconsin, N USA
Greeneville *21 E1* Tennessee, S USA
Greenland *60 D3 Dan.* Grønland, *Inuit* Kalaallit Nunaat. *Danish self-governing territory* NE North America
Greenland Sea *61 F2 sea* Arctic Ocean
Green Mountains *19 G2 mountain range* Vermont, NE USA
Green Mountain State *see* Vermont
Greenock *66 C4* W Scotland, United Kingdom
Green River *22 B3* Wyoming, C USA
Green River *18 C5 river* Kentucky, C USA
Green River *22 B4 river* Utah, W USA
Greensboro *21 F1* North Carolina, SE USA
Greenville *20 B2* Mississippi, S USA
Greenville *21 F1* North Carolina, SE USA
Greenville *21 E1* South Carolina, SE USA
Greenville *27 G2* Texas, SW USA
Greenwich *67 B8* United Kingdom
Greenwood *20 B2* Mississippi, S USA
Greenwood *21 E2* South Carolina, SE USA
Gregory Range *126 C3 mountain range* Queensland, E Australia
Greifenberg/Greifenberg in Pommern *see* Gryfice
Greifswald *72 D2* Mecklenburg-Vorpommern, NE Germany
Grenada *20 C2* Mississippi, S USA
Grenada *33 G5 country* SE West Indies
Grenadines, The *33 H4 island group* Grenada/St Vincent and the Grenadines
Grenoble *69 D5 anc.* Cularo, Gratianopolis. Isère, E France
Gresham *24 B3* Oregon, NW USA
Grevená *82 B4* Dytikí Makedonía, N Greece
Grevenmacher *65 E8* Grevenmacher, E Luxembourg
Greymouth *129 B5* West Coast, South Island, New Zealand
Grey Range *127 C5 mountain range* New South Wales/Queensland, E Australia
Greytown *see* San Juan del Norte
Griffin *20 D2* Georgia, SE USA
Grimari *54 C4* Ouaka, C Central African Republic
Grimsby *67 E5 prev.* Great Grimsby. E England, United Kingdom
Grobin *see* Grobiņa
Grobiņa *84 B3 Ger.* Grobin. Liepāja, W Latvia
Gródek Jagielloński *see* Horodok
Grodno *see* Hrodna
Grodzisk Wielkopolski *76 B3* Wielkopolskie, C Poland
Groesbeek *64 D4* Gelderland, SE Netherlands
Grójec *76 D4* Mazowieckie, C Poland
Groningen *64 E1* Groningen, NE Netherlands
Grønland *see* Greenland
Groote Eylandt *126 B2 island* Northern Territory, N Australia
Grootfontein *56 B3* Otjozondjupa, N Namibia
Groot Karasberge *56 B4 mountain range* S Namibia
Gros Islet *33 F1* N Saint Lucia
Grossa, Isola *see* Dugi Otok
Grossbetschkerek *see* Zrenjanin
Grosse Morava *see* Velika Morava
Grosser Sund *see* Suur Väin
Grosseto *74 B4* Toscana, C Italy
Grossglockner *73 C7 mountain* W Austria
Grosskanizsa *see* Nagykanizsa
Gross-Karol *see* Carei
Grosskikinda *see* Kikinda
Grossmichel *see* Michalovce
Gross-Schlatten *see* Abrud
Grosswardein *see* Oradea
Groznyy *89 B8* Chechenskaya Respublika, SW Russia
Grudovo *see* Sredets
Grudziądz *76 C3 Ger.* Graudenz. Kujawsko-pomorskie, C Poland
Grums *63 B6* Värmland, C Sweden
Grünberg/Grünberg in Schlesien *see* Zielona Góra
Grüneberg *see* Zielona Góra
Gruzinskaya SSR/Gruziya *see* Georgia
Gryazi *89 B6* Lipetskaya Oblast', W Russia
Gryfice *76 B2 Ger.* Greifenberg, Greifenberg in Pommern. Zachodnio-pomorskie, NW Poland
Guabito *31 E4* Bocas del Toro, NW Panama
Guadalajara *28 D4* Jalisco, C Mexico
Guadalajara *71 E3 Ar.* Wad Al-Hajarah; *anc.* Arriaca. Castilla-La Mancha, C Spain
Guadalcanal *122 C3 island* C Solomon Islands
Guadalquivir *70 D4 river* W Spain
Guadalupe *28 D3* Zacatecas, C Mexico
Guadalupe Peak *26 D3 mountain* Texas, SW USA
Guadalupe River *27 G4 river* SW USA
Guadarrama, Sierra de *71 E2 mountain range* C Spain
Guadeloupe *33 H3 French overseas department* E West Indies
Guadiana *70 C4 river* Portugal/Spain
Guadix *71 E4* Andalucía, S Spain
Guaimaca *30 C2* Francisco Morazán, C Honduras
Guajira, Península de la *36 B1 peninsula* N Colombia
Gualaco *30 D2* Olancho, C Honduras
Gualán *30 B2* Zacapa, C Guatemala
Gualdicciolo *74 D1* NW San Marino
Gualeguaychú *42 D4* Entre Ríos, E Argentina
Guam *122 B1 US unincorporated territory* W Pacific Ocean
Guamúchil *28 C3* Sinaloa, C Mexico
Guanabacoa *32 B2* La Habana, W Cuba
Guanajuato *29 E4* Guanajuato, C Mexico
Guanare *36 C2* Portuguesa, N Venezuela
Guanare, Río *36 D2 river* W Venezuela
Guangdong *106 C6 var.* Guangdong Sheng, Kuang-tung, Kwangtung, Yue. *province* S China
Guangdong Sheng *see* Guangdong
Guangju *see* Gwangju
Guangxi *see* Guangxi Zhuangzu Zizhiqu
Guangxi Zhuangzu Zizhiqu *106 C6 var.* Guangxi, Gui, Kuang-hsi, Kwangsi, *Eng.* Kwangsi Chuang Autonomous Region. *autonomous region* S China
Guangyuan *106 B5 var.* Kuang-yuan, Kwangyuan. Sichuan, C China
Guangzhou *106 C6 var.* Kuang-chou, Kwangchow, *Eng.* Canton. *province capital* Guangdong, S China
Guantánamo *32 D3* Guantánamo, SE Cuba
Guantánamo, Bahia de *32 D3 Eng.* Guantanamo Bay. *US military base* SE Cuba
Guantanamo Bay *see* Guantánamo, Bahia de
Guaporé, Rio *40 D3 var.* Río Iténez. *river* Bolivia/Brazil
Guarda *70 C3* Guarda, N Portugal
Guarumal *31 F5* Veraguas, S Panama
Guasave *28 C3* Sinaloa, C Mexico
Guatemala *30 A2 off.* Republic of Guatemala. *country* Central America
Guatemala Basin *13 B7 undersea basin* E Pacific Ocean
Guatemala City *see* Ciudad de Guatemala
Guatemala, Republic of *see* Guatemala
Guaviare *34 B2 off.* Comisaría Guaviare. *province* S Colombia
Guaviare, Comisaría *see* Guaviare
Guaviare, Río *36 D3 river* E Colombia
Guayanas, Macizo de las *see* Guiana Highlands
Guayaquil *38 A2 var.* Santiago de Guayaquil. Guayas, SW Ecuador
Guayaquil, Golfo de *38 A2 var.* Gulf of Guayaquil. *gulf* SW Ecuador
Guayaquil, Gulf of *see* Guayaquil, Golfo de
Guaymas *28 B2* Sonora, NW Mexico
Gubadag *100 C2 Turkm.* Tel'man; *prev.* Tel'mansk. Daşoguz Welaýaty, N Turkmenistan
Guben *72 D4 var.* Wilhelm-Pieck-Stadt. Brandenburg, E Germany
Gudara *see* Ghŭdara
Gudauta *95 E1* NW Georgia
Guéret *68 C4* Creuse, C France
Guernsey *67 D8 British Crown Dependency* Channel Islands, NW Europe
Guerrero Negro *28 A2* Baja California Sur, NW Mexico
Gui *see* Guangxi Zhuangzu Zizhiqu
Guiana *see* French Guiana
Guiana Highlands *40 D1 var.* Macizo de las Guayanas. *mountain range* N South America
Guiba *see* Juba
Guidder *see* Guider
Guider *54 B4 var.* Guidder. Nord, N Cameroon
Guidimouni *53 G3* Zinder, S Niger
Guildford *67 D7* SE England, United Kingdom
Guilin *106 C6 var.* Kuei-lin, Kweilin. Guangxi Zhuangzu Zizhiqu, S China
Guimarães *70 B2 var.* Guimaráes. Braga, N Portugal
Guimaráes *see* Guimarães
Guinea *52 C4 off.* Republic of Guinea, *var.* Guinée; *prev.* French Guinea, People's Revolutionary Republic of Guinea. *country* W Africa
Guinea Basin *47 A5 undersea basin* E Atlantic Ocean
Guinea-Bissau *52 B4 off.* Republic of Guinea-Bissau, *Fr.* Guinée-Bissau, *Port.* Guiné-Bissau; *prev.* Portuguese Guinea. *country* W Africa
Guinea-Bissau, Republic of *see* Guinea-Bissau
Guinea, Gulf of *46 B4 Fr.* Golfe de Guinée. *gulf* E Atlantic Ocean
Guinea, People's Revolutionary Republic of *see* Guinea
Guinea, Republic of *see* Guinea
Guiné-Bissau *see* Guinea-Bissau
Guinée *see* Guinea
Guinée-Bissau *see* Guinea-Bissau
Guinée, Golfe de *see* Guinea, Gulf of
Güiria *37 E1* Sucre, NE Venezuela
Guiyang *106 B6 var.* Kuei-Yang, Kuei-yang, Kueyang, Kweiyang; *prev.* Kweichu. *province capital* Guizhou, S China
Guizhou *106 B6* Guangdong, SE China
Gujarāt *112 C4 var.* Gujerat. *cultural region* W India
Gujerat *see* Gujarāt
Gujrānwāla *112 D2* Punjab, NE Pakistan
Gujrāt *112 D2* Punjab, E Pakistan
Gulbarga *see* Kalaburagi
Gulbene *84 D3 Ger.* Alt-Schwanenburg. Gulbene, NE Latvia
Gulf of Liaotung *see* Liaodong Wan
Gulfport *20 C3* Mississippi, S USA
Gulf, The *see* Persian Gulf
Gulistan *see* Guliston
Guliston *101 E2 Rus.* Gulistan. Sirdaryo Viloyati, E Uzbekistan
Gulja *see* Yining
Gulkana *14 D3* Alaska, USA
Gulu *51 B6* N Uganda
Gulyantsi *82 C1* Pleven, N Bulgaria
Guma *see* Pishan
Gumbinnen *see* Gusev
Gumpolds *see* Humpolec
Gümülcine/Gümüljina *see* Komotiní
Gümüşane *see* Gümüşhane
Gümüşhane *95 E3 var.* Gümüşane, Gumushkhane. Gümüşhane, NE Turkey
Gumushkhane *see* Gümüşhane
Güney Doğu Toroslar *95 E4 mountain range* SE Turkey
Gunnbjørn Fjeld *60 D4 var.* Gunnbjörns Bjerge. *mountain* C Greenland
Gunnbjörns Bjerge *see* Gunnbjørn Fjeld
Gunnedah *127 D6* New South Wales, SE Australia
Gunnison *22 C5* Colorado, C USA
Gurbansoltan Eje *100 C2 prev.* Ýylanly, *Rus.* Il'yaly. Daşoguz Welaýaty, N Turkmenistan
Gurbantünggüt Shamo *104 B2 desert* W China
Gurgan *see* Gorgān
Guri, Embalse de *37 E2 reservoir* E Venezuela
Gurkfeld *see* Krško
Gurktaler Alpen *73 D7 mountain range* S Austria
Gürün *94 D3* Sivas, C Turkey
Gur'yev/Gur'yevskaya Oblast' *see* Atyrau
Gusau *53 G4* Zamfara, NW Nigeria
Gusev *84 B4 Ger.* Gumbinnen. Kaliningradskaya Oblast', W Russia
Gustavus *14 D4* Alaska, USA
Güstrow *72 C3* Mecklenburg-Vorpommern, NE Germany
Guta/Gúta *see* Kolárovo
Gütersloh *72 B4* Nordrhein-Westfalen, W Germany
Gutta *see* Kolárovo
Guttstadt *see* Dobre Miasto
Guwāhāti *113 G3 prev.* Gauhāti. Assam, NE India
Guyana *37 F3 off.* Co-operative Republic of Guyana; *prev.* British Guiana. *country* N South America
Guyana, Co-operative Republic of *see* Guyana
Guyane *see* French Guiana
Guymon *27 E1* Oklahoma, C USA
Güzelyurt Körfezi *80 C5 Gk.* Kólpos Mórfu, Morphou. W Cyprus
Gvardeysk *84 A4 Ger.* Tapaiu. Kaliningradskaya Oblast', W Russia
Gwādar *112 A3 var.* Gwadur. Baluchistān, SW Pakistan
Gwadur *see* Gwādar
Gwalior *112 D3* Madhya Pradesh, C India
Gwanda *56 D3* Matabeleland South, SW Zimbabwe
Gwangju *107 E4 off.* Gwangju Gwang-yeoksi, *prev.* Kwangju, *var.* Guangju, Kwangchu, *Jap.* Kōshū. SW South Korea
Gwangju Gwang-yeoksi *see* Gwangju
Gwy *see* Wye
Gyandzha *see* Gäncä
Gyangzê *104 C5* Xizang Zizhiqu, W China
Gyaring Co *104 C5 lake* W China
Gyêgu *see* Yushu
Gyergyószentmiklós *see* Gheorgheni
Gyixong *see* Gonggar
Gympie *127 E5* Queensland, E Australia
Gyomaendrőd *77 D7* Békés, SE Hungary
Gyöngyös *77 D6* Heves, NE Hungary
Győr *77 C6 Ger.* Raab, *Lat.* Arrabona. Győr-Moson-Sopron, NW Hungary
Gýtheio *83 B6 var.* Githio; *prev.* Ýíthion. Pelopónnisos, S Greece
Gyulafehérvár *see* Alba Iulia
Gyumri *95 F2 var.* Giumri, *Rus.* Kumayri; *prev.* Aleksandropol', Leninakan. W Armenia
Gyzyrlabat *see* Serdar

H

Haabai *see* Ha'apai Group
Haacht *65 C6* Vlaams Brabant, C Belgium
Haaksbergen *64 E3* Overijssel, E Netherlands
Ha'apai Group *123 F4 var.* Haabai. *island group* C Tonga
Haapsalu *84 D2 Ger.* Hapsal. Läänemaa, W Estonia
Ha'Arava *see* 'Arabah, Wādī al
Haarlem *64 C3 prev.* Harlem. Noord-Holland, W Netherlands
Haast *129 B6* West Coast, South Island, New Zealand

Hachijo-jima – Houlton

Hachijo-jima 109 D6 island Izu-shotō, SE Japan
Hachinohe 108 D3 Aomori, Honshū, C Japan
Ḩadabat al Jilf al Kabīr 50 A2 var. Gilf Kebir Plateau. plateau SW Egypt
Hadama see Nazrēt
Hadejia 53 G4 Jigawa, N Nigeria
Hadejia 53 G3 river N Nigeria
Hadera 97 A6 var. Khadera; prev. Ḩadera. Haifa, C Israel
Ḩadera see Hadera
Hadhdhunmathi Atoll 110 A5 atoll S Maldives
Ha Đông 114 D3 var. Hadong. Ha Tây, N Vietnam
Hadong see Ha Đông
Hadramaut see Ḩaḑramawt
Ḩaḑramawt 99 C6 Eng. Hadramaut. mountain range S Yemen
Hadrianopolis see Edirne
Haerbin/Haerhpin/Ha-erh-pin see Harbin
Hafnia see Denmark
Hafnia see København
Hafren see Severn
Hafun, Ras see Xaafuun, Raas
Hagåtña 122 B1 , var. Agaña. dependent territory capital (Guam) NW Guam
Hagerstown 19 E4 Maryland, NE USA
Ha Giang 114 D3 Ha Giang, N Vietnam
Hagios Evstrátios see Ágios Efstrátios
HaGolan see Golan Heights
Hagondange 68 D3 Moselle, NE France
Haguenau 68 E3 Bas-Rhin, NE France
Haibowan see Wuhai
Haicheng 106 D3 Liaoning, NE China
Haidarabad see Hyderābād
Haifa see Hefa
Haifa, Bay of see Mifrats Hefa
Haifong see Hai Phong
Haikou 106 C7 var. Hai-k'ou, Hoihow, Fr. Hoï-Hao. province capital Hainan, S China
Hai-k'ou see Haikou
Ḩā'il 98 B4 Ḩā'il, NW Saudi Arabia
Hailuoto 62 D4 Swe. Karlö. island W Finland
Hainan 106 B7 var. Hainan Sheng, Qiong. province S China
Hainan Dao 106 C7 island S China
Hainan Sheng see Hainan
Hainasch see Ainaži
Haines 14 D4 Alaska, USA
Hainichen 72 D4 Sachsen, E Germany
Hai Phong 114 D3 var. Haifong, Haiphong. N Vietnam
Haiphong see Hai Phong
Haiti 32 D3 off. Republic of Haiti. country C West Indies
Haiti, Republic of see Haiti
Haiya 50 C3 Red Sea, NE Sudan
Hajdúhadház 77 D6 Hajdú-Bihar, E Hungary
Hajine see Abū Ḩardān
Hajnówka 76 E3 Ger. Hermhausen. Podlaskie, NE Poland
Hakodate 108 D3 Hokkaidō, NE Japan
Hal see Halle
Ḩalab 96 B2 Eng. Aleppo, Fr. Alep; anc. Beroea. Ḩalab, NW Syria
Hala'ib Triangle 50 C3 region Egypt/Sudan
Ḩalāniyāt, Juzur al 99 D6 var. Jazā'ir Bin Ghalfān, Eng. Kuria Muria Islands. island group S Oman
Halberstadt 72 C4 Sachsen-Anhalt, C Germany
Halden 63 B6 prev. Fredrikshald. Østfold, S Norway
Halfmoon Bay 129 A8 var. Oban. Stewart Island, Southland, New Zealand
Haliacmon see Aliákmonas
Halifax 17 F4 province capital Nova Scotia, SE Canada
Halkida see Chalkída
Halle 65 B6 Fr. Hal. Vlaams Brabant, C Belgium
Halle 72 C4 var. Halle an der Saale. Sachsen-Anhalt, C Germany
Halle an der Saale see Halle
Halle-Neustadt 72 C4 Sachsen-Anhalt, C Germany
Halley 132 B2 UK research station Antarctica
Hall Islands 120 B2 island group C Micronesia
Halls Creek 124 C3 Western Australia
Halmahera, Laut 117 F3 Eng. Halmahera Sea; sea E Indonesia
Halmahera, Pulau 117 F3 prev. Djailolo, Gilolo, Jailolo. island E Indonesia
Halmahera Sea see Halmahera, Laut
Halmstad 63 B7 Halland, S Sweden
Ha Long 114 E3 prev. Hông Gai; var. Hon Gai, Hongay. Quang Ninh, N Vietnam
Hälsingborg see Helsingborg
Hamada 109 B6 Shimane, Honshū, SW Japan
Hamadān 98 C3 anc. Ecbatana. Hamadān, W Iran
Ḩamāh 96 B3 var. Hama; anc. Epiphania, Bibl. Hamath. Ḩamāh, W Syria
Hamamatsu 109 D6 var. Hamamatu. Shizuoka, Honshū, S Japan
Hamamatu see Hamamatsu
Hamar 63 B5 prev. Storhammer. Hedmark, S Norway
Hamath see Ḩamāh
Hamburg 72 B3 Hamburg, N Germany
Hamd, Wadi al 98 A4 dry watercourse W Saudi Arabia
Hämeenlinna 63 D5 Swe. Tavastehus. Kanta-Häme, S Finland
HaMela h, Yam see Dead Sea
Hamersley Range 124 A4 mountain range Western Australia
Hamhŭng 107 E3 C North Korea
Hami 104 C3 var. Ha-mi, Uigh. Kumul, Qomul. Xinjiang Uygur Zizhiqu, NW China
Ha-mi see Hami
Hamilton 20 A5 dependent territory capital (Bermuda) C Bermuda
Hamilton 16 D5 Ontario, S Canada
Hamilton 128 D3 Waikato, North Island, New Zealand
Hamilton 66 C4 S Scotland, United Kingdom
Hamilton 20 C2 Alabama, S USA
Hamim, Wadi al 49 G2 river NE Libya
Hamïs Musait see Khamīs Mushayt
Hamm 72 B4 var. Hamm in Westfalen. Nordrhein-Westfalen, W Germany
Ḩammāmāt, Khalīj al see Hammamet, Golfe de
Hammamet, Golfe de 80 D3 Ar. Khalīj al Ḩammāmāt. gulf NE Tunisia
Ḩammār, Hawr al 98 C3 lake SE Iraq
Hamm in Westfalen see Hamm
Hampden 129 B7 Otago, South Island, New Zealand
Hampstead 67 A7 Maryland, USA
Ħamrun 80 B5 C Malta
Hāmūn, Daryācheh-ye see Şāberī, Hāmūn-e/Sīstān, Daryācheh-ye
Hamwih see Southampton
Hâncești see Hînceşti
Hancewicze see Hantsavichy
Handan 106 C4 var. Han-tan. Hebei, E China
Haneda 108 A2 (Tōkyō) Tōkyō, Honshū, S Japan
HaNegev 97 A7 Eng. Negev. desert S Israel
Hanford 25 C6 California, W USA
Hangayn Nuruu 104 D2 mountain range C Mongolia
Hang-chou/Hangchow see Hangzhou
Hangö see Hanko
Hangzhou 106 D5 var. Hang-chou, Hangchow. province capital Zhejiang, SE China
Hania see Chaniá
Hanka, Lake see Khanka, Lake
Hanko 63 D6 Swe. Hangö. Uusimaa, SW Finland
Han-kou/Han-k'ou/Hankow see Wuhan
Hanmer Springs 129 C5 Canterbury, South Island, New Zealand
Hannibal 23 G4 Missouri, C USA
Hannover 72 B3 Eng. Hanover. Niedersachsen, NW Germany
Hanöbukten 63 B7 bay S Sweden
Ha Nôi 114 D3 Eng. Hanoi, Fr. Hanoï. country capital (Vietnam) N Vietnam
Hanover see Hannover
Han Shui 105 E4 river C China
Han-tan see Handan
Hantsavichy 85 B6 Pol. Hancewicze, Rus. Gantsevichi. Brestskaya Voblasts', SW Belarus
Hanyang see Wuhan
Hanzhong 106 B5 Shaanxi, C China
Hāora 113 F4 prev. Howrah. West Bengal, NE India
Haparanda 62 D4 Norrbotten, N Sweden
Hapsal see Haapsalu
Haradok 85 E5 Rus. Gorodok. Vitsyebskaya Voblasts', N Belarus
Haradzyets 85 B6 Rus. Gorodets. Brestskaya Voblasts', SW Belarus
Haramachi 108 D4 Fukushima, Honshū, E Japan
Harany 85 D5 Rus. Gorany. Vitsyebskaya Voblasts', N Belarus
Harare 56 D3 prev. Salisbury. country capital (Zimbabwe) Mashonaland East, NE Zimbabwe
Harbavichy 85 E6 Rus. Gorbovichi. Mahilyowskaya Voblasts', E Belarus
Harbel 52 C5 W Liberia
Harbin 107 E2 var. Haerbin, Ha-erh-pin, Kharbin; prev. Haerhpin, Pingkiang, Pinkiang. province capital Heilongjiang, NE China
Hardangerfjorden 63 A6 fjord S Norway
Hardangervidda 63 A6 plateau S Norway
Hardenberg 64 E3 Overijssel, E Netherlands
Harelbeke 65 A6 var. Harlebeke. West-Vlaanderen, W Belgium
Harem see Ḩārim
Haren 64 E2 Groningen, NE Netherlands
Hārer 51 D5 E Ethiopia
Hargeisa see Hargeysa
Hargeysa 51 D5 var. Hargeisa. Woqooyi Galbeed, NW Somalia
Hariana see Haryāna
Hari, Batang 116 B4 prev. Djambi. river Sumatera, W Indonesia
Ḩārim 96 B2 var. Harem. Idlib, W Syria
Harima-nada 109 B6 sea S Japan
Harirud 101 E4 var. Tedzhen, Turkm. Tejen. river Afghanistan/Iran
Harlan 23 F3 Iowa, C USA
Harlebeke see Harelbeke
Harlem see Haarlem
Harlingen 64 D2 Fris. Harns. Friesland, N Netherlands
Harlingen 27 G5 Texas, SW USA
Harlow 67 E6 E England, United Kingdom
Harney Basin 24 B4 basin Oregon, NW USA
Härnösand 63 C5 var. Hernösand. Västernorrland, C Sweden
Harns see Harlingen
Harper 52 D5 var. Cape Palmas. NE Liberia
Harricana 16 D3 river Québec, SE Canada
Harris 66 B3 physical region NW Scotland, United Kingdom
Harrisburg 19 E4 state capital Pennsylvania, NE USA
Harrisonburg 19 E4 Virginia, NE USA
Harrison, Cape 17 F2 headland Newfoundland and Labrador, E Canada
Harris Ridge see Lomonosov Ridge
Harrogate 67 D5 N England, United Kingdom
Hârșova 86 D5 prev. Hîrşova. Constanţa, SE Romania
Harstad 62 C2 Troms, N Norway
Hartford 19 G3 state capital Connecticut, NE USA
Hartlepool 67 D5 N England, United Kingdom
Harunabad see Eslāmābād
Har Us Gol 104 C2 lake Hovd, W Mongolia
Har Us Nuur 104 C2 lake NW Mongolia
Harwich 67 E6 E England, United Kingdom
Haryāna 112 D2 var. Hariana. cultural region N India
Hashemite Kingdom of Jordan see Jordan
Hasselt 65 C6 Limburg, NE Belgium
Hassetché see Al Ḩasakah
Hasta Colonia/Hasta Pompeia see Asti
Hastings 128 E4 Hawke's Bay, North Island, New Zealand
Hastings 67 E7 SE England, United Kingdom
Hastings 23 E4 Nebraska, C USA
Hațeg 86 B4 Ger. Wallenthal, Hung. Hátszeg; prev. Hatzeg, Hötzing. Hunedoara, SW Romania
Hátszeg see Hațeg
Hattem 64 D3 Gelderland, E Netherlands
Hatteras, Cape 21 G1 headland North Carolina, SE USA
Hatteras Plain 13 D6 abyssal plain Atlantic Ocean
Hattiesburg 20 C3 Mississippi, S USA
Hatton Bank see Hatton Ridge
Hatton Ridge 58 B2 var. Hatton Bank. undersea ridge N Atlantic Ocean
Hat Yai 115 C7 var. Ban Hat Yai. Songkhla, SW Thailand
Hatzeg see Hațeg
Hatzfeld see Jimbolia
Haugesund 63 A6 Rogaland, S Norway
Haukeligrend 63 A6 Telemark, S Norway
Haukivesi 63 E5 lake SE Finland
Hauraki Gulf 128 D2 gulf North Island, N New Zealand
Hauroko, Lake 129 A7 lake South Island, New Zealand
Haut Atlas 48 C2 Eng. High Atlas. mountain range C Morocco
Hautes Fagnes 65 D6 Ger. Hohes Venn. mountain range E Belgium
Hauts Plateaux 48 D2 plateau Algeria/Morocco
Hauzenberg 73 D6 Bayern, SE Germany
Havana 13 D6 Illinois, N USA
Havana see La Habana
Havant 67 D7 S England, United Kingdom
Havelock 21 F1 North Carolina, SE USA
Havelock North 128 E4 Hawke's Bay, North Island, New Zealand
Haverfordwest 67 C6 SW Wales, United Kingdom
Havířov 77 C5 Moravskoslezský Kraj, E Czech Republic (Czechia)
Havre 22 C1 Montana, NW USA
Havre see le Havre
Havre-St-Pierre 17 F3 Québec, E Canada
Hawaii 25 A8 off. State of Hawaii, also known as Aloha State, Paradise of the Pacific, var. Hawai'i. state USA, C Pacific Ocean
Hawai'i 25 B8 var. Hawaii. island Hawai'ian Islands, USA, C Pacific Ocean
Hawai'ian Islands 130 D2 prev. Sandwich Islands. island group Hawaii, USA
Hawaiian Ridge 130 H4 undersea ridge N Pacific Ocean
Hawea, Lake 129 B6 lake South Island, New Zealand
Hawera 128 D4 Taranaki, North Island, New Zealand
Hawick 66 C4 SE Scotland, United Kingdom
Hawke Bay 128 E4 bay North Island, New Zealand
Hawkeye State see Iowa
Hawlēr see Arbil
Hawthorne 25 C6 Nevada, W USA
Hay 127 C6 New South Wales, SE Australia
HaYarden see Jordan
Hayastani Hanrapetut'yun see Armenia
Hayes 16 B2 river Manitoba, C Canada
Hay River 15 E4 Northwest Territories, W Canada
Hays 23 E5 Kansas, C USA
Haysyn 86 D3 Rus. Gaysin. Vinnyts'ka Oblast', C Ukraine
Hazar 100 B2 prev. Rus. Cheleken. Balkan Welaýaty, W Turkmenistan
Heard and McDonald Islands 119 B7 Australian external territory S Indian Ocean
Hearst 16 C4 Ontario, S Canada
Heart of Dixie see Alabama
Heathrow 67 A8 (London) SE England, United Kingdom
Hebei 106 C4 var. Hebei Sheng, Hopeh, Hopei, Ji; prev. Chihli. province E China
Hebei Sheng see Hebei
Hebron 97 A6 var. Al Khalil, El Khalil, Heb. Hevron; anc. Kiriath-Arba. S West Bank
Heemskerk 64 C3 Noord-Holland, W Netherlands
Heerde 64 D3 Gelderland, E Netherlands
Heerenveen 64 D2 Fris. It Hearrenfean. Friesland, N Netherlands
Heerhugowaard 64 C2 Noord-Holland, NW Netherlands
Heerlen 65 D6 Limburg, SE Netherlands
Heerwegen see Polkowice
Hefa 97 A5 var. Haifa, hist. Caiffa, Caiphas; anc. Sycaminum. Haifa, N Israel
Hefa, Mifraz see Mifrats Hefa
Hefei 106 D5 var. Hofei, hist. Luchow. province capital Anhui, E China
Hegang 107 E2 Heilongjiang, NE China
Hei see Heilongjiang
Heide 72 B2 Schleswig-Holstein, N Germany
Heidelberg 73 B5 Baden-Württemberg, SW Germany
Heidenheim see Heidenheim an der Brenz
Heidenheim an der Brenz 73 B6 var. Heidenheim. Baden-Württemberg, S Germany
Hei-ho see Nagqu
Heilbronn 73 B6 Baden-Württemberg, SW Germany
Heiligenbeil see Mamonovo
Heilongjiang 106 D2 var. Hei, Heilongjiang Sheng, Hei-lung-chiang, Heilungkiang. province NE China
Heilong Jiang see Amur
Heilongjiang Sheng see Heilongjiang
Heiloo 64 C3 Noord-Holland, NW Netherlands
Heilsberg see Lidzbark Warmiński
Hei-lung-chiang/Heilungkiang see Heilongjiang
Heimdal 63 B5 Sør-Trøndelag, S Norway
Heinaste see Ainaži
Hekimhan 94 D3 Malatya, C Turkey
Helena 22 B2 state capital Montana, NW USA
Helensville 128 D2 Auckland, North Island, New Zealand
Helgoland Bay see Helgoländer Bucht
Helgoländer Bucht 72 A2 var. Helgoland Bay, Heligoland Bight. bay NW Germany
Heligoland Bight see Helgoländer Bucht
Heliopolis see Baalbek
Hellas see Greece
Hellenic Republic see Greece
Hellevoetsluis 64 B4 Zuid-Holland, SW Netherlands
Hellín 71 E4 Castilla-La Mancha, C Spain
Darya-ye Helmand 100 D5 var. Rūd-e Hirmand. river Afghanistan/Iran
Helmantica see Salamanca
Helmond 65 D5 Noord-Brabant, S Netherlands
Helsingborg 63 B7 prev. Hälsingborg. Skåne, S Sweden
Helsingfors see Helsinki
Helsinki 63 D6 Swe. Helsingfors. country capital (Finland) Etelä-Suomi, S Finland
Heltau see Cisnădie
Helvetia see Switzerland
Henan 106 C5 var. Henan Sheng, Honan, Yu. province C China
Henderson 18 B5 Kentucky, S USA
Henderson 25 D7 Nevada, W USA
Henderson 27 H3 Texas, SW USA
Hendū Kosh see Hindu Kush
Hengchow see Hengyang
Hengduan Shan 106 A5 mountain range SW China
Hengelo 64 E3 Overijssel, E Netherlands
Hengnan see Hengyang
Hengyang 106 C6 var. Hengnan, Heng-yang; prev. Hengchow. Hunan, S China
Heng-yang see Hengyang
Heniches'k 87 F4 Rus. Genichesk. Khersons'ka Oblast', S Ukraine
Hennebont 68 A3 Morbihan, NW France
Henrique de Carvalho see Saurimo
Henzada see Hinthada
Herakleion see Irákleio
Herāt 100 D4 var. Herat; anc. Aria. Herāt, W Afghanistan
Heredia 31 E4 Heredia, C Costa Rica
Hereford 27 E2 Texas, SW USA
Herford 72 B4 Nordrhein-Westfalen, NW Germany
Héristal see Herstal
Herk-de-Stad 65 C6 Limburg, NE Belgium
Herlen Gol/Herlen He see Kerulen
Hermannstadt see Sibiu
Hermansverk 63 A5 Sogn Og Fjordane, S Norway
Hermhausen see Hajnówka
Hermiston 24 C2 Oregon, NW USA
Hermon, Mount 97 B5 Ar. Jabal ash Shaykh. mountain S Syria
Hermosillo 28 B2 Sonora, NW Mexico
Hermoupolis see Ermoúpoli
Hernösand see Härnösand
Herrera del Duque 70 D3 Extremadura, W Spain
Herselt 65 C5 Antwerpen, C Belgium
Herstal 65 D6 Fr. Héristal. Liège, E Belgium
Herzogenbusch see 's-Hertogenbosch
Hesse see Hessen
Hessen 73 B5 Eng./Fr. Hesse. state C Germany
Hevron see Hebron
Heydebrech see Kędzierzyn-Kozle
Heydekrug see Šilutė
Heywood Islands 124 C3 island group Western Australia
Hibbing 23 F1 Minnesota, N USA
Hibernia see Ireland
Hidalgo del Parral 28 C2 var. Parral. Chihuahua, N Mexico
Hida-sanmyaku 109 C5 mountain range Honshū, S Japan
Hierosolyma see Jerusalem
Hierro 48 A3 var. Ferro. island Islas Canarias, Spain, NE Atlantic Ocean
High Atlas see Haut Atlas
High Plains see Great Plains
High Point 21 E1 North Carolina, SE USA
Hiiumaa 84 C2 Ger. Dagden, Swe. Dagö. island W Estonia
Hikurangi 128 D2 Northland, North Island, New Zealand
Hildesheim 72 B4 Niedersachsen, N Germany
Hilla see Al Ḩillah
Hillaby, Mount 33 G1 mountain N Barbados
Hill Bank 30 C1 Orange Walk, N Belize
Hillegom 64 C3 Zuid-Holland, W Netherlands
Hilo 25 B8 Hawaii, USA, C Pacific Ocean
Hilton Head Island 21 E2 South Carolina, SE USA
Hilversum 64 C3 Noord-Holland, C Netherlands
Himalaya/Himalaya Shan see Himalayas
Himalayas 113 E2 var. Himalaya, Chin. Himalaya Shan. mountain range S Asia
Himeji 109 C6 var. Himezi. Hyōgo, Honshū, SW Japan
Himezi see Himeji
Ḩimş 96 B4 var. Homs; anc. Emesa. Ḩimş, C Syria
Hînceşti 86 D4 var. Hânceşti; prev. Kotovsk. C Moldova
Hinchinbrook Island 126 D3 island Queensland, NE Australia
Hinds 129 C6 Canterbury, South Island, New Zealand
Hindu Kush 101 F4 Per. Hendū Kosh. mountain range Afghanistan/Pakistan
Hinesville 21 E3 Georgia, SE USA
Hinnøya 62 C3 Lapp. Iinnasuolu. island C Norway
Hinson Bay 20 A5 bay W Bermuda, W Atlantic Ocean
Hinthada 114 B4 var. Henzada. Ayeyarwady, SW Myanmar (Burma)
Hios see Chíos
Hirfanlı Barajı 94 C3 reservoir C Turkey
Hirmand, Rūd-e see Helmand, Daryā-ye
Hirosaki 108 D3 Aomori, Honshū, C Japan
Hiroshima 109 B6 var. Hirosima. Hiroshima, Honshū, SW Japan
Hirschberg/Hirschberg im Riesengebirge/Hirschberg in Schlesien see Jelenia Góra
Hirson 68 D3 Aisne, N France
Hîrşova see Hârşova
Hispalis see Sevilla
Hispana/Hispania see Spain
Hispaniola 34 B1 island Dominion Republic/Haiti
Hitachi 109 D5 var. Hitati. Ibaraki, Honshū, S Japan
Hitati see Hitachi
Hitra 62 A4 prev. Hitteren. island S Norway
Hitteren see Hitra
Hjälmaren 63 C6 Eng. Lake Hjalmar. lake C Sweden
Hjalmar, Lake see Hjälmaren
Hjørring 63 B7 Nordjylland, N Denmark
Hkakabo Razi 114 B1 mountain Myanmar (Burma)/China
Hlobyne 87 F2 Rus. Globino. Poltavs'ka Oblast', NE Ukraine
Hlukhiv 87 F1 Rus. Glukhov. Sums'ka Oblast', NE Ukraine
Hlybokaye 85 D5 Rus. Glubokoye. Vitsyebskaya Voblasts', N Belarus
Hoa Binh 114 D3 Hoa Binh, N Vietnam
Hoang Lien Son 114 D3 mountain range N Vietnam
Hobart 127 C8 prev. Hobarton, Hobart Town. state capital Tasmania, SE Australia
Hobarton/Hobart Town see Hobart
Hobbs 27 E3 New Mexico, SW USA
Hobro 63 A7 Nordjylland, N Denmark
Hô Chi Minh 115 E6 var. Ho Chi Minh City; prev. Saigon. S Vietnam
Ho Chi Minh City see Hô Chi Minh
Hodeida see Al Ḩudaydah
Hódmezővásárhely 77 D7 Csongrád, SE Hungary
Hodna, Chott El 80 C4 var. Chott el-Hodna, Ar. Shatt al-Hodna. salt lake N Algeria
Hodna, Chott el-/Hodna, Shatt al- see Hodna, Chott El
Hodonín 77 C5 Ger. Göding. Jihomoravský Kraj, SE Czech Republic (Czechia)
Hoei see Huy
Hoey see Huy
Hof 73 C5 Bayern, SE Germany
Hofei see Hefei
Hōfu 109 B7 Yamaguchi, Honshū, SW Japan
Hofuf see Al Hufūf
Hogoley Islands see Chuuk Islands
Hohensalza see Inowrocław
Hohenstadt see Zábřeh
Hohes Venn see Hautes Fagnes
Hohe Tauern 73 C7 mountain range W Austria
Hohhot 105 F3 var. Huhehot, Huhohaote, Mong. Kukukhoto; prev. Kweisui, Kwesui. Nei Mongol Zizhiqu, N China
Hôi An 115 E5 prev. Faifo. Quang Nam-Đa Nâng, C Vietnam
Hoï-Hao/Hoihow see Haikou
Hokianga Harbour 128 C2 inlet SE Tasman Sea
Hokitika 129 B5 West Coast, South Island, New Zealand
Hokkaido 108 C2 prev. Ezo, Yeso, Yezo. island NE Japan
Hola Prystan' 87 E4 Rus. Golaya Pristan. Khersons'ka Oblast', S Ukraine
Holbrook 26 B2 Arizona, SW USA
Holetown 33 G1 prev. Jamestown. W Barbados
Holguín 32 C2 Holguín, SE Cuba
Hollabrunn 73 E6 Niederösterreich, NE Austria
Holland see Netherlands
Hollandia see Jayapura
Holly Springs 20 C1 Mississippi, S USA
Holman 15 E3 Victoria Island, Northwest Territories, N Canada
Holmsund 62 D4 Västerbotten, N Sweden
Holon 97 A6 var. Kholon; prev. Ḩolon. Tel Aviv, C Israel
Ḩolon see Holon
Holovanivs'k 87 E3 Rus. Golovanevsk. Kirovohrads'ka Oblast', C Ukraine
Holstebro 63 A7 Ringkøbing, W Denmark
Holsteinborg/Holsteinsborg/Holstenborg/Holstensborg see Sisimiut
Holyhead 67 C5 Wel. Caer Gybi. NW Wales, United Kingdom
Hombori 53 E3 Mopti, S Mali
Homs see Al Khums, Libya
Homs see Ḩimş
Homyel' 85 D7 Rus. Gomel'. Homyel'skaya Voblasts', SE Belarus
Honan see Luoyang, China
Honan see Henan, China
Hondo 27 F4 Texas, SW USA
Hondo see Honshū
Honduras 30 C2 off. Republic of Honduras. country Central America
Honduras, Golfo de see Honduras, Gulf of
Honduras, Gulf of 30 C2 Sp. Golfo de Honduras. gulf W Caribbean Sea
Honduras, Republic of see Honduras
Hønefoss 63 B6 Buskerud, S Norway
Honey Lake 25 B5 lake California, W USA
Hon Gai see Ha Long
Hongay see Ha Long
Hông Gai see Ha Long
Hông Hà, Sông see Red River
Hong Kong 106 A1 Hong Kong, S China
Hong Kong Island 106 B2 island S China Asia
Honiara 122 C3 country capital (Solomon Islands) Guadalcanal, C Solomon Islands
Honjō 108 D4 var. Honzyô, Yurihonjō. Akita, Honshū, C Japan
Honolulu 25 A8 state capital O'ahu, Hawaii, USA, C Pacific Ocean
Honshu 109 E5 var. Hondo, Honsyû. island SW Japan
Honsyû see Honshū
Honte see Westerschelde
Honzyô see Honjō
Hoogeveen 64 E2 Drenthe, NE Netherlands
Hoogezand-Sappemeer 64 E2 Groningen, NE Netherlands
Hoorn 64 C2 Noord-Holland, NW Netherlands
Hoosier State see Indiana
Hopa 95 E2 Artvin, NE Turkey
Hope 14 C3 British Columbia, SW Canada
Hopedale 17 F2 Newfoundland and Labrador, NE Canada
Hopeh/Hopei see Hebei
Hopkinsville 18 B5 Kentucky, S USA
Horasan 95 F3 Erzurum, NE Turkey
Horizon Deep 130 D4 trench W Pacific Ocean
Horki 85 E6 Rus. Gorki. Mahilyowskaya Voblasts', E Belarus
Horlivka 87 G3 Rom. Adâncata, Rus. Gorlovka. Donets'ka Oblast', E Ukraine
Hormoz, Tangeh-ye see Hormuz, Strait of
Hormuz, Strait of 98 D4 var. Strait of Ormuz, Per. Tangeh-ye Hormoz. strait Iran/Oman
Cape Horn 43 C8 Eng. Cape Horn. headland S Chile
Horn, Cape see Hornos, Cabo de
Hornsby 126 E1 New South Wales, SE Australia
Horodnya 87 E1 Rus. Gorodnya. Chernihivs'ka Oblast', NE Ukraine
Horodok 86 B2 Pol. Gródek Jagielloński, Rus. Gorodok, Gorodok Yagellonski. L'vivs'ka Oblast', NW Ukraine
Horodyshche 87 E2 Rus. Gorodishche. Cherkas'ka Oblast', C Ukraine
Horoshiri-dake 108 D2 var. Horosiri Dake. mountain Hokkaidō, N Japan
Horosiri Dake see Horoshiri-dake
Horsburgh Atoll 110 A4 var. Goidhoo Atoll. atoll N Maldives
Horseshoe Bay 20 A5 bay W Bermuda W Atlantic Ocean
Horseshoe Seamounts 58 A4 seamount range E Atlantic Ocean
Horsham 127 B7 Victoria, SE Australia
Horst 65 D5 Limburg, SE Netherlands
Horten 63 B6 Vestfold, S Norway
Horyn' 85 B7 Rus. Goryn. river NW Ukraine
Hosingen 65 D7 Diekirch, NE Luxembourg
Hospitalet see L'Hospitalet de Llobregat
Hotan 104 B4 var. Khotan, Chin. Ho-t'ien. Xinjiang Uygur Zizhiqu, NW China
Ho-t'ien see Hotan
Hoting 62 C4 Jämtland, C Sweden
Hot Springs 20 B1 Arkansas, C USA
Hötzing see Hațeg
Houayxay 114 C3 var. Ban Houayxay. Bokèo, N Laos
Houghton 18 B1 Michigan, N USA
Houilles 69 B5 Yvelines, Île-de-France, N France Europe
Houlton 19 H1 Maine, NE USA

Houm – Jaipur

Houma 20 B3 Louisiana, S USA
Houston 27 H4 Texas, SW USA
Hovd 104 C2 var. Khovd, Kobdo; prev. Jirgalanta. Hovd, W Mongolia
Hove 67 E7 SE England, United Kingdom
Hoverla, Hora 86 C3 Rus. Gora Goverla. mountain W Ukraine
Hovsgol, Lake see Hövsgöl Nuur
Hövsgöl Nuur 104 D1 var. Lake Hovsgol. lake N Mongolia
Howa, Ouadi see Howar, Wādi
Howar, Wadi 50 A3 var. Ouadi Howa. river Chad/Sudan
Howrah see Hāora
Hoy 66 C2 island N Scotland, United Kingdom
Hoyerswerda 72 D4 Lus. Wojerecy. Sachsen, E Germany
Hpa-an 114 B4 var. Pa-an. Kayin State, S Myanmar (Burma)
Hpyu see Phyu
Hradec Králové 77 B5 Ger. Königgrätz. Královéhradecký Kraj, N Czech Republic (Czechia)
Hrandzichy 85 B5 Rus. Grandichi. Hrodzyenskaya Voblasts', W Belarus
Hranice 77 C5 Ger. Mährisch-Weisskirchen. Olomoucký Kraj, E Czech Republic (Czechia)
Hrebinka 87 E2 Rus. Grebenka. Poltavs'ka Oblast', NE Ukraine
Hrodna 85 B5 Pol. Grodno. Hrodzyenskaya Voblasts', W Belarus
Hrvatska see Croatia
Hsia-men see Xiamen
Hsiang-t'an see Xiangtan
Hsi Chiang see Xi Jiang
Hsing-K'ai Hu see Khanka, Lake
Hsi-ning/Hsining see Xining
Hsinking see Changchun
Hsin-yang see Xinyang
Hsu-chou see Xuzhou
Htawei see Dawei
Huacho 38 C4 Lima, W Peru
Hua Hin see Ban Hua Hin
Huaihua 106 C5 Hunan, S China
Huailai 106 C3 var. Shacheng. Hebei, E China
Huainan 106 D5 var. Huai-nan, Hwainan. Anhui, E China
Huai-nan see Huainan
Huajuapan 29 F5 var. Huajuapan de León. Oaxaca, SE Mexico
Huajuapan de León see Huajuapan
Hualapai Peak 26 A2 mountain Arizona, SW USA
Huallaga, Río 38 C3 river N Peru
Huambo 56 B2 Port. Nova Lisboa. Huambo, C Angola
Huancavelica 38 D4 Huancavelica, SW Peru
Huancayo 38 D3 Junín, C Peru
Huang Hai see Yellow Sea
Huang He 106 C4 var. Yellow River. river C China
Huangshi 106 C5 var. Huang-shih, Hwangshih. Hubei, C China
Huang-shih see Huangshi
Huanta 38 D4 Ayacucho, C Peru
Huánuco 38 C3 Huánuco, C Peru
Huanuni 39 F4 Oruro, W Bolivia
Huaral 38 C4 Lima, W Peru
Huarás see Huaraz
Huaraz 38 C3 var. Huarás. Ancash, W Peru
Huarmey 38 C3 Ancash, W Peru
Huatabampo 28 C2 Sonora, NW Mexico
Hubballi 102 B2 prev. Hubli. Karnātaka, SW India
Hubli see Hubballi
Huddersfield 67 D5 N England, United Kingdom
Hudiksvall 63 C5 Gävleborg, C Sweden
Hudson Bay 15 G4 bay NE Canada
Hudson, Détroit d' see Hudson Strait
Hudson Strait 15 H3 Fr. Détroit d'Hudson. strait Northwest Territories/Québec, NE Canada
Hudur see Xuddur
Huê 114 E4 Th,a Thiên-Huê, C Vietnam
Huehuetenango 30 A2 Huehuetenango, W Guatemala
Huelva 70 C4 anc. Onuba. Andalucía, SW Spain
Huesca 71 F2 anc. Osca. Aragón, NE Spain
Huéscar 71 E4 Andalucía, S Spain
Hughenden 126 C3 Queensland, NE Australia
Hugo 27 G2 Oklahoma, C USA
Huhehot/Huhohaote see Hohhot
Huíla Plateau 56 B2 plateau S Angola
Huixtla 29 G5 Chiapas, SE Mexico
Hulingol 105 G2 prev. Huolin Gol. Nei Mongol Zizhiqu, N China
Hull 16 D4 Québec, SE Canada
Hull see Kingston upon Hull
Hull Island see Orona
Hulst 65 B5 Zeeland, SW Netherlands
Hulun see Hulun Buir
Hulun Buir 105 F1 var. Hailar; prev. Hulun. Nei Mongol Zizhiqu, N China
Hu-lun Ch'ih see Hulun Nur
Hulun Nur 105 F1 var. Hu-lun Ch'ih; prev. Dalai Nor. lake NE China
Humaitá 40 D2 Amazonas, N Brazil
Humboldt River 25 C5 river Nevada, W USA
Humphreys Peak 26 B1 mountain Arizona, SW USA
Humpolec 77 B5 Ger. Gumpolds, Humpoletz. Vysočina, C Czech Republic (Czechia)
Humpoletz see Humpolec
Hunan 106 C6 var. Hunan Sheng, Xiang. province S China
Hunan Sheng see Hunan
Hunedoara 86 B4 Ger. Eisenmarkt, Hung. Vajdahunyad. Hunedoara, SW Romania
Hünfeld 73 B5 Hessen, C Germany
Hungarian People's Republic see Hungary
Hungary 77 C6 off. Hungary, Ger. Ungarn, Hung. Magyarország, Rom. Ungaria, SCr. Madarska, Ukr. Uhorshchyna; prev. Hungarian People's Republic. country C Europe
Hungary, Plain of see Great Hungarian Plain
Hunjiang see Baishan
Hunter Island 127 B8 island Tasmania, SE Australia
Huntington 18 D4 West Virginia, NE USA
Huntington Beach 25 B8 California, W USA
Huntly 128 D3 Waikato, North Island, New Zealand
Huntsville 20 D1 Alabama, S USA
Huntsville 27 G3 Texas, SW USA
Huolin Gol see Hulingol
Hurghada see Al Ghurdaqah
Huron 23 E3 South Dakota, N USA
Huron, Lake 18 D2 lake Canada/USA
Hurukawa see Furukawa
Hurunui 129 C5 river South Island, New Zealand
Húsavík 61 E4 Nordhurland Eystra, NE Iceland
Husté see Khust
Husum 72 B2 Schleswig-Holstein, N Germany
Huszt see Khust
Hutchinson 23 E5 Kansas, C USA
Hutchinson Island 21 F4 island Florida, SE USA
Huy 65 C6 Dut. Hoei, Hoey. Liège, E Belgium
Huzi see Fuji
Hvannadalshnúkur 61 E5 volcano S Iceland
Hvar 78 B4 It. Lesina; anc. Pharus. island S Croatia
Hwainan see Huainan
Hwange 56 D3 prev. Wankie. Matabeleland North, W Zimbabwe
Hwang-Hae see Yellow Sea
Hwangshih see Huangshi
Hyargas Nuur 104 C2 lake NW Mongolia
Hyderābād 112 D5 var. Haidarabad. state capital Telangana, C India
Hyderābād 112 B3 var. Haidarabad. Sind, SE Pakistan
Hyères 69 D6 Var, SE France
Hyères, Îles d' 69 D6 island group S France
Hypanis see Kuban'
Hyrcania see Gorgān
Hyvinge see Hyvinkää
Hyvinkää 63 D5 Swe. Hyvinge. Uusimaa, S Finland

I

Iader see Zadar
Ialomiţa 86 C5 river SE Romania
Iaşi 86 D3 Ger. Jassy. Iaşi, NE Romania
Ibadan 53 F5 Oyo, SW Nigeria
Ibagué 36 B3 Tolima, C Colombia
Ibar 78 D4 Alb. Ibër. river C Serbia
Ibarra 38 B1 var. San Miguel de Ibarra. Imbabura, N Ecuador
Ibër see Ibar
Iberia see Spain
Iberian Mountains see Ibérico, Sistema
Iberian Peninsula 58 B4 physical region Portugal/Spain
Iberian Plain 58 B4 abyssal plain E Atlantic Ocean
Ibérica, Cordillera see Ibérico, Sistema
Ibérico, Sistema 71 E2 var. Cordillera Ibérica, Eng. Iberian Mountains. mountain range NE Spain
Ibiza see Eivissa
Ibo see Sassandra
Ica 38 D4 Ica, SW Peru
Icaria see Ikaría
Içá, Rio see Putumayo, Río
İçel see Mersin
Iceland 61 E4 off. Republic of Iceland, Dan. Island, Icel. Ísland. country N Atlantic Ocean
Iceland Basin 58 B1 undersea basin N Atlantic Ocean
Icelandic Plateau see Iceland Plateau
Iceland Plateau 133 B6 var. Icelandic Plateau. undersea plateau S Greenland Sea
Iceland, Republic of see Iceland
Iconium see Konya
Iculisma see Angoulême
Idabel 27 H2 Oklahoma, C USA
Idaho 24 D3 off. State of Idaho, also known as Gem of the Mountains, Gem State. state NW USA
Idaho Falls 24 E3 Idaho, NW USA
Idensalmi see Iisalmi
Idfū 50 B2 var. Edfu. SE Egypt
Idi Amin, Lac see Edward, Lake
Idîni 52 B2 Trarza, W Mauritania
Idlib 96 B3 Idlib, NW Syria
Idre 63 B5 Dalarna, C Sweden
Iecava 84 C3 Bauska, S Latvia
Ieper 65 A6 Fr. Ypres. West-Vlaanderen, W Belgium
Ierápetra 83 D8 Kríti, Greece, E Mediterranean Sea
Ierisós see Ierissós
Ierissós 82 C4 var. Ierisós. Kentrikí Makedonía, N Greece
Iferouâne 53 G2 Agadez, N Niger
Ifôghas, Adrar des 53 E2 var. Adrar des Iforas. mountain range NE Mali
Iforas, Adrar des see Ifôghas, Adrar des
Igarka 92 D3 Krasnoyarskiy Kray, N Russia
Igaunija see Estonia
Iglau/Iglawa/Iglawa see Jihlava
Iglesias 75 A5 Sardegna, Italy, C Mediterranean Sea
Igloolik 15 G2 Nunavut, N Canada
Igoumenítsa 82 A4 Ípeiros, W Greece
Iguaçu, Rio 41 E4 Sp. Río Iguazú. river Argentina/Brazil
Iguaçu, Saltos do 41 E4 Sp. Cataratas del Iguazú; prev. Victoria Falls. waterfall Argentina/Brazil
Iguala 29 E4 var. Iguala de la Independencia. Guerrero, S Mexico
Iguala de la Independencia see Iguala
Iguazú, Cataratas del see Iguaçu, Saltos do
Iguazú, Río see Iguaçu, Rio
Iguid, Erg see Iguîdi, 'Erg
Iguîdi, 'Erg 48 C3 var. Erg Iguid. desert Algeria/Mauritania
Ihavandhippolhu Atoll 110 A3 var. Ihavandiffulu Atoll. atoll N Maldives
Ihavandiffulu Atoll see Ihavandhippolhu Atoll
Ihosy 57 F4 Fianarantsoa, S Madagascar
Iinnasuolu see Hinnøya
Iisalmi 62 E4 var. Idensalmi. Pohjois-Savo, C Finland
IJmuiden 64 C3 Noord-Holland, W Netherlands
IJssel 64 D3 var. Yssel. river Netherlands
IJsselmeer 64 C2 prev. Zuider Zee. lake N Netherlands
IJsselmuiden 64 D3 Overijssel, E Netherlands
Ijzer 65 A6 river W Belgium
Ikaahuk see Sachs Harbour
Ikaluktutiak see Cambridge Bay
Ikaría 83 D6 var. Kariot, Nicaria, Nikaria; anc. Icaria. island Dodekánisa, Greece, Aegean Sea
Ikela 55 D6 Equateur, C Dem. Rep. Congo
Iki 109 A7 island SW Japan
Ilagan 117 E1 Luzon, N Philippines
Ilave 39 E4 Puno, S Peru
Iława 76 D3 Ger. Deutsch-Eylau. Warmińsko-Mazurskie, NE Poland
Ilebo 55 C6 prev. Port-Francqui. Kasai-Occidental, W Dem. Rep. Congo
Île-de-France 68 C3 cultural region N France
Ilerda see Lleida
Ilfracombe 67 C7 SW England, United Kingdom
Ílhavo 70 B2 Aveiro, N Portugal
Iliamna Lake 14 C3 lake Alaska, USA
Ilici see Elche
Iligan 117 E2 off. Iligan City. Mindanao, S Philippines
Iligan City see Iligan
Illapel 42 B4 Coquimbo, C Chile
Illichivs'k see Chornomors'k, Ukraine
Illicis see Elche
Illinois 18 A4 off. State of Illinois, also known as Prairie State, Sucker State. state C USA
Illinois River 18 B4 river Illinois, N USA
Illurco see Lorca
Illuro see Mataró
Ilo 39 E4 Moquegua, SW Peru
Iloilo 117 E2 off. Iloilo City. Panay Island, C Philippines
Iloilo City see Iloilo
Ilorin 53 F4 Kwara, W Nigeria
Ilovlya 89 B6 Volgogradskaya Oblast', SW Russia
Iluh see Batman
Il'yaly see Gurbansoltan Eje
Imatra 63 E5 Etelä-Kariala, SE Finland
Imbros see Gökçeada
Imishli see İmişli
İmişli 95 H3 Rus. Imishli. C Azerbaijan
Imola 74 C3 Emilia-Romagna, N Italy
Imperatriz 41 F2 Maranhão, NE Brazil
Imperia 74 A3 Liguria, NW Italy
Impfondo 55 C5 Likouala, NE Congo
Imphāl 113 H3 state capital Manipur, NE India
Imroz Adası see Gökçeada
Inagua Islands see Little Inagua
Inagua Islands see Great Inagua
Inarijärvi 62 D2 Lapp. Aanaarjävri, Swe. Enareträsk. lake N Finland
Inău see Ineu
Inawashiro-ko 109 D5 var. Inawasiro Ko. lake Honshū, C Japan
Inawasiro Ko see Inawashiro-ko
İncesu 94 D3 Kayseri, Turkey Asia
Incheon 107 E4 off. Incheon Gwang-yeoksi, prev. Inch'ŏn, Jap. Jinsen; prev. Chemulpo. NW South Korea
Incheon-Gwang-yeoksi see Incheon
Inch'ŏn see Incheon
Incudine, Monte 69 E7 mountain Corse, France, C Mediterranean Sea
Indefatigable Island see Santa Cruz, Isla
Independence 23 F4 Missouri, C USA
Independence Fjord 61 E1 fjord N Greenland
Independence Island see Malden Island
Independence Mountains 24 C4 mountain range Nevada, W USA
India 102 B3 off. Republic of India, var. Indian Union, Union of India, Hind. Bhārat. country S Asia
India see Indija
Indiana 18 B4 off. State of Indiana, also known as Hoosier State. state N USA
Indianapolis 18 C4 state capital Indiana, N USA
Indian Church 30 C1 Orange Walk, N Belize
Indian Desert see Thar Desert
Indianola 23 F4 Iowa, C USA
Indian Union see India
India, Republic of see India
India, Union of see India
Indigirka 93 F2 river NE Russia
Indija 78 D3 Hung. India; prev. Indjija. Vojvodina, N Serbia
Indira Point 110 G3 headland Andaman and Nicobar Island, India, NE Indian Ocean
Indjija see Indija
Indomed Fracture Zone 119 B6 tectonic feature SW Indian Ocean
Indonesia 116 B4 off. Republic of Indonesia, Ind. Republik Indonesia; prev. Dutch East Indies, Netherlands East Indies, United States of Indonesia. country SE Asia
Indonesian Borneo see Kalimantan
Indonesia, Republic of see Indonesia
Indonesia, Republik see Indonesia
Indonesia, United States of see Indonesia
Indore 112 D4 Madhya Pradesh, C India
Indreville see Châteauroux
Indus 112 C2 Chin. Yindu He; prev. Yin-tu Ho. river S Asia
Indus Cone see Indus Fan
Indus Fan 90 C5 var. Indus Cone. undersea fan N Arabian Sea
Indus, Mouths of the 112 B4 delta S Pakistan
İnebolu 94 C2 Kastamonu, N Turkey
Ineu 86 A4 Hung. Borosjenő; prev. Inău. Arad, W Romania
Infiernillo, Presa del 29 E4 reservoir S Mexico
Inglewood 24 D2 California, W USA
Ingolstadt 73 C6 Bayern, S Germany
Ingulets see Inhulets'
Inguri see Enguri
Inhambane 57 E4 Inhambane, SE Mozambique
Inhulets' 87 F3 Rus. Ingulets. Dnipropetrovs'ka Oblast', E Ukraine
I-ning see Yining
Inis see Ennis
Inis Ceithleann see Enniskillen
Inn 73 C6 river C Europe
Innaanganeq 60 C1 var. Kap York. headland NW Greenland
Inner Hebrides 66 B4 island group W Scotland, United Kingdom
Inner Islands 57 H1 var. Central Group. island group NE Seychelles
Innisfail 126 D3 Queensland, NE Australia
Inniskilling see Enniskillen
Innsbruch see Innsbruck
Innsbruck 73 C7 var. Innsbruch. Tirol, W Austria
Inoucdjouac see Inukjuak
Inowrazlaw see Inowrocław
Inowrocław 76 C3 Ger. Hohensalza; prev. Inowrazlaw. Kujawski-pomorskie, C Poland
I-n-Sakane, 'Erg 53 E2 desert N Mali
I-n-Salah 48 D3 var. In Salah. C Algeria
Insterburg see Chernyakhovsk
Insula see Lille
Inta 88 E3 Respublika Komi, NW Russia
Interamna see Teramo
Interamna Nahars see Terni
International Falls 23 F1 Minnesota, N USA
Inukjuak 16 D2 var. Inoucdjouac; prev. Port Harrison. Québec, NE Canada
Inuuvik see Inuvik
Inuvik 14 D3 var. Inuuvik. Northwest Territories, NW Canada
Invercargill 129 A7 Southland, South Island, New Zealand
Inverness 66 C3 N Scotland, United Kingdom
Investigator Ridge 119 D5 undersea ridge E Indian Ocean
Investigator Strait 127 B7 strait South Australia
Inyangani 56 D3 mountain NE Zimbabwe
Ioánnina 82 A4 var. Janina, Yannina. Ípeiros, W Greece
Iola 23 F5 Kansas, C USA
Ionia Basin see Ionian Basin
Ionian Basin 58 D5 var. Ionia Basin. undersea basin Ionian Sea, C Mediterranean Sea
Ionian Islands see Iónia Nisiá
Ionian Sea 81 E3 Gk. Iónio Pélagos, It. Mar Ionio. sea C Mediterranean Sea
Iónia Nisiá 83 A5 Eng. Ionian Islands. island group W Greece
Ionio, Mar/Iónio Pélagos see Ionian Sea
Íos 83 D6 var. Nio. island Kykládes, Greece, Aegean Sea
Ioulída 83 C6 prev. Ioulís, Kykládes, Greece, Aegean Sea
Iowa 23 F3 off. State of Iowa, also known as Hawkeye State. state C USA
Iowa City 23 G3 Iowa, C USA
Iowa Falls 23 G3 Iowa, C USA
Ipel' 77 C6 var. Ipoly, Ger. Eipel. river Hungary/Slovakia
Ipiales 36 A4 Nariño, SW Colombia
Ipoh 116 B3 Perak, Peninsular Malaysia
Ipoly see Ipel'
Ippy 54 C4 Ouaka, C Central African Republic
Ipswich 127 E5 Queensland, E Australia
Ipswich 67 E6 hist. Gipeswic. E England, United Kingdom
Iqaluit 15 H3 prev. Frobisher Bay. province capital Baffin Island, Nunavut, NE Canada
Iquique 42 B1 Tarapacá, N Chile
Iquitos 38 C1 Loreto, N Peru
Irákleio 83 D7 var. Herakleion, Eng. Candia; prev. Iráklion. Kríti, Greece, E Mediterranean Sea
Iráklion see Irákleio
Iran 98 C3 off. Islamic Republic of Iran; prev. Persia. country SW Asia
Iranian Plateau 98 D3 var. Plateau of Iran. plateau N Iran
Iran, Islamic Republic of see Iran
Iran, Plateau of see Iranian Plateau
Irapuato 29 E4 Guanajuato, C Mexico
Iraq 98 B3 off. Republic of Iraq, Ar. 'Irāq. country SW Asia
'Irāq see Iraq
Iraq, Republic of see Iraq
Irbid 97 B5 Irbid, N Jordan
Irbīl see Arbīl
Ireland 67 A5 off. Republic of Ireland, Ir. Éire. country NW Europe
Ireland 58 C3 Lat. Hibernia. island Ireland/United Kingdom
Ireland, Republic of see Ireland
Irian see New Guinea
Irian Barat see Papua
Irian Jaya see Papua
Irian, Teluk see Cenderawasih, Teluk
Iringa 51 C7 Iringa, C Tanzania
Iriomote-jima 108 A4 island Sakishima-shotō, SW Japan
Iriona 30 D2 Colón, NE Honduras
Irish Sea 67 C5 Ir. Muir Éireann. sea C British Isles
Irkutsk 93 E4 Irkutskaya Oblast', S Russia
Irminger Basin see Reykjanes Basin
Iroise 68 A3 sea NW France
Iron Mountain 18 B2 Michigan, N USA
Ironwood 18 B1 Michigan, N USA
Irrawaddy 114 B2 var. Ayeyarwady. river W Myanmar (Burma)
Irrawaddy, Mouths of the 115 A5 delta SW Myanmar (Burma)
Irtish see Irtysh
Irtysh 92 C4 var. Irtish, Kaz. Ertis. river C Asia
Irun 71 E1 Cast. Irún. País Vasco, N Spain
Irún see Irun
Iruña see Pamplona
Isabela, Isla 38 A5 var. Albemarle Island. island Galápagos Islands, Ecuador, E Pacific Ocean
Isaccea 86 D4 Tulcea, E Romania
Isachsen 15 F1 Ellef Ringnes Island, Nunavut, N Canada
Ísafjördur 61 E4 Vestfirdhir, NW Iceland
Isbarta see Isparta
Isca Damnoniorum see Exeter
Ise 109 C6 Mie, Honshū, SW Japan
Iseghem see Izegem
Isère 69 D5 river E France
Isernia 75 D5 var. Æsernia. Molise, C Italy
Ise-wan 109 C6 bay S Japan
Isfahan see Eşfahān
Isha Baydhabo see Baydhabo
Ishigaki-jima 108 A4 island Sakishima-shotō, SW Japan
Ishikari-wan 108 C2 bay Hokkaidō, NE Japan
Ishim 92 C4 Tyumenskaya Oblast', C Russia
Ishim 92 C4 Kaz. Esil. river Kazakhstan/Russia
Ishinomaki 108 D4 var. Isinomaki. Miyagi, Honshū, C Japan
Ishkashim see Ishkoshim
Ishkoshim 101 F3 Rus. Ishkashim. S Tajikistan
Isinomaki see Ishinomaki
Isiro 55 E5 Orientale, NE Dem. Rep. Congo
Iskăr see Iskŭr
İskenderun 94 D4 Eng. Alexandretta. Hatay, S Turkey
İskenderun Körfezi 96 A2 Eng. Gulf of Alexandretta. gulf S Turkey
Iskur 82 C2 var. Iskăr. river NW Bulgaria
Yazovir Iskur 82 B2 prev. Yazovir Stalin. reservoir W Bulgaria
Isla Cristina 70 C4 Andalucía, S Spain
Isla de León see San Fernando
Islāmābād 112 C1 country capital (Pakistan) Federal Capital Territory Islāmābād, NE Pakistan
Island/Ísland see Iceland
Islay 66 B4 island SW Scotland, United Kingdom
Isle 69 B5 river W France
Isle of Man 67 B5 British Crown Dependency NW Europe
Isles of Scilly 67 B8 island group SW England, United Kingdom
Ismailia see Al Ismā'īlīya
Ismâ'ilîya see Al Ismā'īlīya
Ismid see İzmit
Isnā 50 B2 var. Esna. SE Egypt
Isoka 56 D1 Northern, NE Zambia
Isparta 94 B4 var. Isbarta. Isparta, SW Turkey
İspir 95 E3 Erzurum, NE Turkey
Israel 97 A7 off. State of Israel, var. Medinat Israel, Heb. Yisrael, Yisra'el. country SW Asia
Israel, State of see Israel
Issa see Vis
Issiq Köl see Issyk-Kul', Ozero
Issoire 69 C5 Puy-de-Dôme, C France
Issyk-Kul' see Balykchy
Issyk-Kul', Ozero 101 G2 var. Issiq Köl, Kir. Ysyk-Köl. lake E Kyrgyzstan
İstanbul 94 B2 Bul. Tsarigrad, Eng. Istanbul, prev. Constantinople; anc. Byzantium. Istanbul, NW Turkey
İstanbul Boğazi 94 B2 var. Bosporus Thracius, Eng. Bosphorus, Bosporus, Turk. Karadeniz Boğazi. strait NW Turkey
Istarska Županija see Istra
Istra 78 A3 off. Istarska Županija. province NW Croatia
Istra 78 A3 Eng. Istria, Ger. Istrien. cultural region NW Croatia
Istria/Istrien see Istra
Itabuna 41 G3 Bahia, E Brazil
Itagüí 36 B3 Antioquia, W Colombia
Itaipú, Represa de 41 E4 reservoir Brazil/Paraguay
Itaituba 41 E2 Pará, NE Brazil
Italia/Italiana, Republica/Italian Republic, The see Italy
Italian Somaliland see Somalia
Italy 74 C3 off. The Italian Republic, It. Italia, Repubblica Italiana. country S Europe
Iténez, Río see Guaporé, Rio
Ithaca 19 E3 New York, NE USA
It Hearrenfean see Heerenveen
Itoigawa 109 C5 Niigata, Honshū, C Japan
Itseqqortoormiit see Ittoqqortoormiit
Ittoqqortoormiit 61 E3 var. Itseqqortoormiit, Dan. Scoresbysund, Eng. Scoresby Sound. Tunu, C Greenland
Iturup, Ostrov 108 E1 island Kuril'skiye Ostrova, SE Russia
Itzehoe 72 B2 Schleswig-Holstein, N Germany
Ivalo 62 D2 Lapp. Avveel, Avvil. Lappi, N Finland
Ivanava 85 B7 Pol. Janów, Janów Poleski, Rus. Ivanovo. Brestskaya Voblasts', SW Belarus
Ivangrad see Berane
Ivanhoe 127 C6 New South Wales, SE Australia
Ivano-Frankivs'k 86 C2 Ger. Stanislau, Pol. Stanisławów, Rus. Ivano-Frankovsk; prev. Stanislav. Ivano-Frankivs'ka Oblast', W Ukraine
Ivano-Frankovsk see Ivano-Frankivs'k
Ivanovo 89 B5 Ivanovskaya Oblast', W Russia
Ivanovo see Ivanava
Ivantsevichi/Ivatsevichi see Ivatsevichy
Ivatsevichy 85 B6 Pol. Iwacewicze, Rus. Ivantsevichi, Ivatsevichi. Brestskaya Voblasts', SW Belarus
Ivigtut see Ivittuut
Ivittuut 60 B4 var. Ivigtut. Kitaa, S Greenland
Iviza see Eivissa\Ibiza
Ivory Coast 52 D4 off. République de la Côte d'Ivoire. country W Africa
Ivujivik 16 D1 Québec, NE Canada
Iwacewicze see Ivatsevichy
Iwaki 109 D5 Fukushima, Honshū, N Japan
Iwakuni 109 B7 Yamaguchi, Honshū, SW Japan
Iwanai 108 C2 Hokkaidō, NE Japan
Iwate 108 D3 Iwate, Honshū, N Japan
Ixtapa 29 E5 Guerrero, S Mexico
Ixtepec 29 F5 Oaxaca, SE Mexico
Iyo-nada 109 B7 sea S Japan
Izabal, Lago de 30 B2 prev. Golfo Dulce. lake E Guatemala
Īzad Khvāst 98 D3 Fārs, C Iran
Izegem 65 A6 prev. Iseghem. West-Vlaanderen, W Belgium
Izhevsk 89 D5 prev. Ustinov. Udmurtskaya Respublika, NW Russia
Izmail see Izmayil
Izmayil 86 D4 Rus. Izmail. Odes'ka Oblast', SW Ukraine
İzmir 94 A3 prev. Smyrna. İzmir, W Turkey
İzmit 94 B2 var. Ismid; anc. Astacus. Kocaeli, NW Turkey
İznik Gölü 94 B3 lake NW Turkey
Izu-hanto 109 D6 peninsula Honshū, S Japan
Izu Shichito see Izu-shotō
Izu-shoto 109 D6 var. Izu Shichito. island group S Japan
Izvor 82 B2 Pernik, W Bulgaria
Izyaslav 86 C2 Khmel'nyts'ka Oblast', W Ukraine
Izyum 87 G2 Kharkivs'ka Oblast', E Ukraine

J

Jabal ash Shifa 98 A4 desert NW Saudi Arabia
Jabalpur 113 E4 prev. Jubbulpore. Madhya Pradesh, C India
Jabbūl, Sabkhat al 96 B2 sabkha NW Syria
Jablah 96 A3 var. Jeble, Fr. Djéblé. Al Lādhiqīyah, W Syria
Jaca 71 F1 Aragón, NE Spain
Jacaltenango 30 A2 Huehuetenango, W Guatemala
Jackson 20 B2 state capital Mississippi, S USA
Jackson 23 H5 Missouri, C USA
Jackson 20 C1 Tennessee, S USA
Jackson Head 129 A6 headland South Island, New Zealand
Jacksonville 21 E3 Florida, SE USA
Jacksonville 18 B4 Illinois, N USA
Jacksonville 21 F1 North Carolina, SE USA
Jacksonville 27 G3 Texas, SW USA
Jacmel 32 D3 var. Jaquemel. S Haiti
Jacob see Nkayi
Jacobābād 112 B3 Sind, SE Pakistan
Jadotville see Likasi
Jadransko More/Jadransko Morje see Adriatic Sea
Jaén 38 B2 Cajamarca, N Peru
Jaén 70 D4 Andalucía, SW Spain
Jaffna 110 D3 Northern Province, N Sri Lanka
Jagannath see Puri
Jagdalpur 113 E5 Chhattīsgarh, C India
Jagdaqi 105 G1 Nei Mongol Zizhiqu, N China
Jagodina 78 D4 prev. Svetozarevo. Serbia, C Serbia
Jahra see Al Jahrā'
Jailolo see Halmahera, Pulau
Jaipur 112 D3 prev. Jeypore. state capital Rājasthān, N India

Jaisalmer – Karakax

Jaisalmer 112 C3 Rājasthān, NW India
Jajce 78 B3 Federacija Bosna I Hercegovina, W Bosnia and Herzegovina
Jakarta 116 C5 prev. Djakarta, Dut. Batavia. country capital (Indonesia) Jawa, C Indonesia
Jakobstad 62 D4 Fin. Pietarsaari. Österbotten, W Finland
Jakobstadt see Jēkabpils
Jalālābād 101 F4 var. Jalalabad, Jelalabad. Nangarhār, E Afghanistan
Jalal-Abad see Dzhalal-Abad, Dzhalal-Abadskaya Oblast', Kyrgyzstan
Jalandhar 112 D2 prev. Jullundur. Punjab, N India
Jalapa 30 D3 Nueva Segovia, NW Nicaragua
Jalpa 28 D4 Zacatecas, C Mexico
Jālū 49 G3 var. Jūlā. NE Libya
Jaluit Atoll 122 D2 var. Jālwōj. atoll Ralik Chain, S Marshall Islands
Jālwōj see Jaluit Atoll
Jamaame 51 D6 It. Giamame; prev. Margherita. Jubbada Hoose, S Somalia
Jamaica 32 A4 country W West Indies
Jamaica 34 A1 island W West Indies
Jamaica Channel 32 D3 channel Haiti/Jamaica
Jamālpur 113 F3 Bihār, NE India
Jambi 116 B4 var. Telanaipura; prev. Djambi. Sumatera, W Indonesia
Jamdena see Yamdena, Pulau
James Bay 16 C3 bay Ontario/Québec, E Canada
James River 23 E2 river North Dakota/South Dakota, N USA
James River 19 E5 river Virginia, NE USA
Jamestown 19 E3 New York, NE USA
Jamestown 23 E2 North Dakota, N USA
Jamestown see Holetown
Jammu 112 D2 prev. Jummoo. state capital Jammu and Kashmir, NW India
Jammu and Kashmir 112 D1 disputed region India/Pakistan
Jāmnagar 112 C4 prev. Navanagar. Gujarāt, W India
Jamshedpur 113 F4 Jhārkhand, NE India
Jamuna see Brahmaputra
Janaúba 41 F3 Minas Gerais, SE Brazil
Janesville 18 B3 Wisconsin, N USA
Janina see Ioánnina
Janischken see Joniškis
Jankovac see Jánoshalma
Jan Mayen 61 F4 constituent part of Norway. island N Atlantic Ocean
Jánoshalma 77 C7 SCr. Jankovac. Bács-Kiskun, S Hungary
Janów see Ivanava, Belarus
Janow/Janów see Jonava, Lithuania
Janów Poleski see Ivanava
Japan 108 C4 var. Nippon, Jap. Nihon. country E Asia
Japan, Sea of 108 A4 var. East Sea, Rus. Yapanskoye More. sea NW Pacific Ocean
Japan Trench 103 F1 trench NW Pacific Ocean
Japen see Yapen, Pulau
Japiim 40 C2 var. Máncio Lima. Acre, W Brazil
Japurá, Rio 40 C2 var. Río Caquetá, Yapurá. river Brazil/Colombia
Japurá, Rio see Caquetá, Río
Jaqué 31 G5 Darién, SE Panama
Jaquemel see Jacmel
Jarablos see Jarābulus
Jarābulus 96 C2 var. Jarablos, Jerablus, Fr. Djérablous. Ḩalab, N Syria
Jarbah, Jazīrat see Jerba, Île de
Jardines de la Reina, Archipiélago de los 32 B2 island group C Cuba
Jarīd, Shaṭṭ al see Jerid, Chott el
Jarocin 76 C4 Wielkopolskie, C Poland
Jaroslau see Jarosław
Jarosław 77 E5 Ger. Jaroslau, Rus. Yaroslav. Podkarpackie, SE Poland
Jarqo'rg'on 101 E3 Rus. Dzharkurgan. Surkhondaryo Viloyati, S Uzbekistan
Jarvis Island 123 G2 US unincorporated territory C Pacific Ocean
Jasło 77 D5 Podkarpackie, SE Poland
Jastrzębie-Zdrój 77 C5 Śląskie, S Poland
Jataí 41 E3 Goiás, C Brazil
Jativa see Xàtiva
Jauf see Al Jawf
Jaunpiebalga 84 D3 Gulbene, NE Latvia
Jaunpur 113 E3 Uttar Pradesh, N India
Java 130 A3 South Dakota, N USA
Javalambre 71 E3 mountain E Spain
Java Sea 116 D4 Ind. Laut Jawa. sea W Indonesia
Java Trench 102 D5 var. Sunda Trench. trench E Indian Ocean
Jawa, Laut see Java Sea
Jawhar 51 D6 var. Jowhar, It. Giohar. Shabeellaha Dhexe, S Somalia
Jaworów see Yavoriv
Jaya, Puncak 117 G4 prev. Puntjak Carstensz, Puntjak Sukarno. mountain Papua, E Indonesia
Jayapura 117 H4 var. Djajapura, Dut. Hollandia; prev. Kotabaru, Sukarnapura. Papua, E Indonesia
Jay Dairen see Dalian
Jayhawker State see Kansas
Jaz Murian, Hamun-e 98 E4 lake SE Iran
Jebba 53 F4 Kwara, W Nigeria
Jebel, Bahr el see White Nile
Jeble see Jablah
Jedda see Jiddah
Jędrzejów 76 D4 Ger. Endersdorf. Świętokrzyskie, C Poland
Jefferson City 23 G5 state capital Missouri, C USA
Jega 53 F4 Kebbi, NW Nigeria
Jehol see Chengde
Jeju-do 107 E4 Jap. Saishū; prev. Cheju-do, Quelpart. island S South Korea
Jeju Strait 107 E4 var. Jeju-haehyŏp; prev. Cheju-Strait. strait S South Korea
Jeju-haehyŏp see Jeju Strait
Jēkabpils 84 D4 Ger. Jakobstadt. Jēkabpils, S Latvia
Jelalabad see Jalālābād
Jelenia Góra 76 B4 Ger. Hirschberg, Hirschberg im Riesengebirge, Hirschberg in Riesengebirge, Hirschberg in Schlesien. Dolnośląskie, SW Poland
Jelgava 84 C3 Ger. Mitau. Jelgava, C Latvia
Jemappes 65 B6 Hainaut, S Belgium
Jember 116 D5 prev. Djember. Jawa, C Indonesia
Jena 72 C4 Thüringen, C Germany
Jengish Chokusu see Tömür Feng
Jenin 97 A6 N West Bank
Jerablus see Jarābulus
Jerada 48 D2 NE Morocco
Jérémie 32 D3 SW Haiti
Jerez see Jeréz de la Frontera, Spain
Jerez de la Frontera 70 C5 var. Jerez; prev. Xeres. Andalucía, SW Spain
Jerez de los Caballeros 70 C4 Extremadura, W Spain
Jericho see Arīḩā
Jerid, Chott el 49 E2 var. Shaṭṭ al Jarīd. salt lake SW Tunisia
Jersey 67 D8 British Crown Dependency Channel Islands, NW Europe
Jerusalem 81 H4 Ar. Al Quds, Al Quds ash Sharīf, Heb. Yerushalayim; anc. Hierosolyma. country capital (Israel) Jerusalem, NE Israel
Jesenice 73 D7 Ger. Assling. NW Slovenia
Jesselton see Kota Kinabalu
Jessore 113 G4 Khulna, W Bangladesh
Jesús María 42 C3 Córdoba, C Argentina
Jeypore see Jaipur, Rājasthān, India
Jhānsi 112 D3 Uttar Pradesh, N India
Jhārkhand 113 F4 cultural region NE India
Jhelum 112 C2 Punjab, NE Pakistan
Ji see Hebei, China
Ji see Jilin, China
Jiangmen 106 C6 Guangdong, S China
Jiangsu 106 D4 var. Chiang-su, Jiangsu Sheng, Kiangsu, Su. province E China
Jiangsu see Nanjing
Jiangsu Sheng see Jiangsu
Jiangxi 106 C6 var. Chiang-hsi, Gan, Jiangxi Sheng, Kiangsi. province S China
Jiangxi Sheng see Jiangxi
Jiaxing 106 D5 Zhejiang, SE China
Jiayi see Chiayi
Jibuti see Djibouti
Jiddah 99 A5 Eng. Jedda. Makkah, W Saudi Arabia
Jih-k'a-tse see Xigazê
Jihlava 77 B5 Ger. Iglau, Pol. Igława. Vysocina, S Czech Republic (Czechia)
Jilib 51 D6 It. Gelib. Jubbada Dhexe, S Somalia
Jilin 106 E3 var. Chi-lin, Girin, Kirin; prev. Yungki, Yunki. Jilin, NE China
Jilin 106 E3 var. Chi-lin, Girin, Ji, Jilin Sheng, Kirin. province NE China
Jilin Sheng see Jilin
Jilong see Keelung
Jīma 51 C5 var. Jimma, It. Gimma. Oromīya, C Ethiopia
Jimbolia 86 A4 Ger. Hatzfeld, Hung. Zsombolya. Timiș, W Romania
Jiménez 28 D2 Chihuahua, N Mexico
Jimma see Jīma
Jimsar 104 C3 Xinjiang Uygur Zizhiqu, NW China
Jin see Shanxi
Jin see Tianjin Shi
Jinan 106 C4 var. Chinan, Chi-nan, Tsinan. province capital Shandong, E China
Jingdezhen 106 C5 Jiangxi, S China
Jinghong 106 A6 var. Yunjinghong. Yunnan, SW China
Jinhua 106 D5 Zhejiang, SE China
Jining 105 F3 Shandong, E China
Jinja 51 C6 S Uganda
Jinotega 30 D3 Jinotega, NW Nicaragua
Jinotepe 30 D3 Carazo, SW Nicaragua
Jinsen see Incheon
Jinzhong 106 C4 var. Yuci. Shanxi, C China
Jinzhou 106 D3 var. Chin-chou, Chinchow; prev. Chinhsien. Liaoning, NE China
Jirgalanta see Hovd
Jisr ash Shadadi see Ash Shadādah
Jiu 86 B5 Ger. Schil, Schyl, Hung. Zsil, Zsily. river S Romania
Jiujiang 106 C5 Jiangxi, S China
Jixi 107 E2 Heilongjiang, NE China
Jīzān 99 B6 var. Qīzān. Jīzān, SW Saudi Arabia
Jizzax 101 E2 Rus. Dzhizak. Jizzax Viloyati, C Uzbekistan
João Belo see Xai-Xai
João Pessoa 41 G2 prev. Paraíba. state capital Paraíba, E Brazil
Joazeiro see Juazeiro
Job'urg see Johannesburg
Jo-ch'iang see Ruoqiang
Jodhpur 112 C3 Rājasthān, NW India
Joensuu 63 E5 Pohjois-Karjala, SE Finland
Jōetsu 109 C5 var. Zyôetu. Niigata, Honshū, C Japan
Jogjakarta see Yogyakarta
Johannesburg 56 D4 var. Egoli, Erautini, Gauteng, abbrev. Job'urg. Gauteng, NE South Africa
Johannisburg see Pisz
John Day River 24 C3 river Oregon, NW USA
John o'Groats 66 C2 N Scotland, United Kingdom
Johnston Atoll 121 E1 US unincorporated territory C Pacific Ocean
Johor Baharu see Johor Bahru
Johor Bahru 116 B3 var. Johor Baharu, Johore Bahru. Johor, Peninsular Malaysia
Johore Bahru see Johor Bahru
Johore Strait 116 A1 strait Johor, Peninsular Malaysia, Malaysia/Singapore Asia Andaman Sea/ South China Sea
Joinvile see Joinville
Joinville 41 E4 var. Joinvile. Santa Catarina, S Brazil
Jokkmokk 62 C3 Lapp. Dálvvadis. Norrbotten, N Sweden
Jokyakarta see Yogyakarta
Joliet 18 B3 Illinois, N USA
Jonava 84 B4 Ger. Janow, Pol. Janów. Kaunas, C Lithuania
Jonesboro 20 B1 Arkansas, C USA
Joniškis 84 C3 Ger. Janischken. Šiaulai, N Lithuania
Jönköping 63 B7 Jönköping, S Sweden
Jonquière 17 E4 Québec, SE Canada
Joplin 23 F5 Missouri, C USA
Jordan 97 B6 off. Hashemite Kingdom of Jordan, Ar. Al Mamlaka al Urduniya al Hashemīyah, Al Urdunn; prev. Transjordan. country SW Asia
Jordan 97 B5 Ar. Urdunn, Heb. HaYarden. river SW Asia
Jorhāt 113 H3 Assam, NE India
Jos 53 G4 Plateau, C Nigeria
Joseph Bonaparte Gulf 124 D2 gulf N Australia
Jos Plateau 53 G4 plateau C Nigeria
Jotunheimen 63 A5 mountain range S Norway
Joûnié 96 A4 var. Juniyah. W Lebanon
Joure 64 D2 Fris. De Jouwer. Friesland, N Netherlands
Joutseno 63 E5 Etelä-Kariala, SE Finland
Jowhar see Jawhar
J.Storm Thurmond Reservoir see Clark Hill Lake
Juan Aldama 28 D3 Zacatecas, C Mexico
Juan de Fuca, Strait of 24 A1 strait Canada/USA
Juan Fernández, Islas 35 A6 Eng. Juan Fernandez Islands. island group W Chile
Juan Fernandez Islands see Juan Fernández, Islas
Juazeiro 41 G2 prev. Joazeiro. Bahia, E Brazil
Juazeiro do Norte 41 G2 Ceará, E Brazil
Juba 51 B5 var. Jūbā. country capital (South Sudan) Bahr el Gabel, S South Sudan
Juba 51 D6 Amh. Genalē Wenz, It. Guiba, Som. Ganaane, Webi Jubba. river Ethiopia/Somalia
Jubba, Webi see Juba
Jubbulpore see Jabalpur
Júcar 71 E3 var. Jucar. river C Spain
Juchitán 29 F5 var. Juchitán de Zaragosa. Oaxaca, SE Mexico
Juchitán de Zaragosa see Juchitán
Judayyidat Hāmir 98 B3 Al Anbār, S Iraq
Judenburg 73 D7 Steiermark, C Austria
Jugoslavija see Serbia
Juigalpa 30 D3 Chontales, S Nicaragua
Juiz de Fora 41 F4 Minas Gerais, SE Brazil
Jujuy see San Salvador de Jujuy
Jūlā see Jālū, Libya
Julia Beterrae see Béziers
Juliaca 39 E4 Puno, SE Peru
Juliana Top 37 G3 mountain C Suriname
Julianehåb see Qaqortoq
Julio Briga see Bragança
Juliobriga see Logroño
Juliomagus see Angers
Jullundur see Jalandhar
Jumilla 71 E4 Murcia, SE Spain
Jummoo see Jammu
Jumna see Yamuna
Jumporn see Chumphon
Junction City 23 F4 Kansas, C USA
Juneau 14 D4 state capital Alaska, USA
Junín 42 C4 Buenos Aires, E Argentina
Juniyah see Joûnié
Junkseylon see Phuket
Jur 51 B5 river C Sudan
Jura 68 D4 cultural region E France
Jura 73 A7 var. Jura Mountains. mountain range France/Switzerland
Jura 66 B4 island SW Scotland, United Kingdom
Jura Mountains see Jura
Jurbarkas 84 B4 Ger. Georgenburg, Jurburg. Tauragė, W Lithuania
Jurburg see Jurbarkas
Jūrmala 84 C3 Rīga, C Latvia
Juruá, Rio 40 C2 var. Río Yuruá. river Brazil/Peru
Juruena, Rio 40 D3 river W Brazil
Jutiapa 30 B2 Jutiapa, S Guatemala
Juticalpa 30 D2 Olancho, C Honduras
Jutland 63 A7 Den. Jylland. peninsula W Denmark
Juvavum see Salzburg
Juventud, Isla de la 32 A2 var. Isla de Pinos, Eng. Isle of Youth; prev. The Isle of the Pines. island W Cuba
Južna Morava 79 E5 Ger. Südliche Morava. river SE Serbia
Jwaneng 56 C4 Southern, S Botswana
Jylland see Jutland
Jyrgalan see Dzhergalan
Jyväskylä 63 D5 Keski-Suomi, C Finland

K

K2 104 A4 Chin. Qogir Feng, Eng. Mount Godwin Austen. mountain China/Pakistan
Kaafu Atoll see Male' Atoll
Kaaimanston 37 G3 Sipaliwini, N Suriname
Kaakhka see Kaka
Kaala see Caála
Kaamanen 62 D2 Lapp. Gámas. Lappi, N Finland
Kaapstad see Cape Town
Kaaresuvanto 62 C3 Lapp. Gárassavon. Lappi, N Finland
Kabale 51 B6 SW Uganda
Kabinda 55 D7 Kasai-Oriental, SE Dem. Rep. Congo
Kabinda see Cabinda
Kābol see Kābul
Kabompo 56 C2 river W Zambia
Kābul 101 E4 prev. Kābol. country capital (Afghanistan) Kābul, E Afghanistan
Kabul 101 E4 var. Daryā-ye Kābul. river Afghanistan/Pakistan
Kābul, Daryā-ye see Kabul
Kabwe 56 D2 Central, C Zambia
Kachchh, Gulf of 112 B4 var. Gulf of Cutch, Gulf of Kutch. gulf W India
Kachchh, Rann of 112 B4 var. Rann of Kachh, Rann of Kutch. salt marsh India/Pakistan
Kachh, Rann of see Kachchh, Rann of
Kadan Kyun 115 B5 prev. King Island. island Myeik Archipelago, S Myanmar (Burma)
Kadavu 123 E4 prev. Kandavu. island S Fiji
Kadiyivka 87 H3 Rus. Stakhanov. Luhans'ka Oblast', E Ukraine
Kadoma 56 D3 prev. Gatooma. Mashonaland West, C Zimbabwe
Kadugli 50 B4 Southern Kordofan, S Sudan
Kaduna 53 G4 Kaduna, C Nigeria
Kadzhi-Say 101 G2 Kir. Kajisay. Issyk-Kul'skaya Oblast', NE Kyrgyzstan
Kaédi 52 C3 Gorgol, S Mauritania
Kaffa see Feodosiya
Kafue 56 D2 Lusaka, SE Zambia
Kafue 56 C2 river C Zambia
Kaga Bandoro 54 C4 prev. Fort-Crampel. Nana-Grébizi, C Central African Republic
Kagan see Kogon
Kâghet 52 D1 var. Karet. physical region N Mauritania
Kagi see Chiayi
Kagoshima 109 B8 var. Kagosima. Kagoshima, Kyūshū, SW Japan
Kagoshima-wan 109 A8 bay SW Japan
Kagosima see Kagoshima
Kagul see Cahul
Darya-ye Kahmard 101 E4 prev. Darya-i-surkhab. river NE Afghanistan
Kahramanmaraş 94 D4 var. Kahraman Maraş, Maraş, Marash. Kahramanmaraş, S Turkey
Kaiapoi 129 C6 Canterbury, South Island, New Zealand
Kaifeng 106 C4 Henan, C China
Kai, Kepulauan 117 F4 prev. Kei Islands. island group Maluku, SE Indonesia
Kaikohe 128 C2 Northland, North Island, New Zealand
Kaikoura 129 C5 Canterbury, South Island, New Zealand
Kaikoura Peninsula 129 C5 peninsula South Island, New Zealand
Kainji Lake see Kainji Reservoir
Kainji Reservoir 53 F4 var. Kainji Lake. reservoir W Nigeria
Kaipara Harbour 128 C2 harbour North Island, New Zealand
Kairouan 49 E2 var. Al Qayrawān. E Tunisia
Kaisaria see Kayseri
Kaiserslautern 73 A5 Rheinland-Pfalz, SW Germany
Kaišiadorys 85 B5 Kaunas, S Lithuania
Kaitaia 128 C2 Northland, North Island, New Zealand
Kajaani 62 E4 Swe. Kajana. Kainuu, C Finland
Kajan see Kayan, Sungai
Kajana see Kajaani
Kajisay see Kadzhi-Say
Kaka 100 C2 Rus. Kaakhka. Ahal Welaýaty, S Turkmenistan
Kake 14 D4 Kupreanof Island, Alaska, USA
Kakhovka 87 F4 Khersons'ka Oblast', S Ukraine
Kakhovs'ke Vodoskhovyshche 87 F4 Rus. Kakhovskoye Vodokhranilishche. reservoir SE Ukraine
Kakhovskoye Vodokhranilishche see Kakhovs'ke Vodoskhovyshche
Kākināda 110 D1 prev. Cocanada. Andhra Pradesh, E India
Kakshaal-Too, Khrebet see Kokshaal-Tau
Kaktovik 14 D2 Alaska, USA
Kalaallit Nunaat see Greenland
Kalaburagi 110 C1 prev. Gulbarga. Karnātaka, C India
Kalahari Desert 56 B4 desert Southern Africa
Kalaikhum see Qal'aikhum
Kalámai see Kalámata
Kalamariá 82 B4 Kentrikí Makedonía, N Greece
Kalamás 82 A4 var. Thiamis; prev. Thýamis. river W Greece
Kalámata 83 B6 prev. Kalámai. Pelopónnisos, S Greece
Kalamazoo 18 C3 Michigan, N USA
Kalambaka see Kalampáka
Kálamos 83 C5 Attikí, C Greece
Kalampáka 82 B4 var. Kalambaka. Thessalía, C Greece
Kalanchak 87 F4 Khersons'ka Oblast', S Ukraine
Kalarash see Călărași
Kalasin 114 D4 var. Muang Kalasin. Kalasin, E Thailand
Kālat 112 B2 var. Kelat, Khelat. Baluchistān, SW Pakistan
Kalāt see Qalāt
Kalbarri 125 A5 Western Australia
Kalecik 94 C3 Ankara, N Turkey
Kalemie 55 E6 prev. Albertville. Katanga, SE Dem. Rep. Congo
Kale Sultanie see Çanakkale
Kalgan see Zhangjiakou
Kalgoorlie 125 B6 Western Australia
Kalima 55 D6 Maniema, E Dem. Rep. Congo
Kalimantan 116 D4 Eng. Indonesian Borneo. geopolitical region Borneo, C Indonesia
Kalinin see Tver'
Kaliningrad 84 A4 Kaliningradskaya Oblast', W Russia
Kaliningrad see Kaliningradskaya Oblast'
Kaliningradskaya Oblast' 84 B4 var. Kaliningrad. province and enclave W Russia
Kalinkavichy 85 C7 Rus. Kalinkovichi. Homyel'skaya Voblasts', SE Belarus
Kalinkovichi see Kalinkavichy
Kalisch/Kalish see Kalisz
Kalispell 22 B1 Montana, NW USA
Kalisz 76 C4 Ger. Kalisch, Rus. Kalish; anc. Calisia. Wielkopolskie, C Poland
Kalix 62 D4 Norrbotten, N Sweden
Kalixälven 62 D3 river N Sweden
Kallaste 84 E3 Ger. Krasnogor. Tartumaa, SE Estonia
Kallavesi 63 E5 lake SE Finland
Kalloní 83 D5 Lésvos, E Greece
Kalmar 63 C7 var. Calmar. Kalmar, S Sweden
Kalmthout 65 C5 Antwerpen, N Belgium
Kalpáki 82 A4 Ípeiros, W Greece
Kalpeni Island 110 B3 island Lakshadweep, India, N Indian Ocean
Kaltdorf see Pruszków
Kaluga 89 B5 Kaluzhskaya Oblast', W Russia
Kalush 86 C2 Pol. Kałusz. Ivano-Frankivs'ka Oblast', W Ukraine
Kałusz see Kalush
Kalutara 110 D4 Western Province, SW Sri Lanka
Kalvarija 85 B5 Pol. Kalwaria. Marijampolė, S Lithuania
Kalwaria see Kalvarija
Kalyān 112 C5 Mahārāshtra, W India
Kálymnos 83 D6 var. Kálimnos. island Dodekánisa, Greece, Aegean Sea
Kama 88 D4 river NW Russia
Kamarang 37 F3 W Guyana
Kambryk see Cambrai
Kamchatka see Kamchatka, Poluostrov
Kamchatka, Poluostrov 93 G3 Eng. Kamchatka. peninsula E Russia
Kamenets-Podol'skiy see Kam"yanets'-Podil's'kyy
Kamenka Dneprovskaya see Kam"yanka-Dniprovs'ka
Kamenskoye see **Kam"yans'ke**
Kamensk-Shakhtinskiy 89 B6 Rostovskaya Oblast', SW Russia
Kamina 55 D7 Katanga, S Dem. Rep. Congo
Kamishli see Al Qāmishlī
Kamloops 15 E5 British Columbia, SW Canada
Kammu Seamount 130 C2 guyot N Pacific Ocean
Kampala 51 B6 country capital (Uganda) S Uganda
Kampong Cham 115 D6 Khmer. Kâmpóng Cham. Kampong Cham, C Cambodia
Kâmpóng Cham see Kampong Cham
Kampong Chhang 115 D6 Khmer. Kâmpóng Chhnăng. Kampong Chhang, C Cambodia
Kâmpóng Chhnăng see Kampong Chhang
Kampong Speu 115 D6 Khmer. Kâmpóng Spoe. Kampong Speu, S Cambodia
Kâmpóng Spoe see Kampong Speu
Kampong Thom 115 D5 Khmer. Kâmpóng Thum, prev. Trâpeăng Vêng. Kampong Thom, C Cambodia
Kâmpóng Thum see Kampong Thom
Kâmpóng Trâbêk 115 D5 prev. Phumĭ Kâmpóng Trâbêk, Phum Kompong Trabek. Kampong Thom, C Cambodia
Kampot 115 D6 Khmer. Kâmpôt. Kampot, SW Cambodia
Kâmpôt see **Kampot**
Kampuchea see Cambodia
Kampuchea, Democratic see Cambodia
Kampuchea, People's Democratic Republic of see Cambodia
Kam"yanets'-Podil's'kyy 86 C3 Rus. Kamenets-Podol'skiy. Khmel'nyts'ka Oblast', W Ukraine
Kam"yanka-Dniprovs'ka 87 F3 Rus. Kamenka Dneprovskaya. Zaporiz'ka Oblast', SE Ukraine
Kam"yans'ke 87 F3 Rus. Dniprodzerzhyns'k, prev. Kamenskoye. Dnipropetrovs'ka Oblast', E Ukraine
Kamyshin 89 B6 Volgogradskaya Oblast', SW Russia
Kanaky see New Caledonia
Kananga 55 D6 prev. Luluabourg. Kasai-Occidental, S Dem. Rep. Congo
Kanara see Karnātaka
Kanash 89 C5 Chuvashskaya Respublika, W Russia
Kanazawa 109 C5 Ishikawa, Honshū, SW Japan
Kanbe 114 B4 Yangon, SW Myanmar (Burma)
Kānchīpuram 110 C2 prev. Conjeeveram. Tamil Nādu, SE India
Kandahār 101 E5 Per. Qandahār. Kandahār, S Afghanistan
Kandalakša see Kandalaksha
Kandalaksha 88 B2 var. Kandalakša, Fin. Kantalahti. Murmanskaya Oblast', NW Russia
Kandangan 116 D4 Borneo, C Indonesia
Kandau see Kandava
Kandava 84 C3 Ger. Kandau. Tukums, W Latvia
Kandavu see Kadavu
Kandi 53 F4 N Benin
Kandy 110 D3 Central Province, C Sri Lanka
Kane Fracture Zone 44 B4 fracture zone NW Atlantic Ocean
Kāne'ohe 25 A8 var. Kaneohe. O'ahu, Hawaii, USA, C Pacific Ocean
Kanestron, Akrotírio see Palioúri, Akrotírio
Kanëv see Kaniv
Kanevskoye Vodokhranilishche see Kanivs'ke Vodoskhovyshche
Kangān see Bandar-e Kangān
Kangaroo Island 127 A7 island South Australia
Kangertittivaq 61 E4 Dan. Scoresby Sund. fjord E Greenland
Kangikajik 61 E4 var. Kap Brewster. headland E Greenland
Kaniv 87 E2 Rus. Kanëv. Cherkas'ka Oblast', C Ukraine
Kanivs'ke Vodoskhovyshche 87 E2 Rus. Kanevskoye Vodokhranilishche. reservoir C Ukraine
Kanjiža 78 D2 Ger. Altkanischa, Hung. Magyarkanizsa, Ókanizsa; prev. Stara Kanjiža. Vojvodina, N Serbia
Kankaanpää 63 D5 Satakunta, SW Finland
Kankakee 18 B3 Illinois, N USA
Kankan 52 D4 E Guinea
Kannur 110 B2 var. Cannanore. Kerala, SW India
Kano 53 G4 Kano, N Nigeria
Kānpur 113 E3 Eng. Cawnpore. Uttar Pradesh, N India
Kansas 23 F5 off. State of Kansas, also known as Jayhawker State, Sunflower State. state C USA
Kansas City 23 F4 Kansas, C USA
Kansas City 23 F4 Missouri, C USA
Kansas River 23 F5 river Kansas, C USA
Kansk 93 E4 Krasnoyarskiy Kray, S Russia
Kansu see Gansu
Kantalahti see Kandalaksha
Kántanos 83 C7 Kríti, Greece, E Mediterranean Sea
Kantemirovka 89 B6 Voronezhskaya Oblast', W Russia
Kantipur see Kathmandu
Kanton 123 F3 var. Abariringa, Canton Island; prev. Mary Island. atoll Phoenix Islands, C Kiribati
Kanye 56 C4 Southern, SE Botswana
Kaohsiung 106 D6 var. Gaoxiong, Jap. Takao, Takow. S Taiwan
Kaolack 52 B3 var. Kaolak. W Senegal
Kaolak see Kaolack
Kaolan see Lanzhou
Kaoma 56 C2 Western, W Zambia
Kapelle 65 B5 Zeeland, SW Netherlands
Kapellen 65 C5 Antwerpen, N Belgium
Kapka, Massif du 54 C2 mountain range E Chad
Kaplangky, Plato see Gaplañgyr Platosy
Kapoeas see Kapuas, Sungai
Kapoeta 51 C5 E Equatoria, SE South Sudan
Kaposvár 77 C7 Somogy, SW Hungary
Kappeln 72 B2 Schleswig-Holstein, N Germany
Kapronca see Koprivnica
Kapstad see Cape Town
Kapsukas see Marijampolė
Kaptsevichy 85 C7 Rus. Koptsevichi. Homyel'skaya Voblasts', SE Belarus
Kapuas, Sungai 116 C4 prev. Kapoeas. river Borneo, C Indonesia
Kapuskasing 16 C4 Ontario, S Canada
Kapyl' 85 C6 Rus. Kopyl'. Minskaya Voblasts', C Belarus
Kara-Balta 101 F2 Chuyskaya Oblast', N Kyrgyzstan
Karabil', Vozvyshennost' see Garabil Belentligi
Kara-Bogaz-Gol, Zaliv see Garabogaz Aylagy
Karabük 94 C2 Karabük, NW Turkey
Karāchi 112 B3 Sind, SE Pakistan
Karácsonkő see Piatra-Neamț
Karadeniz see Black Sea
Karadeniz Boğazi see İstanbul Boğazı
Karaferiye see Véroia
Karagandy 92 C4 prev. Karaganda, Kaz. Qaraghandy. Karagandy, C Kazakhstan
Karaginskiy, Ostrov 93 H2 island E Russia
Karagumskiy Kanal see Garagum Kanaly
Karak see Al Karak
Kara-Kala see Magtymguly
Karakax see Moyu

Kitakyūshū – Kuril Islands

Kitakyūshū 109 A7 var. Kitakyûsyû. Fukuoka, Kyūshū, SW Japan
Kitakyûsyû see Kitakyūshū
Kitami 108 D2 Hokkaidō, NE Japan
Kitchener 16 C5 Ontario, S Canada
Kíthnos see Kýthnos
Kitimat 14 D4 British Columbia, SW Canada
Kitinen 62 D3 river N Finland
Kitob 101 E3 Rus. Kitab. Qashqadaryo Viloyati, S Uzbekistan
Kitwe 56 D2 var. Kitwe-Nkana. Copperbelt, C Zambia
Kitwe-Nkana see Kitwe
Kitzbüheler Alpen 73 C7 mountain range W Austria
Kivalina 14 C2 Alaska, USA
Kivalo 62 D3 ridge C Finland
Kivertsi 86 C1 Pol. Kiwerce, Rus. Kivertsy. Volyns'ka Oblast', NW Ukraine
Kivertsy see Kivertsi
Kivu, Lac see Kivu, Lake
Kivu, Lake 55 E6 Fr. Lac Kivu. lake Rwanda/Dem. Rep. Congo
Kiwerce see Kivertsi
Kiyev see Kyiv
Kiyevskoye Vodokhranilishche see Kyyivs'ke Vodoskhovyshche
Kızıl Irmak 94 C3 river C Turkey
Kizil Kum see Kyzyl Kum
Kizyl-Arvat see Serdar
Kjølen see Kölen
Kladno 77 A5 Středočeský, NW Czech Republic (Czechia)
Klagenfurt 73 D7 Slvn. Celovec. Kärnten, S Austria
Klaipėda 84 B3 Ger. Memel. Klaipeda, NW Lithuania
Klamath Falls 24 B4 Oregon, NW USA
Klamath Mountains 24 A4 mountain range California/Oregon, W USA
Klang 116 B3 var. Kelang; prev. Port Swettenham. Selangor, Peninsular Malaysia
Klarälven 63 B6 river Norway/Sweden
Klatovy 77 A5 Ger. Klattau. Plzeňský Kraj, W Czech Republic (Czechia)
Klattau see Klatovy
Klausenburg see Cluj-Napoca
Klazienaveen 64 E2 Drenthe, NE Netherlands
Kleines Ungarisches Tiefland see Little Alföld
Klein Karas 56 B4 Karas, S Namibia
Kleinwardein see Kisvárda
Kleisoúra 83 A5 Ípeiros, W Greece
Klerksdorp 56 D4 North-West, N South Africa
Klimavichy 85 E7 Rus. Klimovichi. Mahilyowskaya Voblasts', E Belarus
Klimovichi see Klimavichy
Klintsy 89 A5 Bryanskaya Oblast', W Russia
Klisura 82 C2 Plovdiv, C Bulgaria
Ključ 78 B3 Federacija Bosna I Hercegovina, NW Bosnia and Herzegovina
Kłobuck 76 C4 Śląskie, S Poland
Klosters 73 B7 Graubünden, SE Switzerland
Kluang see Keluang
Kluczbork 76 C4 Ger. Kreuzburg, Kreuzburg in Oberschlesien. Opolskie, S Poland
Klyuchevskaya Sopka, Vulkan 93 H3 volcano E Russia
Knin 78 B4 Šibenik-Knin, S Croatia
Knjaževac 78 E4 Serbia, E Serbia
Knokke-Heist 65 A5 West-Vlaanderen, NW Belgium
Knoxville 20 D1 Tennessee, S USA
Knud Rasmussen Land 60 D1 physical region N Greenland
Kobdo see Hovd
Kōbe 109 C6 Hyōgo, Honshū, SW Japan
København 63 B7 Eng. Copenhagen; anc. Hafnia. country capital (Denmark) Sjælland, København, E Denmark
Kobenni 52 D3 Hodh el Gharbi, S Mauritania
Koblenz 73 A5 prev. Coblenz, Fr. Coblence; anc. Confluentes. Rheinland-Pfalz, W Germany
Kobrin see Kobryn
Kobryn 85 A6 Rus. Kobrin. Brestskaya Voblasts', SW Belarus
Kobuleti 95 F2 prev. K'obulet'i. W Georgia
K'obulet'i see Kobuleti
Kočani 79 E6 NE Macedonia
Kočevje 73 D8 Ger. Gottschee. S Slovenia
Koch Bihār 113 G3 West Bengal, NE India
Kochchi see Kochi
Kochi 110 C3 var. Cochin, Kochchi. Kerela SW India
Kōchi 109 B7 var. Kôti. Kōchi, Shikoku, SW Japan
Kochiu see Gejiu
Kodiak 14 C3 Kodiak Island, Alaska, USA
Kodiak Island 14 C3 island Alaska, USA
Koedoes see Kudus
Koeln see Köln
Koepang see Kupang
Ko-erh-mu see Golmud
Koetai see Mahakam, Sungai
Koetaradja see Bandaaceh
Kōfu 109 D5 var. Kôhu. Yamanashi, Honshū, S Japan
Kogarah 126 E2 New South Wales, E Australia
Kogon 100 D2 Rus. Kagan. Buxoro Viloyati, C Uzbekistan
Kōhalom see Rupea
Kohima 113 H3 state capital Nāgāland, E India
Koh I Noh see Büyükağrı Dağı
Kohtla-Järve 84 E2 Ida-Virumaa, NE Estonia
Kôhu see Kōfu
Kokand see Qo'qon
Kokchetav see Kokshetau
Kokkola 62 D4 Swe. Karleby; prev. Swe. Gamlakarleby. Österbotten, W Finland
Koko 53 F4 Kebbi, W Nigeria
Kokomo 18 C4 Indiana, N USA
Koko Nor see Qinghai, China
Koko Nor see Qinghai Hu, China
Kokrines 14 C2 Alaska, USA
Kokshaal-Tau 101 G2 Rus. Khrebet Kakshaal-Too. mountain range China/Kyrgyzstan
Kokshetau 92 C4 Kaz. Kökshetaū; prev. Kokchetav. Kokshetau, N Kazakhstan
Kökshetaū see Kokshetau
Koksijde 65 A5 West-Vlaanderen, W Belgium
Koksoak 16 D2 river Québec, E Canada
Kokstad 56 D5 KwaZulu/Natal, E South Africa
Kolaka 117 E4 Sulawesi, C Indonesia
K'o-la-ma-i see Karamay
Kola Peninsula see Kol'skiy Poluostrov
Kolari 62 D3 Lappi, NW Finland
Kolárovo 77 C6 Ger. Gutta; prev. Guta, Hung. Gúta. Nitriansky Kraj, SW Slovakia
Kolberg see Kołobrzeg
Kolda 52 C3 S Senegal
Kolding 63 A7 Vejle, C Denmark
Kölen 59 E1 Nor. Kjølen. mountain range Norway/Sweden
Kolguyev, Ostrov 88 C2 island NW Russia
Kolhāpur 110 B1 Mahārāshtra, SW India
Kolhumadulu 110 A5 var. Thaa Atoll. atoll S Maldives
Kolín 77 B5 Ger. Kolin. Střední Čechy, C Czech Republic (Czechia)
Kolka 84 C2 Talsi, NW Latvia
Kolkasrags 84 C2 prev. Eng. Cape Domesnes. headland NW Latvia
Kolkata 113 G4 prev. Calcutta. West Bengal, N India
Kollam 110 C3 var. Quilon. Kerala, SW India
Kolmar see Colmar
Köln 72 A4 var. Koeln, Eng./Fr. Cologne, prev. Cöln; anc. Colonia Agrippina, Oppidum Ubiorum. Nordrhein-Westfalen, W Germany
Koło 76 C3 Wielkopolskie, C Poland
Kołobrzeg 76 B2 Ger. Kolberg. Zachodnio-pomorskie, NW Poland
Kolokani 52 D3 Koulikoro, W Mali
Kolomea see Kolomyya
Kolomna 89 B5 Moskovskaya Oblast', W Russia
Kolomyya 86 C3 Ger. Kolomea. Ivano-Frankivs'ka Oblast', W Ukraine
Kolosjoki see Nikel'
Kolozsvár see Cluj-Napoca
Kolpa 78 A2 Ger. Kulpa, SCr. Kupa. river Croatia/Slovenia
Kolpino 88 B4 Leningradskaya Oblast', NW Russia
Kólpos Mórfu see Güzelyurt Körfezi
Kol'skiy Poluostrov 88 C2 Eng. Kola Peninsula. peninsula NW Russia
Kolwezi 55 D7 Katanga, S Dem. Rep. Congo
Kolyma 93 G2 river NE Russia
Komatsu 109 C5 var. Komatu. Ishikawa, Honshū, SW Japan
Komatu see Komatsu
Kommunizm, Qullai see Ismoili Somonī, Qullai
Komoé 53 E4 var. Komoé Fleuve. river E Ivory Coast
Komoé Fleuve see Komoé
Komotau see Chomutov
Komotiní 82 D3 var. Gümüljina, Turk. Gümülcine. Anatolikí Makedonía kai Thráki, NE Greece
Kompong Som see Sihanoukville
Komrat see Comrat
Komsomolets, Ostrov 93 E1 island Severnaya Zemlya, N Russia
Komsomol'sk-na-Amure 93 G4 Khabarovskiy Kray, SE Russia
Kondolovo 82 E3 Burgas, E Bulgaria
Kondopoga 88 B3 Respublika Kareliya, NW Russia
Kondoz see Kunduz
Köneürgenç 100 C2 var. Köneürgench, Rus. Këneurgench; prev. Kunya-Urgench. Daşoguz Welaýaty, N Turkmenistan
Kong Christian IX Land 60 D4 Eng. King Christian IX Land. physical region SE Greenland
Kong Frederik IX Land 60 C3 physical region SW Greenland
Kong Frederik VIII Land 61 E2 Eng. King Frederik VIII Land. physical region NE Greenland
Kong Frederik VI Kyst 60 C4 Eng. King Frederik VI Coast. physical region SE Greenland
Kong Karls Land 61 G2 Eng. King Charles Islands. island group SE Svalbard
Kongo see Congo (river)
Kongolo 55 D6 Katanga, E Dem. Rep. Congo
Kongor 51 B5 Jonglei, E South Sudan
Kong Oscar Fjord 61 E3 fjord E Greenland
Kongsberg 63 B6 Buskerud, S Norway
Kông, Tônle 115 E5 var. Xê Kong. river Cambodia/Laos
Kong, Xê see Kông, Tônle
Königgrätz see Hradec Králové
Königshütte see Chorzów
Konin 76 C3 Ger. Kuhnau. Weilkopolskie, C Poland
Koninkrijk der Nederlanden see Netherlands
Konispol 79 C7 var. Konispoli. Vlorë, S Albania
Konispoli see Konispol
Kónitsa 82 A4 Ípeiros, W Greece
Konitz see Chojnice
Konjic 78 C4 Federacija Bosna I Hercegovina, S Bosnia and Herzegovina
Konosha 88 C4 Arkhangel'skaya Oblast', NW Russia
Konotop 87 F1 Sums'ka Oblast', NE Ukraine
Konstantinovka see Kostyantynivka
Konstanz 73 B7 var. Constanz, Eng. Constance, hist. Kostnitz; anc. Constantia. Baden-Württemberg, S Germany
Konstanza see Constanța
Konya 94 C4 var. Konieh, prev. Konia; anc. Iconium. Konya, C Turkey
Kopaonik 79 D5 mountain range S Serbia
Kopar see Koper
Koper 73 D8 It. Capodistria; prev. Kopar. SW Slovenia
Köpetdag Gershi 100 C3 mountain range Iran/Turkmenistan
Köpetdag Gershi/Kopetdag, Khrebet see Koppeh Dāgh
Koppeh Dagh 98 D2 Rus. Khrebet Kopetdag, Turkm. Köpetdag Gershi. mountain range Iran/Turkmenistan
Kopreinitz see Koprivnica
Koprivnica 78 B2 Ger. Kopreinitz, Hung. Kaproncza. Koprivnica-Križevci, N Croatia
Köprülü see Veles
Koptsevichi see Kaptsevichy
Kopyl' see Kapyl'
Korat see Nakhon Ratchasima
Korat Plateau 114 D4 plateau E Thailand
Korba 113 E4 Chhattisgarh, C India
Korça see Korçë
Korçë 79 D6 var. Korça, Gk. Korytsa, It. Corriza; prev. Koritsa. Korçë, SE Albania
Korčula 78 B4 It. Curzola; anc. Corcyra Nigra. island S Croatia
Korea Bay 105 G3 bay China/North Korea
Korea, Democratic People's Republic of see North Korea
Korea, Republic of see South Korea
Korea Strait 109 A7 Jap. Chōsen-kaikyō, Kor. Taehan-haehyŏp. channel Japan/South Korea
Korhogo 52 D4 N Ivory Coast
Kórinthos 83 B6 anc. Corinthus Eng. Corinth. Pelopónnisos, S Greece
Korinthiakós Kólpos 83 B5 Eng. Gulf of Corinth; anc. Corinthiacus Sinus. gulf C Greece
Koritsa see Korçë
Kōriyama 109 D5 Fukushima, Honshū, C Japan
Korla 104 C3 Chin. K'u-erh-lo. Xinjiang Uygur Zizhiqu, NW China
Körmend 77 B7 Vas, W Hungary
Koróni 83 B6 Pelopónnisos, S Greece
Koror 122 A2 (Palau) Oreor, N Palau
Kőrös see Križevci
Korosten' 86 D1 Zhytomyrs'ka Oblast', NW Ukraine
Koro Toro 54 C2 Borkou-Ennedi-Tibesti, N Chad
Korsovka see Kārsava
Kortrijk 65 A6 Fr. Courtrai. West-Vlaanderen, W Belgium
Koryak Range 93 H2 var. Koryakskiy Khrebet, Eng. Koryak Range. mountain range NE Russia
Koryak Range see Koryakskoye Nagor'ye
Koryakskiy Khrebet see Koryakskoye Nagor'ye
Koryazhma 88 C4 Arkhangel'skaya Oblast', NW Russia
Korytsa see Korçë
Kos 83 E6 Kos, Dodekánisa, Greece, Aegean Sea
Kos 83 E6 It. Coo; anc. Cos. island Dodekánisa, Greece, Aegean Sea
Ko-saki 109 A7 headland Nagasaki, Tsushima, SW Japan
Kościan 76 B4 Ger. Kosten. Wielkopolskie, C Poland
Kościerzyna 76 C2 Pomorskie, NW Poland
Kosciusko, Mount see Kosciuszko, Mount
Kosciuszko, Mount 127 C7 prev. Mount Kosciusko. mountain New South Wales, SE Australia
K'o-shih see Kashi
Koshikijima-retto 109 A8 var. Kosikizima Rettô. island group SW Japan
Kōshū see Gwangju
Košice 77 D6 Ger. Kaschau, Hung. Kassa. Košický Kraj, E Slovakia
Kosikizima Rettô see Koshikijima-rettō
Köslin see Koszalin
Koson 101 E3 Rus. Kasan. Qashqadaryo Viloyati, S Uzbekistan
Kosovo 79 D5 prev. Autonomous Province of Kosovo and Metohija. country (not fully recognised) SE Europe
Kosovo and Metohija, Autonomous Province of see Kosovo
Kosovo Polje see Fushë Kosovë
Kosovska Mitrovica see Mitrovicë
Kosrae 122 C2 prev. Kusaie. island Caroline Islands, E Micronesia
Kossou, Lac de 52 D5 lake C Ivory Coast
Kostanay 92 C4 var. Kustanay, Kaz. Qostanay. Kostanay, N Kazakhstan
Kosten see Kościan
Kostenets 82 C2 prev. Georgi Dimitrov. Sofiya, W Bulgaria
Kostnitz see Konstanz
Kostroma 88 B4 Kostromskaya Oblast', NW Russia
Kostyantynivka 87 G3 Rus. Konstantinovka. Donets'ka Oblast', SE Ukraine
Kostyukovichi see Kastsyukovichy
Kostyukowka see Kastsyukowka
Koszalin 76 B2 Ger. Köslin. Zachodnio-pomorskie, NW Poland
Kota 112 D3 prev. Kotah. Rājasthān, N India
Kota Baharu see Kota Bharu
Kota Bahru see Kota Bharu
Kotabaru see Jayapura
Kota Bharu 116 B3 var. Kota Baharu, Kota Bahru. Kelantan, Peninsular Malaysia
Kotaboemi see Kotabumi
Kotabumi 116 B4 prev. Kotaboemi. Sumatera, W Indonesia
Kotah see Kota
Kota Kinabalu 116 D3 prev. Jesselton. Sabah, East Malaysia
Kotel'nyy, Ostrov 93 E2 island Novosibirskiye Ostrova, N Russia
Kotka 63 E5 Kymenlaakso, S Finland
Kotlas 88 C4 Arkhangel'skaya Oblast', NW Russia
Kotonu see Cotonou
Kotor 79 C5 It. Cattaro. SW Montenegro
Kotovs'k see Podil's'k
Kotovsk see Hînceşti
Kottbus see Cottbus
Kotto 54 D4 river Central African Republic/Dem. Rep. Congo
Kotuy 93 E2 river N Russia
Koudougou 53 E4 C Burkina
Koulamoutou 55 B6 Ogooué-Lolo, C Gabon
Koulikoro 52 D3 Koulikoro, SW Mali
Koumra 54 C4 Moyen-Chari, S Chad
Kourou 37 H3 N French Guiana
Kousseir see Al Quşayr
Kousséri 54 B3 prev. Fort-Foureau. Extrême-Nord, NE Cameroon
Koutiala 52 D4 Sikasso, S Mali
Kouvola 63 E5 Kymenlaakso, S Finland
Kovel' 86 C1 Pol. Kowel. Volyns'ka Oblast', NW Ukraine
Kovno see Kaunas
Koweit see Kuwait
Kowel see Kovel'
Kowloon 106 A2 Hong Kong, S China
Kowno see Kaunas
Kozáni 82 B4 Dytikí Makedonía, N Greece
Kozara 78 B3 mountain range NW Bosnia and Herzegovina
Kozarska Dubica see Bosanska Dubica
Kozhikode 110 C2 var. Calicut. Kerala, SW India
Kozu-shima 109 D6 island E Japan
Kozyatyn 86 D2 Rus. Kazatin. Vinnyts'ka Oblast', C Ukraine
Kpalimé 53 E5 var. Palimé. SW Togo
Krâchéh see Kratie
Kragujevac 78 D4 Serbia, C Serbia
Krainburg see Kranj
Kra, Isthmus of 115 B6 isthmus Malaysia/Thailand
Krakau see Kraków
Kraków 77 D5 Eng. Cracow, Ger. Krakau; anc. Cracovia. Małopolskie, S Poland
Krâlănh 115 D5 Siĕmréab, NW Cambodia
Kralendijk 33 E5 dependent territory capital (Bonaire) Lesser Antilles, S Caribbean Sea
Kraljevo 78 D4 prev. Rankovićevo. Serbia, C Serbia
Kramators'k 87 G3 Rus. Kramatorsk. Donets'ka Oblast', SE Ukraine
Kramatorsk see Kramators'k
Kramfors 63 C5 Västernorrland, C Sweden
Kranéa see Kraniá
Kraniá 82 B4 var. Kranéa. Dytikí Makedonía, N Greece
Kranj 73 D7 Ger. Krainburg. NW Slovenia
Kranz see Zelenogradsk
Krāslava 84 D4 Krāslava, SE Latvia
Krasnaye 85 C5 Rus. Krasnoye. Minskaya Voblasts', C Belarus
Krasnoarmeysk 89 C6 Saratovskaya Oblast', W Russia
Krasnodar 89 A7 prev. Ekaterinodar, Yekaterinodar. Krasnodarskiy Kray, SW Russia
Krasnodon see Sorokyne
Krasnogor see Kallaste
Krasnogvardeyskoye see Krasnohvardiys'ke
Krasnohvardiys'ke 87 F4 Rus. Krasnogvardeyskoye. Respublika Krym, S Ukraine
Krasnokamensk 93 F4 Chitinskaya Oblast', S Russia
Krasnokamsk 89 D5 Permskaya Oblast', W Russia
Krasnoperekops'k see Yany Kapu
Krasnostav see Krasnystaw
Krasnovodsk see Türkmenbaşy
Krasnovodskiy Zaliv see Türkmenbasy Aylagy
Krasnowodsk Aylagy see Türkmenbasy Aylagy
Krasnoyarsk 92 D4 Krasnoyarskiy Kray, S Russia
Krasnoye see Krasnaye
Krasnystaw 76 E4 Rus. Krasnostav. Lubelskie, SE Poland
Krasnyy Kut 89 C6 Saratovskaya Oblast', W Russia
Krasnyy Luch see Khrustal'nyy
Kratie 115 D6 Khmer. Krâchéh. Kratie, E Cambodia
Krâvanh, Chuŏr Phnum 115 C6 Eng. Cardamom Mountains, Fr. Chaîne des Cardamomes. mountain range W Cambodia
Krefeld 72 A4 Nordrhein-Westfalen, W Germany
Kreisstadt see Krosno Odrzańskie
Kremenchug see Kremenchuk
Kremenchugskoye Vodokhranilishche/Kremenchuk Reservoir see Kremenchuts'ke Vodoskhovyshche
Kremenchuk 87 F2 Rus. Kremenchug. Poltavs'ka Oblast', NE Ukraine
Kremenchuk Reservoir 87 F2 Eng. Kremenchuk Reservoir, Rus. Kremenchugskoye Vodokhranilishche. reservoir C Ukraine
Kremenets' 86 C2 Pol. Krzemieniec, Rus. Kremenets. Ternopil's'ka Oblast', W Ukraine
Kremennaya see Kreminna
Kreminna 87 G2 Rus. Kremennaya. Luhans'ka Oblast', E Ukraine
Kresena see Kresna
Kresna 82 C3 var. Kresena. Blagoevgrad, SW Bulgaria
Kretikon Delagos see Kritikó Pélagos
Kretinga 84 B3 Ger. Krottingen. Klaipėda, NW Lithuania
Kreutz see Cristuru Secuiesc
Kreuz see Križevci, Croatia
Kreuz see Risti, Estonia
Kreuzburg/Kreuzburg in Oberschlesien see Kluczbork
Krichëv see Krychaw
Krievija see Russia
Krindachevka see Khrustal'nyy
Krishna 110 C1 prev. Kistna. river C India
Krishnagiri 110 C2 Tamil Nādu, SE India
Kristiania see Oslo
Kristiansand 63 A6 var. Christiansand. Vest-Agder, S Norway
Kristianstad 63 B7 Skåne, S Sweden
Kristiansund 62 A4 var. Christiansund. Møre og Romsdal, S Norway
Kríti 83 C7 Eng. Crete. island Greece, Aegean Sea
Kritikó Pélagos 83 D7 var. Kretikon Delagos, Eng. Sea of Crete; anc. Mare Creticum. sea Greece, Aegean Sea
Krivoy Rog see Kryvyy Rih
Križevci 78 B2 Ger. Kreuz, Hung. Kőrös. Varaždin, NE Croatia
Krk 78 A3 It. Veglia; anc. Curieta. island NW Croatia
Kroatien see Croatia
Krolevets' 87 F1 Rus. Krolevets. Sums'ka Oblast', NE Ukraine
Krolevets see Krolevets'
Królewska Huta see Chorzów
Kronach 73 C5 Bayern, E Germany
Kronstadt see Brașov
Kroonstad 56 D4 Free State, C South Africa
Kropotkin 89 A7 Krasnodarskiy Kray, SW Russia
Kropyvnyts'kyy 87 E3 Rus. Kirovohrad; prev. Kirovo, Yelizavetgrad, Zinov'yevsk. Kirovohrads'ka Oblast', C Ukraine
Krosno 77 D5 Ger. Krossen. Podkarpackie, SE Poland
Krosno Odrzańskie 76 B3 Ger. Crossen, Kreisstadt. Lubuskie, W Poland
Krossen see Krosno
Krottingen see Kretinga
Krško 73 E8 Ger. Gurkfeld; prev. Videm-Krško. E Slovenia
Krugloye see Kruhlaye
Kruhlaye 85 D6 Rus. Krugloye. Mahilyowskaya Voblasts', E Belarus
Kruja see Krujë
Krujë 79 C6 var. Kruja, It. Croia. Durrës, C Albania
Krummau see Český Krumlov
Krung Thep, Ao 115 C5 var. Bight of Bangkok. bay S Thailand
Krung Thep Mahanakhon see Ao Krung Thep
Krupki 85 D6 Minskaya Voblasts', C Belarus
Krušné Hory see Erzgebirge
Krychaw 85 E7 Rus. Krichëv. Mahilyowskaya Voblasts', E Belarus
Kryms'ki Hory 87 F5 mountain range S Ukraine
Kryms'kyy Pivostriv 87 F5 Eng. Crimea. (Ukrainian territory annexed by Russia since 2014). peninsula S Ukraine
Krynica 77 D5 Ger. Tannenhof. Małopolskie, S Poland
Kryve Ozero 87 E3 Odes'ka Oblast', SW Ukraine
Kryvyy Rih 87 F3 Rus. Krivoy Rog. Dnipropetrovs'ka Oblast', SE Ukraine
Krzemieniec see Kremenets'
Ksar al Kabir see Ksar-el-Kebir
Ksar al Soule see Er-Rachidia
Ksar-el-Kebir 48 C2 var. Alcázar, Ksar al Kabir, Ksar-el-Kébir, Ar. Al-Kasr al-Kebir, Al-Qsar al-Kbir, Sp. Alcazarquivir. NW Morocco
Ksar-el-Kébir see Ksar-el-Kebir
Kuala Dungun see Dungun
Kuala Lumpur 116 B3 country capital (Malaysia) Kuala Lumpur, Peninsular Malaysia
Kuala Terengganu 116 B3 var. Kuala Trengganu. Terengganu, Peninsular Malaysia
Kualatungkal 116 B4 Sumatera, W Indonesia
Kuang-chou see Guangzhou
Kuang-hsi see Guangxi Zhuangzu Zizhiqu
Kuang-tung see Guangdong
Kuang-yuan see Guangyuan
Kuantan 116 B3 Pahang, Peninsular Malaysia
Kuba see Quba
Kuban' 87 G5 var. Hypanis. river SW Russia
Kubango see Cubango/Okavango
Kuching 116 C3 prev. Sarawak. Sarawak, East Malaysia
Kūchnay Darwāshān 100 D5 prev. Kūchnay Darweyshān. Helmand, S Afghanistan
Kūchnay Darweyshān see Kūchnay Darwāshān
Kuçova see Kuçovë
Kuçovë 79 C6 var. Kuçova; prev. Qyteti Stalin. Berat, C Albania
Kudara see Ghŭdara
Kudus 116 C5 prev. Koedoes. Jawa, C Indonesia
Kuei-lin see Guilin
Kuei-Yang/Kuei-yang see Guiyang
K'u-erh-lo see Korla
Kueyang see Guiyang
Kugaaruk 15 G3 prev. Pelly Bay. Nunavut, N Canada
Kugluktuk 31 E3 var. Qurlurtuuq; prev. Coppermine. Nunavut, NW Canada
Kuhmo 62 E4 Kainuu, E Finland
Kuhnau see Konin
Kühnö see Kihnu
Kuibyshev see Kuybyshevskoye Vodokhranilishche
Kuito 56 B2 Port. Silva Porto. Bié, C Angola
Kuji 108 D3 var. Kuzi. Iwate, Honshū, C Japan
Kukës 79 D5 var. Kukësi. Kukës, NE Albania
Kukësi see Kukës
Kukong see Shaoguan
Kukukhoto see Hohhot
Kula Kangri 113 G3 var. Kulhakangri. mountain Bhutan/China
Kuldīga 84 B3 Ger. Goldingen. Kuldīga, W Latvia
Kuldja see Yining
Kulhakangri see Kula Kangri
Kullorsuaq 60 D2 var. Kuvdlorssuak. Kitaa, C Greenland
Kulm see Chełmno
Kulmsee see Chełmża
Kŭlob 101 F3 Rus. Kulyab. SW Tajikistan
Kulpa see Kolpa
Kulu 94 C3 Konya, W Turkey
Kulunda Steppe 92 C4 Kaz. Qulyndy Zhazyghy, Rus. Kulundinskaya Ravnina. grassland Kazakhstan/Russia
Kulundinskaya Ravnina see Kulunda Steppe
Kulyab see Kŭlob
Kum see Qom
Kuma 89 B7 river SW Russia
Kumamoto 109 A7 Kumamoto, Kyūshū, SW Japan
Kumanova see Kumanovo
Kumanovo 79 E5 Turk. Kumanova. N Macedonia
Kumasi 53 E5 prev. Coomassie. C Ghana
Kumayri see Gyumri
Kumba 55 A5 Sud-Ouest, W Cameroon
Kumertau 89 D6 Respublika Bashkortostan, W Russia
Kumillā see Comilla
Kumo 53 G4 Gombe, E Nigeria
Kumon Range 114 B2 mountain range N Myanmar (Burma)
Kumul see Hami
Kunashiri see Kunashir, Ostrov
Kunashir, Ostrov 108 E1 var. Kunashiri. island Kuril'skiye Ostrova, SE Russia
Kunda 84 E2 Lääne-Virumaa, NE Estonia
Kunduz 101 E3 prev. Kondoz. NE Afghanistan
Kunene 47 C6 var. Kunene. river Angola/Namibia
Kunene see Cunene
Kungsbacka 63 B7 Halland, S Sweden
Kungur 89 D5 Permskaya Oblast', NW Russia
Kunlun Mountains see Kunlun Shan
Kunlun Shan 104 B4 Eng. Kunlun Mountains. mountain range NW China
Kunming 106 B6 var. K'un-ming; prev. Yunnan. province capital Yunnan, SW China
K'un-ming see Kunming
Kununurra 124 D3 Western Australia
Kunya-Urgench see Köneürgenç
Kuopio 63 E5 Pohjois-Savo, C Finland
Kupa see Kolpa
Kupang 117 E5 prev. Koepang. Timor, C Indonesia
Kup"yans'k 87 G2 Rus. Kupyansk. Kharkivs'ka Oblast', E Ukraine
Kupyansk see Kup"yans'k
Kür see Kura
Kura 95 H3 Az. Kür, Geor. Mtkvari, Turk. Kura Nehri. river SW Asia
Kura Nehri see Kura
Kurashiki 109 B6 var. Kurasiki. Okayama, Honshū, SW Japan
Kurasiki see Kurashiki
Kurdistan 95 F4 cultural region SW Asia
Kŭrdzhali 82 D3 var. Kirdzhali. Kŭrdzhali, S Bulgaria
Kure 109 B7 Hiroshima, Honshū, SW Japan
Küre Dağları 94 C2 mountain range N Turkey
Kuressaare 84 C2 Ger. Arensburg; prev. Kingissepp. Saaremaa, W Estonia
Kureyka 90 D2 river N Russia
Kurgan-Tyube see Qŭrghonteppa
Kuria Muria Islands see Ḩalāniyāt, Juzur al
Kuril'skiye Ostrova 93 H4 Eng. Kuril Islands. island group SE Russia
Kuril Islands see Kuril'skiye Ostrova

Kuril-Kamchatka Depression – Lewis, Isle of

Kuril-Kamchatka Depression see Kuril-Kamchatka Trench
Kuril-Kamchatka Trench 91 F3 var. Kuril-Kamchatka Depression. trench NW Pacific Ocean
Kuril'sk 108 E1 Jap. Shana. Kuril'skiye Ostrova, Sakhalinskaya Oblats', SE Russia
Ku-ring-gai 126 E1 New South Wales, E Australia
Kurisches Haff see Courland Lagoon
Kurkund see Kilingi-Nõmme
Kurnool 110 C1 var. Karnul. Andhra Pradesh, S India
Kursk 89 A6 Kurskaya Oblast', W Russia
Kurskiy Zaliv see Courland Lagoon
Kuršumlija 79 D5 Serbia, S Serbia
Kurtbunar see Tervel
Kurtitsch/Kürtös see Curtici
Kuruktag 104 C3 mountain range NW China
Kurume 109 A7 Fukuoka, Kyūshū, SW Japan
Kurupukari 37 F3 C Guyana
Kusaie see Kosrae
Kushiro 108 D2 var. Kusiro. Hokkaidō, NE Japan
Kushka see Serhetabat
Kusiro see Kushiro
Kuskokwim Mountains 14 C3 mountain range Alaska, USA
Kustanay see Kostanay
Küstence/Küstendje see Constanța
Kütahya 94 B3 prev. Kutaia. Kütahya, W Turkey
Kutai see Mahakam, Sungai
Kutaisi 95 F2 prev. K'ut'aisi 95 F2 W Georgia W Georgia
K'ut'aisi see Kutaisi
Küt al 'Amārah see Al Küt
Kut al Imara see Al Küt
Kutaradja/Kutaraja see Bandaaceh
Kutch, Gulf of see Kachchh, Gulf of
Kutch, Rann of see Kachchh, Rann of
Kutina 78 B3 Sisak-Moslavina, NE Croatia
Kutno 76 C3 Łódzkie, C Poland
Kuujjuaq 17 E2 prev. Fort-Chimo. Québec, E Canada
Kuusamo 62 E3 Pohjois-Pohjanmaa, E Finland
Kuvango see Cubango
Kuvdlorssuak see Kullorsuaq
Kuwait 98 C4 off. State of Kuwait, var. Dawlat al Kuwait, Koweit, Kuweit. country SW Asia
Kuwait see Al Kuwayt
Kuwait City see Al Kuwayt
Kuwait, Dawlat al see Kuwait
Kuwait, State of see Kuwait
Kuwajleen see Kwajalein Atoll
Kuwayt 98 C3 Maysān, E Iraq
Kuweit see Kuwait
Kuybyshev see Samara
Kuybyshev Reservoir see Kuybyshevskoye Vodokhranilishche
Kuybyshevskoye Vodokhranilishche 89 C5 var. Kuibyshev, Eng. Kuybyshev Reservoir. reservoir W Russia
Kuytun 104 B2 Xinjiang Uygur Zizhiqu, NW China
Kuzi see Kuji
Kuznetsk 89 B6 Penzenskaya Oblast', W Russia
Kuźnica 76 E2 Białystok, NE Poland Europe
Kvaløya 62 C2 island N Norway
Kvarnbergsvattnet 62 B4 var. Frostviken. lake N Sweden
Kvarner 78 A3 var. Carnaro, It. Quarnero. gulf W Croatia
Kvitøya 61 G1 island NE Svalbard
Kwajalein Atoll 122 C1 var. Kuwajleen. atoll Ralik Chain, C Marshall Islands
Kwando see Cuando
Kwangchow see Guangzhou
Kwangchu see Gwangju
Kwangju see Gwangju
Kwango 55 C7 Port. Cuango. river Angola/Dem. Rep. Congo
Kwangsi/Kwangsi Chuang Autonomous Region see Guangxi Zhuangzu Zizhiqu
Kwangtung see Guangdong
Kwangyuan see Guangyuan
Kwanza see Cuanza
Kweichu see Guiyang
Kweilin see Guilin
Kweisui see Hohhot
Kweiyang see Guiyang
Kwekwe 56 D3 prev. Que Que. Midlands, C Zimbabwe
Kwesui see Hohhot
Kwidzyń 76 C2 Ger. Marienwerder. Pomorskie, N Poland
Kwigillingok 14 C3 Alaska, USA
Kwilu 55 C6 river W Dem. Rep. Congo
Kwito see Cuito
Kyabé 54 C4 Moyen-Chari, S Chad
Kyaikkami 115 B5 prev. Amherst. Mon State, S Myanmar (Burma)
Kyaiklat 114 B4 Ayeyarwady, SW Myanmar (Burma)
Kyaikto 114 B4 Mon State, S Myanmar (Burma)
Kyakhta 93 E5 Respublika Buryatiya, S Russia
Kyaukse 114 B3 Mandalay, C Myanmar (Burma)
Kyiv 87 E2 var. Kyyiv, Eng. Kiev, Rus. Kiyev. country capital (Ukraine) Kyyivs'ka Oblast', N Ukraine
Kyjov 77 C5 Ger. Gaya. Jihomoravský Kraj, SE Czech Republic (Czechia)
Kykládes 83 D6 var. Kikládhes, Eng. Cyclades. island group SE Greece
Kými 83 C5 prev. Kími. Évvoia, C Greece
Kyŏngsŏng see Seoul
Kyōto 109 C6 Kyōto, Honshū, SW Japan
Kyparissía 83 B6 var. Kiparissía. Pelopónnisos, S Greece
Kypros see Cyprus
Kyrá Panagía 83 C5 island Vóreies Sporádes, Greece, Aegean Sea
Kyrenia see Girne
Kyrgyz Republic see Kyrgyzstan
Kyrgyzstan 101 F2 off. Kyrgyz Republic, var. Kirghizia; prev. Kirgizskaya SSR, Kirghiz SSR, Republic of Kyrgyzstan. country C Asia
Kyrgyzstan, Republic of see Kyrgyzstan
Kythira 83 C7 var. Kíthira, It. Cerigo, Lat. Cythera. island S Greece
Kýthnos 83 C6 Kńythnos, Kykládes, Greece, Aegean Sea
Kythnos 83 C6 var. Kíthnos, Thermiá, It. Termia; anc. Cythnos. island Kykládes, Greece, Aegean Sea
Kythréa see Değirmenlik
Kyushu 109 B7 var. Kyūsyū. island SW Japan
Kyushu-Palau Ridge 103 F3 var. Kyusyu-Palau Ridge. undersea ridge W Pacific Ocean
Kyustendil 82 B2 anc. Pautalia. Kyustendil, W Bulgaria
Kyûsyû see Kyûshû
Kyusyu-Palau Ridge see Kyushu-Palau Ridge
Kyyiv see Kyiv
Kyyivs'ke Vodoskhovyshche 87 E1 Eng. Kiev Reservoir, Rus. Kiyevskoye Vodokhranilishche. reservoir N Ukraine
Kyzyl 92 D4 Respublika Tyva, C Russia
Kyzyl Kum 100 D2 var. Kizil Kum, Qizil Qum, Uzb. Qizilqum. desert Kazakhstan/Uzbekistan
Kyzylrabot see Qizilrabot
Kyzyl-Suu 101 G2 prev. Pokrovka. Issyk-Kul'skaya Oblast', NE Kyrgyzstan
Kzylorda 92 B5 var. Kzyl-Orda, Qizil Orda, Qyzylorda; prev. Perovsk. Kyzylorda, S Kazakhstan
Kzyl-Orda see Kzylorda

L

Laaland see Lolland
La Algaba 70 C4 Andalucía, S Spain
Laarne 65 B5 Oost-Vlaanderen, NW Belgium
La Asunción 37 E1 Nueva Esparta, NE Venezuela
Laatokka see Ladozhskoye, Ozero
Laâyoune 48 B3 var. Aaiún. country capital (Western Sahara) NW Western Sahara
La Banda Oriental see Uruguay
la Baule-Escoublac 68 A4 Loire-Atlantique, NW France
Labé 52 C4 NW Guinea
Labe see Elbe
Laborca see Laborec
Laborec 77 E5 Hung. Laborca. river E Slovakia
Labrador 17 F2 cultural region Newfoundland and Labrador, SW Canada
Labrador Basin 12 E3 var. Labrador Sea Basin. undersea basin Labrador Sea
Labrador Sea 60 A4 sea NW Atlantic Ocean
Labrador Sea Basin see Labrador Basin
Labudalin see Ergun
Labutta 115 A5 Ayeyarwady, SW Myanmar (Burma)
Laç 79 C6 var. Laci. Lezhë, C Albania
La Calera 42 B4 Valparaíso, C Chile
La Carolina 70 D4 Andalucía, S Spain
Laccadive Islands 110 A3 Eng. Laccadive Islands. island group India, N Indian Ocean
Laccadive Islands/Laccadive Minicoy and Amindivi Islands, the see Lakshadweep
La Ceiba 30 D2 Atlántida, N Honduras
Lachanás 82 B3 Kentrikí Makedonía, N Greece
La Chaux-de-Fonds 73 A7 Neuchâtel, W Switzerland
Lachlan River 127 C6 river New South Wales, SE Australia
Laci see Laç
la Ciotat 69 D6 anc. Citharista. Bouches-du-Rhône, SE France
Lacobriga see Lagos
La Concepción 31 E5 var. Concepción. Chiriquí, W Panama
La Concepción 36 C1 Zulia, NW Venezuela
La Condamine 69 C8 W Monaco
Laconia 19 G2 New Hampshire, NE USA
La Crosse 18 A2 Wisconsin, N USA
La Cruz 30 D4 Guanacaste, NW Costa Rica
Ladoga, Lake see Ladozhskoye, Ozero
Ladozhskoye, Ozero 88 B3 Eng. Lake Ladoga, Fin. Laatokka. lake NW Russia
Ladysmith 18 B2 Wisconsin, N USA
Lae 122 B3 Morobe, W Papua New Guinea
La Esperanza 30 C2 Intibucá, SW Honduras
Lafayette 18 C4 Indiana, N USA
Lafayette 20 B3 Louisiana, S USA
La Fé 32 A2 Pinar del Río, W Cuba
Lafia 53 G4 Nassarawa, C Nigeria
la Flèche 68 B4 Sarthe, NW France
Lagdo, Lac de 54 B4 lake N Cameroon
Laghouat 48 D2 N Algeria
Lagos 53 F5 Lagos, SW Nigeria
Lagos 70 B5 anc. Lacobriga. Faro, S Portugal
Lagos de Moreno 29 E4 Jalisco, SW Mexico
Lagouira 48 A4 SW Western Sahara
La Grande 24 C3 Oregon, NW USA
La Guaira 44 B4 Distrito Federal, N Venezuela
Lagunas 42 B1 Tarapacá, N Chile
Lagunillas 39 G4 Santa Cruz, SE Bolivia
La Habana 32 B2 var. Havana. country capital (Cuba) Ciudad de La Habana, W Cuba
Lahat 116 B4 Sumatera, W Indonesia
La Haye see 's-Gravenhage
Laholm 63 B7 Halland, S Sweden
Lahore 112 D2 Punjab, NE Pakistan
Lahr 73 A6 Baden-Württemberg, S Germany
Lahti 63 D5 Swe. Lahtis. Päijät-Häme, S Finland
Lahtis see Lahti
Laï 54 B4 prev. Behagle, De Behagle. Tandjilé, S Chad
Laibach see Ljubljana
Lai Châu 114 D3 Lai Châu, N Vietnam
Laila see Laylā
La Junta 22 D5 Colorado, C USA
Lake Charles 20 A3 Louisiana, S USA
Lake City 21 E3 Florida, SE USA
Lake District 67 C5 physical region NW England, United Kingdom
Lake Havasu City 26 A2 Arizona, SW USA
Lake Jackson 27 H4 Texas, SW USA
Lakeland 21 E4 Florida, SE USA
Lakeside 25 C8 California, W USA
Lake State see Michigan
Lakewood 22 D4 Colorado, C USA
Lakhnau see Lucknow
Lakonikós Kólpos 83 B7 gulf S Greece
Lakselv 62 D2 Lapp. Leavdnja. Finnmark, N Norway
la Laon see Laon
Lalibela 50 C4 Āmara, Ethiopia
La Libertad 30 B1 Petén, N Guatemala
La Ligua 42 B4 Valparaíso, C Chile
Lalín 70 C1 Galicia, NW Spain
Lalitpur 113 F3 Central, C Nepal
La Louvière 65 B6 Hainaut, S Belgium
la Maddalena 74 A4 Sardegna, Italy, C Mediterranean Sea
la Manche see English Channel
Lamar 22 D5 Colorado, C USA
La Marmora, Punta 75 A5 mountain Sardegna, Italy, C Mediterranean Sea
La Massana 69 A8 La Massana, W Andorra Europe
Lambaréné 55 A6 Moyen-Ogooué, W Gabon
Lamego 70 C2 Viseu, N Portugal
Lamesa 27 E3 Texas, SW USA
Lamezia Terme 75 D6 Calabria, SE Italy
Lamía 83 B5 Stereá Ellás, C Greece
Lamoni 23 F4 Iowa, C USA
Lampang 114 C4 var. Muang Lampang. Lampang, NW Thailand
Lámpeia 83 B6 Dytikí Elláς, S Greece
Lanbi Kyun 115 B6 prev. Sullivan Island. island Myeik Archipelago, S Myanmar (Burma)
Lancang Jiang see Mekong
Lancaster 67 D5 NW England, United Kingdom
Lancaster 25 C7 California, W USA
Lancaster 19 F4 Pennsylvania, NE USA
Lancaster Sound 15 F2 sound Nunavut, N Canada
Lan-chou/Lan-chow/Lanchow see Lanzhou
Landao see Lantau Island
Landen 65 C6 Vlaams Brabant, C Belgium
Lander 22 C3 Wyoming, C USA
Landerneau 68 A3 Finistère, NW France
Landes 69 B5 cultural region SW France
Land of Enchantment see New Mexico
The Land of Opportunity see Arkansas
Land of Steady Habits see Connecticut
Land of the Midnight Sun see Alaska
Landsberg see Gorzów Wielkopolski, Lubuskie, Poland
Landsberg an der Warthe see Gorzów Wielkopolski
Land's End 67 B8 headland SW England, United Kingdom
Landshut 73 C6 Bayern, SE Germany
Langar 101 E2 Rus. Lyangar. Navoiy Viloyati, C Uzbekistan
Langfang 106 D4 Hebei, E China
Langkawi, Pulau 115 B7 island Peninsular Malaysia
Langres 68 D4 Haute-Marne, N France
Langsa 116 A3 Sumatera, W Indonesia
Lang Shan 105 E3 mountain range N China
Lang Sơn 114 D3 var. Langson. Lang Sơn, N Vietnam
Langson see Lang Sơn
Lang Suan 115 B6 Chumphon, SW Thailand
Languedoc 69 C6 cultural region S France
Länkäran 95 H3 Rus. Lenkoran'. S Azerbaijan
Lansing 18 C3 state capital Michigan, N USA
Lanta, Ko 115 B7 island S Thailand
Lantau Island 106 A2 Cant. Tai Yue Shan, Chin. Landao. island Hong Kong, S China
Lan-ts'ang Chiang see Mekong
Lantung, Gulf of see Liaodong Wan
Lanzarote 48 B3 island Islas Canarias, Spain, NE Atlantic Ocean
Lanzhou 106 B4 var. Lan-chou, Lanchow, Lan-chow; prev. Kaolan. province capital Gansu, C China
Lao Cai 114 D3 Lao Cai, N Vietnam
Laodicea/Laodicea ad Mare see Al Lādhiqīyah
Laoet see Laut, Pulau
Laojunmiao 106 A3 prev. Yumen. Gansu, N China
Laon 68 D3 var. la Laon; anc. Laudunum. Aisne, N France
Lao People's Democratic Republic see Laos
La Orchila, Isla 36 D1 island N Venezuela
La Oroya 38 C3 Junín, C Peru
Laos 114 D4 off. Lao People's Democratic Republic. country SE Asia
La Palma 31 G5 Darién, SE Panama
La Palma 48 A3 island Islas Canarias, Spain, NE Atlantic Ocean
La Paz 39 F4 var. La Paz de Ayacucho. country capital (Bolivia-seat of government) La Paz, W Bolivia
La Paz 28 B3 Baja California Sur, NW Mexico
La Paz, Bahía de 28 B3 bay NW Mexico
La Paz de Ayacucho see La Paz
La Pérouse Strait 108 D1 Jap. Sōya-kaikyō, Rus. Proliv Laperuza. strait Japan/Russia
Laperuza, Proliv see La Pérouse Strait
Lápithos see Lapta
Lapland 62 D3 Fin. Lappi, Swe. Lappland. cultural region N Europe
La Plata 42 D4 Buenos Aires, E Argentina
La Plata see Sucre
La Pola 70 D1 var. Pola de Lena. Asturias, N Spain
Lappeenranta 63 E5 Swe. Villmanstrand. Etelä-Karjala, SE Finland
Lappi/Lappland see Lapland
Lappo see Lapua
Lapta 80 C5 Gk. Lápithos. NW Cyprus
Laptev Sea see Laptevykh, More
Laptevykh, More 93 E2 Eng. Laptev Sea. sea Arctic Ocean
Lapua 63 D5 Swe. Lappo. Etelä-Pohjanmaa, W Finland
Lapurdum see Bayonne
Łapy 76 E3 Podlaskie, NE Poland
La Quiaca 42 C2 Jujuy, N Argentina
L'Aquila 74 C4 var. Aquila, Aquila degli Abruzzi. Abruzzo, C Italy
Laracha 70 B1 Galicia, NW Spain
Laramie 22 C4 Wyoming, C USA
Laramie Mountains 22 C3 mountain range Wyoming, C USA
Laredo 71 E1 Cantabria, N Spain
Laredo 27 F5 Texas, SW USA
La Réunion see Réunion
Largeau see Faya
Largo 21 E4 Florida, SE USA
Largo, Cayo 32 B2 island W Cuba
Lario see Como, Lago di
La Rioja 42 C3 La Rioja, NW Argentina
La Rioja 71 E2 autonomous community N Spain
Lárisa 82 B4 var. Larissa. Thessalía, C Greece
Larissa see Lárisa
Lārkāna 112 B3 var. Larkhana. Sind, SE Pakistan
Larkhana see Lārkāna
Larnaca see Lárnaka
Lárnaka 80 C5 var. Larnaca, Larnax. SE Cyprus
Larnax see Lárnaka
la Rochelle 68 B4 anc. Rupella. Charente-Maritime, W France
la Roche-sur-Yon 68 B4 prev. Bourbon Vendée, Napoléon-Vendée. Vendée, NW France
La Roda 71 E3 Castilla-La Mancha, C Spain
La Romana 33 E3 E Dominican Republic
Larvotto 69 C8 N Monaco Europe
La-sa see Lhasa
Las Cabezas de San Juan 70 C5 Andalucía, S Spain
Las Cruces 26 D3 New Mexico, SW USA
La See d'Urgel see La Seu d'Urgell
La Serena 42 B3 Coquimbo, C Chile
La Seu d'Urgell 71 G1 prev. La See d'Urgel, Seo de Urgel. Cataluña, NE Spain
la Seyne-sur-Mer 69 D6 Var, SE France
Lashio 114 B3 Shan State, E Myanmar (Burma)
Lashkar Gāh 100 D5 var. Lash-Kar-Gar'. Helmand, S Afghanistan
Lash-Kar-Gar' see Lashkar Gāh
La Sila 75 D6 mountain range SW Italy
La Sirena 30 D3 Región Autónoma Atlántico Sur, E Nicaragua
Łask 76 C4 Łódzkie, C Poland
Las Lomitas 42 D2 Formosa, N Argentina
La Solana 71 E4 Castilla-La Mancha, C Spain
Las Palmas 48 A3 var. Las Palmas de Gran Canaria. Gran Canaria, Islas Canarias, Spain, NE Atlantic Ocean
Las Palmas de Gran Canaria see Las Palmas
La Spezia 74 B3 Liguria, NW Italy
Lassa see Lhasa
Las Tablas 31 F5 Los Santos, S Panama
Last Frontier, The see Alaska
Las Tunas 32 C2 var. Victoria de las Tunas. Las Tunas, E Cuba
La Suisse see Switzerland
Las Vegas 25 D7 Nevada, W USA
Latacunga 38 B1 Cotopaxi, C Ecuador
Latakia see Al Lādhiqīyah
la Teste 69 B5 Gironde, SW France
Latina 75 C5 prev. Littoria. Lazio, C Italy
La Tortuga, Isla 37 E1 var. Isla Tortuga. island N Venezuela
La Tuque 17 E4 Québec, SE Canada
Latvia 84 C3 off. Republic of Latvia, Ger. Lettland, Latv. Latvija, Latvijas Republika; prev. Latvian SSR, Rus. Latviyskaya SSR. country NE Europe
Latvian SSR/Latvija/Latvijas Republika/Latviyskaya SSR see Latvia
Latvia, Republic of see Latvia
Laudunum see Laon
Laudus see St-Lô
Lauenburg/Lauenburg in Pommern see Lębork
Lau Group 123 E4 island group E Fiji
Lauis see Lugano
Launceston 127 C8 Tasmania, SE Australia
La Unión 30 C2 Olancho, C Honduras
La Unión 71 F4 Murcia, SE Spain
Laurel 20 C3 Mississippi, S USA
Laurel 22 C2 Montana, NW USA
Laurentian Highlands see Laurentian Mountains
Laurentian Mountains 17 E3 var. Laurentian Highlands, Fr. Les Laurentides. plateau Newfoundland and Labrador/Québec, Canada
Laurentides, Les see Laurentian Mountains
Lauria 75 D6 Basilicata, S Italy
Laurinburg 21 F1 North Carolina, SE USA
Lauru see Choiseul
Lausanne 73 A7 It. Losanna. Vaud, SW Switzerland
Laut, Pulau 116 D4 prev. Laoet. island Borneo, C Indonesia
Laval 16 D4 Québec, SE Canada
Laval 68 B3 Mayenne, NW France
La Vall d'Uixó 71 F3 var. Vall D'Uxó. País Valenciano, E Spain
La Vega 33 E3 var. Concepción de la Vega. C Dominican Republic
La Vila Joíosa see Villajoyosa
Lávrio 83 C6 prev. Lávrion. Attikí, C Greece
Lávrion see Lávrio
Lawrence 19 G3 Massachusetts, NE USA
Lawrenceburg 20 C1 Tennessee, S USA
Lawton 27 F2 Oklahoma, C USA
La Yarada 39 E4 Tacna, SW Peru
Laylā 99 C5 var. Laila. Ar Riyāḍ, C Saudi Arabia
Lazarev Sea 132 B1 sea Antarctica
Lázaro Cárdenas 29 E5 Michoacán, SW Mexico
Leal see Lihula
Leamhcán see Lucan
Leamington 16 C5 Ontario, S Canada
Leavdnja see Lakselv
Lebak 117 E3 Mindanao, S Philippines
Lebanese Republic see Lebanon
Lebanon 23 G5 Missouri, C USA
Lebanon 19 G2 New Hampshire, NE USA
Lebanon 24 B3 Oregon, NW USA
Lebanon 96 A4 off. Lebanese Republic, Ar. Al Lubnān, Fr. Liban. country SW Asia
Lebanon, Mount see Liban, Jebel
Lebap 100 D2 Lebapskiy Velayat, NE Turkmenistan
Lebedin see Lebedyn
Lebedyn 87 F2 Rus. Lebedin. Sums'ka Oblast', NE Ukraine
Lębork 76 C2 var. Lębórk, Ger. Lauenburg, Lauenburg in Pommern. Pomorskie, N Poland
Lebrija 70 C5 Andalucía, S Spain
Lebu 43 A5 Bío Bío, C Chile
le Cannet 69 D6 Alpes-Maritimes, SE France
Le Cap see Cap-Haïtien
Lecce 75 E6 Puglia, SE Italy
Lechainá 83 B6 var. Lehena, Lekhainá. Dytikí Elláς, S Greece
Ledo Salinarius see Lons-le-Saunier
Leduc 15 E5 Alberta, SW Canada
Leech Lake 23 F2 lake Minnesota, N USA
Leeds 67 D5 N England, United Kingdom
Leek 64 E2 Groningen, NE Netherlands
Leer 72 A3 Niedersachsen, NW Germany
Leeuwarden 64 D1 Fris. Ljouwert. Friesland, N Netherlands
Leeuwin, Cape 120 A5 headland Western Australia
Leeward Islands 33 G3 island group E West Indies
Leeward Islands see Sotavento, Ilhas de
Lefkáda 83 A5 prev. Levkás. Lefkáda, Iónia Nisiá, Greece, C Mediterranean Sea
Lefkáda 83 A5 It. Santa Maura, prev. Levkás; anc. Leucas. island Iónia Nisiá, Greece, C Mediterranean Sea
Lefká Óri 83 C7 mountain range Kríti, Greece, E Mediterranean Sea
Lefkímmi 83 A5 var. Levkímmi. Kérkyra, Iónia Nisiá, Greece, C Mediterranean Sea
Lefkosía/Lefkoşa see Nicosia
Legaceaster see Chester
Legaspi see Legazpi City
Legazpi City 117 E2 var. Legaspi. C Philippines
Leghorn see Livorno
Legnica 76 B4 Ger. Liegnitz. Dolnośląskie, SW Poland
le Havre 68 B3 Eng. Havre; prev. le Havre-de-Grâce. Seine-Maritime, N France
le Havre-de-Grâce see le Havre
Lehena see Lechainá
Leicester 67 D6 Lat. Batae Coritanorum. C England, United Kingdom
Leiden 64 B3 prev. Leyden; anc. Lugdunum Batavorum. Zuid-Holland, W Netherlands
Leie 68 D2 Fr. Lys. river Belgium/France
Leinster 67 B6 Ir. Cúige Laighean. cultural region E Ireland
Leipsic see Leipzig
Leipsoí 83 E6 island Dodekánisa, Greece, Aegean Sea
Leipzig 72 C4 Pol. Lipsk, hist. Leipsic; anc. Lipsia. Sachsen, E Germany
Leiria 70 B3 anc. Collipo. Leiria, C Portugal
Leirvik 63 A6 Hordaland, S Norway
Lek 64 C4 river SW Netherlands
Lekhainá see Lechainá
Lekhchevo see Boychinovtsi
Leksand 63 C5 Dalarna, C Sweden
Lel'chitsy see Lyel'chytsy
le Léman see Geneva, Lake
Lelystad 64 D3 Flevoland, C Netherlands
Léman, Lac see Geneva, Lake
le Mans 68 B3 Sarthe, NW France
Lemberg see L'viv
Lemesós 80 C5 var. Limassol. SW Cyprus
Lemhi Range 24 D3 mountain range Idaho, C USA North America
Lemnos see Límnos
Lemovices see Limoges
Lena 93 F3 river NE Russia
Lena Tablemount 119 B7 seamount S Indian Ocean
Len Dao 106 C8 island S Spratly Islands
Lengshuitan see Yongzhou
Leninabad see Khujand
Leninakan see Gyumri
Lenine 87 G5 Rus. Lenino. Respublika Krym, S Ukraine
Leningor see Ridder
Leningrad see Sankt-Peterburg
Leningradskaya 132 B4 Russian research station Antarctica
Lenino see Lenine, Ukraine
Leninobod see Khujand
Leninogorsk see Ridder
Leninpol' 101 F2 Talasskaya Oblast', NW Kyrgyzstan
Lenin-Turkmenski see Türkmenabat
Lenkoran' see Länkäran
Lenti 77 B7 Zala, SW Hungary
Lentia see Linz
Leoben 73 E7 Steiermark, C Austria
León 29 E4 var. León de los Aldamas. Guanajuato, C Mexico
León 30 C3 León, NW Nicaragua
León 70 D1 Castilla-León, NW Spain
León de los Aldamas see León
Leonídi see Leonídio
Leonídio 83 B6 var. Leonídi. Pelopónnisos, S Greece
Léopold II, Lac see Mai-Ndombe, Lac
Léopoldville see Kinshasa
Lepe 70 C4 Andalucía, S Spain
Lepel' see Lyepyel'
le Portel 68 C2 Pas-de-Calais, N France
Le Puglie see Puglia
le Puy 69 C5 prev. le Puy-en-Velay, hist. Anicium, Podium Anicensis. Haute-Loire, C France
le Puy-en-Velay see le Puy
Léré 54 B4 Mayo-Kébbi, SW Chad
Lérida see Lleida
Lerma 70 D2 Castilla-León, N Spain
Leros 83 D6 island Dodekánisa, Greece, Aegean Sea
Lerrnayin Gharabakh see Nagornyy-Karabakh
Lerwick 66 D1 NE Scotland, United Kingdom
Lesbos see Lésvos
Les Cayes see Cayes
Les Gonaïves see Gonaïves
Leshan 106 B5 Sichuan, C China
les Herbiers 68 B4 Vendée, NW France
Lesh/Leshi see Lezhë
Lesina see Hvar
Leskovac 79 E5 Serbia, SE Serbia
Lesnoy 92 C3 Sverdlovskaya Oblast', C Russia
Lesotho 56 D4 off. Kingdom of Lesotho; prev. Basutoland. country S Africa
Lesotho, Kingdom of see Lesotho
les Sables-d'Olonne 68 B4 Vendée, NW France
Lesser Antarctica see West Antarctica
Lesser Antilles 33 G4 island group E West Indies
Lesser Caucasus 95 F2 Rus. Malyy Kavkaz. mountain range SW Asia
Lesser Khingan Range see Xiao Hinggan Ling
Lesser Sunda Islands 117 E5 Eng. Lesser Sunda Islands. island group C Indonesia
Lesser Sunda Islands see Nusa Tenggara
Lésvos 94 A3 anc. Lesbos. island E Greece
Leszno 76 B4 Ger. Lissa. Wielkopolskie, C Poland
Lethbridge 15 E5 Alberta, SW Canada
Lethem 37 F3 S Guyana
Leti, Kepulauan 117 F5 island group E Indonesia
Letpadan 114 B4 Bago, SW Myanmar (Burma)
Letsôk-aw Kyun 115 B6 var. Letsutan Island; prev. Domel Island. island Myeik Archipelago, S Myanmar (Burma)
Letsutan Island see Letsôk-aw Kyun
Lettland see Latvia
Lëtzebuerg see Luxembourg
Leucas see Lefkáda
Leuven 65 C6 Fr. Louvain, Ger. Löwen. Vlaams Brabant, C Belgium
Leuze see Leuze-en-Hainaut
Leuze-en-Hainaut 65 B6 var. Leuze. Hainaut, SW Belgium
Léva see Levice
Levanger 62 B4 Nord-Trøndelag, C Norway
Levelland 27 E2 Texas, SW USA
Leverkusen 72 A4 Nordrhein-Westfalen, W Germany
Levice 77 C6 Ger. Lewentz, Hung. Léva, Lewenz. Nitriansky Kraj, SW Slovakia
Levin 128 D4 Manawatu-Wanganui, North Island, New Zealand
Levkás see Lefkáda
Levkímmi see Lefkímmi
Lewentz/Lewenz see Levice
Lewis, Isle of 66 B2 island NW Scotland, United Kingdom

Lewis Range – Lyusino

Lewis Range 22 B1 *mountain range* Montana, NW USA
Lewiston 24 C2 Idaho, NW USA
Lewiston 19 G2 Maine, NE USA
Lewistown 22 C1 Montana, NW USA
Lexington 18 C5 Kentucky, S USA
Lexington 23 E4 Nebraska, C USA
Leyden *see* Leiden
Leyte 117 F2 *island* C Philippines
Leżajsk 77 E5 Podkarpackie, SE Poland
Lezha *see* Lezhë
Lezhë 79 C6 *var.* Lezha; *prev.* Lesh, Leshi. Lezhë, NW Albania
Lhasa 104 C5 *var.* La-sa, Lassa. Xizang Zizhiqu, W China
Lhaviyani Atoll *see* Faadhippolhu Atoll
Lhazê 104 C5 *var.* Quxar. Xizang Zizhiqu, China E Asia
L'Hospitalet de Llobregat 71 G2 *var.* Hospitalet. Cataluña, NE Spain
Liancourt Rocks 109 A5 *Jap.* Takeshima, *Kor.* Dokdo. *island group* Japan/South Korea
Lianyungang 106 D4 *var.* Xinpu. Jiangsu, E China
Liao *see* Liaoning
Liaodong Wan 105 G3 *Eng.* Gulf of Lantung, Gulf of Liaotung. *gulf* NE China
Liao He 103 E1 *river* NE China
Liaoning 106 D3 *var.* Liao, Liaoning Sheng, Shengking, *hist.* Fengtien, Shenking. *province* NE China
Liaoning Sheng *see* Liaoning
Liaoyuan 107 E3 *var.* Dongliao, Shuang-liao, *Jap.* Chengchiatun. Jilin, NE China
Liard *see* Fort Liard
Liban *see* Lebanon
Liban, Jebel 96 B4 *Ar.* Jabal al Gharbt, Jabal Lubnān, *Eng.* Mount Lebanon. *mountain range* C Lebanon
Libau *see* Liepāja
Libby 22 A1 Montana, NW USA
Liberal 23 E5 Kansas, C USA
Liberalitas Julia *see* Évora
Liberec 76 B4 *Ger.* Reichenberg. Liberecký Kraj, N Czech Republic (Czechia)
Liberia 30 D4 Guanacaste, NW Costa Rica
Liberia 52 C5 *off.* Republic of Liberia. *country* W Africa
Liberia, Republic of *see* Liberia
Libian Desert *see* Libyan Desert
Lībīyah, Aş Şahrā' al *see* Libyan Desert
Libourne 69 B5 Gironde, SW France
Libreville 55 A5 *country capital* (Gabon) Estuaire, NW Gabon
Libya 49 F3 *off.* Libya, *Ar.* Al Jamāhīrīyah al 'Arabīyah al Lībīyah ash Sha'bīyah al Ishtirākiy; *prev.* Libyan Arab Republic, Great Socialist People's Libyan Arab Jamahiriya. *country* N Africa
Libyan Arab Republic *see* Libya
Libyan Desert 49 H4 *var.* Libian Desert, *Ar.* Aş Şahrā' al Lībīyah. *desert* N Africa
Libyan Plateau 81 F4 *var.* Aḍ Diffah. *plateau* Egypt/Libya
Lichtenfels 73 C5 Bayern, SE Germany
Lichtenvoorde 64 E4 Gelderland, E Netherlands
Lichuan 106 C5 Hubei, C China
Lida 85 B5 Hrodzyenskaya Voblasts', W Belarus
Lidhorikíon *see* Lidoríki
Lidköping 63 B6 Västra Götaland, S Sweden
Lidokhorikion *see* Lidoríki
Lidoríki 83 B5 *prev.* Lidhorikíon, Lidokhorikion. Stereá Ellás, C Greece
Lidzbark Warmiński 76 D2 *Ger.* Heilsberg. Olsztyn, N Poland
Liechtenstein 72 D1 *off.* Principality of Liechtenstein. *country* C Europe
Liechtenstein, Principality of *see* Liechtenstein
Liège 65 D6 *Dut.* Luik, *Ger.* Lüttich. Liège, E Belgium
Liegnitz *see* Legnica
Lienz 73 D7 Tirol, W Austria
Liepāja 84 B3 *Ger.* Libau. Liepāja, W Latvia
Lietuva *see* Lithuania
Lievenhof *see* Līvāni
Liezen 73 D7 Steiermark, C Austria
Liffey 67 B6 *river* Ireland
Lifou 122 D5 *island* Îles Loyauté, E New Caledonia
Liger *see* Loire
Ligure, Appennino 74 A2 *Eng.* Ligurian Mountains. *mountain range* NW Italy
Ligure, Mar *see* Ligurian Sea
Ligurian Mountains *see* Ligure, Appennino
Ligurian Sea 74 A3 *Fr.* Mer Ligurienne, *It.* Mar Ligure. *sea* N Mediterranean Sea
Ligurienne, Mer *see* Ligurian Sea
Līhu'e 25 A7 *var.* Lihue. Kaua'i, Hawaii, USA
Lihue *see* Līhu'e
Lihula 84 D2 *Ger.* Leal. Läänemaa, W Estonia
Liivi Laht *see* Riga, Gulf of
Likasi 55 D7 *prev.* Jadotville. Shaba, SE Dem. Rep. Congo
Liknes 63 A6 Vest-Agder, S Norway
Lille 68 C2 *var.* l'Isle, *Dut.* Rijssel, *Flem.* Ryssel, *prev.* Lisle; *anc.* Insula. Nord, N France
Lillehammer 63 B5 Oppland, S Norway
Lillestrøm 63 B6 Akershus, S Norway
Lilongwe 57 E2 *country capital* (Malawi) Central, W Malawi
Lilybaeum *see* Marsala
Lima 38 C4 *country capital* (Peru) Lima, W Peru
Limanowa 77 D5 Małopolskie, S Poland
Limassol *see* Lemesós
Limerick 67 A6 *Ir.* Luimneach. Limerick, SW Ireland
Límín Vathéos *see* Sámos
Límnos 81 F3 *anc.* Lemnos. *island* E Greece
Limoges 69 C5 *anc.* Augustoritum Lemovicensium, Lemovices. Haute-Vienne, C France
Limón 31 E4 *var.* Puerto Limón. Limón, E Costa Rica
Limón 30 D2 Colón, NE Honduras
Limonum *see* Poitiers
Limousin 69 C5 *cultural region* C France
Limoux 69 C6 Aude, S France
Limpopo 56 D3 *var.* Crocodile. *river* S Africa
Linares 42 B4 Maule, C Chile
Linares 29 E3 Nuevo León, NE Mexico
Linares 70 D4 Andalucía, S Spain
Lincoln 67 D5 *anc.* Lindum, Lindum Colonia. E England, United Kingdom
Lincoln 19 H2 Maine, NE USA
Lincoln 23 F4 *state capital* Nebraska, C USA
Lincoln Sea 12 D2 *sea* Arctic Ocean
Linden 37 F3 E Guyana
Líndhos *see* Líndos
Lindi 51 D8 Lindi, SE Tanzania
Líndos 83 E7 *var.* Líndhos. Ródos, Dodekánisa, Greece, Aegean Sea
Lindum/Lindum Colonia *see* Lincoln
Line Islands 123 G3 *island group* E Kiribati
Lingeh *see* Bandar-e Lengeh
Lingen 72 A3 *var.* Lingen an der Ems. Niedersachsen, NW Germany
Lingen an der Ems *see* Lingen
Lingga, Kepulauan 116 B4 *island group* W Indonesia
Linköping 63 C6 Östergötland, S Sweden
Linz 73 D6 *anc.* Lentia. Oberösterreich, N Austria
Lion, Golfe du 69 C7 *Eng.* Gulf of Lion, Gulf of Lions; *anc.* Sinus Gallicus. *gulf* S France
Lion, Gulf of/Lions, Gulf of *see* Lion, Golfe du
Liozno *see* Lyozna
Lipari 75 D6 *island* Isole Eolie, S Italy
Lipari Islands/Lipari, Isole *see* Eolie, Isole
Lipetsk 89 B5 Lipetskaya Oblast', W Russia
Lipno 76 C3 Kujawsko-pomorskie, C Poland
Lipova 86 A4 *Hung.* Lippa. Arad, W Romania
Lipovets *see* Lypovets'
Lippa *see* Lipova
Lipsia/Lipsk *see* Leipzig
Lira 51 B6 N Uganda
Lisala 55 C5 Equateur, N Dem. Rep. Congo
Lisboa 70 B4 *Eng.* Lisbon; *anc.* Felicitas Julia, Olisipo. *country capital* (Portugal) Lisboa, W Portugal
Lisbon *see* Lisboa
Lisichansk *see* Lysychans'k
Lisieux 68 B3 *anc.* Noviomagus. Calvados, N France
Liski 89 B6 *prev.* Georgiu-Dezh. Voronezhskaya Oblast', W Russia
Lisle/l'Isle *see* Lille
Lismore 127 E5 New South Wales, SE Australia
Lissa *see* Vis, Croatia
Lissa *see* Leszno, Poland
Lisse 64 C3 Zuid-Holland, W Netherlands
Litang 106 A5 *var.* Gaocheng. Sichuan, C China
Lītani, Nahr el 97 B5 *var.* Nahr al Litant. *river* C Lebanon
Litant, Nahr al *see* Lītani, Nahr el
Litauen *see* Lithuania
Lithgow 127 D6 New South Wales, SE Australia
Lithuania 84 B4 *off.* Republic of Lithuania, *Ger.* Litauen, *Lith.* Lietuva, *Pol.* Litwa, *Rus.* Litva; *prev.* Lithuanian SSR, *Rus.* Litovskaya SSR. *country* NE Europe
Lithuanian SSR *see* Lithuania
Lithuania, Republic of *see* Lithuania
Litóchoro 82 B4 *var.* Litohoro, Litókhoron. Kentrikí Makedonía, N Greece
Litohoro/Litókhoron *see* Litóchoro
Litovskaya SSR *see* Lithuania
Little Alföld 77 C6 *Ger.* Kleines Ungarisches Tiefland, *Hung.* Kisalföld, *Slvk.* Podunajská Rovina. *plain* Hungary/Slovakia
Little Andaman 111 F2 *island* Andaman Islands, India, NE Indian Ocean
Little Barrier Island 128 D2 *island* N New Zealand
Little Bay 71 H5 *bay* Alboran Sea, Mediterranean Sea
Little Cayman 32 B3 *island* E Cayman Islands
Little Falls 23 F2 Minnesota, N USA
Littlefield 27 E2 Texas, SW USA
Little Inagua 32 D2 *var.* Inagua Islands. *island* S The Bahamas
Little Minch, The 66 B3 *strait* NW Scotland, United Kingdom
Little Missouri River 22 D2 *river* NW USA
Little Nicobar 111 G3 *island* Nicobar Islands, India, NE Indian Ocean
Little Rhody *see* Rhode Island
Little Rock 20 B1 *state capital* Arkansas, C USA
Little Saint Bernard Pass 69 D5 *Fr.* Col du Petit St-Bernard, *It.* Colle del Piccolo San Bernardo. *pass* France/Italy
Little Sound 20 A5 *bay* Bermuda, NW Atlantic Ocean
Littleton 22 D4 Colorado, C USA
Littoria *see* Latina
Litva/Litwa *see* Lithuania
Liu-chou/Liuchow *see* Liuzhou
Liuzhou 106 C6 *var.* Liu-chou, Liuchow. Guangxi Zhuangzu Zizhiqu, S China
Livanátai *see* Livanátes
Livanátes 83 B5 *prev.* Livanátai. Stereá Ellás, C Greece
Līvāni 84 D4 *Ger.* Lievenhof. Preiļi, SE Latvia
Liverpool 17 F5 Nova Scotia, SE Canada
Liverpool 67 C5 NW England, United Kingdom
Livingston 22 B2 Montana, NW USA
Livingston 27 H3 Texas, SW USA
Livingstone 56 C3 *var.* Maramba. Southern, S Zambia
Livingstone Mountains 129 A7 *mountain range* South Island, New Zealand
Livno 78 B4 Federicija Bosna I Hercegovina, SW Bosnia and Herzegovina
Livojoki 62 D4 *river* C Finland
Livonia 18 D3 Michigan, N USA
Livorno 74 B3 *Eng.* Leghorn. Toscana, C Italy
Lixian Jiang *see* Black River
Lixoúri 83 A5 *prev.* Lixoúrion. Kefallinía, Iónia Nisiá, Greece, C Mediterranean Sea
Lixoúrion *see* Lixoúri
Lizarra *see* Estella
Ljouwert *see* Leeuwarden
Ljubelj *see* Loibl Pass
Ljubljana 73 D7 *Ger.* Laibach, *It.* Lubiana; *anc.* Aemona, Emona. *country capital* (Slovenia) C Slovenia
Ljungby 63 B7 Kronoberg, S Sweden
Ljusdal 63 C5 Gävleborg, C Sweden
Ljusnan 63 C5 *river* C Sweden
Llanelli 67 C6 *prev.* Llanelly. SW Wales, United Kingdom
Llanelly *see* Llanelli
Llanes 70 D1 Asturias, N Spain
Llanos 36 D2 *physical region* Colombia/Venezuela
Lleida 71 F2 *Cast.* Lérida; *anc.* Ilerda. Cataluña, NE Spain
Llucmajor 71 G3 Mallorca, Spain, W Mediterranean Sea
Loaita Island 106 C8 *island* W Spratly Islands
Loanda *see* Luanda
Lobamba 56 D4 *country capital* (Swaziland- royal and legislative) NW Swaziland
Lobatse 56 C4 *var.* Lobatsi. Kgatleng, SE Botswana
Lobatsi *see* Lobatse
Löbau 72 D4 Sachsen, E Germany
Lobito 56 B2 Benguela, W Angola
Lob Nor *see* Lop Nur
Lobositz *see* Lovosice
Loburi *see* Lop Buri
Locarno 73 B8 *Ger.* Luggarus. Ticino, S Switzerland
Lochem 64 E3 Gelderland, E Netherlands
Lockport 19 E3 New York, NE USA
Lodja 55 D6 Kasai-Oriental, C Dem. Rep. Congo
Lodwar 51 C6 Rift Valley, NW Kenya
Łódź 76 D4 *Rus.* Lodz. Łódź, C Poland
Loei 114 C4 *var.* Loey, Muang Loei. Loei, C Thailand
Loey *see* Loei
Lofoten 62 B3 *var.* Lofoten Islands. *island group* C Norway
Lofoten Islands *see* Lofoten
Logan 22 B3 Utah, W USA
Logan, Mount 14 D3 *mountain* Yukon, W Canada
Logroño 71 E1 *anc.* Vareia, *Lat.* Juliobriga. La Rioja, N Spain
Loibl Pass 73 D7 *Ger.* Loiblpass, *Slvn.* Ljubelj. *pass* Austria/Slovenia
Loiblpass *see* Loibl Pass
Loikaw 114 B4 Kayah State, C Myanmar (Burma)
Loire 68 B4 *var.* Liger. *river* C France
Loja 38 B2 Loja, S Ecuador
Lokitaung 51 C5 Rift Valley, NW Kenya
Lokoja 53 G4 Kogi, C Nigeria
Loksa 84 E2 *Ger.* Loxa. Harjumaa, NW Estonia
Lolland 63 B8 *prev.* Laaland. *island* S Denmark
Lom 82 C1 *prev.* Lom-Palanka. Montana, NW Bulgaria
Lomami 55 D6 *river* C Dem. Rep. Congo
Lomas 38 D4 Arequipa, SW Peru
Lomas de Zamora 42 D4 Buenos Aires, E Argentina
Lombamba 56 D4 *country capital* (Swaziland - royal and legislative) NW Swaziland
Lombardia 74 B2 *Eng.* Lombardy. *region* N Italy
Lombardy *see* Lombardia
Lombok, Pulau 116 D5 *island* Nusa Tenggara, C Indonesia
Lomé 53 F5 *country capital* (Togo) S Togo
Lomela 55 D6 Kasai-Oriental, C Dem. Rep. Congo
Lommel 65 C5 Limburg, N Belgium
Lomond, Loch 66 B4 *lake* C Scotland, United Kingdom
Lomonosov Ridge 133 B3 *var.* Harris Ridge, *Rus.* Khrebet Homonsova. *undersea ridge* Arctic Ocean
Lomonsova, Khrebet *see* Lomonosov Ridge
Lom-Palanka *see* Lom
Lompoc 25 B7 California, W USA
Lom Sak 114 C4 *var.* Muang Lom Sak. Phetchabun, C Thailand
Łomża 76 D3 *Rus.* Lomzha. Podlaskie, NE Poland
Lomzha *see* Łomża
Loncoche 43 B5 Araucanía, C Chile
Londinium *see* London
London 67 A7 *anc.* Augusta, *Lat.* Londinium. *country capital* (United Kingdom) SE England, United Kingdom
London 16 C5 Ontario, S Canada
London 18 C5 Kentucky, S USA
Londonderry 66 B4 *var.* Derry, *Ir.* Doire. NW Northern Ireland, United Kingdom
Londonderry, Cape 124 C2 *cape* Western Australia
Londrina 41 E4 Paraná, S Brazil
Lone Star State *see* Texas
Long Bay 21 F2 *bay* W Jamaica
Long Beach 25 C7 California, W USA
Longford 67 B5 *Ir.* An Longfort. Longford, C Ireland
Long Island 32 D2 *island* C The Bahamas
Long Island 19 G4 *island* New York, NE USA
Longlac 16 C3 Ontario, S Canada
Longmont 22 D4 Colorado, C USA
Longreach 126 C4 Queensland, E Australia
Long Strait 93 G1 *Eng.* Long Strait. *strait* NE Russia
Long Strait *see* Longa, Proliv
Longview 27 H3 Texas, SW USA
Longview 24 B2 Washington, NW USA
Long Xuyên 115 D6 *var.* Longxuyen. An Giang, S Vietnam
Longxuyen *see* Long Xuyên
Longyan 106 D6 Fujian, SE China
Longyearbyen 61 G2 *dependent territory capital* (Svalbard) Spitsbergen, W Svalbard
Lons-le-Saunier 68 D4 *anc.* Ledo Salinarius. Jura, E France
Lop Buri 115 C5 *var.* Loburi. Lop Buri, C Thailand
Lop Nor *see* Lop Nur
Lop Nur 104 C3 *var.* Lob Nor, Lop Nor, Lo-pu Po. *seasonal lake* NW China
Loppersum 64 E1 Groningen, NE Netherlands
Lo-pu Po *see* Lop Nur
Lorca 71 E4 *Ar.* Lurka; *anc.* Eliocroca, *Lat.* Illurco. Murcia, S Spain
Lord Howe Island 120 C4 *island* E Australia
Lord Howe Rise 120 C4 *undersea rise* SW Pacific Ocean
Loreto 28 B3 Baja California Sur, NW Mexico
Lorient 68 A3 *prev.* l'Orient. Morbihan, NW France
l'Orient *see* Lorient
Lorn, Firth of 66 B4 *inlet* W Scotland, United Kingdom
Loro Sae *see* East Timor
Lörrach 73 A7 Baden-Württemberg, S Germany
Lorraine 68 D3 *cultural region* NE France
Los Alamos 26 C1 New Mexico, SW USA
Los Amates 30 B2 Izabal, E Guatemala
Los Ángeles 43 B5 Bío Bío, C Chile
Los Angeles 25 C7 California, W USA
Losanna *see* Lausanne
Lošinj 78 A3 *Ger.* Lussin, *It.* Lussino. *island* W Croatia
Loslau *see* Wodzisław Śląski
Los Mochis 28 C3 Sinaloa, C Mexico
Losonc/Losontz *see* Lučenec
Los Roques, Islas 36 D1 *island group* N Venezuela
Lot 69 B5 *cultural region* S France
Lot 69 B5 *river* S France
Lotagipi Swamp 51 C5 *wetland* Kenya/ South Sudan
Lötzen *see* Giżycko
Loualaba *see* Lualaba
Louangnamtha 114 C3 *var.* Luong Nam Tha. Louang Namtha, N Laos
Louangphabang 102 D3 *var.* Louangphrabang, Luang Prabang. Louangphabang, N Laos
Louangphrabang *see* Louangphabang
Loubomo *see* Dolisie
Loudéac 68 A3 Côtes d'Armor, NW France
Loudi 106 C5 Hunan, S China
Louga 52 B3 NW Senegal
Louisiade Archipelago 122 B4 *island group* SE Papua New Guinea
Louisiana 20 A2 *off.* State of Louisiana, *also known as* Creole State, Pelican State. *state* S USA
Louisville 18 C5 Kentucky, S USA
Louisville Ridge 121 E4 *undersea ridge* S Pacific Ocean
Loup River 23 E4 *river* Nebraska, C USA
Lourdes 69 B6 Hautes-Pyrénées, S France
Lourenço Marques *see* Maputo
Louth 67 E5 E England, United Kingdom
Loutrá 82 C4 Kentrikí Makedonía, N Greece
Louvain *see* Leuven
Louvain-la Neuve 65 C6 Walloon Brabant, C Belgium
Louviers 68 C3 Eure, N France
Lovech 82 C2 Lovech, N Bulgaria
Loveland 22 D4 Colorado, C USA
Lovosice 76 A4 *Ger.* Lobositz. Ústecký Kraj, NW Czech Republic (Czechia)
Lóvua 56 C1 Moxico, E Angola
Lowell 19 G3 Massachusetts, NE USA
Löwen *see* Leuven
Lower California *see* Baja California
Lower Hutt 129 D5 Wellington, North Island, New Zealand
Lower Lough Erne 67 A5 *lake* SW Northern Ireland, United Kingdom
Lower Red Lake 23 F1 *lake* Minnesota, N USA
Lower Rhine *see* Neder Rijn
Lower Tunguska *see* Nizhnyaya Tunguska
Lowestoft 67 E6 E England, United Kingdom
Loxa *see* Loksa
Lo-yang *see* Luoyang
Loyauté, Îles 122 D5 *island group* S New Caledonia
Loyev *see* Loyew
Loyew 85 D8 *Rus.* Loyev. Homyel'skaya Voblasts', SE Belarus
Loznica 78 C3 Serbia, W Serbia
Lu *see* Shandong, China
Lualaba 55 D6 *Fr.* Loualaba. *river* SE Dem. Rep. Congo
Luanda 56 A1 *var.* Loanda, *Port.* São Paulo de Loanda. *country capital* (Angola) Luanda, NW Angola
Luang Prabang *see* Louangphabang
Luang, Thale 115 C7 *lagoon* S Thailand
Luangua, Rio *see* Luangwa
Luangwa 51 B8 *var.* Aruângua, Rio Luangua. *river* Mozambique/Zambia
Luanshya 56 D2 Copperbelt, C Zambia
Luarca 70 C1 Asturias, N Spain
Lubaczów 77 E5 *var.* Lúbaczów. Podkarpackie, SE Poland
L'uban' 76 B4 Leningradskaya Oblast', Russia
Lubānas Ezers *see* Lubāns
Lubango 56 B2 *Port.* Sá da Bandeira. Huíla, SW Angola
Lubāns 84 D4 *var.* Lubānas Ezers. *lake* E Latvia
Lubao 55 D6 Kasai-Oriental, C Dem. Rep. Congo
Lübben 72 D4 Brandenburg, E Germany
Lübbenau 72 D4 Brandenburg, E Germany
Lubbock 27 E2 Texas, SW USA
Lübeck 72 C2 Schleswig-Holstein, N Germany
Lubelska, Wyżyna 76 E4 *plateau* SE Poland
Lüben *see* Lubin
Lubiana *see* Ljubljana
Lubin 76 B4 *Ger.* Lüben. Dolnośląskie, SW Poland
Lublin 76 E4 *Rus.* Lyublin. Lubelskie, E Poland
Lubliniec 76 C4 Śląskie, S Poland
Lubnān, Jabal *see* Liban, Jebel
Lubny 87 F2 Poltavs'ka Oblast', NE Ukraine
Lubsko 76 B4 *Ger.* Sommerfeld. Lubuskie, W Poland
Lubumbashi 55 E8 *prev.* Élisabethville. Shaba, SE Dem. Rep. Congo
Lubutu 55 D6 Maniema, E Dem. Rep. Congo
Luca *see* Lucca
Lucan 67 B5 *Ir.* Leamhcán. Dublin, E Ireland
Lucanian Mountains *see* Lucano, Appennino
Lucano, Appennino 75 D5 *Eng.* Lucanian Mountains. *mountain range* S Italy
Lucapa 56 C1 *var.* Lukapa. Lunda Norte, NE Angola
Lucca 74 B3 *anc.* Luca. Toscana, C Italy
Lucea 32 A4 W Jamaica
Lucena 117 E1 *off.* Lucena City. Luzon, N Philippines
Lucena 70 D4 Andalucía, S Spain
Lucena City *see* Lucena
Lučenec 77 D6 *Ger.* Losontz, *Hung.* Losonc. Banskobystrický Kraj, C Slovakia
Lucentum *see* Alicante
Lucerna/Lucerne *see* Luzern
Luchow *see* Hefei
Łuck *see* Luts'k
Lucknow 113 E3 *var.* Lakhnau. *state capital* Uttar Pradesh, N India
Lüda *see* Dalian
Luda Kamchiya 82 D2 *river* E Bulgaria
Ludasch *see* Luduş
Lüderitz 56 B4 *prev.* Angra Pequena. Karas, SW Namibia
Ludhiāna 112 D2 Punjab, N India
Ludington 18 C2 Michigan, N USA
Ludsan *see* Ludza
Luduş 86 B4 *Ger.* Ludasch, *Hung.* Marosludas. Mureş, C Romania
Ludvika 63 C6 Dalarna, C Sweden
Ludwigsburg 73 B6 Baden-Württemberg, SW Germany
Ludwigsfelde 72 D3 Brandenburg, NE Germany
Ludwigshafen 73 B5 *var.* Ludwigshafen am Rhein. Rheinland-Pfalz, W Germany
Ludwigshafen am Rhein *see* Ludwigshafen
Ludwigslust 72 C3 Mecklenburg-Vorpommern, N Germany
Ludza 84 D4 *Ger.* Ludsan. Ludza, E Latvia
Luebo 55 C6 Kasai-Occidental, SW Dem. Rep. Congo
Luena 56 C2 *var.* Lwena, *Port.* Luso. Moxico, E Angola
Lufira 55 E7 *river* SE Dem. Rep. Congo
Lufkin 27 H3 Texas, SW USA
Luga 88 A4 Leningradskaya Oblast', NW Russia
Lugano 73 B8 *Ger.* Lauis. Ticino, S Switzerland
Lugansk *see* Luhans'k
Lugdunum *see* Lyon
Lugdunum Batavorum *see* Leiden
Lugenda, Rio 57 E2 *river* N Mozambique
Luggarus *see* Locarno
Lugh Ganana *see* Luuq
Lugo 70 C1 *anc.* Lugus Augusti. Galicia, NW Spain
Lugoj 86 A4 *Ger.* Lugosch, *Hung.* Lugos. Timiş, W Romania
Lugos/Lugosch *see* Lugoj
Lugus Augusti *see* Lugo
Luguvallium/Luguvallum *see* Carlisle
Luhans'k 87 H3 *Rus.* Lugansk; *prev.* Voroshilovgrad. Luhans'ka Oblast', E Ukraine
Luimneach *see* Limerick
Lukapa *see* Lucapa
Lukenie 55 C6 *river* C Dem. Rep. Congo
Lukovit 82 C2 Lovech, N Bulgaria
Łuków 76 E4 *Ger.* Bogendorf. Lubelskie, E Poland
Lukuga 55 D7 *river* SE Dem. Rep. Congo
Luleå 62 D4 Norrbotten, N Sweden
Luleälven 62 C3 *river* N Sweden
Lulonga 55 C5 *river* NW Dem. Rep. Congo
Lulua 55 D7 *river* S Dem. Rep. Congo
Luluabourg *see* Kananga
Lumber State *see* Maine
Lumbo 57 F2 Nampula, NE Mozambique
Lumsden 129 A7 Southland, South Island, New Zealand
Lund 63 B7 Skåne, S Sweden
Lüneburg 72 C3 Niedersachsen, N Germany
Lunga, Isola *see* Dugi Otok
Lungkiang *see* Qiqihar
Lungué-Bungo 56 C2 *var.* Lungwebungu. *river* Angola/Zambia
Lungwebungu *see* Lungué-Bungo
Luninets *see* Luninyets
Łuniniec *see* Luninyets
Luninyets 85 B7 *Pol.* Łuniniec, *Rus.* Luninets. Brestskaya Voblasts', SW Belarus
Lunteren 64 D4 Gelderland, C Netherlands
Luong Nam Tha *see* Louangnamtha
Luoyang 106 C4 *var.* Honan, Lo-yang. Henan, C China
Lupatia *see* Altamura
Lúrio 57 F2 Nampula, NE Mozambique
Lúrio, Rio 57 E2 *river* NE Mozambique
Lurka *see* Lorca
Lusaka 56 D2 *country capital* (Zambia) Lusaka, SE Zambia
Lushnja *see* Lushnjë
Lushnjë 79 C6 *var.* Lushnja. Fier, C Albania
Luso *see* Luena
Lussin/Lussino *see* Lošinj
Lūt, Baḩrat/Lut, Bahret *see* Dead Sea
Lut, Dasht-e 98 D3 *var.* Kavīr-e Lūt. *desert* E Iran
Lutetia/Lutetia Parisiorum *see* Paris
Lūt, Kavīr-e *see* Lūt, Dasht-e
Luton 67 D6 E England, United Kingdom
Łutselk'e 15 F4 *prev.* Snowdrift. Northwest Territories, W Canada
Luts'k 86 C1 *Pol.* Łuck, *Rus.* Lutsk. Volyns'ka Oblast', NW Ukraine
Lutsk *see* Luts'k
Lüttich *see* Liège
Lutzow-Holm Bay 132 C2 *var.* Lutzow-Holm Bay. *bay* Antarctica
Lutzow-Holm Bay *see* Lützow Holmbukta
Luuq 51 D6 *It.* Lugh Ganana. Gedo, SW Somalia
Luvua 55 D7 *river* SE Dem. Rep. Congo
Luwego 51 C8 *river* S Tanzania
Luxembourg 65 D8 *country capital* (Luxembourg) Luxembourg, S Luxembourg
Luxembourg 65 D8 *off.* Grand Duchy of Luxembourg, *var.* Lëtzebuerg, Luxemburg. *country* NW Europe
Luxemburg *see* Luxembourg
Luxor 50 B2 *Ar.* Al Uqşur. E Egypt
Luza 88 C4 Kirovskaya Oblast', NW Russia
Luz, Costa de la 70 C5 *coastal region* SW Spain
Luzern 73 B7 *Fr.* Lucerne, *It.* Lucerna. Luzern, C Switzerland
Luzon 117 E1 *island* N Philippines
Luzon Strait 103 E3 *strait* Philippines/Taiwan
L'viv 86 B2 *Ger.* Lemberg, *Pol.* Lwów, *Rus.* L'vov. L'vivs'ka Oblast', W Ukraine
L'vov *see* L'viv
Lwena *see* Luena
Lwów *see* L'viv
Lyakhavichy 85 B6 *Rus.* Lyakhovichi. Brestskaya Voblasts', SW Belarus
Lyakhovichi *see* Lyakhavichy
Lyallpur *see* Faisalābād
Lyangar *see* Langar
Lyck *see* Ełk
Lycksele 62 C4 Västerbotten, N Sweden
Lycopolis *see* Asyūţ
Lyel'chytsy 85 C7 *Rus.* Lel'chitsy. Homyel'skaya Voblasts', SE Belarus
Lyepyel' 85 D5 *Rus.* Lepel'. Vitsyebskaya Voblasts', N Belarus
Lyme Bay 67 C7 *bay* S England, United Kingdom
Lynchburg 19 E5 Virginia, NE USA
Lynn *see* King's Lynn
Lynn Lake 15 F4 Manitoba, C Canada
Lynn Regis *see* King's Lynn
Lyon 69 D5 *Eng.* Lyons; *anc.* Lugdunum. Rhône, E France
Lyons *see* Lyon
Lyozna 85 E6 *Rus.* Liozno. Vitsyebskaya Voblasts', NE Belarus
Lypovets' 86 D2 *Rus.* Lipovets. Vinnyts'ka Oblast', C Ukraine
Lys *see* Leie
Lysychans'k 87 H3 *Rus.* Lisichansk. Luhans'ka Oblast', E Ukraine
Lyttelton 129 C6 South Island, New Zealand
Lyublin *see* Lublin
Lyubotin *see* Lyubotyn
Lyubotyn 87 G2 *Rus.* Lyubotin. Kharkivs'ka Oblast', E Ukraine
Lyulyakovo 82 E2 *prev.* Keremitlik. Burgas, E Bulgaria
Lyusina 85 B6 *Rus.* Lyusino. Brestskaya Voblasts', SW Belarus
Lyusino *see* Lyusina

Maale – Marshall Islands

M

Maale 110 B4 var. Male'. country capital (Maldives) Male' Atoll, C Maldive
Ma'ān 97 B7 Ma'ān, SW Jordan
Maardu 84 D2 Ger. Maart. Harjumaa, NW Estonia
Ma'aret-en-Nu'man see Ma'arrat an Nu'mān
Ma'arrat an Nu'mān 96 B3 var. Ma'aret-en-Nu'man, Fr. Maarret enn Naamâne. Idlib, NW Syria
Maarret enn Naamâne see Ma'arrat an Nu'mān
Maart see Maardu
Maas see Meuse
Maaseik 65 D5 prev. Maeseyck. Limburg, NE Belgium
Maastricht 65 D6 var. Maestricht; anc. Traiectum ad Mosam, Traiectum Tungorum. Limburg, SE Netherlands
Macao 107 C6 Port. Macau. Guangdong, SE China
Macapá 41 E1 state capital Amapá, N Brazil
Macarsca see Makarska
Macassar see Makassar
Macău see Makó, Hungary
Macau see Macao
MacCluer Gulf see Berau, Teluk
Macdonnell Ranges 124 D4 mountain range Northern Territory, C Australia
Macedonia 79 D6 off. Republic of Macedonia, var. the former Yugoslav Republic of Macedonia (used by UN), Mac. Makedonija, abbrev. FYR Macedonia, FYROM. country SE Europe
Macedonia, FYR see Macedonia
Macedonia, Republic of see Macedonia
Macedonia, the former Yugoslav Republic of see Macedonia
Maceió 41 G3 state capital Alagoas, E Brazil
Machachi 38 B1 Pichincha, C Ecuador
Machala 38 B2 El Oro, SW Ecuador
Machanga 57 E3 Sofala, E Mozambique
Machilipatnam 110 D1 var. Bandar Masulipatnam. Andhra Pradesh, E India
Machiques 36 C2 Zulia, NW Venezuela
Macías Nguema Biyogo see Bioco, Isla de
Măcin 86 D5 Tulcea, SE Romania
Mackay 126 D4 Queensland, NE Australia
Mackay, Lake 124 C4 salt lake Northern Territory/Western Australia
Mackenzie 15 E3 river Northwest Territories, NW Canada
Mackenzie Bay 132 D3 bay Antarctica
Mackenzie Mountains 14 D3 mountain range Northwest Territories, NW Canada
Macleod, Lake 124 A4 lake Western Australia
Macomb 18 A4 Illinois, N USA
Macomer 75 A5 Sardegna, Italy, C Mediterranean Sea
Mâcon 69 D5 anc. Matisco, Matisco Ædourum. Saône-et-Loire, C France
Macon 20 D2 Georgia, SE USA
Macon 23 G4 Missouri, C USA
Macquarie Ridge 132 C5 undersea ridge SW Pacific Ocean
Macuspana 29 G4 Tabasco, SE Mexico
Ma'dabā 97 B6 var. Mādabā, Madeba; anc. Medeba. Ma'dabā, NW Jordan
Mādabā see Ma'dabā
Madagascar 57 F3 off. Democratic Republic of Madagascar, Malg. Madagasikara; prev. Malagasy Republic. country W Indian Ocean
Madagascar 57 F3 island W Indian Ocean
Madagascar Basin 47 E7 undersea basin W Indian Ocean
Madagascar, Democratic Republic of see Madagascar
Madagascar Plateau 47 E7 var. Madagascar Ridge, Madagascar Rise, Rus. Madagaskarskiy Khrebet. undersea plateau W Indian Ocean
Madagascar Rise/Madagascar Ridge see Madagascar Plateau
Madagasikara see Madagascar
Madagaskarskiy Khrebet see Madagascar Plateau
Madang 122 B3 Madang, N Papua New Guinea
Madaniyin see Médenine
Mađarska see Hungary
Made 64 C4 Noord-Brabant, S Netherlands
Madeba see Ma'dabā
Madeira 48 A2 var. Ilha de Madeira. island Madeira, Portugal, NE Atlantic Ocean
Madeira, Ilha de see Madeira
Madeira Plain 44 C3 abyssal plain E Atlantic Ocean
Madeira, Rio 40 D2 var. Río Madera. river Bolivia/Brazil
Madeleine, Îles de la 17 F4 Eng. Magdalen Islands. island group Québec, E Canada
Madera 25 B6 California, W USA
Madera, Río see Madeira, Rio
Madhya Pradesh 113 E4 prev. Central Provinces and Berar. cultural region C India
Madinat ath Thawrah 96 C2 var. Ath Thawrah. Ar Raqqah, N Syria
Madioen see Madiun
Madison 23 F3 South Dakota, N USA
Madison 18 B3 state capital Wisconsin, N USA
Madiun 116 D5 prev. Madioen. Jawa, C Indonesia
Madoera see Madura, Pulau
Madona 84 D4 Ger. Modohn. Madona, E Latvia
Madras see Chennai
Madras see Tamil Nādu
Madre de Dios, Río 39 E3 river Bolivia/Peru
Madre del Sur, Sierra 29 E5 mountain range S Mexico
Madre, Laguna 29 F3 lagoon NE Mexico
Madre, Laguna 27 G5 lagoon Texas, SW USA
Madre Occidental, Sierra 28 C3 var. Western Sierra Madre. mountain range C Mexico
Madre Oriental, Sierra 29 E3 var. Eastern Sierra Madre. mountain range C Mexico
Madre, Sierra 30 B2 var. Sierra de Soconusco. mountain range Guatemala/Mexico
Madrid 70 D3 country capital (Spain) Madrid, C Spain
Madura see Madurai
Madurai 110 C3 prev. Madura, Mathurai. Tamil Nādu, S India
Madura, Pulau 116 D5 prev. Madoera. island C Indonesia
Maebashi 109 D5 var. Maebasi, Mayebashi. Gunma, Honshū, S Japan
Maebasi see Maebashi
Mae Nam Khong see Mekong
Mae Nam Nan 114 C4 river NW Thailand
Mae Nam Yom 114 C4 river W Thailand
Maeseyck see Maaseik
Maestricht see Maastricht
Maéwo 122 D4 prev. Aurora. island C Vanuatu
Mafia 51 D7 island E Tanzania
Mafraq/Muḥāfaẓat al Mafraq see Al Mafraq
Magadan 93 G3 Magadanskaya Oblast', E Russia
Magallanes see Punta Arenas
Magallanes, Estrecho de see Magellan, Strait of
Magangué 36 B2 Bolívar, N Colombia
Magdalena 39 F3 Beni, N Bolivia
Magdalena 28 B1 Sonora, NW Mexico
Magdalena, Isla 28 B3 island NW Mexico
Magdalena, Río 36 B2 river C Colombia
Magdalen Islands see Madeleine, Îles de la
Magdeburg 72 C4 Sachsen-Anhalt, C Germany
Magelang 116 C5 Jawa, C Indonesia
Magellan, Strait of 43 B8 Sp. Estrecho de Magallanes. strait Argentina/Chile
Magerøy see Magerøya
Magerøya 62 D1 var. Magerøy, Lapp. Máhkarávju. island N Norway
Maggiore, Lago see Maggiore, Lake
Maggiore, Lake 74 B1 It. Lago Maggiore. lake Italy/Switzerland
Maglaj 78 C3 Federacija Bosna I Hercegovina, N Bosnia and Herzegovina
Maglie 75 E6 Puglia, SE Italy
Magna 22 B4 Utah, W USA
Magnesia see Manisa
Magnitogorsk 92 B4 Chelyabinskaya Oblast', C Russia
Magnolia State see Mississippi
Magta' Lahjar 52 C3 var. Magta Lahjar, Magta' Lahjar, Magtá Lahjar. Brakna, SW Mauritania
Magtymguly 100 C3 prev. Garrygala; Rus. Kara-gala. W Turkmenistan
Magway 114 A3 var. Magwe. Magway, W Myanmar (Burma)
Magyar-Becse see Bečej
Magyarkanizsa see Kanjiža
Magyarország see Hungary
Mahajanga 57 F2 var. Majunga. Mahajanga, NW Madagascar
Mahakam, Sungai 116 D4 var. Koetai, Kutai. river Borneo, C Indonesia
Mahalapye 56 D3 var. Mahalatswe. Central, SE Botswana
Mahalatswe see Mahalapye
Māhān 98 D3 Kermān, E Iran
Mahanādi 113 F4 river E India
Mahārāshtra 112 D5 cultural region W India
Mahbés see El Mahbas
Mahbūbnagar 112 D5 Telangana, C India
Mahdia 49 F2 var. Al Mahdīyah, Mehdia. NE Tunisia
Mahé 57 H1 island Inner Islands, NE Seychelles
Mahia Peninsula 128 E4 peninsula North Island, New Zealand
Mahilyow 85 D6 Rus. Mogilëv. Mahilyowskaya Voblasts', E Belarus
Máhkarávju see Magerøya
Mahmūd-e 'Erāqī see Maḥmūd-e Rāqī
Maḥmūd-e Rāqī 101 E4 var. Mahmūd-e 'Erāqī. Kāpīsā, NE Afghanistan
Mahón see Maó
Mähren see Moravia
Mährisch-Weisskirchen see Hranice
Maicao 36 C1 La Guajira, N Colombia
Mai Ceu/Mai Chio see Maych'ew
Maidān Shahr 101 E4 prev. Meydān Shahr. Vardak, E Afghanistan
Maidstone 67 E7 SE England, United Kingdom
Maiduguri 53 H4 Borno, NE Nigeria
Mailand see Milano
Maimanah 100 D3 var. Meymaneh, Maymana. Fāryāb, NW Afghanistan
Main 73 B5 river C Germany
Mai-Ndombe, Lac 55 C6 prev. Lac Léopold II. lake W Dem. Rep. Congo
Maine 19 G2 off. State of Maine, also known as Lumber State, Pine Tree State. state NE USA
Maine 68 B3 cultural region NW France
Maine, Gulf of 19 H2 gulf NE USA
Mainland 66 C2 island N Scotland, United Kingdom
Mainland 66 D1 island NE Scotland, United Kingdom
Mainz 73 B5 Fr. Mayence. Rheinland-Pfalz, SW Germany
Maio 52 A3 var. Mayo. island Ilhas de Sotavento, SE Cape Verde
Maisur see Mysūru, India
Maisur see Karnātaka, India
Maitri 132 C2 Indian research station Antarctica
Maizhokunggar 104 C5 Xizang Zizhiqu, W China
Majorca see Mallorca
Mājro see Majuro Atoll
Majunga see Mahajanga
Majuro Atoll 122 D2 var. Mājro. atoll and capital (Marshall Islands) Ratak Chain, SE Marshall Islands
Makale see Mek'elē
Makarov Basin 133 B3 undersea basin Arctic Ocean
Makarska 78 B4 It. Macarsca. Split-Dalmacija, SE Croatia
Makasar see Makassar
Makasar, Selat see Makassar Straits
Makassar 117 E4 var. Macassar, Makasar; prev. Ujungpandang. Sulawesi, C Indonesia
Makassar Straits 116 D4 Ind. Makasar Selat. strait C Indonesia
Makay 57 F3 var. Massif du Makay. mountain range SW Madagascar
Makay, Massif du see Makay
Makedonija see Macedonia
Makeni 52 C4 C Sierra Leone
Makeyevka see Makiyivka
Makhachkala 92 A4 prev. Petrovsk-Port. Respublika Dagestan, SW Russia
Makin 122 D2 prev. Pitt Island. atoll Tungaru, W Kiribati
Makira see San Cristobal
Makiyivka 87 G3 Rus. Makeyevka; prev. Dmitriyevsk. Donets'ka Oblast', E Ukraine
Makkah 99 A5 Eng. Mecca. Makkah, W Saudi Arabia
Makkovik 17 F2 Newfoundland and Labrador, NE Canada
Makó 77 D7 Rom. Macău. Csongrád, SE Hungary
Makoua 55 B5 Cuvette, C Congo
Makran Coast 98 E4 coastal region SE Iran
Makrany 85 A6 Rus. Mokrany. Brestskaya Voblasts', SW Belarus
Mākū 98 B2 Āżarbāyjān-e Gharbī, NW Iran
Makurdi 53 G4 Benue, C Nigeria
Mala see Malaita, Solomon Islands
Malabār Coast 110 B3 coast SW India
Malabo 55 A5 prev. Santa Isabel. country capital (Equatorial Guinea) Isla de Bioco, NW Equatorial Guinea
Malaca see Málaga
Malacca, Strait of 116 B3 Ind. Selat Malaka. strait Indonesia/Malaysia
Malacka see Malacky
Malacky 77 C6 Hung. Malacka. Bratislavský Kraj, W Slovakia
Maladzyechna 85 C5 Pol. Molodeczno, Rus. Molodechno. Minskaya Voblasts', C Belarus
Málaga 70 D5 anc. Malaca. Andalucía, S Spain
Malagarasi River 51 B7 river W Tanzania Africa
Malagasy Republic see Madagascar
Malaita 122 C3 var. Mala. island N Solomon Islands
Malakal 51 B5 Upper Nile, NE South Sudan
Malakula see Malekula
Malang 116 D5 Jawa, C Indonesia
Malange see Malanje
Malanje 56 B1 var. Malange. Malanje, NW Angola
Mälaren 63 C6 lake C Sweden
Malatya 95 E4 anc. Melitene. Malatya, SE Turkey
Mala Vyska 87 E3 Rus. Malaya Viska. Kirovohrads'ka Oblast', S Ukraine
Malawi 57 E1 off. Republic of Malaŵi; prev. Nyasaland, Nyasaland Protectorate. country S Africa
Malawi, Lake see Nyasa, Lake
Malaŵi, Republic of see Malawi
Malaya Viska see Mala Vyska
Malay Peninsula 102 D4 peninsula Malaysia/Thailand
Malaysia 116 B3 off. Malaysia, var. Federation of Malaysia; prev. the separate territories of Federation of Malaya, Sarawak and Sabah (North Borneo) and Singapore. country SE Asia
Malaysia, Federation of see Malaysia
Malbork 76 C2 Ger. Marienburg, Marienburg in Westpreussen. Pomorskie, N Poland
Malchin 72 C3 Mecklenburg-Vorpommern, N Germany
Malden 23 H5 Missouri, C USA
Malden Island 123 G3 prev. Independence Island. atoll E Kiribati
Maldives 110 A4 off. Maldivian Divehi, Republic of Maldives. country N Indian Ocean
Maldives, Republic of see Maldives
Maldivian Divehi see Maldives
Male' see Maale
Male' Atoll 110 B4 var. Kaafu Atoll. atoll C Maldives
Malekula 122 D4 var. Malakula; prev. Mallicolo. island W Vanuatu
Malesína 83 C5 Stereá Ellás, E Greece
Malheur Lake 24 C3 lake Oregon, NW USA
Mali 53 E3 off. Republic of Mali, Fr. République du Mali; prev. French Sudan, Sudanese Republic. country W Africa
Malik, Wadi al see Milk, Wadi el
Mali Kyun 115 B5 var. Tavoy Island. island Myeik Archipelago, S Myanmar (Burma)
Malin see Malyn
Malindi 51 D7 Coast, SE Kenya
Malines see Mechelen
Mali, Republic of see Mali
Mali, République du see Mali
Malkiye see Al Mālikīyah
Malko Tŭrnovo 82 E3 Burgas, E Bulgaria
Mallaig 66 B3 N Scotland, United Kingdom
Mallawī 50 B2 var. Mallawi. C Egypt
Mallawi see Mallawī
Mallicolo see Malekula
Mallorca 71 G3 Eng. Majorca; anc. Baleares Major. island Islas Baleares, Spain, W Mediterranean Sea
Malmberget 62 C3 Lapp. Malmivaara. Norrbotten, N Sweden
Malmédy 65 D6 Liège, E Belgium
Malmivaara see Malmberget
Malmö 63 B7 Skåne, S Sweden
Maloelap see Maloelap Atoll
Maloelap Atoll 122 D1 var. Maloelap. atoll E Marshall Islands
Małopolska, Wyżyna 76 D4 plateau S Poland
Malozemel'skaya Tundra 88 D3 physical region NW Russia
Malta 84 D4 Rēzekne, SE Latvia
Malta 22 C1 Montana, NW USA
Malta 75 C8 off. Republic of Malta. country C Mediterranean Sea
Malta 75 C8 island Malta, C Mediterranean Sea
Malta, Canale di see Malta Channel
Malta Channel 75 C8 It. Canale di Malta. strait Italy/Malta
Malta, Republic of see Malta
Maluku 117 F4 Dut. Molukken, Eng. Moluccas; prev. Spice Islands. island group E Indonesia
Maluku, Laut see Molucca Sea
Malung 63 B6 Dalarna, C Sweden
Malventum see Benevento
Malvina, Isla Gran see West Falkland
Malvinas, Islas see Falkland Islands
Malyn 86 D2 Rus. Malin. Zhytomyrs'ka Oblast', N Ukraine
Malyy Kavkaz see Lesser Caucasus
Mamberamo, Sungai 117 H4 river Papua, E Indonesia
Mambij see Manbij
Mamonovo 84 A4 Ger. Heiligenbeil. Kaliningradskaya Oblast', W Russia
Mamoré, Rio 39 F3 river Bolivia/Brazil
Mamou 52 C4 W Guinea
Mamoudzou 57 F2 dependent territory capital (Mayotte) C Mayotte
Mamuno 56 C3 Ghanzi, W Botswana
Manacor 71 G3 Mallorca, Spain, W Mediterranean Sea
Manado 117 F3 prev. Menado. Sulawesi, C Indonesia
Managua 30 D3 country capital (Nicaragua) Managua, W Nicaragua
Managua, Lake 30 C3 var. Xolotlán. lake W Nicaragua
Manakara 57 G4 Fianarantsoa, SE Madagascar
Manama see Al Manāmah
Mananjary 57 G3 Fianarantsoa, SE Madagascar
Manáos see Manaus
Manapouri, Lake 129 A7 lake South Island, New Zealand
Manar see Mannar
Manas, Gora 101 E2 mountain Kyrgyzstan/Uzbekistan
Manaus 40 D2 prev. Manáos. state capital Amazonas, NW Brazil
Manavgat 94 B4 Antalya, SW Turkey
Manbij 96 C2 var. Mambij, Fr. Membidj. Ḥalab, N Syria
Manchester 67 D5 Lat. Mancunium. NW England, United Kingdom
Manchester 19 G3 New Hampshire, NE USA
Man-chou-li see Manzhouli
Manchurian Plain 103 E1 plain NE China
Máncio Lima see Japiim
Mancunium see Manchester
Mand see Mand, Rūd-e
Mandalay 114 B3 Mandalay, C Myanmar (Burma)
Mandan 23 E2 North Dakota, N USA
Mandeville 32 B5 C Jamaica
Mándra 83 C6 Attikí, C Greece
Rud-e Mand 98 D4 var. Mand. river S Iran
Mandurah 125 A6 Western Australia
Manduria 75 E5 Puglia, SE Italy
Mandya 110 C2 Karnātaka, C India
Manfredonia 75 D5 Puglia, SE Italy
Mangai 55 C6 Bandundu, W Dem. Rep. Congo
Mangaia 123 G5 island group S Cook Islands
Mangalia 86 D5 anc. Callatis. Constanța, SE Romania
Mangalmé 54 C3 Guéra, SE Chad
Mangalore see Mangalūru
Mangalūru 110 B2 prev. Mangalore. Karnātaka, W India
Mangaung see Bloemfontein
Mango see Sansanné-Mango, Togo
Mangoky 57 F3 river W Madagascar
Manhattan 23 F4 Kansas, C USA
Manicouagan, Réservoir 16 D3 lake Québec, E Canada
Manihiki 123 G4 atoll N Cook Islands
Manihiki Plateau 121 E3 undersea plateau C Pacific Ocean
Maniitsoq 60 C3 var. Manitsoq, Dan. Sukkertoppen. Kitaa, S Greenland
Manila 117 E1 off. City of Manila. country capital (Philippines) Luzon, N Philippines
Manila, City of see Manila
Manisa 94 A3 var. Manissa, prev. Saruhan; anc. Magnesia. Manisa, W Turkey
Manissa see Manisa
Manitoba 15 F5 province S Canada
Manitoba, Lake 15 F5 lake Manitoba, S Canada
Manitoulin Island 16 C4 island Ontario, S Canada
Manītsoq see Maniitsoq
Manizales 36 B3 Caldas, W Colombia
Manjimup 125 A7 Western Australia
Mankato 23 F3 Minnesota, N USA
Manlleu 71 G2 Cataluña, NE Spain
Manly 126 E1 Iowa, C USA
Manmād 112 C5 Mahārāshtra, W India
Mannar 110 C3 var. Manar. Northern Province, NW Sri Lanka
Mannar, Gulf of 110 C3 gulf India/Sri Lanka
Mannheim 73 B5 Baden-Württemberg, SW Germany
Manokwari 117 G4 New Guinea, E Indonesia
Manono 55 E7 Shaba, SE Dem. Rep. Congo
Manosque 69 D6 Alpes-de-Haute-Provence, SE France
Manra 123 F3 prev. Sydney Island. atoll Phoenix Islands, C Kiribati
Mansa 56 D2 prev. Fort Rosebery. Luapula, N Zambia
Mansel Island 15 G3 island Nunavut, NE Canada
Mansfield 18 D4 Ohio, N USA
Manta 38 A2 Manabí, W Ecuador
Manteca 25 B6 California, W USA
Mantoue see Mantova
Mantova 74 B2 Eng. Mantua, Fr. Mantoue. Lombardia, NW Italy
Mantua see Mantova
Manuae 123 G4 island S Cook Islands
Manukau see Manurewa
Manurewa 128 D3 var. Manukau. Auckland, North Island, New Zealand
Manzanares 71 E3 Castilla-La Mancha, C Spain
Manzanillo 32 C3 Granma, E Cuba
Manzanillo 28 D4 Colima, SW Mexico
Manzhouli 105 F1 var. Man-chou-li. Nei Mongol Zizhiqu, N China
Mao 54 B3 Kanem, W Chad
Maó 71 H3 Cast. Mahón, Eng. Port Mahon; anc. Portus Magonis. Menorca, Spain, W Mediterranean Sea
Maoke, Pegunungan 117 H4 Dut. Sneeuw-gebergte, Eng. Snow Mountains. mountain range Papua, E Indonesia
Maoming 106 C6 Guangdong, S China
Mapmaker Seamounts 103 H2 seamount range N Pacific Ocean
Maputo 56 D4 prev. Lourenço Marques. country capital (Mozambique) Maputo, S Mozambique
Marabá 41 F2 Pará, NE Brazil
Maracaibo 36 C1 Zulia, NW Venezuela
Maracaibo, Gulf of see Venezuela, Golfo de
Maracaibo, Lago de 36 C2 var. Lake Maracaibo. inlet NW Venezuela
Maracaibo, Lake see Maracaibo, Lago de
Maracay 36 D2 Aragua, N Venezuela
Marada see Marādah
Marādah 49 G3 var. Marada. N Libya
Maradi 53 G3 Maradi, S Niger
Maragha see Marāgheh
Marāgheh 98 C2 var. Maragha. Āżarbāyjān-e Khāvarī, NW Iran
Marajó, Baía de 41 F1 bay N Brazil
Marajó, Ilha de 41 E1 island N Brazil
Marakesh see Marrakech
Maramba see Livingstone
Maranhão 41 F2 off. Estado do Maranhão. state/region E Brazil
Maranhão, Estado do see Maranhão
Marañón, Río 38 B2 river N Peru
Marathon 16 C4 Ontario, S Canada
Marathón see Marathónas
Marathónas 83 C5 prev. Marathón. Attikí, C Greece
Marbella 70 D5 Andalucía, S Spain
Marble Bar 124 B4 Western Australia
Marburg see Marburg an der Lahn, Germany
Marburg see Maribor, Slovenia
Marburg an der Lahn 72 B4 hist. Marburg. Hessen, W Germany
March see Morava
Marche 74 C3 Eng. Marches. region C Italy
Marche 69 C5 cultural region C France
Marche-en-Famenne 65 C7 Luxembourg, SE Belgium
Marchena, Isla 38 B5 var. Bindloe Island. island Galápagos Islands, Ecuador, E Pacific Ocean
Marches see Marche
Mar Chiquita, Laguna 42 C3 lake C Argentina
Marcounda see Markounda
Mardān 112 C1 North-West Frontier Province, N Pakistan
Mar del Plata 43 D5 Buenos Aires, E Argentina
Mardin 95 E4 Mardin, SE Turkey
Maré 122 D5 island Îles Loyauté, E New Caledonia
Marea Neagră see Black Sea
Mareeba 126 D3 Queensland, NE Australia
Marek see Dupnitsa
Marganets see Marhanets'
Margarita, Isla de 37 E1 island N Venezuela
Margate 67 E7 prev. Mergate. SE England, United Kingdom
Margherita see Jamaame
Margherita, Lake 51 C5 Eng. Lake Margherita, It. Abbaia. lake SW Ethiopia
Margherita, Lake see Ābaya Hāyk'
Marghita 86 B3 Hung. Margitta. Bihor, NW Romania
Margitta see Marghita
Marhanets' 87 F3 Rus. Marganets. Dnipropetrovs'ka Oblast', E Ukraine
María Cleofas, Isla 28 C4 island C Mexico
Maria Island 127 C8 island Tasmania, SE Australia
María Madre, Isla 28 C4 island C Mexico
María Magdalena, Isla 28 C4 island C Mexico
Mariana Trench 103 G4 trench W Pacific Ocean
Mariánské Lázně 77 A5 Ger. Marienbad. Karlovarský Kraj, W Czech Republic (Czechia)
Marías, Islas 28 C4 island group C Mexico
Maria-Theresiopel see Subotica
Maribor 73 E7 Ger. Marburg. NE Slovenia
Marica see Maritsa
Maridi 51 B5 W Equatoria, S South Sudan
Marie Byrd Land 132 A3 physical region Antarctica
Marie-Galante 33 G4 var. Ceyre to the Caribs. island SE Guadeloupe
Marienbad see Mariánské Lázně
Marienburg see Alūksne, Latvia
Marienburg see Malbork, Poland
Marienburg in Westpreussen see Malbork
Marienhausen see Viļaka
Mariental 56 B4 Hardap, SW Namibia
Marienwerder see Kwidzyń
Mariestad 63 B6 Västra Götaland, S Sweden
Marietta 20 D2 Georgia, SE USA
Marijampolė 84 B4 prev. Kapsukas. Marijampolė, S Lithuania
Marília 41 E4 São Paulo, S Brazil
Marín 70 B1 Galicia, NW Spain
Mar"ina Gorka see Mar"ina Horka
Mar"ina Horka 85 C6 Rus. Mar'ina Gorka. Minskaya Voblasts', C Belarus
Maringá 41 E4 Paraná, S Brazil
Marion 23 G3 Iowa, C USA
Marion 18 D4 Ohio, N USA
Marion, Lake 21 E2 reservoir South Carolina, SE USA
Mario Zucchelli 132 C4 Italian research station Antarctica
Mariscal Estigarribia 42 D2 Boquerón, NW Paraguay
Maritsa 82 D3 var. Marica, Gk. Évros, Turk. Meriç; anc. Hebrus. river SW Europe
Maritzburg see Pietermaritzburg
Mariupol' 87 G4 prev. Zhdanov. Donets'ka Oblast', SE Ukraine
Marka 51 D6 var. Merca. Shabeellaha Hoose, S Somalia
Markham, Mount 132 B4 mountain Antarctica
Markounda 54 C4 var. Marcounda. Ouham, NW Central African Republic
Marktredwitz 73 C5 Bayern, E Germany
Marlborough 126 D4 Queensland, E Australia
Marmanda see Marmande
Marmande 69 B5 anc. Marmanda. Lot-et-Garonne, SW France
Sea of Marmara 94 A2 Eng. Sea of Marmara. sea NW Turkey
Marmara, Sea of see Marmara Denizi
Marmaris 94 A4 Muğla, SW Turkey
Marne 68 C3 cultural region N France
Marne 68 D3 river N France
Maro 54 C4 Moyen-Chari, S Chad
Maroantsetra 57 G2 Toamasina, NE Madagascar
Maromokotro 57 G2 mountain N Madagascar
Maroni 37 G3 Dut. Marowijne. river French Guiana/Suriname
Marosheviz see Toplița
Marosludas see Luduș
Marosvásárhely see Târgu Mureș
Marotiri 121 F4 var. Îlots de Bass, Morotiri. island group Îles Australes, SW French Polynesia
Maroua 54 B3 Extrême-Nord, N Cameroon
Marowijne see Maroni
Marquesas Fracture Zone 131 E3 fracture zone E Pacific Ocean
Marquette 18 B1 Michigan, N USA
Marrakech 48 C2 var. Marakesh, Eng. Marrakesh; prev. Morocco. W Morocco
Marrakesh see Marrakech
Marrawah 127 C8 Tasmania, SE Australia
Marree 127 B5 South Australia
Marsá al Burayqah 49 G3 var. Al Burayqah. N Libya
Marsabit 51 C6 Eastern, N Kenya
Marsala 75 B7 anc. Lilybaeum. Sicilia, Italy, C Mediterranean Sea
Marsberg 72 B4 Nordrhein-Westfalen, W Germany
Marseille 69 D6 Eng. Marseilles; anc. Massilia. Bouches-du-Rhône, SE France
Marseilles see Marseille
Marshall 23 F2 Minnesota, N USA
Marshall 27 H2 Texas, SW USA
Marshall Islands 122 C1 off. Republic of the Marshall Islands. country W Pacific Ocean

Marshall Islands, Republic of the – Misti, Volcán

Marshall Islands, Republic of the *see* Marshall Islands
Marshall Seamounts *103 H3 seamount range* SW Pacific Ocean
Marsh Harbour *32 C1* Great Abaco, W The Bahamas
Martaban *see* Mottama
Martha's Vineyard *19 G3 island* Massachusetts, NE USA
Martigues *69 D6* Bouches-du-Rhône, SE France
Martin *77 C5 Ger.* Sankt Martin, *Hung.* Turócszentmárton; *prev.* Turčianský Svätý Martin. Žilinský Kraj, N Slovakia
Martinique *33 G4 French overseas department* E West Indies
Martinique Channel *see* Martinique Passage
Martinique Passage *33 G4 var.* Dominica Channel, Martinique Channel. *channel* Dominica/Martinique
Marton *128 D4* Manawatu-Wanganui, North Island, New Zealand
Martos *70 D4* Andalucía, S Spain
Marungu *55 E7 mountain range* SE Dem. Rep. Congo
Mary *100 D3 prev.* Merv. Mary Welaýaty, S Turkmenistan
Maryborough *127 D4* Queensland, E Australia
Maryborough *see* Port Laoise
Mary Island *see* Kanton
Maryland *19 E5 off.* State of Maryland, *also known as* America in Miniature, Cockade State, Free State, Old Line State. *state* NE USA
Maryland, State of *see* Maryland
Maryville *23 F4* Missouri, C USA
Maryville *20 D1* Tennessee, S USA
Masai Steppe *51 C7 grassland* NW Tanzania
Masaka *51 B6* SW Uganda
Masallı *95 H3 Rus.* Masally. S Azerbaijan
Masally *see* Masallı
Masasi *51 C8* Mtwara, SE Tanzania
Masawa/Massawa *see* Mits'iwa
Masaya *30 D3* Masaya, W Nicaragua
Mascarene Basin *119 B5 undersea basin* W Indian Ocean
Mascarene Islands *57 H4 island group* W Indian Ocean
Mascarene Plain *119 B5 abyssal plain* W Indian Ocean
Mascarene Plateau *119 B5 undersea plateau* W Indian Ocean
Maseru *56 D4 country capital* (Lesotho) W Lesotho
Mas-ha *97 D7* W West Bank Asia
Mashhad *98 E2 var.* Meshed. Khorāsān-Razavī, NE Iran
Masindi *51 B6* W Uganda
Masira *see* Maşīrah, Jazīrat
Masira, Gulf of *see* Maşīrah, Khalīj
Maşīrah, Jazīrat *99 E5 var.* Masīra. *island* E Oman
Maşīrah, Khalīj *99 E5 var.* Gulf of Masira. *bay* E Oman
Masis *see* Büyükağrı Dağı
Maskat *see* Masqaţ
Mason City *23 F3* Iowa, C USA
Masqaţ *99 E5 var.* Maskat, *Eng.* Muscat. *country capital* (Oman) NE Oman
Massa *74 B3* Toscana, C Italy
Massachusetts *19 G3 off.* Commonwealth of Massachusetts, *also known as* Bay State, Old Bay State, Old Colony State. *state* NE USA
Massenya *54 B3* Chari-Baguirmi, SW Chad
Massif Central *69 C5 plateau* C France
Massilia *see* Marseille
Massoukou *see* Franceville
Mastanli *see* Momchilgrad
Masterton *129 D5* Wellington, North Island, New Zealand
Masty *85 B5 Rus.* Mosty. Hrodzyenskaya Voblasts', W Belarus
Masuda *109 B6* Shimane, Honshū, SW Japan
Masuku *see* Franceville
Masvingo *56 D3 prev.* Fort Victoria, Nyanda, Victoria. Masvingo, SE Zimbabwe
Maşyāf *96 B3 Fr.* Misiaf. Ḩamāh, C Syria
Matadi *55 B6* Bas-Congo, W Dem. Rep. Congo
Matagalpa *30 D3* Matagalpa, C Nicaragua
Matale *110 D3* Central Province, C Sri Lanka
Matam *52 C3* NE Senegal
Matamata *128 D3* Waikato, North Island, New Zealand
Matamoros *28 D3* Coahuila, NE Mexico
Matamoros *29 E2* Tamaulipas, C Mexico
Matane *17 E4* Québec, SE Canada
Matanzas *32 B2* Matanzas, NW Cuba
Matara *110 D4* Southern Province, S Sri Lanka
Mataram *116 D5* Pulau Lombok, C Indonesia
Mataró *71 G2 anc.* Illuro. Cataluña, E Spain
Mataura *129 B7* Southland, South Island, New Zealand
Mataura *129 B7 river* South Island, New Zealand
Mata Uta *see* Matā'utu
Matā'utu *123 E4 var.* Mata Uta. *dependent territory capital* (Wallis and Futuna) Île Uvea, Wallis and Futuna
Matera *75 E5* Basilicata, S Italy
Mathurai *see* Madurai
Matianus *see* Orūmīyeh, Daryācheh-ye
Matías Romero *29 F5* Oaxaca, SE Mexico
Matisco/Matisco Ædourum *see* Mâcon
Mato Grosso *41 E3 off.* Estado de Mato Grosso; *prev.* Matto Grosso. *state/region* W Brazil
Mato Grosso do Sul *41 E4 off.* Estado de Mato Grosso do Sul. *state/region* S Brazil
Mato Grosso do Sul, Estado de *see* Mato Grosso do Sul
Mato Grosso, Estado de *see* Mato Grosso
Mato Grosso, Planalto de *34 C4 plateau* C Brazil
Matosinhos *70 B2 prev.* Matozinhos. Porto, NW Portugal
Matozinhos *see* Matosinhos
Matsue *109 B6 var.* Matsuye, Matue. Shimane, Honshū, SW Japan
Matsumoto *109 C5 var.* Matumoto. Nagano, Honshū, S Japan
Matsuyama *109 B7 var.* Matuyama. Ehime, Shikoku, SW Japan
Matsuye *see* Matsue
Matterhorn *73 A8 It.* Monte Cervino. *mountain* Italy/Switzerland
Matthews Ridge *37 F2* N Guyana
Matthew Town *32 D2* Great Inagua, S The Bahamas
Matto Grosso *see* Mato Grosso
Matucana *38 C4* Lima, W Peru
Matue *see* Matsue
Matumoto *see* Matsumoto
Maturín *37 E2* Monagas, NE Venezuela
Matuyama *see* Matsuyama
Mau *113 E3 var.* Maunāth Bhanjan. Uttar Pradesh, N India
Maui *25 B8 island* Hawaii, USA, C Pacific Ocean
Maun *56 C3* North-West, C Botswana
Maunāth Bhanjan *see* Mau
Mauren *72 E1* NE Liechtenstein Europe
Maurice *see* Mauritius
Mauritania *52 C2 off.* Islamic Republic of Mauritania, *Ar.* Mūrītānīyah. *country* W Africa
Mauritania, Islamic Republic of *see* Mauritania
Mauritius *57 H3 off.* Republic of Mauritius, *Fr.* Maurice. *country* W Indian Ocean
Mauritius *119 B5 island* W Indian Ocean
Mauritius, Republic of *see* Mauritius
Mawlamyaing *see* Mawlamyine
Mawlamyine *114 B4 var.* Mawlamyaing, Moulmein. Mon State, S Myanmar (Burma)
Mawson *132 D2 Australian research station* Antarctica
Mayadin *see* Al Mayādīn
Mayaguana *32 D2 island* SE The Bahamas
Mayaguana Passage *32 D2 passage* SE The Bahamas
Mayagüez *33 F3* W Puerto Rico
Mayamey *98 D2* Semnān, N Iran
Maya Mountains *30 B2 Sp.* Montañas Mayas. *mountain range* Belize/Guatemala
Mayas, Montañas *see* Maya Mountains
Maych'ew *50 C4 var.* Mai Chio, *It.* Mai Ceu. Tigray, N Ethiopia
Mayebashi *see* Maebashi
Mayence *see* Mainz
Mayfield *129 B6* Canterbury, South Island, New Zealand
Maykop *89 A7* Respublika Adygeya, SW Russia
Maymana *see* Maīmanah
Maymyo *see* Pyin-Oo-Lwin
Mayo *see* Maio
Mayor Island *128 D3 island* NE New Zealand
Mayor Pablo Lagerenza *see* Capitán Pablo Lagerenza
Mayotte *57 F2 French territorial collectivity* E Africa
May Pen *32 B5* C Jamaica
Mayyit, Al Baḩr al *see* Dead Sea
Mazabuka *56 D2* Southern, S Zambia
Mazaca *see* Kayseri
Mazagan *see* El-Jadida
Mazār-e Sharīf *101 E3 var.* Mazār-i Sharif. Balkh, N Afghanistan
Mazār-i Sharif *see* Mazār-e Sharīf
Mazatlán *28 C3* Sinaloa, C Mexico
Mažeikiai *84 B3* Telšiai, NW Lithuania
Mazirbe *84 C2* Talsi, NW Latvia
Mazra'a *see* Al Mazra'ah
Mazury *76 D3 physical region* NE Poland
Mazyr *85 C7 Rus.* Mozyr'. Homyel'skaya Voblasts', SE Belarus
Mbabane *56 D4 country capital* (Swaziland - administrative) NW Swaziland
Mbacké *see* Mbaké
Mbaïki *55 C5 var.* M'Baiki. Lobaye, SW Central African Republic
M'Baiki *see* Mbaïki
Mbaké *52 B3 var.* Mbacké. W Senegal
Mbala *56 D1 prev.* Abercorn. Northern, NE Zambia
Mbale *51 C6* E Uganda
Mbandaka *55 C5 prev.* Coquilhatville. Equateur, NW Dem. Rep. Congo
M'Banza Congo *56 B1 var.* Mbanza Congo; *prev.* São Salvador, São Salvador do Congo. Dem. Rep. Congo, NW Angola
Mbanza-Ngungu *55 B6* Bas-Congo, W Dem. Rep. Congo
Mbarara *51 B6* SW Uganda
Mbé *54 B4* Nord, N Cameroon
Mbeya *51 C7* Mbeya, SW Tanzania
Mbomou/M'Bomu/Mbomu *see* Bomu
Mbour *52 B3* W Senegal
Mbuji-Mayi *55 D7 prev.* Bakwanga. Kasai-Oriental, S Dem. Rep. Congo
McAlester *27 G2* Oklahoma, C USA
McAllen *27 G5* Texas, SW USA
McCamey *27 E3* Texas, SW USA
M'Clintock Channel *15 F2 channel* Nunavut, N Canada
McComb *20 B3* Mississippi, S USA
McCook *23 E4* Nebraska, C USA
McKean Island *123 E3 island* Phoenix Islands, C Kiribati
McKinley, Mount *see* Denali
McKinley Park *14 C3* Alaska, USA
McMinnville *24 B3* Oregon, NW USA
McMurdo *132 B4 US research station* Antarctica
McPherson *23 E5* Kansas, C USA
McPherson *see* Fort McPherson
Mdantsane *56 D5* Eastern Cape, SE South Africa
Mead, Lake *25 D6 reservoir* Arizona/Nevada, W USA
Mecca *see* Makkah
Mechelen *65 C5 Eng.* Mechlin, *Fr.* Malines. Antwerpen, C Belgium
Mechlin *see* Mechelen
Mecklenburger Bucht *72 C2 bay* N Germany
Mecsek *77 C7 mountain range* SW Hungary
Medan *116 B3* Sumatera, E Indonesia
Medeba *see* Ma'dabā
Medellín *36 B3* Antioquia, NW Colombia
Médenine *49 F2 var.* Madanīyīn. SE Tunisia
Medeshamstede *see* Peterborough
Medford *24 B4* Oregon, NW USA
Medgidia *86 D5* Constanţa, SE Romania
Medgyes *see* Mediaş
Mediaş *86 B4 Ger.* Mediasch, *Hung.* Medgyes. Sibiu, C Romania
Mediasch *see* Mediaş
Medicine Hat *15 F5* Alberta, SW Canada
Medina *see* Al Madīnah
Medinaceli *71 E2* Castilla-León, N Spain
Medina del Campo *70 D2* Castilla-León, N Spain
Medinat Israel *see* Israel
Mediolanum *see* Saintes, France
Mediolanum *see* Milano, Italy
Mediomatrica *see* Metz
Mediterranean Sea *80 D3 Fr.* Mer Méditerranée. *sea* Africa/Asia/Europe
Méditerranée, Mer *see* Mediterranean Sea
Médoc *69 B5 cultural region* SW France
Medvezh'yegorsk *88 B3* Respublika Kareliya, NW Russia
Meekatharra *125 B5* Western Australia
Meemu Atoll *see* Mulakatholhu
Meerssen *65 D6 var.* Mersen. Limburg, SE Netherlands
Meerut *112 D2* Uttar Pradesh, N India
Megáli Préspa, Límni *see* Prespa, Lake
Meghālaya *91 G3 cultural region* NE India
Mehdia *see* Mahdia
Meheso *see* Mī'ēso
Me Hka *see* Nmai Hka
Mehrīz *98 D3* Yazd, C Iran
Mehtar Lām *101 F4 var.* Mehtarlām, Meterlam, Methariam, Metharlam. Laghmān, E Afghanistan
Mehtarlām *see* Mehtar Lām
Meiktila *114 B3* Mandalay, C Myanmar (Burma)
Méjico *see* Mexico
Mejillones *42 B2* Antofagasta, N Chile
Mek'elē *50 C4 var.* Makale. Tigray, N Ethiopia
Mékhé *52 B3* NW Senegal
Mekong *102 D3 var.* Lan-ts'ang Chiang, *Cam.* Mékôngk, *Chin.* Lancang Jiang, *Lao.* Mènam Khong, *Th.* Mae Nam Khong, *Tib.* Dza Chu, *Vtn.* Sông Tiên Giang. *river* SE Asia
Mékôngk *see* Mekong
Mekong, Mouths of the *115 E6 delta* S Vietnam
Melaka *116 B3 var.* Malacca. Melaka, Peninsular Malaysia
Melaka, Selat *see* Malacca, Strait of
Melanesia *122 D3 island group* W Pacific Ocean
Melanesian Basin *120 C2 undersea basin* W Pacific Ocean
Melbourne *127 C7 state capital* Victoria, SE Australia
Melbourne *21 E4* Florida, SE USA
Meleda *see* Mljet
Melghir, Chott *49 E2 var.* Chott Melrhir. *salt lake* E Algeria
Melilla *58 B5 anc.* Rusaddir, Russadir. Melilla, Spain, N Africa
Melilla *48 D2 enclave* Spain, N Africa
Melita *15 F5* Manitoba, S Canada
Melita *see* Mljet
Melitene *see* Malatya
Melitopol' *87 F4* Zaporiz'ka Oblast', SE Ukraine
Melle *65 B5* Oost-Vlaanderen, NW Belgium
Mellerud *63 B6* Västra Götaland, S Sweden
Mellieħa *80 B5* E Malta
Mellizo Sur, Cerro *43 A7 mountain* S Chile
Melo *42 E4* Cerro Largo, NE Uruguay
Melodunum *see* Melun
Melrhir, Chott *see* Melghir, Chott
Melsungen *72 B4* Hessen, C Germany
Melun *68 C3 anc.* Melodunum. Seine-et-Marne, N France
Melville Bay/Melville Bugt *see* Qimusseriarsuaq
Melville Island *124 D2 island* Northern Territory, N Australia
Melville Island *15 E2 island* Parry Islands, Northwest Territories, NW Canada
Melville, Lake *17 F2 lake* Newfoundland and Labrador, E Canada
Melville Peninsula *15 G3 peninsula* Nunavut, NE Canada
Melville Sound *see* Viscount Melville Sound
Membidj *see* Manbij
Memel *see* Neman, NE Europe
Memel *see* Klaipeda, Lithuania
Memmingen *73 B6* Bayern, S Germany
Memphis *20 C1* Tennessee, S USA
Menaam *see* Menaldum
Menado *see* Manado
Ménaka *53 F3* Goa, E Mali
Menaldum *64 D1 Fris.* Menaam. Friesland, N Netherlands
Mènam Khong *see* Mekong
Mendaña Fracture Zone *131 F4 fracture zone* E Pacific Ocean
Mende *69 C5 anc.* Mimatum. Lozère, S France
Mendeleyev Ridge *133 B2 undersea ridge* Arctic Ocean
Mendocino Fracture Zone *130 D2 fracture zone* NE Pacific Ocean
Mendoza *42 B4* Mendoza, W Argentina
Menemen *94 A3* İzmir, W Turkey
Menengiyn Tal *105 F2 plain* E Mongolia
Menongue *56 B2 var.* Vila Serpa Pinto, *Port.* Serpa Pinto. Cuando Cubango, C Angola
Menorca *71 H3 Eng.* Minorca; *anc.* Balearis Minor. *island* Islas Baleares, Spain, W Mediterranean Sea
Mentawai, Kepulauan *116 A4 island group* W Indonesia
Meppel *64 D2* Drenthe, NE Netherlands
Meran *see* Merano
Merano *74 C1 Ger.* Meran. Trentino-Alto Adige, N Italy
Merca *see* Marka
Mercedes *42 D3* Corrientes, NE Argentina
Mercedes *42 D4* Soriano, SW Uruguay
Meredith, Lake *27 E1 reservoir* Texas, SW USA
Merefa *87 G2* Kharkivs'ka Oblast', E Ukraine
Mergate *see* Margate
Mergui *see* Myeik
Mergui Archipelago *see* Myeik Archipelago
Mérida *29 H3* Yucatán, SW Mexico
Mérida *70 C4 anc.* Augusta Emerita. Extremadura, W Spain
Mérida *36 C2* Mérida, W Venezuela
Meridian *20 C2* Mississippi, S USA
Mérignac *69 B5* Gironde, SW France
Merín, Laguna *see* Mirim Lagoon
Merkulovichi *see* Myerkulavichy
Merowe *50 B3* Northern, N Sudan
Merredin *125 B6* Western Australia
Mersen *see* Meerssen
Mersey *67 D5 river* NW England, United Kingdom
Mersin *94 C4 var.* İçel. İçel, S Turkey
Mērsrags *84 C3* Talsi, NW Latvia
Meru *51 C6* Eastern, C Kenya
Merv *see* Mary
Merzifon *94 D2* Amasya, N Turkey
Merzig *73 A5* Saarland, SW Germany
Mesa *26 B2* Arizona, SW USA
Meseritz *see* Międzyrzecz
Meshed *see* Mashhad
Mesopotamia *35 C5 var.* Mesopotamia Argentina. *physical region* NE Argentina
Mesopotamia Argentina *see* Mesopotamia
Messalo, Rio *57 E2 var.* Mualo. *river* NE Mozambique
Messana/Messene *see* Messina
Messina *see* Musina
Messina, Strait of *see* Messina, Stretto di
Messina, Stretto di *75 D7 Eng.* Strait of Messina. *strait* SW Italy
Messíni *83 B6* Pelopónnisos, S Greece
Mesta *see* Néstos
Mestghanem *see* Mostaganem
Mestia *95 F1 var.* Mestiya. N Georgia
Mestiya *see* Mestia
Mestre *74 C2* Veneto, NE Italy
Metairie *20 B3* Louisiana, S USA
Metán *42 C2* Salta, N Argentina
Metapán *30 B2* Santa Ana, NW El Salvador
Meta, Río *36 D3 river* Colombia/Venezuela
Meterlam *see* Mehtar Lām
Methariam/Metharlam *see* Mehtar Lām
Metis *see* Metz
Metković *78 B4* Dubrovnik-Neretva, SE Croatia
Métsovo *82 B4 prev.* Métsovon. Ípeiros, C Greece
Métsovon *see* Métsovo
Metz *68 D3 anc.* Divodurum Mediomatricum, Mediomatrica, Metis. Moselle, NE France
Meulaboh *116 A3* Sumatera, W Indonesia
Meuse *65 C6 Dut.* Maas. *river* W Europe
Mexcala, Río *see* Balsas, Río
Mexicali *28 A1* Baja California Norte, NW Mexico
Mexicanos, Estados Unidos *see* Mexico
México *29 E4 var.* Ciudad de México, *Eng.* Mexico City. *country capital* (Mexico) México, C Mexico
Mexico *23 G4* Missouri, C USA
Mexico *28 C3 off.* United Mexican States, *var.* Méjico, México, *Sp.* Estados Unidos Mexicanos. *country* N Central America
México *see* Mexico
Mexico City *see* México
México, Golfo de *see* Mexico, Gulf of
Mexico, Gulf of *29 F2 Sp.* Golfo de México. *gulf* W Atlantic Ocean
Meyadine *see* Al Mayādīn
Meydān Shahr *see* Maidān Shahr
Meymaneh *see* Maīmanah
Mezen' *88 D3 river* NW Russia
Mezőtúr *77 D7* Jász-Nagykun-Szolnok, E Hungary
Mġarr *80 A5* Gozo, N Malta
Miadzioł Nowy *see* Myadzyel
Miahuatlán *29 F5 var.* Miahuatlán de Porfirio Díaz. Oaxaca, SE Mexico
Miahuatlán de Porfirio Díaz *see* Miahuatlán
Miami *21 F5* Florida, SE USA
Miami *27 G1* Oklahoma, C USA
Miami Beach *21 F5* Florida, SE USA
Miāneh *98 C2 var.* Miyāneh. Āżarbāyjān-e Sharqī, NW Iran
Mianyang *106 B5* Sichuan, C China
Miastko *76 C2 Ger.* Rummelsburg in Pommern. Pomorskie, N Poland
Mi Chai *see* Nong Khai
Michalovce *77 E5 Ger.* Grossmichel, *Hung.* Nagymihály. Košický Kraj, E Slovakia
Michigan *18 C1 off.* State of Michigan, *also known as* Great Lakes State, Lake State, Wolverine State. *state* N USA
Michigan, Lake *18 C2 lake* N USA
Michurin *see* Tsarevo
Michurinsk *89 B5* Tambovskaya Oblast', W Russia
Micoud *33 F2* SE Saint Lucia
Micronesia *122 B1 off.* Federated States of Micronesia. *country* W Pacific Ocean
Micronesia *122 C1 island group* W Pacific Ocean
Micronesia, Federated States of *see* Micronesia
Mid-Atlantic Cordillera *see* Mid-Atlantic Ridge
Mid-Atlantic Ridge *44 C3 var.* Mid-Atlantic Cordillera, Mid-Atlantic Rise, Mid-Atlantic Swell. *undersea ridge* Atlantic Ocean
Mid-Atlantic Rise/Mid-Atlantic Swell *see* Mid-Atlantic Ridge
Middelburg *65 B5* Zeeland, SW Netherlands
Middelharnis *64 B4* Zuid-Holland, SW Netherlands
Middelkerke *65 A5* West-Vlaanderen, W Belgium
Middle America Trench *13 B7 trench* E Pacific Ocean
Middle Andaman *111 F2 island* Andaman Islands, India, NE Indian Ocean
Middle Atlas *48 C2 Eng.* Middle Atlas. *mountain range* N Morocco
Middle Atlas *see* Moyen Atlas
Middleburg Island *see* 'Eua
Middle Congo *see* Congo (Republic of)
Middlesboro *18 C5* Kentucky, S USA
Middlesbrough *67 D5* N England, United Kingdom
Middletown *19 F4* New Jersey, NE USA
Middletown *19 F3* New York, NE USA
Mid-Indian Basin *119 C5 undersea basin* N Indian Ocean
Mid-Indian Ridge *119 C5 var.* Central Indian Ridge. *undersea ridge* C Indian Ocean
Midland *16 D5* Ontario, S Canada
Midland *18 C3* Michigan, N USA
Midland *27 E3* Texas, SW USA
Mid-Pacific Mountains *130 C2 var.* Mid-Pacific Seamounts. *seamount range* NW Pacific Ocean
Mid-Pacific Seamounts *see* Mid-Pacific Mountains
Midway Islands *130 D2 US unincorporated territory* C Pacific Ocean
Miechów *77 D5* Małopolskie, S Poland
Międzyrzec Podlaski *76 E3* Lubelskie, E Poland
Międzyrzecz *76 B3 Ger.* Meseritz. Lubuskie, W Poland
Mielec *77 D5* Podkarpackie, SE Poland
Miercurea-Ciuc *86 C4 Ger.* Szeklerburg, *Hung.* Csíkszereda. Harghita, C Romania
Mieres del Camín *70 D1 var.* Mieres del Camino. Asturias, NW Spain
Mieres del Camino *see* Mieres del Camín
Mī'ēso *51 D5 var.* Meheso, Miesso. Oromīya, C Ethiopia
Miesso *see* Mī'ēso
Mifrats Hefa *97 A5 Eng.* Bay of Haifa; *prev.* MifrazHefa. *bay* N Israel
Miguel Asua *28 D3 var.* Miguel Auza. Zacatecas, C Mexico
Miguel Auza *see* Miguel Asua
Mijdrecht *64 C3* Utrecht, C Netherlands
Mikashevichi *see* Mikashevichy
Mikashevichy *85 C7 Pol.* Mikaszewicze, *Rus.* Mikashevichi. Brestskaya Voblasts', SW Belarus
Mikaszewicze *see* Mikashevichy
Mikhaylovgrad *see* Montana
Mikhaylovka *89 B6* Volgogradskaya Oblast', SW Russia
Míkonos *see* Mýkonos
Mikre *82 C2* Lovech, N Bulgaria
Mikun' *88 D4* Respublika Komi, NW Russia
Mikuni-sanmyaku *109 D5 mountain range* Honshū, N Japan Asia
Mikura-jima *109 D6 island* E Japan
Milagro *38 B2* Guayas, SW Ecuador
Milan *see* Milano
Milange *57 E2* Zambézia, NE Mozambique
Milano *74 B2 Eng.* Milan, *Ger.* Mailand; *anc.* Mediolanum. Lombardia, N Italy
Milas *94 A4* Muğla, SW Turkey
Milashavichy *85 C7 Rus.* Milashevichi. Homyel'skaya Voblasts', SE Belarus
Milashevichi *see* Milashavichy
Mildura *127 C6* Victoria, SE Australia
Mile *see* Mili Atoll
Miles *127 D5* Queensland, E Australia
Miles City *22 C2* Montana, NW USA
Milford *see* Milford Haven
Milford Haven *67 C6 prev.* Milford. SW Wales, United Kingdom
Milford Sound *129 A6* Southland, South Island, New Zealand
Milford Sound *129 A6 inlet* South Island, New Zealand
Milḩ, Baḩr al *see* Razāzah, Buḩayrat ar
Mili Atoll *122 D2 var.* Mile. *atoll* Ratak Chain, SE Marshall Islands
Mil'kovo *93 H3* Kamchatskaya Oblast', E Russia
Milk River *15 E5* Alberta, SW Canada
Milk River *22 C1 river* Montana, NW USA
Milk, Wadi el *66 B4 var.* Wadi al Malik. *river* C Sudan
Milledgeville *21 E2* Georgia, SE USA
Mille Lacs Lake *23 F2 lake* Minnesota, N USA
Millennium Island *121 F3 prev.* Caroline Island, Thornton Island. *atoll* Line Islands, E Kiribati
Millerovo *89 B6* Rostovskaya Oblast', SW Russia
Mílos *83 C7 island* Kykládes, Greece, Aegean Sea
Milton *129 B7* Otago, South Island, New Zealand
Milton Keynes *67 D6* SE England, United Kingdom
Milwaukee *18 B3* Wisconsin, N USA
Mimatum *see* Mende
Min *see* Fujian
Mīnā' Qābūs *118 B3* NE Oman
Minas Gerais *41 F3 off.* Estado de Minas Gerais. *state/region* E Brazil
Minas Gerais, Estado de *see* Minas Gerais
Minatitlán *29 F4* Veracruz-Llave, E Mexico
Minbu *114 A3* Magway, W Myanmar (Burma)
Minch, The *66 B3 var.* North Minch. *strait* NW Scotland, United Kingdom
Mindanao *117 F2 island* S Philippines
Mindanao Sea *see* Bohol Sea
Mindelheim *73 C6* Bayern, S Germany
Mindello *see* Mindelo
Mindelo *52 A2 var.* Mindello; *prev.* Porto Grande. São Vicente, N Cape Verde
Minden *72 B4 anc.* Minthun. Nordrhein-Westfalen, NW Germany
Mindoro *117 E2 island* N Philippines
Mindoro Strait *117 E2 strait* W Philippines
Mineral Wells *27 F2* Texas, SW USA
Mingäçevir *95 G2 Rus.* Mingechaur, Mingechevir. C Azerbaijan
Mingechaur/Mingechevir *see* Mingäçevir
Mingora *see* Saidu Sharif
Minho *70 B2 former province* N Portugal
Minho *70 B2 Sp.* Miño. *river* Portugal/Spain
Minho, Rio *see* Miño
Minicoy Island *110 B3 island* SW India
Minius *see* Miño
Minna *53 G4* Niger, C Nigeria
Minneapolis *23 F2* Minnesota, N USA
Minnesota *23 F2 off.* State of Minnesota, *also known as* Gopher State, New England of the West, North Star State. *state* N USA
Miño *70 B2 var.* Mino, Minius, *Port.* Rio Minho. *river* Portugal/Spain
Miño *see* Minho, Rio
Minorca *see* Menorca
Minot *23 E1* North Dakota, N USA
Minsk *85 C6 country capital* (Belarus) Minskaya Voblasts', C Belarus
Minskaya Wzvyshsha *85 C6 mountain range* C Belarus
Mińsk Mazowiecki *76 D3 var.* Nowo-Minsk. Mazowieckie, C Poland
Minthun *see* Minden
Minto, Lac *16 D2 lake* Québec, C Canada
Minya *see* Al Minyā
Miraflores *28 C3* Baja California Sur, NW Mexico
Miranda de Ebro *71 E1* La Rioja, N Spain
Mirgorod *see* Myrhorod
Miri *116 D3* Sarawak, East Malaysia
Mirim Lagoon *41 E5 var.* Lake Mirim, *Sp.* Laguna Merín. *lagoon* Brazil/Uruguay
Mirim, Lake *see* Mirim Lagoon
Mírina *see* Mýrina
Mīrjāveh *98 E4* Sīstān va Balūchestān, SE Iran
Mirny *132 C3 Russian research station* Antarctica
Mirnyy *93 F3* Respublika Sakha (Yakutiya), NE Russia
Mirpur Khās *112 B3* Sind, SE Pakistan
Mirtoan Sea *see* Mirtóo Pélagos
Mirtóo Pélagos *83 C6 Eng.* Mirtoan Sea; *anc.* Myrtoum Mare. *sea* S Greece
Misiaf *see* Maşyāf
Miskito Coast *see* Mosquito Coast
Miskitos, Cayos *31 E2 island group* NE Nicaragua
Miskolc *77 D6* Borsod-Abaúj-Zemplén, NE Hungary
Misool, Pulau *117 F4 island* Maluku, E Indonesia
Mişrātah *49 F2 var.* Misurata. NW Libya
Mission *27 G5* Texas, SW USA
Mississippi *20 B2 off.* State of Mississippi, *also known as* Bayou State, Magnolia State. *state* SE USA
Mississippi Delta *20 B4 delta* Louisiana, S USA
Mississippi River *13 C6 river* C USA
Missoula *22 B1* Montana, NW USA
Missouri *23 F5 off.* State of Missouri, *also known as* Bullion State, Show Me State. *state* C USA
Missouri River *23 E3 river* C USA
Mistassini, Lac *16 D3 lake* Québec, SE Canada
Mistelbach an der Zaya *73 E6* Niederösterreich, NE Austria
Misti, Volcán *39 E4 volcano* S Peru

Misurata – Nablus

Misurata *see* Mişrātah
Mitau *see* Jelgava
Mitchell *127 D5* Queensland, E Australia
Mitchell *23 E3* South Dakota, N USA
Mitchell, Mount *21 E1 mountain* North Carolina, SE USA
Mitchell River *126 C2 river* Queensland, NE Australia
Mi Tho *see* My Tho
Mitilíni *see* Mytilíni
Mito *109 D5* Ibaraki, Honshū, S Japan
Mitrovica *see* Mitrovicë
Mitrovica/Mitrowitz *see* Sremska Mitrovica, Serbia
Mitrovicë *79 D5 Serb.* Mitrovica, *prev.* Kosovska Mitrovica, Titova Mitrovica. N Kosovo
Mits'iwa *50 C4 var.* Masawa, Massawa. E Eritrea
Mitspe Ramon *97 A7 prev.* Mizpe Ramon. Southern, S Israel
Mittelstadt *see* Baia Sprie
Mitú *36 C4* Vaupés, SE Colombia
Mitumba, Chaîne des/Mitumba Range *see* Mitumba, Monts
Mitumba Monts *55 E7 var.* Chaîne des Mitumba, Mitumba Range. *mountain range* E Dem. Rep. Congo
Miueru Wantipa, Lake *55 E7 lake* N Zambia
Miyake-jima *109 D6 island* Sakishima-shotō, SW Japan
Miyako *108 D4* Iwate, Honshū, C Japan
Miyakonojō *109 B8 var.* Miyakonzyô. Miyazaki, Kyūshū, SW Japan
Miyakonzyô *see* Miyakonojō
Miyāneh *see* Mīāneh
Miyazaki *109 B8* Miyazaki, Kyūshū, SW Japan
Mizil *86 C5* Prahova, SE Romania
Miziya *82 C1* Vratsa, NW Bulgaria
Mizpe Ramon *see* Mitspe Ramon
Mjøsa *63 B6 var.* Mjøsen. *lake* S Norway
Mjøsen *see* Mjøsa
Mladenovac *78 D4* Serbia, C Serbia
Mława *76 D3* Mazowieckie, C Poland
Mljet *79 B5 It.* Meleda; *anc.* Melita. *island* S Croatia
Mmabatho *56 C4* North-West, N South Africa
Moab *22 B5* Utah, W USA
Moa Island *126 C1 island* Queensland, NE Australia
Moanda *55 B6 var.* Mouanda. Haut-Ogooué, SE Gabon
Moba *55 E7* Katanga, E Dem. Rep. Congo
Mobay *see* Montego Bay
Mobaye *55 C5* Basse-Kotto, S Central African Republic
Moberly *23 G4* Missouri, C USA
Mobile *20 C3* Alabama, S USA
Mobutu Sese Seko, Lac *see* Albert, Lake
Moçâmedes *see* Namibe
Mochudi *56 C4* Kgatleng, SE Botswana
Mocímboa da Praia *57 F2 var.* Vila de Mocímboa da Praia. Cabo Delgado, N Mozambique
Môco *56 B2 var.* Morro de Môco. *mountain* W Angola
Mocoa *36 A4* Putumayo, SW Colombia
Môco, Morro de *see* Môco
Mocuba *57 E3* Zambézia, NE Mozambique
Modena *74 B3 anc.* Mutina. Emilia-Romagna, N Italy
Modesto *25 B6* California, W USA
Modica *75 C7 anc.* Motyca. Sicilia, Italy, C Mediterranean Sea
Modimolle *56 D4 prev.* Nylstroom. Limpopo, NE South Africa
Modohn *see* Madona
Modriča *78 C3* Republika Srpska, N Bosnia and Herzegovina
Moe *127 C7* Victoria, SE Australia
Möen *see* Møn, Denmark
Moero, Lac *see* Mweru, Lake
Moeskroen *see* Mouscron
Mogadiscio/Mogadishu *see* Muqdisho
Mogador *see* Essaouira
Mogilëv *see* Mahilyow
Mogilev-Podol'skiy *see* Mohyliv-Podil's'kyy
Mogilno *76 C3* Kujawsko-pomorskie, C Poland
Moḥammadābād-e Rīgān *98 E4* Kermān, SE Iran
Mohammedia *48 C2 prev.* Fédala. NW Morocco
Mohave, Lake *25 D7 reservoir* Arizona/Nevada, W USA
Mohawk River *19 F3 river* New York, NE USA
Mohéli *see* Mwali
Mohns Ridge *61 F3 undersea ridge* Greenland Sea/Norwegian Sea
Moho *39 E4* Puno, SE Peru
Mohoro *51 C7* Pwani, E Tanzania
Mohyliv-Podil's'kyy *86 D3 Rus.* Mogilev-Podol'skiy. Vinnyts'ka Oblast', C Ukraine
Moi *63 A6* Rogaland, S Norway
Moili *see* Mwali
Mo i Rana *62 C3* Nordland, C Norway
Mõisaküla *84 D3 Ger.* Moiseküll. Viljandimaa, S Estonia
Moiseküll *see* Mõisaküla
Moissac *69 B6* Tarn-et-Garonne, S France
Mojácar *71 E5* Andalucía, S Spain
Mojave Desert *25 D7 plain* California, W USA
Mokrany *see* Makrany
Moktama *see* Mottama
Mol *65 C5 prev.* Moll. Antwerpen, N Belgium
Moldavia *see* Moldova
Moldavian SSR/Moldavskaya SSR *see* Moldova
Molde *63 A5* Møre og Romsdal, S Norway
Moldotau, Khrebet *see* Moldo-Too, Khrebet
Moldo-Too, Khrebet *101 G2 prev.* Khrebet Moldotau. *mountain range* C Kyrgyzstan
Moldova *86 D3 off.* Republic of Moldova, *var.* Moldavia; *prev.* Moldavian SSR, *Rus.* Moldavskaya SSR. *country* SE Europe
Moldova Nouă *86 A4 Ger.* Neumoldowa, *Hung.* Újmoldova. Caraş-Severin, SW Romania
Moldova, Republic of *see* Moldova
Moldoveanul *see* Vârful Moldoveanu
Molfetta *75 E5* Puglia, SE Italy
Moll *see* Mol
Mollendo *39 E4* Arequipa, SW Peru
Mölndal *63 B7* Västra Götaland, S Sweden
Molochans'k *87 G4 Rus.* Molochansk. Zaporiz'ka Oblast', SE Ukraine
Molodechno/Molodeczno *see* Maladzyechna
Moloka'i *25 B8 var.* Molokai. *island* Hawai'ian Islands, Hawaii, USA
Molokai Fracture Zone *131 E2 tectonic feature* NE Pacific Ocean
Molopo *56 C4 seasonal river* Botswana/South Africa
Mólos *83 B5* Stereá Elláś, C Greece
Molotov *see* Severodvinsk, Arkhangel'skaya Oblast', Russia
Molotov *see* Perm', Permskaya Oblast', Russia
Moluccas *see* Maluku
Molucca Sea *117 F4 Ind.* Laut Maluku. *sea* E Indonesia
Molukken *see* Maluku
Mombasa *51 D7* Coast, SE Kenya
Mombetsu *see* Monbetsu
Momchilgrad *82 D3 prev.* Mastanli. Kŭrdzhali, S Bulgaria
Møn *63 B8 prev.* Möen. *island* SE Denmark
Mona, Canal de la *see* Mona Passage
Monaco *69 C7 var.* Monaco-Ville; *anc.* Monoecus. *country capital* (Monaco) S Monaco
Monaco *69 E6 off.* Principality of Monaco. *country* W Europe
Monaco *see* München
Monaco, Port de *69 C8 bay* S Monaco W Mediterranean Sea
Monaco, Principality of *see* Monaco
Monaco-Ville *see* Monaco
Monahans *27 E3* Texas, SW USA
Mona, Isla *33 E3 island* W Puerto Rico
Mona Passage *33 E3 Sp.* Canal de la Mona. *channel* Dominican Republic/Puerto Rico
Monastir *see* Bitola
Monbetsu *108 D2 var.* Mombetsu, Monbetu. Hokkaidō, NE Japan
Monbetu *see* Monbetsu
Moncalieri *74 A2* Piemonte, NW Italy
Monchegorsk *88 C2* Murmanskaya Oblast', NW Russia
Monclova *28 D2* Coahuila, NE Mexico
Moncton *17 F4* New Brunswick, SE Canada
Mondovì *74 A2* Piemonte, NW Italy
Monfalcone *74 D2* Friuli-Venezia Giulia, NE Italy
Monforte de Lemos *70 C1* Galicia, NW Spain
Mongo *54 C3* Guéra, C Chad
Mongolia *104 C2 Mong.* Mongol Uls. *country* E Asia
Mongolia, Plateau of *102 D1 plateau* E Mongolia
Mongol Uls *see* Mongolia
Mongora *see* Saidu Sharif
Mongos, Chaîne des *see* Bongo, Massif des
Mongu *56 C2* Western, W Zambia
Monkchester *see* Newcastle upon Tyne
Monkey Bay *57 E2* Southern, SE Malawi
Monkey River *see* Monkey River Town
Monkey River Town *30 C2 var.* Monkey River. Toledo, SE Belize
Monoecus *see* Monaco
Mono Lake *25 C6 lake* California, W USA
Monostor *see* Beli Manastir
Monóvar *71 F4 Cat.* Monòver. País Valenciano, E Spain
Monover *see* Monovar
Monroe *20 B2* Louisiana, S USA
Monrovia *52 C5 country capital* (Liberia) W Liberia
Mons *65 B6 Dut.* Bergen. Hainaut, S Belgium
Monselice *74 C2* Veneto, NE Italy
Montana *82 C2 prev.* Ferdinand, Mikhaylovgrad. Montana, NW Bulgaria
Montana *22 B1 off.* State of Montana, *also known as* Mountain State, Treasure State. *state* NW USA
Montargis *68 C4* Loiret, C France
Montauban *69 B6* Tarn-et-Garonne, S France
Montbéliard *68 D4* Doubs, E France
Mont Cenis, Col du *69 D5 pass* E France
Mont-de-Marsan *69 B6* Landes, SW France
Monteagudo *39 G4* Chuquisaca, S Bolivia
Montecarlo *69 C8* Misiones, NE Argentina
Monte Caseros *42 D3* Corrientes, NE Argentina
Monte Cristi *32 D3 var.* San Fernando de Monte Cristi. NW Dominican Republic
Monte Croce Carnico, Passo di *see* Plöcken Pass
Montegiardino *74 E2* SE San Marino
Montego Bay *32 A4 var.* Mobay. W Jamaica
Montélimar *69 D5 anc.* Acunum Acusio, Montilium Adhemari. Drôme, E France
Montemorelos *29 E3* Nuevo León, NE Mexico
Montenegro *79 C5 Serb.* Crna Gora. *country* SW Europe
Monte Patria *42 B3* Coquimbo, N Chile
Monterey *25 B6* California, W USA
Monterey *see* Monterrey
Monterey Bay *25 A6 bay* California, W USA
Montería *36 B2* Córdoba, NW Colombia
Montero *39 G4* Santa Cruz, C Bolivia
Monterrey *29 E3 var.* Monterey. Nuevo León, NE Mexico
Montes Claros *41 F3* Minas Gerais, SE Brazil
Montevideo *42 D4 country capital* (Uruguay) Montevideo, S Uruguay
Montevideo *23 F2* Minnesota, N USA
Montgenèvre, Col de *69 D5 pass* France/Italy
Montgomery *20 D2 state capital* Alabama, S USA
Montgomery *see* Sāhīwāl
Monthey *73 A7* Valais, SW Switzerland
Montilium Adhemari *see* Montélimar
Montluçon *68 C4* Allier, C France
Montoro *70 D4* Andalucía, S Spain
Montpelier *19 G2 state capital* Vermont, NE USA
Montpellier *69 C6* Hérault, S France
Montréal *17 E4 Eng.* Montreal. Québec, SE Canada
Montrose *66 D3* E Scotland, United Kingdom
Montrose *22 C5* Colorado, C USA
Montserrat *33 G3 var.* Emerald Isle. *UK dependent territory* E West Indies
Monywa *114 B3* Sagaing, C Myanmar (Burma)
Monza *74 B2* Lombardia, N Italy
Monze *56 D2* Southern, S Zambia
Monzón *71 F2* Aragón, NE Spain
Moonie *127 D5* Queensland, E Australia
Moon-Sund *see* Väinameri
Moora *125 A6* Western Australia
Moore *27 G1* Oklahoma, C USA
Moore, Lake *125 B6 lake* Western Australia
Moorhead *23 F2* Minnesota, N USA
Moose *16 C3 river* Ontario, S Canada
Moosehead Lake *19 G1 lake* Maine, NE USA
Moosonee *16 C3* Ontario, SE Canada
Mopti *53 E3* Mopti, C Mali
Moquegua *39 E4* Moquegua, SE Peru
Mora *63 C5* Dalarna, C Sweden
Morales *30 C2* Izabal, E Guatemala
Morant Bay *32 B5* E Jamaica
Moratalla *71 E4* Murcia, SE Spain
Morava *77 C5 var.* March. *river* C Europe
Morava *see* Moravia, Czech Republic (Czechia)
Morava *see* Velika Morava, Serbia
Moravia *77 B5 Cz.* Morava, *Ger.* Mähren. *cultural region* E Czech Republic (Czechia)
Moray Firth *66 C3 inlet* N Scotland, United Kingdom
Morea *see* Pelopónnisos
Moreau River *22 D2 river* South Dakota, N USA
Moree *127 D5* New South Wales, SE Australia
Morelia *29 E4* Michoacán, S Mexico
Morena, Sierra *70 C4 mountain range* S Spain
Moreni *86 C5* Dâmboviţa, S Romania
Morgan City *20 B3* Louisiana, S USA
Morghab, Darya-ye *100 D3 Rus.* Murgab, Murghab, *Turkm.* Murgap, Murgap Deryasy. *river* Afghanistan/Turkmenistan
Morioka *108 D4* Iwate, Honshū, C Japan
Morlaix *68 A3* Finistère, NW France
Mormon State *see* Utah
Mornington Abyssal Plain *45 A7 abyssal plain* SE Pacific Ocean
Mornington Island *126 B2 island* Wellesley Islands, Queensland, N Australia
Morocco *48 B3 off.* Kingdom of Morocco, *Ar.* Al Mamlakah. *country* N Africa
Morocco *see* Marrakech
Morocco, Kingdom of *see* Morocco
Morogoro *51 C7* Morogoro, E Tanzania
Moro Gulf *117 E3 gulf* S Philippines
Morón *32 C2* Ciego de Ávila, C Cuba
Mörön *104 D2* Hövsgöl, N Mongolia
Morondava *57 F3* Toliara, W Madagascar
Moroni *57 F2 country capital* (Comoros) Grande Comore, NW Comoros
Morotai, Pulau *117 F3 island* Maluku, E Indonesia
Morotiri *see* Marotiri
Morphou *see* Güzelyurt
Morrinsville *128 D3* Waikato, North Island, New Zealand
Morris *23 F2* Minnesota, N USA
Morris Jesup, Kap *61 E1 headland* N Greenland
Morvan *68 D4 physical region* C France
Moscow *24 C2* Idaho, NW USA
Moscow *see* Moskva
Mosel *73 A5 Fr.* Moselle. *river* W Europe
Mosel *see* Moselle
Moselle *65 E8 Ger.* Mosel. *river* W Europe
Moselle *see* Mosel
Mosgiel *129 B7* Otago, South Island, New Zealand
Moshi *51 C7* Kilimanjaro, NE Tanzania
Mosjøen *62 B4* Nordland, C Norway
Moskovskiy *see* Moskva
Moskva *89 B5 Eng.* Moscow. *country capital* (Russia) Gorod Moskva, W Russia
Moskva *101 E3 Rus.* Moskovskiy; *prev.* Chubek. SW Tajikistan
Moson and Magyaróvár *see* Mosonmagyaróvár
Mosonmagyaróvár *77 C6 Ger.* Wieselburg-Ungarisch-Altenburg; *prev.* Moson and Magyaróvár, *Ger.* Wieselburg and Ungarisch-Altenburg. Győr-Moson-Sopron, NW Hungary
Mosquito Coast *31 E3 var.* Miskito Coast, *Eng.* Mosquito Coast. *coastal region* E Nicaragua
Mosquito Coast *see* La Mosquitia
Mosquito Gulf *31 F4 Eng.* Mosquito Gulf. *gulf* N Panama
Mosquito Gulf *see* Mosquitos, Golfo de los
Moss *63 B6* Østfold, S Norway
Mossâmedes *see* Namibe
Mosselbaai *56 C5 var.* Mosselbai, *Eng.* Mossel Bay. Western Cape, SW South Africa
Mosselbai/Mossel Bay *see* Mosselbaai
Mossendjo *55 B6* Niari, SW Congo
Mossoró *41 G2* Rio Grande do Norte, NE Brazil
Most *76 A4 Ger.* Brüx. Ústecký Kraj, NW Czech Republic (Czechia)
Mosta *80 B5 var.* Musta. C Malta
Mostaganem *48 D2 var.* Mestghanem. NW Algeria
Mostar *78 C4* Federacija Bosna I Hercegovina, S Bosnia and Herzegovina
Mosty *see* Masty
Mosul *see* Al Mawşil
Mota del Cuervo *71 E3* Castilla-La Mancha, C Spain
Motagua, Río *30 B2 river* Guatemala/Honduras
Mother of Presidents/Mother of States *see* Virginia
Motril *70 D5* Andalucía, S Spain
Motru *86 B4* Gorj, SW Romania
Mottama *114 B4 prev.* Martaban; *var.* Moktama. Mon State, S Myanmar (Burma)
Motueka *129 C5* Tasman, South Island, New Zealand
Motul *29 H3 var.* Motul de Felipe Carrillo Puerto. Yucatán, SE Mexico
Motul de Felipe Carrillo Puerto *see* Motul
Motyca *see* Modica
Mouanda *see* Moanda
Mouhoun *see* Black Volta
Mouila *55 A6* Ngounié, C Gabon
Moukden *see* Shenyang
Mould Bay *15 E2* Prince Patrick Island, Northwest Territories, N Canada
Moulins *68 C4* Allier, C France
Moulmein *see* Mawlamyine
Moundou *54 B4* Logone-Occidental, SW Chad
Moŭng Roessei *115 D5* Battambang, W Cambodia
Moun Hou *see* Black Volta
Mountain Home *20 B1* Arkansas, C USA
Mountain State *see* Montana
Mountain State *see* West Virginia
Mount Cook *129 B6* Canterbury, South Island, New Zealand
Mount Desert Island *19 H2 island* Maine, NE USA
Mount Gambier *127 B7* South Australia
Mount Isa *126 B3* Queensland, C Australia
Mount Magnet *125 B5* Western Australia
Mount Pleasant *23 G4* Iowa, C USA
Mount Pleasant *18 C3* Michigan, N USA
Mount Vernon *18 B5* Illinois, N USA
Mount Vernon *24 B1* Washington, NW USA
Mourdi, Dépression du *54 C2 desert lowland* Chad/Sudan
Mouscron *65 A6 Dut.* Moeskroen. Hainaut, W Belgium
Mouse River *see* Souris River
Moussoro *54 B3* Kanem, W Chad
Moyen-Congo *see* Congo (Republic of)
Mo'ynoq *100 C1 Rus.* Muynak. Qoraqalpog'iston Respublikasi, NW Uzbekistan
Moyobamba *38 B2* San Martín, NW Peru
Moyu *104 B3 var.* Karakax. Xinjiang Uygur Zizhiqu, NW China
Moyynkum, Peski *101 F1 Kaz.* Moyynqum. *desert* S Kazakhstan
Moyynqum *see* Moyynkum, Peski
Mozambika, Lakandranon' i *see* Mozambique Channel
Mozambique *57 E3 off.* Republic of Mozambique; *prev.* People's Republic of Mozambique, Portuguese East Africa. *country* S Africa
Mozambique Basin *see* Natal Basin
Mozambique, Canal de *see* Mozambique Channel
Mozambique Channel *57 E3 Fr.* Canal de Mozambique, *Mal.* Lakandranon' i Mozambika. *strait* W Indian Ocean
Mozambique, People's Republic of *see* Mozambique
Mozambique Plateau *47 D7 var.* Mozambique Rise. *undersea plateau* SW Indian Ocean
Mozambique, Republic of *see* Mozambique
Mozambique Rise *see* Mozambique Plateau
Mozyr' *see* Mazyr
Mpama *55 B6 river* C Congo
Mpika *56 D2* Northern, NE Zambia
Mqinvartsveri *see* Kazbek
Mrągowo *76 D2 Ger.* Sensburg. Warmińsko-Mazurskie, NE Poland
Mthatha *56 D5 prev.* Umtata. Eastern Cape, SE South Africa
Mtkvari *see* Kura
Mtwara *51 D8* Mtwara, SE Tanzania
Mualo *see* Messalo, Rio
Muang Chiang Rai *see* Chiang Rai
Muang Kalasin *see* Kalasin
Muang Khammouan *see* Thakhèk
Muang Không *115 D5* Champasak, S Laos
Muang Khôngxédôn *115 D5 var.* Khong Sedone. Salavan, S Laos
Muang Khon Kaen *see* Khon Kaen
Muang Lampang *see* Lampang
Muang Loei *see* Loei
Muang Lom Sak *see* Lom Sak
Muang Nakhon Sawan *see* Nakhon Sawan
Muang Namo *114 C3* Oudômxai, N Laos
Muang Nan *see* Nan
Muang Phalan *114 D4 var.* Muang Phalane. Savannakhét, S Laos
Muang Phalane *see* Muang Phalan
Muang Phayao *see* Phayao
Muang Phitsanulok *see* Phitsanulok
Muang Phrae *see* Phrae
Muang Roi Et *see* Roi Et
Muang Sakon Nakhon *see* Sakon Nakhon
Muang Samut Prakan *see* Samut Prakan
Muang Sing *114 C3* Louang Namtha, N Laos
Muang Ubon *see* Ubon Ratchathani
Muang Xaignabouri *see* Xaignabouli
Muar *116 B3 var.* Bandar Maharani. Johor, Peninsular Malaysia
Mucojo *57 F2* Cabo Delgado, N Mozambique
Mudanjiang *107 E3 var.* Mu-tan-chiang. Heilongjiang, NE China
Mudon *115 B5* Mon State, S Myanmar (Burma)
Muenchen *see* München
Muenster *see* Münster
Mufulira *56 D2* Copperbelt, C Zambia
Mughla *see* Muğla
Muğla *94 A4 var.* Mughla. Muğla, SW Turkey
Mūḩ, Sabkhat al *96 C3 lake* C Syria
Muhu Väin *see* Väinameri
Muisne *38 A1* Esmeraldas, NW Ecuador
Mukacheve *86 B3 Hung.* Munkács, *Rus.* Mukachevo. Zakarpats'ka Oblast', W Ukraine
Mukachevo *see* Mukacheve
Mukalla *see* Al Mukallā
Mukden *see* Shenyang
Mula *71 E4* Murcia, SE Spain
Mulakatholhu *110 B4 var.* Meemu Atoll, Mulaku Atoll. *atoll* C Maldives
Mulaku Atoll *see* Mulakatholhu
Muleshoe *27 E2* Texas, SW USA
Mulhacén *71 E5 var.* Cerro de Mulhacén. *mountain* S Spain
Mulhacén, Cerro de *see* Mulhacén
Mülhausen *see* Mulhouse
Mülheim *73 A6 var.* Mulheim an der Ruhr. Nordrhein-Westfalen, W Germany
Mulheim an der Ruhr *see* Mülheim
Mulhouse *68 E4 Ger.* Mülhausen. Haut-Rhin, NE France
Müller-gerbergte *see* Muller, Pegunungan
Muller, Pegunungan *116 D4 Dut.* Müller-gerbergte. *mountain range* Borneo, C Indonesia
Mull, Isle of *66 B4 island* W Scotland, United Kingdom
Mulongo *55 D7* Katanga, SE Dem. Rep. Congo
Multān *112 C2* Punjab, E Pakistan
Mumbai *112 C5 prev.* Bombay. *state capital* Mahārāshtra, W India
Munamägi *see* Suur Munamägi
Münchberg *73 C5* Bayern, E Germany
München *73 C6 var.* Muenchen, *Eng.* Munich, *It.* Monaco. Bayern, SE Germany
Muncie *18 C4* Indiana, N USA
Mungbere *55 E5* Orientale, NE Dem. Rep. Congo
Mu Nggava *see* Rennell
Munich *see* München
Munkács *see* Mukacheve
Münster *72 A4 var.* Muenster, Münster in Westfalen. Nordrhein-Westfalen, W Germany
Munster *67 A6 Ir.* Cúige Mumhan. *cultural region* S Ireland
Münster in Westfalen *see* Münster
Muong Xiang Ngeun *114 C4 var.* Xieng Ngeun. Louangphabang, N Laos
Muonio *62 D3* Lappi, N Finland
Muonioälv/Muoniojoki *see* Muonionjoki
Muonionjoki *62 D3 var.* Muoniojoki, *Swe.* Muonioälv. *river* Finland/Sweden
Muqāţ *97 C5* Al Mafraq, E Jordan
Muqdisho *51 D6 Eng.* Mogadishu, *It.* Mogadiscio. *country capital* (Somalia) Banaadir, S Somalia
Mur *73 E7 SCr.* Mura. *river* C Europe
Mura *see* Mur
Muradiye *95 F3* Van, E Turkey
Murapara *see* Murupara
Murata *74 E2* S San Marino
Murchison River *125 A5 river* Western Australia
Murcia *71 F4* Murcia, SE Spain
Murcia *71 E4 autonomous community* SE Spain
Mureş *86 A4 river* Hungary/Romania
Murfreesboro *20 D1* Tennessee, S USA
Murgab *see* Morghāb, Daryā-yeb
Murgap *100 D3* Mary Welaýaty, S Turkmenistan
Murgap *see* Morghāb, Daryā-ye
Murgap Deryasy *see* Morghāb, Daryā-ye
Murghab *see* Morghāb, Daryā-ye
Murghob *101 F3 Rus.* Murgab. SE Tajikistan
Murgon *127 E5* Queensland, E Australia
Mūrītānīyah *see* Mauritania
Müritz *72 C3 var.* Müritzee. *lake* NE Germany
Müritzee *see* Müritz
Murmansk *88 C2* Murmanskaya Oblast', NW Russia
Murmashi *88 C2* Murmanskaya Oblast', NW Russia
Murom *89 B5* Vladimirskaya Oblast', W Russia
Muroran *108 D3* Hokkaidō, NE Japan
Muros *70 B1* Galicia, NW Spain
Murray Fracture Zone *131 E2 fracture zone* NE Pacific Ocean
Murray Range *see* Murray Ridge
Murray Ridge *90 C5 var.* Murray Range. *undersea ridge* N Arabian Sea
Murray River *127 B6 river* SE Australia
Murrumbidgee River *127 C6 river* New South Wales, SE Australia
Murska Sobota *73 E7 Ger.* Olsnitz. NE Slovenia
Murupara *128 E3 var.* Murapara. Bay of Plenty, North Island, New Zealand
Murviedro *see* Sagunto
Murwāra *113 E4* Madhya Pradesh, N India
Murwillumbah *127 E5* New South Wales, SE Australia
Murzuq, Edeyin *see* Murzuq, Idhān
Murzuq, Idhan *49 F4 var.* Edeyin Murzuq. *desert* SW Libya
Mürzzuschlag *73 E7* Steiermark, E Austria
Muş *95 F3 var.* Mush. Muş, E Turkey
Musa, Gebel *50 C2 var.* Gebel Mûsa. *mountain* NE Egypt
Mûsa, Gebel *see* Mūsá, Jabal
Musala *82 B3 mountain* W Bulgaria
Muscat *see* Masqaţ
Muscat and Oman *see* Oman
Muscatine *23 G3* Iowa, C USA
Musgrave Ranges *125 D5 mountain range* South Australia
Musina *75 D7 var.* Messana, Messene; *anc.* Zancle. Sicilia, Italy, C Mediterranean Sea
Musina *56 D3 prev.* Messina. Limpopo, NE South Africa
Muskegon *18 C3* Michigan, N USA
Muskogean *see* Tallahassee
Muskogee *27 G1* Oklahoma, C USA
Musoma *51 C6* Mara, N Tanzania
Musta *see* Mosta
Mustafa-Pasha *see* Svilengrad
Musters, Lago *43 B6 lake* S Argentina
Muswellbrook *127 D6* New South Wales, SE Australia
Mut *94 C4* İçel, S Turkey
Mu-tan-chiang *see* Mudanjiang
Mutare *56 D3 var.* Mutari; *prev.* Umtali. Manicaland, E Zimbabwe
Mutari *see* Mutare
Mutina *see* Modena
Mutsu-wan *108 D3 bay* N Japan
Muttonbird Islands *129 A8 island group* SW New Zealand
Mu Us Shadi *105 E3 var.* Ordos Desert; *prev.* Mu Us Shamo. *desert* N China
Mu Us Shamo *see* Mu Us Shadi
Muy Muy *30 D3* Matagalpa, C Nicaragua
Muynak *see* Mo'ynoq
Mužlja *78 D3 Hung.* Felsőmuzslya; *prev.* Gornja Mužlja. Vojvodina, N Serbia
Mwali *57 F2 var.* Moili, *Fr.* Mohéli. *island* S Comoros
Mwanza *51 B6* Mwanza, NW Tanzania
Mweka *55 C6* Kasai-Occidental, C Dem. Rep. Congo
Mwene-Ditu *55 D7* Kasai-Oriental, S Dem. Rep. Congo
Mweru, Lake *55 D7 var.* Lac Moero. *lake* Dem. Rep. Congo/Zambia
Myadel' *see* Myadzyel
Myadzyel *85 C5 Pol.* Miadzioł Nowy, *Rus.* Myadel'. Minskaya Voblasts', N Belarus
Myanaung *114 B4* Ayeyarwady, SW Myanmar (Burma)
Myanmar *114 A3 off.* Union of Myanmar; *var.* Burma, Myanmar. *country* SE Asia
Myaungmya *114 A4* Ayeyarwady, SW Myanmar (Burma)
Myaydo *see* Aunglan
Myeik *115 B6 var.* Mergui. Tanintharyi, S Myanmar (Burma)
Myeik Archipelago *115 B6 prev.* Mergui Archipelago. *island group* S Myanmar (Burma)
Myerkulavichy *85 D7 Rus.* Merkulovichi. Homyel'skaya Voblasts', SE Belarus
Myingyan *114 B3* Mandalay, C Myanmar (Burma)
Myitkyina *114 B2* Kachin State, N Myanmar (Burma)
Mykolayiv *87 E4 Rus.* Nikolayev. Mykolayivs'ka Oblast', S Ukraine
Mykonos *83 D6 var.* Míkonos. *island* Kykládes, Greece, Aegean Sea
Myrhorod *87 F2 Rus.* Mirgorod. Poltavs'ka Oblast', NE Ukraine
Mýrina *82 D4 var.* Mírina. Límnos, SE Greece
Myrtle Beach *21 F2* South Carolina, SE USA
Mýrtos *83 D8* Kriti, Greece, E Mediterranean Sea
Myrtoum Mare *see* Mirtóo Pélagos
Myślibórz *76 B3* Zachodnio-pomorskie, NW Poland
Mysore *see* Mysūru
Mysore *see* Karnātaka
Mysūru *110 C2 prev.* Mysore, *var.* Maisur. Karnātaka, W India
My Tho *115 E6 var.* Mi Tho. Tiền Giang, S Vietnam
Mytilene *see* Mytilíni
Mytilíni *83 D5 var.* Mitilíni; *anc.* Mytilene. Lésvos, E Greece
Mzuzu *57 E2* Northern, N Malawi

N

Naberezhnyye Chelny *89 D5 prev.* Brezhnev. Respublika Tatarstan, W Russia
Nablus *97 A6 var.* Nābulus, *Heb.* Shekhem; *anc.* Neapolis, *Bibl.* Shechem. N West Bank

Ommen – Parral

Ommen 64 E3 Overijssel, E Netherlands
Omsk 92 C4 Omskaya Oblast', C Russia
Ōmuta 109 A7 Fukuoka, Kyūshū, SW Japan
Onda 71 F3 País Valenciano, E Spain
Ondjiva see N'Giva
Öndörhaan 105 E2 var. Undur Khan; prev. Tsetsen Khan. Hentiy, E Mongolia
Onega 88 C3 Arkhangel'skaya Oblast', NW Russia
Onega 88 B4 river NW Russia
Onega, Lake see Onezhskoye Ozero
Onex 73 A7 Genève, SW Switzerland
Onezhskoye Ozero 88 B4 Eng. Lake Onega. lake NW Russia
Ongole 110 D1 Andhra Pradesh, E India
Onitsha 53 G5 Anambra, S Nigeria
Onon Gol 105 E2 river N Mongolia
Onslow 124 A4 Western Australia
Onslow Bay 21 F1 bay North Carolina, E USA
Ontario 16 B3 province S Canada
Ontario, Lake 19 E3 lake Canada/USA
Onteniente see Ontinyent
Ontinyent 71 F4 var. Onteniente. País Valenciano, E Spain
Ontong Java Rise 103 H4 undersea feature W Pacific Ocean
Onuba see Huelva
Oodeypore see Udaipur
Oos-Londen see East London
Oostakker 65 B5 Oost-Vlaanderen, NW Belgium
Oostburg 65 B5 Zeeland, SW Netherlands
Oostende 65 A5 Eng. Ostend, Fr. Ostende. West-Vlaanderen, NW Belgium
Oosterbeek 64 D4 Gelderland, SE Netherlands
Oosterhout 64 C4 Noord-Brabant, S Netherlands
Opatija 78 A2 It. Abbazia. Primorje-Gorski Kotar, NW Croatia
Opava 77 C5 Ger. Troppau. Moravskoslezský Kraj, E Czech Republic (Czechia)
Ópazova see Stara Pazova
Opelika 20 D2 Alabama, S USA
Opelousas 20 B3 Louisiana, S USA
Ophiusa see Formentera
Opmeer 64 C2 Noord-Holland, NW Netherlands
Opochka 88 A4 Pskovskaya Oblast', W Russia
Opole 76 C4 Ger. Oppeln. Opolskie, S Poland
Oporto see Porto
Opotiki 128 E3 Bay of Plenty, North Island, New Zealand
Oppeln see Opole
Oppidum Ubiorum see Köln
Oqtosh 101 E2 Rus. Aktash. Samarqand Viloyati, C Uzbekistan
Oradea 86 B3 prev. Oradea Mare, Ger. Grosswardein, Hung. Nagyvárad. Bihor, NW Romania
Oradea Mare see Oradea
Orahovac see Rahovec
Oral see Ural'sk
Oran 48 D2 var. Ouahran, Wahran. NW Algeria
Orange 127 D6 New South Wales, SE Australia
Orange 69 D6 anc. Arausio. Vaucluse, SE France
Orangeburg 21 E2 South Carolina, SE USA
Orange Cone see Orange Fan
Orange Fan 47 C7 var. Orange Cone. undersea feature SW Indian Ocean
Orange Mouth/Orangemund see Oranjemund
Orange River 56 B4 Afr. Oranjerivier. river S Africa
Orange Walk 30 C1 Orange Walk, N Belize
Oranienburg 72 D3 Brandenburg, NE Germany
Oranjemund 56 B4 var. Orangemund; prev. Orange Mouth. Karas, SW Namibia
Oranjerivier see Orange River
Oranjestad 33 E5 dependent territory capital (Aruba) Lesser Antilles, S Caribbean Sea
Orany see Varėna
Orașul Stalin see Brașov
Oravicabánya see Oravița
Oravița 86 A4 Ger. Orawitza, Hung. Oravicabánya. Caraș-Severin, SW Romania
Orawitza see Oravița
Orbetello 74 B4 Toscana, C Italy
Orcadas 132 A1 Argentinian research station South Orkney Islands, Antarctica
Orchard Homes 22 B1 Montana, NW USA
Ordino 69 A8 Ordino, NW Andorra Europe
Ordos Desert see Mu Us Shadi
Ordu 94 D2 anc. Cotyora. Ordu, N Turkey
Ordzhonikidze see Pokrov, Ukraine
Ordzhonikidze see Vladikavkaz, Russia
Ordzhonikidze see Yenakiyeve, Ukraine
Orealla 37 G3 E Guyana
Örebro 63 C6 Örebro, C Sweden
Oregon 24 B3 off. State of Oregon, also known as Beaver State, Sunset State, Valentine State, Webfoot State. state NW USA
Oregon City 24 B3 Oregon, NW USA
Oregon, State of see Oregon
Orekhov see Orikhiv
Orël 89 B5 Orlovskaya Oblast', W Russia
Orem 22 B4 Utah, W USA
Ore Mountains see Erzgebirge/Krušné Hory
Orenburg 89 D6 prev. Chkalov. Orenburgskaya Oblast', W Russia
Orense see Ourense
Orestiáda 82 D3 prev. Orestiás. Anatolikí Makedonía kai Thráki, NE Greece
Orestiás see Orestiáda
Organ Peak 26 D3 mountain New Mexico, SW USA
Orgeyev see Orhei
Orhei 86 D3 var. Orheiu, Rus. Orgeyev. N Moldova
Orheiu see Orhei
Oriental, Cordillera 38 D3 mountain range Bolivia/Peru
Oriental, Cordillera 36 B3 mountain range C Colombia
Orihuela 71 F4 País Valenciano, E Spain
Orikhiv 87 G3 Rus. Orekhov. Zaporiz'ka Oblast', SE Ukraine
Orinoco, Río 37 E2 river Colombia/Venezuela
Orinoquía 36 C3 region NE Colombia
Orissa see Odisha
Orissaar see Orissaare
Orissaare 84 C2 Ger. Orissaar. Saaremaa, W Estonia
Oristano 75 A5 Sardegna, Italy, C Mediterranean Sea
Orito 36 A4 Putumayo, SW Colombia
Orizaba, Volcán Pico de 13 C7 var. Citlaltépetl. mountain S Mexico
Orkney see Orkney Islands
Orkney Islands 66 C2 var. Orkney, Orkneys. island group N Scotland, United Kingdom
Orkneys see Orkney Islands
Orlando 21 E4 Florida, SE USA
Orléanais 68 C4 cultural region C France
Orléans 68 C4 anc. Aurelianum. Loiret, C France
Orléansville see Chlef
Orly 68 E2 (Paris) Essonne, N France
Orlya 85 B6 Hrodzyenskaya Voblasts', W Belarus
Ormsö see Vormsi
Ormuz, Strait of see Hormuz, Strait of
Örnsköldsvik 63 C5 Västernorrland, C Sweden
Orolaunum see Arlon
Orol Dengizi see Aral Sea
Oromocto 17 F4 New Brunswick, SE Canada
Orona 123 F3 prev. Hull Island. atoll Phoenix Islands, C Kiribati
Oropeza see Cochabamba
Orosirá Rodhópis see Rhodope Mountains
Orpington 67 B8 United Kingdom
Orschowa see Orșova
Orsha 85 E6 Vitsyebskaya Voblasts', NE Belarus
Orsk 92 B4 Orenburgskaya Oblast', W Russia
Orșova 86 A4 Ger. Orschowa, Hung. Orsova. Mehedinți, SW Romania
Ortelsburg see Szczytno
Orthez 69 B6 Pyrénées-Atlantiques, SW France
Ortona 74 D4 Abruzzo, C Italy
Oruba see Aruba
Orūmīyeh, Daryācheh-ye 99 C2 var. Matianus, Sha Hi, Urumi Yeh, Eng. Lake Urmia; prev. Daryācheh-ye Rezā'īyeh. lake NW Iran
Oruro 39 F4 Oruro, W Bolivia
Oryokko see Yalu
Oss 64 D4 Noord-Brabant, S Netherlands
Ōsaka 109 C6 hist. Naniwa. Ōsaka, Honshū, SW Japan
Ōsaki see Furukawa
Osa, Península de 31 E5 peninsula S Costa Rica
Osborn Plateau 119 D5 undersea feature E Indian Ocean
Osca see Huesca
Ösel see Saaremaa
Osh 101 F2 Oshskaya Oblast', SW Kyrgyzstan
Oshawa 16 D5 Ontario, SE Canada
Oshikango 56 B3 Ohangwena, N Namibia
O-shima 109 D6 island S Japan
Oshkosh 18 B2 Wisconsin, N USA
Oshmyany see Ashmyany
Osiek see Osijek
Osijek 78 C3 prev. Osiek, Osjek, Ger. Esseg, Hung. Eszék. Osijek-Baranja, E Croatia
Osipenko see Berdyans'k
Osipovichi see Asipovichy
Osjek see Osijek
Oskaloosa 23 G4 Iowa, C USA
Oskarshamn 63 C7 Kalmar, S Sweden
Öskemen see Ust'-Kamenogorsk
Oskol 87 G2 Rus. Oskil. river Russia/Ukraine
Oskil see Oskol
Oslo 63 B6 prev. Christiania, Kristiania. country capital (Norway) Oslo, S Norway
Osmaniye 94 D4 Osmaniye, S Turkey
Osnabrück 72 A3 Niedersachsen, NW Germany
Osogov Mountains 82 B3 var. Osogovske Planine, Osogovski Planina, Mac. Osogovski Planini. mountain range Bulgaria/Macedonia
Osogovske Planine/Osogovski Planina/ Osogovski Planini see Osogov Mountains
Oșorhei see Târgu Mureș
Osorno 43 B5 Los Lagos, C Chile
Ossa, Serra d' 70 C4 mountain range SE Portugal
Ossora 93 H2 Koryakskiy Avtonomnyy Okrug, E Russia
Ostee see Baltic Sea
Ostend/Ostende see Oostende
Oster 87 E1 Chernihivs'ka Oblast', N Ukraine
Östermyra see Seinäjoki
Osterode/Osterode in Ostpreussen see Ostróda
Österreich see Austria
Östersund 63 C5 Jämtland, C Sweden
Ostia Aterni see Pescara
Ostiglia 74 C2 Lombardia, N Italy
Ostrava 77 C5 Moravskoslezský Kraj, E Czech Republic (Czechia)
Ostróda 76 D3 Ger. Osterode, Osterode in Ostpreussen. Warmińsko-Mazurskie, NE Poland
Ostrołęka 76 D3 Ger. Wiesenhof, Rus. Ostrolenka. Mazowieckie, C Poland
Ostrolenka see Ostrołęka
Ostrov 88 A4 Latv. Austrava. Pskovskaya Oblast', W Russia
Ostrovets see Ostrowiec Świętokrzyski
Ostrovnoy 88 C2 Murmanskaya Oblast', NW Russia
Ostrów see Ostrów Wielkopolski
Ostrowiec see Ostrowiec Świętokrzyski
Ostrowiec Świętokrzyski 76 D4 var. Ostrowiec, Rus. Ostrovets. Świętokrzyskie, C Poland
Ostrów Mazowiecka 76 D3 var. Ostrów Mazowiecki. Mazowieckie, NE Poland
Ostrów Mazowiecki see Ostrów Mazowiecka
Ostrowo see Ostrów Wielkopolski
Ostrów Wielkopolski 76 C4 var. Ostrów, Ger. Ostrowo. Wielkopolskie, C Poland
Ostyako-Voguls'k see Khanty-Mansiysk
Osum see Osumit, Lumi i
Osumi-shoto 109 A8 island group Kagoshima, Nansei-shotō, SW Japan Asia East China Sea Pacific Ocean
Osumit, Lumi i 79 D7 var. Osum. river SE Albania
Osuna 70 D4 Andalucía, S Spain
Oswego 19 F2 New York, NE USA
Otago Peninsula 129 B7 peninsula South Island, New Zealand
Otaki 128 D4 Wellington, North Island, New Zealand
Otaru 108 C2 Hokkaidō, NE Japan
Otavalo 38 B1 Imbabura, N Ecuador
Otavi 56 B3 Otjozondjupa, N Namibia
Oțelu Roșu 86 B4 Ger. Ferdinandsberg, Hung. Nándorhgy. Caras-Severin, SW Romania
Otepää 84 D3 Ger. Odenpäh. Valgamaa, SE Estonia
Oti 53 E4 river N Togo
Otira 129 C6 West Coast, South Island, New Zealand
Otjiwarongo 56 B3 Otjozondjupa, N Namibia
Otorohanga 128 D3 Waikato, North Island, New Zealand
Otranto, Canale d' see Otranto, Strait of
Otranto, Strait of 79 C6 It. Canale d'Otranto. strait Albania/Italy
Otrokovice 77 C5 Ger. Otrokowitz. Zlínský Kraj, E Czech Republic (Czechia)
Otrokowitz see Otrokovice
Ōtsu 109 C6 var. Ōtu. Shiga, Honshū, SW Japan
Ottawa 16 D5 country capital (Canada) Ontario, SE Canada
Ottawa 18 B3 Illinois, N USA
Ottawa 23 F5 Kansas, C USA
Ottawa 19 E2 Fr. Outaouais. river Ontario/Québec, SE Canada
Ottawa Islands 16 C1 island group Nunavut, C Canada
Ottignies 65 C6 Wallon Brabant, C Belgium
Ottumwa 23 G4 Iowa, C USA
Ōtu see Ōtsu
Ouachita Mountains 20 A1 mountain range Arkansas/Oklahoma, C USA
Ouachita River 20 B2 river Arkansas/Louisiana, C USA
Ouagadougou 53 E4 var. Wagadugu. country capital (Burkina) C Burkina
Ouahigouya 53 E3 NW Burkina
Ouahran see Oran
Oualâta 52 D3 var. Oualata. Hodh ech Chargui, SE Mauritania
Ouanary 37 H3 E French Guiana
Ouanda Djallé 54 D4 Vakaga, NE Central African Republic
Ouarâne 52 D2 desert C Mauritania
Ouargla 49 E2 var. Wargla. NE Algeria
Ouarzazate 48 C3 S Morocco
Oubangui 55 C5 Fr. Oubangui. river C Africa
Oubangui see Ubangi
Oubangui-Chari see Central African Republic
Oubangui-Chari, Territoire de l' see Central African Republic
Oudjda see Oujda
Ouessant, Île d' 68 A3 Eng. Ushant. island NW France
Ouésso 55 B5 Sangha, NW Congo
Oujda 48 D2 Ar. Oudjda, Ujda. NE Morocco
Oujeft 52 C2 Adrar, C Mauritania
Oulu 62 D4 Swe. Uleåborg. Pohjois-Pohjanmaa, C Finland
Oulujärvi 62 D4 Swe. Uleträsk. lake C Finland
Oulujoki 62 D4 Swe. Uleälv. river C Finland
Ounasjoki 62 D3 river N Finland
Ounianga Kébir 54 C2 Borkou-Ennedi-Tibesti, N Chad
Oup see Auob
Oupeye 65 D6 Liège, E Belgium
Our 65 D6 river NW Europe
Ourense 70 C1 Cast. Orense, Lat. Aurium. Galicia, NW Spain
Ourique 70 B4 Beja, S Portugal
Ours, Grand Lac de l' see Great Bear Lake
Ourthe 65 D7 river E Belgium
Ouse 67 D5 river N England, United Kingdom
Outaouais see Ottawa
Outer Hebrides 66 B3 var. Western Isles. island group NW Scotland, United Kingdom
Outer Islands 57 G1 island group SW Seychelles Africa W Indian Ocean
Outes 70 B1 Galicia, NW Spain
Ouvéa 122 D5 island Îles Loyauté, NE New Caledonia
Ouyen 127 C6 Victoria, SE Australia
Ovalle 42 B3 Coquimbo, N Chile
Ovar 70 B2 Aveiro, N Portugal
Overflakkee 64 B4 island SW Netherlands
Overijse 65 C6 Vlaams Brabant, C Belgium
Oviedo 70 C1 anc. Asturias. Asturias, NW Spain
Ovilava see Wels
Ovruch 86 D1 Zhytomyrs'ka Oblast', N Ukraine
Owando 55 B5 prev. Fort Rousset. Cuvette, C Congo
Owase 109 C6 Mie, Honshū, SW Japan
Owatonna 23 F3 Minnesota, N USA
Owen Fracture Zone 118 B4 tectonic feature W Arabian Sea
Owen, Mount 129 C5 mountain South Island, New Zealand
Owensboro 18 B5 Kentucky, S USA
Owen Stanley Range 122 B3 mountain range S Papua New Guinea
Owerri 53 G5 Imo, S Nigeria
Owo 53 F5 Ondo, SW Nigeria
Owyhee River 24 C4 river Idaho/Oregon, NW USA
Oxford 129 C6 Canterbury, South Island, New Zealand
Oxford 67 D6 Lat. Oxonia. S England, United Kingdom
Oxkutzcab 29 H4 Yucatán, SE Mexico
Oxnard 25 B7 California, W USA
Oxonia see Oxford
Oxus see Amu Darya
Oyama 109 D5 Tochigi, Honshū, S Japan
Oyem 55 B5 Woleu-Ntem, N Gabon
Oyo 55 B6 Cuvette, C Congo
Oyo 53 F4 Oyo, W Nigeria
Ozark 20 D3 Alabama, S USA
Ozark Plateau 23 G5 plain Arkansas/Missouri, C USA
Ozarks, Lake of the 23 F5 reservoir Missouri, C USA
Ozbourn Seamount 130 D4 undersea feature W Pacific Ocean
Ózd 77 D6 Borsod-Abaúj-Zemplén, NE Hungary
Ozieri 75 A5 Sardegna, Italy, C Mediterranean Sea

P

Paamiut 60 B4 var. Pâmiut, Dan. Frederikshåb. S Greenland
Pa-an see Hpa-an
Pabianice 76 C4 Łódzski, Poland
Pabna 113 G4 Rajshahi, W Bangladesh
Pacaraima, Sierra/Pacaraim, Serra see Pakaraima Mountains
Pachuca 29 E4 var. Pachuca de Soto. Hidalgo, C Mexico
Pachuca de Soto see Pachuca
Pacific-Antarctic Ridge 132 B5 undersea feature S Pacific Ocean
Pacific Ocean 130 D3 ocean
Padalung see Phatthalung
Padang 116 B4 Sumatera, W Indonesia
Paderborn 72 B4 Nordrhein-Westfalen, NW Germany
Padma see Brahmaputra
Padma see Ganges
Padova 74 C2 Eng. Padua; anc. Patavium. Veneto, NE Italy
Padre Island 27 G5 island Texas, SW USA
Padua see Padova
Paducah 18 B5 Kentucky, S USA
Paeroa 128 D3 Waikato, North Island, New Zealand
Páfos 80 C5 var. Paphos. W Cyprus
Pag 78 A3 It. Pago. island Zadar, C Croatia
Page 26 B1 Arizona, SW USA
Pago see Pag
Pago Pago 123 F4 dependent territory capital (American Samoa) Tutuila, W American Samoa
Pahiatua 128 D4 Manawatu-Wanganui, North Island, New Zealand
Pahsien see Chongqing
Paide 84 D2 Ger. Weissenstein. Järvamaa, N Estonia
Paihia 128 D2 Northland, North Island, New Zealand
Päijänne 63 D5 lake S Finland
Paine, Cerro 43 A7 mountain S Chile
Painted Desert 26 B1 desert Arizona, SW USA
Paisance see Piacenza
Paisley 66 C4 W Scotland, United Kingdom
País Valenciano 71 F3 var. Valencia, Cat. València; anc. Valentia. autonomous community NE Spain
País Vasco 71 E1 cultural region N Spain
Paita 38 B3 Piura, NW Peru
Pakanbaru see Pekanbaru
Pakaraima Mountains 37 E3 var. Serra Pacaraim, Sierra Pacaraima. mountain range N South America
Pakistan 112 A2 off. Islamic Republic of Pakistan, var. Islami Jamhuriya e Pakistan. country S Asia
Pakistan, Islamic Republic of see Pakistan
Pakistan, Islami Jamhuriya e see Pakistan
Paknam see Samut Prakan
Pakokku 114 A3 Magway, C Myanmar (Burma)
Pak Phanang 115 C7 var. Ban Pak Phanang. Nakhon Si Thammarat, SW Thailand
Pakruojis 84 C4 Šiaulai, N Lithuania
Paks 77 C7 Tolna, S Hungary
Paksé see Pakxé
Pakxé 115 D5 var. Paksé. Champasak, S Laos
Palafrugell 71 G2 Cataluña, NE Spain
Palagruža 79 B5 It. Pelagosa. island SW Croatia
Palaiá Epídavros 83 C6 Pelopónnisos, S Greece
Palaiseau 68 D2 Essonne, N France
Palamós 71 G2 Cataluña, NE Spain
Palamuse 84 E2 Ger. Sankt-Bartholomäi. Jõgevamaa, E Estonia
Palanka see Bačka Palanka
Pālanpur 112 C4 Gujarāt, W India
Palantia see Palencia
Palapye 56 D3 Central, SE Botswana
Palau 122 A2 var. Belau. country W Pacific Ocean
Palawan 117 E2 island W Philippines
Palawan Passage 116 D2 passage W Philippines
Paldiski 84 D2 prev. Baltiski, Eng. Baltic Port, Ger. Baltischport. Harjumaa, NW Estonia
Palembang 116 B4 Sumatera, W Indonesia
Palencia 70 D2 anc. Palantia, Pallantia. Castilla-León, NW Spain
Palerme see Palermo
Palermo 75 C7 Fr. Palerme; anc. Panhormus, Panormus. Sicilia, Italy, C Mediterranean Sea
Palestine, State of see West Bank; Gaza Strip
Pāli 112 C3 Rājasthān, N India
Palikir 122 C2 country capital (Micronesia) Pohnpei, E Micronesia
Palimé see Kpalimé
Palioúri, Akrotírio 82 C4 var. Akrotírio Kanestron. headland N Greece
Palk Strait 110 C3 strait India/Sri Lanka
Pallantia see Palencia
Palliser, Cape 129 D5 headland North Island, New Zealand
Palma 71 G3 var. Palma de Mallorca. Mallorca, Spain, W Mediterranean Sea
Palma del Río 70 D4 Andalucía, S Spain
Palma de Mallorca see Palma
Palmar Sur 31 E5 Puntarenas, SE Costa Rica
Palma Soriano 32 C3 Santiago de Cuba, E Cuba
Palm Beach 126 E1 New South Wales, E Australia
Palmer 132 A2 US research station Antarctica
Palmer Land 132 A3 physical region Antarctica
Palmerston 123 F4 island S Cook Islands
Palmerston see Darwin
Palmerston North 128 D4 Manawatu-Wanganui, North Island, New Zealand
Palmetto State, The see South Carolina
Palmi 75 D7 Calabria, SW Italy
Palmira 36 B3 Valle del Cauca, W Colombia
Palm Springs 25 D7 California, W USA
Palmyra see Tudmur
Palmyra Atoll 123 G2 US privately owned unincorporated territory C Pacific Ocean
Palo Alto 25 B6 California, W USA
Paloe see Denpasar, Bali, C Indonesia
Paloe see Palu
Palu 117 E4 prev. Paloe. Sulawesi, C Indonesia
Pamiers 69 B6 Ariège, S France
Pamir 101 F3 var. Daryā-ye Pāmīr, Taj. Dar"yoi Pomir. river Afghanistan/Tajikistan
Pāmīr, Daryā-ye see Pamir
Pamir/Pāmīr, Daryā-ye see Pamirs
Pamirs 101 F3 Pash. Daryā-ye Pāmīr, Rus. Pamir. mountain range C Asia
Pâmiut see Paamiut
Pamlico Sound 21 G1 sound North Carolina, SE USA
Pampa 27 E1 Texas, SW USA
Pampa Aullagas, Lago see Poopó, Lago
Pampas 42 C4 plain C Argentina
Pampeluna see Pamplona
Pamplona 36 C2 Norte de Santander, N Colombia
Pamplona 71 E1 Basq. Iruña, prev. Pampeluna; anc. Pompaelo. Navarra, N Spain
Panaji 110 B1 var. Pangim, Panjim, New Goa. state capital Goa, W India
Panamá 31 G4 var. Ciudad de Panama, Eng. Panama City. country capital (Panama) Panamá, C Panama
Panama 31 G5 off. Republic of Panama. country Central America
Panama Basin 13 C8 undersea feature E Pacific Ocean
Panama Canal 31 F4 canal E Panama
Panama City 20 D3 Florida, SE USA
Panama City see Panamá
Panamá, Golfo de 31 G5 var. Gulf of Panama. gulf S Panama
Panama, Gulf of see Panamá, Golfo de
Panama, Isthmus of see Panama, Istmo de
Panama, Istmo de 31 G4 Eng. Isthmus of Panama; prev. Isthmus of Darien. isthmus E Panama
Panama, Republic of see Panama
Panay Island 117 E2 island C Philippines
Pančevo 78 D3 Ger. Pantschowa, Hung. Pancsova. Vojvodina, N Serbia
Pancsova see Pančevo
Paneas see Bāniyās
Panevėžys 84 C4 Panevėžys, C Lithuania
Pangim see Panaji
Pangkalpinang 116 C4 Pulau Bangka, W Indonesia
Pang-Nga see Phang-Nga
Panhormus see Palermo
Panjim see Panaji
Panopolis see Akhmīm
Pánormos 83 C7 Kríti, Greece, E Mediterranean Sea
Panormus see Palermo
Pantanal 41 E3 var. Pantanalmato-Grossense. swamp SW Brazil
Pantanalmato-Grossense see Pantanal
Pantelleria, Isola di 75 B7 island SW Italy
Pantschowa see Pančevo
Pánuco 29 E3 Veracruz-Llave, E Mexico
Pao-chi/Paoki see Baoji
Paola 80 B5 E Malta
Pao-shan see Baoshan
Pao-t'ou/Paotow see Baotou
Papagayo, Golfo de 30 C4 gulf NW Costa Rica
Papakura 128 D3 Auckland, North Island, New Zealand
Papantla 29 F4 var. Papantla de Olarte. Veracruz-Llave, E Mexico
Papantla de Olarte see Papantla
Papeete 123 H4 dependent territory capital (French Polynesia) Tahiti, W French Polynesia
Paphos see Páfos
Papilė 84 B3 Šiaulai, NW Lithuania
Papillion 23 F4 Nebraska, C USA
Papua 117 H4 var. Irian Barat, West Irian, West New Guinea, West Papua; prev. Dutch New Guinea, Irian Jaya, Netherlands New Guinea. province E Indonesia
Papua and New Guinea, Territory of see Papua New Guinea
Papua, Gulf of 122 B3 gulf S Papua New Guinea
Papua New Guinea 122 B3 off. Independent State of Papua New Guinea; prev. Territory of Papua and New Guinea. country NW Melanesia
Papua New Guinea, Independent State of see Papua New Guinea
Papuk 78 C3 mountain range NE Croatia
Pará 41 E2 off. Estado do Pará. state/region NE Brazil
Pará see Belém
Paracel Islands 103 E3 disputed territory SE Asia
Paraćin 78 D4 Serbia, C Serbia
Paradise of the Pacific see Hawaii
Pará, Estado do see Pará
Paragua, Río 37 E3 river SE Venezuela
Paraguay 42 C2 country C South America
Paraguay 42 D2 var. Río Paraguay. river C South America
Paraguay, Río see Paraguay
Parahiba/Parahyba see Paraíba
Paraíba 41 G2 off. Estado da Paraíba; prev. Parahiba, Parahyba. state/region E Brazil
Paraíba see João Pessoa
Paraíba, Estado da see Paraíba
Parakou 53 F4 C Benin
Paramaribo 37 G3 country capital (Suriname) Paramaribo, N Suriname
Paramushir, Ostrov 93 H3 island SE Russia
Paraná 41 E4 Entre Ríos, E Argentina
Paraná 41 E5 off. Estado do Paraná. state/region S Brazil
Paraná 35 C5 var. Alto Paraná. river C South America
Paraná, Estado do see Paraná
Paranéstì 82 C3 var. Paranéstio. Anatolikí Makedonía kai Thráki, NE Greece
Paranéstio see Paranésti
Paraparaumu 129 D5 Wellington, North Island, New Zealand
Parchim 72 C3 Mecklenburg-Vorpommern, N Germany
Parczew 76 E4 Lubelskie, E Poland
Pardubice 77 B5 Ger. Pardubitz. Pardubický Kraj, C Czech Republic (Czechia)
Pardubitz see Pardubice
Parechcha 85 B5 Pol. Porzecze, Rus. Porech'ye. Hrodzyenskaya Voblasts', W Belarus
Parecis, Chapada dos 40 D3 var. Serra dos Parecis. mountain range W Brazil
Parecis, Serra dos see Parecis, Chapada dos
Parenzo see Poreč
Parepare 117 E4 Sulawesi, C Indonesia
Párga 83 A5 Ípeiros, W Greece
Paria, Golfo de see Paria, Gulf of
Paria, Gulf of 37 E1 var. Golfo de Paria. gulf Trinidad and Tobago/Venezuela
Parika 37 F2 NE Guyana
Parikiá 83 D6 Kykládes, Greece, Aegean Sea
Paris 68 D1 anc. Lutetia, Lutetia Parisiorum, Parisii. country capital (France) Paris, N France
Paris 27 G2 Texas, SW USA
Parisii see Paris
Parkersburg 18 D4 West Virginia, NE USA
Parkes 127 D6 New South Wales, SE Australia
Parkhar see Farkhor
Parma 74 B2 Emilia-Romagna, N Italy
Parnahyba see Parnaíba
Parnaíba 41 F2 var. Parnahyba. Piauí, E Brazil
Pärnu 84 D2 Ger. Pernau, Latv. Pērnava; prev. Rus. Pernov. Pärnumaa, SW Estonia
Pärnu 84 D2 var. Parnu Jõgi, Ger. Pernau. river SW Estonia
Pärnu-Jaagupi 84 D2 Ger. Sankt-Jakobi. Pärnumaa, SW Estonia
Parnu Jõgi see Pärnu
Pärnu Laht 84 D2 Ger. Pernauer Bucht. bay SW Estonia
Paroikiá see Páros
Paropamisus Range see Sefīd Kūh, Selseleh-ye
Páros 83 D6 var. Paroikiá, Greece
Páros 83 D6 island Kykládes, Greece, Aegean Sea
Parral 42 B4 Maule, C Chile
Parral see Hidalgo del Parral

Ponziane Island – Quezaltenango

Ponziane Island *75 C5* *island* C Italy
Poole *67 D7* S England, United Kingdom
Poona *see* Pune
Poopó, Lago *39 F4* *var.* Lago Pampa Aullagas. *lake* W Bolivia
Popayán *36 B4* Cauca, SW Colombia
Poperinge *65 A6* West-Vlaanderen, W Belgium
Poplar Bluff *23 G5* Missouri, C USA
Popocatépetl *29 E4* *volcano* S Mexico
Popper *see* Poprad
Poprad *77 D5* *Ger.* Deutschendorf, *Hung.* Poprád. Prešovský Kraj, E Slovakia
Poprád *77 D5* *Ger.* Popper, *Hung.* Poprád. *river* Poland/Slovakia
Porbandar *112 B4* Gujarāt, W India
Porcupine Plain *58 B3* *undersea feature* E Atlantic Ocean
Pordenone *74 C2* *anc.* Portenau. Friuli-Venezia Giulia, NE Italy
Poreč *78 A2* *It.* Parenzo. Istra, NW Croatia
Porech'ye *see* Parechcha
Pori *63 D5* *Swe.* Björneborg. Satakunta, SW Finland
Porirua *129 D5* Wellington, North Island, New Zealand
Porkhov *88 A4* Pskovskaya Oblast', W Russia
Porlamar *37 E1* Nueva Esparta, NE Venezuela
Póros *83 C6* Póros, S Greece
Póros *83 A5* Kefallinía, Iónia Nisiá, Greece, C Mediterranean Sea
Pors *see* Porsangenfijorden
Porsangenfijorden *62 D2* *Lapp.* Pors. *fjord* N Norway
Porsgrunn *63 B6* Telemark, S Norway
Portachuelo *39 G4* Santa Cruz, C Bolivia
Portadown *67 B5* *Ir.* Port An Dúnáin. S Northern Ireland, United Kingdom
Portalegre *70 C3* *anc.* Ammaia, Amoea. Portalegre, E Portugal
Port Alexander *14 D4* Baranof Island, Alaska, USA
Port Alfred *56 D5* Eastern Cape, S South Africa
Port Amelia *see* Pemba
Port An Dúnáin *see* Portadown
Port Angeles *24 B1* Washington, NW USA
Port Antonio *32 B5* NE Jamaica
Port Arthur *27 H4* Texas, SW USA
Port Augusta *127 B6* South Australia
Port-au-Prince *32 D3* *country capital* (Haiti) C Haiti
Port Blair *111 F2* Andaman and Nicobar Islands, SE India
Port Charlotte *21 E4* Florida, SE USA
Port Darwin *see* Darwin
Port d'Envalira *69 B8* E Andorra Europe
Port Douglas *126 D3* Queensland, NE Australia
Port Elizabeth *56 C5* Eastern Cape, S South Africa
Portenau *see* Pordenone
Porterville *25 C7* California, W USA
Port-Étienne *see* Nouâdhibou
Port Florence *see* Kisumu
Port-Francqui *see* Ilebo
Port-Gentil *55 A6* Ogooué-Maritime, W Gabon
Port Harcourt *53 G5* Rivers, S Nigeria
Port Hardy *14 D5* Vancouver Island, British Columbia, SW Canada
Port Harrison *see* Inukjuak
Port Hedland *124 B4* Western Australia
Port Huron *18 D3* Michigan, N USA
Portimão *70 B4* *var.* Vila Nova de Portimão. Faro, S Portugal
Port Jackson *126 E1* *harbour* New South Wales, E Australia
Portland *127 B7* Victoria, SE Australia
Portland *19 G2* Maine, NE USA
Portland *24 B3* Oregon, NW USA
Portland *27 G4* Texas, SW USA
Portland Bight *32 B5* *bay* S Jamaica
Portlaoighise *see* Port Laoise
Port Laoise *67 B6* *var.* Portlaoise, *Ir.* Portlaoighise; *prev.* Maryborough. C Ireland
Portlaoise *see* Port Laoise
Port Lavaca *27 G4* Texas, SW USA
Port Lincoln *127 A6* South Australia
Port Louis *57 H3* *country capital* (Mauritius) NW Mauritius
Port-Lyautey *see* Kénitra
Port Macquarie *127 E6* New South Wales, SE Australia
Port Mahon *see* Mahón
Portmore *32 B5* C Jamaica
Port Moresby *122 B3* *country capital* (Papua New Guinea) Central/National Capital District, SW Papua New Guinea
Port Natal *see* Durban
Porto *70 B2* *Eng.* Oporto; *anc.* Portus Cale. Porto, NW Portugal
Porto Alegre *41 E5* *var.* Pôrto Alegre. *state capital* Rio Grande do Sul, S Brazil
Porto Alegre *54 E2* São Tomé, S Sao Tome and Principe, Africa
Porto Alexandre *see* Tombua
Porto Amélia *see* Pemba
Porto Bello *see* Portobelo
Portobelo *31 G4* *var.* Porto Bello, Puerto Bello. Colón, N Panama
Port O'Connor *27 G4* Texas, SW USA
Porto Edda *see* Sarandë
Portoferraio *74 B4* Toscana, C Italy
Port of Spain *33 H5* *country capital* (Trinidad and Tobago) Trinidad, Trinidad and Tobago
Porto Grande *see* Mindelo
Portogruaro *74 C2* Veneto, NE Italy
Porto-Novo *53 F5* *country capital* (Benin - official) S Benin
Porto Rico *see* Puerto Rico
Porto Santo *48 A2* *var.* Ilha do Porto Santo. *island* Madeira, Portugal, NE Atlantic Ocean
Porto Santo, Ilha do *see* Porto Santo
Porto Torres *75 A5* Sardegna, Italy, C Mediterranean Sea
Porto Velho *40 D2* *var.* Velho. *state capital* Rondônia, W Brazil
Portoviejo *38 A2* *var.* Puertoviejo. Manabí, W Ecuador
Port Pirie *127 B6* South Australia
Port Rex *see* East London
Port Said *see* Būr Sa'īd
Portsmouth *67 D7* S England, United Kingdom
Portsmouth *19 G3* New Hampshire, NE USA
Portsmouth *18 D4* Ohio, N USA
Portsmouth *19 F5* Virginia, NE USA
Port Stanley *see* Stanley
Port Sudan *50 C3* Red Sea, NE Sudan
Port Swettenham *see* Klang/Pelabuhan Klang
Port Talbot *67 C7* S Wales, United Kingdom
Portugal *70 B3* *off.* Portuguese Republic. *country* SW Europe
Portuguese East Africa *see* Mozambique
Portuguese Guinea *see* Guinea-Bissau
Portuguese Republic *see* Portugal
Portuguese Timor *see* East Timor
Portuguese West Africa *see* Angola
Portus Cale *see* Porto
Portus Magnus *see* Almería
Portus Magonis *see* Mahón
Port-Vila *122 D4* *var.* Vila. *country capital* (Vanuatu) Éfaté, C Vanuatu
Porvenir *39 E3* Pando, NW Bolivia
Porvenir *43 B8* Magallanes, S Chile
Porvoo *63 E6* *Swe.* Borgå. Uusimaa, S Finland
Porzecze *see* Parechcha
Posadas *42 D3* Misiones, NE Argentina
Poschega *see* Požega
Posen *see* Poznań
Posnania *see* Poznań
Postavy/Postawy *see* Pastavy
Posterholt *65 D5* Limburg, SE Netherlands
Postojna *73 D8* *Ger.* Adelsberg, *It.* Postumia. SW Slovenia
Postumia *see* Postojna
Pöstyén *see* Piešťany
Potamós *83 C7* Antikýthira, S Greece
Potentia *see* Potenza
Potenza *75 D5* *anc.* Potentia. Basilicata, S Italy
Poti *95 F2* *prev.* P'ot'i. W Georgia
P'ot'i *see* Poti
Potiskum *53 G4* Yobe, NE Nigeria
Potomac River *19 E5* *river* NE USA
Potosí *39 F4* Potosí, S Bolivia
Potsdam *72 D3* Brandenburg, NE Germany
Potwar Plateau *112 C2* *plateau* NE Pakistan
Poŭthĭsăt *see* Pursat
Po, Valle del *see* Po Valley
Po Valley *74 C2* *It.* Valle del Po. *valley* N Italy
Považská Bystrica *77 C5* *Ger.* Waagbistritz, *Hung.* Vágbeszterce. Trenčiansky Kraj, W Slovakia
Poverty Bay *128 E4* *inlet* North Island, New Zealand
Póvoa de Varzim *70 B2* Porto, NW Portugal
Powder River *22 D2* *river* Montana/Wyoming, NW USA
Powell *22 C2* Wyoming, C USA
Powell, Lake *22 B5* *lake* Utah, W USA
Požarevac *78 D4* *Ger.* Passarowitz. Serbia, NE Serbia
Poza Rica *29 F4* *var.* Poza Rica de Hidalgo. Veracruz-Llave, E Mexico
Poza Rica de Hidalgo *see* Poza Rica
Požega *78 D4* *prev.* Slavonska Požega, *Ger.* Poschega, *Hung.* Pozsega. Požega-Slavonija, NE Croatia
Požega *78 D4* Serbia
Poznań *76 C3* *Ger.* Posen, Posnania. Wielkolpolskie, C Poland
Pozoblanco *70 D4* Andalucía, S Spain
Pozsega *see* Požega
Pozsony *see* Bratislava
Pozzallo *75 C8* Sicilia, Italy, C Mediterranean Sea
Prachatice *77 A5* *Ger.* Prachatitz. Jihočeský Kraj, S Czech Republic (Czechia)
Prachatitz *see* Prachatice
Prado del Ganso *see* Goose Green
Prae *see* Phrae
Prag/Praga/Prague *see* Praha
Praha *77 A5* *Eng.* Prague, *Ger.* Prag, *Pol.* Praga. *country capital* (Czech Republic (Czechia)) Středočeský Kraj, NW Czech Republic (Czechia)
Praia *52 A3* *country capital* (Cape Verde) Santiago, S Cape Verde
Prairie State *see* Illinois
Prathet Thai *see* Thailand
Prato *74 B3* Toscana, C Italy
Pratt *23 E5* Kansas, C USA
Prattville *20 D2* Alabama, S USA
Pravda *see* Glavinitsa
Pravia *70 C1* Asturias, N Spain
Preăh Seihânŭ *see* Sihanoukville
Preny *see* Prienai
Prenzlau *72 D3* Brandenburg, NE Germany
Prerau *see* Přerov
Přerov *77 C5* *Ger.* Prerau. Olomoucký Kraj, E Czech Republic (Czechia)
Preschau *see* Prešov
Prescott *26 B2* Arizona, SW USA
Preševo *79 D5* Serbia, SE Serbia
Presidente Epitácio *41 E4* São Paulo, S Brazil
Presidente Stroessner *see* Ciudad del Este
Prešov *77 D5* *var.* Preschau, *Ger.* Eperies, *Hung.* Eperjes. Prešovský Kraj, E Slovakia
Prespa, Lake *79 D6* *Alb.* Liqen i Prespës, *Gk.* Límni Megáli Préspa, Limni Prespa, *Mac.* Prespansko Ezero, *Serb.* Prespansko Jezero. *lake* SE Europe
Prespa, Limni/Prespansko Ezero/Prespansko Jezero/Prespës, Liqen i *see* Prespa, Lake
Presque Isle *19 H1* Maine, NE USA
Pressburg *see* Bratislava
Preston *67 D5* NW England, United Kingdom
Prestwick *66 C4* W Scotland, United Kingdom
Pretoria *56 D4* *var.* Epitoli. *country capital* (South Africa-administrative capital) Gauteng, NE South Africa
Preussisch Eylau *see* Bagrationovsk
Preußisch Holland *see* Pasłęk
Preussisch-Stargard *see* Starogard Gdański
Préveza *83 A5* Ípeiros, W Greece
Pribilof Islands *14 A3* *island group* Alaska, USA
Priboj *78 C4* Serbia, W Serbia
Price *22 B4* Utah, W USA
Prichard *20 C3* Alabama, S USA
Priekulė *84 B3* *Ger.* Prökuls. Klaipėda, W Lithuania
Prienai *85 B5* *Pol.* Preny. Kaunas, S Lithuania
Prieska *56 C4* Northern Cape, C South Africa
Prijedor *78 B3* Republika Srpska, NW Bosnia and Herzegovina
Prijepolje *78 D4* Serbia, W Serbia
Prikaspiyskaya Nizmennost' *see* Caspian Depression
Prilep *79 D6* *Turk.* Perlepe. S Macedonia
Priluki *see* Pryluky
Primorsk *84 A4* *Ger.* Fischhausen. Kaliningradskaya Oblast', W Russia
Primorsko *82 E2* *prev.* Keupriya. Burgas, E Bulgaria
Primorsk/Primorskoye *see* Prymors'k
Prince Albert *15 F5* Saskatchewan, S Canada
Prince Edward Island *17 F4* *Fr.* Île-du Prince-Édouard. *province* SE Canada
Prince Edward Islands *47 E8* *island group* S South Africa
Prince George *15 E5* British Columbia, SW Canada
Prince of Wales Island *126 B1* *island* Queensland, E Australia
Prince of Wales Island *15 F2* *island* Queen Elizabeth Islands, Nunavut, NW Canada
Prince of Wales Island *see* Pinang, Pulau
Prince Patrick Island *15 E2* *island* Parry Islands, Northwest Territories, NW Canada
Prince Rupert *14 D4* British Columbia, SW Canada
Prince's Island *see* Príncipe
Princess Charlotte Bay *126 C2* *bay* Queensland, NE Australia
Princess Elizabeth Land *132 C3* *physical region* Antarctica
Príncipe *55 A5* *var.* Príncipe Island, *Eng.* Prince's Island. *island* N Sao Tome and Principe
Príncipe Island *see* Príncipe
Prinzapolka *31 E3* Región Autónoma Atlántico Norte, NE Nicaragua
Pripet *85 C7* *Bel.* Prypyats', *Ukr.* Pryp"yat'. *river* Belarus/Ukraine
Pripet Marshes *85 B7* *wetland* Belarus/Ukraine
Prishtinë *79 D5* *Eng.* Pristina, *Serb.* Priština. C Kosovo
Pristina *see* Prishtinë
Priština *see* Prishtinë
Privas *69 D5* Ardèche, E France
Privolzhskaya Vozvyshennost' *59 G3* *var.* Volga Uplands. *mountain range* W Russia
Prizren *79 D5* S Kosovo
Probolinggo *116 D5* Jawa, C Indonesia
Probstberg *see* Wyszków
Progreso *29 H3* Yucatán, SE Mexico
Prokhladnyy *89 B8* Kabardino-Balkarskaya Respublika, SW Russia
Prokletije *see* North Albanian Alps
Prökuls *see* Priekulė
Prokuplje *79 D5* Serbia, SE Serbia
Prome *see* Pyay
Promyshlennyy *88 E3* Respublika Komi, NW Russia
Prościejów *see* Prostějov
Proskurov *see* Khmel 'nyts'kyy
Prossnitz *see* Prostějov
Prostějov *77 C5* *Ger.* Prossnitz, *Pol.* Prościejów. Olomoucký Kraj, E Czech Republic (Czechia)
Provence *69 D6* *cultural region* SE France
Providence *19 G3* *state capital* Rhode Island, NE USA
Providence *see* Fort Providence
Providencia, Isla de *31 F3* *island* NW Colombia, Caribbean Sea
Provideniya *133 B1* Chukotskiy Avtonomnyy Okrug, NE Russia
Provo *22 B4* Utah, W USA
Prudhoe Bay *14 D2* Alaska, USA
Prusa *see* Bursa
Pruszków *76 D3* *Ger.* Kaltdorf. Mazowieckie, C Poland
Prut *86 D4* *Ger.* Pruth. *river* E Europe
Pruth *see* Prut
Pružana *see* Pruzhany
Pruzhany *85 B6* *Pol.* Pružana. Brestskaya Voblasts', SW Belarus
Prychornomor'ska Nyzovyna *see* Black Sea Lowland
Prydniprovs'ka Nyzovyna/Prydnyaprowskaya Nizina *see* Dnieper Lowland
Prydz Bay *132 D3* *bay* Antarctica
Pryluky *87 E2* *Rus.* Priluki. Chernihivs'ka Oblast', NE Ukraine
Prymors'k *87 G4* *Rus.* Primorsk; *prev.* Primorskoye. Zaporiz'ka Oblast', SE Ukraine
Pryp"yat'/Prypyats' *see* Pripet
Przemyśl *77 E5* *Rus.* Peremyshl. Podkarpackie, C Poland
Przheval'sk *see* Karakol
Psará *83 D5* *island* E Greece
Psel *87 F2* *Rus.* Psël. *river* Russia/Ukraine
Psël *see* Psel
Pskov *92 B2* *Ger.* Pleskau, *Latv.* Pleskava. Pskovskaya Oblast', W Russia
Pskov, Lake *84 E3* *Est.* Pihkva Järv, *Ger.* Pleskauer See, *Rus.* Pskovskoye Ozero. *lake* Estonia/Russia
Pskovskoye Ozero *see* Pskov, Lake
Ptich' *see* Ptsich
Ptsich *85 C7* *Rus.* Ptich'. Homyel'skaya Voblasts', SE Belarus
Ptsich *85 C7* *Rus.* Ptich'. *river* SE Belarus
Ptuj *73 E7* *Ger.* Pettau; *anc.* Poetovio. NE Slovenia
Pucallpa *38 C3* Ucayali, C Peru
Puck *76 C2* Pomorskie, N Poland
Pudasjärvi *62 D4* Pohjois-Pohjanmaa, C Finland
Puebla *29 F4* *var.* Puebla de Zaragoza. Puebla, S Mexico
Puebla de Zaragoza *see* Puebla
Pueblo *22 D5* Colorado, C USA
Puerto Acosta *39 E4* La Paz, W Bolivia
Puerto Aisén *43 B6* Aisén, S Chile
Puerto Ángel *29 F5* Oaxaca, SE Mexico
Puerto Argentino *see* Stanley
Puerto Ayacucho *36 D3* Amazonas, SW Venezuela
Puerto Baquerizo Moreno *38 B5* *var.* Baquerizo Moreno. Galápagos Islands, Ecuador, E Pacific Ocean
Puerto Barrios *30 C2* Izabal, E Guatemala
Puerto Bello *see* Portobelo
Puerto Berrío *36 B2* Antioquia, C Colombia
Puerto Cabello *36 D1* Carabobo, N Venezuela
Puerto Cabezas *31 E2* *var.* Bilwi. Región Autónoma Atlántico Norte, NE Nicaragua
Puerto Carreño *36 D3* Vichada, E Colombia
Puerto Cortés *30 C2* Cortés, NW Honduras
Puerto Cumarebo *36 C1* Falcón, N Venezuela
Puerto Deseado *43 C7* Santa Cruz, SE Argentina
Puerto Escondido *29 F5* Oaxaca, SE Mexico
Puerto Francisco de Orellana *38 B1* *var.* Coca. NE Ecuador
Puerto Gallegos *see* Río Gallegos
Puerto Inírida *36 D3* *var.* Obando. Guainía, E Colombia
Puerto La Cruz *37 E1* Anzoátegui, NE Venezuela
Puerto Lempira *31 E2* Gracias a Dios, E Honduras
Puerto Limón *see* Limón
Puertollano *70 D4* Castilla-La Mancha, C Spain
Puerto López *36 C1* La Guajira, N Colombia
Puerto Maldonado *39 E3* Madre de Dios, E Peru
Puerto México *see* Coatzacoalcos
Puerto Montt *43 B5* Los Lagos, C Chile
Puerto Natales *43 B7* Magallanes, S Chile
Puerto Obaldía *31 H5* Kuna Yala, NE Panama
Puerto Plata *33 E3* *var.* San Felipe de Puerto Plata. N Dominican Republic
Puerto Presidente Stroessner *see* Ciudad del Este
Puerto Princesa *117 E2* *off.* Puerto Princesa City. Palawan, W Philippines
Puerto Princesa City *see* Puerto Princesa
Puerto Príncipe *see* Camagüey
Puerto Rico *33 F3* *off.* Commonwealth of Puerto Rico; *prev.* Porto Rico. *US commonwealth territory* C West Indies
Puerto Rico *34 B1* *island* C West Indies
Puerto Rico, Commonwealth of *see* Puerto Rico
Puerto Rico Trench *34 B1* *trench* NE Caribbean Sea
Puerto San José *see* San José
Puerto San Julián *43 B7* *var.* San Julián. Santa Cruz, SE Argentina
Puerto Suárez *39 H4* Santa Cruz, E Bolivia
Puerto Vallarta *28 D4* Jalisco, SW Mexico
Puerto Varas *43 B5* Los Lagos, C Chile
Puerto Viejo *31 E4* Heredia, NE Costa Rica
Puertoviejo *see* Portoviejo
Puget Sound *24 B1* *sound* Washington, NW USA
Puglia *75 E5* *var.* Le Puglie, *Eng.* Apulia. *region* SE Italy
Pukaki, Lake *129 B6* *lake* South Island, New Zealand
Pukekohe *128 D3* Auckland, North Island, New Zealand
Puket *see* Phuket
Pukhavichy *85 C6* *Rus.* Pukhovichi. Minskaya Voblasts', C Belarus
Pukhovichi *see* Pukhavichy
Pula *78 A3* *It.* Pola; *prev.* Pulj. Istra, NW Croatia
Pulaski *18 D5* Virginia, NE USA
Puławy *76 D4* *Ger.* Neu Amerika. Lubelskie, E Poland
Pul-e Khumrī *101 E4* *prev.* Pol-e Khomrī, *var.* Pul-i-Khumri. Baghlān, NE Afghanistan
Pul-i-Khumri *see* Pul-e Khumrī
Pulj *see* Pula
Pullman *24 C2* Washington, NW USA
Pułtusk *76 D3* Mazowieckie, C Poland
Puná, Isla *38 A2* *island* SW Ecuador
Pune *112 C5* *prev.* Poona. Mahārāshtra, W India
Punjab *112 C2* *prev.* West Punjab, Western Punjab. *province* E Pakistan
Puno *39 E4* Puno, SE Peru
Punta Alta *43 C5* Buenos Aires, E Argentina
Punta Arenas *43 B8* *prev.* Magallanes. Magallanes, S Chile
Punta Gorda *30 C2* Toledo, SE Belize
Punta Gorda *31 E4* Región Autónoma Atlántico Sur, SE Nicaragua
Puntarenas *30 D4* Puntarenas, W Costa Rica
Punto Fijo *36 C1* Falcón, N Venezuela
Pupuya, Nevado *39 E4* *mountain* W Bolivia
Puri *113 F5* *var.* Jagannath. Odisha, E India
Puriramya *see* Buriram
Purmerend *64 C3* Noord-Holland, C Netherlands
Pursat *115 D6* *Khmer.* Poŭthĭsăt. Pursat, W Cambodia
Purus, Rio *40 C2* *var.* Río Purús. *river* Brazil/Peru
Pusan *see* Busan
Pushkino *see* Biläsuvar
Püspökladány *77 D6* Hajdú-Bihar, E Hungary
Putorana, Gory/Putorana Mountains *see* Putorana, Plato
Putorana Mountains *93 E3* *var.* Gory Putorana, *Eng.* Putorana Mountains. *mountain range* N Russia
Putrajaya *116 B3* *administrative capital* (Malaysia) Kuala Lumpur, Peninsular Malaysia
Puttalam *110 C3* North Western Province, W Sri Lanka
Puttgarden *72 C2* Schleswig-Holstein, N Germany
Putumayo, Río *36 B5* *var.* Içá, Rio. *river* NW South America
Puurmani *84 D2* *Ger.* Talkhof. Jõgevamaa, E Estonia
Pyatigorsk *89 B7* Stavropol'skiy Kray, SW Russia
Pyatikhatki *see* P"yatykhatky
P"yatykhatky *87 F3* *Rus.* Pyatikhatki. Dnipropetrovs'ka Oblast', E Ukraine
Pyay *114 B4* *var.* Prome, Pye. Bago, C Myanmar (Burma)
Pye *see* Pyay
Pyetrykaw *85 C7* *Rus.* Petrikov. Homyel'skaya Voblasts', SE Belarus
Pyin-Oo-Lwin *114 B3* *var.* Maymyo. Mandalay, C Myanmar (Burma)
Pýlos *83 B6* *var.* Pilos. Pelopónnisos, S Greece
P'yŏngyang *107 E3* *var.* P'yŏngyang-si, *Eng.* Pyongyang. *country capital* (North Korea) SW North Korea
P'yŏngyang-si *see* P'yŏngyang
Pyramid Lake *25 C5* *lake* Nevada, W USA
Pyrenaei Montes *see* Pyrenees
Pyrenees *80 B2* *Fr.* Pyrénées, *Sp.* Pirineos; *anc.* Pyrenaei Montes. *mountain range* SW Europe
Pýrgos *83 B6* *var.* Pírgos. Dytikí Elláς, S Greece
Pyritz *see* Pyrzyce
Pyryatyn *87 E2* *Rus.* Piryatin. Poltavs'ka Oblast', NE Ukraine
Pyrzyce *76 B3* *Ger.* Pyritz. Zachodnio-pomorskie, NW Poland
Pyu *see* Phyu
Pyuntaza *114 B4* Bago, SW Myanmar (Burma)

Q

Qā' al Jafr *97 C7* *lake* S Jordan
Qaanaaq *60 D1* *var.* Qânâq, *Dan.* Thule. Avannaarsua, N Greenland
Qabātiya *97 E6* N West Bank Asia
Qābis *see* Gabès
Qābis, Khalīj *see* Gabès, Golfe de
Qacentina *see* Constantine
Qafşah *see* Gafsa
Qagan Us *see* Dulan
Qahremānshahr *see* Kermānshāh
Qaidam Pendi *104 C4* *basin* C China
Qal'aikhum *101 F3* *Rus.* Kalaikhum. S Tajikistan
Qalāt *101 E5* *Per.* Kalāt. Zābol, S Afghanistan
Qal'at Bīshah *99 B5* 'Asīr, SW Saudi Arabia
Qalqiliya *see* Qalqīlya
Qalqīlya *97 D6* *var.* Qalqiliya. Central, W West Bank, Asia
Qamdo *104 D5* Xizang Zizhiqu, W China
Qamishly *see* Al Qāmishlī
Qânâq *see* Qaanaaq
Qaqortoq *60 C4* *Dan.* Julianehåb. Kitaa, S Greenland
Qaraghandy/Qaraghandy Oblysy *see* Karagandy
Qara Qum *see* Garagum
Qarataŭ *see* Karatau, Zhambyl, Kazakhstan
Qarkilik *see* Ruoqiang
Qarokŭl *101 F3* *Rus.* Karakul'. E Tajikistan
Qarqannah, Juzur *see* Kerkenah, Îles de
Qars *see* Kars
Qarshi *101 E3* *Rus.* Karshi; *prev.* Bek-Budi. Qashqadaryo Viloyati, S Uzbekistan
Qasigianguit *see* Qasigiannguit
Qasigiannguit *60 C3* *var.* Qasigianguit, *Dan.* Christianshåb. Kitaa, C Greenland
Qaşr al Farāfirah *50 B2* *var.* Qasr Farâfra. W Egypt
Qasr Farâfra *see* Qaşr al Farāfirah
Qaţanā *97 B5* *var.* Katana. Dimashq, S Syria
Qatar *98 C4* *off.* State of Qatar, *Ar.* Dawlat Qaţar. *country* SW Asia
Qatar, State of *see* Qatar
Qattara Depression *see* Qaţţārah, Munkhafaḑ al
Qaţţāra, Monkhafad el *see* Qaţţārah, Munkhafaḑ al
Qaţţārah, Munkhafaḑ al *50 A1* *var.* Monkhafad el Qaţţāra, *Eng.* Qattara Depression. *desert* NW Egypt
Qausuittuq *see* Resolute
Qazaqstan/Qazaqstan Respublikasy *see* Kazakhstan
Qazimämmäd *95 H3* *Rus.* Kazi Magomed. SE Azerbaijan
Qazris *see* Cáceres
Qazvīn *98 C2* *var.* Kazvin. Qazvīn, N Iran
Qena *see* Qinā
Qeqertarssuaq *see* Qeqertarsuaq
Qeqertarsuaq *60 C3* *var.* Qeqertarssuaq, *Dan.* Godhavn. Kitaa, S Greenland
Qeqertarsuaq *60 C3* *island* W Greenland
Qeqertarsuup Tunua *60 C3* *Dan.* Disko Bugt. *inlet* W Greenland
Qerveh *see* Qorveh
Qeshm *98 D4* *var.* Jazīreh-ye Qeshm, Qeshm Island. *island* S Iran
Qeshm Island/Qeshm, Jazīreh-ye *see* Qeshm
Qilian Shan *104 D3* *var.* Kilien Mountains. *mountain range* N China
Qimusseriarsuaq *60 C2* *Dan.* Melville Bugt, *Eng.* Melville Bay. *bay* NW Greenland
Qinā *50 B2* *var.* Qena; *anc.* Caene, Caenepolis. E Egypt
Qing *see* Qinghai
Qingdao *106 D4* *var.* Ching-Tao, Ch'ing-tao, Tsingtao, Tsintao, *Ger.* Tsingtau. Shandong, E China
Qinghai *104 C4* *var.* Chinghai, Koko Nor, Qing, Qinghai Sheng, Tsinghai. *province* C China
Qinghai Hu *104 D4* *var.* Ch'ing Hai, Tsing Hai, *Mong.* Koko Nor. *lake* C China
Qinghai Sheng *see* Qinghai
Qingzang Gaoyuan *104 B4* *var.* Xizang Gaoyuan, *Eng.* Plateau of Tibet. *plateau* W China
Qinhuangdao *106 D3* Hebei, E China
Qinzhou *106 B6* Guangxi Zhuangzu Zizhiqu, S China
Qiong *see* Hainan
Qiqihar *106 D2* *var.* Ch'i-ch'i-ha-erh, Tsitsihar; *prev.* Lungkiang. Heilongjiang, NE China
Qira *104 B4* Xinjiang Uygur Zizhiqu, NW China
Qita Ghazzah *see* Gaza Strip
Qitai *104 C3* Xinjiang Uygur Zizhiqu, NW China
Qīzān *see* Jīzān
Qizil Orda *see* Kzylorda
Qizil Qum/Qizilqum *see* Kyzyl Kum
Qizilrabot *101 G3* *Rus.* Kyzylrabot. SE Tajikistan
Qogir Feng *see* K2
Qom *98 C3* *var.* Kum, Qum. Qom, N Iran
Qomolangma Feng *see* Everest, Mount
Qomul *see* Hami
Qo'qon *101 F2* *var.* Khokand, *Rus.* Kokand. Farg'ona Viloyati, E Uzbekistan
Qorveh *98 C3* *var.* Qerveh, Qurveh. Kordestān, W Iran
Qostanay/Qostanay Oblysy *see* Kostanay
Qoubaïyât *96 B4* *var.* Al Qubayyāt. N Lebanon
Qoussantîna *see* Constantine
Quang Ngai *115 E5* *var.* Quangngai, Quang Nghia. Quang Ngai, C Vietnam
Quangngai *see* Quang Ngai
Quang Nghia *see* Quang Ngai
Quan Long *see* Ca Mau
Quanzhou *106 D6* *var.* Ch'uan-chou, Tsinkiang; *prev.* Chin-chiang. Fujian, SE China
Quanzhou *106 C6* Guangxi Zhuangzu Zizhiqu, S China
Qu'Appelle *15 F5* *river* Saskatchewan, S Canada
Quarles, Pegunungan *117 E4* *mountain range* Sulawesi, C Indonesia
Quarnero *see* Kvarner
Quartu Sant' Elena *75 A6* Sardegna, Italy, C Mediterranean Sea
Quba *95 H2* *Rus.* Kuba. N Azerbaijan
Qubba *see* Ba'qūbah
Québec *17 E4* *var.* Quebec. *province capital* Québec, SE Canada
Québec *16 D3* *var.* Quebec. *province* SE Canada
Queen Charlotte Islands *14 C5* *Fr.* Îles de la Reine-Charlotte. *island group* British Columbia, SW Canada
Queen Charlotte Sound *14 C5* *sea area* British Columbia, W Canada
Queen Elizabeth Islands *15 E1* *Fr.* Îles de la Reine-Élisabeth. *island group* Nunavut, N Canada
Queensland *126 B4* *state* N Australia
Queenstown *129 B7* Otago, South Island, New Zealand
Queenstown *56 D5* Eastern Cape, S South Africa
Quelimane *57 E3* *var.* Kilimane, Kilmain, Quilimane. Zambézia, NE Mozambique
Quelpart *see* Jeju-do
Quepos *31 E4* Puntarenas, S Costa Rica
Que Que *see* Kwekwe
Quera *see* Chur
Querétaro *29 E4* Querétaro de Arteaga, C Mexico
Quesada *31 E4* *var.* Ciudad Quesada, San Carlos. Alajuela, N Costa Rica
Quetta *112 B2* Baluchistān, SW Pakistan
Quetzalcoalco *see* Coatzacoalcos
Quezaltenango *30 A2* *var.* Quetzaltenango. Quezaltenango, W Guatemala

Rovinj – San Juan Mountains

Rovinj 78 A3 It. Rovigno. Istra, NW Croatia
Rovno see Rivne
Rovuma, Rio 57 F2 var. Ruvuma. river Mozambique/Tanzania
Rovuma, Rio see Ruvuma
Równe see Rivne
Roxas City 117 E2 Panay Island, C Philippines
Royale, Isle 18 B1 island Michigan, N USA
Royan 69 B5 Charente-Maritime, W France
Rozdol'ne 87 F4 Rus. Razdolnoye. Respublika Krym, S Ukraine
Rožňava 77 D6 Ger. Rosenau, Hung. Rozsnyó. Košický Kraj, E Slovakia
Rózsahegy see Ružomberok
Rozsnyó see Râşnov, Romania
Rozsnyó see Rožňava, Slovakia
Ruanda see Rwanda
Ruapehu, Mount 128 D4 volcano North Island, New Zealand
Ruapuke Island 129 B8 island SW New Zealand
Ruatoria 128 E3 Gisborne, North Island, New Zealand
Ruawai 128 D2 Northland, North Island, New Zealand
Rubezhnoye see Rubizhne
Rubizhne 87 H3 Rus. Rubezhnoye. Luhans'ka Oblast', E Ukraine
Ruby Mountains 25 D5 mountain range Nevada, W USA
Rucava 84 B3 Liepāja, SW Latvia
Rudensk see Rudzyensk
Rūdiškės 85 B5 Vilnius, S Lithuania
Rudnik see Dolni Chiflik
Rudny see Rudnyy
Rudnyy 92 C4 var. Rudny. Kostanay, N Kazakhstan
Rudolf, Lake see Turkana, Lake
Rudolfswert see Novo mesto
Rudzyensk 85 C6 Rus. Rudensk. Minskaya Voblasts', C Belarus
Rufiji 51 C7 river E Tanzania
Rufino 42 C4 Santa Fe, C Argentina
Rugāji 84 D4 Balvi, E Latvia
Rügen 72 D2 headland NE Germany
Ruggell 72 E1 N Liechtenstein Europe
Ruhja see Rūjiena
Ruhnu 84 C2 var. Ruhnu Saar, Swe. Runö. island SW Estonia
Ruhnu Saar see Ruhnu
Rujen see Rūjiena
Rūjiena 84 D3 Est. Ruhja, Ger. Rujen. Valmiera, N Latvia
Rukwa, Lake 51 B7 lake SE Tanzania
Rum see Rhum
Ruma 78 D3 Vojvodina, N Serbia
Rumadiya see Ar Ramādī
Rumania/Rumänien see Romania
Rumbek 51 B5 El Buhayrat, C South Sudan
Rum Cay 32 D2 island C The Bahamas
Rumia 76 C2 Pomorskie, N Poland
Rummah, Wādī ar see Rimah, Wādī ar
Rummelsburg in Pommern see Miastko
Rumuniya/Rumŭniya/Rumunjska see Romania
Runanga 129 B5 West Coast, South Island, New Zealand
Rundu 56 C3 var. Runtu. Okavango, NE Namibia
Runö see Ruhnu
Runtu see Rundu
Ruoqiang 104 C3 var. Jo-ch'iang, Uigh. Charkhlik, Charkhliq, Qarkilik. Xinjiang Uygur Zizhiqu, NW China
Rupea 86 C4 Ger. Reps, Hung. Kőhalom; prev. Cohalm. Braşov, C Romania
Rupel 65 B5 river N Belgium
Rupella see la Rochelle
Rupert, Rivière de 16 D3 river Québec, C Canada
Rusaddir see Melilla
Ruschuk/Rusçuk see Ruse
Ruse 82 D1 var. Ruschuk, Rustchuk, Turk. Rusçuk. Ruse, N Bulgaria
Russadir see Melilla
Russellville 20 A1 Arkansas, C USA
Russia 90 D2 off. Russian Federation, Latv. Krievija, Rus. Rossiyskaya Federatsiya. country Asia/Europe
Russian America see Alaska
Russian Federation see Russia
Rustaq see Ar Rustāq
Rustavi 95 G2 prev. Rust'avi. SE Georgia
Rust'avi see Rustavi
Rustchuk see Ruse
Ruston 20 B2 Louisiana, S USA
Rutanzige, Lake see Edward, Lake
Rutba see Ar Ruṭbah
Rutlam see Ratlām
Rutland 19 F2 Vermont, NE USA
Rutog 104 A4 var. Rutög, Rutok. Xizang Zizhiqu, W China
Rutok see Rutog
Ruvuma see Rovuma, Rio
Ruwenzori 55 E5 mountain range Dem. Rep. Congo/Uganda
Ruzhany 85 B6 Brestskaya Voblasts', SW Belarus
Ružomberok 77 C5 Ger. Rosenberg, Hung. Rózsahegy. Žilinský Kraj, N Slovakia
Rwanda 51 B6 off. Rwandese Republic; prev. Ruanda. country C Africa
Rwandese Republic see Rwanda
Ryazan' 89 B5 Ryazanskaya Oblast', W Russia
Rybach'ye see Balykchy
Rybinsk 88 B4 prev. Andropov. Yaroslavskaya Oblast', W Russia
Rybnik 77 C5 Śląskie, S Poland
Rybnitsa see Rîbniţa
Ryde 126 E1 United Kingdom
Ryki 76 D4 Lubelskie, E Poland
Rykovo see Yenakiyeve
Rypin 76 C3 Kujawsko-pomorskie, C Poland
Ryssel see Lille
Rysy 77 C5 mountain S Poland
Ryukyu Islands see Nansei-shotō
Ryukyu Trench 103 F3 var. Nansei Syotō Trench. trench S East China Sea
Rzeszów 77 E5 Podkarpackie, SE Poland
Rzhev 88 B4 Tverskaya Oblast', W Russia

S

Saale 72 C4 river C Germany
Saalfeld 73 C5 var. Saalfeld an der Saale. Thüringen, C Germany
Saalfeld an der Saale see Saalfeld
Saarbrücken 73 A6 Fr. Sarrebruck. Saarland, SW Germany
Sääre 84 C2 var. Sjar. Saaremaa, W Estonia
Saare see Saaremaa
Saaremaa 84 C2 Ger. Oesel, Ösel; prev. Saare. island W Estonia
Saariselkä 62 D2 Lapp. Suoločielgi. Lappi, N Finland
Sab' Ābār 96 C4 var. Sab'a Biyar, Sa'b Bi'ār. Ḩimş, C Syria
Sab'a Biyar see Sab' Ābār
Šabac 78 D3 Serbia, W Serbia
Sabadell 71 G2 Cataluña, E Spain
Sabah 116 D3 prev. British North Borneo, North Borneo. state East Malaysia
Sabanalarga 36 B1 Atlántico, N Colombia
Sabaneta 36 C1 Falcón, N Venezuela
Sabaria see Szombathely
Sab'atayn, Ramlat as 99 C6 desert C Yemen
Sabaya 39 F4 Oruro, S Bolivia
Sa'b Bi'ār see Sab' Ābār
Saberi, Hamun-e 100 C5 var. Daryācheh-ye Hāmun, Daryācheh-ye Sīstān. lake Afghanistan/Iran
Sabhā 49 F3 C Libya
Sabi see Save
Sabinas 29 E2 Coahuila, NE Mexico
Sabinas Hidalgo 29 E2 Nuevo León, NE Mexico
Sabine River 27 H3 river Louisiana/Texas, SW USA
Sabkha see As Sabkhah
Sable, Cape 21 E5 headland Florida, SE USA
Sable Island 17 G4 island Nova Scotia, SE Canada
Şabyā 99 B6 Jīzān, SW Saudi Arabia
Sabzawar see Sabzevār
Sabzevār 98 D2 var. Sabzawar. Khorāsān-Razavī, NE Iran
Sachsen 72 D4 Eng. Saxony, Fr. Saxe. state E Germany
Sachs Harbour 15 E2 var. Ikaahuk. Banks Island, Northwest Territories, N Canada
Sächsisch-Reen/Sächsisch-Regen see Reghin
Sacramento 25 B5 state capital California, W USA
Sacramento Mountains 26 D2 mountain range New Mexico, SW USA
Sacramento River 25 B5 river California, W USA
Sacramento Valley 25 B5 valley California, W USA
Sá da Bandeira see Lubango
Şa'dah 99 B6 NW Yemen
Sado 109 C5 var. Sadoga-shima. island C Japan
Sadoga-shima see Sado
Saena Julia see Siena
Safad see Tsefat
Şafāqis see Sfax
Şafāshahr 98 D3 var. Deh Bīd. Fārs, C Iran
Safed see Tsefat
Säffle 63 B6 Värmland, C Sweden
Safford 26 C3 Arizona, SW USA
Safi 48 B2 W Morocco
Selseleh-ye Safid Kuh 100 D4 Eng. Paropamisus Range. mountain range W Afghanistan
Sagaing 114 B3 Sagaing, C Myanmar (Burma)
Sagami-nada 109 D6 inlet SW Japan
Sagan see Żagań
Sāgar 112 D4 prev. Saugor. Madhya Pradesh, C India
Sagarmāthā see Everest, Mount
Sagebrush State see Nevada
Saghez see Saqqez
Saginaw 18 C3 Michigan, N USA
Saginaw Bay 18 D2 lake bay Michigan, N USA
Sagua la Grande 32 B2 Villa Clara, C Cuba
Sagunto 71 F3 Cat. Sagunt, Ar. Murviedro; anc. Saguntum. País Valenciano, E Spain
Sagunt/Saguntum see Sagunto
Sahara 46 B3 desert Libya/Algeria
Sahara el Gharbîya see Şaḩrā' al Gharbīyah
Saharan Atlas see Atlas Saharien
Sahel 52 D3 physical region C Africa
Sāḩilīyah, Jibāl as 96 B3 mountain range NW Syria
Sāhīwāl 112 C2 prev. Montgomery. Punjab, E Pakistan
Şaḩrā' al Gharbīyah 50 B2 var. Sahara el Gharbîya, Eng. Western Desert. desert C Egypt
Şaḩrā' ash Sharqīyah 81 H5 Eng. Arabian Desert, Eastern Desert. desert E Egypt
Saïda 97 A5 var. Şaydā, Sayida; anc. Sidon. W Lebanon
Sa'īdābād see Sīrjān
Saidpur 113 G3 var. Syedpur. Rajshahi, NW Bangladesh
Saidu Sharif 112 C1 var. Mingora, Mongora. North-West Frontier Province, N Pakistan
Saigon see Hô Chi Minh
Saimaa 63 E5 lake SE Finland
St Albans 67 E6 anc. Verulamium. E England, United Kingdom
Saint Albans 18 D5 West Virginia, NE USA
St Andrews 66 C4 E Scotland, United Kingdom
Saint Anna Trough see Svyataya Anna Trough
St. Ann's Bay 32 B4 C Jamaica
St. Anthony 17 G3 Newfoundland and Labrador, SE Canada
Saint Augustine 21 E3 Florida, SE USA
St Austell 67 C7 SW England, United Kingdom
St.Botolph's Town see Boston
St-Brieuc 68 A3 Côtes d'Armor, NW France
St. Catharines 16 D5 Ontario, S Canada
St-Chamond 69 D5 Loire, E France
Saint Christopher and Nevis, Federation of see Saint Kitts and Nevis
Saint Christopher-Nevis see Saint Kitts and Nevis
Saint Clair, Lake 18 D3 var. Lac à L'Eau Claire. lake Canada/USA
St-Claude 69 D5 anc. Condate. Jura, E France
Saint Cloud 23 F2 Minnesota, N USA
Saint Croix 33 F3 island S Virgin Islands (US)
Saint Croix River 18 A2 river Minnesota/Wisconsin, N USA
St David's Island 20 B5 island E Bermuda
St-Denis 57 G4 dependent territory capital (Réunion) NW Réunion
St-Dié 68 E4 Vosges, NE France
St-Egrève 69 D5 Isère, E France
Sainte Marie, Cap see Vohimena, Tanjona
Saintes 69 B5 anc. Mediolanum. Charente-Maritime, W France
St-Étienne 69 D5 Loire, E France
St-Flour 69 C5 Cantal, C France
St-Gall/Saint Gall/St. Gallen see Sankt Gallen
St-Gaudens 69 B6 Haute-Garonne, S France
Saint George 127 D5 Queensland, E Australia
St George 20 B4 N Bermuda
Saint George 22 A5 Utah, W USA
St. George's 33 G5 country capital (Grenada) SW Grenada
St-Georges 17 E4 Québec, SE Canada
St-Georges 37 H3 E French Guiana
Saint George's Channel 67 B6 channel Ireland/Wales, United Kingdom
St George's Island 20 B4 island E Bermuda
St Helena see St Helena, Ascension and Tristan da Cunha
St Helena, Ascension and Tristan da Cunha 47 A6 UK overseas territory C Atlantic Ocean
St Helier 67 D8 dependent territory capital (Jersey) S Jersey, Channel Islands
St.Iago de la Vega see Spanish Town
Saint Ignace 18 C2 Michigan, N USA
St-Jean, Lac 17 E4 lake Québec, SE Canada
Saint Joe River 24 D2 river Idaho, NW USA North America
St. John 17 F4 New Brunswick, SE Canada
Saint-John see Saint John
Saint John 19 H1 Fr. Saint-John. river Canada/USA
St John's 33 G3 country capital (Antigua and Barbuda) Antigua, Antigua and Barbuda
St. John's 17 H3 province capital Newfoundland and Labrador, E Canada
Saint Joseph 23 F4 Missouri, C USA
St Julian's see San Giljan
St Kilda 66 A3 island NW Scotland, United Kingdom
Saint Kitts and Nevis 33 F3 off. Federation of Saint Christopher and Nevis, var. Saint Christopher-Nevis. country E West Indies
St-Laurent see St-Laurent-du-Maroni
St-Laurent-du-Maroni 37 H3 var. St-Laurent. NW French Guiana
St-Laurent, Fleuve see St. Lawrence
St. Lawrence 17 E4 Fr. Fleuve St-Laurent. river Canada/USA
St. Lawrence, Gulf of 17 F3 gulf NW Atlantic Ocean
Saint Lawrence Island 14 B2 island Alaska, USA
St-Lô 68 B3 anc. Briovera, Laudus. Manche, N France
St-Louis 68 E4 Haut-Rhin, NE France
Saint Louis 52 B3 NW Senegal
Saint Louis 23 G4 Missouri, C USA
Saint Lucia 33 E1 country SE West Indies
Saint Lucia Channel 33 H4 channel Martinique/Saint Lucia
St-Malo 68 B3 Ille-et-Vilaine, NW France
St-Malo, Golfe de 68 A3 gulf NW France
Saint Martin see Sint Maarten
St.Matthew's Island see Zadetkyi Kyun
St.Matthias Group 122 B3 island group NE Papua New Guinea
St. Moritz 73 B7 Ger. Sankt Moritz, Rmsch. San Murezzan. Graubünden, SE Switzerland
St-Nazaire 68 A4 Loire-Atlantique, NW France
Saint Nicholas see São Nicolau
Saint-Nicolas see Sint-Niklaas
St-Omer 68 C2 Pas-de-Calais, N France
Saint Paul 23 F2 state capital Minnesota, N USA
St-Paul, Île 119 C6 var. St.Paul Island. island Île St-Paul, NE French Southern and Antarctic Lands Antarctica Indian Ocea
St.Paul Island see St-Paul, Île
St Peter Port 67 D8 dependent territory capital (Guernsey) C Guernsey, Channel Islands
Saint Petersburg 21 E4 Florida, SE USA
Saint Petersburg see Sankt-Peterburg
St-Pierre and Miquelon 17 G4 Fr. Îles St-Pierre et Miquelon. French territorial collectivity NE North America
St-Quentin 68 C3 Aisne, N France
Saint Thomas see São Tomé, Sao Tome and Principe
Saint Thomas see Charlotte Amalie, Virgin Islands (US)
Saint Ubes see Setúbal
Saint Vincent 33 G4 island N Saint Vincent and the Grenadines
Saint Vincent see São Vicente
Saint Vincent and the Grenadines 33 H4 country SE West Indies
Saint Vincent, Cape see São Vicente, Cabo de
Saint Vincent Passage 33 H4 passage Saint Lucia/Saint Vincent and the Grenadines
Saint Yves see Setúbal
Saipan 120 B1 island/country capital (Northern Mariana Islands) S Northern Mariana Islands
Saishū see Jeju-do
Sajama, Nevado 39 F4 mountain W Bolivia
Sajószentpéter 77 D6 Borsod-Abaúj-Zemplén, NE Hungary
Sakākah 98 B4 Al Jawf, NW Saudi Arabia
Sakakawea, Lake 22 D1 reservoir North Dakota, N USA
Sak'art'velo see Georgia
Sakata 108 D4 Yamagata, Honshū, C Japan
Sakhalin 93 G4 var. Sakhalin. island SE Russia
Sakhalin see Sakhalin, Ostrov
Sakhon Nakhon see Sakon Nakhon
Şäki 95 G2 Rus. Sheki; prev. Nukha. NW Azerbaijan
Saki see Saky
Sakishima-shotō 108 A3 var. Sakisima Syotô. island group SW Japan
Sakisima Syotô see Sakishima-shotō
Sakiz see Saqqez
Sakiz-Adasi see Chios
Sakon Nakhon 114 D4 var. Muang Sakon Nakhon, Sakhon Nakhon. Sakon Nakhon, E Thailand
Saky 87 F5 Rus. Saki. Respublika Krym, S Ukraine
Sal 52 A3 island Ilhas de Barlavento, NE Cape Verde
Sala 63 C6 Västmanland, C Sweden
Salacgrīva 84 C3 Est. Salatsi. Limbaži, N Latvia
Sala Consilina 75 D5 Campania, S Italy
Salado, Río 40 D5 river E Argentina
Salado, Río 42 C3 river C Argentina
Şalālah 99 D6 SW Oman
Salamá 30 B2 Baja Verapaz, C Guatemala
Salamanca 42 B4 Coquimbo, C Chile
Salamanca 70 D2 anc. Helmantica, Salmantica. Castilla-León, NW Spain
Salamīyah 96 B3 var. As Salamīyah. Ḩamāh, W Syria
Salang see Phuket
Salantai 84 B3 Klaipėda, NW Lithuania
Salatsi see Salacgrīva
Salavan 115 D5 var. Saravan, Saravane. Salavan, S Laos
Salavat 89 D6 Respublika Bashkortostan, W Russia
Sala y Gomez 131 F4 island Chile, E Pacific Ocean
Sala y Gomez Fracture Zone see Sala y Gomez Ridge
Sala y Gomez Ridge 131 G4 var. Sala y Gomez Fracture Zone. fracture zone SE Pacific Ocean
Salazar see N'Dalatando
Šalčininkai 85 C5 Vilnius, SE Lithuania
Salduba see Zaragoza
Saldus 84 B3 Ger. Frauenburg. Saldus, W Latvia
Sale 127 C7 Victoria, SE Australia
Salé 48 C2 NW Morocco
Salekhard 92 D3 prev. Obdorsk. Yamalo-Nenetskiy Avtonomnyy Okrug, N Russia
Salem 110 C2 Tamil Nādu, SE India
Salem 24 B3 state capital Oregon, NW USA
Salerno 75 D5 anc. Salernum. Campania, S Italy
Salerno, Gulf of 75 C5 Eng. Gulf of Salerno. gulf S Italy
Salerno, Gulf of see Salerno, Golfo di
Salernum see Salerno
Salihorsk 85 C7 Rus. Soligorsk. Minskaya Voblasts', S Belarus
Salima 57 E2 Central, C Malawi
Salina 23 E5 Kansas, C USA
Salina Cruz 29 F5 Oaxaca, SE Mexico
Salinas 38 A2 Guayas, W Ecuador
Salinas 25 B6 California, W USA
Salisbury 67 D7 var. New Sarum. S England, United Kingdom
Salisbury see Harare
Sällan see Sørøya
Salliq see Coral Harbour
Sallyana see Şalyān
Salmantica see Salamanca
Salmon River 24 D3 river Idaho, NW USA
Salmon River Mountains 24 D3 mountain range Idaho, NW USA
Salo 63 D6 Länsi-Suomi, SW Finland
Salon-de-Provence 69 D6 Bouches-du-Rhône, SE France
Salonica/Salonika see Thessaloníki
Salonta 86 A3 Hung. Nagyszalonta. Bihor, W Romania
Sal'sk 89 B7 Rostovskaya Oblast', SW Russia
Salt see As Salṭ
Salta 42 C2 Salta, NW Argentina
Saltash 67 C7 SW England, United Kingdom
Saltillo 29 E3 Coahuila, NE Mexico
Salt Lake City 22 B4 state capital Utah, W USA
Salto 42 D4 Salto, N Uruguay
Salton Sea 25 D8 lake California, W USA
Salvador 41 G3 prev. São Salvador. state capital Bahia, E Brazil
Salween 102 C2 Bur. Thanlwin, Chin. Nu Chiang, Nu Jiang. river SE Asia
Şalyān 113 E3 var. Sallyana. Mid Western, W Nepal
Salzburg 73 D6 anc. Juvavum. Salzburg, N Austria
Salzgitter 72 C4 prev. Watenstedt-Salzgitter. Niedersachsen, C Germany
Salzwedel 72 C3 Sachsen-Anhalt, N Germany
Šamac see Bosanski Šamac
Samakhixai see Attapu
Samalayuca 28 C1 Chihuahua, N Mexico
Samar 117 F2 island C Philippines
Samara 92 B3 prev. Kuybyshev. Samarskaya Oblast', W Russia
Samarang see Semarang
Samarinda 116 D4 Borneo, C Indonesia
Samarkand see Samarqand
Samarkandski/Samarkandskoye see Temirtau
Samarobriva see Amiens
Samarqand 101 E2 Rus. Samarkand. Samarqand Viloyati, C Uzbekistan
Samawa see As Samāwah
Şamaxı 95 H2 E Azerbaijan
Sambalpur 113 F4 Odisha, E India
Sambava 57 G2 Antsiranana, NE Madagascar
Sambir 86 B2 Rus. Sambor. L'vivs'ka Oblast', NW Ukraine
Sambor see Sambir
Sambre 68 D2 river Belgium/France
Samfya 56 D2 Luapula, N Zambia
Saminatal 72 E2 valley Austria/Liechtenstein, Europe
Samnān see Semnān
Sam Neua see Xam Nua
Samoa 123 E4 off. Independent State of Western Samoa, var. Sāmoa; prev. Western Samoa. country W Polynesia
Sāmoa see Samoa
Samoa Basin 121 E3 undersea basin W Pacific Ocean
Samobor 78 A2 Zagreb, N Croatia
Sámos 83 E6 prev. Limín Vathéos. Sámos, Dodekánisa, Greece, Aegean Sea
Sámos 83 D6 island Dodekánisa, Greece, Aegean Sea
Samothrace see Samothráki
Samothráki 82 D4 Samothráki, NE Greece
Samothráki 82 C4 anc. Samothrace. island NE Greece
Sampit 116 C4 Borneo, C Indonesia
Sâmraông 115 D5 prev. Phumĭ Sâmraông, Phum Samrong. Siĕmréab, NW Cambodia
Samsun 94 D2 anc. Amisus. Samsun, N Turkey
Samt'redia 95 F2 W Georgia
Samui, Ko 115 C6 island SW Thailand
Samut Prakan 115 C5 var. Muang Samut Prakan, Paknam. Samut Prakan, C Thailand
San 52 D3 Ségou, C Mali
San 77 E5 river SE Poland
Şan'ā' 99 B6 Eng. Sanaa. country capital (Yemen) W Yemen
Sana 78 B3 river NW Bosnia and Herzegovina
Sanaa see Şan'ā'
Sanae 132 B2 South African research station Antarctica
Sanaga 55 B5 river C Cameroon
San Ambrosio, Isla 35 A5 Eng. San Ambrosio Island. island W Chile
San Ambrosio Island see San Ambrosio, Isla
Sanandaj 98 C3 prev. Sinneh. Kordestān, W Iran
San Andrés, Isla de 31 F3 island NW Colombia, Caribbean Sea
San Andrés Tuxtla 29 F4 var. Tuxtla. Veracruz-Llave, E Mexico
San Angelo 27 F3 Texas, SW USA
San Antonio 30 B2 Toledo, S Belize
San Antonio 42 B4 Valparaíso, C Chile
San Antonio 27 F4 Texas, SW USA
San Antonio Oeste 43 C5 Río Negro, E Argentina
San Antonio River 27 G4 river Texas, SW USA
Sanāw 99 C6 var. Sanaw. NE Yemen
San Benedicto, Isla 28 B4 island W Mexico
San Benito 30 B1 Petén, N Guatemala
San Benito 27 G5 Texas, SW USA
San Bernardino 25 C7 California, W USA
San Blas 28 C3 Sinaloa, C Mexico
San Blas, Cape 20 D3 headland Florida, SE USA
San Blas, Cordillera de 31 G4 mountain range NE Panama
San Carlos 30 D4 Río San Juan, S Nicaragua
San Carlos 26 B2 Arizona, SW USA
San Carlos see Quesada, Costa Rica
San Carlos de Ancud see Ancud
San Carlos de Bariloche 43 B5 Río Negro, SW Argentina
San Carlos del Zulia 36 C2 Zulia, W Venezuela
San Clemente Island 25 B8 island Channel Islands, California, W USA
San Cristóbal 36 C2 Táchira, W Venezuela
San Cristóbal 122 C4 var. Makira. island SE Solomon Islands
San Cristóbal see San Cristóbal de Las Casas
San Cristóbal de Las Casas 29 G5 var. San Cristóbal. Chiapas, SE Mexico
San Cristóbal, Isla 38 B5 var. Chatham Island. island Galápagos Islands, Ecuador, E Pacific Ocean
Sancti Spíritus 32 B2 Sancti Spíritus, C Cuba
Sandakan 116 D3 Sabah, East Malaysia
Sandalwood Island see Sumba, Pulau
Sandanski 82 C3 prev. Sveti Vrach. Blagoevgrad, SW Bulgaria
Sanday 66 D2 island NE Scotland, United Kingdom
Sanders 26 C2 Arizona, SW USA
Sand Hills 22 D3 mountain range Nebraska, C USA
San Diego 25 C8 California, W USA
Sandnes 63 A6 Rogaland, S Norway
Sandomierz 76 D4 Rus. Sandomir. Świętokrzyskie, C Poland
Sandomir see Sandomierz
Sandoway see Thandwe
Sandpoint 24 C1 Idaho, NW USA
Sand Springs 27 G1 Oklahoma, C USA
Sandusky 18 D3 Ohio, N USA
Sandvika 63 A6 Akershus, S Norway
Sandviken 63 C6 Gävleborg, C Sweden
Sandwich Island see Efate
Sandwich Islands see Hawaiian Islands
Sandy Bay 71 H5 Saskatchewan, C Canada
Sandy City 22 B4 Utah, W USA
Sandy Lake 16 B3 lake Ontario, C Canada
San Esteban 30 D2 Olancho, C Honduras
San Eugenio/San Eugenio del Cuareim see Artigas
San Felipe 36 D1 Yaracuy, NW Venezuela
San Felipe de Puerto Plata see Puerto Plata
San Félix, Isla 35 A5 Eng. San Felix Island. island W Chile
San Felix Island see San Félix, Isla
San Fernando 70 C5 prev. Isla de León. Andalucía, S Spain
San Fernando 33 H5 Trinidad, Trinidad and Tobago
San Fernando 24 D1 California, W USA
San Fernando 36 D2 var. San Fernando de Apure. Apure, C Venezuela
San Fernando de Apure see San Fernando
San Fernando del Valle de Catamarca 42 C3 var. Catamarca. Catamarca, NW Argentina
San Fernando de Monte Cristi see Monte Cristi
San Francisco 25 B6 California, W USA
San Francisco del Oro 28 C2 Chihuahua, N Mexico
San Francisco de Macorís 33 E3 C Dominican Republic
San Fructuoso see Tacuarembó
San Gabriel 38 B1 Carchi, N Ecuador
San Gabriel Mountains 24 E1 mountain range California, USA
Sangihe, Kepulauan see Sangir, Kepulauan
San Giljan 80 B5 var. St Julian's. N Malta
Sangir, Kepulauan 117 F3 var. Kepulauan Sangihe. island group N Indonesia
Sāngli 110 B1 Mahārāshtra, W India
Sangmélima 55 B5 Sud, S Cameroon
Sangre de Cristo Mountains 26 D1 mountain range Colorado/New Mexico, C USA
San Ignacio 30 B1 prev. Cayo, El Cayo. Cayo, W Belize
San Ignacio 39 F3 Beni, N Bolivia
San Ignacio 28 B2 Baja California Sur, NW Mexico
San Joaquin Valley 25 B7 valley California, W USA
San Jorge, Golfo 43 C6 var. Gulf of San Jorge. gulf S Argentina
San Jorge, Gulf of see San Jorge, Golfo
San José 31 E4 country capital (Costa Rica) San José, C Costa Rica
San José 39 G3 var. San José de Chiquitos. Santa Cruz, E Bolivia
San José 30 B3 var. Puerto San José. Escuintla, S Guatemala
San Jose 25 B6 California, W USA
San José see San José del Guaviare, Colombia
San José de Chiquitos see San José
San José de Cúcuta see Cúcuta
San José del Guaviare 36 C4 var. San José. Guaviare, S Colombia
San Juan 42 B4 San Juan, W Argentina
San Juan 33 F3 dependent territory capital (Puerto Rico) NE Puerto Rico
San Juan see San Juan de los Morros
San Juan Bautista 42 D3 Misiones, S Paraguay
San Juan Bautista see Villahermosa
San Juan Bautista Tuxtepec see Tuxtepec
San Juan de Alicante see Sant Joan d'Alacant
San Juan del Norte 31 E4 var. Greytown. Río San Juan, SE Nicaragua
San Juan de los Morros 36 D2 var. San Juan. Guárico, N Venezuela
San Juanito, Isla 28 C4 island C Mexico
San Juan Mountains 26 D1 mountain range Colorado, C USA

San Juan, Río – Severn

San Juan, Río 31 E4 river Costa Rica/Nicaragua
San Juan River 26 C1 river Colorado/Utah, W USA
San Julián see Puerto San Julián
Sankt-Bartholomäi see Palamuse
Sankt Gallen 73 B7 var. St. Gallen, Eng. Saint Gall, Fr. St-Gall. Sankt Gallen, NE Switzerland
Sankt-Georgen see Sfântu Gheorghe
Sankt-Jakobi see Pärnu-Jaagupi, Pärnumaa, Estonia
Sankt Martin see Martin
Sankt Moritz see St. Moritz
Sankt-Peterburg 88 B4 prev. Leningrad, Petrograd, Eng. Saint Petersburg, Fin. Pietari. Leningradskaya Oblast', NW Russia
Sankt Pölten 73 E6 Niederösterreich, N Austria
Sankt Veit am Flaum see Rijeka
Sankuru 55 D6 river C Dem. Rep. Congo
Şanlıurfa 95 E4 prev. Sanli Urfa, Urfa; anc. Edessa. şanlıurfa, S Turkey
Sanli Urfa see Şanlıurfa
San Lorenzo 39 G5 Tarija, S Bolivia
San Lorenzo 38 A1 Esmeraldas, N Ecuador
San Lorenzo, Isla 38 C4 island W Peru
Sanlúcar de Barrameda 70 C5 Andalucía, S Spain
San Luis 42 C4 San Luis, C Argentina
San Luis 30 B2 Petén, NE Guatemala
San Luis see San Luis Río Colorado
San Luis Obispo 25 B7 California, W USA
San Luis Potosí 29 E3 San Luis Potosí, C Mexico
San Luis Río Colorado 28 A1 var. San Luis Río Colorado. Sonora, NW Mexico
San Marcos 30 A2 San Marcos, W Guatemala
San Marcos 27 G4 Texas, SW USA
San Marcos de Arica see Arica
San Marino 74 E1 country capital (San Marino) C San Marino
San Marino 74 D1 off. Republic of San Marino. country S Europe
San Marino, Republic of see San Marino
San Martín 132 A2 Argentinian research station Antarctica
San Mateo 37 E2 Anzoátegui, NE Venezuela
San Matías 39 H3 Santa Cruz, E Bolivia
San Matías, Gulf of 43 C5 var. Gulf of San Matías. gulf E Argentina
San Matías, Gulf of see San Matías, Golfo
Sanmenxia 106 C4 var. Shan Xian. Henan, C China
Sânmiclăuş Mare see Sânnicolau Mare
San Miguel 30 C3 San Miguel, SE El Salvador
San Miguel 28 D2 Coahuila, N Mexico
San Miguel de Ibarra see Ibarra
San Miguel de Tucumán 42 C3 var. Tucumán. Tucumán, N Argentina
San Miguelito 31 G4 Panamá, C Panama
San Miguel, Río 39 G3 river E Bolivia
San Murezzan see St. Moritz
Sannār see Sennar
Sânnicolaul-Mare see Sânnicolau Mare
Sânnicolau Mare 86 A4 var. Sânnicolaul-Mare, Hung. Nagyszentmiklós; prev. Sânmiclăuş Mare, Sînnicolau Mare. Timiş, W Romania
Sanok 77 E5 Podkarpackie, SE Poland
San Pablo 39 F5 Potosí, S Bolivia
San Pedro 30 C1 Corozal, NE Belize
San-Pédro 52 D5 S Ivory Coast
San Pedro 28 D3 var. San Pedro de las Colonias. Coahuila, NE Mexico
San Pedro de la Cueva 28 C2 Sonora, NW Mexico
San Pedro de las Colonias see San Pedro
San Pedro de Lloc 38 B3 La Libertad, NW Peru
San Pedro Mártir, Sierra 28 A1 mountain range NW Mexico
San Pedro Sula 30 C2 Cortés, NW Honduras
San Rafael 42 B4 Mendoza, W Argentina
San Rafael Mountains 25 C7 mountain range California, W USA
San Ramón de la Nueva Orán 42 C2 Salta, N Argentina
San Remo 74 A3 Liguria, NW Italy
San Salvador 30 B3 country capital (El Salvador) San Salvador, SW El Salvador
San Salvador 32 D2 prev. Watlings Island. island E The Bahamas
San Salvador de Jujuy 42 C2 var. Jujuy. Jujuy, N Argentina
San Salvador, Isla 38 A4 island Ecuador
Sansanné-Mango 53 E4 var. Mango. N Togo
San Sebastián 71 E1 País Vasco, N Spain see also Donostia
Sansepolcro 74 C3 Toscana, C Italy
San Severo 75 D5 Puglia, SE Italy
Santa Ana 39 F3 Beni, N Bolivia
Santa Ana 30 B3 Santa Ana, NW El Salvador
Santa Ana 24 D2 California, W USA
Santa Ana de Coro see Coro
Santa Ana Mountains 24 E2 mountain range California, W USA
Santa Barbara 28 C2 Chihuahua, N Mexico
Santa Barbara 25 C7 California, W USA
Santa Catalina de Armada 70 B1 var. Santa Comba. Galicia, NW Spain
Santa Catalina Island 25 B8 island Channel Islands, California, W USA
Santa Catarina 41 E5 off. Estado de Santa Catarina. state/region S Brazil
Santa Catarina, Estado de see Santa Catarina
Santa Clara 32 B2 Villa Clara, C Cuba
Santa Clarita 24 D1 California, USA
Santa Comba see Santa Cataliña de Armada
Santa Cruz 54 E2 São Tomé, S Sao Tome and Principe, Africa
Santa Cruz 25 B6 California, W USA
Santa Cruz 39 G4 department E Bolivia
Santa Cruz Barillas see Barillas
Santa Cruz del Quiché 30 B2 Quiché, W Guatemala
Santa Cruz de Tenerife 48 A3 Tenerife, Islas Canarias, Spain, NE Atlantic Ocean
Santa Cruz, Isla 38 B5 var. Indefatigable Island, Isla Chávez. island Galápagos Islands, Ecuador, E Pacific Ocean
Santa Cruz Islands 122 D3 island group E Solomon Islands
Santa Cruz, Río 43 B7 river S Argentina
Santa Elena 30 B1 Cayo, W Belize
Santa Fe 42 C4 Santa Fe, C Argentina
Santa Fe 26 D1 state capital New Mexico, SW USA
Santa Fe see Bogotá
Santa Fe de Bogotá see Bogotá
Santa Genoveva 28 B3 mountain NW Mexico
Santa Isabel 122 C3 var. Bughotu. island N Solomon Islands
Santa Isabel see Malabo
Santa Lucia Range 25 B7 mountain range California, W USA
Santa Margarita, Isla 28 B3 island NW Mexico
Santa Maria 41 E5 Rio Grande do Sul, S Brazil
Santa Maria 25 B7 California, USA
Santa Maria 70 A5 island Azores, Portugal, NE Atlantic Ocean
Santa Maria del Buen Aire see Buenos Aires
Santa María, Isla 38 A5 var. Isla Floreana, Charles Island. island Galápagos Islands, Ecuador, E Pacific Ocean
Santa Marta 36 B1 Magdalena, N Colombia
Santa Maura see Lefkáda
Santa Monica 24 D1 California, W USA
Santana 54 E2 São Tomé, C Sao Tome and Principe, Africa
Santander 70 D1 Cantabria, N Spain
Santarém 41 E2 Pará, N Brazil
Santarém 70 B3 anc. Scalabis. Santarém, W Portugal
Santa Rosa 42 C4 La Pampa, C Argentina
Santa Rosa see Santa Rosa de Copán
Santa Rosa de Copán 30 C2 var. Santa Rosa. Copán, W Honduras
Santa Rosa Island 25 B8 island California, W USA
Santa Uxía de Ribeira 70 B1 var. Ribeira. NW Spain
Sant Carles de la Rápida see Sant Carles de la Ràpita
Sant Carles de la Ràpita 71 F3 var. Sant Carles de la Rápida. Cataluña, NE Spain
Santiago 42 B4 var. Gran Santiago. country capital (Chile) Santiago, C Chile
Santiago 33 E3 var. Santiago de los Caballeros. N Dominican Republic
Santiago 31 F5 Veraguas, S Panama
Santiago see Santiago de Compostela
Santiago 52 A3 var. São Tiago. island Ilhas de Sotavento, S Cape Verde
Santiago see Santiago de Cuba, Cuba
Santiago de Compostela 70 B1 var. Santiago, Eng. Compostella; anc. Campus Stellae. Galicia, NW Spain
Santiago de Cuba 32 C3 var. Santiago. Santiago de Cuba, E Cuba
Santiago de Guayaquil see Guayaquil
Santiago del Estero 42 C3 Santiago del Estero, C Argentina
Santiago de los Caballeros see Santiago, Dominican Republic
Santiago de los Caballeros see Ciudad de Guatemala, Guatemala
Santiago Pinotepa Nacional see Pinotepa Nacional
Santiago, Río 38 B2 river N Peru
Santi Quaranta see Sarandë
Santissima Trinidad see Keelung
San Joan d'Alacant 71 F4 Cast. San Juan de Alicante. País Valenciano, E Spain
Sant Julià de Lòria 69 A8 Sant Julià de Lòria, SW Andorra Europe
Santo see Espiritu Santo
Santo Antão 52 A2 island Ilhas de Barlavento, N Cape Verde
Santo António 54 E1 Príncipe, N Sao Tome and Principe, Africa
Santo Domingo 33 D3 prev. Ciudad Trujillo. country capital (Dominican Republic) SE Dominican Republic
Santo Domingo de los Colorados 38 B1 Pichincha, NW Ecuador
Santo Domingo Tehuantepec see Tehuantepec
Santorini see Thíra
Santorini 83 D7 island Kykládes, Greece, Aegean Sea
San Tomé de Guayana see Ciudad Guayana
Santos 41 F4 São Paulo, S Brazil
Santos Plateau 35 D5 undersea plateau SW Atlantic Ocean
Santo Tomé 42 D3 Corrientes, NE Argentina
Santo Tomé de Guayana see Ciudad Guayana
San Valentín, Cerro 43 A6 mountain S Chile
San Vicente 30 C3 San Vicente, C El Salvador
São Francisco, Rio 41 F3 river E Brazil
Sao Hill 51 C7 Iringa, S Tanzania
São João da Madeira 70 B2 Aveiro, N Portugal
São Jorge 70 A5 island Azores, Portugal, NE Atlantic Ocean
São Luís 41 F2 state capital Maranhão, NE Brazil
São Mandol see São Manuel, Rio
São Manuel, Rio 41 E3 var. São Mandol, Teles Pirés. river C Brazil
São Marcos, Baía de 41 F1 bay N Brazil
São Miguel 70 A5 island Azores, Portugal, NE Atlantic Ocean
Saona, Isla 33 E3 island SE Dominican Republic
Saône 69 D5 river E France
São Nicolau 52 A3 Eng. Saint Nicholas. island Ilhas de Barlavento, N Cape Verde
São Paulo 41 E4 state capital São Paulo, S Brazil
São Paulo 41 E4 off. Estado de São Paulo. state/region S Brazil
São Paulo de Loanda see Luanda
São Paulo, Estado de see São Paulo
São Pedro do Rio Grande do Sul see Rio Grande
São Roque, Cabo de 41 G2 headland E Brazil
São Salvador see Salvador, Brazil
São Salvador/São Salvador do Congo see M'Banza Congo, Angola
São Tiago see Santiago
São Tomé 55 A5 country capital (Sao Tome and Principe) São Tomé, S Sao Tome and Principe
São Tomé 54 E2 Eng. Saint Thomas. island S Sao Tome and Principe
Sao Tome and Principe 54 D1 off. Democratic Republic of Sao Tome and Principe. country E Atlantic Ocean
Sao Tome and Principe, Democratic Republic of see Sao Tome and Principe
São Tomé, Pico de 54 D2 mountain São Tomé, C Sao Tome and Principe, Africa
São Vicente 52 A3 Eng. Saint Vincent. island Ilhas de Barlavento, N Cape Verde
São Vicente, Cabo de 70 B5 Eng. Cape Saint Vincent, Port. Cabo de São Vicente. cape S Portugal
São Vicente, Cabo de see São Vicente, Cabo de
Sápai see Sápes
Sapele 53 F5 Delta, S Nigeria
Sápes 82 D3 var. Sápai. Anatolikí Makedonía kai Thráki, NE Greece
Sapir 97 B7 prev. Sappir. Southern, S Israel
Sa Pobla 71 G3 Mallorca, Spain, W Mediterranean Sea
Sappir see Sapir
Sapporo 108 D2 Hokkaidō, NE Japan
Sapri 75 D6 Campania, S Italy
Sapulpa 27 G1 Oklahoma, C USA
Saqqez 98 C2 var. Saghez, Sakiz, Saqqiz. Kordestān, NW Iran
Saqqiz see Saqqez
Sara Buri 115 C5 var. Saraburi. Saraburi, C Thailand
Saraburi see Sara Buri
Saragossa see Zaragoza
Saragt see Sarahs
Saraguro 38 B2 Loja, S Ecuador
Sarahs 100 D3 var. Saragt, Rus. Serakhs. Ahal Welaýaty, S Turkmenistan
Sarajevo 78 C4 country capital (Bosnia and Herzegovina) Federacija Bosna I Hercegovina, SE Bosnia and Herzegovina
Sarakhs 98 E2 Khorāsān-Razavī, NE Iran
Saraktash 89 D6 Orenburgskaya Oblast', W Russia
Saran' 92 C4 Kaz. Saran. Karagandy, C Kazakhstan
Saranda see Sarandë
Sarandë 79 C7 var. Saranda, It. Porto Edda; prev. Santi Quaranta. Vlorë, S Albania
Saransk 89 C5 Respublika Mordoviya, W Russia
Sarasota 21 E4 Florida, SE USA
Saratov 92 B3 Saratovskaya Oblast', W Russia
Saravan/Saravane see Salavan
Sarawak 116 D3 state East Malaysia
Sarawak see Kuching
Sarcelles 68 D1 Val-d'Oise, Île-de-France, N France Europe
Sardegna 75 A5 Eng. Sardinia. island Italy, C Mediterranean Sea
Sardinia see Sardegna
Sarera, Teluk see Cenderawasih, Teluk
Sargasso Sea 44 B4 sea W Atlantic Ocean
Sargodha 112 C2 Punjab, NE Pakistan
Sarh 54 C4 prev. Fort-Archambault. Moyen-Chari, S Chad
Sārī 98 D2 var. Sari, Sāri. Māzandarān, N Iran
Saría 83 E7 island SE Greece
Sarıkamış 95 F3 Kars, NE Turkey
Sarikol Range 101 G3 Rus. Sarykol'skiy Khrebet. mountain range China/Tajikistan
Sark 67 D8 Fr. Sercq. island Channel Islands
Şarkışla 94 D3 Sivas, C Turkey
Sarmiento 43 B6 Chubut, S Argentina
Sarnia 16 C5 Ontario, S Canada
Sarny 86 C1 Rivnens'ka Oblast', NW Ukraine
Sarochyna 85 D5 Rus. Sorochino. Vitsyebskaya Voblasts', N Belarus
Sarov 89 C5 prev. Sarova. Respublika Mordoviya, SW Russia
Sarova see Sarov
Sarpsborg 63 B6 Østfold, S Norway
Sarrebruck see Saarbrücken
Sartène 69 E7 Corse, France, C Mediterranean Sea
Sarthe 68 B4 river N France
Sárti 82 C4 Kentrikí Makedonía, N Greece
Saruhan see Manisa
Saryarqa see Kazakhskiy Melkosopochnik
Sarykol'skiy Khrebet see Sarikol Range
Sary-Tash 101 F2 Oshskaya Oblast', SW Kyrgyzstan
Saryyesik-Atyrau, Peski 101 G1 desert E Kazakhstan
Sasebo 109 A7 Nagasaki, Kyūshū, SW Japan
Saskatchewan 15 F5 province SW Canada
Saskatchewan 15 F5 river Manitoba/ Saskatchewan, C Canada
Saskatoon 15 F5 Saskatchewan, S Canada
Sasovo 89 B5 Ryazanskaya Oblast', W Russia
Sassandra 52 D5 S Ivory Coast
Sassandra 52 D5 var. Ibo, Sassandra Fleuve. river S Ivory Coast
Sassandra Fleuve see Sassandra
Sassari 75 A5 Sardegna, Italy, C Mediterranean Sea
Sassenheim 64 C3 Zuid-Holland, W Netherlands
Sassnitz 72 D2 Mecklenburg-Vorpommern, NE Germany
Sathmar see Satu Mare
Sátoraljaújhely 77 D6 Borsod-Abaúj-Zemplén, NE Hungary
Satpura Range 112 D4 mountain range C India
Satsuma-Sendai 109 A8 Kagoshima, Kyūshū, SW Japan
Satsunan-shoto 108 A3 island group Nansei-shotō, SW Japan Asia
Sattanen 62 D3 Lappi, NE Finland
Satu Mare 86 B3 Ger. Sathmar, Hung. Szatmárnémeti. Satu Mare, NW Romania
Sau see Sava
Saudi Arabia 99 B5 off. Kingdom of Saudi Arabia, Al 'Arabīyah as Su'ūdīyah, Ar. Al Mamlakah al 'Arabīyah as Su'ūdīyah. country SW Asia
Saudi Arabia, Kingdom of see Saudi Arabia
Sauer see Sûre
Saugor see Sāgar
Saulkrasti 84 C3 Rīga, C Latvia
Sault Sainte Marie 16 C4 Michigan, N USA
Sault Sainte Marie 18 C1 Michigan, N USA
Sault Ste. Marie 16 C4 Ontario, S Canada
Saumur 68 B4 Maine-et-Loire, NW France
Saurimo 56 C1 Port. Henrique de Carvalho, Vila Henrique de Carvalho. Lunda Sul, NE Angola
Sava 85 E6 Mahilyowskaya Voblasts', E Belarus
Sava 78 B3 Eng. Save, Ger. Sau, Hung. Száva. river SE Europe
Savá 30 D2 Colón, N Honduras
Savai'i 123 E4 island NW Samoa
Savannah 21 E2 Georgia, SE USA
Savannah River 21 E2 river Georgia/South Carolina, SE USA
Savannakhét see Khanthabouli
Savanna-La-Mar 32 A5 W Jamaica
Savaria see SzombathelySave see Sava
Save, Rio 57 E3 var. Sabi. river Mozambique/ Zimbabwe
Saverne 68 E3 var. Zabern; anc. Tres Tabernae. Bas-Rhin, NE France
Savigliano 74 A2 Piemonte, NW Italy
Savigsivik see Savissivik
Savinski see Savinskiy
Savinskiy 88 C3 var. Savinski. Arkhangel'skaya Oblast', NW Russia
Savissivik 60 D1 var. Savigsivik. Avannaarsua, N Greenland
Savoie 69 D5 cultural region E France
Savona 74 A2 Liguria, NW Italy
Savu Sea 117 E5 Ind. Laut Sawu. sea S Indonesia
Sawakin see Suakin
Sawdirī see Sodiri
Sawhāj 50 B2 var. Sawhāj var. Sohâg, Suliag. C Egypt
Sawhāj see Sawhāj
Şawqirah 99 D6 var. Suqrah. S Oman
Sawu, Laut see Savu Sea
Saxe see Sachsen
Saxony see Sachsen
Sayaboury see Xaignabouli
Sayanskiy Khrebet 90 D3 mountain range S Russia
Saýat 100 D3 Rus. Sayat. Lebap Welaýaty, E Turkmenistan
Sayaxché 30 B2 Petén, N Guatemala
Şaydā/Sayida see Saïda
Sayḩūt 99 D6 E Yemen
Saynshand 105 E2 Dornogovĭ, SE Mongolia
Sayre 19 E3 Pennsylvania, NE USA
Say'ūn 99 C6 var. Saywūn. C Yemen
Saywūn see Say'ūn
Scalabis see Santarém
Scandinavia 44 D2 geophysical region NW Europe
Scarborough 67 D5 N England, United Kingdom
Scarpanto see Kárpathos
Scebeli see Shebeli
Schaan 72 E1 W Liechtenstein Europe
Schaerbeek 65 C6 Brussels, C Belgium
Schaffhausen 73 B7 Fr. Schaffhouse. Schaffhausen, N Switzerland
Schaffhouse see Schaffhausen
Schagen 64 C2 Noord-Holland, NW Netherlands
Schaulen see Šiauliai
Schebschi Mountains see Shebshi Mountains
Scheessel 72 B3 Niedersachsen, NW Germany
Schefferville 17 E2 Québec, E Canada
Schelde see Scheldt
Scheldt 65 B5 Dut. Schelde, Fr. Escaut. river W Europe
Schell Creek Range 25 D5 mountain range Nevada, W USA
Schenectady 19 F3 New York, NE USA
Schertz 27 G4 Texas, SW USA
Schiermonnikoog 64 D1 Fris. Skiermûntseach. island Waddeneilanden, N Netherlands
Schijndel 64 D4 Noord-Brabant, S Netherlands
Schil see Jiu
Schiltigheim 68 E3 Bas-Rhin, NE France
Schivelbein see Świdwin
Schleswig 72 B2 Schleswig-Holstein, N Germany
Schleswig-Holstein 72 B2 state N Germany
Schlettstadt see Sélestat
Schlochau see Człuchów
Schneekoppe see Sněžka
Schneidemühl see Piła
Schoden see Skuodas
Schönebeck 72 C4 Sachsen-Anhalt, C Germany
Schönlanke see Trzcianka
Schooten see Schoten
Schoten 65 C5 var. Schooten. Antwerpen, N Belgium
Schouwen 64 B4 island SW Netherlands
Schwabenalb see Schwäbische Alb
Schwäbische Alb 73 B6 var. Schwabenalb, Eng. Swabian Jura. mountain range S Germany
Schwandorf 73 C5 Bayern, SE Germany
Schwaz 73 C7 Tirol, W Austria
Schweidnitz see Świdnica
Schweinfurt 73 B5 Bayern, SE Germany
Schweiz see Switzerland
Schwerin 72 C3 Mecklenburg-Vorpommern, N Germany
Schwertberg see Świecie
Schwiebus see Świebodzin
Schwyz 73 B7 var. Schwiz. Schwyz, C Switzerland
Schyl see Jiu
Scio see Chíos
Scoresby Sound/Scoresbysund see Ittoqqortoormiit
Scoresby Sund see Kangertittivaq
Scotia Sea 35 C8 sea SW Atlantic Ocean
Scotland 66 C3 cultural region Scotland, U K
Scott Base 132 B4 NZ research station Antarctica
Scott Island 132 B5 island Antarctica
Scottsbluff 22 D3 Nebraska, C USA
Scottsboro 20 D1 Alabama, S USA
Scottsdale 26 B2 Arizona, SW USA
Scranton 19 F3 Pennsylvania, NE USA
Scrobesbyrig' see Shrewsbury
Scupi see Skopje
Scutari see Shkodër
Scutari, Lake 79 C5 Alb. Liqeni i Shkodrës, SCr. Skadarsko Jezero. lake Albania/Montenegro
Scyros see Skýros
Searcy 20 B1 Arkansas, C USA
Seattle 24 B2 Washington, NW USA
Sébaco 30 D3 Matagalpa, W Nicaragua
Sebaste/Sebastia see Sivas
Sebastián Vizcaíno, Bahía 28 A2 bay NW Mexico
Sebastopol see Sevastopol'
Sebenico see Šibenik
Sechura, Bahía de 38 A3 bay NW Peru
Secunderābād 112 D5 var. Sikandarabad. Telangana, C India
Sedan 68 D3 Ardennes, N France
Seddon 129 D5 Marlborough, South Island, New Zealand
Seddonville 129 C5 West Coast, South Island, New Zealand
Sédhiou 52 B3 SW Senegal
Sedlez see Siedlce
Sedona 26 B2 Arizona, SW USA
Sedunum see Sion
Seeland see Sjælland
Seenu Atoll see Addu Atoll
Seesen 72 B4 Niedersachsen, C Germany
Segestica see Sisak
Segezha 88 B3 Respublika Kareliya, NW Russia
Seghedin see Szeged
Segna see Senj
Segodunum see Rodez
Ségou 52 D3 var. Segu. Ségou, C Mali
Segovia 70 D2 Castilla-León, C Spain
Segoviao Wangkí see Coco, Río
Segu see Ségou
Séguédine 53 H2 Agadez, NE Niger
Seguin 27 G4 Texas, SW USA
Segura 71 E4 river S Spain
Seinäjoki 63 D5 Swe. Östermyra. Etelä-Pohjanmaa, W Finland
Seine 68 D1 river N France
Seine, Baie de la 68 B3 bay N France
Sejong City 106 E4 administrative capital (South Korea) C South Korea
Sekondi see Sekondi-Takoradi
Sekondi-Takoradi 53 E5 var. Sekondi. S Ghana
Selânik see Thessaloníki
Selenga 105 E1 Mong. Selenge Mörön. river Mongolia/Russia
Selenge Mörön see Selenga
Sélestat 68 E4 Ger. Schlettstadt. Bas-Rhin, NE France
Seleucia see Silifke
Selfoss 61 E5 Sudhurland, SW Iceland
Sélibabi 52 C3 var. Sélibaby. Guidimaka, S Mauritania
Sélibaby see Sélibabi
Selma 25 C6 California, W USA
Selway River 24 D2 river Idaho, NW USA North America
Selwyn Range 126 B3 mountain range Queensland, C Australia
Selzaete see Zelzate
Semarang 116 C5 var. Samarang. Jawa, C Indonesia
Sembé 55 B5 Sangha, NW Congo
Semendria see Smederevo
Semey 92 D4 prev. Semipalatinsk. Vostochnyy Kazakhstan, E Kazakhstan
Semechevo see Syemyezhava
Seminole 27 E3 Texas, SW USA
Seminole, Lake 20 D3 reservoir Florida/Georgia, SE USA
Semipalatinsk see Semey
Semnān 98 D3 var. Samnān. Semnān, N Iran
Semois 65 C8 river SE Belgium
Sendai 108 D4 Miyagi, Honshū, C Japan
Sendai-wan 108 D4 bay E Japan
Senec 77 C6 Ger. Wartberg, Hung. Szenc; prev. Szempcz. Bratislavský Kraj, W Slovakia
Senegal 52 B3 off. Republic of Senegal, Fr. Sénégal. country W Africa
Senegal 52 C3 Fr. Sénégal. river W Africa
Senegal, Republic of see Senegal
Senftenberg 72 D4 Brandenburg, E Germany
Senia see Senj
Senica 77 C6 Ger. Senitz, Hung. Szenice. Trnavský Kraj, W Slovakia
Seniça see Sjenica
Senitz see Senica
Senj 78 A3 Ger. Zengg, It. Segna; anc. Senia. Lika-Senj, NW Croatia
Senja 62 C2 prev. Senjen. island N Norway
Senjen see Senja
Senkaku-shoto 108 A3 island group SW Japan
Senlis 68 C3 Oise, N France
Sennar 50 C4 var. Sannār. Sinnar, C Sudan
Senones see Sens
Sens 68 C3 anc. Agendicum, Senones. Yonne, C France
Sensburg see Mrągowo
Sên, Stœng 115 D5 river C Cambodia
Senta 78 D3 Hung. Zenta. Vojvodina, N Serbia
Seo de Urgel see La See d'Urgel
Seoul 107 E4 off. Seoul Teukbyeolsi, prev. Sŏul, Jap. Keijō; prev. Kyŏngsŏng. country capital (South Korea) NW South Korea
Seoul Teukbyeolsi see Seoul
Şepşi-Sângeorz/Sepsiszentgyőrgy see Sfântu Gheorghe
Sept-Îles 17 E3 Québec, SE Canada
Seraing 65 D6 Liège, E Belgium
Serakhs see Sarahs
Seram, Laut 117 F4 Eng. Ceram Sea. sea E Indonesia
Pulau Seram 117 F4 var. Serang, Eng. Ceram. island Maluku, E Indonesia
Serang 116 C5 Jawa, C Indonesia
Serang see Seram, Pulau
Serasan, Selat 116 C3 strait Indonesia/Malaysia
Serbia 78 D4 off. Federal Republic of Serbia; prev. Yugoslavia, SCr. Jugoslavija. country SE Europe
Serbia, Federal Republic of see Serbia
Sercq see Sark
Serdar 100 C2 prev. Rus. Gyzyrlabat, Kizyl-Arvat. Balkan Welaýaty, W Turkmenistan
Serdica see Sofiya
Serdobol' see Sortavala
Serenje 56 D2 Central, E Zambia
Seres see Sérres
Seret/Sereth see Siret
Serhetabat 100 D4 prev. Rus. Gushgy, Kushka. Mary Welaýaty, S Turkmenistan
Sérifos 83 C6 anc. Seriphos. island Kykládes, Greece, Aegean Sea
Seriphos see Sérifos
Serov 92 C3 Sverdlovskaya Oblast', C Russia
Serowe 56 D3 Central, SE Botswana
Serpa Pinto see Menongue
Serpent's Mouth, The 37 F2 Sp. Boca de la Serpiente. strait Trinidad and Tobago/Venezuela
Serpiente, Boca de la see Serpent's Mouth, The
Serpukhov 89 B5 Moskovskaya Oblast', W Russia
Sérrai see Sérres
Serrana, Cayo de 31 F2 island group NW Colombia South America
Serranilla, Cayo de 31 F2 island group NW Colombia South America Caribbean Sea
Serravalle 74 E1 N San Marino
Sérres 82 C3 var. Seres; prev. Sérrai. Kentrikí Makedonía, NE Greece
Sesdlets see Siedlce
Sesto San Giovanni 74 B2 Lombardia, N Italy
Sesvete 78 B2 Zagreb, N Croatia
Setabis see Xàtiva
Sète 69 C6 prev. Cette. Hérault, S France
Setesdal 63 A6 valley S Norway
Sétif 49 E2 var. Stif. N Algeria
Setté Cama 55 A6 Ogooué-Maritime, SW Gabon
Setúbal 70 B4 Eng. Saint Ubes, Saint Yves. Setúbal, W Portugal
Setúbal, Baía de 70 B4 bay W Portugal
Seul, Lac 16 B3 lake Ontario, S Canada
Sevan 95 G2 C Armenia
Sevana Lich 95 G3 Eng. Lake Sevan, Rus. Ozero Sevan. lake E Armenia
Sevan, Lake/Sevan, Ozero see Sevana Lich
Sevastopol' 87 F5 Eng. Sebastopol. Respublika Krym, S Ukraine
Severn 16 B2 river Ontario, S Canada
Severn 67 D6 Wel. Hafren. river England/Wales, United Kingdom

Severnaya Dvina – Sognefjorden

Severnaya Dvina 88 C4 var. Northern Dvina. river NW Russia
Severnaya Zemlya 93 E2 var. Nicholas II Land. island group N Russia
Severnyy 88 E3 Respublika Komi, NW Russia
Severodonetsk see Syeverodonets'k
Severodvinsk 88 C3 prev. Molotov, Sudostroy. Arkhangel'skaya Oblast', NW Russia
Severomorsk 88 C2 Murmanskaya Oblast', NW Russia
Seversk 92 D4 Tomskaya Oblast', C Russia
Sevier Lake 22 A4 lake Utah, W USA
Sevilla 70 C4 Eng. Seville; anc. Hispalis. Andalucía, SW Spain
Seville see Sevilla
Sevlievo 82 D2 Gabrovo, N Bulgaria
Sevluš/Sevlyush see Vynohradiv
Seward's Folly see Alaska
Seychelles 57 G1 off. Republic of Seychelles. country W Indian Ocean
Seychelles, Republic of see Seychelles
Seyðisfjörður 61 E5 Austurland, E Iceland
Seýdi 100 D2 Rus. Seydi; prev. Neftezavodsk. Lebap Welaýaty, E Turkmenistan
Seyhan see Adana
Sfákia see Chóra Sfakíon
Sfântu Gheorghe 86 C4 Ger. Sankt-Georgen, Hung. Sepsiszentgyörgy; prev. Şepşi-Sângeorz, Sfîntu Gheorghe. Covasna, C Romania
Sfax 49 F2 Ar. Şafāqis. E Tunisia
Sfîntu Gheorghe see Sfântu Gheorghe
's-Gravenhage 64 B4 var. Den Haag, Eng. The Hague, Fr. La Haye. country capital (Netherlands-seat of government) Zuid-Holland, W Netherlands
's-Gravenzande 64 B4 Zuid-Holland, W Netherlands
Shaan/Shaanxi Sheng see Shaanxi
Shaanxi 106 B5 var. Shaan, Shaanxi Sheng, Shan-hsi, Shenshi, Shensi. province C China
Shabani see Zvishavane
Shabeelle, Webi see Shebeli
Shache 104 A3 var. Yarkant. Xinjiang Uygur Zizhiqu, NW China
Shacheng see Huailai
Shackleton Ice Shelf 132 D3 ice shelf Antarctica
Shaddādī see Ash Shadādah
Shāhābād see Eslāmābād
Sha Hi see Orūmīyeh, Daryācheh-ye
Shahjahanabad see Delhi
Shahr-e Kord 98 C3 var. Shahr Kord. Chahār Maḩall va Bakhtīārī, C Iran
Shahr Kord see Shahr-e Kord
Shāhrūd 98 D2 prev. Emāmrūd, Emāmshahr. Semnān, N Iran
Shalkar 92 B4 var. Chelkar. Aktyubinsk, W Kazakhstan
Shām, Bādiyat ash see Syrian Desert
Shana see Kuril'sk
Shandī see Shendi
Shandong 106 D4 var. Lu, Shandong Sheng, Shantung. province E China
Shandong Sheng see Shandong
Shanghai 106 D5 var. Shang-hai. Shanghai Shi, E China
Shangrao 106 D5 Jiangxi, S China
Shan-hsi see Shaanxi, China
Shan-hsi see Shanxi, China
Shannon 67 A6 Ir. An tSionainn. river W Ireland
Shan Plateau 114 B3 plateau E Myanmar (Burma)
Shansi see Shanxi
Shantar Islands see Shantarskiye Ostrova
Shantarskiye Ostrova 93 G3 Eng. Shantar Islands. island group E Russia
Shantou 106 D6 var. Shan-t'ou, Swatow. Guangdong, S China
Shan-t'ou see Shantou
Shantung see Shandong
Shanxi 106 C4 var. Jin, Shan-hsi, Shansi, Shanxi Sheng. province C China
Shan Xian see Sanmenxia
Shanxi Sheng see Shanxi
Shaoguan 106 C6 var. Shao-kuan, Cant. Kukong; prev. Ch'u-chiang. Guangdong, S China
Shao-kuan see Shaoguan
Shaqrā' 98 B4 Ar Riyāḑ, C Saudi Arabia
Shaqrā see Shuqrah
Shar 92 D5 var. Charsk. Vostochnyy Kazakhstan, E Kazakhstan
Shari 108 D2 Hokkaidō, NE Japan
Shari see Chari
Sharjah see Ash Shāriqah
Shark Bay 125 A5 bay Western Australia
Sharqī, Al Jabal ash/Sharqi, Jebel esh see Anti-Lebanon
Shashe 56 D3 var. Shashi. river Botswana/Zimbabwe
Shashi see Shashe
Shatskiy Rise 103 G1 undersea rise N Pacific Ocean
Shawnee 27 G1 Oklahoma, C USA
Shaykh, Jabal ash see Hermon, Mount
Shchadryn 85 D7 Rus. Shchedrin. Homyel'skaya Voblasts', SE Belarus
Shchedrin see Shchadryn
Shcheglovsk see Kemerovo
Shchëkino 89 B5 Tul'skaya Oblast', W Russia
Shchors see Snovs'k
Shchuchin see Shchuchyn
Shchuchinsk 92 C4 prev. Shchuchye. Akmola, N Kazakhstan
Shchuchye see Shchuchinsk
Shchuchyn 85 B5 Pol. Szczuczyn Nowogródzki, Rus. Shchuchin. Hrodzyenskaya Voblasts', W Belarus
Shebekino 89 A6 Belgorodskaya Oblast', W Russia
Shebelē Wenz, Wabē see Shebeli
Shebeli 51 D5 Amh. Wabē Shebelē Wenz, It. Scebeli, Som. Webi Shabeelle. river Ethiopia/Somalia
Sheberghān see Shibirghān
Sheboygan 18 B2 Wisconsin, N USA
Shebshi Mountains 54 A4 var. Schebschi Mountains. mountain range E Nigeria
Shechem see Nablus
Shedadi see Ash Shadādah
Sheffield 67 D5 N England, United Kingdom
Shekhem see Nablus
Sheki see Şäki
Shelby 22 B1 Montana, NW USA
Sheldon 23 F3 Iowa, C USA
Shelekhov Gulf see Shelikhova, Zaliv
Shelikhova, Zaliv 93 G2 Eng. Shelekhov Gulf. gulf E Russia
Shendi 50 C4 var. Shandī. River Nile, NE Sudan
Shengking see Liaoning
Shenking see Liaoning
Shenshi/Shensi see Shaanxi
Shenyang 106 D3 Chin. Shen-yang, Eng. Moukden, Mukden; prev. Fengtien. province capital Liaoning, NE China
Shen-yang see Shenyang
Shepetivka 86 D2 Rus. Shepetovka. Khmel'nyts'ka Oblast', NW Ukraine
Shepetovka see Shepetivka
Shepparton 127 C7 Victoria, SE Australia
Sherbrooke 17 E4 Québec, SE Canada
Shereik 50 C3 River Nile, N Sudan
Sheridan 22 C2 Wyoming, C USA
Sherman 27 G2 Texas, SW USA
's-Hertogenbosch 64 C4 Fr. Bois-le-Duc, Ger. Herzogenbusch. Noord-Brabant, S Netherlands
Shetland Islands 66 D1 island group NE Scotland, United Kingdom
Shevchenko see Aktau
Shiberghan/Shiberghān see Shibirghān
Shibirghān 101 E3 var. Sheberghān, Shiberghan, Shiberghān. Jowzjān, N Afghanistan
Shibetsu 108 D2 var. Sibetu. Hokkaidō, NE Japan
Shibh Jazirat Sīnā' see Sinai
Shibushi-wan 109 B8 bay SW Japan
Shigatse see Xigazê
Shih-chia-chuang/Shihmen see Shijiazhuang
Shihezi 104 C2 Xinjiang Uygur Zizhiqu, NW China
Shiichi see Shyichy
Shijiazhuang 106 C4 var. Shih-chia-chuang; prev. Shihmen. province capital Hebei, E China
Shikārpur 112 B3 Sind, S Pakistan
Shikoku 109 C7 var. Sikoku. island SW Japan
Shikoku Basin 103 F2 var. Sikoku Basin. undersea basin N Philippine Sea
Shikotan, Ostrov 108 E2 Jap. Shikotan-tō. island NE Russia
Shikotan-tō see Shikotan, Ostrov
Shilabo 51 D5 Sumalē, E Ethiopia
Shiliguri 113 F3 prev. Siliguri. West Bengal, NE India
Shilka 93 F4 river S Russia
Shillong 113 G3 state capital Meghālaya, NE India
Shimanto see Nakamura
Shimbir Berris see Shimbiris
Shimbiris 50 E4 var. Shimbir Berris. mountain N Somalia
Shimoga see Shivamogga
Shimonoseki 109 A7 var. Simonoseki, hist. Akamagaseki, Bakan. Yamaguchi, Honshū, SW Japan
Shinano-gawa 109 C5 var. Sinano Gawa. river Honshū, C Japan
Shīndand 100 D4 prev. Shīndand. Herāt, W Afghanistan
Shīndand see Shīndand
Shingū 109 C6 var. Singû. Wakayama, Honshū, SW Japan
Shinjō 108 D4 var. Sinzyô. Yamagata, Honshū, C Japan
Shinyanga 51 C7 Shinyanga, NW Tanzania
Shiprock 26 C1 New Mexico, SW USA
Shīrāz 98 D4 var. Shīrāz. Fārs, S Iran
Shishchitsy see Shyshchytsy
Shivamogga 110 C2 prev. Shimoga. Karnātaka, W India
Shivpuri 112 D3 Madhya Pradesh, C India
Shizugawa 108 D4 Miyagi, Honshū, NE Japan
Shizuoka 109 D6 var. Sizuoka. Shizuoka, Honshū, S Japan
Shklov see Shklow
Shklow 85 D6 Rus. Shklov. Mahilyowskaya Voblasts', E Belarus
Shkodër 79 C5 var. Shkodra, It. Scutari, SCr. Skadar. Shkodër, NW Albania
Shkodra see Shkodër
Shkodrës, Liqeni i see Scutari, Lake
Shkumbinit, Lumi i 79 C6 var. Shkumbi, Shkumbin. river C Albania
Shkumbî/Shkumbin see Shkumbinit, Lumi i
Sholāpur see Solāpur
Shostka 87 F1 Sums'ka Oblast', NE Ukraine
Show Low 26 B2 Arizona, SW USA
Show Me State see Missouri
Shpola 87 E3 Cherkas'ka Oblast', N Ukraine
Shqipëria/Shqipërisë, Republika e see Albania
Shreveport 20 A2 Louisiana, S USA
Shrewsbury 67 D6 hist. Scrobesbyrig'. W England, United Kingdom
Shu 92 C5 Kaz. Shū. Zhambyl, SE Kazakhstan
Shuang-liao see Liaoyuan
Shū, Kazakhstan see Shu
Shumagin Islands 14 B3 island group Alaska, USA
Shumen 82 D2 Shumen, NE Bulgaria
Shumilina 85 E5 Rus. Shumilino. Vitsyebskaya Voblasts', NE Belarus
Shumilino see Shumilina
Shunsen see Chuncheon
Shuqrah 99 B7 var. Shaqrā. SW Yemen
Shwebo 114 B3 Sagaing, C Myanmar (Burma)
Shyichy 85 C7 Rus. Shiichi. Homyel'skaya Voblasts', SE Belarus
Shymkent 92 B5 prev. Chimkent. Yuzhnyy Kazakhstan, S Kazakhstan
Shyshchytsy 85 C6 Rus. Shishchitsy. Minskaya Voblasts', C Belarus
Siam see Thailand
Siam, Gulf of see Thailand, Gulf of
Sian see Xi'an
Siang see Brahmaputra
Siangtan see Xiangtan
Šiauliai 84 B4 Ger. Schaulen. Šiauliai, N Lithuania
Siazan' see Siyäzän
Sibay 89 D6 Respublika Bashkortostan, W Russia
Šibenik 78 B4 It. Sebenico. Šibenik-Knin, S Croatia
Siberia see Sibir'
Siberoet see Siberut, Pulau
Siberut, Pulau 116 A4 prev. Siberoet. island Kepulauan Mentawai, W Indonesia
Sibi 112 B2 Baluchistān, SW Pakistan
Sibir' 93 E3 var. Siberia. physical region NE Russia
Sibiti 55 B6 Lékoumou, S Congo
Sibiu 86 B4 Ger. Hermannstadt, Hung. Nagyszeben. Sibiu, C Romania
Sibolga 116 B3 Sumatera, W Indonesia
Sibu 116 D3 Sarawak, East Malaysia
Sibut 54 C4 prev. Fort-Sibut. Kémo, S Central African Republic
Sibuyan Sea 117 E2 sea W Pacific Ocean
Sichon 115 C6 var. Ban Sichon, Si Chon. Nakhon Si Thammarat, SW Thailand
Si Chon see Sichon
Sichuan 106 B5 var. Chuan, Sichuan Sheng, Ssu-ch'uan, Szechuan, Szechwan. province C China
Sichuan Pendi 106 B5 basin C China
Sichuan Sheng see Sichuan
Sicilian Channel see Sicily, Strait of
Sicily 75 C7 Eng. Sicily; anc. Trinacria. island Italy, C Mediterranean Sea
Sicily, Strait of 75 B7 var. Sicilian Channel. strait C Mediterranean Sea
Sicuani 39 E4 Cusco, S Peru
Sidári 82 A4 Kérkyra, Iónia Nisiá, Greece, C Mediterranean Sea
Sidas 116 C4 Borneo, C Indonesia
Siderno 75 D7 Calabria, SW Italy
Sidhirókastron see Sidirókastro
Sidi Barrâni 50 A1 NW Egypt
Sidi Bel Abbès 48 D2 var. Sidi bel Abbès, Sidi-Bel-Abbès. NW Algeria
Sidirókastro 82 C3 prev. Sidhirókastron. Kentrikí Makedonía, NE Greece
Sidley, Mount 132 B4 mountain Antarctica
Sidney 22 D1 Montana, NW USA
Sidney 22 D4 Nebraska, C USA
Sidney 18 C4 Ohio, N USA
Sidon see Saïda
Sidra see Surt
Sidra/Sidra, Gulf of see Surt, Khalīj, N Libya
Siebenbürgen see Transylvania
Siedlce 76 E3 Ger. Sedlez, Rus. Sesdlets. Mazowieckie, C Poland
Siegen 72 B4 Nordrhein-Westfalen, W Germany
Siemiatycze 76 E3 Podlaskie, NE Poland
Siena 74 B3 Fr. Sienne; anc. Saena Julia. Toscana, C Italy
Sienne see Siena
Sieradz 76 C4 Sieradz, C Poland
Sierpc 76 D3 Mazowieckie, C Poland
Sierra Leone 52 C4 off. Republic of Sierra Leone. country W Africa
Sierra Leone Basin 44 C4 undersea basin E Atlantic Ocean
Sierra Leone, Republic of see Sierra Leone
Sierra Leone Ridge see Sierra Leone Rise
Sierra Leone Rise 44 C4 var. Sierra Leone Ridge, Sierra Leone Schwelle. undersea rise E Atlantic Ocean
Sierra Leone Schwelle see Sierra Leone Rise
Sierra Vista 26 B3 Arizona, SW USA
Sífnos 83 C6 anc. Siphnos. island Kykládes, Greece, Aegean Sea
Sigli 116 A3 Sumatera, W Indonesia
Siglufjörður 61 E4 Nordhurland Vestra, N Iceland
Signal Peak 26 A2 mountain Arizona, SW USA
Signan see Xi'an
Signy 132 A2 UK research station South Orkney Islands, Antarctica
Siguatepeque 30 C2 Comayagua, W Honduras
Siguiri 52 D4 NE Guinea
Sihanoukville 115 D6 Khmer. Preăh Seihânŭ; prev. Kompong Som. Sihanoukville, SW Cambodia
Siilinjärvi 62 E4 Pohjois-Savo, C Finland
Siirt 95 F4 var. Sert; anc. Tigranocerta. Siirt, SE Turkey
Sikandarabad see Secunderābād
Sikasso 52 D4 Sikasso, S Mali
Sikeston 23 H5 Missouri, C USA
Sikhote-Alin', Khrebet 93 G4 mountain range SE Russia
Siking see Xi'an
Siklós 77 C7 Baranya, SW Hungary
Sikoku see Shikoku
Sikoku Basin see Shikoku Basin
Šilalė 84 B4 Tauragė, W Lithuania
Silchar 113 G3 Assam, NE India
Silesia 76 B4 physical region SW Poland
Silifke 94 C4 anc. Seleucia. İçel, S Turkey
Siliguri see Shiliguri
Siling Co 104 C5 lake W China
Silinhot see Xilinhot
Silistra 82 E1 var. Silistria; anc. Durostorum. Silistra, NE Bulgaria
Silistria see Silistra
Sillamäe 84 E2 Ger. Sillamäggi. Ida-Virumaa, NE Estonia
Sillamäggi see Sillamäe
Sillein see Žilina
Šilutė 84 B4 Ger. Heydekrug. Klaipėda, W Lithuania
Silvan 95 E4 Dıyarbakır, SE Turkey
Silva Porto see Kuito
Silver State see Colorado
Silver State see Nevada
Simanichy 85 C7 Rus. Simonichi. Homyel'skaya Voblasts', SE Belarus
Simav 94 B3 Kütahya, W Turkey
Simav Çayı 94 A3 river NW Turkey
Simbirsk see Ul'yanovsk
Simeto 75 C7 river Sicilia, Italy, C Mediterranean Sea
Simeulue, Pulau 116 A3 island NW Indonesia
Simferopol' 87 F5 Respublika Krym, S Ukraine
Simitla 82 C3 Blagoevgrad, SW Bulgaria
Şimlăul Silvaniei/Şimleul Silvaniei see Şimleu Silvaniei
Şimleu Silvaniei 86 B3 Hung. Szilágysomlyó; prev. Şimlăul Silvaniei, Şimleul Silvaniei. Sălaj, NW Romania
Simonichi see Simanichy
Simonoseki see Shimonoseki
Simpelveld 65 D6 Limburg, SE Netherlands
Simplon Pass 73 B8 pass S Switzerland
Simpson see Fort Simpson
Simpson Desert 126 B4 desert Northern Territory/South Australia
Sinai 50 C2 var. Sinai Peninsula, Ar. Shibh Jazirat Sīnā', Sīnā'. physical region NE Egypt
Sinaia 86 C4 Prahova, SE Romania
Sinano Gawa see Shinano-gawa
Sīnā/Sinai Peninsula see Sinai
Sincelejo 36 B2 Sucre, NW Colombia
Sind 112 B3 var. Sindh. province SE Pakistan
Sindelfingen 73 B6 Baden-Württemberg, SW Germany
Sindh see Sind
Sindi 84 D2 Ger. Zintenhof. Pärnumaa, SW Estonia
Sines 70 B4 Setúbal, S Portugal
Singan see Xi'an
Singapore 116 B3 country capital (Singapore) S Singapore
Singapore 116 A1 off. Republic of Singapore. country SE Asia
Singapore, Republic of see Singapore
Singen 73 B6 Baden-Württemberg, S Germany
Singida 51 C7 Singida, C Tanzania
Singkang 117 E4 Sulawesi, C Indonesia
Singkawang 116 C3 Borneo, C Indonesia
Singora see Songkhla
Singû see Shingū
Sining see Xining
Siniscola 75 A5 Sardegna, Italy, C Mediterranean Sea
Sinj 78 B4 Split-Dalmacija, SE Croatia
Sinkiang/Sinkiang Uighur Autonomous Region see Xinjiang Uygur Zizhiqu
Sinnamarie see Sinnamary
Sinnamary 37 H3 var. Sinnamarie. N French Guiana
Sinneh see Sanandaj
Sînnicolau Mare see Sânnicolau Mare
Sinoe, Lacul see Sinoie, Lacul
Sinoie, Lacul 86 D5 prev. Lacul Sinoe. lagoon SE Romania
Sinop 94 D2 anc. Sinope. Sinop, N Turkey
Sinope see Sinop
Sinsheim 73 B6 Baden-Württemberg, SW Germany
Sint Maarten 33 G3 Eng. Saint Martin. island Lesser Antilles
Sint-Michielsgestel 64 C4 Noord-Brabant, S Netherlands
Sîn-Miclăuş see Gheorgheni
Sint-Niklaas 65 B5 Fr. Saint-Nicolas. Oost-Vlaanderen, N Belgium
Sint-Pieters-Leeuw 65 B6 Vlaams Brabant, C Belgium
Sintra 70 B3 prev. Cintra. Lisboa, W Portugal
Sinujiif 51 E5 Nugaal, NE Somalia
Sinus Aelaniticus see Aqaba, Gulf of
Sinus Gallicus see Lion, Golfe du
Sinyang see Xinyang
Sinzyô see Shinjō
Sion 73 A7 Ger. Sitten; anc. Sedunum. Valais, SW Switzerland
Sioux City 23 F3 Iowa, C USA
Sioux Falls 23 F3 South Dakota, N USA
Sioux State see North Dakota
Siphnos see Sífnos
Siping 106 D3 var. Ssu-p'ing, Szeping; prev. Ssu-p'ing-chieh. Jilin, NE China
Siple, Mount 132 A4 mountain Siple Island, Antarctica
Siquirres 31 E4 Limón, E Costa Rica
Siracusa 75 D7 Eng. Syracuse. Sicilia, Italy, C Mediterranean Sea
Sir Edward Pellew Group 126 B2 island group Northern Territory, NE Australia
Siret 86 C3 var. Siretul, Ger. Sereth, Rus. Seret. river Romania/Ukraine
Siretul see Siret
Siria see Syria
Sirikit Reservoir 114 C4 lake N Thailand
Sīrjān 98 D4 prev. Sa'īdābād. Kermān, S Iran
Sirna see Sýrna
Şırnak 95 F4 Şırnak, SE Turkey
Síros see Sýros
Sirte see Surt
Sirti, Gulf of see Surt, Khalīj
Sisak 78 B3 var. Siscia, Ger. Sissek, Hung. Sziszek; anc. Segestica. Sisak-Moslavina, C Croatia
Siscia see Sisak
Sisimiut 60 C3 var. Holsteinborg, Holsteinsborg, Holstenborg, Holstensborg. Kitaa, S Greenland
Sissek see Sisak
Sīstān, Daryācheh-ye see Şāberī, Hāmūn-e
Sitaş Cristuru see Cristuru Secuiesc
Siteía 83 D8 var. Sitía. Kríti, Greece, E Mediterranean Sea
Sitges 71 G2 Cataluña, NE Spain
Sitía see Siteía
Sittang see Sittoung
Sittard 65 D5 Limburg, SE Netherlands
Sitten see Sion
Sittoung 114 B4 var. Sittang. river S Myanmar (Burma)
Sittwe 114 A3 var. Akyab. Rakhine State, W Myanmar (Burma)
Siuna 30 D3 Región Autónoma Atlántico Norte, NE Nicaragua
Siut see Asyūţ
Sivas 94 D3 anc. Sebastia, Sebaste. Sivas, C Turkey
Siverek 95 E4 Şanlıurfa, S Turkey
Siwa see Sīwah
Sīwah 50 A2 var. Siwa. NW Egypt
Six Counties, The see Northern Ireland
Six-Fours-les-Plages 69 D6 Var, SE France
Siyäzän 95 H2 Rus. Siazan'. NE Azerbaijan
Sjar see Sääre
Sjenica 79 D5 Turk. Seniça. Serbia, SW Serbia
Skadar see Shkodër
Skadarsko Jezero see Scutari, Lake
Skagerak see Skagerrak
Skagerrak 63 A6 var. Skagerak. channel N Europe
Skagit River 24 B1 river Washington, NW USA
Skalka 62 C3 lake N Sweden
Skarżysko-Kamienna 76 D4 Świętokrzyskie, C Poland
Skaudvilė 84 B4 Tauragė, SW Lithuania
Skegness 67 E6 E England, United Kingdom
Skellefteå 62 D4 Västerbotten, N Sweden
Skellefteälven 62 C4 river N Sweden
Ski 63 B6 Akershus, S Norway
Skíathos 83 C5 Skíathos, Vóreies Sporádes, Greece, Aegean Sea
Skidal' 85 B5 Rus. Skidel'. Hrodzyenskaya Voblasts', W Belarus
Skidel' see Skidal'
Skiermûntseach see Schiermonnikoog
Skierniewice 76 D3 Łódzkie, C Poland
Skiftet 84 C1 strait Finland Atlantic Ocean Baltic Sea Gulf of Bothnia/Gulf of Finland
Skíros see Skýros
Skópelos 83 C5 Skópelos, Vóreies Sporádes, Greece, Aegean Sea
Skopje 79 D6 var. Üsküb, Turk. Üsküp, prev. Skoplje; anc. Scupi. country capital (FYR Macedonia) N FYR Macedonia
Skoplje see Skopje
Skovorodino 93 F4 Amurskaya Oblast', SE Russia
Skudnesfjorden 63 A6 fjord S Norway
Skuodas 84 B3 Ger. Schoden, Pol. Szkudy. Klaipėda, NW Lithuania
Skye, Isle of 66 B3 island NW Scotland, United Kingdom
Skylge see Terschelling
Skýros 83 C5 var. Skíros. Skńyros, Vóreies Sporádes, Greece, Aegean Sea
Skýros 83 C5 var. Skíros; anc. Scyros. island Vóreies Sporádes, Greece, Aegean Sea
Slagelse 63 B7 Vestsjælland, E Denmark
Slatina 78 C3 Hung. Szlatina; prev. Podravska Slatina. Virovitica-Podravina, NE Croatia
Slatina 86 B5 Olt, S Romania
Slavgorod see Slawharad
Slavonski Brod 78 C3 Ger. Brod, Hung. Bród; prev. Brod, Brod na Savi. Brod-Posavina, NE Croatia
Slavuta 86 C2 Khmel'nyts'ka Oblast', NW Ukraine
Slavyansk see Slov"yans'k
Slawharad 85 E7 Rus. Slavgorod. Mahilyowskaya Voblasts', E Belarus
Sławno 76 C2 Zachodnio-pomorskie, NW Poland
Slēmānī see As Sulaymānīyah
Sliema 80 B5 N Malta
Sligo 67 A5 Ir. Sligeach. Sligo, NW Ireland
Sliven 82 D2 var. Slivno. Sliven, C Bulgaria
Slivnitsa 82 B2 Sofiya, W Bulgaria
Slivno see Sliven
Slobozia 86 C5 Ialomiţa, SE Romania
Slonim 85 B6 Pol. Słonim. Hrodzyenskaya Voblasts', W Belarus
Słonim see Slonim
Slovakia 77 C6 off. Slovenská Republika, Ger. Slowakei, Hung. Szlovákia, Slvk. Slovensko. country C Europe
Slovak Ore Mountains see Slovenské rudohorie
Slovenia 73 D8 off. Republic of Slovenia, Ger. Slowenien, Slvn. Slovenija. country SE Europe
Slovenia, Republic of see Slovenia
Slovenija see Slovenia
Slovenská Republika see Slovakia
Slovenské rudohorie 77 D6 Eng. Slovak Ore Mountains, Ger. Slowakisches Erzgebirge, Ungarisches Erzgebirge. mountain range C Slovakia
Slovensko see Slovakia
Slov"yans'k 87 G3 Rus. Slavyansk. Donets'ka Oblast', E Ukraine
Slowakei see Slovakia
Slowakisches Erzgebirge see Slovenské rudohorie
Slowenien see Slovenia
Słubice 76 B3 Ger. Frankfurt. Lubuskie, W Poland
Sluch 86 D1 river NW Ukraine
Słupsk 76 C2 Ger. Stolp. Pomorskie, N Poland
Slutsk 85 C6 Minskaya Voblasts', S Belarus
Smallwood Reservoir 17 F2 lake Newfoundland and Labrador, S Canada
Smara 48 B3 var. Es Semara. N Western Sahara
Smarhon' 85 C5 Pol. Smorgonie, Rus. Smorgon'. Hrodzyenskaya Voblasts', W Belarus
Smederevo 78 D4 Ger. Semendria. Serbia, N Serbia
Smederevska Palanka 78 D4 Serbia, C Serbia
Smela see Smila
Smila 87 E2 Rus. Smela. Cherkas'ka Oblast', C Ukraine
Smilten see Smiltene
Smiltene 84 D3 Ger. Smilten. Valka, N Latvia
Smøla 62 A4 island W Norway
Smolensk 89 A5 Smolenskaya Oblast', W Russia
Smorgon'/Smorgonie see Smarhon'
Smyrna see İzmir
Snake 12 B4 river Yukon, NW Canada
Snake River 24 C3 river NW USA
Snake River Plain 24 D4 plain Idaho, NW USA
Sneek 64 D2 Friesland, N Netherlands
Sneeuw-gebergte see Maoke, Pegunungan
Sněžka 76 B4 Ger. Schneekoppe, Pol. Śnieżka. mountain N Czech Republic (Czechia)/Poland
Śniardwy, Jezioro 76 D2 Ger. Spirdingsee. lake NE Poland
Sniečkus see Visaginas
Śnieżka see Sněžka
Snina 77 E5 Hung. Szinna. Prešovský Kraj, E Slovakia
Snovs'k 87 E1 Rus. Shchors. Chernihivs'ka Oblast', N Ukraine
Snowdonia 67 C6 mountain range NW Wales, United Kingdom
Snowdrift see Łutselk'e
Snow Mountains see Maoke, Pegunungan
Snyder 27 F3 Texas, SW USA
Sobradinho, Barragem de see Sobradinho, Represa de
Sobradinho, Represa de 41 F2 var. Barragem de Sobradinho. reservoir E Brazil
Sochi 89 A7 Krasnodarskiy Kray, SW Russia
Société, Îles de la/Society Islands see Société, Archipel de la
Society Islands 123 G4 var. Archipel de Tahiti, Îles de la Société, Eng. Society Islands. island group W French Polynesia
Soconusco, Sierra de see Madre, Sierra
Socorro 26 D2 New Mexico, SW USA
Socorro, Isla 28 B5 island W Mexico
Socotra see Suquţrā
Soc Trăng 115 D6 var. Khanh Hung. Soc Trăng, S Vietnam
Socuéllamos 71 E3 Castilla-La Mancha, C Spain
Sodankylä 62 D3 Lappi, N Finland
Sodari see Sodiri
Söderhamn 63 C5 Gävleborg, C Sweden
Södertälje 63 C6 Stockholm, C Sweden
Sodiri 50 B4 var. Sawdirī, Sodari. Northern Kordofan, C Sudan
Soekaboemi see Sukabumi
Soemba see Sumba, Pulau
Soengaipenoeh see Sungaipenuh
Soerabaja see Surabaya
Soerakarta see Surakarta
Sofia see Sofiya
Sofiya 82 C2 var. Sophia, Eng. Sofia, Lat. Serdica. country capital (Bulgaria) (Bulgaria) Sofiya-Grad, W Bulgaria
Sogamoso 36 B3 Boyacá, C Colombia
Sognefjorden 63 A5 fjord NE North Sea

Tessalit - Tsangpo

Tessalit *53 E2* Kidal, NE Mali
Tessaoua *53 G3* Maradi, S Niger
Tessenderlo *65 C5* Limburg, NE Belgium
Tessenei *see* Teseney
Testigos, Islas los *37 E1 island group* N Venezuela
Tete *57 E2* Tete, NW Mozambique
Teterow *72 C3* Mecklenburg-Vorpommern, NE Germany
Tétouan *48 C2 var.* Tetouan, Tetuán. N Morocco
Tetovo *79 D5* Razgrad, N Bulgaria
Tetschen *see* Děčín
Tetuán *see* Tétouan
Teverya *see* Tverya
Te Waewae Bay *129 A7 bay* South Island, New Zealand
Texarkana *20 A2* Arkansas, C USA
Texarkana *27 H2* Texas, SW USA
Texas *27 F3 off.* State of Texas, *also known as* Lone Star State. *state* S USA
Texas City *27 H4* Texas, SW USA
Texel *64 C2 island* Waddeneilanden, NW Netherlands
Texoma, Lake *27 G2 reservoir* Oklahoma/ Texas, C USA
Teziutlán *29 F4* Puebla, S Mexico
Thaa Atoll *see* Kolhumadulu
Thai, Ao *see* Thailand, Gulf of
Thai Binh *114 D3* Thai Binh, N Vietnam
Thailand *115 C5 off.* Kingdom of Thailand, *Th.* Prathet Thai; *prev.* Siam. *country* SE Asia
Thailand, Gulf of *115 C6 var.* Gulf of Siam, *Th.* Ao Thai, *Vtn.* Vinh Thai Lan. *gulf* SE Asia
Thailand, Kingdom of *see* Thailand
Thai Lan, Vinh *see* Thailand, Gulf of
Thai Nguyên *114 D3* Bắc Thai, N Vietnam
Thakhèk *114 D4 var.* Muang Khammouan. Khammouan, C Laos
Thamarīd *see* Thamarīt
Thamarīt *99 D6 var.* Thamarīd, Thumrayt. SW Oman
Thames *128 D3* Waikato, North Island, New Zealand
Thames *67 B8 river* S England, United Kingdom
Thandwe *114 A4 var.* Sandoway. Rakhine State, W Myanmar (Burma)
Thanh Hoa *114 D3* Thanh Hoa, N Vietnam
Thanintari Taungdan *see* Bilauktaung Range
Thanlwin *see* Salween
Thar Desert *112 C3 var.* Great Indian Desert, Indian Desert. *desert* India/Pakistan
Tharthar, Buhayrat ath *98 B3 lake* C Iraq
Thásos *82 C4* Thásos, E Greece
Thásos *82 C4 island* E Greece
Thaton *114 B4* Mon State, S Myanmar (Burma)
Thayetmyo *114 A4* Magway, C Myanmar (Burma)
The Crane *33 H2 var.* Crane. S Barbados
The Dalles *24 B3* Oregon, NW USA
The Flatts Village *see* Flatts Village
The Hague *see* 's-Gravenhage
Theodosia *see* Feodosiya
The Pas *15 F5* Manitoba, C Canada
Therezina *see* Teresina
Thérma *83 D6* Ikaría, Dodekánisa, Greece, Aegean Sea
Thermaic Gulf/Thermaicus Sinus *see* Thermaïkós Kólpos
Thermaïkós Kólpos *82 B4 Eng.* Thermaic Gulf; *anc.* Thermaicus Sinus. *gulf* N Greece
Thermiá *see* Kýthnos
Thérmo *83 B5* Dytikí Ellás, C Greece
The Rock *71 H4* New South Wales, SE Australia
The Sooner State *see* Oklahoma
Thessaloníki *82 C3 Eng.* Salonica, Salonika, *SCr.* Solun, *Turk.* Selânik. Kentrikí Makedonía, N Greece
The Valley *33 G3 dependent territory capital* (Anguilla) E Anguilla
The Village *27 G1* Oklahoma, C USA
The Volunteer State *see* Tennessee
Thiamis *see* Kalamás
Thian Shan *see* Tien Shan
Thibet *see* Xizang Zizhiqu
Thief River Falls *23 F1* Minnesota, N USA
Thienen *see* Tienen
Thiers *69 C5* Puy-de-Dôme, C France
Thiès *52 B3* W Senegal
Thikombia *see* Cikobia
Thimbu *see* Thimphu
Thimphu *113 G3 var.* Thimbu; *prev.* Tashi Chho Dzong. *country capital* (Bhutan) W Bhutan
Thionville *68 D3 Ger.* Diedenhofen. Moselle, NE France
Thíra *83 D7 var.* Santorini. Kykládes, Greece, Aegean Sea
Thiruvananthapuram *110 C3 var.* Trivandrum, Tiruvantapuram. *state capital* Kerala, SW India
Thitu Island *106 C8 island* NW Spratly Islands
Tholen *64 B4 island* SW Netherlands
Thomasville *20 D3* Georgia, SE USA
Thompson *15 F4* Manitoba, C Canada
Thonon-les-Bains *69 D5* Haute-Savoie, E France
Thorenburg *see* Turda
Thorlákshöfn *61 E5* Sudhurland, SW Iceland
Thorn *see* Toruń
Thornton Island *see* Millennium Island
Thorshavn *see* Tórshavn
Thospitis *see* Van Gölü
Thouars *68 B4* Deux-Sèvres, W France
Thoune *see* Thun
Thracian Sea *82 D4 Gk.* Thrakikó Pélagos; *anc.* Thracium Mare. *sea* Greece/Turkey
Thracium Mare/Thrakikó Pélagos *see* Thracian Sea
Three Gorges Reservoir *107 C5 reservoir* C China
Three Kings Islands *128 C1 island group* N New Zealand
Thrissur *110 C3 var.* Trichūr. Kerala, SW India
Thuin *65 B7* Hainaut, S Belgium
Thule *see* Qaanaaq
Thumrayt *see* Thamarīt
Thun *73 A7 Fr.* Thoune. Bern, W Switzerland
Thunder Bay *16 B4* Ontario, S Canada
Thuner See *73 A7 lake* C Switzerland
Thung Song *115 C7 var.* Cha Mai. Nakhon Si Thammarat, SW Thailand
Thurso *66 C2* N Scotland, United Kingdom
Thýamis *see* Kalamás
Tianjin *106 D4 var.* Tientsin. Tianjin Shi, E China
Tianjin *see* Tianjin Shi
Tianjin Shi *106 D4 var.* Jin, Tianjin, T'ien-ching, Tientsin. *municipality* E China
Tian Shan *see* Tien Shan
Tianshui *106 B4* Gansu, C China
Tiba *see* Chiba
Tiber *74 C4 Eng.* Tiber. *river* C Italy
Tiber *see* Tevere, Italy
Tiber *see* Tivoli, Italy
Tiberias *see* Tverya
Tiberias, Lake *97 B5 var.* Chinnereth, Sea of Bahr Tabariya, Sea of Galilee, *Ar.* Bahrat Tabariya, *Heb.* Yam Kinneret. *lake* N Israel
Tibesti *54 C2 var.* Tibesti Massif, *Ar.* Tibistī. *mountain range* N Africa
Tibesti Massif *see* Tibesti
Tibet *see* Xizang Zizhiqu
Tibetan Autonomous Region *see* Xizang Zizhiqu
Tibet, Plateau of *see* Qingzang Gaoyuan
Tibistī *see* Tibesti
Tibnī *see* At Tibnī
Tiburón, Isla *28 B2 var.* Isla del Tiburón. *island* NW Mexico
Tiburón, Isla del *see* Tiburón, Isla
Tichau *see* Tychy
Tichît *52 D2 var.* Tichitt. Tagant, C Mauritania
Tichitt *see* Tichît
Ticinum *see* Pavia
Ticul *29 H3* Yucatán, SE Mexico
Tidjikdja *see* Tidjikja
Tidjikja *52 C2 var.* Tidjikdja; *prev.* Fort-Cappolani. Tagant, C Mauritania
T'ien-ching *see* Tianjin Shi
Tienen *65 C6 var.* Thienen, *Fr.* Tirlemont. Vlaams Brabant, C Belgium
Tiên Giang, Sông *see* Mekong
Tien Shan *104 B3 Chin.* Thian Shan, Tian Shan, T'ien Shan, *Rus.* Tyan'-Shan'. *mountain range* C Asia
Tientsin *see* Tianjin
Tierp *63 C6* Uppsala, C Sweden
Tierra del Fuego *43 B8 island* Argentina/Chile
Tiflis *see* T'bilisi
Tifton *20 D3* Georgia, SE USA
Tifu *117 F4* Pulau Buru, E Indonesia
Tighina *86 D4 Rus.* Bendery; *prev.* Bender. E Moldova
Tigranocerta *see* Siirt
Tigris *98 B2 Ar.* Dijlah, *Turk.* Dicle. *river* Iraq/Turkey
Tiguentourine *49 E3* E Algeria
Ti-hua/Tihwa *see* Ürümqi
Tijuana *28 A1* Baja California Norte, NW Mexico
Tikhoretsk *89 A7* Krasnodarskiy Kray, SW Russia
Tikhvin *88 B4* Leningradskaya Oblast', NW Russia
Tikiarjuaq *see* Whale Cove
Tiki Basin *121 G3 undersea basin* S Pacific Ocean
Tikinsso *52 C4 river* C Guinea
Tiksi *93 F2* Respublika Sakha (Yakutiya), NE Russia
Tilburg *64 C4* Noord-Brabant, S Netherlands
Tilimsen *see* Tlemcen
Tilio Martius *see* Toulon
Tillabéri *53 F3 var.* Tillabéry. Tillabéri, W Niger
Tillabéry *see* Tillabéri
Tílos *83 E7 island* Dodekánisa, Greece, Aegean Sea
Timan Ridge *see* Timanskiy Kryazh
Timanskiy Kryazh *88 D3 Eng.* Timan Ridge. *ridge* NW Russia
Timaru *129 B6* Canterbury, South Island, New Zealand
Timbaki/Timbákion *see* Tympáki
Timbedgha *52 D3 var.* Timbédra. Hodh ech Chargui, SE Mauritania
Timbédra *see* Timbedgha
Timbuktu *see* Tombouctou
Timiş *86 A4 county* SW Romania
Timişoara *86 A4 Ger.* Temeschwar, Temeswar, *Hung.* Temesvár; *prev.* Temeschburg. Timiş, W Romania
Timmins *16 C4* Ontario, S Canada
Timor *103 F5 island* Nusa Tenggara, C Indonesia
Timor Sea *103 F5 sea* E Indian Ocean
Timor Timur *see* East Timor
Timor Trench *see* Timor Trough
Timor Trough *103 F5 var.* Timor Trench. *trough* NE Timor Sea
Timrå *63 C5* Västernorrland, C Sweden
Tindouf *48 C3* W Algeria
Tineo *70 C1* Asturias, N Spain
Tingis *see* Tanger
Tingo María *38 C3* Huánuco, C Peru
Tingréla *see* Tengréla
Tinhosa Grande *54 E2 island* N Sao Tome and Principe, Africa, E Atlantic Ocean
Tinhosa Pequena *54 E1 island* N Sao Tome and Principe, Africa, E Atlantic Ocean
Tinian *122 B1 island* S Northern Mariana Islands
Tínos *83 D6* Tínos, Kykládes, Greece, Aegean Sea
Tínos *83 D6 anc.* Tenos. *island* Kykládes, Greece, Aegean Sea
Tip *79 E6* Papua, E Indonesia
Tipitapa *30 D3* Managua, W Nicaragua
Tip Top Mountain *16 C4 mountain* Ontario, S Canada
Tirana *see* Tiranë
Tiranë *79 C6 var.* Tirana. *country capital* (Albania) Tiranë, C Albania
Tiraspol *86 D4 Rus.* Tiraspol'. E Moldova
Tiraspol' *see* Tiraspol
Tiree *66 B3 island* W Scotland, United Kingdom
Tirgovişte *see* Târgovişte
Tirgu Jiu *see* Targu Jiu
Tirgu Mures *see* Târgu Mureş
Tirgu-Neamţ *see* Târgu-Neamţ
Tirgu Ocna *see* Târgu Ocna
Tirgu Secuiesc *see* Târgu Secuiesc
Tirlemont *see* Tienen
Tírnavos *see* Týrnavos
Tirnovo *see* Veliko Tŭrnovo
Tirol *73 C7 off.* Land Tirol, *var.* Tyrol, *It.* Tirolo. *state* W Austria
Tirol, Land *see* Tirol
Tirolo *see* Tirol
Tirreno, Mare *see* Tyrrhenian Sea
Tiruchchirāppalli *110 C3 prev.* Trichinopoly. Tamil Nādu, SE India
Tiruppattūr *110 C2* Tamil Nādu, SE India
Tiruvantapuram *see* Thiruvananthapuram
Tisa *see* Tisza
Tisza *81 F1 Ger.* Theiss, *Rom./Slvn./SCr.* Tisa, *Rus.* Tissa, *Ukr.* Tysa. *river* SE Europe
Tiszakécske *77 D7* Bács-Kiskun, C Hungary
Titano, Monte *74 E1 mountain* C San Marino
Titicaca, Lake *39 E4 lake* Bolivia/Peru
Titograd *see* Podgorica
Titose *see* Chitose
Titova Mitrovica *see* Mitrovicë
Titovo Užice *see* Užice
Titu *86 C5* Dâmboviţa, S Romania
Titule *55 D5* Orientale, N Dem. Rep. Congo
Tiverton *67 C7* SW England, United Kingdom
Tivoli *74 C4 anc.* Tiber. Lazio, C Italy
Tizimín *29 H3* Yucatán, SE Mexico
Tizi Ouzou *49 E1 var.* Tizi-Ouzou. N Algeria
Tizi-Ouzou *see* Tizi Ouzou
Tiznit *48 B3* SW Morocco
Tjilatjap *see* Cilacap
Tjirebon *see* Cirebon
Tlaquepaque *28 D4* Jalisco, C Mexico
Tlascala *see* Tlaxcala
Tlaxcala *29 F4 var.* Tlascala, Tlaxcala de Xicohténcatl. Tlaxcala, C Mexico
Tlaxcala de Xicohténcatl *see* Tlaxcala
Tlemcen *48 D2 var.* Tilimsen, Tlemsen. NW Algeria
Tlemsen *see* Tlemcen
Toamasina *57 G3 var.* Tamatave. Toamasina, E Madagascar
Toba, Danau *116 B3 lake* Sumatera, W Indonesia
Tobago *33 H5 island* NE Trinidad and Tobago
Toba Kakar Range *112 B2 mountain range* NW Pakistan
Tobol *92 C4 Kaz.* Tobyl. *river* Kazakhstan/Russia
Tobol'sk *92 C3* Tyumenskaya Oblast', C Russia
Tobruch/Tobruk *see* Ţubruq
Tobyl *see* Tobol
Tocantins *41 E3 off.* Estado do Tocantins. *state/ region* C Brazil
Tocantins, Estado do *see* Tocantins
Tocantins, Rio *41 F2 river* N Brazil
Tocoa *30 D2* Colón, N Honduras
Tocopilla *42 B2* Antofagasta, N Chile
Todi *74 C4* Umbria, C Italy
Todos os Santos, Baía de *41 G3 bay* E Brazil
Toetoes Bay *129 B8 bay* South Island, New Zealand
Tofua *123 E4 island* Ha'apai Group, C Tonga
Togo *53 E4 off.* Togolese Republic; *prev.* French Togoland. *country* W Africa
Togolese Republic *see* Togo
Tojikiston, Jumhurii *see* Tajikistan
Tokanui *129 B7* Southland, South Island, New Zealand
Tokar *50 C3 var.* Ţawkar. Red Sea, NE Sudan
Tokat *94 D3* Tokat, N Turkey
Tokelau *123 E3 NZ overseas territory* W Polynesia
Tőketerebes *see* Trebišov
Tokio *see* Tōkyō
Tokmak *101 G2 Kir.* Tokmok. Chuyskaya Oblast', N Kyrgyzstan
Tokmak *87 G4 var.* Velykyy Tokmak. Zaporiz'ka Oblast', SE Ukraine
Tokmok *see* Tokmak
Tokoroa *128 D3* Waikato, North Island, New Zealand
Tokounou *52 C4* C Guinea
Tokushima *109 C6 var.* Tokusima. Tokushima, Shikoku, SW Japan
Tōkyō *108 A1 var.* Tokio. *country capital* (Japan) Tōkyō, Honshū, S Japan
Tōkyō-wan *108 A2 bay* S Japan
Tolbukhin *see* Dobrich
Toledo *70 D3 anc.* Toletum. Castilla-La Mancha, C Spain
Toledo *18 D3* Ohio, N USA
Toledo Bend Reservoir *27 G3 reservoir* Louisiana/ Texas, SW USA
Toletum *see* Toledo
Toliara *57 F4 var.* Toliary; *prev.* Tuléar. Toliara, SW Madagascar
Toliary *see* Toliara
Tolmein *see* Tolmin
Tolmin *73 D7 Ger.* Tolmein, *It.* Tolmino. W Slovenia
Tolmino *see* Tolmin
Tolna *77 C7 Ger.* Tolnau. Tolna, S Hungary
Tolnau *see* Tolna
Tolochin *see* Talachyn
Tolosa *71 E1* País Vasco, N Spain
Tolosa *see* Toulouse
Toluca *29 E4 var.* Toluca de Lerdo. México, S Mexico
Toluca de Lerdo *see* Toluca
Tol'yatti *89 C6 prev.* Stavropol'. Samarskaya Oblast', W Russia
Tomah *18 B2* Wisconsin, N USA
Tomakomai *108 D2* Hokkaidō, NE Japan
Tomar *70 B3* Santarém, W Portugal
Tomaschow *see* Tomaszów Mazowiecki
Tomaschow *see* Tomaszów Lubelski
Tomaszów *see* Tomaszów Mazowiecki
Tomaszów Lubelski *76 E4 Ger.* Tomaschow. Lubelskie, E Poland
Tomaszów Mazowiecka *see* Tomaszów Mazowiecki
Tomaszów Mazowiecki *76 D4 var.* Tomaszów Mazowiecka; *prev.* Tomaszów, *Ger.* Tomaschow. Łódzkie, C Poland
Tombigbee River *20 C3 river* Alabama/ Mississippi, S USA
Tombouctou *53 E3 Eng.* Timbuktu. *var.* Taoudenni, N Mali
Tombua *56 A2 Port.* Porto Alexandre. Namibe, SW Angola
Tomelloso *71 E3* Castilla-La Mancha, C Spain
Tomini, Gulf of *see* Teluk Tomini
Tomini, Teluk *117 E4 Eng.* Gulf of Tomini; *prev.* Teluk Gorontalo. *bay* Sulawesi, C Indonesia
Tomsk *92 D4* Tomskaya Oblast', C Russia
Tömür Feng *104 B3 pre. Pik Pobedy; Kyrg.* Jengish Chokusu. *mountain* China/Kyrgyzstan
Tonezh *see* Tonyezh
Tonga *123 E4 off.* Kingdom of Tonga, *var.* Friendly Islands. *country* SW Pacific Ocean
Tonga, Kingdom of *see* Tonga
Tongatapu *123 E5 island* Tongatapu Group, S Tonga
Tongatapu Group *123 E5 island group* S Tonga
Tonga Trench *121 E3 trench* S Pacific Ocean
Tongchuan *106 C4* Shaanxi, C China
Tongeren *65 D6 Fr.* Tongres. Limburg, NE Belgium
Tongking, Gulf of *see* Tonkin, Gulf of
Tongliao *105 G2* Nei Mongol Zizhiqu, N China
Tongres *see* Tongeren
Tongshan *see* Xuzhou, Jiangsu, China
Tongtian He *104 C4 river* C China
Tonj *51 B5* Warab, C South Sudan
Tonkin, Gulf of *106 B7 var.* Tongking, Gulf of, *Chin.* Beibu Wan, *Vtn.* Vinh Bắc Bộ. *gulf* China/Vietnam
Tônlé Sap *115 D5 Eng.* Great Lake. *lake* W Cambodia
Tonopah *25 C6* Nevada, W USA
Tonyezh *85 C7 Rus.* Tonezh. Homyel'skaya Voblasts', SE Belarus
Tooele *22 B4* Utah, W USA
Toowoomba *127 E5* Queensland, E Australia
Topeka *23 F4 state capital* Kansas, C USA
Toplicza *see* Topliţa
Topliţa *86 C3 Ger.* Töplitz, *Hung.* Marosheviz; *prev.* Topliţa Română, *Hung.* Oláh-Toplicza, Toplicza. Harghita, C Romania
Topliţa Română/Töplitz *see* Topliţa
Topoľčany *77 C6 Hung.* Nagytapolcsány. Nitriansky Kraj, W Slovakia
Topolovgrad *82 D3 prev.* Kavakli. Khaskovo, S Bulgaria
Topolya *see* Bačka Topola
Top Springs Roadhouse *124 E3* Northern Territory, N Australia
Tor *132 C2 Norwegian research station* Antarctica
Torda *see* Turda
Torez *see* Chystyakove
Torgau *72 D4* Sachsen, E Germany
Torhout *65 A5* West-Vlaanderen, W Belgium
Torino *74 A2 Eng.* Turin. Piemonte, NW Italy
Tornacum *see* Tournai
Torneå *see* Tornio
Torneträsk *62 C3 lake* N Sweden
Tornio *62 D4 Swe.* Torneå. Lappi, NW Finland
Tornionjoki *62 D3 river* Finland/Sweden
Toro *70 D2* Castilla-León, N Spain
Toronto *16 D5 province capital* Ontario, S Canada
Toros Dağları *94 C4 Eng.* Taurus Mountains. *mountain range* S Turkey
Torquay *67 C7* SW England, United Kingdom
Torrance *24 D2* California, W USA
Torre, Alto da *70 B3 mountain* C Portugal
Torre del Greco *75 D5* Campania, S Italy
Torrejón de Ardoz *71 E3* Madrid, C Spain
Torrelavega *70 D1* Cantabria, N Spain
Torrens, Lake *127 A6 salt lake* South Australia
Torrent *71 F3 Cas.* Torrente, *var.* Torrent de l'Horta. País Valenciano, E Spain
Torrent de l'Horta/Torrente *see* Torrent
Torreón *28 D3* Coahuila, NE Mexico
Torres Strait *126 C1 strait* Australia/Papua New Guinea
Torres Vedras *70 B3* Lisboa, C Portugal
Torrington *22 D3* Wyoming, C USA
Tórshavn *61 F5 Dan.* Thorshavn. *Dependent territory capital* Faroe Islands
To'rtkok'l *100 D2 var.* Türtkül, *Rus.* Turtkul'; *prev.* Petroaleksandrovsk. Qoraqalpog'iston Respublikasi, W Uzbekistan
Tortoise Islands *see* Galápagos Islands
Tortosa *71 F2 anc.* Dertosa. Cataluña, E Spain
Tortue, Montagne *37 H3 mountain range* C French Guiana
Tortuga, Isla *see* La Tortuga, Isla
Toruń *76 C3 Ger.* Thorn. Toruń, Kujawsko-pomorskie, C Poland
Tõrva *84 D3 Ger.* Törwa. Valgamaa, S Estonia
Törwa *see* Tõrva
Torzhok *88 B4* Tverskaya Oblast', W Russia
Tosa-wan *109 B7 bay* SW Japan
Toscana *74 B3 Eng.* Tuscany. *region* C Italy
Toscano, Archipelago *74 B4 Eng.* Tuscan Archipelago. *island group* C Italy
Toshkent *101 E2 Eng./Rus.* Tashkent. *country capital* (Uzbekistan) Toshkent Viloyati, E Uzbekistan
Totana *71 E4* Murcia, SE Spain
Tot'ma *see* Sukhona
Totness *37 G3* Coronie, N Suriname
Tottori *109 B6* Tottori, Honshū, SW Japan
Touâjîl *52 C2* Tiris Zemmour, N Mauritania
Touggourt *49 E2* NE Algeria
Toukoto *52 C3* Kayes, W Mali
Toul *68 D3* Meurthe-et-Moselle, NE France
Toulon *69 D6 anc.* Telo Martius, Tilio Martius. Var, SE France
Toulouse *69 B6 anc.* Tolosa. Haute-Garonne, S France
Toungoo *see* Taungoo
Touraine *68 B4 cultural region* C France
Tourane *see* Đă Nẵng
Tourcoing *68 C2* Nord, N France
Tournai *65 A6 var.* Tournay, *Dut.* Doornik; *anc.* Tornacum. Hainaut, SW Belgium
Tournay *see* Tournai
Tours *68 B4 anc.* Caesarodunum, Turoni. Indre-et-Loire, C France
Tovarkovskiy *89 B5* Tul'skaya Oblast', W Russia
Tower Island *see* Genovesa, Isla
Townsville *126 D3* Queensland, NE Australia
Towoeti Meer *see* Towuti, Danau
Towraghoudī *100 D4* Herāt, NW Afghanistan
Towson *19 F4* Maryland, NE USA
Towuti, Danau *117 E4 Dut.* Towoeti Meer. *lake* Sulawesi, C Indonesia
Toyama *109 C5* Toyama, Honshū, SW Japan
Toyama-wan *109 B5 bay* W Japan
Toyohara *see* Yuzhno-Sakhalinsk
Toyota *109 C6* Aichi, Honshū, SW Japan
Tozeur *49 E2 var.* Tawzar. W Tunisia
Trâblous *see* Tripoli
Trabzon *95 E2 Eng.* Trebizond; *anc.* Trapezus. Trabzon, NE Turkey
Traiectum ad Mosam/Traiectum Tungorum *see* Maastricht
Traiskirchen *73 E6* Niederösterreich, NE Austria
Trajani Portus *see* Civitavecchia
Trajectum ad Rhenum *see* Utrecht
Trakai *85 C5 Ger.* Traken, *Pol.* Troki. Vilnius, SE Lithuania
Traken *see* Trakai
Tralee *67 A6 Ir.* Trá Lí. SW Ireland
Trá Lí *see* Tralee
Tralles Aydin *see* Aydın
Trang *115 C7* Trang, S Thailand
Transantarctic Mountains *132 B3 mountain range* Antarctica
Transilvania *see* Transylvania
Transilvaniei, Alpi *see* Carpaţii Meridionalii
Transjordan *see* Jordan
Transnistria *86 D3 cultural region* NE Moldova
Transsylvanische Alpen/Transylvanian Alps *see* Carpaţii Meridionalii
Transylvania *86 B4 Eng.* Ardeal, Transilvania, *Ger.* Siebenbürgen, *Hung.* Erdély. *cultural region* NW Romania
Trapani *75 B7 anc.* Drepanum. Sicilia, Italy, C Mediterranean Sea
Trâpeăng Vêng *see* Kampong Thom
Trapezus *see* Trabzon
Traralgon *127 C7* Victoria, SE Australia
Trasimenischersee *see* Trasimeno, Lago
Trasimeno, Lago *74 C4 Eng.* Lake of Perugia, *Ger.* Trasimenischersee. *lake* C Italy
Traù *see* Trogir
Traverse City *18 C2* Michigan, N USA
Tra Vinh *115 D6 var.* Phu Vinh. Tra Vinh, S Vietnam
Travis, Lake *27 F3 reservoir* Texas, SW USA
Travnik *78 C4* Federacija Bosna I Hercegovina, C Bosnia and Herzegovina
Trbovlje *73 E7 Ger.* Trifail. C Slovenia
Treasure State *see* Montana
Třebíč *77 B5 Ger.* Trebitsch. Vysočina, C Czech Republic (Czechia)
Trebinje *79 C5* Republika Srpska, S Bosnia and Herzegovina
Trebišov *77 D6 Hung.* Tőketerebes. Košický Kraj, E Slovakia
Trebitsch *see* Třebíč
Trebnitz *see* Trzebnica
Tree Planters State *see* Nebraska
Trélazé *68 B4* Maine-et-Loire, NW France
Trelew *43 C6* Chubut, SE Argentina
Tremelo *65 C5* Vlaams Brabant, C Belgium
Trenčín *77 C5 Ger.* Trentschin, *Hung.* Trencsén. Trenčiansky Kraj, W Slovakia
Trencsén *see* Trenčín
Trengganu, Kuala *see* Kuala Terengganu
Trenque Lauquen *42 C4* Buenos Aires, E Argentina
Trent *see* Trento
Trento *74 C2 Eng.* Trent, *Ger.* Trient; *anc.* Tridentum. Trentino-Alto Adige, N Italy
Trenton *19 F4 state capital* New Jersey, NE USA
Trentschin *see* Trenčín
Tres Arroyos *43 D5* Buenos Aires, E Argentina
Treskavica *78 C4 mountain range* SE Bosnia and Herzegovina
Tres Tabernae *see* Saverne
Treves/Trèves *see* Trier
Treviso *74 C2 anc.* Tarvisium. Veneto, NE Italy
Trichinopoly *see* Tiruchchirāppalli
Trichūr *see* Thrissur
Tridentum/Trient *see* Trento
Trier *73 A5 Eng.* Treves, *Fr.* Trèves; *anc.* Augusta Treverorum. Rheinland-Pfalz, SW Germany
Triesen *72 E2* SW Liechtenstein Europe
Triesenberg *72 E2* SW Liechtenstein
Trieste *74 D2 Slvn.* Trst. Friuli-Venezia Giulia, NE Italy
Trifail *see* Trbovlje
Tríkala *82 B4 prev.* Trikkala. Thessalía, C Greece
Trikkala *see* Tríkala
Trimontium *see* Plovdiv
Trinacria *see* Sicilia
Trincomalee *110 D3 var.* Trinkomali. Eastern Province, NE Sri Lanka
Trindade, Ilha da *45 C5 island* Brazil, W Atlantic Ocean
Trinidad *39 F3* Beni, N Bolivia
Trinidad *42 D4* Flores, S Uruguay
Trinidad *22 D5* Colorado, C USA
Trinidad *33 H5 island* C Trinidad and Tobago
Trinidad and Tobago *33 H5 off.* Republic of Trinidad and Tobago. *country* SE West Indies
Trinidad and Tobago, Republic of *see* Trinidad and Tobago
Trinité, Montagnes de la *37 H3 mountain range* C French Guiana
Trinity River *27 G3 river* Texas, SW USA
Trinkomali *see* Trincomalee
Trípoli *83 B6 prev.* Trípolis. Pelopónnisos, S Greece
Tripoli *96 B4 var.* Tarābulus, Ţarābulus ash Shām, Trâblous; *anc.* Tripolis. N Lebanon
Tripoli *see* Ţarābulus
Trípolis *see* Trípoli, Greece
Tripolis *see* Tripoli, Lebanon
Tristan da Cunha *see* St Helena, Ascension and Tristan da Cunha
Triton Island *106 B7 island* S Paracel Islands
Trivandrum *see* **Thiruvananthapuram**
Trnava *77 C6 Ger.* Tyrnau, *Hung.* Nagyszombat. Trnavský Kraj, W Slovakia
Trnovo *see* Veliko Tŭrnovo
Trogir *78 B4 It.* Traù. Split-Dalmacija, S Croatia
Troglav *78 B4 mountain* Bosnia and Herzegovina/ Croatia
Trois-Rivières *17 E4* Québec, SE Canada
Troki *see* Trakai
Troll *132 C2 Norwegian research station* Antarctica
Trollhättan *63 B6* Västra Götaland, S Sweden
Tromsø *62 C2 Fin.* Tromssa. Troms, N Norway
Tromssa *see* Tromsø
Trondheim *62 B4 Ger.* Drontheim; *prev.* Nidaros, Trondhjem. Sør-Trøndelag, S Norway
Trondheimsfjorden *62 B4 fjord* S Norway
Trondhjem *see* Trondheim
Troódos *80 C5 var.* Troodos Mountains. *mountain range* C Cyprus
Troodos Mountains *see* Troódos
Troppau *see* Opava
Troy *20 D3* Alabama, S USA
Troy *19 F3* New York, NE USA
Troyan *82 C2* Lovech, N Bulgaria
Troyes *68 D3 anc.* Augustobona Tricassium. Aube, N France
Trst *see* Trieste
Trstenik *78 E4* Serbia, C Serbia
Trucial States *see* United Arab Emirates
Trujillo *30 D2* Colón, NE Honduras
Trujillo *38 B3* La Libertad, NW Peru
Trujillo *70 C3* Extremadura, W Spain
Truk Islands *see* Chuuk Islands
Trŭn *82 B2* Pernik, W Bulgaria
Truro *17 F4* Nova Scotia, SE Canada
Truro *67 C7* SW England, United Kingdom
Trzcianka *76 B3 Ger.* Schönlanke. Pila, Wielkopolskie, C Poland
Trzebnica *76 C4 Ger.* Trebnitz. Dolnośląskie, SW Poland
Tsalka *95 F2* S Georgia Asia
Tsamkong *see* Zhanjiang
Tsangpo *see* Brahmaputra

Tsarevo – Vanuatu, Republic of

Tsarevo 82 E2 prev. Michurin. Burgas, E Bulgaria
Tsarigrad see İstanbul
Tsaritsyn see Volgograd
Tschakathurn see Čakovec
Tschaslau see Čáslav
Tschenstochau see Częstochowa
Tsefat 97 B5 var. Safed, Ar. Safad; prev. Ẕefat. Northern, N Israel
Tselinograd see Astana
Tsetsen Khan see Öndörhaan
Tsetserleg 104 D2 Arhangay, C Mongolia
Tshela 55 B6 Bas-Congo, W Dem. Rep. Congo
Tshikapa 55 C7 Kasai-Occidental, SW Dem. Rep. Congo
Tshuapa 55 D6 river C Dem. Rep. Congo
Tsinan see Jinan
Tsing Hai see Qinghai Hu, China
Tsinghai see Qinghai, China
Tsingtao/Tsingtau see Qingdao
Tsinkiang see Quanzhou
Tsintao see Qingdao
Tsitsihar see Qiqihar
Tsu 109 C6 var. Tu. Mie, Honshū, SW Japan
Tsugaru-kaikyo 108 C3 strait N Japan
Tsumeb 56 B3 Otjikoto, N Namibia
Tsuruga 109 C6 var. Turuga. Fukui, Honshū, SW Japan
Tsuruoka 108 D4 var. Turuoka. Yamagata, Honshū, C Japan
Tsushima 109 A7 var. Tsushima-tō, Tusima. island group SW Japan
Tsushima-tō see Tsushima
Tsyerakhowka 85 D8 Rus. Terekhovka. Homyel'skaya Voblasts', SE Belarus
Tsyurupyns'k see Oleshk
Tu see Tsu
Tuamotu, Archipel des see Tuamotu, Îles
Tuamotu Fracture Zone 121 H3 fracture zone E Pacific Ocean
Tuamotu, Îles 123 H4 var. Archipel des Tuamotu, Dangerous Archipelago, Tuamotu Islands. island group N French Polynesia
Tuamotu Islands see Tuamotu, Îles
Tuapi 31 E2 Región Autónoma Atlántico Norte, NE Nicaragua
Tuapse 89 A7 Krasnodarskiy Kray, SW Russia
Tuba City 26 B1 Arizona, SW USA
Tubbergen 64 E3 Overijssel, E Netherlands
Tubeke see Tubize
Tubize 65 B6 Dut. Tubeke. Walloon Brabant, C Belgium
Tubmanburg 52 C5 NW Liberia
Ţubruq 49 H2 Eng. Tobruk, It. Tobruch. NE Libya
Tubuai, Îles/Tubuai Islands see Australes, Îles
Tucker's Town 20 B5 E Bermuda
Tuckum see Tukums
Tucson 26 B3 Arizona, SW USA
Tucumán see San Miguel de Tucumán
Tucumcari 27 E2 New Mexico, SW USA
Tucupita 37 E2 Delta Amacuro, NE Venezuela
Tucuruí, Represa de 41 F2 reservoir NE Brazil
Tudela 71 E2 Basq. Tutera; anc. Tutela. Navarra, N Spain
Tudmur 96 C3 var. Tadmur, Tamar, Gk. Palmyra, Bibl. Tadmor. Ḥimṣ, C Syria
Tuguegarao 117 E1 Luzon, N Philippines
Tuktoyaktuk 15 E3 Northwest Territories, NW Canada
Tukums 84 C3 Ger. Tuckum. Tukums, W Latvia
Tula 89 B5 Tul'skaya Oblast', W Russia
Tulancingo 29 E4 Hidalgo, C Mexico
Tulare Lake Bed 25 C7 salt flat California, W USA
Tulcán 38 B1 Carchi, N Ecuador
Tulcea 86 D5 Tulcea, E Romania
Tul'chin see Tul'chyn
Tul'chyn 86 D3 Rus. Tul'chin. Vinnyts'ka Oblast', C Ukraine
Tuléar see Toliara
Tulia 27 E2 Texas, SW USA
Ţūlkarm 97 D7 West Bank, Israel
Tulle 69 C5 anc. Tutela. Corrèze, C France
Tulln 73 E6 var. Oberhollabrunn. Niederösterreich, NE Austria
Tully 126 D3 Queensland, NE Australia
Tulsa 27 G1 Oklahoma, C USA
Tuluá 36 B3 Valle del Cauca, W Colombia
Tulun 93 E4 Irkutskaya Oblast', S Russia
Tumaco 36 A4 Nariño, SW Colombia
Tumakūru 110 C2 prev. Tumkūr. Karnātaka, W India
Tumba, Lac see Ntomba, Lac
Tumbes 38 A2 Tumbes, NW Peru
Tumkūr see Tumakūru
Tumuc-Humac Mountains 41 E1 var. Serra Tumucumaque. mountain range N South America
Tumucumaque, Serra see Tumuc-Humac Mountains
Tunca Nehri see Tundzha
Tunduru 51 C8 Ruvuma, S Tanzania
Tundzha 82 D3 Turk. Tunca Nehri. river Bulgaria/Turkey
Tungabhadra Reservoir 110 C2 lake S India
Tungaru 123 E2 prev. Gilbert Islands. island group W Kiribati
T'ung-shan see Xuzhou
Tungsten 14 D4 Northwest Territories, W Canada
Tung-t'ing Hu see Dongting Hu
Tunis 49 E1 var. Tūnis. country capital (Tunisia) NE Tunisia
Tunis, Golfe de 80 D3 Ar. Khalīj Tūnis. gulf NE Tunisia
Tunisia 49 F2 off. Republic of Tunisia, Ar. Al Jumhūrīyah at Tūnisīyah, Fr. République Tunisienne. country N Africa
Tunisia, Republic of see Tunisia
Tunisienne, République see Tunisia
Tūnisīyah, Al Jumhūrīyah at see Tunisia
Tūnis, Khalīj see Tunis, Golfe de
Tunja 36 B3 Boyacá, C Colombia
Tuong Buong see Tương Đương
Tương Đương 114 D4 var. Tuong Buong. Nghê An, N Vietnam
Tüp see Tyup
Tupelo 20 C2 Mississippi, S USA
Tupiza 39 G5 Potosí, S Bolivia
Turabah 99 B5 Makkah, W Saudi Arabia
Turangi 128 D4 Waikato, North Island, New Zealand
Turan Lowland 100 C2 var. Turan Plain, Kaz. Turan Oypaty, Rus. Turanskaya Nizmennost', Turk. Turan Pesligi, Uzb. Turan Pasttekisligi. plain C Asia
Turan Oypaty/Turan Pesligi/Turan Plain/Turanskaya Nizmennost' see Turan Lowland
Turan Pasttekisligi see Turan Lowland
Ţurayf 98 A3 Al Ḥudūd ash Shamālīyah, NW Saudi Arabia
Turba see Teruel
Turbat 112 A3 Baluchistān, SW Pakistan
Turčianský Svätý Martin see Martin
Turda 86 B4 Ger. Thorenburg, Hung. Torda. Cluj, NW Romania
Turek 76 C3 Wielkopolskie, C Poland
Turfan see Turpan
Turin see Torino
Turkana, Lake 51 C6 var. Lake Rudolf. lake N Kenya
Turkestan see Turkistan
Turkey 94 B3 off. Republic of Turkey, Turk. Türkiye Cumhuriyeti. country SW Asia
Turkey, Republic of see Turkey
Turkish Republic of Northern Cyprus 80 D5 disputed territory administered by Turkey Cyprus
Turkistan 92 B5 prev. Turkestan. Yuzhnyy Kazakhstan, S Kazakhstan
Türkiye Cumhuriyeti see Turkey
Türkmenabat 100 D3 prev. Rus. Chardzhev, Chardzhou, Chardzhui, Lenin-Turkmenski, Turkm. Chärjew. Lebap Welaýaty, E Turkmenistan
Türkmen Aylagy 100 B2 Rus. Turkmenskiy Zaliv. lake gulf W Turkmenistan
Turkmenbashi see Türkmenbaşy
Türkmenbaşy 100 B2 Rus. Turkmenbashi; prev. Krasnovodsk. Balkan Welaýaty, W Turkmenistan
Türkmenbasy Aylagy 100 A2 prev. Rus. Krasnovodskiy Zaliv, Turkm. Krasnowodsk Aylagy. lake Gulf W Turkmenistan
Turkmenistan 100 B2 ; prev. Turkmenskaya Soviet Socialist Republic. country C Asia
Turkmenskaya Soviet Socialist Republic see Turkmenistan
Turkmenskiy Zaliv see Türkmen Aylagy
Turks and Caicos Islands 33 E2 UK dependent territory N West Indies
Turku 63 D6 Swe. Åbo. Varsinais-Suomi, SW Finland
Turlock 25 B6 California, W USA
Turnagain, Cape 128 D4 headland North Island, New Zealand
Turnau see Turnov
Turnhout 65 C5 Antwerpen, N Belgium
Turnov 76 B4 Ger. Turnau. Liberecký Kraj, N Czech Republic (Czechia)
Tŭrnovo see Veliko Tŭrnovo
Turnu Măgurele 86 B5 var. Turnu-Măgurele. Teleorman, S Romania
Turnu Severin see Drobeta-Turnu Severin
Turócszentmárton see Martin
Turoni see Tours
Turpan 104 C3 var. Turfan. Xinjiang Uygur Zizhiqu, NW China
Turpan Depression see Turpan Pendi
Turpan Pendi 104 C3 Eng. Turpan Depression. depression NW China
Turpentine State see North Carolina
Türtkül/Turtkul' see To'rtkok'l
Turuga see Tsuruga
Turuoka see Tsuruoka
Tuscaloosa 20 C2 Alabama, S USA
Tuscan Archipelago see Toscano, Archipelago
Tuscany see Toscana
Tusima see Tsushima
Tutela see Tulle, France
Tutela see Tudela, Spain
Tutera see Tudela
Tuticorin 110 C3 Tamil Nādu, SE India
Tutrakan 82 D1 Silistra, NE Bulgaria
Tutuila 123 F4 island W American Samoa
Tuvalu 123 E3 prev. Ellice Islands. country SW Pacific Ocean
Tuwayq, Jabal 99 C5 mountain range C Saudi Arabia
Tuxpan 28 D4 Jalisco, C Mexico
Tuxpán 29 F4 var. Tuxpán de Rodríguez Cano. Veracruz-Llave, E Mexico
Tuxpán de Rodríguez Cano see Tuxpán
Tuxtepec 29 F4 var. San Juan Bautista Tuxtepec. Oaxaca, S Mexico
Tuxtla 29 G5 var. Tuxtla Gutiérrez. Chiapas, SE Mexico
Tuxtla see San Andrés Tuxtla
Tuxtla Gutiérrez see Tuxtla
Tuy Hoa 115 E5 Phu Yên, S Vietnam
Tuz, Lake 94 C3 lake C Turkey
Tver' 88 B4 prev. Kalinin. Tverskaya Oblast', W Russia
Tverya 97 B5 var. Tiberias; prev. Teverya. Northern, N Israel
Twin Falls 24 D4 Idaho, NW USA
Tyan'-Shan' see Tien Shan
Tychy 77 D5 Ger. Tichau. Śląskie, S Poland
Tyler 27 G3 Texas, SW USA
Tylos see Bahrain
Tympáki 83 C8 var. Timbaki; prev. Timbákion. Kríti, Greece, E Mediterranean Sea
Tynda 93 F4 Amurskaya Oblast', SE Russia
Tyne 66 D4 river N England, United Kingdom
Tyôsi see Chōshi
Tyras see Dniester
Tyre see Soûr
Tyrnau see Trnava
Týrnavos 82 B4 var. Tírnavos. Thessalía, C Greece
Tyrol see Tirol
Tyros see Bahrain
Tyrrhenian Sea 75 B6 It. Mare Tirreno. sea N Mediterranean Sea
Tyumen' 92 C3 Tyumenskaya Oblast', C Russia
Tyup 101 G2 Kir. Tüp. Issyk-Kul'skaya Oblast', NE Kyrgyzstan
Tywyn 67 C6 W Wales, United Kingdom
Tzekung see Zigong
Tziá see Kéa

U

Uaco Cungo 56 B1 C Angola
UAE see United Arab Emirates
Uanle Uen see Wanlaweyn
Uaupés, Rio see Vaupés, Rio
Ubangi-Shari see Central African Republic
Ube 109 B7 Yamaguchi, Honshū, SW Japan
Ubeda 71 E4 Andalucía, S Spain
Uberaba 41 F4 Minas Gerais, SE Brazil
Uberlândia 41 F4 Minas Gerais, SE Brazil
Ubol Rajadhani/Ubol Ratchathani see Ubon Ratchathani
Ubon Ratchathani 115 D5 var. Muang Ubon, Ubol Rajadhani, Ubol Ratchathani, Udon Ratchathani. Ubon Ratchathani, E Thailand
Ubrique 70 D5 Andalucía, S Spain
Ubsu-Nur, Ozero see Uvs Nuur
Ucayali, Río 38 D3 river C Peru
Uchiura-wan 108 D3 bay NW Pacific Ocean
Uchkuduk see Uchquduq
Uchquduq 100 D2 Rus. Uchkuduk. Navoiy Viloyati, N Uzbekistan
Uchtagan Gumy/Uchtagan, Peski see Uçtagan Gumy
Uçtagan Gumy 100 C2 var. Uchtagan Gumy, Rus. Peski Uchtagan. desert NW Turkmenistan
Udaipur 112 C3 prev. Oodeypore. Rājasthān, N India
Uddevalla 63 B6 Västra Götaland, S Sweden
Udine 74 D2 anc. Utina. Friuli-Venezia Giulia, NE Italy
Udintsev Fracture Zone 132 A5 tectonic feature S Pacific Ocean
Udipi see Udupi
Udon Ratchathani see Ubon Ratchathani
Udon Thani 114 C4 var. Ban Mak Khaeng, Udorndhani. Udon Thani, N Thailand
Udorndhani see Udon Thani
Udupi 110 B2 var. Udipi. Karnātaka, SW India
Uele 55 D5 var. Welle. river NE Dem. Rep. Congo
Uelzen 72 C3 Niedersachsen, N Germany
Ufa 89 D6 Respublika Bashkortostan, W Russia
Ugāle 84 C2 Ventspils, NW Latvia
Uganda 51 B6 off. Republic of Uganda. country E Africa
Uganda, Republic of see Uganda
Uhorshchyna see Hungary
Uhuru Peak see Kilimanjaro
Uíge 56 B1 Port. Carmona, Vila Marechal Carmona. Uíge, NW Angola
Uinta Mountains 22 B4 mountain range Utah, W USA
Uitenhage 56 C5 Eastern Cape, S South Africa
Uithoorn 64 C3 Noord-Holland, C Netherlands
Ujda see Oujda
Ujelang Atoll 122 C1 var. Wujlān. atoll Ralik Chain, W Marshall Islands
Újgradiska see Nova Gradiška
Újmoldova see Moldova Nouă
Ujungpandang see Makassar
Ujung Salang see Phuket
Újvidék see Novi Sad
UK see United Kingdom
Ukhta 92 C3 Respublika Komi, NW Russia
Ukiah 25 B5 California, W USA
Ukmergė 84 C4 Pol. Wiłkomierz. Vilnius, C Lithuania
Ukraine 86 C2 off. Ukraine, Rus. Ukraina, Ukr. Ukrayina; prev. Ukrainian Soviet Socialist Republic, Ukrainskay S.S.R. country SE Europe
Ukraine see Ukraine
Ukrainian Soviet Socialist Republic see Ukraine
Ukrainskay S.S.R/Ukrayina see Ukraine
Ulaanbaatar 105 E2 Eng. Ulan Bator; prev. Urga. country capital (Mongolia) Töv, C Mongolia
Ulaangom 104 C2 Uvs, NW Mongolia
Ulan Bator see Ulaanbaatar
Ulanhad see Chifeng
Ulan-Ude 93 E4 prev. Verkhneudinsk. Respublika Buryatiya, S Russia
Uleälv see Oulujoki
Uleträsk see Oulujärvi
Ulft 64 E4 Gelderland, E Netherlands
Ullapool 66 C3 N Scotland, United Kingdom
Ulm 73 B6 Baden-Württemberg, S Germany
Ulsan 107 E4 Jap. Urusan. SE South Korea
Ulster 67 B5 province Northern Ireland, United Kingdom/Ireland
Ulungur Hu 104 B2 lake NW China
Uluru 125 D5 var. Ayers Rock. monolith Northern Territory, C Australia
Ulyanivka see Blahovishchens'ke
Ul'yanovsk 89 C5 prev. Simbirsk. Ul'yanovskaya Oblast', W Russia
Umán 29 H3 Yucatán, SE Mexico
Uman' 87 E3 Rus. Uman. Cherkas'ka Oblast', C Ukraine
Uman see Uman'
Umanak/Umanaq see Uummannaq
'Umān, Khalīj see Oman, Gulf of
'Umān, Salṭanat see Oman
Umbrian-Machigian Mountains see Umbro-Marchigiano, Appennino
Umbro-Marchigiano, Appennino 74 C3 Eng. Umbrian-Machigian Mountains. mountain range C Italy
Umeå 62 C4 Västerbotten, N Sweden
Umeälven 62 C4 river N Sweden
Umiat 14 D2 Alaska, USA
Umm Buru 50 A4 Western Darfur, W Sudan
Umm Durmān see Omdurman
Umm Ruwaba 50 C4 var. Umm Ruwābah, Um Ruwāba. Northern Kordofan, C Sudan
Umm Ruwābah see Umm Ruwaba
Umnak Island 14 A3 island Aleutian Islands, Alaska, USA
Um Ruwāba see Umm Ruwaba
Umtali see Mutare
Umtata see Mthatha
Una 78 B3 river Bosnia and Herzegovina/Croatia
Unac 78 B3 river W Bosnia and Herzegovina
Unalaska Island 14 A3 island Aleutian Islands, Alaska, USA
'Unayzah 98 B4 var. Anaiza. Al Qaṣīm, C Saudi Arabia
Unci see Almería
Uncía 39 F4 Potosí, C Bolivia
Uncompahgre Peak 22 B5 mountain Colorado, C USA
Undur Khan see Öndörhaan
Ungaria see Hungary
Ungarisches Erzgebirge see Slovenské rudohorie
Ungarn see Hungary
Ungava Bay 17 E1 bay Québec, E Canada
Ungava Peninsula see Ungava, Péninsule d'
Ungava, Péninsule d' 16 D1 Eng. Ungava Peninsula. peninsula Québec, SE Canada
Ungeny see Ungheni
Ungheni 86 D3 Rus. Ungeny. W Moldova
Unguja see Zanzibar
Üngüz Angyrsyndaky Garagum 100 C2 Rus. Zaunguzskiye Garagumy. desert N Turkmenistan
Ungvár see Uzhhorod
Unimak Island 14 B3 island Aleutian Islands, Alaska, USA
Union 21 E1 South Carolina, SE USA
Union City 20 C1 Tennessee, S USA
Union of Myanmar see Myanmar
United Arab Emirates 99 C5 Ar. Al Imārāt al 'Arabīyah al Muttaḥidah, abbrev. UAE; prev. Trucial States. country SW Asia
United Arab Republic see Egypt
United Kingdom 67 B5 off. United Kingdom of Great Britain and Northern Ireland, abbrev. UK. country NW Europe
United Kingdom of Great Britain and Northern Ireland see United Kingdom
United Mexican States see Mexico
United Provinces see Uttar Pradesh
United States of America 13 B5 off. United States of America, var. America, The States, abbrev. U.S., USA. country North America
Unst 66 D1 island NE Scotland, United Kingdom
Ünye 94 D2 Ordu, W Turkey
Upala 30 D4 Alajuela, NW Costa Rica
Upata 37 E2 Bolívar, E Venezuela
Upemba, Lac 55 D7 lake SE Dem. Rep. Congo
Upernavik 60 C2 var. Upernivik. Kitaa, C Greenland
Upernivik see Upernavik
Upington 56 C4 Northern Cape, W South Africa
'Upolu 123 F4 island SE Samoa
Upper Klamath Lake 24 A4 lake Oregon, NW USA
Upper Lough Erne 67 A5 lake SW Northern Ireland, United Kingdom
Upper Red Lake 23 F1 lake Minnesota, N USA
Upper Volta see Burkina
Uppsala 63 C6 Uppsala, C Sweden
Uqsuqtuuq see Gjoa Haven
Ural 90 B3 Kaz. Zhayyk. river Kazakhstan/Russia
Ural Mountains see Ural'skiye Gory
Ural'sk 92 B3 Kaz. Oral. Zapadnyy Kazakhstan, NW Kazakhstan
Ural'skiye Gory 92 C3 var. Ural'skiy Khrebet, Eng. Ural Mountains. mountain range Kazakhstan/Russia
Ural'skiy Khrebet see Ural'skiye Gory
Uraricoera 40 D1 Roraima, N Brazil
Ura-Tyube see Ŭroteppa
Urbandale 23 F3 Iowa, C USA
Urdunn see Jordan
Uren' 89 C5 Nizhegorodskaya Oblast', W Russia
Urga see Ulaanbaatar
Urganch 100 D2 Rus. Urgench; prev. Novo-Urgench. Xorazm Viloyati, W Uzbekistan
Urgench see Urganch
Urgut 101 E3 Samarqand Viloyati, C Uzbekistan
Urmia, Lake see Orūmīyeh, Daryācheh-ye
Uroševac see Ferizaj
Ŭroteppa 101 E2 Rus. Ura-Tyube. NW Tajikistan
Uruapan 29 E4 var. Uruapan del Progreso. Michoacán, SW Mexico
Uruapan del Progreso see Uruapan
Uruguai, Rio see Uruguay
Uruguay 42 D4 off. Oriental Republic of Uruguay; prev. La Banda Oriental. country E South America
Uruguay 42 D3 var. Rio Uruguai, Río Uruguay. river E South America
Uruguay, Oriental Republic of see Uruguay
Uruguay, Río see Uruguay
Urumchi see Ürümqi
Urumi Yeh see Orūmīyeh, Daryācheh-ye
Ürümqi 104 C3 var. Tihwa, Urumchi, Urumqi, Urumtsi, Wu-lu-k'o-mu-shi, Wu-lu-mu-ch'i; prev. Ti-hua. Xinjiang Uygur Zizhiqu, NW China
Urumtsi see Ürümqi
Urundi see Burundi
Urup, Ostrov 93 H4 island Kuril'skiye Ostrova, SE Russia
Urusan see Ulsan
Urziceni 86 C5 Ialomiţa, SE Romania
Usa 88 E3 river NW Russia
Uşak 94 B3 prev. Ushak. Uşak, W Turkey
Ushak see Uşak
Ushant see Ouessant, Île d'
Ushuaia 43 B8 Tierra del Fuego, S Argentina
Usinsk 88 E3 Respublika Komi, NW Russia
Üsküb/Üsküp see Skopje
Usmas Ezers 84 B3 lake NW Latvia
Usol'ye-Sibirskoye 93 E4 Irkutskaya Oblast', C Russia
Ussel 69 C5 Corrèze, C France
Ussuriysk 93 G5 prev. Nikol'sk, Nikol'sk-Ussuriyskiy, Voroshilov. Primorskiy Kray, SE Russia
Ustica 75 B6 island S Italy
Ust'-Ilimsk 93 E4 Irkutskaya Oblast', C Russia
Ústí nad Labem 76 A4 Ger. Aussig. Ústecký Kraj, NW Czech Republic (Czechia)
Ustinov see Izhevsk
Ustka 76 C2 Ger. Stolpmünde. Pomorskie, N Poland
Ust'-Kamchatsk 93 H2 Kamchatskaya Oblast', E Russia
Ust'-Kamenogorsk 92 D5 Kaz. Öskemen. Vostochnyy Kazakhstan, E Kazakhstan
Ust'-Kut 93 E4 Irkutskaya Oblast', C Russia
Ust'-Olenëk 93 E3 Respublika Sakha (Yakutiya), NE Russia
Ustrzyki Dolne 77 E5 Podkarpackie, SE Poland
Ust'-Sysol'sk see Syktyvkar
Ust Urt see Ustyurt Plateau
Ustyurt Plateau 100 B1 var. Ust Urt, Uzb. Ustyurt Platosi. plateau Kazakhstan/Uzbekistan
Ustyurt Platosi see Ustyurt Plateau
Usulután 30 C3 Usulután, SE El Salvador
Usumacinta, Río 30 B1 river Guatemala/Mexico
Usumbura see Bujumbura
U.S./USA see United States of America
Utah 22 B4 off. State of Utah, also known as Beehive State, Mormon State. state W USA
Utah Lake 22 B4 lake Utah, W USA
Utena 84 C4 Utena, E Lithuania
Utica 19 F3 New York, NE USA
Utina see Udine
Utrecht 64 C4 Lat. Trajectum ad Rhenum. Utrecht, C Netherlands
Utsunomiya 109 D5 var. Utunomiya. Tochigi, Honshū, S Japan
Uttarakhand 113 E2 cultural region N India
Uttar Pradesh 113 E3 prev. United Provinces, United Provinces of Agra and Oudh. cultural region N India
Utunomiya see Utsunomiya
Uulu 84 D2 Pärnumaa, SW Estonia
Uummannaq 60 C3 var. Umanak, Umanaq. Kitaa, C Greenland
Uummannarsuaq see Nunap Isua
Uvalde 27 F4 Texas, SW USA
Uvaravichy 85 D7 Rus. Uvarovichi. Homyel'skaya Voblasts', SE Belarus
Uvarovichi see Uvaravichy
Uvea, Île 123 E4 island N Wallis and Futuna
Uvs Nuur 104 C1 var. Ozero Ubsu-Nur. lake Mongolia/Russia
'Uwaynāt, Jabal al 66 A3 var. Jebel Uweinat. mountain Libya/Sudan
Uweinat, Jebel see 'Uwaynāt, Jabal al
Uyo 53 G5 Akwa Ibom, S Nigeria
Uyuni 39 F5 Potosí, W Bolivia
Uzbekistan 100 D2 off. Republic of Uzbekistan. country C Asia
Uzbekistan, Republic of see Uzbekistan
Uzhgorod see Uzhhorod
Uzhhorod 86 B2 Rus. Uzhgorod; prev. Ungvár. Zakarpats'ka Oblast', W Ukraine
Užice 78 D4 prev. Titovo Užice. Serbia, W Serbia

V

Vaal 56 D4 river C South Africa
Vaals 65 D6 Limburg, SE Netherlands
Vaasa 63 D5 Swe. Vasa; prev. Nikolainkaupunki. Österbotten, W Finland
Vaassen 64 D3 Gelderland, E Netherlands
Vác 77 C6 Ger. Waitzen. Pest, N Hungary
Vadodara 112 C4 prev. Baroda. Gujarāt, W India
Vaduz 72 E2 country capital (Liechtenstein) W Liechtenstein
Vág see Váh
Vágbeszterce see Považská Bystrica
Váh 77 C5 Ger. Waag, Hung. Vág. river W Slovakia
Vähtjer see Gällivare
Väinameri 84 C2 prev. Muhu Väin, Ger. Moon-Sund. sea E Baltic Sea
Vajdahunyad see Hunedoara
Valachia see Wallachia
Valday 88 B4 Novgorodskaya Oblast', W Russia
Valdecañas, Embalse de 70 D3 reservoir W Spain
Valdepeñas 71 E4 Castilla-La Mancha, C Spain
Valdez 14 C3 Alaska, USA
Valdia see Weldiya
Valdivia 43 B5 Los Lagos, C Chile
Val-d'Or 16 D4 Québec, SE Canada
Valdosta 21 E3 Georgia, SE USA
Valence 69 D5 anc. Valentia, Valentia Julia, Ventia. Drôme, E France
Valencia 71 F3 País Valenciano, E Spain
Valencia 24 D1 California, USA
Valencia 36 D1 Carabobo, N Venezuela
Valencia, Gulf of 71 F3 var. Gulf of Valencia. gulf E Spain
Valencia, Gulf of see Valencia, Golfo de
Valencia/València see País Valenciano
Valenciennes 68 D2 Nord, N France
Valentia see Valence, France
Valentia see País Valenciano
Valentia Julia see Valence
Valentine State see Oregon
Valera 36 C2 Trujillo, NW Venezuela
Valetta see Valletta
Valga 84 D3 Ger. Walk, Latv. Valka. Valgamaa, S Estonia
Valira 69 A8 river Andorra/Spain Europe
Valjevo 78 C4 Serbia, W Serbia
Valjok see Válljohka
Valka 84 D3 Ger. Walk. Valka, N Latvia
Valka see Valga
Valkenswaard 65 D5 Noord-Brabant, S Netherlands
Valladolid 29 H3 Yucatán, SE Mexico
Valladolid 70 D2 Castilla-León, NW Spain
Vall D'Uxó see La Vall d'Uixó
Valle de La Pascua 36 D2 Guárico, N Venezuela
Valledupar 36 B1 Cesar, N Colombia
Vallenar 42 B3 Atacama, N Chile
Valletta 75 C8 prev. Valetta. country capital (Malta) E Malta
Valley City 23 E2 North Dakota, N USA
Válljohka 62 D2 var. Valjok. Finnmark, N Norway
Valls 71 G2 Cataluña, NE Spain
Valmiera 84 D3 Est. Volmari, Ger. Wolmar. Valmiera, N Latvia
Valona see Vlorë
Valozhyn 85 C5 Pol. Wołożyn, Rus. Volozhin. Minskaya Voblasts', C Belarus
Valparaíso 42 B4 Valparaíso, C Chile
Valparaiso 18 C3 Indiana, N USA
Valverde del Camino 70 C4 Andalucía, S Spain
Van 95 F3 Van, E Turkey
Vanadzor 95 F2 prev. Kirovakan. N Armenia
Vancouver 14 D5 British Columbia, SW Canada
Vancouver 24 B3 Washington, NW USA
Vancouver Island 14 D5 island British Columbia, SW Canada
Vanda see Vantaa
Van Diemen Gulf 124 D2 gulf Northern Territory, N Australia
Van Diemen's Land see Tasmania
Vaner, Lake see Vänern
Vänern 63 B6 Eng. Lake Vaner; prev. Lake Vener. lake S Sweden
Vangaindrano 57 G4 Fianarantsoa, SE Madagascar
Van Gölü 95 F3 Eng. Lake Van; anc. Thospitis. salt lake E Turkey
Van Horn 26 D3 Texas, SW USA
Van, Lake see Van Gölü
Vannes 68 A3 anc. Dariorigum. Morbihan, NW France
Vantaa 63 D6 Swe. Vanda. Uusimaa, S Finland
Vanua Levu 123 E4 island N Fiji
Vanuatu 122 C4 off. Republic of Vanuatu; prev. New Hebrides. country SW Pacific Ocean
Vanuatu, Republic of see Vanuatu

Wanlaweyn – Yerevan

Wanlaweyn *51 D6 var.* Wanle Weyn, *It.* Uanle Uen. Shabeellaha Hoose, SW Somalia
Wanle Weyn *see* Wanlaweyn
Wanxian *see* Wanzhou
Wanzhou *106 B5 var.* Wanxian. Chongqing, C China
Warangal *113 E5* Telangana, C India
Warasdin *see* Varaždin
Warburg *72 B4* Nordrhein-Westfalen, W Germany
Ware *15 E4* British Columbia, W Canada
Waremme *65 C6* Liège, E Belgium
Waren *72 C3* Mecklenburg-Vorpommern, NE Germany
Wargla *see* Ouargla
Warkworth *128 D2* Auckland, North Island, New Zealand
Warnemünde *72 C2* Mecklenburg-Vorpommern, NE Germany
Warner *27 G1* Oklahoma, C USA
Warnes *39 G4* Santa Cruz, C Bolivia
Warrego River *127 C5 seasonal river* New South Wales/Queensland, E Australia
Warren *18 D3* Michigan, N USA
Warren *18 D3* Ohio, N USA
Warren *19 E3* Pennsylvania, NE USA
Warri *53 F5* Delta, S Nigeria
Warrnambool *127 B7* Victoria, SE Australia
Warsaw/Warschau *see* Warszawa
Warszawa *76 D3 Eng.* Warsaw, *Ger.* Warschau, *Rus.* Varshava. *country capital* (Poland) Mazowieckie, C Poland
Warta *76 B3 Ger.* Warthe. *river* W Poland
Wartberg *see* Senec
Warthe *see* Warta
Warwick *127 E5* Queensland, E Australia
Wasa *132 B2 Swedish research station* Antarctica
Washington *22 A2* NE England, United Kingdom
Washington D.C. *19 E4 country capital* (USA) District of Columbia, NE USA
Washington Island *see* Teraina
Washington, Mount *19 G2 mountain* New Hampshire, NE USA
Wash, The *67 E6 inlet* E England, United Kingdom
Wasiliszki *see* Vasilishki
Waspam *31 E2 var.* Waspán. Región Autónoma Atlántico Norte, NE Nicaragua
Waspán *see* Waspam
Watampone *117 E4 var.* Bone. Sulawesi, C Indonesia
Watenstedt-Salzgitter *see* Salzgitter
Waterbury *19 F3* Connecticut, NE USA
Waterford *67 B6 Ir.* Port Láirge. Waterford, S Ireland
Waterloo *23 G3* Iowa, C USA
Watertown *19 F2* New York, NE USA
Watertown *23 F2* South Dakota, N USA
Waterville *19 G2* Maine, NE USA
Watford *67 A7* E England, United Kingdom
Watlings Island *see* San Salvador
Watsa *55 E5* Orientale, NE Dem. Rep. Congo
Watts Bar Lake *20 D1 reservoir* Tennessee, S USA
Wau *51 B5 var.* Wāw. Western Bahr el Ghazal, C South Sudan
Waukegan *18 B3* Illinois, N USA
Waukesha *18 B3* Wisconsin, N USA
Wausau *18 B2* Wisconsin, N USA
Waverly *23 G3* Iowa, C USA
Wavre *65 C6* Walloon Brabant, C Belgium
Wāw *see* Wau
Wawa *16 C4* Ontario, S Canada
Waycross *21 E3* Georgia, SE USA
Wearmouth *see* Sunderland
Webfoot State *see* Oregon
Webster City *23 F3* Iowa, C USA
Weddell Plain *132 A2 abyssal plain* SW Atlantic Ocean
Weddell Sea *132 A2 sea* SW Atlantic Ocean
Weener *72 A3* Niedersachsen, NW Germany
Weert *65 D5* Limburg, SE Netherlands
Weesp *64 C3* Noord-Holland, C Netherlands
Węgorzewo *76 D2 Ger.* Angerburg. Warmińsko-Mazurskie, NE Poland
Weichsel *see* Wisła
Weimar *72 C4* Thüringen, C Germany
Weissenburg *see* Alba Iulia, Romania
Weissenburg in Bayern *73 C6* Bayern, SE Germany
Weissenstein *see* Paide
Weisskirchen *see* Bela Crkva
Weiswampach *65 D7* Diekirch, N Luxembourg
Wejherowo *76 C2* Pomorskie, NW Poland
Welchman Hall *33 G1* C Barbados
Weldiya *50 C4 var.* Waldia, *It.* Valdia. Āmara, N Ethiopia
Welkom *56 D4* Free State, C South Africa
Welle *see* Uele
Wellesley Islands *126 B2 island group* Queensland, N Australia
Wellington *129 D5 country capital* (New Zealand) Wellington, North Island, New Zealand
Wellington *23 F5* Kansas, C USA
Wellington *see* Wellington, Isla
Wellington, Isla *43 A7 var.* Wellington. *island* S Chile
Wells *24 D4* Nevada, W USA
Wellsford *128 D2* Auckland, North Island, New Zealand
Wells, Lake *125 C5 lake* Western Australia
Wels *73 D6 anc.* Ovilava. Oberösterreich, N Austria
Wembley *67 A8* Alberta, W Canada
Wemmel *65 B6* Vlaams Brabant, C Belgium
Wenatchee *24 B2* Washington, NW USA
Wenchi *53 E4* W Ghana
Wen-chou/Wenchow *see* Wenzhou
Wendau *see* Võnnu
Wenden *see* Cēsis
Wenzhou *106 D5 var.* Wen-chou, Wenchow. Zhejiang, SE China
Werda *56 C4* Kgalagadi, S Botswana
Werder *see* Virtsu
Werenów *see* Voranava
Werkendam *64 C4* Noord-Brabant, S Netherlands
Werowitz *see* Virovitica
Werro *see* Võru
Werschetz *see* Vršac
Wesenberg *see* Rakvere
Weser *72 B3 river* NW Germany
Wessel Islands *126 B1 island group* Northern Territory, N Australia
West Antarctica *132 A3 var.* Lesser Antarctica. *physical region* Antarctica
West Australian Basin *see* Wharton Basin
West Bank *97 A6 disputed region* SW Asia
West Bend *18 B3* Wisconsin, N USA
West Bengal *113 F4 cultural region* NE India
West Cape *129 A7 headland* South Island, New Zealand
West Des Moines *23 F3* Iowa, C USA
Westerland *72 B2* Schleswig-Holstein, N Germany
Western Australia *124 B4 state* W Australia
Western Bug *see* Bug
Western Carpathians *77 E7 mountain range* W Romania Europe
Western Desert *see* Ṣaḩrā' al Gharbīyah
Western Dvina *63 E7 Bel.* Dzvina, *Ger.* Düna, *Latv.* Daugava, *Rus.* Zapadnaya Dvina. *river* W Europe
Western Ghats *112 C5 mountain range* SW India
Western Isles *see* Outer Hebrides
Western Punjab *see* Punjab
Western Sahara *48 B3 disputed territory administered by Morocco* N Africa
Western Samoa *see* Samoa
Western Samoa, Independent State of *see* Samoa
Western Sayans *see* Zapadnyy Sayan
Western Scheldt *see* Westerschelde
Western Sierra Madre *see* Madre Occidental, Sierra
Westerschelde *65 B5 Eng.* Western Scheldt; *prev.* Honte. *inlet* S North Sea
West Falkland *43 C7 var.* Gran Malvina, Isla Gran Malvina. *island* W Falkland Islands
West Fargo *23 F2* North Dakota, N USA
West Frisian Islands *see* Waddeneilanden
West Irian *see* Papua
Westliche Morava *see* Zapadna Morava
West Mariana Basin *120 B1 var.* Perece Vela Basin. *undersea feature* W Pacific Ocean
West Memphis *20 B1* Arkansas, C USA
West New Guinea *see* Papua
Weston-super-Mare *67 D7* SW England, United Kingdom
West Palm Beach *21 F4* Florida, SE USA
West Papua *see* Papua
Westport *129 C5* West Coast, South Island, New Zealand
West Punjab *see* Punjab
West River *see* Xi Jiang
West Siberian Plain *see* Zapadno-Sibirskaya Ravnina
West Virginia *18 D4 off.* State of West Virginia, *also known as* Mountain State. *state* NE USA
Wetar, Pulau *117 F5 island* Kepulauan Damar, E Indonesia
Wetzlar *73 B5* Hessen, W Germany
Wevok *14 C2 var.* Wewuk. Alaska, USA
Wewuk *see* Wevok
Wexford *67 B6 Ir.* Loch Garman. SE Ireland
Weyburn *15 F5* Saskatchewan, S Canada
Weymouth *67 D7* S England, United Kingdom
Wezep *64 D3* Gelderland, E Netherlands
Whakatane *128 E3* Bay of Plenty, North Island, New Zealand
Whale Cove *15 G3 var.* Tikiarjuaq. Nunavut, C Canada
Whangarei *128 D2* Northland, North Island, New Zealand
Wharton Basin *119 D5 var.* West Australian Basin. *undersea feature* E Indian Ocean
Whataroa *129 B6* West Coast, South Island, New Zealand
Wheatland *22 D3* Wyoming, C USA
Wheeler Peak *26 D1 mountain* New Mexico, SW USA
Wheeling *18 D4* West Virginia, NE USA
Whitby *67 D5* N England, United Kingdom
Whitefish *22 B1* Montana, NW USA
Whitehaven *67 C5* NW England, United Kingdom
Whitehorse *14 D4 territory capital* Yukon, W Canada
White Nile *50 B4 Ar.* Al Baḩr al Abyaḑ, An Nīl al Abyaḑ, Bahr el Jebel. *river* C South Sudan
White River *22 D3 river* South Dakota, N USA
White Sea *see* Beloye More
White Volta *53 E4 var.* Nakambé, *Fr.* Volta Blanche. *river* Burkina/Ghana
Whitianga *128 D2* Waikato, North Island, New Zealand
Whitney, Mount *25 C6 mountain* California, W USA
Whitsunday Group *126 D3 island group* Queensland, E Australia
Whyalla *127 B6* South Australia
Wichita *23 F5* Kansas, C USA
Wichita Falls *27 F2* Texas, SW USA
Wichita River *27 F2 river* Texas, SW USA
Wickenburg *26 B2* Arizona, SW USA
Wicklow *67 B6 Ir.* Cill Mhantáin. *county* E Ireland
Wicklow Mountains *67 B6 Ir.* Sléibhte Chill Mhantáin. *mountain range* E Ireland
Wieliczka *77 D5* Małopolskie, S Poland
Wieluń *76 C4* Sieradz, C Poland
Wien *73 E6 Eng.* Vienna, *Hung.* Bécs, *Slvk.* Vídeň, *Slvn.* Dunaj; *anc.* Vindobona. *country capital* (Austria) Wien, NE Austria
Wiener Neustadt *73 E6* Niederösterreich, E Austria
Wierden *64 E3* Overijssel, E Netherlands
Wiesbaden *73 B5* Hessen, W Germany
Wieselburg and Ungarisch-Altenburg/ Wieselburg-Ungarisch-Altenburg *see* Mosonmagyaróvár
Wiesenhof *see* Ostrołęka
Wight, Isle of *67 D7 island* United Kingdom
Wigorna Ceaster *see* Worcester
Wijchen *64 D4* Gelderland, SE Netherlands
Wijk bij Duurstede *64 D4* Utrecht, C Netherlands
Wilcannia *127 C6* New South Wales, SE Australia
Wilejka *see* Vilyeyka
Wilhelm, Mount *122 B3 mountain* C Papua New Guinea
Wilhelm-Pieck-Stadt *see* Guben
Wilhelmshaven *72 B3* Niedersachsen, NW Germany
Wilia/Wilja *see* Neris
Wilkes Barre *19 F3* Pennsylvania, NE USA
Wilkes Land *132 C4 physical region* Antarctica
Wilkomierz *see* Ukmergė
Willard *26 D2* New Mexico, SW USA
Willcox *26 C3* Arizona, SW USA
Willebroek *65 B5* Antwerpen, C Belgium
Willemstad *33 E5 dependent territory capital* (Curaçao) Lesser Antilles, S Caribbean Sea
Williston *22 D1* North Dakota, N USA
Wilmington *19 F4* Delaware, NE USA
Wilmington *21 F2* North Carolina, SE USA
Wilmington *18 C4* Ohio, N USA
Wilna/Wilno *see* Vilnius
Wilrijk *65 C5* Antwerpen, N Belgium
Winchester *67 D7 hist.* Wintanceaster, *Lat.* Venta Belgarum. S England, United Kingdom
Winchester *19 E4* Virginia, NE USA
Windau *see* Ventspils, Latvia
Windau *see* Venta, Latvia/Lithuania
Windhoek *56 B3 Ger.* Windhuk. *country capital* (Namibia) Khomas, C Namibia
Windhuk *see* Windhoek
Windorah *126 C4* Queensland, C Australia
Windsor *16 C5* Ontario, S Canada
Windsor *67 D7* S England, United Kingdom
Windsor *19 G3* Connecticut, NE USA
Windward Islands *33 H4 island group* E West Indies
Windward Islands *see* Barlavento, Ilhas de, Cape Verde
Windward Passage *32 D3 Sp.* Paso de los Vientos. *channel* Cuba/Haiti
Winisk *16 C2 river* Ontario, C Canada
Winkowitz *see* Vinkovci
Winnebago, Lake *18 B2 lake* Wisconsin, N USA
Winnemucca *25 C5* Nevada, W USA
Winnipeg *15 G5 province capital* Manitoba, S Canada
Winnipeg, Lake *15 G5 lake* Manitoba, C Canada
Winnipegosis, Lake *16 A3 lake* Manitoba, C Canada
Winona *23 G3* Minnesota, N USA
Winschoten *64 E2* Groningen, NE Netherlands
Winsen *72 B3* Niedersachsen, N Germany
Winston Salem *21 E1* North Carolina, SE USA
Winsum *64 D1* Groningen, NE Netherlands
Wintanceaster *see* Winchester
Winterswijk *64 E4* Gelderland, E Netherlands
Winterthur *73 B7* Zürich, NE Switzerland
Winton *126 C4* Queensland, E Australia
Winton *129 A7* Southland, South Island, New Zealand
Wisby *see* Visby
Wisconsin *18 A2 off.* State of Wisconsin, *also known as* Badger State. *state* N USA
Wisconsin Rapids *18 B2* Wisconsin, N USA
Wisconsin River *18 B3 river* Wisconsin, N USA
Wisłany, Zalew *see* Vistula Lagoon
Wismar *72 C2* Mecklenburg-Vorpommern, N Germany
Wittenberge *72 C3* Brandenburg, N Germany
Wittlich *73 A5* Rheinland-Pfalz, SW Germany
Wittstock *72 C3* Brandenburg, NE Germany
W. J. van Blommesteinmeer *37 G3 reservoir* E Suriname
Władysławowo *76 C2* Pomorskie, N Poland
Włocławek *76 C3 Ger./Rus.* Vlotslavsk. Kujawsko-pomorskie, C Poland
Włodawa *76 E4 Rus.* Vlodava. Lubelskie, SE Poland
Włodzimierz *see* Volodymyr-Volyns'kyy
Wlotzkasbaken *56 B3* Erongo, W Namibia
Wodonga *127 C7* Victoria, SE Australia
Wodzisław Śląski *77 C5 Ger.* Loslau. Śląskie, S Poland
Wojerecy *see* Hoyerswerda
Wōjjā *see* Wotje Atoll
Wojwodina *see* Vojvodina
Woking *67 D7* SE England, United Kingdom
Wolf, Isla *38 A4 island* Galápagos Islands, W Ecuador South America
Wolfsberg *73 D7* Kärnten, SE Austria
Wolfsburg *72 C3* Niedersachsen, N Germany
Wolgast *72 D2* Mecklenburg-Vorpommern, NE Germany
Wołkowysk *see* Vawkavysk
Wöllan *see* Velenje
Wollaston Lake *15 F4* Saskatchewan, C Canada
Wollongong *127 D6* New South Wales, SE Australia
Wolmar *see* Valmiera
Wołożyn *see* Valozhyn
Wolvega *64 D2 Fris.* Wolvegea. Friesland, N Netherlands
Wolvegea *see* Wolvega
Wolverhampton *67 D6* C England, United Kingdom
Wolverine State *see* Michigan
Wŏnsan *107 E3* SE North Korea
Woodburn *24 B3* Oregon, NW USA
Woodland *25 B5* California, W USA
Woodruff *18 B2* Wisconsin, N USA
Woods, Lake of the *16 A3 Fr.* Lac des Bois. *lake* Canada/USA
Woodville *128 D4* Manawatu-Wanganui, North Island, New Zealand
Woodward *27 F1* Oklahoma, C USA
Worcester *56 C5* Western Cape, SW South Africa
Worcester *67 D6 hist.* Wigorna Ceaster. W England, United Kingdom
Worcester *19 G3* Massachusetts, NE USA
Workington *67 C5* NW England, United Kingdom
Worland *22 C3* Wyoming, C USA
Wormatia *see* Worms
Worms *73 B5 anc.* Augusta Vangionum, Borbetomagus, Wormatia. Rheinland-Pfalz, SW Germany
Worms *see* Vormsi
Worthington *23 F3* Minnesota, N USA
Wotje Atoll *122 D1 var.* Wōjjā. *atoll* Ratak Chain, E Marshall Islands
Woudrichem *64 C4* Noord-Brabant, S Netherlands
Wrangel Island *93 F1 Eng.* Wrangel Island. *island* NE Russia
Wrangel Island *see* Vrangelya, Ostrov
Wrangel Plain *133 B2 undersea feature* Arctic Ocean
Wrocław *76 C4 Eng./Ger.* Breslau. Dolnośląskie, SW Poland
Września *76 C3* Wielkopolskie, C Poland
Wsetin *see* Vsetín
Wuchang *see* Wuhan
Wuday'ah *99 C6 spring/well* S Saudi Arabia
Wuhai *105 E3 var.* Haibowan. Nei Mongol Zizhiqu, N China
Wuhan *106 C5 var.* Han-kou, Han-k'ou, Hanyang, Wuchang, Wu-han; *prev.* Hankow. *province capital* Hubei, C China
Wu-han *see* Wuhan
Wuhsien *see* Suzhou
Wuhsi/Wu-his *see* Wuxi
Wuhu *106 D5 var.* Wu-na-mu. Anhui, E China
Wujlān *see* Ujelang Atoll
Wukari *53 G4* Taraba, E Nigeria
Wuliang Shan *106 A6 mountain range* SW China
Wu-lu-k'o-mu-shi/Wu-lu-mu-ch'i *see* Ürümqi
Wu-na-mu *see* Wuhu
Wuppertal *72 A4 prev.* Barmen-Elberfeld. Nordrhein-Westfalen, W Germany
Würzburg *73 B5* Bayern, SW Germany
Wusih *see* Wuxi
Wuxi *106 D5 var.* Wuhsi, Wu-hsi, Wusih. Jiangsu, E China
Wuyi Shan *103 E3 mountain range* SE China
Wye *67 C6 Wel.* Gwy. *river* England/Wales, United Kingdom
Wyłkowyszki *see* Vilkaviškis
Wyndham *124 D3* Western Australia
Wyoming *18 C3* Michigan, N USA
Wyoming *22 B3 off.* State of Wyoming, *also known as* Equality State. *state* C USA
Wyszków *76 D3 Ger.* Probstberg. Mazowieckie, NE Poland

X

Xaafuun, Raas *50 E4 var.* Ras Hafun. *cape* NE Somalia
Xaçmaz *95 H2 Rus.* Khachmas. N Azerbaijan
Xaignabouli *114 C4 prev.* Muang Xaignabouri, *Fr.* Sayaboury. Xaignabouli, N Laos
Xai-Xai *57 E4 prev.* João Belo, Vila de João Belo. Gaza, S Mozambique
Xalapa *29 F4* Veracruz-Llave, Mexico
Xam Nua *114 D3 var.* Sam Neua. Houaphan, N Laos
Xankändi *95 G3 Rus.* Khankendi; *prev.* Stepanakert. SW Azerbaijan
Xánthi *82 C3* Anatolikí Makedonía kai Thráki, NE Greece
Xàtiva *71 F3 Cas.* Xátiva; *anc.* Setabis, *var.* Jativa. País Valenciano, E Spain
Xauen *see* Chefchaouen
Xäzär Dänizi *see* Caspian Sea
Xeres *see* Jeréz de la Frontera
Xiaguan *see* Dali
Xiamen *106 D6 var.* Hsia-men; *prev.* Amoy. Fujian, SE China
Xi'an *106 C4 var.* Changan, Sian, Signan, Siking, Singan, Xian. *province capital* Shaanxi, C China
Xiang *see* Hunan
Xiangkhoang *see* Phônsaven
Xiangtan *106 C5 var.* Hsiang-t'an, Siangtan. Hunan, S China
Xiao Hinggan Ling *106 D2 Eng.* Lesser Khingan Range. *mountain range* NE China
Xichang *106 B5* Sichuan, C China
Xieng Khouang *see* Phônsaven
Xieng Ngeun *see* Muong Xiang Ngeun
Xigazê *104 C5 var.* Jih-k'a-tse, Shigatse, Xigaze. Xizang Zizhiqu, W China
Xi Jiang *102 D3 var.* Hsi Chiang, *Eng.* West River. *river* S China
Xilinhot *105 F2 var.* Silinhot. Nei Mongol Zizhiqu, N China
Xilokastro *see* Xylókastro
Xin *see* Xinjiang Uygur Zizhiqu
Xingkai Hu *see* Khanka, Lake
Xingu, Rio *41 E2 river* C Brazil
Xingxingxia *104 D3* Xinjiang Uygur Zizhiqu, NW China
Xining *105 E4 var.* Hsining, Hsi-ning, Sining. *province capital* Qinghai, C China
Xinjiang *see* Xinjiang Uygur Zizhiqu
Xinjiang Uygur Zizhiqu *104 B3 var.* Sinkiang, Sinkiang Uighur Autonomous Region, Xin, Xinjiang. *autonomous region* NW China
Xinpu *see* Lianyungang
Xinxiang *106 C4* Henan, C China
Xinyang *106 C5 var.* Hsin-yang, Sinyang. Henan, C China
Xinzo de Limia *70 C2* Galicia, NW Spain
Xiqing Shan *102 D2 mountain range* C China
Xiva *100 D2 Rus.* Khiva, Khiwa. Xorazm Viloyati, W Uzbekistan
Xixón *see* Gijón
Xizang *see* Xizang Zizhiqu
Xizang Gaoyuan *see* Qingzang Gaoyuan
Xizang Zizhiqu *104 B4 var.* Thibet, Tibetan Autonomous Region, Xizang, *Eng.* Tibet. *autonomous region* W China
Xolotlán *see* Managua, Lago de
Xucheng *see* Xuwen
Xuddur *51 D5 var.* Hudur, *It.* Oddur. Bakool, SW Somalia
Xuwen *106 C7 var.* Xucheng. Guangdong, S China
Xuzhou *106 D4 var.* Hsu-chou, Suchow, Tongshan; *prev.* T'ung-shan. Jiangsu, E China
Xylókastro *83 B5 var.* Xilokastro. Pelopónnisos, S Greece

Y

Ya'an *106 B5 var.* Yaan. Sichuan, C China
Yabēlo *51 C5* Oromīya, C Ethiopia
Yablis *31 E2* Región Autónoma Atlántico Norte, NE Nicaragua
Yablonovyy Khrebet *93 F4 mountain range* S Russia
Yabrai Shan *105 E3 mountain range* NE China
Yafran *49 F2* NW Libya
Yaghan Basin *45 B7 undersea feature* S E Pacific Ocean
Yagotin *see* Yahotyn
Yahotyn *87 E2 Rus.* Yagotin. Kyyivs'ka Oblast', N Ukraine
Yahualica *28 D4* Jalisco, SW Mexico
Yakima *24 B2* Washington, NW USA
Yakima River *24 B2 river* Washington, NW USA
Yakoruda *82 C3* Blagoevgrad, SW Bulgaria
Yaku-shima *109 B8 island* Nansei-shotō, SW Japan
Yakutat *14 D4* Alaska, USA
Yakutsk *93 F3* Respublika Sakha (Yakutiya), NE Russia
Yala *115 C7* Yala, SW Thailand
Yalizava *85 D6 Rus.* Yelizovo. Mahilyowskaya Voblasts', E Belarus
Yalong Jiang *106 A5 river* C China
Yalova *94 B3* Yalova, NW Turkey
Yalpug, Ozero *see* Yalpuh, Ozero
Yalpuh, Ozero *86 D4 Rus.* Ozero Yalpug. *lake* SW Ukraine
Yalta *87 F5* Respublika Krym, S Ukraine
Yalu *103 E2 Chin.* Yalu Jiang, *Jap.* Oryokko, *Kor.* Amnok-kang. *river* China/North Korea
Yalu Jiang *see* Yalu
Yamaguchi *109 B7 var.* Yamaguti. Yamaguchi, Honshū, SW Japan
Yamal, Poluostrov *92 D2 peninsula* N Russia
Yamanīyah, Al Jumhūrīyah al *see* Yemen
Yambio *51 B5 var.* Yambiyo. Western Equatoria, S South Sudan
Yambiyo *see* Yambio
Yambol *82 D2 Turk.* Yanboli. Yambol, E Bulgaria
Yamdena, Pulau *117 G5 prev.* Jamdena. *island* Kepulauan Tanimbar, E Indonesia
Yamoussoukro *52 D5 country capital* (Ivory Coast) C Ivory Coast
Yamuna *112 D3 prev.* Jumna. *river* N India
Yana *93 F2 river* NE Russia
Yanboli *see* Yambol
Yanbu 'al Baḩr *99 A5* Al Madīnah, W Saudi Arabia
Yangambi *55 D5* Orientale, N Dem. Rep. Congo
Yangchow *see* Yangzhou
Yangiyo'l *101 E2 Rus.* Yangiyul'. Toshkent Viloyati, E Uzbekistan
Yangiyul' *see* Yangiyo'l
Yangku *see* Taiyuan
Yangon *114 B4 Eng.* Rangoon. Yangon, S Myanmar (Burma)
Yangtze *106 B5 var.* Yangtze Kiang, *Eng.* Yangtze. *river* C China
Yangtze *see* Chang Jiang
Yangtze Kiang *see* Chang Jiang
Yangzhou *106 D5 var.* Yangchow. Jiangsu, E China
Yankton *23 E3* South Dakota, N USA
Yany Kapu *87 F4 Rus.* Krasnoperekops'k. Respublika Krym, S Ukraine
Yannina *see* Ioánnina
Yanskiy Zaliv *91 F2 bay* N Russia
Yantai *106 D4 var.* Yan-t'ai; *prev.* Chefoo, Chih-fu. Shandong, E China
Yaoundé *55 B5 var.* Yaunde. *country capital* (Cameroon) Centre, S Cameroon
Yap *122 A1 island* Caroline Islands, W Micronesia
Yapanskoye More East Sea/Japan, Sea of
Yapen, Pulau *117 G4 prev.* Japen. *island* E Indonesia
Yap Trench *120 B2 var.* Yap Trough. *undersea feature* SE Philippine Sea
Yap Trough *see* Yap Trench
Yapurá *see* Caquetá, Río, Brazil/Colombia
Yapurá *see* Japurá, Rio, Brazil/Colombia
Yaqui, Río *28 C2 river* NW Mexico
Yaransk *89 C5* Kirovskaya Oblast', NW Russia
Yarega *88 D4* Respublika Komi, NW Russia
Yaren *122 D2 de facto country capital* (Nauru) Nauru, SW Pacific
Yarkant *see* Shache
Yarlung Zangbo Jiang *see* Brahmaputra
Yarmouth *17 F5* Nova Scotia, SE Canada
Yarmouth *see* Great Yarmouth
Yaroslav *see* Jarosław
Yaroslavl' *88 B4* Yaroslavskaya Oblast', W Russia
Yarumal *36 B2* Antioquia, NW Colombia
Yasyel'da *85 B7 river* Brestskaya Voblasts', SW Belarus Europe
Yatsushiro *109 A7 var.* Yatusiro. Kumamoto, Kyūshū, SW Japan
Yatusiro *see* Yatsushiro
Yaunde *see* Yaoundé
Yavarí *see* Javari, Rio
Rio Yavari *40 C2 var.* Yavarí. *river* Brazil/Peru
Yaviza *31 H5* Darién, SE Panama
Yavoriv *86 B2 Pol.* Jaworów, *Rus.* Yavorov. L'vivs'ka Oblast', NW Ukraine
Yavorov *see* Yavoriv
Yazd *98 D3 var.* Yezd. Yazd, C Iran
Yazoo City *20 B2* Mississippi, S USA
Yding Skovhøj *63 A7 hill* C Denmark
Ýdra *83 C6 var.* Ídhra. *island* Ýdra, S Greece
Ye *115 B5* Mon State, S Myanmar (Burma)
Yecheng *104 A3 var.* Kargilik. Xinjiang Uygur Zizhiqu, NW China
Yefremov *89 B5* Tul'skaya Oblast', W Russia
Yekaterinburg *92 C3 prev.* Sverdlovsk. Sverdlovskaya Oblast', C Russia
Yekaterinodar *see* Krasnodar
Yekaterinoslav *see* Dnipro
Yelets *89 B5* Lipetskaya Oblast', W Russia
Yelisavetpol *see* Gäncä
Yelizavetgrad *see* Kropyvnyts'kyy
Yelizovo *see* Yalizava
Yell *66 D1 island* NE Scotland, United Kingdom
Yellowhammer State *see* Alabama
Yellowknife *15 E4 territory capital* Northwest Territories, W Canada
Yellow River *see* Huang He
Yellow Sea *106 D4 Chin.* Huang Hai, *Kor.* Hwang-Hae. *sea* E Asia
Yellowstone River *22 C2 river* Montana/Wyoming, NW USA
Yel'sk *85 C7* Homyel'skaya Voblasts', SE Belarus
Yelwa *53 F4* Kebbi, W Nigeria
Yemen *99 C7 off.* Republic of Yemen, *Ar.* Al Jumhuriyah al Yamaniyah, Al Yaman. *country* SW Asia
Yemen, Republic of *see* Yemen
Yemva *88 D4 prev.* Zheleznodorozhnyy. Respublika Komi, NW Russia
Yenakiyeve *87 G3 Rus.* Yenakiyevo; *prev.* Ordzhonikidze, Rykovo. Donets'ka Oblast', E Ukraine
Yenakiyevo *see* Yenakiyeve
Yenangyaung *114 A3* Magway, W Myanmar (Burma)
Yendi *53 E4* NE Ghana
Yengisar *104 A3* Xinjiang Uygur Zizhiqu, NW China
Yenierenköy *80 D4 var.* Yialousa, *Gk.* Agialoúsa. NE Cyprus
Yenipazar *see* Novi Pazar
Yenisey *92 D3 river* Mongolia/Russia
Yenping *see* Nanping
Yeovil *67 D7* SW England, United Kingdom
Yeppoon *126 D4* Queensland, E Australia
Yerevan *95 F3 Eng.* Erivan. *country capital* (Armenia) C Armenia

Yeriḥo see Jericho
Yerushalayim see Jerusalem
Yeso see Hokkaidō
Yeu, Île d' 68 A4 island NW France
Yevlakh see Yevlax
Yevlax 95 G2 Rus. Yevlakh. C Azerbaijan
Yevpatoriya 87 F5 Respublika Krym, S Ukraine
Yeya 87 H4 river SW Russia
Yezerishche see Yezyaryshcha
Yezo see Hokkaidō
Yezyaryshcha 85 E5 Rus. Yezerishche. Vitsyebskaya Voblasts', NE Belarus
Yialousa see Yenierenköy
Yiannitsá see Giannitsá
Yichang 106 C5 Hubei, C China
Yıldızeli 94 D3 Sivas, N Turkey
Yinchuan 106 B4 var. Yinch'uan, Yin-ch'uan, Yinchwan. province capital Ningxia, N China
Yinchwan see Yinchuan
Yindu He see Indus
Yin-hsien see Ningbo
Yining 104 B2 var. I-ning, Uigh. Gulja, Kuldja. Xinjiang Uygur Zizhiqu, NW China
Yin-tu Ho see Indus
Yisrael/Yisra'el see Israel
Yíthion see Gýtheio
Yogyakarta 116 C5 prev. Djokjakarta, Jogjakarta, Jokyakarta. Jawa, C Indonesia
Yokohama 109 D5 Aomori, Honshū, C Japan
Yokohama 108 A2 Kanagawa, Honshū, S Japan
Yokote 108 D4 Akita, Honshū, C Japan
Yola 53 H4 Adamawa, E Nigeria
Yonago 109 B6 Tottori, Honshū, SW Japan
Yong'an 106 D6 var. Yongan. Fujian, SE China
Yongzhou 107 C6 var. Lengshuitan. Hunan, S China
Yonkers 19 F3 New York, NE USA
Yonne 68 C4 river C France
Yopal 36 C3 var. El Yopal. Casanare, C Colombia
York 67 D5 anc. Eboracum, Eburacum. N England, United Kingdom
York 23 E4 Nebraska, C USA
York, Cape 126 C1 headland Queensland, NE Australia
York, Kap see Innaanganeq
Yorkton 15 F5 Saskatchewan, S Canada
Yoro 30 C2 Yoro, C Honduras
Yoshkar-Ola 89 C5 Respublika Mariy El, W Russia
Yösönbulag see Altay
Youngstown 18 D4 Ohio, N USA
Youth, Isle of see Juventud, Isla de la
Ypres see Ieper
Yreka 24 B4 California, W USA
Yrendagué see General Eugenio A. Garay
Yssel see IJssel
Ysyk-Köl see Issyk-Kul', Ozero
Ysyk-Köl see Balykchy
Yu see Henan
Yuan see Red River
Yuan Jiang see Red River
Yuba City 25 B5 California, W USA
Yucatán, Canal de see Yucatan Channel
Yucatan Channel 29 H3 Sp. Canal de Yucatán. channel Cuba/Mexico
Yucatan Peninsula see Yucatán, Península de
Yucatán, Península de 13 C7 Eng. Yucatan Peninsula. peninsula Guatemala/Mexico
Yuci see Jinzhong
Yue see Guangdong
Yue Shan, Tai see Lantau Island
Yueyang 106 C5 Hunan, S China
Yugoslavia see Serbia
Yukhavichy 85 D5 Rus. Yukhovichi. Vitsyebskaya Voblasts', N Belarus
Yukhovichi see Yukhavichy
Yukon 14 D3 prev. Yukon Territory, Fr. Territoire du Yukon. territory NW Canada
Yukon River 14 C2 river Canada/USA
Yukon, Territoire du see Yukon
Yukon Territory see Yukon
Yulin 106 C6 Guangxi Zhuangzu Zizhiqu, S China
Yuma 26 A2 Arizona, SW USA

Yun see Yunnan
Yungki see Jilin
Yung-ning see Nanning
Yunjinghong see Jinghong
Yunki see Jilin
Yunnan 106 A6 var. Yun, Yunnan Sheng, Yünnan, Yun-nan. province SW China
Yunnan see Kunming
Yunnan Sheng see Yunnan
Yünnan/Yun-nan see Yunnan
Yurev see Tartu
Yurihonjō see Honjō
Yuruá, Río see Juruá, Rio
Yury'ev see Tartu
Yushu 104 D4 var. Gyêgu. Qinghai, C China
Yuty 42 D3 Caazapá, S Paraguay
Yuzhno-Sakhalinsk 93 H4 Jap. Toyohara; prev. Vladimirovka. Ostrov Sakhalin, Sakhalinskaya Oblast', SE Russia
Yuzhnyy Bug see Pivdennyy Buh
Yuzhou see Chongqing
Ýlanly see Gurbansoltan Eje

Z

Zaandam see Zaanstad
Zaanstad 64 C3 prev. Zaandam. Noord-Holland, C Netherlands
Zabaykal'sk 93 F5 Chitinskaya Oblast', S Russia
Zabern see Saverne
Zabid 99 B7 W Yemen
Żabinka see Zhabinka
Ząbkowice see Ząbkowice Śląskie
Ząbkowice Śląskie 76 B4 var. Ząbkowice, Ger. Frankenstein, Frankenstein in Schlesien. Dolnośląskie, SW Poland
Zábřeh 77 C5 Ger. Hohenstadt. Olomoucký Kraj, E Czech Republic (Czechia)
Zacapa 30 B2 Zacapa, E Guatemala
Zacatecas 28 D3 Zacatecas, C Mexico
Zacatepec 29 E4 Morelos, S Mexico
Zacháro 83 B6 var. Zaharo, Zakháro. Dytikí Elláś, S Greece
Zadar 78 A3 It. Zara; anc. Iader. Zadar, SW Croatia
Zadetkyi Kyun 115 B6 var. St.Matthew's Island. island Myeik Archipelago, S Myanmar (Burma)
Zafra 70 C4 Extremadura, W Spain
Żagań 76 B4 var. Zagań, Żegań, Ger. Sagan. Lubuskie, W Poland
Zagazig see Az Zaqāzīq
Zágráb see Zagreb
Zagreb 78 B2 Ger. Agram, Hung. Zágráb. country capital (Croatia) Zagreb, N Croatia
Zagros Mountains 98 C3 Eng. Zagros Mountains. mountain range W Iran
Zagros Mountains see Zāgros, Kūhhā-ye
Zaharo see Zacháro
Zāhedān 98 E4 var. Zahidan; prev. Duzdab. Sīstān va Balūchestān, SE Iran
Zahidan see Zāhedān
Zaḥlah see Zahlé
Zahlé 96 B4 var. Zaḥlah. C Lebanon
Záhony 77 E6 Szabolcs-Szatmár-Bereg, NE Hungary
Zaire see Congo (river)
Zaire see Congo (Democratic Republic of)
Zaječar 78 E4 Serbia, E Serbia
Zakataly see Zaqatala
Zakháro see Zacháro
Zakhidnyy Buh/Zakhodni Buh see Bug
Zākhō 98 B2 var. Zākhū. Dahūk, N Iraq
Zākhū see Zākhō
Zakopane 77 D5 Małopolskie, S Poland
Zákynthos 83 A6 var. Zákinthos, It. Zante. island Iónia Nísoi, Greece, C Mediterranean Sea
Zalaegerszeg 77 B7 Zala, W Hungary
Zalău 86 B3 Ger. Waltenberg, Hung. Zilah; prev. Ger. Zillenmarkt. Sălaj, NW Romania
Zalim 99 B5 Makkah, W Saudi Arabia
Zambesi/Zambeze see Zambezi
Zambezi 56 C2 North Western, W Zambia
Zambezi 56 D2 var. Zambesi, Port. Zambeze. river S Africa

Zambia 56 C2 off. Republic of Zambia; prev. Northern Rhodesia. country S Africa
Zambia, Republic of see Zambia
Zamboanga 117 E3 off. Zamboanga City. Mindanao, S Philippines
Zamboanga City see Zamboanga
Zambrów 76 E3 Łomża, E Poland
Zamora 70 D2 Castilla-León, NW Spain
Zamora de Hidalgo 28 D4 Michoacán, SW Mexico
Zamość 76 E4 Rus. Zamoste. Lubelskie, E Poland
Zamoste see Zamość
Zancle see Messina
Zanda 104 A4 Xizang Zizhiqu, W China
Zanesville 18 D4 Ohio, N USA
Zanjān 98 C2 var. Zenjan, Zinjan. Zanjān, NW Iran
Zante see Zákynthos
Zanthus 125 C6 Western Australia, S Australia Oceania
Zanzibar 51 D7 Zanzibar, E Tanzania
Zanzibar 51 C7 Swa. Unguja. island E Tanzania
Zaozhuang 106 D4 Shandong, E China
Zapadna Morava 78 D4 Ger. Westliche Morava. river C Serbia
Zapadnaya Dvina 88 A4 Tverskaya Oblast', W Russia
Zapadnaya Dvina see Western Dvina
Zapadno-Sibirskaya Ravnina 92 C3 Eng. West Siberian Plain. plain C Russia
Zapadnyy Bug see Bug
Zapadnyy Sayan 92 D4 Eng. Western Sayans. mountain range S Russia
Zapala 43 B5 Neuquén, W Argentina
Zapiola Ridge 45 B6 undersea feature SW Atlantic Ocean
Zapolyarnyy 88 C2 Murmanskaya Oblast', NW Russia
Zaporizhzhya 87 F3 Rus. Zaporozh'ye; prev. Aleksandrovsk. Zaporiz'ka Oblast', SE Ukraine
Zaporozh'ye see Zaporizhzhya
Zapotiltic 28 D4 Jalisco, SW Mexico
Zaqatala 95 G2 Rus. Zakataly. NW Azerbaijan
Zara 94 D3 Sivas, C Turkey
Zara see Zadar
Zarafshan see Zarafshon
Zarafshon 100 D2 Rus. Zarafshan. Navoiy Viloyati, N Uzbekistan
Zarafshon see Zeravshan
Zaragoza 71 F2 Eng. Saragossa; anc. Caesaraugusta, Salduba. Aragón, NE Spain
Zarand 98 D3 Kermān, C Iran
Zaranj 100 D5 Nīmrūz, SW Afghanistan
Zarasai 84 C4 Utena, E Lithuania
Zárate 42 D4 prev. General José F.Uriburu. Buenos Aires, E Argentina
Zarautz 71 E1 var. Zara
Zelenogradsk 84 A4 Ger. Cranz, Kranz. Kaliningradskaya Oblast', W Russia
Zelle see Celle
Zel'va 85 B6 Pol. Zelwa. Hrodzyenskaya Voblasts', W Belarus
Zelwa see Zel'va
Zelzate 65 B5 var. Selzaete. Oost-Vlaanderen, NW Belgium
Žemaičių Aukštumas 84 B4 physical region W Lithuania
Zemst 65 C5 Vlaams Brabant, C Belgium
Zemun 78 D3 Serbia, N Serbia
Zengg see Senj
Zenica 78 C4 Federacija Bosna I Hercegovina, C Bosnia and Herzegovina
Zenta see Senta
Zeravshan 101 E3 Taj./Uzb. Zarafshon. river Tajikistan/Uzbekistan
Zevenaar 64 D4 Gelderland, SE Netherlands
Zevenbergen 64 C4 Noord-Brabant, S Netherlands
Zeya 91 E3 river SE Russia
Zgierz see Zgierz
Zgierz 76 C4 Ger. Neuhof, Rus. Zgerzh. Łódź, C Poland
Zgorzelec 76 B4 Ger. Görlitz. Dolnośląskie, SW Poland
Zhabinka 85 A6 Pol. Żabinka. Brestskaya Voblasts', SW Belarus

Zhambyl see Taraz
Zhanaozen 92 A4 Kaz. Zhangaözen; prev. Novyy Uzen'. Mangistau, W Kazakhstan
Zhangaözen see Zhanaozen
Zhangaqazaly see Ayteke Bi
Zhang-chia-k'ou see Zhangjiakou
Zhangdian see Zibo
Zhangjiakou 106 C3 var. Changkiakow, Zhang-chia-k'ou, Eng. Kalgan; prev. Wanchuan. Hebei, E China
Zhangzhou 106 D6 Fujian, SE China
Zhanjiang 106 C7 var. Chanchiang, Chan-chiang, Cant. Tsamkong, Fr. Fort-Bayard. Guangdong, S China
Zhaoqing 106 C6 Guangdong, S China
Zhayyk see Ural
Zhdanov see Mariupol'
Zhe see Zhejiang
Zhejiang 106 D5 var. Che-chiang, Chekiang, Zhe, Zhejiang Sheng. province SE China
Zhejiang Sheng see Zhejiang
Zheleznodorozhnyy 84 A4 Kaliningradskaya Oblast', W Russia
Zheleznodorozhnyy see Yemva
Zheleznogorsk 89 A5 Kurskaya Oblast', W Russia
Zhëltyye Vody see Zhovti Vody
Zhengzhou 106 C4 var. Ch'eng-chou, Chengchow; prev. Chenghsien. province capital Henan, C China
Zhezkazgan 92 C4 Kaz. Zhezqazghan; prev. Dzhezkazgan. Karagandy, C Kazakhstan
Zhezqazghan see Zhezkazgan
Zhidachov see Zhydachiv
Zhitkovichi see Zhytkavichy
Zhitomir see Zhytomyr
Zhlobin 85 D7 Homyel'skaya Voblasts', SE Belarus
Zhmerinka see Zhmerynka
Zhmerynka 86 D2 Rus. Zhmerinka. Vinnyts'ka Oblast', C Ukraine
Zhodino see Zhodzina
Zhodzina 85 D6 Rus. Zhodino. Minskaya Voblasts', C Belarus
Zholkev/Zholkva see Zhovkva
Zhonghua Renmin Gongheguo see China
Zhosaly 92 B4 prev. Dzhusaly. Kzylorda, SW Kazakhstan
Zhovkva 86 B2 Pol. Żółkiew, Rus. Zholkev, Zholkva; prev. Nesterov. L'vivs'ka Oblast', NW Ukraine
Zhovti Vody 87 F3 Rus. Zhëltyye Vody. Dnipropetrovs'ka Oblast', E Ukraine
Zhovtneve 87 E4 Rus. Zhovtnevoye. Mykolayivs'ka Oblast', S Ukraine
Zhovtnevoye see Zhovtneve
Zhydachiv 86 B2 Pol. Żydaczów, Rus. Zhidachov. L'vivs'ka Oblast', NW Ukraine
Zhytkavichy 85 C7 Rus. Zhitkovichi. Homyel'skaya Voblasts', SE Belarus
Zhytomyr 86 D2 Rus. Zhitomir. Zhytomyrs'ka Oblast', NW Ukraine
Zibo 106 D4 var. Zhangdian. Shandong, E China
Zichenau see Ciechanów
Zielona Góra 76 B4 Ger. Grünberg, Grünberg in Schlesien, Grüneberg. Lubuskie, W Poland
Zierikzee 64 B4 Zeeland, SW Netherlands
Zigong 106 B5 var. Tzekung. Sichuan, C China
Ziguinchor 52 B3 SW Senegal
Zilah see Zalău
Žilina 77 C5 Ger. Sillein, Hung. Zsolna. Žilínský Kraj, N Slovakia
Zillenmarkt see Zalău
Zimbabwe 56 D3 off. Republic of Zimbabwe; prev. Rhodesia. country S Africa
Zimbabwe, Republic of see Zimbabwe
Zimnicea 86 C5 Teleorman, S Romania
Zimovniki 89 B7 Rostovskaya Oblast', SW Russia
Zinder 53 G3 Zinder, S Niger
Zinov'yevsk see Kropyvnyts'kyy
Zintenhof see Sindi
Zipaquirá 36 B3 Cundinamarca, C Colombia
Zittau 72 D4 Sachsen, E Germany
Zlatni Pyasŭtsi 82 E2 Dobrich, NE Bulgaria
Zlín 77 C5 prev. Gottwaldov. Zlínský Kraj, E Czech Republic (Czechia)

Złoczów see Zolochev
Złotów 76 C3 Wielkopolskie, C Poland
Znamenka see Znam''yanka
Znam''yanka 87 F3 Rus. Znamenka. Kirovohrads'ka Oblast', C Ukraine
Żnin 76 C3 Kujawsko-pomorskie, C Poland
Zoetermeer 64 C4 Zuid-Holland, W Netherlands
Zółkiew see Zhovkva
Zolochev 86 C2 Pol. Złoczów, Rus. Zolochiv. L'vivs'ka Oblast', W Ukraine
Zolochiv 87 G2 Rus. Zolochev. Kharkivs'ka Oblast', E Ukraine
Zolochiv see Zolochev
Zolote 87 H3 Rus. Zolotoye. Luhans'ka Oblast', E Ukraine
Zolotonosha 87 E2 Cherkas'ka Oblast', C Ukraine
Zolotoye see Zolote
Zólyom see Zvolen
Zomba 57 E2 Southern, S Malawi
Zombor see Sombor
Zongo 55 C5 Equateur, N Dem. Rep. Congo
Zonguldak 94 C2 Zonguldak, NW Turkey
Zonhoven 65 D6 Limburg, NE Belgium
Zoppot see Sopot
Żory 77 C5 var. Zory, Ger. Sohrau. Śląskie, S Poland
Zouar 54 C2 Borkou-Ennedi-Tibesti, N Chad
Zouérat 52 C2 var. Zouérate, Zouîrât. Tiris Zemmour, N Mauritania
Zouérate see Zouérat
Zouîrât see Zouérat
Zrenjanin 78 D3 prev. Petrovgrad, Veliki Bečkerek, Ger. Grossbetschkerek, Hung. Nagybecskerek. Vojvodina, N Serbia
Zsil/Zsily see Jiu
Zsolna see Žilina
Zsombolya see Jimbolia
Zsupanya see Županja
Zubov Seamount 45 D5 undersea feature E Atlantic Ocean
Zueila see Zawilah
Zug 73 B7 Fr. Zoug. Zug, C Switzerland
Zugspitze 73 C7 mountain S Germany
Zuid-Beveland 65 B5 var. South Beveland. island SW Netherlands
Zuider Zee see IJsselmeer
Zuidhorn 64 E1 Groningen, NE Netherlands
Zuidlaren 64 E2 Drenthe, NE Netherlands
Zula 50 C4 E Eritrea
Züllichau see Sulechów
Zumbo see Vila do Zumbo
Zundert 65 C5 Noord-Brabant, S Netherlands
Zunyi 106 B5 Guizhou, S China
Županja 78 C3 Hung. Zsupanya. Vukovar-Srijem, E Croatia
Zürich 73 B7 Eng./Fr. Zurich, It. Zurigo. Zürich, N Switzerland
Zurich, Lake see Zürichsee
Zürichsee 73 B7 Eng. Lake Zurich. lake NE Switzerland
Zurigo see Zürich
Zutphen 64 D3 Gelderland, E Netherlands
Zuwārah 49 F2 NW Libya
Zuwaylah see Zawilah
Zuyevka 89 D5 Kirovskaya Oblast', NW Russia
Zvenigorodka see Zvenyhorodka
Zvenyhorodka 87 E2 Rus. Zvenigorodka. Cherkas'ka Oblast', C Ukraine
Zvishavane 56 D3 prev. Shabani. Matabeleland South, S Zimbabwe
Zvolen 77 C6 Ger. Altsohl, Hung. Zólyom. Banskobystrický Kraj, C Slovakia
Zvornik 78 C4 E Bosnia and Herzegovina
Zwedru 52 D5 var. Tchien. E Liberia
Zwettl 73 E6 Wien, NE Austria
Zwevegem 65 A6 West-Vlaanderen, W Belgium
Zwickau 73 C5 Sachsen, E Germany
Zwolle 64 D3 Overijssel, E Netherlands
Żydaczów see Zhydachiv
Żýoetu see Jōetsu
Żyrardów 76 D3 Mazowieckie, C Poland
Zyryanovsk 92 D5 Vostochnyy Kazakhstan, E Kazakhstan